Springer Proceedings in Mathematics & Statistics

Volume 222

Springer Proceedings in Mathematics & Statistics

This book series features volumes composed of selected contributions from workshops and conferences in all areas of current research in mathematics and statistics, including operation research and optimization. In addition to an overall evaluation of the interest, scientific quality, and timeliness of each proposal at the hands of the publisher, individual contributions are all refereed to the high quality standards of leading journals in the field. Thus, this series provides the research community with well-edited, authoritative reports on developments in the most exciting areas of mathematical and statistical research today.

More information about this series at http://www.springer.com/series/10533

Raimundo Nonato Araújo dos Santos
Aurélio Menegon Neto · David Mond
Marcelo J. Saia · Jawad Snoussi
Editors

Singularities and Foliations. Geometry, Topology and Applications

BMMS 2/NBMS 3, Salvador, Brazil, 2015

Springer

Editors
Raimundo Nonato Araújo dos Santos
ICMC
University of São Paulo
São Carlos, São Paulo
Brazil

Marcelo J. Saia
ICMC
University of São Paulo
São Carlos, São Paulo
Brazil

Aurélio Menegon Neto
Department of Mathematics
Federal University of Paraíba
João Pessoa, Paraíba
Brazil

Jawad Snoussi
Instituto de Matemáticas, Unidad Cuernavaca
Universidad Nacional Autónoma de México
Cuernavaca, Morelos
Mexico

David Mond
Mathematics Institute
University of Warwick
Coventry
UK

ISSN 2194-1009 ISSN 2194-1017 (electronic)
Springer Proceedings in Mathematics & Statistics
ISBN 978-3-030-08826-2 ISBN 978-3-319-73639-6 (eBook)
https://doi.org/10.1007/978-3-319-73639-6

Mathematics Subject Classification (2010): 58K05, 58K15, 58K35, 58K60, 32S05, 32S15, 32S55

Printed on acid-free paper

This Springer imprint is published by the registered company Springer International Publishing AG part of Springer Nature
The registered company address is: Gewerbestrasse 11, 6330 Cham, Switzerland

Preface

The last mathematical proceedings of an event held at the Institute of Mathematics of the Federal University of Bahia were the famous volume of the "SALVADOR SYMPOSIUM ON DYNAMICAL SYSTEMS," organized in 1971 by Jacob Palis, Elon L. Lima, and Mauricio Matos Peixoto and published in 1973.

As is well known, several of the papers published there have played an important role in the areas of Dynamical Systems, Singularity Theory, and Foliation Theory.

Forty-six years on (almost a half-century), it would be wonderful to present some mathematical developments following on from those foundations.

This volume of proceedings comes as an initiative in that direction.

In fact, this book is a result of the joint organization, by the Brazilian and Mexican research groups in Singularities and Foliations, of two international mathematical meetings in the Brazilian Northeast: the 3rd Singularity Theory Meeting of Northeast region (or 3º ENSINO), from July 8 to 11, 2015, and the Brazil-Mexico 2nd Meeting on Singularities, from July 13 to 17, 2015.

The choice of the city of Salvador was intended to promote the development of research in Singularities and Foliations in the Northeast of Brazil.

The organization from the Brazilian side was carried out by the Singularities group of ICMC-USP/São Carlos, the groups from the Federal Universities of Paraiba and Ceará, and Simone Moraes, Kleiber Cunha and Evandro Carlos dos Santos from the Institute of Mathematics of the Federal University of Bahia.

From the Mexican side, the organizers were F. Aroca, E. Rosales, J. Seade, and J. Snoussi from Instituto de Matemáticas UNAM, together with X. Gómez-Mont from CIMAT and A. Giles Flores from Universidad Autónoma de Aguascalientes.

The meetings brought some of the leading researchers from Brazil and Mexico and from elsewhere, to the Northeast of Brazil, to discuss the latest developments in Singularity Theory, Foliations, and associated areas.

Since the 1980s, there has been a strong collaboration between Mexican and Brazilian researchers in Singularities and Complex Dynamical Systems. Current interactions involve researchers and students from São Carlos, João Pessoa, and Fortaleza, on the Brazilian side, and from the Institute of Mathematics of Cuernavaca (UNAM), CIMAT, and the Faculty of Science of the UNAM in

Mexico. The 1st Brazil/Mexico meeting on Singularities took place in Querétaro, Mexico, from August 1 to 3, 2013, with the participation of 30 researchers and students from both sides.

During the 3rd Singularity Theory Meeting of Northeast region, a school was offered for young researchers and graduate students at master's and Ph.D. levels, with minicourses delivered by Nicolas Dutertre (Université Aix-Marseille—France), Terence Gaffney (Northeastern University—USA), David Mond (University of Warwick—England), and Juan J. Nuño-Ballesteros (Universitat de Valencia—Spain).

During the Brazil-Mexico 2nd Meeting on Singularities, plenary talks were delivered by participants from Brazil and Mexico, as well as by invited mathematicians from different countries, as described below.

Brazilian team: Abramo Hefez (UFF), Alexandre Fernandes (UFC), Alice Libardi (UNESP), Aurélio Menegon (UFPB), Jonny Ardila (UFRJ), Leonardo Câmara (UEFS), Marcelo Escudeiro (UEM), Marcio Soares (UFMG), Marcos Craizer (PUC), Maria Elenice (UEM), M. Salarinoghabi (ICMC), N. Thuy (UNESP), Roberta Wik Atique (ICMC), Rodrigo Mendes (UFC).

Mexican team: Agustín Romano (UNAM), Edwin Leon (CONACyT/CIMAT), Ernesto Rosales (UNAM), Federico Sánchez-Bringas (UNAM), Fuensanta Aroca (UNAM), Jessica Jáurez (UNAM), José-Luis Cisneros-Molina (UNAM), Mirna Gómez (UNAM), Daniel Duarte (CONACyT/UAZ), José Seade (UNAM), Jawad Snoussi (UNAM), L. Nunes (CIMAT).

Invited Speakers from the International Math Community:

Alexey Remizov (Russia), Carles Bivià-Ausina (Spain), Maria del Carmen Romero Fuster (Spain), Christopher Eyral (Poland), Jean-Paul Brasselet (France), Kasuto Takao (Japan), Lê Dũng Tráng (France), Mutsuo Oka (Japan), Osamu Saeki (Japan).

Beside the speakers listed above, there were poster sessions presented by postdoctoral researchers and master's and Ph.D. students from all over Brazil, from Mexico, and from many others countries. Further information on the minicourses, plenary talks, and poster presentation sessions and a list of participants may be found via the link

http://bramexsing2015.icmc.usp.br/index.php/programme.

We hope this initiative does not stop here and that in the near future it will motivate many other meetings in Singularity Theory and Foliation Theory, contributing to their development and to the development of mathematics in the Northeast of Brazil.

The Brazil/Mexico 3rd Meeting on Singularities was held in Cuernavaca, Mexico, from August 7 to 11, 2017. Further information may be found in the link http://www.matcuer.unam.mx/3rdMeetingOnSingularities/.

São Carlos, Brazil — Raimundo Nonato Araújo dos Santos
João Pessoa, Brazil — Aurélio Menegon Neto
Coventry, UK — David Mond
São Carlos, Brazil — Marcelo J. Saia
Cuernavaca, Mexico — Jawad Snoussi

Acknowledgements

Many people and institutions contributed to the realization of these meetings. Without their help it would certainly have been impossible.

Special thanks go to

1. the Brazilian scientific agencies CNPq-ARC for grant number 466390/2014-9, to CAPES-PAEP for grant number 1052/2015, and to FAPESP-São Paulo for grant number 2015/06135-0, for their support;
2. the Mexican scientific agency CONACYT, who supported this meeting through a special grant DADC 259611, and through their Redes Temáticas; the Instituto de Matemáticas of UNAM offered support throughout the organization of the meeting; the last editor would especially like to thank UNAM-DGAPA-PAPIIT 113817;
3. everyone from the International Math Community (England, France, Japan, Russia, Spain and USA) for their mini-courses and talks.

These contributions were fundamental for the richness of the meetings. Thank you so much!

We also thank:

4. the ICMC-USP for their technical and material support. Specially we thank the Director Prof. Alexandre Nolasco de Carvalho, the coordinator of the Math Graduate Program Prof. Sergio Monari, and all members of the Singularities Groups of ICMC/USP-São Carlos, UFPB-João Pessoa and UFC-Fortaleza, for their help and engagement in these meetings.
5. the international section of Bahiatursa for all technical support and information, specially, Ms. Rizia Falcão, of the Coordination of International Relationship Bureau, for her kindness and for the informative folders, and also for providing the nice Bahian characters during the opening ceremony of the meeting.
6. Ms. Terezinha and her nice staff who provided the delicious juice and cakes for the coffee breaks, giving us the opportunity to experience a piece of Bahian

cuisine. Thanks for representing so well one of the nicest part of Bahia's culture. Your kindness, love and smiles make everything possible.

Last but not least, we thank Prof. Evandro Carlos Ferreira dos Santos, Director of the Institute of Mathematics of UFBA for making available the IM installations during the meetings. Also, Profs. Antonio Andrade do E. Santo and Maria Amelia B. Holenwerger of CETEC/UFRB-Bahia, for believing and working together with us from the beginning all the way up to the realization of the meetings.

Thank You All Very Much !!!

Contents

Participants of the Second Brazil-Mexico Meeting on Singularities and Third Northeastern Meeting on Singularities. Photo taken at the Federal University of Bahia, Salvador, Brazil, in July 2015. *Source*: Paulo César Soares de Oliveira/ICMC-USP

Part I
Lecture Notes: Geometry, Topology, and Algebraic Aspects of Singularities

Combinatorial Models in the Topological Classification of Singularities of Mappings

J. J. Nuño-Ballesteros

Abstract The topological classification of finitely determined map germs $f : (\mathbb{R}^n, 0) \to (\mathbb{R}^p, 0)$ is discrete (by a theorem due to R. Thom), hence we want to obtain combinatorial models which codify all the topological information of the map germ f. According to Fukuda's work, the topology of such germs is determined by the link, which is obtained by taking the intersection of the image of f with a small enough sphere centered at the origin. If $f^{-1}(0) = \{0\}$, then the link is a topologically stable map $\gamma : S^{n-1} \to S^{p-1}$ (or stable if (n, p) are nice dimensions) and f is topologically equivalent to the cone of γ. When $f^{-1}(0) \neq \{0\}$, the situation is more complicated. The link is a topologically stable map $\gamma : N \to S^{p-1}$, where N is a manifold with boundary of dimension $n - 1$. However, in this case, we have to consider a generalized version of the cone, so that f is again topologically equivalent to the cone of the link diagram. We analyze some particular cases in low dimensions, where the combinatorial models are provided by objects which are well known in Computational Geometry, for instance, the Gauss word or the Reeb graph.

Keywords Finite determinacy · Topological classification · Gauss word · Reeb graph

2000 Mathmematics Subject Classification Primary 58K15 · Secondary 58K40 · 58K65

This work has been partially supported by DGICYT Grant MTM2015–64013–P and by PVE - CAPES Grant 88881.062217/2014–01.

J. J. Nuño-Ballesteros (✉)
Departament de Geometria i Topologia, Universitat de València,
Campus de Burjassot, 46100 Burjassot, Spain
e-mail: Juan.Nuno@uv.es

R. N. Araújo dos Santos et al. (eds.), *Singularities and Foliations. Geometry, Topology and Applications*, Springer Proceedings in Mathematics & Statistics 222, https://doi.org/10.1007/978-3-319-73639-6_1

1 Introduction

René Thom showed in [39] that the topological classification of finitely determined map germs $f : (\mathbb{R}^n, 0) \to (\mathbb{R}^p, 0)$ is discrete and hence, there are no moduli. The same assertion is not true if we consider the C^∞ classification by $\mathscr{A}$-equivalence (for instance, consider the 1-parameter family $f_t(x, y) = xy(x + y)(x - ty)$) or if we remove the finite determinacy assumption. In fact, Thom himself found a 1-parameter family of germs $f_t : (\mathbb{R}^3, 0) \to (\mathbb{R}^3, 0)$ with the property that any two distinct members of the family are not topologically equivalent (see [38]). Since the classification problem is discrete, a natural open question is to find a good combinatorial model which codifies the topological information of the map germ.

In [9], T. Fukuda proved that if the map germ is finitely determined and has isolated zeros (i.e., if $f^{-1}(0) = \{0\}$), then f has a cone structure on its link. The link is obtained by intersecting the image of f with a small enough sphere centered at the origin in $\mathbb{R}^p$. The main result is that the link turns out to be a mapping between spheres $\gamma : S^{n-1} \to S^{p-1}$ which is topologically stable (in fact, stable if (n, p) are nice dimensions in Mather's sense). Moreover, f is topologically equivalent to the cone of its link. Thus, the topological classification of germs can be deduced from the topological classification of topological stable mappings between spheres of one dimension less. We remark that the condition of isolated zeros is automatically satisfied when $n \leq p$. We review the proof of Fukuda's cone structure theorem for germs with isolated zeros in Sect. 4.

In a later paper [10], Fukuda also considered the case of non isolated zeros (i.e., $f^{-1}(0) \neq \{0\}$). The classification problem in this case is much more complicated. He showed the link is a mapping $\gamma : N \to S^{p-1}$ from a manifold with boundary N which is again topologically stable (or stable in nice dimensions). However, the germ f has not a cone structure on its link in the usual sense. We introduce in Sect. 7 the notion of generalized cone (following [5]) and also give an adapted version of the cone theorem for the case of non isolated zeros, by using this generalized version of the cone.

In low dimensions, the topological classification of finitely determined map germs has been widely developed by the author and other collaborators. We have studied the cases of map germs $f : (\mathbb{R}^2, 0) \to (\mathbb{R}^3, 0)$ in [22–24], map germs $f : (\mathbb{R}^2, 0) \to (\mathbb{R}^2, 0)$ in [30, 31, 33] and map germs $f : (\mathbb{R}^3, 0) \to (\mathbb{R}^3, 0)$ in [32, 34, 35]. In all these cases, the combinatorial model used for the topological classification is provided by the Gauss words.

More recently, we have also considered the case of map germs $f : (\mathbb{R}^3, 0) \to (\mathbb{R}^2, 0)$ in [2, 4] where we use the Reeb graph as a good combinatorial model for the singularity and the case of map germs $f : (\mathbb{R}^2, 0) \to (\mathbb{R}^4, 0)$ in [25], where we find a connection with Knot Theory. In [5], we also consider the topological classification of map germs with respect to the contact equivalence $\mathscr{K}$ instead of the right-left equivalence $\mathscr{A}$.

Gauss words and Reeb graphs are well known objects in Computational Geometry. We will explain in Sects. 5, 6 how to construct these models as well as the main results for the case of map germs $f : (\mathbb{R}^2, 0) \to (\mathbb{R}^3, 0)$ and map germs $f : (\mathbb{R}^3, 0) \to (\mathbb{R}^2, 0)$ with isolated zeros, respectively.

In Sects. 2 and 3 we review the basic concepts of stability and finite determinacy that we need for the course. There are no proofs for all the results in these sections, but we provide precise references which can be found basically in the celebrated six papers about stability by Mather [16–21], the survey papers by Wall [40, 41] and the book by Gibson et al. [12].

2 Stability

Along the text, we use the following notation:

$$\begin{aligned} \mathscr{E}_n &= \text{set of } C^\infty \text{ function germs } h : (\mathbb{R}^n, 0) \to \mathbb{R}, \\ \mathscr{E}(n, p) &= \text{set of } C^\infty \text{ map germs } f : (\mathbb{R}^n, 0) \to (\mathbb{R}^p, 0), \\ \mathscr{R}_n &= \text{set of } C^\infty \text{ diffeomorphism germs } \phi : (\mathbb{R}^n, 0) \to (\mathbb{R}^n, 0). \end{aligned}$$

Definition 2.1 We say that two germs $f, g \in \mathscr{E}(n, p)$ are *$\mathscr{A}$-equivalent* if there exist $\phi \in \mathscr{R}_n$ and $\psi \in \mathscr{R}_p$ such that $g = \psi \circ f \circ \phi^{-1}$. That is, the following diagram is commutative, where the columns are diffeomorphisms:

$$\begin{array}{ccc} (\mathbb{R}^n, 0) & \xrightarrow{\ f\ } & (\mathbb{R}^p, 0) \\ \downarrow{\scriptstyle \phi} & & \downarrow{\scriptstyle \psi} \\ (\mathbb{R}^n, 0) & \xrightarrow{\ g\ } & (\mathbb{R}^p, 0) \end{array}$$

In the case that ϕ, ψ are homeomorphisms instead of diffeomorphisms, then we say that f, g are *C^0-$\mathscr{A}$-equivalent.*

We can characterize the $\mathscr{A}$-equivalence through the group action $\mathscr{R}_n \times \mathscr{R}_p$ on $\mathscr{E}(n, p)$ given by $(\phi, \psi) \cdot f = \psi \circ f \circ \phi^{-1}$. Then $f, g \in \mathscr{E}(n, p)$ are $\mathscr{A}$-equivalent if they are in the same orbit.

Given $f \in \mathscr{E}(n, p)$, an r-parameter *unfolding* of f is another germ $F \in \mathscr{E}(r + n, r + p)$ of the form $F(u; x) = (u; f_u(x))$ and such that $f_0 = f$.

Definition 2.2 Two unfolding F, G of f are *$\mathscr{A}$-equivalent* (as unfoldings) if there exist diffeomorphisms $\Phi \in \mathscr{R}_{r+n}$ and $\Psi \in \mathscr{R}_{r+p}$ unfoldings of the identity maps in $(\mathbb{R}^n, 0)$ and $(\mathbb{R}^p, 0)$ respectively, such that $G = \Psi \circ F \circ \Phi^{-1}$.

We say that an unfolding F of f is *trivial* if F is $\mathscr{A}$-equivalent to the constant unfolding $G = \mathrm{id} \times f$. If $F(u; x) = (u; f_u(x))$, $\Phi(u; x) = (u; \phi_u(x))$ and $\Psi(u; y) = (u; \psi_u(y))$, we have

$$\psi_u \circ f_u \circ \phi_u^{-1} = f.$$

Thus, the germ of f_u at the point $\phi_u^{-1}(0)$ is $\mathscr{A}$-equivalent to f, but in general we do not have $\phi_u^{-1}(0) = 0$.

We say that $f \in \mathscr{E}(n, p)$ is *stable* if any unfolding F of f is trivial.

The above definition is known as stability by deformations or by homotopies when we consider mappings instead of germs. An immediate consequence of the definition is that the property that a germ is stable is invariant under $\mathscr{A}$-equivalence. Therefore, the definition can be extended without problem by taking coordinate charts for smooth map germs between smooth manifolds $f : (N, x) \to (P, y)$.

Example 2.3 From the definition, one deduces easily that if f is regular (i.e., the differential df_0 has maximal rank), then f is stable.

The set of germs $\mathscr{E}_n$ has a structure of commutative and unit local $\mathbb{R}$-algebra, whose maximal ideal $\mathfrak{m}_n$ is given by the germs $h \in \mathscr{E}_n$ such that $h(0) = 0$. Any $f \in \mathscr{E}(n, p)$ induces an $\mathbb{R}$-algebra homomorphism $f^* : \mathscr{E}_p \to \mathscr{E}_n$ through $f^*(h) = h \circ f$. Moreover, $\mathscr{E}(n, p)$ has a structure of $\mathscr{E}_n$-module and of $\mathscr{E}_p$-module via f^*.

Given $f \in \mathscr{E}(n, p)$, we denote by $\theta(f)$ the set of C^∞ germs of vector fields $\eta : (\mathbb{R}^n, 0) \to T\mathbb{R}^p$ along f, that is, such that $\pi \circ \eta = f$ where $\pi : T\mathbb{R}^p \to \mathbb{R}^p$ is the canonical projection. A generic element of $\theta(f)$ is written in a unique way as

$$\eta = \sum_{i=1}^{p} g_i \left(\frac{\partial}{\partial y_i} \circ f \right), \quad g_i \in \mathscr{E}_n,$$

where $y_1, \ldots, y_p$ are the coordinates in $\mathbb{R}^p$. In this way, $\theta(f)$ has a structure of $\mathscr{E}_n$-module isomorphic to $(\mathscr{E}_n)^p$, after identification of η with the p-tuple $(g_1, \ldots, g_p)$. In case that f is the germ of the identity map in $(\mathbb{R}^n, 0)$ or $(\mathbb{R}^p, 0)$, we denote $\theta(f)$ by θ_n or θ_p respectively.

For each $f \in \mathscr{E}(n, p)$ we can define two module homomorphisms. First, we have an $\mathscr{E}_n$-module homomorphism:

$$\begin{aligned} tf : \theta_n &\to \theta(f) \\ \xi &\mapsto df \circ \xi \end{aligned},$$

where df is the differential of f. On the other hand, we have an $\mathscr{E}_p$-module homomorphism:

$$\begin{aligned} wf : \theta_p &\to \theta(f) \\ \eta &\mapsto \eta \circ f \end{aligned},$$

where now in $\theta(f)$ we consider the $\mathscr{E}_p$-module structure induced by $f^* : \mathscr{E}_p \to \mathscr{E}_n$.

Definition 2.4 Given $f \in \mathscr{E}(n, p)$, the *$\mathscr{A}$-tangent space* (of f) and the *extended $\mathscr{A}$-tangent space* (of f) are defined respectively as

$$T\mathscr{A}f = tf(\mathfrak{m}_n\theta_n) + wf(\mathfrak{m}_p\theta_p),$$
$$T\mathscr{A}_e f = tf(\theta_n) + wf(\theta_p).$$

The $\mathscr{A}$*-codimension* and the $\mathscr{A}_e$*-codimension* are defined as

$$\mathscr{A} - \operatorname{codim}(f) = \dim_{\mathbb{R}} \frac{\mathfrak{m}_n \theta(f)}{T\mathscr{A} f},$$
$$\mathscr{A}_e - \operatorname{codim}(f) = \dim_{\mathbb{R}} \frac{\theta(f)}{T\mathscr{A}_e f}.$$

The following theorem is known as the infinitesimal stability criterion of Mather. The proof can be found in [12, 2.2].

Theorem 2.5 *A germ* $f \in \mathscr{E}(n, p)$ *is stable if and only if its* $\mathscr{A}_e$*-codimension is zero.*

By using the above identification of $\theta(f)$ with $(\mathscr{E}_n)^p$, the above theorem says that $f \in \mathscr{E}(n, p)$ is stable if and only if for each $\alpha \in (\mathscr{E}_n)^p$ there exist $g \in (\mathscr{E}_n)^n$ and $h \in (\mathscr{E}_p)^p$ such that

$$\alpha = h \circ f + \sum_{i=1}^{n} g_i \frac{\partial f}{\partial x_i}.$$

Example 2.6 We begin with the case of functions $p = 1$, we see that $f \in \mathscr{E}(n, 1)$ has stable singularity if and only f is a Morse function. In fact, if f is a Morse function (i.e., it has non degenerate critical point at the origin) by the Morse lemma, we can assume that f is given (up to coordinate changes) by

$$f(x) = x_1^2 + \cdots + x_s^2 - x_{s+1}^2 - \cdots - x_n^2,$$

in such a way that $\frac{\partial f}{\partial x_i} = \pm 2x_i$. Given $\alpha \in \mathscr{E}_n$, by the Hadamard lemma there exist $g_i \in \mathscr{E}_n$ such that α is written as

$$\alpha = \alpha(0) + \sum_{i=1}^{n} x_i g_i = \alpha(0) \circ f + \sum_{i=1}^{n} \left(\pm \frac{g_i}{2}\right) \frac{\partial f}{\partial x_i},$$

and f is stable by Theorem 2.5.

Conversely, suppose that f is not a Morse function and has a degenerate critical point at the origin. Consider the n-parameter unfolding $F(a, x) = (a, f_a(x))$ given by

$$f_a(x) = f(x) + a_1 x_1 + \cdots + a_n x_n.$$

Fix a representative $F : \mathbb{R}^n \times U \to \mathbb{R}^n \times \mathbb{R}$, where $U \subset \mathbb{R}^n$ is an open neighbourhood of the origin. By the Thom Transversality Theorem [13, Theorem 4.9], for almost any $a \in \mathbb{R}^n$, $f_a : U \to \mathbb{R}$ is a Morse function, and hence, f_a cannot be $\mathscr{A}$-equivalent to f. This shows that the unfolding F is not trivial and f is not stable.

Example 2.7 Let $n = 1$ and $p = 2$. Then $f \in \mathscr{E}(1, 2)$ is stable only in the case that f is an immersion. In fact, if f is singular we can consider the 2-parameter unfolding $F(a, x) = (a, f_a(x))$ given by

$$f_a(x) = f(x) + ax.$$

Again we fix a representative $F : \mathbb{R}^2 \times (-\epsilon, \epsilon) \to \mathbb{R}^2 \times \mathbb{R}^2$. By the Thom Transversality Theorem, for almost any $a \in \mathbb{R}^2$, $f_a : (-\epsilon, \epsilon) \to \mathbb{R}^2$ is an immersion and f_a cannot be $\mathscr{A}$-equivalent to f. Thus, F is not trivial and f is not stable.

Sometimes it can be complicated to see that a certain germ is stable by means of Theorem 2.5. Also, the genericity arguments we have used to check that the germs are the only stable singularities may not work in higher dimensions. We present here a pair of results which give easy methods to check stability and to obtain normal forms for stable germs. Both are related to the concept of contact or $\mathscr{K}$-equivalence. This is another important equivalence introduced by Mather, which is weaker than $\mathscr{A}$-equivalence. We do not include here the definition, details can be found in [19].

Definition 2.8 For each germ $f \in \mathscr{E}(n, p)$, the *$\mathscr{K}$-extended tangent space* is defined as:

$$T\mathscr{K}_e f = tf(\theta_n) + (f^*\mathfrak{m}_p)\theta(f).$$

Note that $\omega f(\mathfrak{m}_p\theta_p) \subset (f^*\mathfrak{m}_p)\theta(f)$, hence ωf induces a well defined morphism:

$$\overline{\omega} f : \mathbb{R}^p \cong \frac{\theta_p}{\mathfrak{m}_p\theta_p} \longrightarrow \frac{\theta(f)}{T\mathscr{K}_e f}.$$

Lemma 2.9 ([19, Proof of Proposition I.6]) *A germ $f \in \mathscr{E}(n, p)$ is stable if and only if $\overline{\omega}(f)$ is an epimorphism.*

An equivalent statement of Lemma 2.9 is that f is stable if and only if $\theta(f)/T\mathscr{K}_e f$ is generated over $\mathbb{R}$ by the classes of the canonical basis $\{e_1, \ldots, e_p\}$ in $\mathbb{R}^p$. Note that $T\mathscr{K}_e f$ is an $\mathscr{E}_n$-module which is finitely generated, in fact, it is generated over $\mathscr{E}_n$ by $\partial f/\partial x_i$, $i = 1, \ldots, n$ and by $f_j e_k$, with $j, k = 1, \ldots, p$. Thus, it is possible to compute it by using some computer algebra system like Singular [14].

Definition 2.10 For each germ $f \in \mathscr{E}(n, p)$, the *local algebra* (of f) is defined as

$$Q(f) = \frac{\mathscr{E}_n}{f^*\mathfrak{m}_p}.$$

Theorem 2.11 ([19]) *Two stable germs are $\mathscr{A}$-equivalent if and only if their local algebras are isomorphic.*

Example 2.12 Let us see that for $n = p = 2$, a singular germ $f \in \mathscr{E}(2, 2)$ is stable if and only if f has a singularity of type fold $f(x, y) = (x, y^2)$ or cusp $f(x, y) = (x, xy + y^3)$.

If f is a fold, we have:

$$\begin{aligned}T\mathcal{K}_e f &= \mathcal{E}_2\left\{\begin{pmatrix}1\\0\end{pmatrix},\begin{pmatrix}0\\2y\end{pmatrix}\right\}+\langle x,y^2\rangle\mathcal{E}_2^2\\ &= \mathcal{E}_2\left\{\begin{pmatrix}1\\0\end{pmatrix},\begin{pmatrix}0\\x\end{pmatrix},\begin{pmatrix}0\\y\end{pmatrix}\right\}.\end{aligned}$$

Thus $\theta(f)/T\mathcal{K}_e f$ is generated over $\mathbb{R}$ by the class of $(0,1)$ and the map $\overline{\omega}f$ is obviously surjective, so f is stable by Lemma 2.9.

In the case of the cusp, we have:

$$\begin{aligned}T\mathcal{K}_e f &= \mathcal{E}_2\left\{\begin{pmatrix}1\\y\end{pmatrix},\begin{pmatrix}0\\3y^2+x\end{pmatrix}\right\}+\langle x,y^3+xy\rangle\mathcal{E}_2^2\\ &= \mathcal{E}_3\left\{\begin{pmatrix}1\\y\end{pmatrix},\begin{pmatrix}0\\x\end{pmatrix},\begin{pmatrix}0\\y^2\end{pmatrix}\right\}.\end{aligned}$$

Now, $\theta(f)/T\mathcal{K}_e f$ is generated over $\mathbb{R}$ by the classes of $\{(1,0),(0,1)\}$. Again $\overline{\omega}f$ is surjective and hence, f is stable.

Assume now that f is stable, so that $\overline{\omega}f$ is surjective. If f has rank 0, then $T\mathcal{K}_e f\subset\mathfrak{m}_2\theta(f)$. Since $\theta(f)/\mathfrak{m}_2\theta(f)$ has dimension 2, we must have necessarily that $T\mathcal{K}_e f=\mathfrak{m}_2\theta(f)$. Moreover, $(f^*\mathfrak{m}_2)\subset\mathfrak{m}_2^2\theta(f)$, the classes of $\partial f/\partial x$ and $\partial f/\partial y$ should generate $\mathfrak{m}_2\theta(f)/\mathfrak{m}_2^2\theta(f)$ over $\mathbb{R}$. But this is not possible, since this space has dimension 4.

Thus, f must have rank 1 and after a coordinate change in the source, we can assume that $f(x,y)=(x,g(x,y))$, for some function $g\in\mathfrak{m}_2^2$. In other words, we see f as an unfolding of $g_0(y)=g(0,y)$. An easy exercise shows that

$$\frac{\theta(f)}{T\mathcal{K}_e(f)}\cong\frac{\theta(g_0)}{T\mathcal{K}_e(g_0)}\cong\frac{\mathcal{E}_1}{\langle g_0'\rangle}.$$

If $g_0\in\mathfrak{m}_1^4$, then $g_0'\in\mathfrak{m}_1^3$ and thus $\dim_{\mathbb{R}}(\mathcal{E}_1/\langle g_0'\rangle)\geq 3$, which is not possible by the surjectivity of $\overline{\omega}f$. Hence, g_0 must have order 2 or 3. But this implies that either $Q(f)\cong\mathcal{E}_1/\langle y^2\rangle$ or $Q(f)\cong\mathcal{E}_1/\langle y^3\rangle$. By Theorem 2.11, f is $\mathcal{A}$-equivalent either to the fold or the cusp, respectively.

Given a germ $f\in\mathcal{E}(n,p)$, for each $k\in\mathbb{N}$ we denote by $j^kf(0)$ the k-jet of f, that is, the Taylor polynomial of degree k of f at the origin. The k-jet space $J^k(n,p)$ is the space of k-jets $j^kf(0)$ of germs $f\in\mathcal{E}(n,p)$. Then $J^k(n,p)$ is identified with the space of polynomial maps $\sigma:\mathbb{R}^n\to\mathbb{R}^p$ of degree less than or equal to k and such that $\sigma(0)=0$. We denote by $L^k(n)\subset J^k(n,n)$ the group of k-jets of diffeomorphism germs with the product defined by the k-jet of the composition. Moreover, we have the action of $L^k(n)\times L^k(p)$ on $J^k(n,p)$ induced by the action of $\mathcal{R}_n\times\mathcal{R}_p$ on $\mathcal{E}(n,p)$.

The k-jet spaces provide a finite-dimensional model of the classification problem by $\mathcal{A}$-equivalence. The jet space $J^k(n,p)$ can be identified with an Euclidean space $\mathbb{R}^N$ and the group $G=L^k(n)\times L^k(p)$ is a Lie group of finite dimension acting on

$J^k(n, p)$ in a semialgebraic way. As a consequence, for each $\sigma \in J^k(n, p)$ the orbit $G \cdot \sigma$ is a semialgebraic submanifold of $J^k(n, p)$. In fact, $G \cdot \sigma$ is a semialgebraic subset which contains at least a regular point. But the orbit is locally diffeomorphic at all of its points because of the group action. Thus, $G \cdot \sigma$ is regular at all of its points.

For each $f \in \mathscr{E}(n, p)$, we have an epimorphism from $\mathfrak{m}_n\theta(f)$ to $J^k(n, p)$ given by $g \mapsto j^k g(0)$ and whose kernel is $\mathfrak{m}_n^{k+1}\theta(f)$. This allows us to identify

$$T_\sigma(J^k(n, p)) \cong \frac{\mathfrak{m}_n\theta(f)}{\mathfrak{m}_n^{k+1}\theta(f)}.$$

Under this identification, the tangent space to the orbit $G \cdot \sigma$ is precisely

$$T_\sigma(G \cdot \sigma) = \frac{T\mathscr{A}f + \mathfrak{m}_n^{k+1}\theta(f)}{\mathfrak{m}_n^{k+1}\theta(f)}.$$

In particular, if $\mathfrak{m}_n^{k+1}\theta(f) \subset T\mathscr{A}f$, we deduce that the codimension of the orbit $G \cdot \sigma$ is equal to the $\mathscr{A}$-codimension.

Given $f \in \mathscr{E}(n, p)$, we denote by $j^k f : (\mathbb{R}^n, 0) \to J^k(n, p)$ the germ of the k-jet extension of f (for each $x \in \mathbb{R}^n$, $j^k f(x)$ is the k-jet of f at the point x after translation to the origin). Then, we have the following result which characterizes the stability in terms of k-jets.

Theorem 2.13 *Let $f \in \mathscr{E}(n, p)$, $k \geq p + 1$ and $\sigma = j^k f(0)$. Then:*

(1) f is stable if and only if $j^k f : (\mathbb{R}^n, 0) \to J^k(n, p)$ is transverse to $G \cdot \sigma$.
(2) If f is stable, then $\mathfrak{m}_n^{k+1}\theta(f) \subset T\mathscr{A}f$ and hence, $\operatorname{codim}(G \cdot \sigma) = \mathscr{A} - \operatorname{codim}(f)$.
(3) If f is stable, then g is $\mathscr{A}$-equivalent to f if and only if $j^k g(0) \in G \cdot \sigma$.

Proof Part (1) can be found in [41, Theorem 15] whilst (2) and (3) follow from the fact that if f is stable then it is $(p + 1)$-determined (see again [41, Theorem 15] and Sect. 3), then use [40, Theorem 1.2]. □

Definition 2.14 Given a stable germ $f \in \mathscr{E}(n, p)$, we denote by $A \subset (\mathbb{R}^n, 0)$ the germ of the subset of points x such that the germ of f at x is $\mathscr{A}$-equivalent to the germ of f at 0. As a consequence of Theorem 2.13, A is the germ of a submanifold in $(\mathbb{R}^n, 0)$ of codimension $\mathscr{A} - \operatorname{codim}(f)$ and the restriction $f|_A : A \to (\mathbb{R}^p, 0)$ is an immersion (unless the trivial case that f is a submersion). We say that A is the *analytic stratum of f in the source*. Note that if f is defined by a polynomial map, then A is a semialgebraic subset of $\mathbb{R}^n$.

We pass now to the study of multi-germs. Given a finite subset $S = \{x_1, \dots, x_r\} \subset \mathbb{R}^n$, we consider multi-germs of C^∞ maps of the form $f : (\mathbb{R}^n, S) \to (\mathbb{R}^p, y)$. The definitions of $\mathscr{A}$-equivalence, unfolding, trivial unfolding and stable germ can be generalized without any problem for multi-germs. Also the definitions of $\theta(f)$ and of extended $\mathscr{A}$-tangent space $T_e\mathscr{A} f$ can be adapted to the case of multi-germs and the infinitesimal stability criterion (Theorem 2.5) is still true. Moreover, next theorem allows to check in a relatively easy way whether a multi-germ is stable.

Given a multi-germ $f : (\mathbb{R}^n, S) \to (\mathbb{R}^p, y)$ with $S = \{x_1, \dots, x_r\}$, we denote the restriction germ by $f_i : (\mathbb{R}^n, x_i) \to (\mathbb{R}^p, y)$ and by A_i the analytic stratum of f_i in the source, $i = 1, \dots, r$.

Theorem 2.15 ([19, 1.6]) *A multi-germ $f : (\mathbb{R}^n, S) \to (\mathbb{R}^p, y)$ is stable if and only if for each $i = 1, \dots, r$, f_i is stable and the subspaces*

$$df_{x_1}(T_{x_1}A_1), \dots, df_{x_r}(T_{x_r}A_r)$$

have regular intersection in $\mathbb{R}^p$.

The regular intersection condition means that the codimension of the intersection is the sum of all the codimensions. Note that if f is a submersion at a point x_i, then $df_{x_i}(T_{x_i}A_i) = \mathbb{R}^p$. In this way, if $\tilde{S} \subset S$ is the subset of critical (i.e., non submersive) points of f, $f : (\mathbb{R}^n, S) \to (\mathbb{R}^p, y)$ is stable if and only if $f : (\mathbb{R}^n, \tilde{S}) \to (\mathbb{R}^p, y)$ is stable. Thus, we can assume without loss of generality that all the points of S are critical.

Example 2.16 Let us see what happens in the above examples when we consider multi-germs.

- Let $f : (\mathbb{R}^n, S) \to (\mathbb{R}, y)$. At each critical point x_i of S, f must be a Morse singularity and the analytic stratum A_i is only the point $\{x_i\}$. Hence, f is stable if and only if S is a single point and f has a Morse singularity at that point.
- Let $f : (\mathbb{R}, S) \to (\mathbb{R}^2, y)$. Then f is stable only in the case that it is an immersion with normal crossings. So, the stable multi-germs are the simple regular point and the transverse double point.
- Let $f : (\mathbb{R}^2, S) \to (\mathbb{R}^2, y)$. At each critical point x_i of S, f must have fold or cusp type. If any of the points x_i has cusp type, then again the analytic stratum is $\{x_i\}$ and necessarily $S = \{x_i\}$. Otherwise, if all the points have fold type, then each A_i is a curve and now the regular intersection condition implies that we can have either simple points or transverse double points. In conclusion, f is stable if and only if S is made of a simple fold, a simple cusp or two transverse folds.

Given a C^∞-mapping $f : N \to P$ between manifolds, we denote by $\Sigma(f) \subset N$ the set of critical points (where f is not submersive) and its image $\Delta(f) = f(\Sigma(f))$ is called the discriminant.

Definition 2.17 Given a stable multi-germ $f : (\mathbb{R}^n, S) \to (\mathbb{R}^p, y)$, we denote by $B \subset (\mathbb{R}^p, y)$ the germ of all points $y' \in \Delta(f)$ such that the multi-germ of f at $S' = f^{-1}(y') \cap \Sigma(f)$ is $\mathscr{A}$-equivalent to the multi-germ of f at S. Then B is a germ of submanifold in $(\mathbb{R}^p, y)$ which results from the intersection of the submanifolds $f_i(A_i)$, where each A_i is the analytic stratum of $f_i : (\mathbb{R}^n, x_i) \to (\mathbb{R}^p, y)$ in the source. We say that B is the *analytic stratum of f in the target*.

If $\dim B = d$, then we say that the multi-germ f represents a d-dimensional stable type. In particular, when $d = 0$, we say that f is a 0-stable type. Again, in the case that f is polynomial, B is a semialgebraic subset of $\mathbb{R}^p$.

Next, we will prove an interesting property which will be used in Sect. 4 to construct the cone structure of finitely determined germs.

Definition 2.18 For each germ $f \in \mathscr{E}(n, p)$, we define $\tau(f)$ as the subspace of $\mathbb{R}^p$ given by the kernel of $\overline{\omega} f : \mathbb{R}^p \to \theta(f)/T\mathscr{K}_e f$.

We will see that if f is stable, then $\tau(f)$ is nothing but $T_0 B$, where B is the analytic stratum of f in the target. The first step is to prove that they have the same dimension.

Lemma 2.19 *If $f \in \mathscr{E}(n, p)$ is stable, then* $\dim B = \dim_{\mathbb{R}} \tau(f)$, *where B is the analytic stratum in the target.*

Proof Assume that $\dim_{\mathbb{R}} \tau(f) = d$. By Lemma 2.9, we have

$$\dim_{\mathbb{R}} \frac{\theta(f)}{T\mathscr{K}_e f} = \dim_{\mathbb{R}} \frac{\mathbb{R}^p}{\tau(f)} = p - d.$$

We use the formulas of [40, 4.5.1, 4.5.2], which give:

$$\mathscr{A} - \operatorname{codim}(f) = p - d + (n - p) = n - d.$$

It follows from Theorem 2.13 that this is equal to the codimension of the analytic straum in the source in $(\mathbb{R}^n, 0)$. Thus, the analytic stratum in the source (and hence in the target) has dimension d. □

Lemma 2.20 *Let $f \in \mathscr{E}(n, p)$ be a stable germ and assume that* $\dim_{\mathbb{R}} \tau(f) = d$. *Then f is $\mathscr{A}$-equivalent to* $\mathrm{id}_{\mathbb{R}^d,0} \times g_0$, *where $g_0 \in \mathscr{E}(n - d, p - d)$ is also a stable germ.*

Proof We choose linear coordinates in $\mathbb{R}^p$ such that $\tau(f) = \mathbb{R}^d \times \{0\}$. Given $v \in \tau(f)$, there exists $\xi \in \theta_p$ such that $\xi_0 = v$ and $\omega f(\xi) = tf(\eta) + \nu$, for some $\eta \in \theta_n$ and $\nu \in f^* \mathfrak{m}_p \theta(f)$ and evaluating at 0, we get $v = df(\eta_0)$. This shows that $\tau(f) \subset df_0(\mathbb{R}^n)$. Hence, we can choose smooth coordinates in $(\mathbb{R}^n, 0)$ such that f is an unfolding of a map germ $g \in \mathscr{E}(n - d, p - d)$, that is,

$$f(u, x) = (u, g_u(x)), \ u \in \mathbb{R}^d, \ x \in \mathbb{R}^{n-d}.$$

Consider the following commutative diagram:

$$\begin{array}{ccc} \dfrac{\mathbb{R}^p}{\tau(f)} & \xrightarrow{\cong} & \dfrac{\theta(f)}{T\mathscr{K}_e(f)} \\ \cong\Big\downarrow & & \Big\downarrow\cong \\ \mathbb{R}^{p-d} & \xrightarrow{\overline{\omega}g_0} & \dfrac{\theta(g_0)}{T\mathscr{K}_e(g_0)} \end{array}$$

The top arrow is an isomorphism induced by $\overline{\omega}f$ and the columns are also isomorphisms: the left arrow is induced by the projection and is an isomorphism because $\tau(f) = \mathbb{R}^d \times \{0\}$ and the right arrow is also an isomorphism because f is an unfolding of g_0. Then, $\overline{\omega}g_0$ is also an isomorphism, so g_0 is stable by Lemma 2.9. By definition of stability, f is $\mathscr{A}$-equivalent to the constant unfolding $\mathrm{id}_{\mathbb{R}^d,0} \times g_0$. □

Corollary 2.21 *If $f \in \mathscr{E}(n, p)$ is stable, then $\tau(f) = T_0B$, where B is the analytic stratum in the target.*

Proof By Lemma 2.20, we can assume that $f = \mathrm{id}_{\mathbb{R}^d,0} \times g_0$, where $g_0 \in \mathscr{E}(n - d, p - d)$ is also a stable germ such that $\tau(g_0) = \{0\}$. We know from Lemma 2.19 that the analytic stratum of g_0 is also equal to $\{0\}$, so

$$\tau(f) = \mathbb{R}^d \times \{0\} = T_0B.$$

□

Proposition 2.22 *Let $f : (\mathbb{R}^n, S) \to (\mathbb{R}^p, y)$ be a stable multi-germ and let B be the analytic stratum in the target. If $P \subset \mathbb{R}^p$ is a submanifold transverse to B at y, then $N = f^{-1}(P)$ is a submanifold of $\mathbb{R}^n$ in a neighbourhood of S and the restriction $f|_{N,S} : (N, S) \to (P, y)$ is stable.*

Proof For each i, we have that $df_{x_i}(\mathbb{R}^n) \supset df_{x_i}(T_{x_i}A_i) \supset T_yB$. If B and P are transverse at y, then f is transverse to P at x_i and N is a submanifold of $\mathbb{R}^n$ in a neighbourhood of x_i. Let us assume for a moment that each restriction $f|_{N,x_i} : (N, x_i) \to (P, y)$ is a stable germ. The analytic stratum in the source is $N \cap A_i$ and the image by the differential of the tangent space is

$$df_{x_i}(T_{x_i}(N \cap A_i)) = df_{x_i}(T_{x_i}A_i) \cap T_yP.$$

Thus, the transversality between B and P at the point y ensures that the images of the tangent spaces of $N \cap A_i$ have regular intersection in T_yP.

It only remains to show that each germ $f|_{N,x_i} : (N, x_i) \to (P, y)$ is stable. We assume, for simplicity, that $x_i = 0$ and $y = 0$. By Lemma 2.20, we can also assume that $f_i = \mathrm{id}_{\mathbb{R}^d,0} \times g$, where $g \in \mathscr{E}(n - d, p - d)$ is also a stable germ and $d = \dim A_i$. The transversality assumption implies that T_0P contains $\{0\} \times \mathbb{R}^{p-d}$ and T_0N contains $\{0\} \times \mathbb{R}^{n-d}$.

Consider the following diagram:

$$\begin{array}{ccc} (N,0) & \xrightarrow{f|_{N,0}} & (P,0) \\ \uparrow{\scriptstyle i} & & \uparrow{\scriptstyle j} \\ (\mathbb{R}^{n-d},0) & \xrightarrow{g} & (\mathbb{R}^{p-d},0), \end{array}$$

where $i(z) = (0, z)$ and $j(w) = (0, w)$. Then we have that i, j are immersions, that j is transverse to $f|_{N,0}$ and that diagram is cartesian (that is, it is commutative and the mapping (i, g) from $\mathbb{R}^{n-d}$ into the submanifold $\{(x, w) \in N \times \mathbb{R}^{p-d} : f(x) = j(w)\}$ is a diffeomorphism). Thus, $f|_{N,0}$ can be seen, after $\mathscr{A}$-equivalence, as an unfolding of g (see [12, III.0.1]). But it is easy to see that if g is stable, then any unfolding of g is also stable. □

If instead of germs of C^∞ maps, we consider germs of analytic maps (real or complex), then all the definitions of $\mathscr{A}$-equivalence, unfoldings, stability, extended $\mathscr{A}$-tangent space, $\mathscr{A}_e$-codimension as well as all the theorems relating these concepts are still valid. In that case, the diffeomorphisms, vector fields and manifolds are considered also of analytic class (real or complex). Moreover, all the showed examples of stable germs or multi-germs work in the same way in the analytic case (real or complex). In fact, we have the following result which gives the relation between the three classes [40, 1.7].

Proposition 2.23 *Let $f : (\mathbb{R}^n, S) \to (\mathbb{R}^p, y)$ be a real analytic multi-germ, then the following statements are equivalent:*

(1) f is stable as a C^∞ multi-germ.
(2) f is stable as a real analytic multi-germ.
(3) The complexification $\hat{f}$ is stable as a complex analytic multi-germ.

We finish this section with the notion of stability of mappings.

Definition 2.24 We say that $f : N \to P$ is *locally stable* if

(1) the restriction $f : \Sigma(f) \to P$ is finite (i.e., finite-to-one and closed),
(2) for any $y \in \Delta(f)$, the multi-germ $f : (N, S) \to (P, y)$ is stable, where $S = f^{-1}(y) \cap \Sigma(f)$.

Example 2.25 We come back to the above examples. Let $f : N \to P$, with $\dim N = n$ and $\dim P = p$.

- If $p = 1$, then f is locally stable if and only if f is a Morse function with distinct critical points (see Fig. 1).
- If $n = 1$ and $p = 2$, then f is locally stable if and only if f is an immersion with transverse double points (see Fig. 2).
- If $n = p = 2$, then f is locally stable if and only if the singularities of f are simple folds, simple cusps of transverse double folds (see Fig. 3).

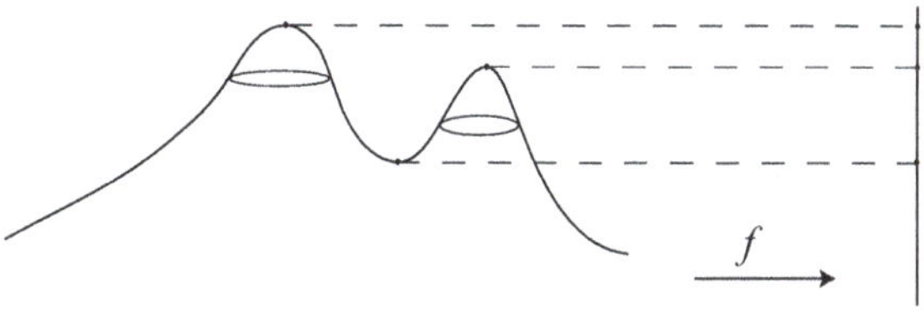

Fig. 1 Example of locally stable function

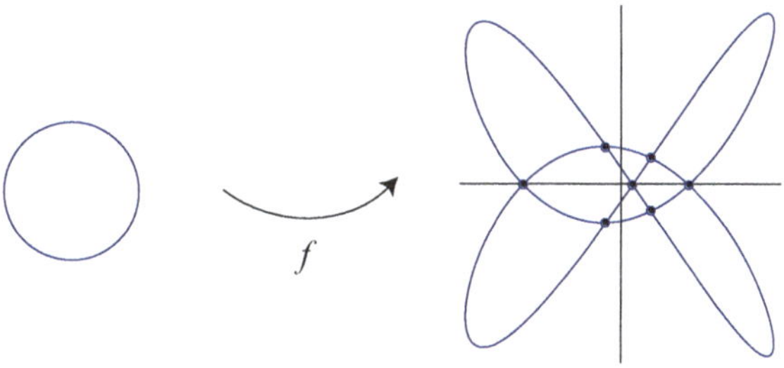

Fig. 2 Example of locally stable plane curve

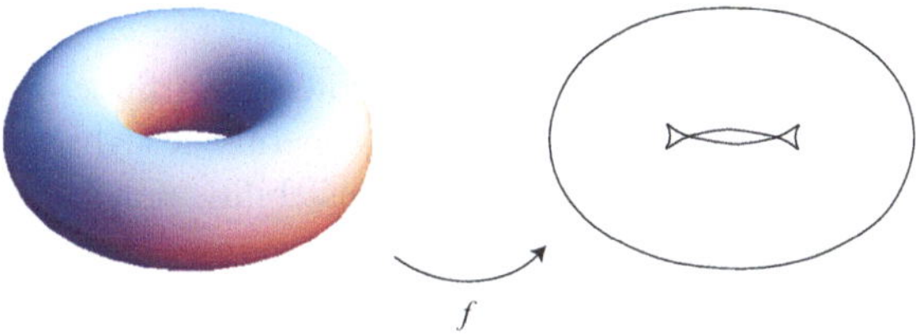

Fig. 3 Example of locally stable map between surfaces

There exists a concept of global stability. We say that a C^∞ mapping $f : N \to P$ between smooth manifolds is *globally stable* if there exists a neighbourhood W of f in $C^\infty(N, P)$ with the Whitney C^∞ topology, such that any $g \in W$ is $\mathscr{A}$-equivalent to f. Mather proved in [18] that if the restriction $f|_{\Sigma(f)}$ is proper, then the local and the global stability coincide. However, this result cannot be used in the real or complex analytic case.

3 Finite Determinacy

We begin this section with the definition of finite determinacy.

Definition 3.1 Given $f \in \mathscr{E}(n, p)$ and $k \in \mathbb{N}$, we say that f is *k-determined* if for any $g \in \mathscr{E}(n, p)$ such that $j^k f(0) = j^k g(0)$, then f, g are $\mathscr{A}$-equivalent. We say that f is *finitely determined* (FD) if it is k-determined for some $k \in \mathbb{N}$.

From the definition we deduce that if f is k-determined then f is $\mathscr{A}$-equivalent to $j^k f(0)$. Thus, when studying FD germs, we can assume without loss of generality that f is the germ of a polynomial mapping. Another consequence of the definition and of Theorem 2.13 is that if $f \in \mathscr{E}(n, p)$ is stable, then it is $(p + 1)$-determined.

Finite determinacy is a very desirable property, but usually it is difficult to check it directly from the definition. By this reason, the criteria of finite determinacy are very important. The following criterion is known as the infinitesimal criterion of finite determinacy. It is due to J. Mather and a proof can be found in [40, 1.2].

Theorem 3.2 *A germ $f \in \mathscr{E}(n, p)$ is FD if and only if its $\mathscr{A}_e$-codimension is finite.*

Next property is analogous to the Proposition 2.23 and it relates the finite determinacy of the three classes of map germs: C^∞, real analytic and complex analytic. The proof is based again on the fact that the $\mathscr{A}_e$-codimension coincides in the three classes [40, 1.7].

Proposition 3.3 *Let $f \in \mathscr{E}(n, p)$ a real analytic germ, then the following statements are equivalent:*

(1) f is FD as a C^∞ germ.
(2) f is FD as a real analytic germ.
(3) The complexification $\hat{f}$ is FD as a complex analytic germ.

We give now the geometric criterion of finite determinacy of Mather-Gaffney which works for complex analytic germs. Roughly speaking, it says that a germ is FD if and only if it has isolated instability at the origin. The proof can be found in [40, Theorem 2.1].

Theorem 3.4 *Let $f : (\mathbb{C}^n, 0) \to (\mathbb{C}^p, 0)$ a complex analytic germ. Then f is FD if and only if there exists a representative $f : U \to V$ where U, V are open neighbourhoods of the origin in $\mathbb{C}^n$ and $\mathbb{C}^p$ respectively, such that $f^{-1}(0) \cap \Sigma(f) = \{0\}$ and the restriction $f : U \setminus f^{-1}(0) \to V \setminus \{0\}$ is a locally stable mapping.*

If $f : (\mathbb{R}^n, 0) \to (\mathbb{R}^p, 0)$ is FD and is defined by polynomials, then we can complexify $\hat{f} : (\mathbb{C}^n, 0) \to (\mathbb{C}^p, 0)$ and apply the geometric criterion to $\hat{f}$. We deduce that there exists a representative $f : U \to V$ where U, V open neighbourhoods of the origin in $\mathbb{R}^n$ and $\mathbb{R}^p$ respectively, such that $f^{-1}(0) \cap \Sigma(f) = \{0\}$ and the restriction $f : U \setminus f^{-1}(0) \to V \setminus \{0\}$ is a locally stable mapping.

The converse is not true in general in the real case. For instance, consider the function $f : (\mathbb{R}^2, 0) \to (\mathbb{R}, 0)$ given by $f(x, y) = (x^2 + y^2)^2$. We have $f^{-1}(0) = \Sigma(f) = \{0\}$ and the restriction $f : \mathbb{R}^2 \setminus \{0\} \to \mathbb{R} \setminus \{0\}$ is regular and hence, locally stable. However, $\hat{f}^{-1}(0) = \Sigma(\hat{f}) = \{(x, y) \in \mathbb{C}^2 : x^2 + y^2 = 0\}$, so f is not FD by Proposition 3.3 and Theorem 3.4.

Since $f^{-1}(0) \cap \Sigma(f) = \{0\}$, after shrinking the neighbourhoods U, V if necessary, we can assume that the restriction $f : \Sigma(f) \to V$ is finite. Moreover, if $f : U \setminus f^{-1}(0) \to V \setminus \{0\}$ is a locally stable mapping, then the 0-stable types are isolated points in $U \setminus \{0\}$. But since these sets are semialgebraic, then by the Curve Selection Lemma [27], we have that the 0-stable types are also isolated points in U. Thus, we can shrink the neighbourhoods U, V in such a way that f has no 0-stable singularities in $U \setminus \{0\}$.

This fact motivates the following definition.

Definition 3.5 We say that a germ $f \in \mathscr{E}(n, p)$ has *isolated instability* (II) if there exists a representative $f : U \to V$ where U, V are open neighbourhoods of the origin in $\mathbb{R}^n$ and $\mathbb{R}^p$ respectively, such that

(1) $f^{-1}(0) \cap \Sigma(f) = \{0\}$,
(2) the restriction $f : \Sigma(f) \to V$ is finite,
(3) the restriction $f : U \setminus f^{-1}(0) \to V \setminus \{0\}$ is a locally stable mapping with no 0-stable singularities.

In such case we also say that $f : U \to V$ is a *good representative* of f. In the case that f is a polynomial mapping, we also add the condition that the open sets U, V are semialgebraic.

It follows from the above remarks that any FD germ has II, but the converse is not true in general.

Other important definitions related to the finite determinacy are the concepts of finite type singularity and of finite germ. These two concepts correspond to the finite determinacy when we consider the groups $\mathscr{K}$ and $\mathscr{C}$ respectively instead of the group $\mathscr{A}$ (see [40, Theorem 1.2]).

Definition 3.6 Given $f \in \mathscr{E}(n, p)$, we say that f has *finite singularity type* if

$$\dim_{\mathbb{R}} \frac{\theta(f)}{T\mathscr{K}_e f} < +\infty,$$

where $T\mathscr{K}_e f$ was defined in Definition 2.8. We say that f is *finite* if

$$\dim_{\mathbb{R}} Q(f) < +\infty.$$

Some properties can be deduced immediately from the definitions:

(1) f is FD $\Longrightarrow f$ has finite singularity type.
(2) f is finite $\Longrightarrow f$ has finite singularity type.
(3) f is finite $\Longrightarrow n \leq p$.
(4) If $n \leq p$, f has finite singularity type $\Longrightarrow f$ is finite.

Properties (1) and (2) are consequence of the fact that the $\mathscr{C}$ and the $\mathscr{A}$-equivalence imply the $\mathscr{K}$-equivalence. Property (3) follows from the fact that if $n > p$, then $f^*\mathfrak{m}_p$ is generated by less that n elements and hence, it cannot have finite codimension. Finally, property (4) is proved in [40, 2.4.(ii)] (note that the case $n < p$ is trivial).

Given a complex analytic germ $f : (\mathbb{C}^n, 0) \to (\mathbb{C}^p, 0)$, we have that f has finite singularity type if and only if $f^{-1}(0) \cap \Sigma(f) = \{0\}$ and f is finite if and only if $f^{-1}(0) = \{0\}$. Both properties are consequence of the Hilbert Nullstellensatz (in the complex analytic version [7, Theorem 3.4.4]).

In the real case we have only one of the implications: if $f \in \mathscr{E}(n, p)$ has finite singularity type then $f^{-1}(0) \cap \Sigma(f) = \{0\}$ and if f is finite then $f^{-1}(0) = \{0\}$. Another two very important properties are stated in the next theorem, the proof can be found in [12, 2.8, 3.1].

Theorem 3.7 *Let $f \in \mathscr{E}(n, p)$.*

(1) f has finite singularity type if and only if there exists a stable unfolding F of f.
(2) If F, G are r-parameter stable unfoldings of f, then F, G are $\mathscr{A}$-equivalent.

Let $f \in \mathscr{E}(n, p)$ be of finite singularity type and let F a stable unfolding of f. Given a stable type represented by the $\mathscr{A}$-class of a stable multi-germ $g : (\mathbb{R}^n, S) \to (\mathbb{R}^p, y)$, we say that F presents the stable type if for any representative $F : U \to V$ there exists $(u; y') \in V$ such that the multi-germ $f_u : (\mathbb{R}^n, S') \to (\mathbb{R}^p, y')$, with $S' = f_u^{-1}(y') \cap \Sigma(f_u)$, is $\mathscr{A}$-equivalent to g.

Definition 3.8 We say that $f \in \mathscr{E}(n, p)$ has *discrete stable type* (DST) if there exists a stable unfolding F of f which only presents a finite number of stable types.

Some cases in which $f \in \mathscr{E}(n, p)$ has DST are:

(1) when (n, p) are nice dimensions or are in the boundary of the nice dimensions in Mather's sense [21];
(2) when f has corank 1.

Definition 3.9 Let $f : U \to V$ be a good representative of a germ $f \in \mathscr{E}(n, p)$ with II and DST. We construct a stratification $(\mathcal{A}, \mathcal{B})$ of f defined as follows:

- The strata B of $\mathcal{B}$ are either $B = \{0\}$, $B = V \setminus \Delta(f)$ or B is the analytic stratum in the target of $f : (\mathbb{R}^n, S) \to (\mathbb{R}^p, y)$ for some $y \in \Delta(f)$ and $S = f^{-1}(y) \cap \Sigma(f)$.
- The strata A of $\mathcal{A}$ are either strata of the form $A = f^{-1}(B) \cap \Sigma(f)$ or strata of the form $A = f^{-1}(B) \setminus \Sigma(f)$, for some $B \in \mathcal{B}$. In particular, we always have the strata $A = \{0\}$ and $A = f^{-1}(0) \setminus \{0\}$ (if $f^{-1}(0) \neq \{0\}$).

We call $(\mathcal{A}, \mathcal{B})$ the *stratification by stable types*. The fact that f has DST guarantees that the stratification is finite. If in addition f is polynomial, then all the strata are semialgebraic sets.

4 The Cone Structure Theorem for Map Germs with Isolated Zeros

In this section, we show the cone structure theorem for FD germs $f \in \mathscr{E}(n, p)$, with $f^{-1}(0) = \{0\}$ and DST, following the arguments of Fukuda in [9]. We fix some notation:

$$D_\epsilon^p = \{y \in \mathbb{R}^p : \|y\|^2 \leq \epsilon\}, \quad S_\epsilon^{p-1} = \{y \in \mathbb{R}^p : \|y\|^2 = \epsilon\}.$$

Given a map germ $f : (\mathbb{R}^n, 0) \to (\mathbb{R}^p, 0)$ we take a representative $f : U \to V$ and put:

$$\tilde{D}_\epsilon^n = f^{-1}(D_\epsilon^p), \quad \tilde{S}_\epsilon^{n-1} = f^{-1}(S_\epsilon^{p-1}).$$

We recall that if $f \in \mathscr{E}(n,p)$ is FD, then after coordinate changes we can assume that it is polynomial and has II.

Theorem 4.1 ([9]) *Let $f : U \to V$ a good representative of a polynomial map germ $f \in \mathscr{E}(n,p)$ with II, DST and such that $f^{-1}(0) = \{0\}$. Then, there exists $\epsilon_0 > 0$ such that for any ϵ with $0 < \epsilon \leq \epsilon_0$ we have:*

(1) $\tilde{S}^{n-1}_\epsilon$ is a smooth submanifold diffeomorphic to S^{n-1},
(2) $f|_{\tilde{S}^{n-1}_\epsilon} : \tilde{S}^{n-1}_\epsilon \to S^{p-1}_\epsilon$ is a stable mapping, whose $\mathscr{A}$-class is independent of ϵ,
(3) $f|_{\tilde{D}^n_\epsilon \setminus \{0\}} : \tilde{D}^n_\epsilon \setminus \{0\} \to D^p_\epsilon \setminus \{0\}$ is $\mathscr{A}$-equivalent to the product map $\mathrm{id} \times f|_{\tilde{S}^{n-1}_\epsilon} : (0,\epsilon] \times \tilde{S}^{n-1}_\epsilon \to (0,\epsilon] \times S^{p-1}_\epsilon$,
(4) By adding the origin, $f|_{\tilde{D}^n_\epsilon} : \tilde{D}^n_\epsilon \to D^p_\epsilon$ is C^0-$\mathscr{A}$-equivalent to the cone of $f|_{\tilde{S}^{n-1}_\epsilon}$.

Proof Let $(\mathcal{A}, \mathcal{B})$ be the stratification by stable types of $f : U \to V$, which has a finite number of semialgebraic strata. We consider the polynomial function $g : U \to \mathbb{R}$ given by $g = \|f\|^2$ and its restriction $g|_{A_i} : A_i \to \mathbb{R}$ to each stratum A_i of $\mathcal{A}$. By the Curve Selection Lemma [27], each $g|_{A_i}$ has a finite number of critical values. Thus, there exists $\epsilon_0 > 0$ such that for any ϵ with $0 < \epsilon \leq \epsilon_0$, ϵ is a regular value of g and $g|_{A_i}$ for all $A_i \in \mathcal{A}$.

Since ϵ is a regular value of g, $\tilde{S}^{n-1}_\epsilon = g^{-1}(\epsilon)$ is a hypersurface in U. Moreover, the condition that ϵ is a regular value of $g|_{A_i}$, for all $A_i \in \mathcal{A}$, is equivalent to that S^{p-1}_ϵ is transverse to all the strata B_i of $\mathcal{B}$. By Proposition 2.22, the restriction $f|_{\tilde{S}^{n-1}_\epsilon} : \tilde{S}^{n-1}_\epsilon \to S^{p-1}_\epsilon$ is stable. Thus, we have showed the first part of (2).

To see (1), we use Reeb's theorem [26, p. 25]. Since $f^{-1}(0) = \{0\}$, 0 is an isolated minimum of g. Then, $\tilde{D}^n_\epsilon = g^{-1}([0,\epsilon])$, is homeomorphic to the closed disk D^n. Thus, $\tilde{S}^{n-1}_\epsilon = \partial \tilde{D}^n_\epsilon$ is homeomorphic (and hence diffeomorphic) to S^{n-1}.

It only remains to show the second part of (2) and (3), since (4) is an immediate consequence of (3). We set $I = (0,\epsilon]$ and consider the following diffeomorphisms:

$$\Phi : \tilde{D}^n_\epsilon \setminus \{0\} \longrightarrow I \times \tilde{S}^{n-1}_\epsilon, \qquad \Psi : D^p_\epsilon \setminus \{0\} \longrightarrow I \times S^{p-1}_\epsilon,$$
$$x \longmapsto (g(x), \phi(x)), \qquad y \longmapsto \left(\|y\|^2, \sqrt{\epsilon}\frac{y}{\|y\|}\right),$$

where $\phi(x)$ is the point of $\tilde{S}^{n-1}_\epsilon$ where the integral curve of the gradient of g passing through x meets $\tilde{S}^{n-1}_\epsilon$. We define $F : I \times \tilde{S}^{n-1}_\epsilon \to I \times S^{p-1}_\epsilon$ as $F = \Psi \circ f \circ \Phi^{-1}$. By construction, we have that $F(\{t\} \times \tilde{S}^{n-1}_\epsilon) \subset \{t\} \times S^{p-1}_\epsilon$, for any $t \in I$. This implies that F can be written in the form $F(t;x) = (t; f_t(x))$, with $f_t : \tilde{S}^{n-1}_\epsilon \to S^{p-1}_\epsilon$ and $t \in I$.

It is obvious that f_t is $\mathscr{A}$-equivalent to $f|_{\tilde{S}^{n-1}_t}$ and thus, f_t is stable. In particular, the unfolding F must be trivial, that is, there exist diffeomorphisms $H : I \times \tilde{S}^{n-1}_\epsilon \to \tilde{I} \times S^{n-1}_\epsilon$ and $K : I \times S^{p-1}_\epsilon \to I \times S^{p-1}_\epsilon$ of the form $H(t;x) = (t; h_t(x))$ and $K(t;y) = (t; k_t(y))$ and such that $K \circ F \circ H^{-1} = \mathrm{id} \times f_\epsilon$. Hence, we have (3). The second part of (2) follows form the fact that $k_t \circ f_t \circ h_t^{-1} = f_\epsilon$. □

Definition 4.2 Let $f : U \to V$ a good representative of a polynomial map germ $f \in \mathscr{E}(n, p)$ with II, DST and such that $f^{-1}(0) = \{0\}$. We say that $\epsilon_0 > 0$ is a *Milnor-Fukuda radius* for f if for any ϵ with $0 < \epsilon \le \epsilon_0$, S_ϵ^{p-1} is transverse to the stratification by stable types of f.

We also say that the mapping $f|_{\tilde{S}_\epsilon^{n-1}} : \tilde{S}_\epsilon^{n-1} \to S_\epsilon^{p-1}$ is the *link* of f and denote it by $L(f)$. It follows from Theorem 4.1 that:

(1) the link is a stable mapping between spheres,
(2) the link is well defined up to $\mathscr{A}$-equivalence,
(3) the germ f is C^0-$\mathscr{A}$-equivalent to the cone of its link.

We include now a couple of important remarks with respect to Theorem 4.1 and the definition of the link.

Remark 4.3 The condition $f^{-1}(0) = \{0\}$ is always satisfied if $f \in \mathscr{E}(n, p)$ is FD and $n \le p$. In the case $n > p$, we may have the two possibilities, either $f^{-1}(0) = \{0\}$ or $f^{-1}(0) \neq \{0\}$. We will give another version of the cone structure theorem for the case $f^{-1}(0) \neq \{0\}$ in the last section.

Remark 4.4 If f is real analytic instead of polynomial, then the theorem is still valid, it is enough to use the semianalytic version of the Curve Selection Lemma. When f is only of class C^∞, if f is FD, f is $\mathscr{A}$-equivalent to a polynomial map and hence, the theorem is valid for a representative of a germ which is $\mathscr{A}$-equivalent to f. If we want to apply the theorem directly to a representative $f : U \to V$ of f, then we have to change the spheres S_ϵ^{p-1} by hypersurfaces $P_\epsilon \subset V$ diffeomorphic to the sphere S^{p-1} (since the diffeomorphisms do not preserve spheres in general).

More exactly, there exists a function called control function $\rho : V \to \mathbb{R}$ with a unique critical point of Morse type in the origin, which plays the role of the function $\|y\|^2$ in the analytic case. We consider $g = \rho \circ f : U \to \mathbb{R}$ and choose $\epsilon_0 > 0$ in such a way that for all ϵ with $0 < \epsilon \le \epsilon_0$, ϵ is a regular value of $g|_{A_i}$ for all $A_i \in \mathcal{A}$. The hypersurfaces P_ϵ are defined as $P_\epsilon = \rho^{-1}(\epsilon)$ and are diffeomorphic to S^{p-1}. Then, the inverse image $N_\epsilon = f^{-1}(P_\epsilon)$ is diffeomorphic to S^{n-1}, the restriction $f|_{N_\epsilon} : N_\epsilon \to P_\epsilon$ is stable and f is topologically equivalent to the cone of $f|_{N_\epsilon}$.

Remark 4.5 If f has no DST, then the theorem is still valid with the only difference that the link $f|_{\tilde{S}_\epsilon^{n-1}} : \tilde{S}_\epsilon^{n-1} \to S_\epsilon^{p-1}$ is C^0-stable instead of stable. The proof in this case can be adapted by using the Mather canonical stratification (see [12]) instead of the stratification by stable types. We leave the details of this construction for the reader.

Next corollary is an immediate consequence of Theorem 4.1.

Corollary 4.6 *Let $f, g \in \mathscr{E}(n, p)$ be two FD germs with $f^{-1}(0) = g^{-1}(0) = \{0\}$. If $L(f)$, $L(g)$ are C^0-$\mathscr{A}$-equivalent, then f, g are C^0-$\mathscr{A}$-equivalent.*

Example 4.7 Let $f \in \mathscr{E}(1, 1)$ be a FD germ, then the link is a mapping $\gamma : S^0 \to S^0$. Since that $S^0 = \{-1, 1\}$, we have only two non equivalent possibilities, namely, either

$\gamma = \text{id}$ or $\gamma = \text{constant}$. In fact, if f is FD, then the infinite jet $j^\infty f(0) \neq 0$. Thus, we have that $j^\infty f(0) = a_k x^k + \cdots$, with $a_k \neq 0$, and f is $\mathscr{A}$-equivalent to x^k. We have

$$\gamma = \begin{cases} \text{id}, & \text{if } k \text{ is odd,} \\ \text{constant}, & \text{if } k \text{ is even.} \end{cases}$$

Basically, this is the well known criterion by the Calculus students for the existence of local maxima, minima or inflections in one variable functions (Fig. 4).

Example 4.8 Given a FD germ $f \in \mathscr{E}(1,2)$, its link is a non constant mapping $\gamma : S^0 \to S^1$. In this case, two non constant mappings $\gamma_1, \gamma_2 : S^0 \to S^1$ are always C^0-$\mathscr{A}$-equivalent, it is enough to take any homeomorphism from S^1 to S^1 which takes two points in other two points. As a consequence, there exists a unique topological class of FD germs $f \in \mathscr{E}(1,2)$ (Fig. 5).

Example 4.9 Given a FD germ $f \in \mathscr{E}(2,1)$ such that $f^{-1}(0) = \{0\}$, the link is a constant mapping $\gamma : S^1 \to S^0$ and again we have only one topological class (Fig. 6).

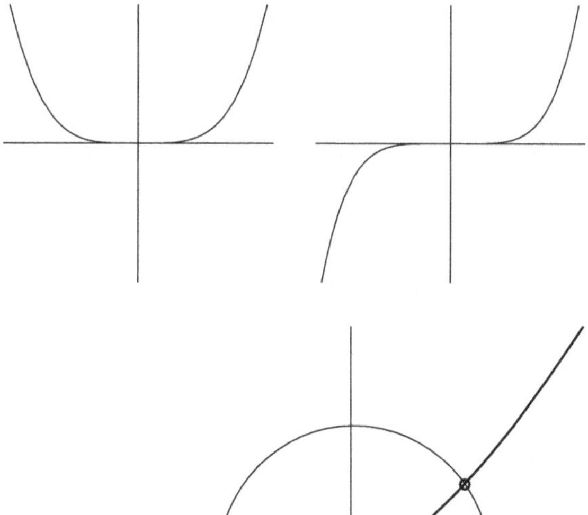

Fig. 4 Graph of $f \in \mathscr{E}(1,1)$ when k even (left) and k odd (right)

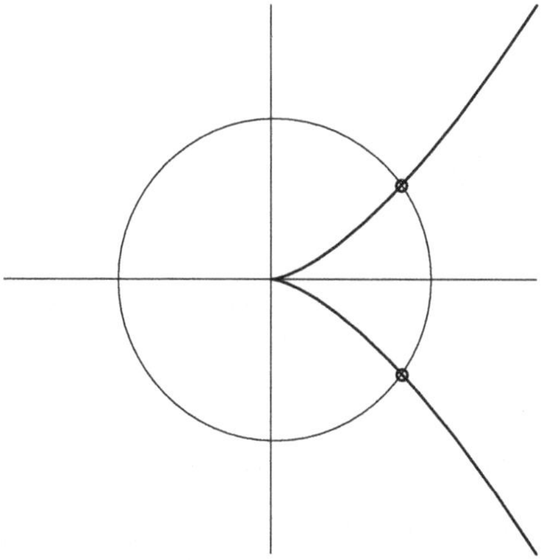

Fig. 5 The link of a FD germ $f \in \mathscr{E}(1,2)$

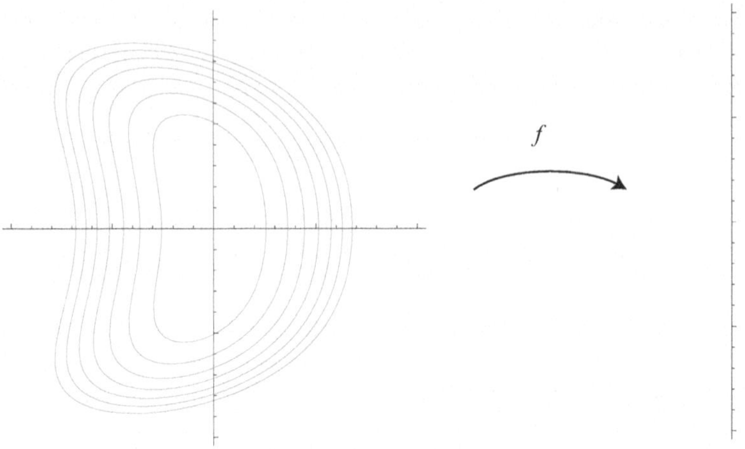

Fig. 6 A FD germ $f \in \mathscr{E}(2,1)$ with $f^{-1}(0)=\{0\}$

We conclude this section with the main open questions related to the topological classification of FD germs $f \in \mathscr{E}(n,p)$ with isolated zeros:

(1) Find a good combinatorial model which codifies all the topological information of a stable mapping $\gamma : S^{n-1} \to S^{p-1}$ (and hence, of the germ f).
(2) Determine the stable mappings $\gamma : S^{n-1} \to S^{p-1}$ which can be realized as the link of a FD germ f, with $f^{-1}(0)=\{0\}$.
(3) Determine if the converse of Corollary 4.6 is true or not, that is, if f, g are C^0-$\mathscr{A}$-equivalent, then does this imply that $L(f)$, $L(g)$ are C^0-$\mathscr{A}$-equivalent?
(4) Find relations between analytic invariants of f (corank, 0-stable invariants, $\mathscr{A}_e$-codimension, etc.) and the topological invariants of the link (number of 0-stable singularities, Vassiliev invariants, etc.).
(5) Study the topological transitions in 1-parameter families of FD germs, in particular, study the topological triviality of the family.

5 Gauss Words

Let $f \in \mathscr{E}(2,3)$ be FD germ. Then the link is a stable map $\gamma : S^1 \to S^2$, that is, γ defines a closed regular curve in S^2 with only transverse double points or crossings. We call such type of curves *doodles*. The topological classification of doodles in the sphere S^2 (or in the plane $\mathbb{R}^2$) is well known since Gauss time [11]. The combinatorial model is given by the so-called "Gauss words". Most the results of this section appear in the paper [22].

Definition 5.1 Let $\gamma : S^1 \to S^2$ be a doodle with r crossings. We choose r letters $a_1, \dots, a_r$ to label the crossings, orientations in S^1 and S^2, and a base point $z_0 \in S^1$.

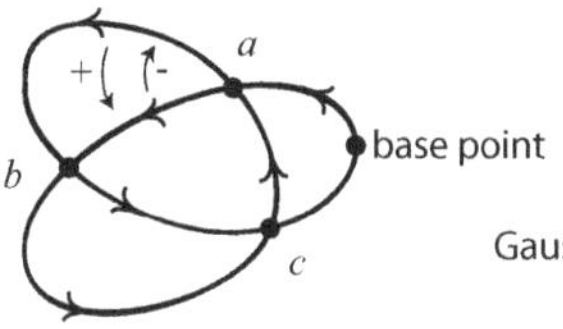

Fig. 7 Gauss word of the trefoil

We define the *Gauss word* as the sequence of crossings starting from the base point and following the orientation of the curve. Each letter a_i appears twice, one with exponent $+1$ and another one with exponent -1, according to the orientation of the two branches near the crossing in the sphere S^2 (see Fig. 7).

It is obvious that the Gauss word is not uniquely defined since it depends on the choice of the labels of the crossings, the base point and the orientations in S^1, S^2. Different choices will produce the following changes in the Gauss word:

(1) permuting the alphabet set $a_1, \dots, a_r$;
(2) cyclically permuting the sequence;
(3) reversing the sequence;
(4) changing all the exponents from $+1$ to -1 and vice versa.

We say that two Gauss words are *equivalent* if they related by means of these four operations. Up to this equivalence, the Gauss word is now well defined. Moreover, the following theorem shows that the Gauss words provide a complete invariant in the topological classification of doodles in the sphere.

Theorem 5.2 (Gauss Theorem) *Two doodles on the sphere are C^0-$\mathscr{A}$-equivalent if and only if their Gauss words are equivalent.*

Proof Let $\gamma, \delta : S^1 \to S^2$ be two doodles which are C^0-$\mathscr{A}$-equivalent. There exist homeomorphisms $\alpha : S^1 \to S^1$ and $\beta : S^2 \to S^2$ such that $\delta = \beta \circ \gamma \circ \alpha^{-1}$. We start with γ and we choose letters $a_1, \dots, a_r$ to label the crossings, a base point $z_0 \in S^1$ and orientations in S^1, S^2, so that we have the Gauss word of γ. Since β takes crossings of γ into crossings of δ, we can choose for each crossing of δ the same letter of the corresponding crossing in γ through β. We also take $\alpha(z_0) \in S^1$ as the base point of δ. Finally, we choose in S^1, S^2 the orientations induced by α, β respectively. With these choices, we have that the Gauss word of δ is equal to the Gauss word of γ.

To see the converse, we first observe that each doodle $\gamma : S^1 \to S^2$ has a natural CW-structure: in S^1 the 0-cells are the inverse images of the crossings and the 1-cells are the connected components of the complement. In S^2, the 0-cells are the crossings, the 1-cells are the edges of the curve joining the crossings and the 2-cells are the connected components of the complement of the curve (this is possible because the curve is a connected graph).

It follows that the CW-structure of S^2 can be read from the Gauss word. In fact, the 0-cells are given by the letters $a_1, \dots, a_r$, each 1-cell is an oriented edge defined by two consecutive letters $a_i^\epsilon a_j^\eta$ in the Gauss word (including the oriented edge joining the last with the first letter) and each 2-cell is a face which is determined by a closed sequence of oriented edges or their inverses.

Assume now that $\gamma, \delta : S^1 \to S^2$ have the same Gauss word. Then the two S^2 are isomorphic as CW-complexes with the CW-structure induced by γ, δ. We choose any cellular homeomorphism $\beta : S^2 \to S^2$. Then we construct another cellular homeomorphism $\alpha : S^1 \to S^1$ such that $\delta = \beta \circ \gamma \circ \alpha^{-1}$. In fact, on each 1-cell E, α is univocally defined as $\alpha|_E = (\delta^{-1} \circ \beta \circ \gamma)|_E$ and then α is extended by continuity to the 0-cells.

If $\gamma, \delta : S^1 \to S^2$ have equivalent Gauss words, then we can take homeomorphisms $\alpha : S^1 \to S^1$ and $\beta : S^2 \to S^2$ such that $\beta \circ \gamma \circ \alpha^{-1}$ and δ have the same Gauss word. Then, we apply the above argument to these two doodles. □

Example 5.3 In the trefoil (see Fig. 7), the CW-structure on the sphere is constructed from the Gauss word $a^{-1}bc^{-1}ab^{-1}c$ as follows:

(1) we have three 0-cells given by a, b and c;
(2) we have six 1-cells given by $a^{-1}b$, bc^{-1}, $c^{-1}a$, ab^{-1}, $b^{-1}c$ and ca^{-1};
(3) there are five 2-cells given by three 2-gons $\{ab^{-1}, ba^{-1}\}$, $\{bc^{-1}, cb^{-1}\}$, $\{ca^{-1}, ac^{-1}\}$ and two triangles $\{a^{-1}b, b^{-1}c, c^{-1}a\}$, $\{a^{-1}c, c^{-1}b, b^{-1}a\}$.

The theorem is not true for doodles in the plane $\mathbb{R}^2$. For instance, the two doodles in Fig. 8 are topologically equivalent on the sphere and have the same Gauss word aa^{-1}, but they are not topologically equivalent on the plane (in fact, they have different Whitney index).

We show in Fig. 9 the classification of doodles in the sphere with up to three crossings. There are 10 non equivalent doodles and their corresponding Gauss words are the following;

(a) $\emptyset$
(b) aa^{-1}
(c) $ab^{-1}ba^{-1}$
(d) $abb^{-1}a^{-1}$
(e) $ab^{-1}cc^{-1}ba^{-1}$
(f) $ab^{-1}c^{-1}cba^{-1}$
(g) $abcc^{-1}b^{-1}a^{-1}$
(h) $ab^{-1}ca^{-1}bc^{-1}$
(i) $aa^{-1}bb^{-1}cc^{-1}$
(j) $aa^{-1}b^{-1}bcc^{-1}$

Gauss was interested in the problem of "planarity" of Gauss words: determine the words which can be realized as the word of a doodle in the sphere (or in the plane). It

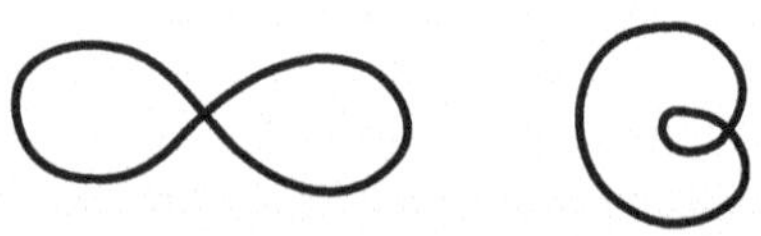

Fig. 8 Two non equivalent doodles in the plane with the same Gauss word aa^{-1}

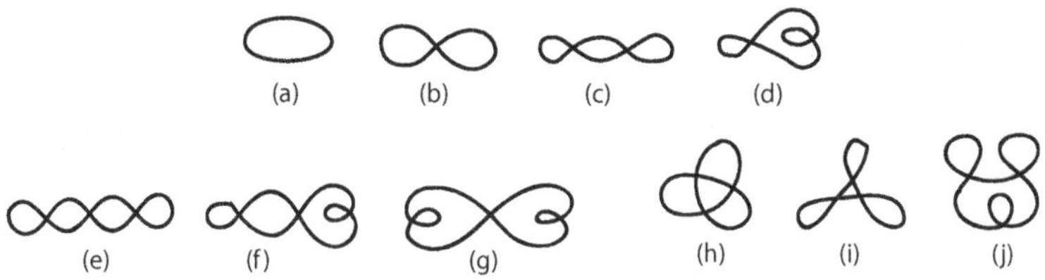

Fig. 9 Doodles with up to three crossings

Fig. 10 A doodle in the torus with Gauss word $aba^{-1}b^{-1}$

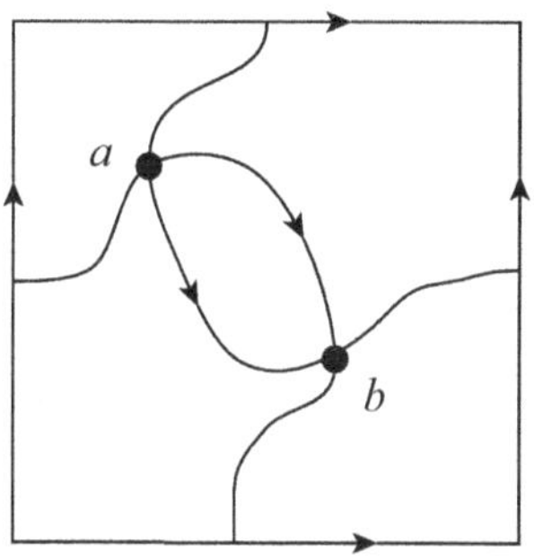

is well known that any Gauss word can be realized as the word of a doodle in some orientable compact surface of genus g. For instance, the word $aba^{-1}b^{-1}$ cannot be realized in the sphere (or the plane), but it can be realized in the torus (see Fig. 10).

Gauss could not solve the planarity problem, but he only was able to find a necessary condition. The planarity problem was completely solved by M. Dehn in 1936 [6]. The planarity problem of Gauss words is of the same nature as the planarity problem of graphs (Kuratowski Theorem). Nowadays, the Gauss words constitute a very active field of research in Computational Geometry.

Definition 5.4 Given a FD germ $f \in \mathscr{E}(2,3)$, we define the *Gauss word* of f as the Gauss word of the doodle of f.

It follows from Gauss Theorem that if two map germs have equivalent Gauss words, then they are C^0-$\mathscr{A}$-equivalent. We will see that the converse is also true. But to this we need to analyze the structure of a FD germ.

We begin with the characterization of stable singularities. We see that a C^∞ mapping $f : N^2 \to P^3$ is stable if and only if it is *semiregular* in the sense of Whitney [42]: f is an immersion with normal crossings, except at isolated points, where f presents singularities of type *cross-cap* or *Whitney umbrella*. At each of this points, the germ of f is $\mathscr{A}$-equivalent to the germ in $\mathscr{E}(2,3)$ given by $(x, y) \mapsto (x, y^2, xy)$ (see Fig. 11).

Theorem 5.5 *The only stable multi-germs from $\mathbb{R}^2$ to $\mathbb{R}^3$ are: regular simple point, transverse double point, transverse triple point and cross-cap.*

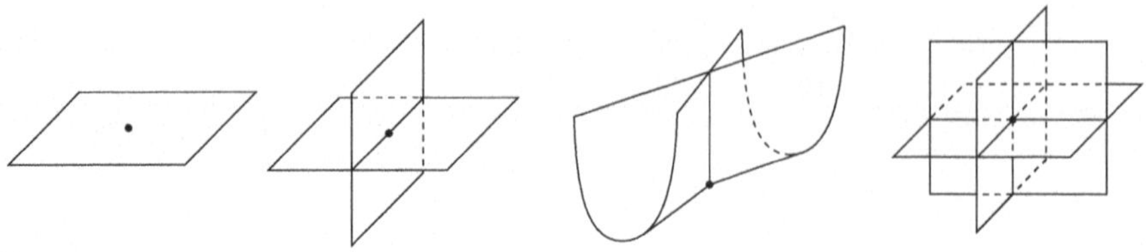

Fig. 11 Stable singularities of surfaces in $\mathbb{R}^3$

Proof We first show that a singular germ $f \in \mathscr{E}(2,3)$ is stable if and only if it has cross-cap type. After coordinate changes in the source and the target, we can assume that f is given by the standard parametrization $f(x,y) = (x, y^2, xy)$, then:

$$T\mathscr{K}_e f = \mathscr{E}_2 \left\{ \begin{pmatrix} 1 \\ 0 \\ y \end{pmatrix}, \begin{pmatrix} 0 \\ 2y \\ x \end{pmatrix} \right\} + \langle x, y^2 \rangle \mathscr{E}_2^2$$
$$= \mathscr{E}_2 \left\{ \begin{pmatrix} 1 \\ 0 \\ y \end{pmatrix}, \begin{pmatrix} 0 \\ x \\ 0 \end{pmatrix}, \begin{pmatrix} 0 \\ y \\ 0 \end{pmatrix}, \begin{pmatrix} 0 \\ 0 \\ x \end{pmatrix}, \begin{pmatrix} 0 \\ 0 \\ y^2 \end{pmatrix} \right\}.$$

We have that $\theta(f)/T\mathscr{K}_e f$ is generated over $\mathbb{R}$ by the classes of the canonical basis $\{e_1, e_2, e_3\}$, hence $\overline{\omega} f$ is surjective and f is stable by Lemma 2.9.

To see the converse, suppose first that f is stable and has rank 0. Then $T\mathscr{K}_e f \subset \mathfrak{m}_2\theta(f)$. Since $\theta(f)/\mathfrak{m}_2\theta(f)$ has dimension 3, we must have necessarily that $T\mathscr{K}_e f = \mathfrak{m}_2\theta(f)$. Moreover, $(f^*\mathfrak{m}_3) \subset \mathfrak{m}_2^2\theta(f)$, hence the classes of $\partial f/\partial x$ and $\partial f/\partial y$ should generate $\mathfrak{m}_2\theta(f)/\mathfrak{m}_2^2\theta(f)$ over $\mathbb{R}$. But this is not possible, since this space has dimension 6.

Thus, if f is stable, it must have rank 1 and after a coordinate change in the source, we can assume that $f(x,y) = (x, g(x,y))$, for some germ $g \in \mathscr{E}(2,2)$. In other words, we see f as an unfolding of $g_0(y) = g(0,y)$. In particular, we have:

$$\frac{\theta(f)}{T\mathscr{K}_e(f)} \cong \frac{\theta(g_0)}{T\mathscr{K}_e(g_0)} \cong \frac{\mathscr{E}_1^2}{\langle g_0' \rangle}.$$

If $g_0 \in \mathfrak{m}_1^3\mathscr{E}_1^2$, then $g_0' \in \mathfrak{m}_1^2\mathscr{E}_1^2$ and thus $\dim_{\mathbb{R}}(\mathscr{E}_1^2/\langle g_0' \rangle) \geq 4$, which is not possible by the surjectivity of $\overline{\omega} f$. Hence, g_0 must have order 2. But this implies that $Q(f) \cong \mathscr{E}_1/\langle y^2 \rangle$, hence f is $\mathscr{A}$-equivalent to the cross-cap by Theorem 2.11.

We consider now multi-germs $f : (\mathbb{R}^2, S) \to (\mathbb{R}^3, y)$, with $S \subset \mathbb{R}^2$ a finite set. If one of the points $x_i \in S$ is singular, then f has cross-cap type at x_i and the analytic stratum is only the point $\{x_i\}$. Thus, the regular intersection condition of Theorem 2.15 implies that $S = \{x_i\}$. Otherwise, if all the points of S are regular, then f is an immersion with normal crossings and we find the remaining types: regular simple point, transverse double point and transverse triple point. □

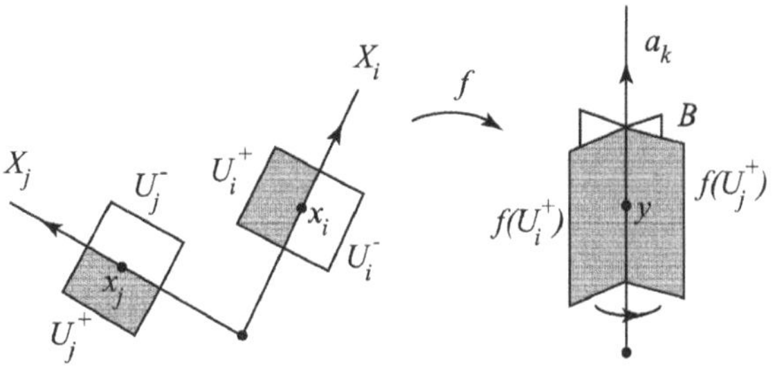

Fig. 12 Orientation of the branches

Assume now that $f \in \mathscr{E}(2,3)$ is FD. The 0-stable types are the cross-caps and the triple points. Thus, a good representative of f is a mapping $f : U \to V$ where $U \subset \mathbb{R}^2$ and $V \subset \mathbb{R}^3$ are open neighbourhoods of the origin such that:

(1) $f^{-1}(0) = \{0\}$,
(2) $f : U \to V$ is proper,
(3) $f : U \setminus \{0\} \to V \setminus \{0\}$ is an immersion with only transverse double points.

An important set associated with f is the *double point curve*, which is defined as

$$D(f) = \{z \in U : f^{-1}(f(z)) \neq \{z\}\} \cup S(f),$$

where $S(f)$ is the singular set. Then $D(f)$ is a closed subset of U. Since f is a good representative, it follows that $S(f) = \{0\}$ and that $D(f) \setminus \{0\}$ is a 1-dimensional submanifold of U. By shrinking the neighbourhoods if necessary, we can assume that all the connected components of $D(f) \setminus \{0\}$ are arcs going from the origin to the boundary of U.

Moreover, f restricted to each connected component is a diffeomorphism, so that the image $f(D(f)) \setminus \{0\}$ is also a 1-dimensional submanifold of V, whose connected components are arcs going from the origin to the boundary of V (since f is proper). Moreover, the restriction $f : D(f) \setminus \{0\} \to f(D(f)) \setminus \{0\}$ is a 2-fold covering. The connected components of $D(f)$ (resp. $f(D(f))$) are called *half-branches* of $D(f)$ (resp. $f(D(f))$).

We claim that we can recover the Gauss word of f just by looking at the relative position of the half-branches of $D(f)$ and $f(D(f))$ and the orientation of the leaves of $f(U)$ at each half-branch. In fact, each half-branch of $f(D(f))$ corresponds to a crossing in the doodle of f. So, we can choose letters $a_1, \dots, a_r$ to label the half-branches. We also choose orientations in U, V and a base point in U. Then, we construct the Gauss word as the sequence of letters according to the relative position of the half-branches of $D(f)$ in U, starting from the base point and following the orientation in U. Moreover, we put the exponent $+1$ if the two leaves of $f(U)$ intersect positively along the half-branch or -1 otherwise. It is obvious that the word obtained with this method is exactly the Gauss word of f (see Fig. 12).

Assume now that we have two FD $f, g \in \mathscr{E}(2,3)$ which are C^0-$\mathscr{A}$-equivalent. Then, the homeomorphisms must preserve the double point sets $D(f)$ and $f(D(f))$.

An argument analogous to that of the proof of Theorem 5.2 gives that f, g have the same Gauss word (up to equivalence). Thus, we have proved the following theorem (see [22, Corollaries 3.4 and 3.8]).

Theorem 5.6 *Let $f, g \in \mathscr{E}(2,3)$ be two FD germs. The following statements are equivalent:*

(1) f, g are C^0-$\mathscr{A}$-equivalent,
(2) the doodles of f, g are C^0-$\mathscr{A}$-equivalent,
(3) f, g have equivalent Gauss words.

Example 5.7 All the doodles with up to three crossings (see Fig. 9) are realizable as the link of a FD map germ $f \in \mathscr{E}(2,3)$:

$$\begin{array}{ll} (a)\ (x, y, 0), & (b)\ (x, y^2, xy), \\ (c)\ (x, y^2, y(x^2-y^2)), & (d)\ (x, xy+y^3, xy^3+\frac{3}{2}y^5), \\ (e)\ (x, y^2, xy(x^2-y^2)), & (f)\ (x, x^4-6x^2y^2+y^4, x^3y-xy^3), \\ (g)\ (x, x^4-6x^2y^2+y^4, x^3y-xy^3), & (h)\ (x, xy+y^3, xy^2+\frac{3}{4}y^4), \\ (i)\ (x, xy+y^3, xy^2+\frac{5}{4}y^4), & (j)\ (x^2, xy+y^3, \frac{1}{2}x^3+\frac{1}{4}x^2y+3xy^3+3y^5). \end{array}$$

To check this, we use a tailor-made computer program `SphereXSurface` by A. Montesinos-Amilibia [29], which pictures the doodle of any map. We remark that all of them except "Mickey" (j) admit a corank 1 realization. We do not know, up to now, if it is also possible to find a corank 1 realization for this doodle.

To finish this section, we see the topological classification of all FD germs $f \in \mathscr{E}(2,3)$ with Boardman type $\Sigma^{1,0}$. We recall the definition of Boardman symbol of order 2.

Definition 5.8 Given $f \in \mathscr{E}(n,p)$, let $M_1, \ldots, M_r$ be the minors of order $n-i+1$ of the Jacobian matrix of f and set $\tilde{f} = (f_1, \ldots, f_p, M_1, \ldots, M_r)$. We say that it has *Boardman type* $\Sigma^{i,j}$ if

$$\dim_{\mathbb{R}} \ker df(0) = i, \quad \dim_{\mathbb{R}} \ker d\tilde{f}(0) = j.$$

The following lemma is due to Mond [28] and gives a prenormal form for all germs with Boardman type $\Sigma^{1,0}$.

Lemma 5.9 *Let $f \in \mathscr{E}(2,3)$ be a germ with Boardman type $\Sigma^{1,0}$. Then f is $\mathscr{A}$-equivalent to a map germ of the form*

$$\tilde{f}(x, y) = (x, y^2, yp(x, y^2)),$$

for some $p \in \mathscr{E}_2$.

Proof The condition that $\dim_{\mathbb{R}} \ker df(0) = 1$ implies that f has corank 1, then after $\mathscr{A}$-equivalence, f can be written in the form

$$f(x, y) = (x, g(x, y), h(x, y)),$$

for some $g, h \in \mathfrak{m}_2^2$. Then the 2-minors of the Jacobian matrix are $g_y, h_y, g_x h_y - g_y h_x$, where the subscripts mean the partial derivatives. Then, an easy computation shows that f has Boardman type $\Sigma^{1,0}$ if and only if either $g_{yy}(0) \neq 0$ or $h_{yy}(0) \neq 0$.

Assume, for instance, that $g_{yy}(0) \neq 0$. Then, we can write

$$f(x, y) = (x, ax^2 + 2bxy + cy^2 + \tilde{g}(x, y), h(x, y)),$$

where $\tilde{g} \in \mathfrak{m}_2^3$ and $c \neq 0$. If $c > 0$, we put

$$ax^2 + 2bxy + cy^2 = (\frac{b}{\sqrt{c}}x + \sqrt{c}y)^2 + (a - \frac{b^2}{c})x^2,$$

then the coordinate change in the source given by $\bar{y} = (b/\sqrt{c})x + \sqrt{c}y$, followed by the coordinate change in the target given by $\bar{Y} = Y - (a - b^2/c)X^2$ transform f into:

$$(x, y) \mapsto (x, y^2 + G(x, y), H(x, y)),$$

for some $G \in \mathfrak{m}_2^3$ and $H \in \mathfrak{m}_2^2$.

Now we use the fact that the fold $(x, y) \to (x, y^2)$ is 2-determined. This implies that there are coordinate changes in the source and the target which transform the above map germ into:

$$(x, y) \mapsto (x, y^2, K(x, y)),$$

for some $K \in \mathfrak{m}_2^2$. Finally, by the Malgrange Preparation Theorem, we split K as

$$K(x, y) = K_1(x, y^2) + yK_2(x, y^2).$$

We take the coordinate change in the target given by $\bar{Z} = Z - K_1(X, Y)$, which now transforms the map germ into

$$(x, y) \mapsto (x, y^2, yK_2(x, y^2)). \qquad \square$$

Theorem 5.10 ([22]) *Any FD germ $f \in \mathscr{E}(2, 3)$ with Boardman type $\Sigma^{1,0}$ has a doodle of type "warm" (see Fig. 13). In particular, two FD germs with Boardman type $\Sigma^{1,0}$ are C^0-$\mathscr{A}$-equivalent if and only if their double point curves have the same number of half-branches.*

Proof We can assume $f(x, y) = (x, y^2, yp(x, y^2))$. We consider $f : U \to V$ a good representative and $\epsilon > 0$ a Milnor-Fukuda radius. The doodle is given by $f|_{\tilde{S}^1_\epsilon} : \tilde{S}^1_\epsilon \to S^2_\epsilon$. We have $D(f) = \{(x, y) : \ p(x, y^2) = 0\}$ and

$$\tilde{S}^1_\epsilon = \{(x, y) : \ x^2 + y^4 + y^2 p(x, y^2)^2 = \epsilon^2\},$$

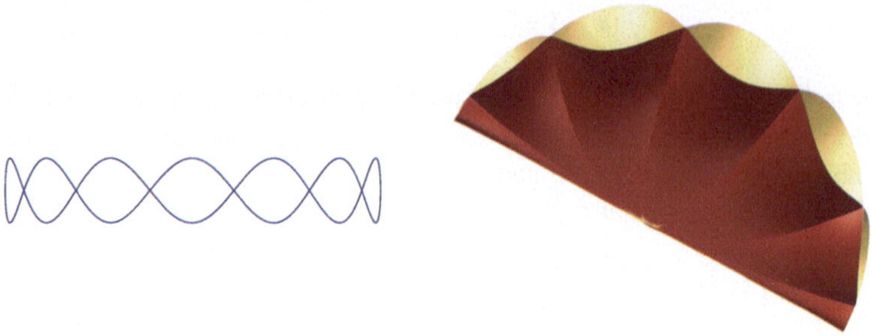

Fig. 13 Singularity of type $\Sigma^{1,0}$ with six crossings

and both sets are symmetric with respect to the x-axis.

We choose $z_0 = (\epsilon, 0)$ as the base point of $\tilde{S}^1_\epsilon$. The crossings of the doodle are determined by $D(f) \cap \tilde{S}^1_\epsilon$, which gives: $z_1, \dots, z_r$ and $\overline{z}_1, \dots, \overline{z}_r$, with

$$z_i = (x_i, y_i), \quad \overline{z}_i = (x_i, -y_i), \quad -\epsilon < x_r \leq \cdots \leq x_1 < \epsilon, \quad y_i \geq 0.$$

This implies that the Gauss word of the doodle (up to the signs) is equal to:

$$a_1 a_2 \cdots a_r a_r \cdots a_2 a_1,$$

where $a_i = f(z_i) = f(\overline{z}_i)$ (see Fig. 14). The doodle has the following properties:

- The doodle is contained in the hemisphere $Y \geq 0$ of S^2_ϵ and intersects the equator $Y = 0$ at the base point $f(z_0)$ and its opposite $f(-z_0)$.
- The doodle is symmetric with respect to the meridian $Z = 0$.
- The doodle intersects the meridian $Z = 0$ only at the double points $a_1, \dots, a_r$, together with $f(z_0)$ and $f(-z_0)$. Moreover, they present the following relative position on the meridian:

$$f(-z_0) < a_r < \cdots < a_1 < f(z_0).$$

The only possible doodles which satisfy these properties are those of type "warm", with Gauss word:

$$a_1 a_2^{-1} \cdots a_r^{\pm 1} a_r^{\mp 1} \cdots a_2 a_1^{-1}. \qquad \square$$

We remark that any doodle of type "warm" with r is crossings is realizable as the link of a FD $f \in \mathscr{E}(2, 3)$. In fact, we consider:

$$f(x, y) = (x, y^2, \Im((x + iy)^{r+1})),$$

where $\Im(z)$ is imaginary part of $z \in \mathbb{C}$. Then, we have

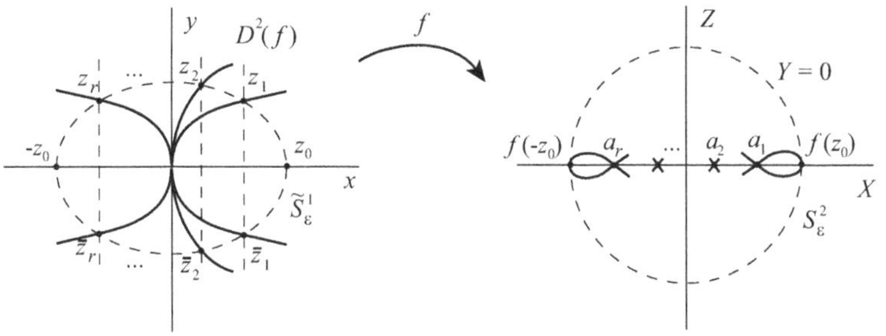

Fig. 14 Configuration of the crossings

$$p(x, y^2) = \Im((x+iy)^{r+1})/y = \prod_{k=1}^{r}\left(-\sin\left(\frac{k\pi}{r+1}\right)x + \cos\left(\frac{k\pi}{r+1}\right)y\right),$$

hence $D(f) = \{(x, y) : \ p(x, y^2) = 0\}$ has exactly $2r$ half-branches.

6 Reeb Graphs

In this section we consider the topological classification of FD germs $f \in \mathscr{E}(3,2)$ with isolated zeros, that is, $f^{-1}(0) = \{0\}$. By Theorem 4.1, the link is a stable mapping $\gamma : S^2 \to S^1$, that is, it has only Morse singularities with distinct critical values. The combinatorial model to describe this type of mappings is given by the Reeb graph. The Reeb graph was introduced by Reeb in [36] and it is well known that it is a complete topological invariant for Morse functions from S^2 to $\mathbb{R}$ (see [1, 37]). In this section we extend the concept of Reeb graph for stable maps from S^2 to S^1. All the results of this section appear in the paper [4].

The following result is probably well known for fibre bundles (that is, locally trivial fibrations), but we include here a elementary proof for the sake of completeness.

Lemma 6.1 *Let $p : E \to B$ be a fibre bundle with fibre F, where B, E, F are all finite CW-complexes. Then,*

$$\chi(E) = \chi(B)\chi(F).$$

Proof After subdivision, we can choose a finite covering $\{U_i\}_{i=1}^k$ of B which trivializes the fibre bundle and such that each U_i is a subcomplex of B. For each i, there exists a homeomorphism $\varphi_i : p^{-1}(U_i) \to U_i \times F$ such that $\pi_1 \circ \varphi_i = p$, where π_1 is the projection onto the first factor. In particular, we have that $p^{-1}(A)$ is homeomorphic to $A \times F$, for any subset $A \subseteq U_i$ and $i = 1, \dots, k$.

Let $B_i = \cup_{j=1}^i U_j$, then we see that $\chi(p^{-1}(B_i)) = \chi(B_i)\chi(F)$ by induction on i. In fact, this is true for $i = 1$ and if we assume it for i, then

$$\begin{aligned}\chi(p^{-1}(B_{i+1})) &= \chi(p^{-1}(B_i)) + \chi(p^{-1}(U_{i+1})) - \chi(p^{-1}(B_i \cap U_{i+1}))\\ &= \chi(B_i)\chi(F) + \chi(U_{i+1})\chi(F) - \chi(B_i \cap U_{i+1})\chi(F)\\ &= (\chi(B_i) + \chi(U_{i+1}) - \chi(B_i \cap U_{i+1}))\chi(F)\\ &= \chi(B_{i+1})\chi(F).\end{aligned}$$

□

Proposition 6.2 *Let $\gamma : S^2 \to S^1$ be a stable map. Then γ is not a regular map.*

Proof Suppose γ is a regular map, then $\gamma(S^2) \subset S^1$ would be an open set. Since $\gamma(S^2)$ is also closed, we get $\gamma(S^2) = S^1$ and hence, γ is surjective. By Ehresmann's fibration theorem [8, p. 31], f is a smooth fibre bundle. In particular, if F is the fiber, we have by Lemma 6.1 that

$$2 = \chi(S^2) = \chi(S^1)\chi(F) = 0,$$

which is an absurd. □

Given a continuous map $f : X \to Y$ between topological spaces, we consider the following equivalence relation on X: $x \sim y$ if $f(x) = f(y)$ and x and y are in the same connected component of $f^{-1}(f(x))$.

Proposition 6.3 *Let $\gamma : S^2 \to S^1$ be a stable map. Then the quotient space $S^2/\sim$ admits the structure of a connected graph in the following way:*

(1) the vertices are the connected components of level curves $\gamma^{-1}(v)$, where $v \in S^1$ is a critical value;
(2) each edge is formed by points that correspond to connected components of level curves $\gamma^{-1}(v)$, where $v \in S^1$ is a regular value.

Proof Since γ is stable we have a finite number of critical values $v_1, \dots, v_r$ and for each $i = 1, \dots, r$, $\gamma^{-1}(v_i)$ has a finite number of connected components. Then,

$$\gamma|S^2 - \gamma^{-1}(\{v_1, \dots, v_r\}) : S^2 - \gamma^{-1}(\{v_1, \dots, v_r\}) \to S^1 - \{v_1, \dots, v_r\}$$

is regular, and the induced map

$$\tilde{\gamma} : (S^2 - \gamma^{-1}(\{v_1, \dots, v_r\}))/\sim \to S^1 - \{v_1, \dots, v_r\}$$

is a local homeomorphism. Each connected component of $S^1 - \{v_1, \dots, v_r\}$ is homeomorphic to an open interval, so each connected component of $(S^2 - \gamma^{-1}(\{v_1, \dots, v_r\}))/\sim$ is also homeomorphic to an open interval. □

Each vertex of the graph can be of three types, depending on if the connected component has a maximum/minimum critical point, a saddle point or just regular points. Then, the possible incidence rules of edges and vertices are given in Fig. 15.

Let $v_1, \dots, v_r \in S^1$ be the critical values of γ. We choose a base point $v_0 \in S^1$ and an orientation. We can reorder the critical values such that $v_0 \le v_1 < \dots < v_r$

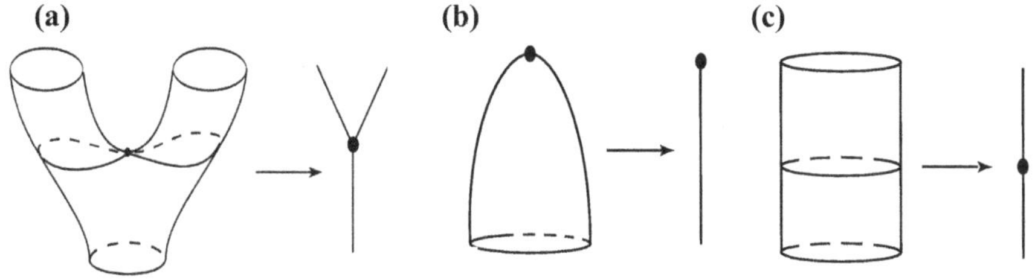

Fig. 15 Incidence rules for the three types of vertices

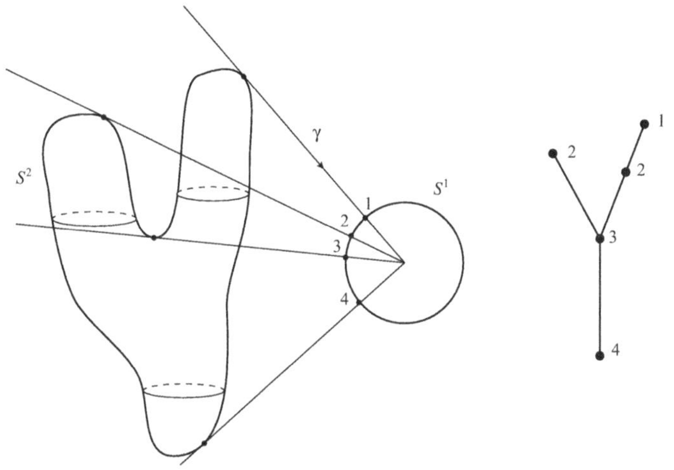

Fig. 16 Example of Reeb graph of a stable map $\gamma : S^2 \to S^1$

and we label each vertex with the index $i \in \{1, \ldots, r\}$, if it corresponds to the critical value v_i.

Definition 6.4 The graph given by $S^2/\!\sim$ together with the labels of the vertices, as previously defined, is said to be the *generalized Reeb graph* associated to $\gamma : S^2 \to S^1$ (see Fig. 16).

For simplicity, from now on we will just call Reeb graph to the generalized Reeb graph, unless otherwise specified.

Proposition 6.5 *Let $\gamma : S^2 \to S^1$ be a stable map. Then the Reeb graph of γ is a tree.*

Proof Let Γ be the Reeb graph of γ. Since Γ is connected, in order to show that Γ is a tree, we only need to prove that its Euler characteristic is $\chi(\Gamma) = 1$. We have that $\chi(\Gamma) = V - E$, where V, E are the number of vertices and edges of Γ, respectively.

On one hand, $V = M + S + I$ where M, S, I are the numbers of vertices of each type: maximum/minimum, saddle or regular, respectively. Note that $V \neq 0$ by Proposition 6.2.

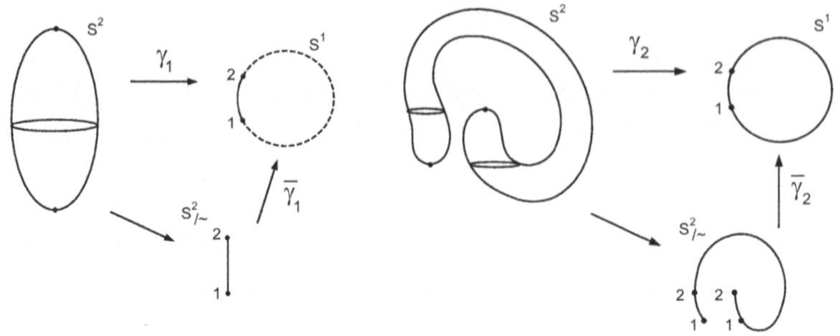

Fig. 17 Two non-equivalent stable maps with the same classical Reeb graph

On the other hand, by Euler's formula $E = \frac{1}{2}\sum \deg(v_i)$ where v_i are the vertices of Γ and $\deg(v_i)$ is the degree of v_i, that is, the number of edges adjacent to v_i. Since γ is stable, the degree of each vertex of maximum/minimum type is 1, while of regular type is 2 and of saddle type is 3 (see Fig. 15). Hence,

$$\chi(\Gamma) = V - E = M + S + I - \frac{1}{2}(M + 2I + 3S) = \frac{M - S}{2} = 1,$$

where the last equality follows from the Morse formula: $M - S = \chi(S^2) = 2$. □

Remark 6.6 The classical Reeb graph is defined in the same way, but the vertices are just the connected components of level curves $\gamma^{-1}(v)$ which contain a critical point. Hence, our generalized Reeb graph contains some extra vertices corresponding to the regular connected components of $\gamma^{-1}(v)$, where v is a critical value. Of course the classical Reeb graph can be obtained from the generalized one just by eliminating the extra vertices and joining the two adjacent edges. But in general, the generalized Reeb graph provides more information.

We present in Fig. 17 two examples of stable maps $\gamma_1, \gamma_2 : S^2 \to S^1$ with their respective generalized Reeb graphs. Both examples share the same classical Reeb graph, but the generalized Reeb graphs are different. The example on the left hand side is a non-surjective map, whilst the map on the right hand side is surjective, therefore the maps are not topologically equivalent. This shows that the classical Reeb graph is not sufficient to distinguish between these two examples.

Notice that if $\gamma : S^2 \to S^1$ is not surjective, then γ may be regarded as a Morse function from S^2 to $\mathbb{R}$ (via stereographic projection). In this case, the generalized Reeb graph can be deduced from the classical one just by adding the extra vertices each time that one passes through a critical value.

It is obvious that labeling of vertices of the Reeb graph is not uniquely determined, since it depends on the chosen orientations and the base points on each S^1. Different choices will produce either a cyclic permutation or a reversion of the labeling in the Reeb graph. This leads us to the following definition of equivalent Reeb graphs.

Let $\gamma, \delta : S^2 \to S^1$ be two stable maps. Let Γ_γ and Γ_δ be their respective Reeb graphs. Consider the induced quotient maps $\bar{\gamma} : \Gamma_\gamma \to S^1_\gamma$ and $\bar{\delta} : \Gamma_\delta \to S^1_\delta$, where S^1_γ, S^1_δ is S^1 with the graph structure whose vertices are the critical values of γ, δ respectively (as illustrated in Fig. 17).

Definition 6.7 We say that Γ_γ *is equivalent to* Γ_δ and we denote it by $\Gamma_\gamma \sim \Gamma_\delta$, if there exist graph isomorphisms $j : \Gamma_\gamma \to \Gamma_\delta$ and $l : S^1_\gamma \to S^1_\delta$, such that the following diagram is commutative:

$$\begin{array}{ccc} V_\gamma & \xrightarrow{\bar{\gamma}|V_\gamma} & \Delta_\gamma \\ {\scriptstyle j|_{V_\gamma}} \downarrow & & \downarrow {\scriptstyle l|_{\Delta_\gamma}} \\ V_\delta & \xrightarrow{\bar{\delta}|V_\delta} & \Delta_\delta \end{array}$$

where $V_\gamma = \{\text{vertices of } \Gamma_\gamma\}$, $V_\delta = \{\text{vertices of } \Gamma_\delta\}$ and Δ_γ and Δ_δ are their respective discriminant sets.

Theorem 6.8 *Let $\gamma, \delta : S^2 \to S^1$ be two stable maps. If γ and δ are C^0-$\mathscr{A}$- equivalent then their respective Reeb graphs are equivalent.*

Proof Since γ and δ are topologically equivalent there exist homeomorphisms $h : S^2 \to S^2$ and $k : S^1 \to S^1$ such that $k \circ \gamma \circ h = \delta$. Then h maps critical points into critical points and k maps critical values into critical values. Hence h induces a graph isomorphism from Γ_γ to Γ_δ and k induces a graph isomorphism from S^1_γ to S^1_δ which gives the equivalence between the Reeb graphs. □

The above theorem allows us to extend the definition of Reeb graph for C^0-stable maps between topological spheres.

Definition 6.9 Let $\gamma : M \to P$ be a continuous map, where M is homeomorphic to S^2 and P is homeomorphic to S^1. We say that γ is C^0*-stable* if there exist a C^∞-stable map $\delta : S^2 \to S^1$ and homeomorphisms $k : M \to S^2$, $h : P \to S^1$ such that the following diagram is commutative

$$\begin{array}{ccc} M & \xrightarrow{\gamma} & P \\ {\scriptstyle k} \downarrow & & \downarrow {\scriptstyle h} \\ S^2 & \xrightarrow{\delta} & S^1 \end{array}$$

We say that $y \in P$ is a *critical value* of γ if $h(y)$ is a critical value of δ. Moreover, $M/\sim$ has a graph structure induced by the Reeb graph of δ. We call this graph the *Reeb graph* of γ and denote it by Γ_γ. The notion of equivalence of graphs given in Definition 6.7 can be also extended for C^0-stable maps in the obvious way. By Theorem 6.8, the Reeb graph Γ_γ is well defined up to equivalence of graphs.

The main result is the following theorem which says that the Reeb graph is a complete invariant for $\mathscr{A}$-equivalence of stable maps from S^2 to S^1. The idea of the proof is that we can "inflate" the Reeb graph and then recover the surface together with the stable map. Near each vertex, we have a Morse singularity and the local normal form is given in Fig. 14. Along the edges, the map is regular, so we have pieces of "tubes" which connect the singularities. The detailed proof, although intuitive, is rather technical and in fact is an adaptation of the proof of [15, Theorem 4.1]. All the details can be found in [4, Theorem 3.8].

Theorem 6.10 *Let $\gamma, \delta : S^2 \to S^1$ be two stable maps such that $\Gamma_\gamma \sim \Gamma_\delta$. Then γ is $\mathscr{A}$-equivalent to δ.*

As we said before, the two Theorems 6.8 and 6.10 together give that the Reeb graph is a complete topological invariant for stable maps from S^2 to S^1. In fact, we have a little bit more, as we can see in the following corollary.

Corollary 6.11 *Let $\gamma, \delta : S^2 \to S^1$ be two stable maps. Then the following statements are equivalent:*

(1) γ, δ are $\mathscr{A}$-equivalent,
(2) γ, δ are C^0-$\mathscr{A}$-equivalent,
(3) $\Gamma_\gamma \sim \Gamma_\delta$.

In the last part of this section, we consider the Reeb graph of the link of a finitely determined map germ with isolated zeros.

Definition 6.12 Given a FD germ $f \in \mathscr{E}(3, 2)$ with $f^{-1}(0) = \{0\}$, we define the *Reeb graph* of f as the Reeb graph of the link of f.

It follows from Theorem 6.10 and Corollary 4.6 that if two FD germs have equivalent Reeb graphs, then they are C^0-$\mathscr{A}$-equivalent. Again in this case we can show the converse. But we need to see how is the structure of a FD germ in this case. The first step is to describe the stable singularities. The characterization of stable singularities of maps from $\mathbb{R}^3$ to $\mathbb{R}^2$ is well known (cf. [13]) and it is given by:

Theorem 6.13 *Let $f : (\mathbb{R}^3, S) \to (\mathbb{R}^2, 0)$ be a C^∞ multi-germ germ such that f is singular at each point of S. Then, f is stable if only if $|S| \leq 2$ and f is $\mathscr{A}$-equivalent to one of the following normal forms:*

(1) For $|S| = 1$:

- $(x, y^2 + z^2)$, *called definite fold D;*
- $(x, y^2 - z^2)$, *called indefinite fold I;*
- $(x, y^3 + xy + z^2)$, *called cusp.*

(2) For $|S| = 2$:

- $(x_1, y_1^2 + z_1^2), (y_2^2 + z_2^2, x_2)$, *called double-fold D&D;*
- $(x_1, y_1^2 + z_1^2), (y_2^2 - z_2^2, x_2)$, *called double-fold D&I;*
- $(x_1, y_1^2 - z_1^2), (y_2^2 - z_2^2, x_2)$, *called double-fold I&I.*

Proof We follow the same arguments as in Example 2.12 and Theorem 5.5. We first consider the mono-germ case $|S| = 1$. If f is a fold (either definite or indefinite), then

$$T\mathscr{K}_e f = \mathscr{E}_3\left\{\begin{pmatrix}1\\0\end{pmatrix}, \begin{pmatrix}0\\2y\end{pmatrix}, \begin{pmatrix}0\\\pm 2z\end{pmatrix}\right\} + \langle x, y^2 \pm z^2\rangle \mathscr{E}_3^2$$
$$= \mathscr{E}_3\left\{\begin{pmatrix}1\\0\end{pmatrix}, \begin{pmatrix}0\\x\end{pmatrix}, \begin{pmatrix}0\\y\end{pmatrix}, \begin{pmatrix}0\\z\end{pmatrix}\right\}.$$

Thus $\theta(f)/T\mathscr{K}_e f$ is generated over $\mathbb{R}$ by the class of $(0, 1)$ and the map $\overline{\omega} f$ is obviously surjective, so f is stable (see Lemma 2.9). In the case of the cusp, we have:

$$T\mathscr{K}_e f = \mathscr{E}_3\left\{\begin{pmatrix}1\\y\end{pmatrix}, \begin{pmatrix}0\\3y^2+x\end{pmatrix}, \begin{pmatrix}0\\\pm 2z\end{pmatrix}\right\} + \langle x, y^3 + xy + z^2\rangle \mathscr{E}_3^2$$
$$= \mathscr{E}_3\left\{\begin{pmatrix}1\\y\end{pmatrix}, \begin{pmatrix}0\\x\end{pmatrix}, \begin{pmatrix}0\\y^2\end{pmatrix}, \begin{pmatrix}0\\z\end{pmatrix}\right\}.$$

Now, $\theta(f)/T\mathscr{K}_e f$ is generated over $\mathbb{R}$ by the classes of $\{(1, 0), (0, 1)\}$. Again $\overline{\omega} f$ is surjective and hence, f is stable.

Assume now that $f \in \mathscr{E}(3, 2)$ is stable. If f has rank 0, then $T\mathscr{K}_e f \subset \mathfrak{m}_3\theta(f)$. Since $\theta(f)/\mathfrak{m}_2\theta(f)$ has dimension 2, we must have necessarily that $T\mathscr{K}_e f = \mathfrak{m}_2\theta(f)$. Moreover, $(f^*\mathfrak{m}_2) \subset \mathfrak{m}_3^2\theta(f)$, hence the classes of $\partial f/\partial x$, $\partial f/\partial y$ and $\partial f/\partial z$ should generate $\mathfrak{m}_3\theta(f)/\mathfrak{m}_3^2\theta(f)$ over $\mathbb{R}$. But this is not possible, since this space has dimension 6.

Thus, if f is stable, it must have rank 1 and after a coordinate change in the source, we can assume that $f(x, y, z) = (x, g(x, y, z))$, for some function $g \in \mathfrak{m}_3^2$. In other words, we see f as an unfolding of $g_0(y, z) = g(0, y, z)$. In particular, we have:

$$\frac{\theta(f)}{T\mathscr{K}_e(f)} \cong \frac{\theta(g_0)}{T\mathscr{K}_e(g_0)} \cong \frac{\mathscr{E}_2}{\left\langle \frac{\partial g_0}{\partial y}, \frac{\partial g_0}{\partial z}, g_0 \right\rangle}.$$

Let $I = \left\langle \frac{\partial g_0}{\partial y}, \frac{\partial g_0}{\partial z}, g_0 \right\rangle$. If $g_0 \in \mathfrak{m}_2^3$, then $I \subset \mathfrak{m}_2^2$ and thus $\dim_{\mathbb{R}}(\mathscr{E}_2/I) \geq 3$, which is not possible by the surjectivity of $\overline{\omega} f$. Hence, the Hessian matrix of g_0 at the origin must have rank ≥ 1. By the splitting lemma, g_0 is $\mathscr{A}$-equivalent to $y^{k+1} \pm z^2$, for some $k \geq 1$. This implies $\dim_{\mathbb{R}}(\mathscr{E}_2/I) = k$, hence we must have necessarily $k \leq 2$. If $k = 1$, then f is a fold (either definite or indefinite) and if $k = 2$, then f is a cusp, by Theorem 2.11.

We consider now multi-germs $f : (\mathbb{R}^3, S) \to (\mathbb{R}^2, y)$, with $S \subset \mathbb{R}^3$ a finite set. If one of the points $x_i \in S$ is a cusp, then the analytic stratum is only the point $\{x_i\}$. Thus, the regular intersection condition of Theorem 2.15 implies that $S = \{x_i\}$. Otherwise, if all the points of S are folds, the analytic stratum at each point is a line. The regular intersection condition now implies that $|S| \leq 2$ and that the two lines are transverse in the plane in the case $|S| = 2$. This implies that f is a double-fold. □

Note that the 0-stable types are the cusps and the double-folds. Hence if $f \in \mathscr{E}(3,2)$ is FD, then there exists a good representative $f : U \to V$ such that

(1) $S(f) \cap f^{-1}(0) = \{0\}$,
(2) the restriction $f : U \setminus f^{-1}(0) \to V \setminus \{0\}$ has only definite and indefinite simple fold singularities.

We have that $S(f)$ and the discriminant $\Delta(f) = f(S(f))$ are curves which are regular outside the origin. After shrinking U, V if necessary, we can assume that $S(f)$, $\Delta(f)$ are made of a finite number of arcs joining the origin with the boundary of U, V, called *half-branches*. Moreover, the restriction $f : S(f) \setminus \{0\} \to \Delta(f) \setminus \{0\}$ is a diffeomorphism. Each half-branch of $\Delta(f)$ corresponds to a critical value of the link of f, which is of type max/min if we are in a half-branch of type definite fold and of type saddle if we are in a half-branch of type indefinite fold. Another important set is

$$X(f) = \overline{f^{-1}(\Delta(f)) \setminus S(f)}.$$

The set $X(f)$ is a regular surface outside the origin and will also assume that the connected components of $X(f) \setminus \{0\}$ are cylinders going from the origin to the boundary of U. Each half-branch of $S(f)$ corresponds to a vertex of the Reeb graph of type max/min if we are in a half-branch of type definite fold and of type saddle if we are in a half-branch of type indefinite fold. Each connected component of $X(f) \setminus \{0\}$ corresponds to a regular vertex of the Reeb graph.

Theorem 6.14 *Let $f, g \in \mathscr{E}(3,2)$ be FD germs such that $f^{-1}(0) = \{0\} = g^{-1}(0)$. If f and g are C^0-$\mathscr{A}$-equivalent then their Reeb graphs are equivalent.*

Proof By hypothesis, there exist two homeomorphisms germs h, k such that the following diagram commutes:

$$(1)\qquad \begin{array}{ccc} (\mathbb{R}^3,0) & \xrightarrow{\ f\ } & (\mathbb{R}^2,0) \\ {\scriptstyle h}\downarrow & & \downarrow{\scriptstyle k} \\ (\mathbb{R}^3,0) & \xrightarrow{\ g\ } & (\mathbb{R}^2,0) \end{array}$$

We take representatives of f, g, h and k and for any small enough $\epsilon > 0$, the next diagram is also commutative:

$$(2)\qquad \begin{array}{ccc} \tilde{S}^2_\epsilon & \xrightarrow{\ \gamma_f\ } & S^1_\epsilon \\ {\scriptstyle h}\downarrow & & \downarrow{\scriptstyle k} \\ M_\epsilon & \xrightarrow{\ g|_{M_\epsilon}\ } & P_\epsilon \end{array}$$

where $M_\epsilon = h(\tilde{S}^2_\epsilon)$ and $P_\epsilon = k(S^1_\epsilon)$.

From the commutativity of diagram (2) follows that $g|M_\epsilon$ is C^0-stable. Choose $\epsilon_0, \epsilon_1 > 0$ such that $\gamma_f : \tilde{S}^2_{\epsilon_0} \to S^1_{\epsilon_0}$ and $\gamma_g : \tilde{S}^2_{\epsilon_1} \to S^1_{\epsilon_1}$ are the links of f and g, respectively, and $S^1_{\epsilon_1} \subset k(D^2_{\epsilon_0})$. By Definition 6.9, let $\Gamma_{g|M_{\epsilon_0}}$ be the Reeb graph associated to $g|M_{\epsilon_0}$. Then, we can conclude that $\Gamma_{g|M_{\epsilon_0}}$ is equivalent to Γ_{γ_f}, where Γ_{γ_f} is the Reeb graph of γ_f.

Consider $A_1, \dots, A_n$ the half branches of the discriminant $\Delta(g)$ ordered in the anti-clockwise orientation. By the cone structure of f (see Theorem 4.1), each half branch A_i intersects P_{ϵ_0} in a unique point v_i so that $v_1, \dots, v_n$ are the critical points of $g|M_{\epsilon_0}$. Analogously, each A_i intersects $S^1_{\epsilon_1}$ in a unique point w_i, where now $w_1, \dots, w_n$ are the critical points of γ_g. We have a graph isomorphism $l : P_{\epsilon_0} \to S^1_{\epsilon_1}$ given by $l(v_i) = w_i, \forall i = 1, \dots, n$.

Let $C_1, \dots, C_r$ be the connected components of

$$g^{-1}(\Delta(g)) \setminus \{0\} = \cup_{i=1}^n g^{-1}(A_i).$$

Again by the cone structure of f, each connected component C_j intersects M_{ϵ_0} in a unique connected component V_j of some $g^{-1}(v_i)$, so that $V_1, \dots, V_r$ are the vertices of $\Gamma_{g|M_{\epsilon_0}}$. Finally, each C_j intersects $\tilde{S}^2_{\epsilon_1}$ in a unique connected component W_j of $g^{-1}(w_i)$, in such a way that $W_1, \dots, W_r$ are now the vertices of Γ_{γ_g}. We have a bijection φ defined by $\varphi(V_j) = W_j, \forall j = 1, \dots, r$. In order to have a graph isomorphism between $\Gamma_{g|M_{\epsilon_0}}$ and Γ_{γ_g} we need to show that φ is edge preserving.

Consider $U = k(D^2_{\epsilon_0}) \setminus (\Delta(g) \cup B^2_{\epsilon_1})$, and let Y_i be one of its connected components limited by two consecutive half branches A_i and A_{i+1}. We denote by α_i and β_i the arcs of $S^1_{\epsilon_1}$ and P_{ϵ_0} respectively, which bound $Y_i, \forall i = 1, \dots, n$ (see Fig. 18). The connected components of $g^{-1}(\alpha_i)$ and $g^{-1}(\beta_i)$ give all the edges of the graphs Γ_{γ_g} and $\Gamma_{g|M_{\epsilon_0}}$, respectively.

Take X any connected component of $f^{-1}(Y_i)$, for some $1 \le i \le n$. Since $g|X : X \to Y_i$ is regular, the induced map $\tilde{g} : X/\sim \to Y_i$ is a local homeomorphism and hence, a covering space. But Y_i is simply connected, so $\tilde{g}$ is in fact a homeomorphism. We deduce that the boundary of $X/\sim$ has two components: one is an edge of Γ_{γ_g} given by the quotient of $X \cap g^{-1}(\alpha_i)$ and the other is an edge of $\Gamma_{g|M_{\epsilon_0}}$ given by the quotient of $X \cap g^{-1}(\beta_i)$.

Notice that all the edges of Γ_{γ_g} and $\Gamma_{g|M_{\epsilon_0}}$ can be obtained in this way, hence we have a bijection between the edges of Γ_{γ_g} and $\Gamma_{g|M_{\epsilon_0}}$ which is compatible with the above bijection φ defined between the vertices. □

Again, Theorem 6.14 together with Corollary 4.6 and Theorem 6.10 show that the Reeb graph is a complete topological invariant for map germs from with isolated zeros.

Corollary 6.15 *Let $f, g \in \mathscr{E}(3,2)$ be FD germs such that $f^{-1}(0) = \{0\} = g^{-1}(0)$. Then the following statements are equivalent:*

(1) *f, g are C^0-$\mathscr{A}$-equivalent,*

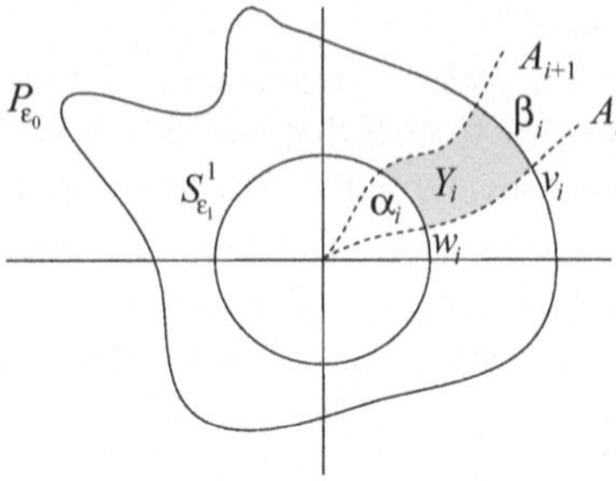

Fig. 18 The graph isomorphism ϕ

(2) *the Reeb graphs of f, g are equivalent,*
(3) *the links of f, g are C^0-$\mathscr{A}$-equivalent.*

As we did in Sect. 5, in the last part of this section, we will describe the topology of FD germs $f \in \mathscr{E}(3,2)$ with Boardman type $\Sigma^{2,1}$. These germs constitute the simplest non trivial class of singular germs. The Boardman type Σ^2 means that f has corank 1 and the next result gives a restriction on the link for this class of germs.

Lemma 6.16 *Let $f \in \mathscr{E}(3,2)$ be a corank 1 FD germ given by $f(x,y,z) = (x, h_x(y,z))$. Then $h_0 : (\mathbb{R}^2, 0) \to (\mathbb{R}, 0)$ is FD.*

Proof Since f is FD, we can assume it is polynomial. Then its complexification $f_{\mathbb{C}}$ is also FD and by the Mather-Gaffney criterion $S(f_{\mathbb{C}}) \cap f_{\mathbb{C}}^{-1}(0) = \{0\}$ (see Theorem 3.4). This implies that $S((h_0)_{\mathbb{C}}) \cap (h_0)_{\mathbb{C}}^{-1}(0) = \{0\}$ and hence h_0 is FD for the contact group $\mathscr{K}$. But for function germs, it is well-known that the FD with respect the contact group $\mathscr{K}$ is equivalent to the FD with respect to the group $\mathscr{A}$ (see again [40, Proposition 2.3]). □

Theorem 6.17 *Let $f \in \mathscr{E}(3,2)$ be a corank 1 FD germ with $f^{-1}(0) = \{0\}$. Then the link of f is not surjective.*

Proof Consider f written by $f(x,y,z) = (x, h_x(y,z))$, where h_0 is also FD and $h_0^{-1}(0) = \{0\}$. By Theorem 4.1, $h_0^{-1}(S^0_\epsilon)$ is diffeomorphic to S^1, for small enough $\epsilon > 0$.

Suppose that associated link of f is surjective. Then $(0,\epsilon)$ and $(0,-\epsilon)$ belong to image of the map $\gamma_f : f^{-1}(S^1_\epsilon) \to S^1_\epsilon$. But

$$\gamma_f^{-1}(\{(0,\epsilon),(0,-\epsilon)\}) = f^{-1}(\{(0,\epsilon),(0,-\epsilon)\}) \simeq h_0^{-1}(\{\epsilon,-\epsilon\}) \simeq S^1,$$

where $\simeq$ indicates homeomorphism of sets. This gives a contradiction because S^1 is connected, $\{(0,\epsilon),(0,-\epsilon)\}$ is not connected and γ_f is a continuous map. □

Remark 6.18 (1) It follows from Theorem 6.17 that the stable map $\gamma : S^2 \to S^1$ presented in the right hand side of Fig. 17 cannot be realized as the link of a corank 1 FD map germ $f \in \mathscr{E}(3,2)$. Up to this moment, we do not know if in fact, this stable map can be realized or not as the link of a corank 2 map germ.

(2) Another consequence of Theorem 6.17 is that if f has corank 1 and $f^{-1}(0) = \{0\}$, then the generalized Reeb graph can obtained from the classical one, since the link is not surjective (see Remark 6.6). From now on in this section, the Reeb graph will be referred to the classical version, unless otherwise specified.

Any corank 1 germ $f \in \mathscr{E}(3, 2)$ may have Boardman type $\Sigma^{2,0}$ or $\Sigma^{2,1}$, $\Sigma^{2,2}$. It is easy to see that if f has type $\Sigma^{2,0}$, then it is $\mathscr{A}$-equivalent to the definite or indefinite fold $(x, y, z) \mapsto (x, y^2 \pm z^2)$, so we do not need to consider this case. From now on, we restrict ourselves to germs of type $\Sigma^{2,1}$.

Lemma 6.19 *Any FD germ $f \in \mathscr{E}(3, 2)$ of Boardman type $\Sigma^{2,1}$ with $f^{-1}(0) = \{0\}$ can be written, up to $\mathscr{A}$-equivalence, as*

$$f(x, y, z) = (x, y^k + a_{k-2}(x)y^{k-2} + \cdots + a_1(x)y + z^2), \tag{3}$$

for some $k \geq 4$ even and functions $a_1, \ldots, a_{k-2} \in \mathscr{E}_1$.

Proof Consider f written by $f(x, y, z) = (x, h_x(y, z))$, where h_0 is also FD and $h_0^{-1}(0) = \{0\}$. The fact f has type $\Sigma^{2,1}$ implies that the Hessian of h_0 has rank 1, hence up to $\mathscr{A}$-equivalence, h_0 is given by $h_0(y, z) = y^k + z^2$, for some $k \geq 4$ even. The mini-versal deformation of h_0 is

$$H(a_1, \ldots, a_{k-2}, y, z) = y^k + a_{k-2}y^{k-2} + \cdots + a_1 y + z^2.$$

Then, there exist functions $a_1, \ldots, a_{k-2} \in \mathscr{E}_1$ such that

$$f(x, y, z) = (x, H(a_1(x), \ldots, a_{k-2}(x), y, z)).$$

□

Definition 6.20 We say that a FD germ $f \in \mathscr{E}(3, 2)$ of Boardman type $\Sigma^{2,1}$ with $f^{-1}(0) = \{0\}$ has *multiplicity* k, if it can be written, up to $\mathscr{A}$-equivalence as in (3).

Let $f \in \mathscr{E}(3, 2)$ be FD germ of Boardman type $\Sigma^{2,1}$ with $f^{-1}(0) = \{0\}$ and multiplicity k given as in (3). We write, for simplicity,

$$h_x(y) = y^k + a_{k-2}(x)y^{k-2} + \cdots + a_1(x)y.$$

We fix a good representative $f : U \to V$ and take $\epsilon > 0$ such that $(-\epsilon, \epsilon) \times (-\epsilon, \epsilon) \times (-\epsilon, \epsilon) \subset U$. The singular points of f are points $(x, y, 0)$ such that $h_x'(y) = 0$. The fact that f has fold type outside the origin implies if $x \neq 0$, then $h_x''(y) \neq 0$ at the singular points. Moreover, f has a definite fold if $h_x''(y) > 0$ and an indefinite fold if $h_x''(y) < 0$. Moreover, all the critical values of have to be distinct.

We deduce that x with $0 < |x| < \epsilon$, the function $h_x : (-\epsilon, \epsilon) \to \mathbb{R}$ is a Morse with distinct critical values. In particular, all the functions h_x with $0 < x < \epsilon$ are $\mathscr{A}$-equivalent and all the functions h_x with $-\epsilon < x < 0$ are also $\mathscr{A}$-equivalent. In both we have a Morsification of x^k and the relative position of the critical values in both functions determine the Reeb graph of f.

Since k is even, h_x will have an odd number of critical points $y_1, \dots, y_r$ with $r \leq k-1$. The points $y_1, y_3, \dots, y_r$ are the local minima and the points $y_2, y_4, \dots, y_{r-1}$ are the local maxima of h_x. If the critical values are $v_1 < \cdots < v_r$, then can associate with h_x a permutation $\sigma \in \Sigma_r$ such that $h_x(y_i) = v_{\sigma(i)}$. We denote by σ^+, σ^- the two permutations of h_x for $x > 0$ and $x < 0$ respectively. Then, the pair (σ^+, σ^-) determines the Reeb graph of f.

Example 6.21 Let $f \in \mathscr{E}(3, 2)$ be FD germ of Boardman type $\Sigma^{2,1}$ with $f^{-1}(0)=\{0\}$ and multiplicity 4. After change of coordinates in the source and target, we can assume f is given by

$$f(x, y, z) = (x, y^4 + a(x)y^2 + b(x)y + z^2).$$

Notice that the bifurcation set $\mathcal{B}$ of the versal unfolding of h_0 in this case is given in the (a, b)-plane by by $b(-4a^3 - 27b^2) = 0$ (see Fig. 19), which permits us to choose appropriate functions $a(x)$ and $b(x)$ such that we can obtain all types of possible configurations.

Then, there are three possibilities for the Reeb graph of the link of f, according to the number of saddles:

- 0 saddle, in this case $(\sigma^+, \sigma^-) = ((1), (1))$, then f is topologically equivalent to $(x, y^4 + x^2y + z^2)$ (see Fig. 20);
- 1 saddle, this corresponds to $(\sigma^+, \sigma^-) = ((1), (1, 3, 2))$, then f is topologically equivalent to $(x, y^4 + xy^2 + 3x^5y + z^2)$ (see Fig. 21);
- 2 saddles, this happens if $(\sigma^+, \sigma^-) = ((1, 3, 2), (1, 3, 2))$ and f is topologically equivalent to $(x, y^4 - x^2y^2 + x^5y + z^2)$. (see Fig. 22).

We remark that the configuration $((1, 3, 2), (2, 3, 1))$ is topologically equivalent to $((1, 3, 2), (2, 3, 1))$ since the corresponding Reeb graphs are equivalent.

7 The Cone Structure Theorem for Map Germs with Non Isolated Zeros

The case of a FD germ $f \in \mathscr{E}(n, p)$ with $f^{-1}(0) \neq \{0\}$ is much more complicated than the case with $f^{-1}(0) = \{0\}$. Fukuda gave in [10] an analogous theorem to Theorem 4.1, which in our notation can be stated as follows(see [10, Theorem 1']).

Theorem 7.1 *Let $f : U \to V$ a good representative of a polynomial map germ $f : (\mathbb{R}^n, 0) \to (\mathbb{R}^p, 0)$ with II, DST and such that $f^{-1}(0) \neq \{0\}$. Then, there exist $\epsilon_0 > 0$ and a strictly increasing smooth function $\delta : [0, \epsilon_0] \to [0, +\infty)$ with $\delta(0) = 0$ such that for any ϵ, δ with $0 < \epsilon \leq \epsilon_0$ and $0 < \delta \leq \delta(\epsilon)$, the following properties hold:*

(1) $f^{-1}(0) \cap S^{n-1}_\epsilon$ is a smooth submanifold of dimension $n - p - 1$, whose diffeomorphic type is independent of ϵ.

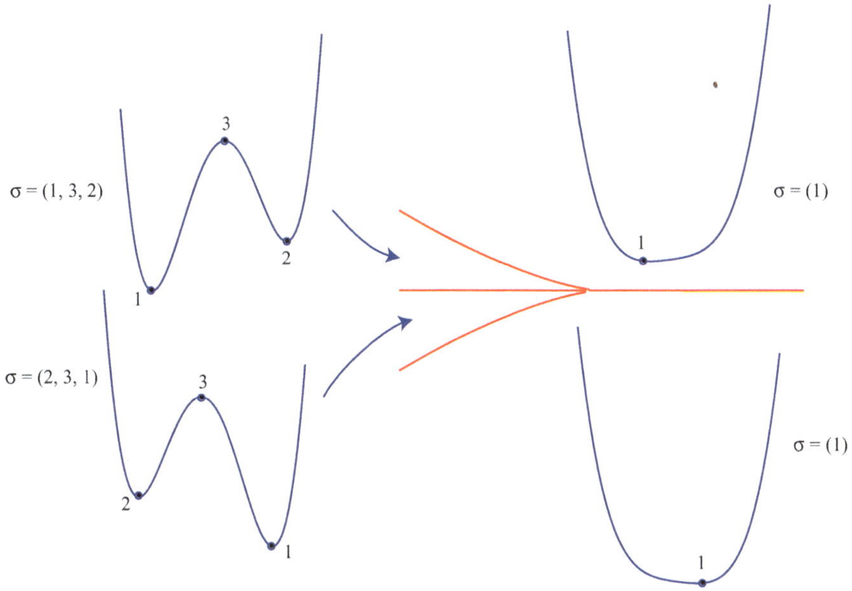

Fig. 19 Morsifications of y^4

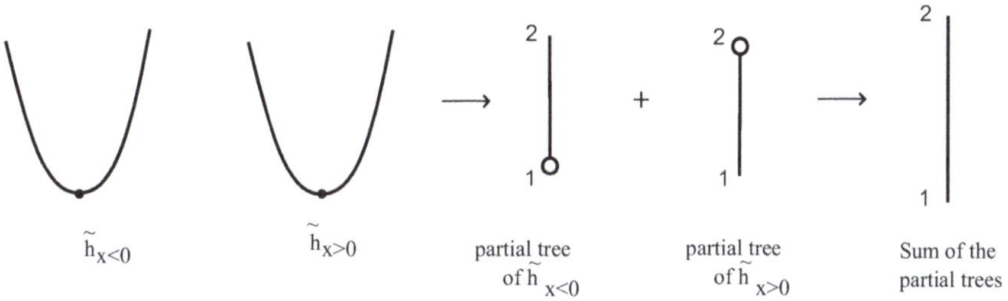

Fig. 20 Reeb graph with no saddles

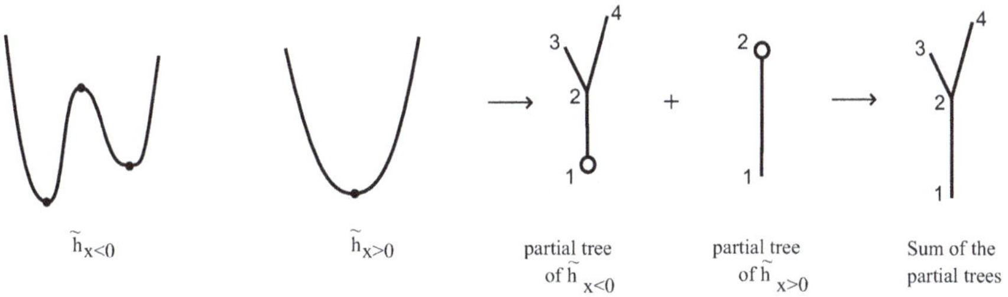

Fig. 21 Reeb graph with one saddle

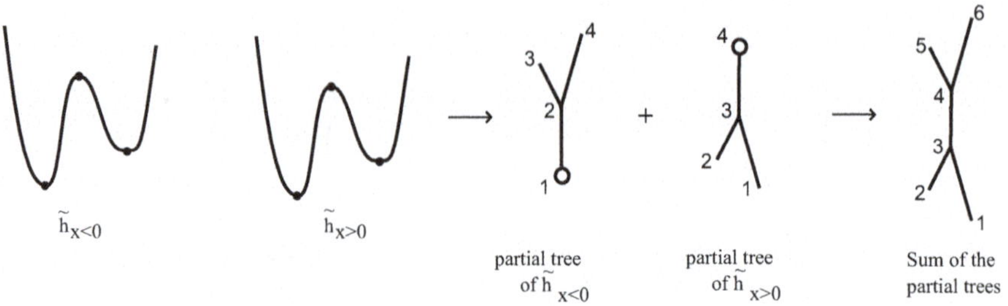

Fig. 22 Reeb graph with two saddles

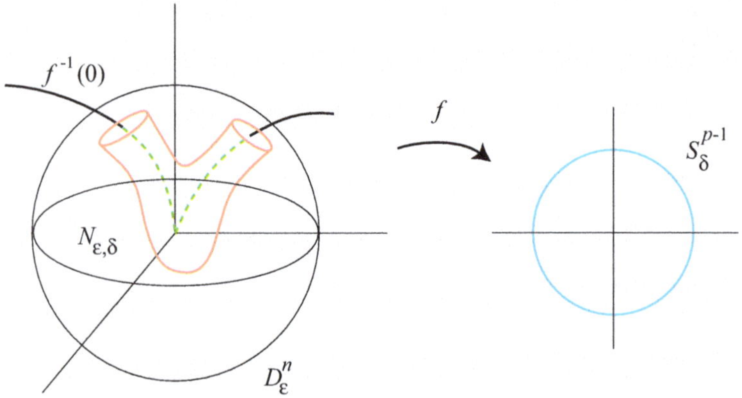

Fig. 23 The map $f|N_{\epsilon,\delta}$

(2) *$N_{\epsilon,\delta} := D^n_\epsilon \cap f^{-1}(S^{p-1}_\delta)$ is a smooth submanifold with boundary of dimension $n-1$, whose diffeomorphic type is independent of ϵ, δ.*

(3) *The restriction $f|_{N_{\epsilon,\delta}} : N_{\epsilon,\delta} \to S^{p-1}_\delta$ is a stable mapping, whose $\mathscr{A}$-class is independent of ϵ, δ.*

The proof of this theorem can be done by using similar arguments to those of the proof of Theorem 4.1 for the case $f^{-1}(0) = \{0\}$. Of course, we can define the link of f as being the stable mapping $f|_{N_{\epsilon,\delta}} : N_{\epsilon,\delta} \to S^{p-1}_\delta$ (Fig. 23). The main problem now is that f is not C^0-$\mathscr{A}$-equivalent to the cone of $f|_{N_{\epsilon,\delta}}$ in the usual sense. In fact, since $N_{\epsilon,\delta}$ is not a sphere, its cone is not a disk. So, we need to introduce a generalized version of the cone in order to solve this. The following construction is given in [5]. We recall that if X, Y are topological spaces and $f : A \to Y$ is a continuous map on $A \subset X$, then the attachment is defined as

$$X \cup_f Y = \frac{X \sqcup Y}{x \sim f(x) : \forall x \in A},$$

where $\sqcup$ means disjoint union and $\sim$ indicates that all points of A are identified with its images.

Definition 7.2 A *link diagram* is a diagram of the form

$$V \xleftarrow{\ r\ } N \xrightarrow{\ \gamma\ } S^{p-1},$$

where N is a manifold with boundary, γ is a continuous map, V is a contractible space and r is a continuous surjective map such that the attachment $(N \times I) \cup_r V$ is homeomorphic to the closed disk D^n (here we identify $N \equiv N \times \{0\} \subset N \times I$).

Definition 7.3 Given a link diagram $V \xleftarrow{\ r\ } N \xrightarrow{\ \gamma\ } S^{p-1}$, the *generalized cone of a link diagram* is the induced map

$$C(\gamma, r) : (N \times I) \cup_r V \to c(S^{p-1}),$$

defined in the obvious way (that is, $[x, t] \mapsto [\gamma(x), t]$ if $(x, t) \in N \times I$ and $[y] \mapsto [0]$ if $y \in V$).

Notice that here we are using the small letter c to the usual notion of cone and the capital letter C to indicate the generalized cone. Also note that in applying the notion of generalized cone of a link diagram for the case $V = \{0\}$, we obtain essentially the usual notion of the cone.

Definition 7.4 We say that two link diagrams

$$V_0 \xleftarrow{\ r_0\ } N_0 \xrightarrow{\ \gamma_0\ } S^{p-1}, \quad V_1 \xleftarrow{\ r_1\ } N_1 \xrightarrow{\ \gamma_1\ } S^{p-1}$$

are *$\mathscr{A}$-equivalent* (resp. *C^0-$\mathscr{A}$-equivalent*) if there are diffeomorphisms (resp. homeomorphisms) $\alpha : V_0 \to V_1$, $\phi : N_0 \to N_1$ and $\psi : S^{p-1} \to S^{p-1}$ such that $r_1 = \alpha \circ r_0 \circ \phi^{-1}$ and $\gamma_1 = \psi \circ \gamma_0 \circ \phi^{-1}$.

The following lemma follows easily from the definitions.

Lemma 7.5 *If two link diagrams are C^0-$\mathscr{A}$-equivalent, then their generalized cones are C^0-$\mathscr{A}$-equivalent.*

We present now the structure cone theorem for map germs with non isolated zeros. Let $f \in \mathscr{E}(n, p)$, in order to simplify the notation, we put $f_{\epsilon,\delta} := f|_{N_{\epsilon,\delta}} : N_{\epsilon,\delta} \to S_\delta^{p-1}$ and $V_\epsilon = f^{-1}(0) \cap D_\epsilon^n$.

Theorem 7.6 ([3]) *Let $f : U \to V$ a good representative of a polynomial map germ $f \in \mathscr{E}(n, p)$ with II, DST and such that $f^{-1}(0) \neq \{0\}$. For each ϵ, δ with $0 < \delta \ll \epsilon \ll 1$, there exists a continuous and surjective mapping $r_{\epsilon,\delta} : N_{\epsilon,\delta} \to V_\epsilon$, such that:*

(1) The link diagram

$$V_\epsilon \xleftarrow{\ r_{\epsilon,\delta}\ } N_{\epsilon,\delta} \xrightarrow{\ f_{\epsilon,\delta}\ } S_\delta^{p-1}$$

is independent of ϵ, δ up to C^0-$\mathscr{A}$-equivalence.

(2) *The restriction* $f|_{D^n_\epsilon\cap f^{-1}(D^p_\delta)}: D^n_\epsilon \cap f^{-1}(D^p_\delta) \to D^p_\delta$ *is* C^0-$\mathscr{A}$*-equivalent to the generalized cone:*

$$C(f_{\epsilon,\delta}, r_{\epsilon,\delta}) : (N_{\epsilon,\delta} \times I) \cup_{r_{\epsilon,\delta}} V_\epsilon \to c(S^{p-1}_\delta),$$

where $I = [0, \delta]$.

Here we give a sketch of the proof of Theorem 7.6, full details of the proof can be found in [3, Theorem 4.4]. Let $(\mathcal{A}, \mathcal{B})$ be the stratification by stable types of f, which is a Thom stratification of f. We choose $\epsilon_0 > 0$ and $0 < \delta_0 \ll \epsilon_0 \ll 1$ small enough and denote by $B^n_{\epsilon_0}$, $B^p_{\delta_0}$ the interiors of $D^n_{\epsilon_0}$, $D^p_{\delta_0}$ respectively. Then $D^n_{\epsilon_0} \cap f^{-1}(B^p_{\delta_0})$ is a manifold with boundary. We consider the mappings

$$D^n_{\epsilon_0} \cap f^{-1}(B^p_{\delta_0}) \xrightarrow{\ f\ } B^p_{\delta_0} \xrightarrow{\ \rho\ } [0, \delta_0),$$

where $\rho(y) = \|y\|^2$. Both are proper and we have that the restriction of $(\mathcal{A}, \mathcal{B})$ is a Thom stratification of f and $(\mathcal{B}, \mathcal{C})$ is a Thom stratification of ρ, where $\mathcal{C} = \{(0, \delta_0), \{0\}\}$. We take stratified vector fields X, Y, T on $D^n_{\epsilon_0} \cap f^{-1}(B^p_{\delta_0})$, $B^p_{\delta_0}$ and $[0, \delta_0)$ respectively, as follows: $T = \frac{d}{dt}$ in $(0, \delta_0)$ and $T_0 = 0$; Y is a lifting of T through ρ and X is a lifting of Y through f. The existence of X, Y is given by [12, Theorem 3.2]. Moreover, since T is globally integrable, then Y, X are also globally integrable, by [12, Lemma 4.8].

Let $0 < \delta_1 < \delta_0$. We define the mapping

$$r : D^n_{\epsilon_0} \cap f^{-1}(D^p_{\delta_1}) \to V_{\epsilon_0},$$

such that $r(x)$ is the point of V_{ϵ_0} where the integral curve of X passing through x meets V_{ϵ_0}. We consider the link diagram:

$$V_{\epsilon_0} \xleftarrow{\ r\ } N_{\epsilon_0,\delta_1} \xrightarrow{\ f\ } S^{p-1}_{\delta_1}$$

We define

$$\Phi : D^n_{\epsilon_0} \cap f^{-1}(D^p_{\delta_1}) \to (N_{\epsilon_0,\delta_1} \times [0, \delta_1]) \cup_r V_{\epsilon_0},$$

as follows:

$$\Phi(x) = \begin{cases} [\phi(x), \|f(x)\|^2], & \text{if } x \notin V_{\epsilon_0}, \\ [r(x), 0], & \text{if } x \in V_{\epsilon_0}, \end{cases}$$

being $\phi(x)$ the point of N_{ϵ_0,δ_1} where the integral curve of X passing through x meets N_{ϵ_0,δ_1}. Analogously, we also define $\Psi : D^p_{\delta_1} \to c(S^{p-1}_{\delta_1})$, as

$$\Psi(y) = \begin{cases} [\psi(y), \|y\|^2], & \text{if } y \neq 0, \\ [y_0, 0], & \text{if } y = 0, \end{cases}$$

being $\psi(y)$ the point of $S^{p-1}_{\delta_1}$ where the integral curve of Y passing through y meets $S^{p-1}_{\delta_1}$ and $y_0 \in S^{p-1}_{\delta_1}$. It is not difficult to see that Φ, Ψ are homeomorphisms which make commutative the following diagram

$$\begin{array}{ccc} D^n_{\epsilon_0} \cap f^{-1}(D^p_{\delta_1}) & \xrightarrow{f} & D^p_{\delta_1} \\ \Phi \downarrow & & \Psi \downarrow \\ (N_{\epsilon_0,\delta_1} \times [0, \delta_1]) \cup_r V_{\epsilon_0} & \xrightarrow{C(f,r)} & c(S^{p-1}_{\delta_1}). \end{array}$$

This proves that f is C^0-$\mathscr{A}$-equivalent to the generalized cone of the link diagram. The construction for other values of ϵ, δ can be done by using Theorem 7.1.

In the case that f has no DST, then the theorem is still valid, but we use the canonical Thom stratification of f instead of the stratification by stable types (see [12, page 32]).

Definition 7.7 Let $f : U \to V$ a good representative of a polynomial map germ $f \in \mathscr{E}(n, p)$ with II and $f^{-1}(0) \neq \{0\}$. The *link diagram* of f is the link diagram

$$V_\epsilon \xleftarrow{r_{\epsilon,\delta}} N_{\epsilon,\delta} \xrightarrow{f_{\epsilon,\delta}} S^{p-1}_\delta$$

given in Theorem 7.6 for $0 < \delta \ll \epsilon \ll 1$. Then, f is C^0-$\mathscr{A}$-equivalent to the generalized cone of its link diagram.

Corollary 7.8 *Let $f, g \in \mathscr{E}(n, p)$ be two FD germs with non isolated zeros. If their link diagrams are C^0-$\mathscr{A}$-equivalent, then f, g are C^0-$\mathscr{A}$-equivalent.*

Example 7.9 Consider a FD function germ $f \in \mathscr{E}(2, 1)$ with $f^{-1}(0) \neq \{0\}$. The FD condition implies that f has isolated critical point in the origin. We fix $0 < \delta \ll \epsilon \ll 1$ as in Theorems 7.1 and 7.6. We can assume f is polynomial, hence $f^{-1}(0)$ is the algebraic curve given by $f(x, y) = 0$. Then, $V_\epsilon = f^{-1}(0) \cap D^2_\epsilon$ is made of a finite an even number $2r$ of half-branches which intersect transversally the boundary S^1_ϵ and separate the disk D^2_ϵ into $2r$ sectors, so that the sign of f alternates on consecutive sectors.

The manifold $N_{\epsilon,\delta}$ is given by the level curves $f(x, y) = \pm\delta$ in D^2_ϵ. It has $2r$ connected components, one in each sector of $D^2_\epsilon \setminus f^{-1}(0)$ and diffeomorphic to a closed interval. Moreover, he retraction map $r : N_{\epsilon,\delta} \to V_\epsilon$, when restricted to each connected component, is a diffeomorphism onto the two half-branches which bound the sector containing the connected component.

Thus, the C^0-$\mathscr{A}$-class only depends on the number of half-branches $2r$. We deduce that two functions f, g are C^0-$\mathscr{A}$-equivalent if and only if the curves $f^{-1}(0)$ and $g^{-1}(0)$ have the same number of half-branches (Fig. 24).

Fig. 24 The link of a FD function $f \in \mathscr{E}(2, 1)$

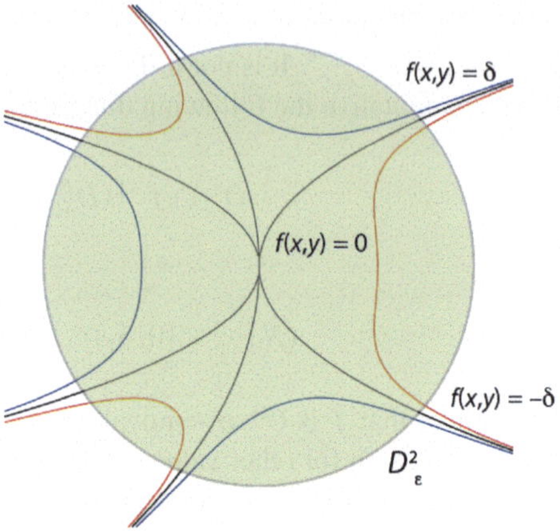

References

1. Arnold, V.I.: Topological classification of Morse functions and generalisations of Hilbert's 16-th problem. Math. Phys. Anal. Geom. **10**, 227–236 (2007)
2. Batista, E.B., Costa, J.C.F., Nuño-Ballesteros, J.J.: The Reeb graph of a map germ from $\mathbb{R}^3$ to $\mathbb{R}^2$ with non isolated zeros. Bull. Braz. Math. Soc. https://doi.org/10.1007/s00574-017-0058-4
3. Batista, E.B., Costa, J.C.F., Nuño-Ballesteros, J.J.: The cone structure theorem for map germs with non isolated zeros. http://www.uv.es/nuno/Preprints/ConeTheorem.pdf
4. Batista, E.B., Costa, J.C.F., Nuño-Ballesteros, J.J.: The Reeb graph of a map germ from $\mathbb{R}^3$ to $\mathbb{R}^2$ with isolated zeros. Proc. Edinb. Math. Soc. **60**(2), 319–348 (2017)
5. Costa, J.C.F., Nuño-Ballesteros, J.J.: Topological $\mathcal{K}$-classification of finitely determined map germs. Geom. Dedicata **166**, 147–162 (2013)
6. Dehn, M.: Über kombinatorische topologie. Acta Math. **67**(1), 123–168 (1936)
7. de Jong, T., Pfister, G.: Local analytic geometry. Basic Theory and Applications. Advanced Lectures in Mathematics. Friedr. Vieweg & Sohn, Braunschweig (2000)
8. Ehresmann, C.: *Les connexions infinitésimales dans un espace fibré différentiable*, Colloque de topologie (espaces fibrés), Bruxelles, pp. 29–55 (1950); Georges Thone, Liége; Masson et Cie., Paris, 1951
9. Fukuda, T.: Local topological properties of differentiable mappings I. Invent. Math. **65**, 227–250 (1981/82)
10. Fukuda, T.: Local topological properties of differentiable mappings II. Tokyo J. Math. **8**(2), 501–520 (1985)
11. Gauss, C.F.: Werke VIII, pp. 271–286. Teubner, Leipzig (1900)
12. Gibson, C.G., Wirthmüller, K., du Plessis, A.A., Looijenga, E.J.N.: Topological stability of smooth mappings. Lecture Notes in Mathematics, vol. 552. Springer, Berlin (1976)
13. Golubitsky, M., Guillemin, V.: Stable mappings and their singularities. Graduate Texts in Mathematics, vol. 14. Springer, New York (1973)
14. Greuel, G.M., Pfister, G.: A Singular Introduction to Commutative Algebra. Springer, Berlin (2008)
15. Izar, S.A.: Funções de Morse e topologia das superfícies II: Classificação das funções de Morse estáveis sobre superfícies, Métrica no. 35, Estudo e Pesquisas em Matemática, IBILCE, Brazil (1989). http://www.ibilce.unesp.br/Home/Departamentos/Matematica/metrica-35.pdf
16. Mather, J.: Stability of C^∞ mappings I: the division theorem. Ann. Math. **87**(2), 89–104 (1968)
17. Mather, J.: Stability of C^∞ mappings III: finitely determined mapgerms. Publ. Math. Inst. Hautes Études Sci. **35**, 279–308 (1968)

18. Mather, J.: Stability of C^∞ mappings II: infinitesimal stability implies stability. Ann. Math. **89**(2), 254–291 (1969)
19. Mather, J.: Stability of C^∞ mappings IV: classification of stable germs by $\mathbb{R}$-algebras. Publ. Math. Inst. Hautes Études Sci. **37**, 223–248 (1969)
20. Mather, J.: Stability of C^∞ mappings V: transversality. Adv. Math. **4**, 301–336 (1970)
21. Mather, J.: Stability of C^∞ mappings VI: the nice dimensions. In: Proceedings of Liverpool Singularities-Symposium, I (1969/70). Lecture Notes in Mathematics, vol. 192, pp. 207–253. Springer, Berlin (1971)
22. Marar, W.L., Nuño-Ballesteros, J.J.: The doodle of a finitely determined map germ from $\mathbb{R}^2$ to $\mathbb{R}^3$. Adv. Math. **221**(4), 1281–1301 (2009)
23. Martins, R., Nuño-Ballesteros, J.J.: Finitely determined singularities of ruled surfaces in $\mathbb{R}^3$. Math. Proc. Camb. Philos. Soc. **147**, 701–733 (2009)
24. Martins, R., Nuño-Ballesteros, J.J.: The link of a ruled frontal surface singularity. Real and Complex Singularities. Contemporary Mathematics, vol. 675, pp. 181–195. American Mathematical Society, Providence (2016)
25. Mendes, R., Nuño-Ballesteros, J.J.: Knots and the topology of singular surfaces in $\mathbb{R}^4$. Real and Complex Singularities. Contemporary Mathematics, vol. 675, pp. 229–239. American Mathematical Society, Providence (2016)
26. Milnor, J.: Morse theory. Annals of Mathematics Studies, vol. 51. Princeton University Press, Princeton (1963)
27. Milnor, J.: Singular points of complex hypersurfaces. Annals of Mathematics Studies, vol. 61. Princeton University Press, Princeton (1968)
28. Mond, D.: On the classification of germs of maps from $\mathbb{R}^2$ to $\mathbb{R}^3$. Proc. Lond. Math. Soc. **50**, 333–369 (1985)
29. Montesinos-Amilibia, A.: SphereXSurface, computer program. http://www.uv.es/montesin
30. Moya-Pérez, J.A., Nuño-Ballesteros, J.J.: The link of finitely determined map germ from $\mathbb{R}^2$ to$\mathbb{R}^2$. J. Math. Soc. Jpn. **62**(4), 1069–1092 (2010)
31. Moya-Pérez, J.A., Nuño-Ballesteros, J.J.: Topological triviality of families of map germs from $\mathbb{R}^2$ to $\mathbb{R}^2$. J. Singul. **6**, 112–123 (2012)
32. Moya-Pérez, J.A., Nuño-Ballesteros, J.J.: Topological classification of corank 1 map germs from $\mathbb{R}^3$ to $\mathbb{R}^3$. Rev. Mat. Complut. **27**(2), 421–445 (2014)
33. Moya-Pérez, J.A., Nuño-Ballesteros, J.J.: Some remarks about the topology of corank 2 map germs from $\mathbb{R}^2$ to $\mathbb{R}^2$. J. Singul. **10**, 200–224 (2014)
34. Moya-Pérez, J.A., Nuño-Ballesteros, J.J.: Gauss words and the topology of map germs from $\mathbb{R}^3$ to $\mathbb{R}^3$. Rev. Mat. Iberoam. **31**(3), 977–988 (2015)
35. Moya-Pérez, J.A., Nuño-Ballesteros, J.J.: Topological triviality of families of map germs from $\mathbb{R}^3$ to $\mathbb{R}^3$. Rocky Mountain J. Math. **46**(5), 1643–1664 (2016)
36. Reeb, G.: Sur les points singuliers d'une forme de Pfaff completement intégrable ou d'une fonction numérique. C. R. Math. Acad. Sci. Paris **222**, 847–849 (1946)
37. Sharko, V.V.: Smooth and topological equivalence of functions on surfaces. Ukr. Math. J. **55**, 832–846 (2003)
38. Thom, R.: La stabilité topologique des applications polynomiales. Enseign. Math. **8**(2), 24–33 (1962)
39. Thom, R.: Local topological properties of differentiable mappings. Colloquium on Differential Analysis (Tata Institute), pp. 191–202. Oxford University Press, Oxford (1964)
40. Wall, C.T.C.: Finite determinacy of smooth map-germs. Bull. Lond. Math. Soc. **13**(6), 481–539 (1981)
41. Wall, C.T.C.: Classification and Stability of Singularities of Smooth Maps Singularity Theory (Trieste, 1991), pp. 920–952. World Scientific Publishing, River Edge (1995)
42. Whitney, H.: The singularities of a smooth n-manifold in $(2n-1)$-space. Ann. Math. **45**(2), 247–293 (1944)

Topology of Real Singularities

Nicolas Dutertre

Abstract In this mini-course, we study the topology of real singularities. After recalling basic notions and classical results of differential topology, we present formulas for topological invariants of semi-analytic or semi-algebraic sets due to several authors.

Keywords Topological degree · Euler characteristic · Real singularities
Real Milnor fibre · Semi-algebraic sets

1 Introduction

This mini-course is aimed at young researchers and graduate students who want to learn basic tools and techniques of real singularity theory.

It starts with well-known notions and results of differential topology: the Brouwer degree, the index of a vector-field, the Poincaré–Hopf theorem, Morse functions. Although theses notions may be very familiar to any researcher experienced in singularity theory, we believe it is worth recalling them here.

In the next chapter, we apply these techniques of differential topology to some real analytic or semi-analytic sets and we get several nice formulas for topological invariants of these sets. In chapter "Degree Formulas and Signature Formulas for the Euler Characteristic of Algebraic Sets and Semi-algebraic Sets", always using the same techniques, we give several methods for the computation of the Euler characteristic of a real algebraic or semi-algebraic set.

We end with a chapter about the real Milnor fibration and the topology of the real Milnor fibre.

The author would like to thank the organizing committee of **The Brazilian Northeastern Meeting on Singularities**, and especially Raimundo Araújo dos Santos, for inviting him to give this mini-course.

N. Dutertre (✉)
CNRS, Centrale Marseille, Aix Marseille Université, Marseille, France
e-mail: nicolas.dutertre@univ-amu.fr

R. N. Araújo dos Santos et al. (eds.), *Singularities and Foliations. Geometry, Topology and Applications*, Springer Proceedings in Mathematics & Statistics 222, https://doi.org/10.1007/978-3-319-73639-6_2

2 The Brouwer Degree, the Poincaré–Hopf Index and Morse Functions

In this chapter, we recall important tools and results from differential topology. We will apply them in the next chapters in order to study the topology of real singularities. We refer to [20, 22, 28, 29] for more details.

2.1 The Brouwer Degree

Let M be a compact oriented manifold without boundary of dimension n and N a connected oriented manifold without boundary of the same dimension. Let $f : M \to N$ be a smooth (i.e. C^∞) map and let x be a regular point of f. The differential map $Df(x) : T_xM \to T_{f(x)}N$ is a linear isomorphism between the two oriented tangent spaces. The "sign" of $Df(x)$ is $+1$ (resp. -1) if $Df(x)$ preserves (resp. reverses) the orientation.

Definition 2.1.1 If $y \in N$ is a regular value of f, we define

$$deg(f, y) = \sum_{x \in f^{-1}(y)} sign\ Df(x).$$

Remark 2.1.2 The integer $deg(f, y)$ is well-defined because $f^{-1}(y)$ is a 0-dimensional submanifold of M, hence a finite union of points since M is compact.

Theorem 2.1.3 ([29], p. 28, Theorem A) *The integer* $\deg(f, y)$ *does not depend on the regular value* y.

Definition 2.1.4 The integer $deg(f, y)$ is called the (Brouwer) degree of f and denoted by $deg\ f$.

Theorem 2.1.5 ([29], p. 28, Theorem B) *If f is smoothly homotopic to g then* $\deg f = \deg g$.

Examples: (1) The map on the left has degree 0 and the map on the right has degree 1.

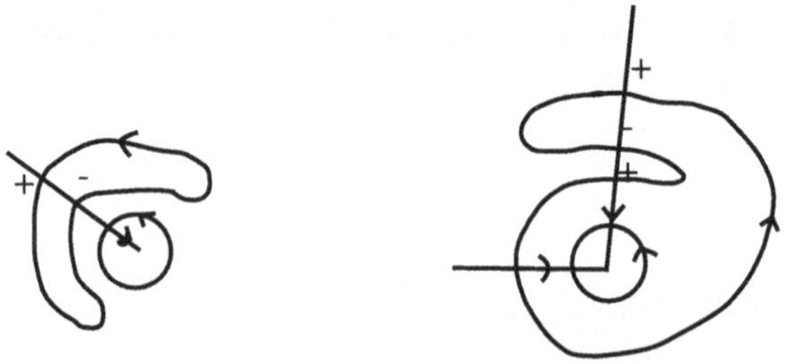

(2) The following map has degree 2.

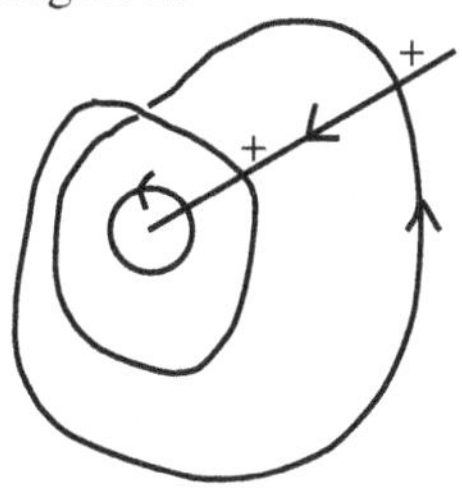

(3) Let $r_i : S^n \to S^n$ be given by

$$r_i(x_1, \ldots, x_{i-1}, x_i, x_{i+1}, \ldots, x_{n+1}) = (x_1, \ldots, x_{i-1}, -x_i, x_{i+1}, \ldots, x_{n+1}).$$

The map r_i has degree -1.

(4) The map $S^1 \subset \mathbb{C} \to S^1 \subset \mathbb{C}$, $z \mapsto z^m$, $m \in \mathbb{Z}^*$, has degree m. It is easy to see because each point in S^1 has exactly m preimages and the map is regular and preserves orientation if $m > 0$ and reverses it if $m < 0$.

Proposition 2.1.6 *Let M, N and L be three smooth oriented manifolds of the same dimension. We assume that M and N are compact and that N and L are connected. If $f : M \to N$ and $g : N \to L$ are two smooth maps then* $\deg(g \circ f) = \deg g \times \deg f$.

Proof It is an easy application of the definition of the degree. □

Example: Let $\sigma : S^n \to S^n$, $x \mapsto -x$ be the antipodal map. It has degree $(-1)^{n+1}$ because $\sigma = r_1 \circ r_2 \circ \cdots \circ r_{n+1}$, where the maps r_i, $i = 1, \ldots, n+1$, are defined on Example 3. Therefore for n even, σ is not smoothly homotopic to the identity.

Theorem 2.1.7 ([29], p. 28, Lemma 1) *If M is the boundary of a compact oriented manifold W and $f : M \to N$ extends to a map $F : W \to N$ then* $\deg f = 0$.

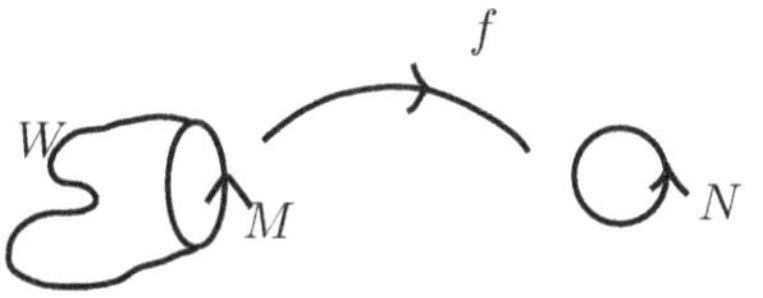

Application: Here is an application of the previous theorem that we will use in the next chapters.

We consider a smooth compact connected hypersurface M in $\mathbb{R}^n$. By the Jordan–Brouwer Separation Theorem, it bounds a connected bounded and open subset $D \subset \mathbb{R}^n$, i.e. $M = \partial\overline{D}$. The canonical orientation on D induces an orientation on M. Let $F : \overline{D} \to \mathbb{R}^n$ be a smooth map which does not vanish on $\partial\overline{D}$. We assume that F has a finite number of zeroes $p_1, \ldots, p_m$ inside D and that $p_1, \ldots, p_m$ are regular points of F. The map $\overline{F} = \frac{F}{|F|} : M \to S^{n-1}$ is well-defined and we have

$$\deg \overline{F} = \sum_{i=1}^{m} \text{sign } \det[DF(p_i)].$$

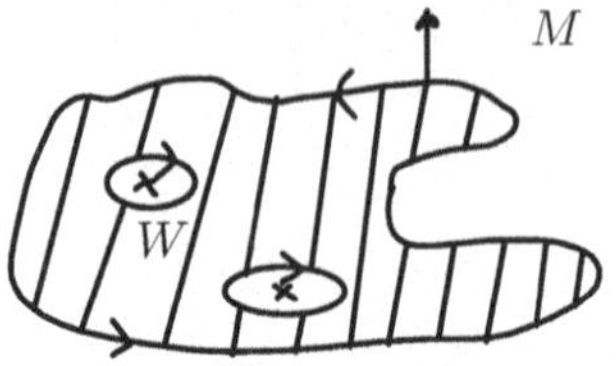

Let us prove briefly this equality. Around each p_i, we remove a small open ball $B(p_i, \epsilon_i)$ and set $W = \overline{D} \setminus \cup_{i=1}^m B(p_i, \epsilon_i)$. It is a manifold with boundary

$$\partial W = M \bigcup \cup_{i=1}^m S(p_i, \epsilon_i).$$

The hypersurface ∂W is oriented by the canonical orientation of the boundary. Let $\overline{F}_{\partial W} : \partial W \to S^{n-1}$ be defined by $\overline{F}_{\partial W}(x) = \frac{F}{|F|}(x)$. Then $\overline{F}_{\partial W}$ extends to W, and so by Theorem 2.1.7, deg $\overline{F}_{\partial W} = 0$. But this degree is also equal to

$$\deg \overline{F} - \sum_{i=1}^{m} \deg \overline{F}_i,$$

where $\overline{F}_i = \frac{F}{|F|} : S(p_i, \epsilon_i) \to S^{n-1}$. The minus sign is explained by the fact that the orientation of $S(p_i, \epsilon_i)$ as a component of the boundary of W is the opposite of the orientation of $S(p_i, \epsilon_i)$ as the boundary of $B(p_i, \epsilon_i)$.

Since p_i is a regular point of F, the degree of $\overline{F}_i$ is equal to the sign of det$[DF(p_i)]$, because $\frac{F}{|F|} : S(p_i, \epsilon_i) \to S^{n-1}$ is homotopic to the map

$$S(p_i, \epsilon_i) \to S^{n-1}, p \mapsto \frac{DF(p_i)(p - p_i)}{|DF(p_i)(p - p_i)|},$$

and so has the same degree as the map

$$S(0, \epsilon_i) \to S^{n-1}, h \mapsto \frac{DF(p_i)(h)}{|DF(p_i)(h)|}.$$

This last map has degree equal to sign det $[DF(p_i)]$, because the map $h \mapsto DF(p_i)(h)$ is homotopic to $\pm Id_{\mathbb{R}^n}$, depending on the sign of det$[DF(p_i)]$ since $GL(n, \mathbb{R})$ has two connected components.

2.2 *The Poincaré–Hopf Index*

Definition 2.2.1 Let M be a smooth manifold. A vector field V on M is a smooth map $V : M \to TM$ such that for all $x \in M$, $pr(V(x)) \in T_x M$, where $pr : TM \to M$ is the natural projection.

Remark 2.2.2 If M is a submanifold of $\mathbb{R}^n$, a vector field is a smooth map $V : M \to \mathbb{R}^n$ such that for all $x \in M$, $V(x) \in T_x M$.

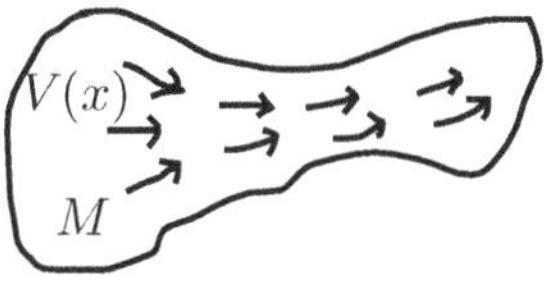

Definition 2.2.3 Let V be a vector field on a manifold M of dimension n and let p be an isolated zero of a vector field V. In local coordinates, V can be seen as a map from a small open set $U \subset \mathbb{R}^n$ to a small open set $U' \subset \mathbb{R}^n$ where $0 \in U$, $0 \in U'$ and 0 is the only zero of V in U. We define the Poincaré–Hopf index of V at p by

$$Ind(V, p) = degree\ of\ \frac{V}{|V|} : S_\epsilon^{n-1} \to S^{n-1},$$

where S_ϵ^{n-1} is a small sphere centered at 0 and included in U.

Examples in $\mathbb{R}^2$:

1. If $V(x, y) = (y, -x)$ (circulation) then $\text{Ind}(V, 0) = +1$.

2. If $V(x, y) = (-x, -y)$ (sink) then $\text{Ind}(V, 0) = +1$.

3. If $V(x, y) = (x, y)$ (source) then $\text{Ind}(V, 0) = +1$.

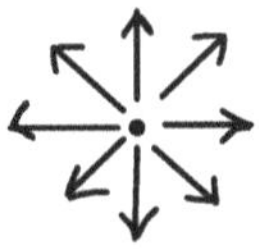

4. If $V(x, y) = (-x, y)$ (saddle) then $\text{Ind}(V, 0) = -1$.

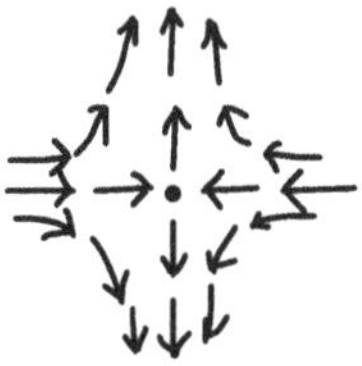

5. If $V(x, y) = (x^2, x + y)$ then $\mathrm{Ind}(V, 0) = 0$.

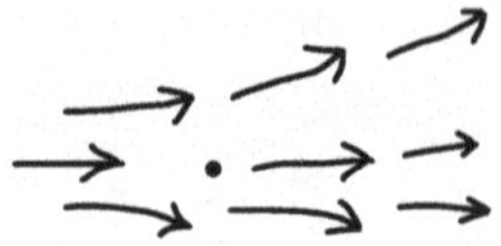

6. If $V(x, y) = (x^2 - y^2, 2xy)$ ($z \mapsto z^2$ in complex coordinates) then $\mathrm{Ind}(V, 0) = +2$.

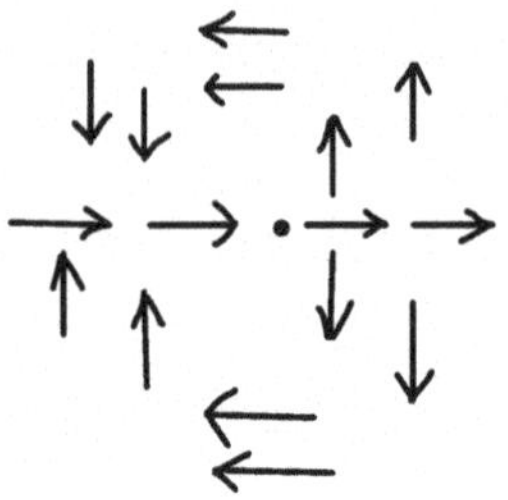

Remark 2.2.4 The definition of $Ind(V, p)$ does not depend on the choice of the local coordinates (see [20, 29] for example).

Theorem 2.2.5 (Poincaré–Hopf theorem) *Let M be a smooth compact manifold. Let V be a smooth vector field on M, with a finite number of zeroes $p_1, \ldots, p_k$. We have*

$$\chi(M) = \sum_{i=1}^{k} \mathrm{Ind}(V, p_i).$$

Proof See [29], p. 35, [20], p. 134 or [22], p. 133. □

2.3 *Morse Functions*

Definition 2.3.1 Let M be a smooth manifold of dimension n, let $p \in M$ and $f : M \to \mathbb{R}$ be a smooth function. Let $(x_1, \ldots, x_n)$ be a local coordinate system around p in M. We say that p is a non-degenerate critical point of f if p is a critical point of f (i.e. $\frac{\partial f}{\partial x_1}(p) = \cdots = \frac{\partial f}{\partial x_n}(p) = 0$) and the Hessian matrix

$$\left[\frac{\partial^2 f}{\partial x_i \partial x_j}(p) \right]_{1 \le i, j \le n},$$

is non-singular.

Remark 2.3.2 The previous definition does not depend on the local coordinates (see [20] or [28]).

Proposition 2.3.3 (Morse lemma) *Let M be a smooth manifold of dimension n and $p \in M$ be a non-degenerate critical point of a smooth function $f : M \to \mathbb{R}$. There exists a local coordinate system $(u_1, \ldots, u_n)$ around p such that*

$$f = f(p) - u_1^2 - \cdots - u_\lambda^2 + u_{\lambda+1}^2 + \cdots + u_n^2.$$

Proof See [28], p. 6, Lemma 2.2. □

Definition 2.3.4 The integer λ is called the (Morse) index of f at p.

Corollary 2.3.5 *Non-degenerate critical points are isolated in the set of critical points.*

Definition 2.3.6 Let M be a smooth manifold. A function $f : M \to \mathbb{R}$ is called a Morse function if it admits only non-degenerate critical points.

Theorem 2.3.7 (Openness and density) *For any manifold M, Morse functions form a dense open set in $C_S^\infty(M, \mathbb{R})$, where $C_S^\infty(M, \mathbb{R})$ denotes the space $C^\infty(M, \mathbb{R})$ equipped with the strong Whitney topology (see [22], Chap. 2).*

Proof It is a consequence of the Thom transversality theorem (see [18, 22] or [1]). □

In the next chapters, we will be interested in semi-analytic or semi-algebraic subsets of euclidian spaces, so from now on, we shall assume that $M \subset \mathbb{R}^N$ and that $\dim M = n$. Let $f : M \to \mathbb{R}$ be a smooth function. The gradient vector field of f on M, denoted by $\nabla_M f$, is defined as follows:

$$\forall p \in M, \forall v \in T_p M, Df(p)(v) = \langle \nabla_M f(p), v \rangle,$$

where $\langle -, - \rangle$ is the usual scalar product. Therefore the critical points are exactly the zeroes of $\nabla_M f$. If p is a Morse critical point of f of index λ, then there is a local coordinate system such that

$$f = f(p) - u_1^2 - \cdots - u_\lambda^2 + u_{\lambda+1}^2 + \cdots + u_n^2,$$

and so

$$\nabla_M f = (-2u_1, \ldots, -2u_\lambda, 2u_{\lambda+1}, \ldots, 2u_n).$$

We see that the Poincaré–Hopf index $\mathrm{Ind}(\nabla_M f, p)$ is equal to $(-1)^\lambda$ because, as already explained above, the mapping $\frac{\nabla_M f}{|\nabla_M f|} : S(p, \epsilon) \to S^{n-1}$ has degree equal to sign $\det[D(\nabla_M f)(p)]$.

Theorem 2.3.8 *Let $M \subset \mathbb{R}^N$ be a smooth compact manifold and let $f : M \to \mathbb{R}$ be a Morse function with critical points $p_1, \ldots, p_k$. We have*

$$\chi(M) = \sum_{i=1}^{k} (-1)^{\lambda(p_i)},$$

where $\lambda(p_i)$ denotes the Morse index of p_i.

In our study of the topology of semi-analytic and semi-algebraic sets, we will need a version of this theorem for manifolds with boundary and manifolds with corners. Let us start with some basic facts on manifolds with corners. Our reference is [7]. A manifold with corners M is defined by an atlas of charts modelled on open subsets of $\mathbb{R}_+^k \times \mathbb{R}^{n-k}$. We write ∂M for its boundary. We will make the additional assumption that the boundary is partitioned into pieces $\partial_i M$, themselves manifolds with corners, such that in each chart, the intersections with the coordinate hyperplanes $\{x_j = 0\}$ correspond to distinct pieces $\partial_i M$ of the boundary. For any set I of suffices, we write $\partial_I M = \cap_{i \in I} \partial_i M$ and we make the convention that $\partial_\emptyset M = M \setminus \partial M$.

Any n-manifold M with corners can be embedded in a n-manifold M^+ without boundary so that the pieces $\partial_i M$ extend to submanifolds $\partial_i M^+$ of codimension 1 in M^+. We will assume that M^+ is provided with a Riemannian metric.

Let M be a manifold with corners and $f : M^+ \to \mathbb{R}$ a smooth map. We consider the points P which are critical points of $f_{|\partial_I M^+}$.

Definition 2.3.9 A critical point P is correct (respectively Morse correct) if, taking $I(P) := \{i \mid P \in \partial_i M\}$, P is a critical (respectively Morse critical) point of $f_{|\partial_{I(P)} M^+}$, and is not a critical point of $f_{|\partial_J M^+}$ for any proper subset J of $I(P)$.

Note that a 0-dimensional corner point P is always a critical point because in that case $\partial_{I(P)} M^+ = \{P\}$, which is a 0-dimensional manifold.

Definition 2.3.10 The maps f with all critical points Morse correct are called Morse correct.

Proposition 2.3.11 *The set of Morse correct functions is dense and open in the space of all maps $M^+ \to \mathbb{R}$.*

Proof It is clear from classical Morse theory, because there is a finite number of pieces $\partial_I M^+$. □

The index $\lambda(P)$ of f at a Morse correct point P is defined to be that of $f_{|\partial_{I(P)} M^+}$. If P is a correct critical point of f, $i \in I(P)$, and J is formed from $I(P)$ by deleting i, then in a chart at P with $\partial_J M$ mapping to $\mathbb{R}_+^p$ and $\partial_{I(P)} M$ to the subset $\{x_1 = 0\}$, the function f is non-critical, but its restriction to $\{x_1 = 0\}$ is. Hence $\partial f / \partial x_1 \neq 0$.

Definition 2.3.12 We say that f is inward at P if, for each $i \in I(P)$, we have $\partial f / \partial x_1 > 0$.

Remark 2.3.13 By our convention, if $I(P) = \emptyset$, then f is inward at P.

Theorem 2.3.14 *If M is compact and f is Morse correct,*

$$\chi(M) = \sum \left\{ (-1)^{\lambda(P)} \mid P \text{ inward Morse critical point} \right\}.$$

Proof This is a consequence of stratified Morse theory [19]. The manifold with corners M is a compact Whitney stratified set of M^+, with stratum the $\partial_I M$'s. The function $f : M \to \mathbb{R}$ is easily seen to be a Morse function in the sense of [19] and so

$$\chi(M) = \sum \{\alpha(f, P) \mid P \text{correct critical point}\},$$

where

$$\alpha(f, P) = 1 - \chi\big(f^{-1}(f(P) - \delta) \cap B(P, \nu)\big),$$

with $0 < \delta \ll \nu \ll 1$. Here $B(P, \nu)$ is the ball centered at P of radius ν in the Riemannian manifold M^+. If P belongs to $\partial_\emptyset M$ then $\alpha(f, P)$ is exactly $(-1)^{\lambda(P)}$. If P belongs to $\partial_I M$, $I \neq \emptyset$, then $\alpha(f, P) = (-1)^{\lambda(P)}.\alpha_{\text{nor}}(f, P)$, where $\alpha_{\text{nor}}(f, P)$ is the normal index of f at P. It is defined as follows. Choose a normal slice V at P, that is a closed submanifold of M^+ of dimension $n - \dim \partial_I M$, which intersects $\partial_I M$ in P transversally, then

$$\alpha_{\text{nor}}(f, P) = 1 - \chi\big(f^{-1}(f(P) - \delta) \cap B(P, \nu) \cap V\big).$$

Let us compute this normal index. We can assume that $f(P) = 0$. We can choose a local chart $(x_1, \ldots, x_n)$ centered at P such that $\partial_I M$ is given by $\{x_1 = \cdots = x_k = 0\}$, V is given by $\{x_{k+1} = \cdots = x_n = 0\}$, $k < n$. Locally M is the set $\{x_1 \geq 0, \ldots, x_k \geq 0\}$. Furthermore, since P is a correct point, $\partial f/\partial x_j(P) \neq 0$ for each $j \in \{1, \ldots, k\}$ and, by an appropriate change of coordinates, the restriction of f to V is just the linear form

$$\sum_{j=1}^{k} \frac{\partial f}{\partial x_j}(P) x_j.$$

It is then straightforward to see that $\alpha_{\text{nor}}(f, P) = 1$ if $\partial f/\partial x_j(P) > 0$ for all $j \in \{1, \ldots, k\}$ and $\alpha_{\text{nor}}(f, P) = 0$ otherwise. This proves the theorem. □

Let us apply this to the case of manifolds with boundary. Let $(M, \partial M) \subset \mathbb{R}^N$ be a manifold with boundary. Let $q \in \partial M$, then $T_q\partial M$ is a hyperplane in $T_q M$ and $T_q M$ decomposes in the following way:

$$T_q M = T_q \partial M \sqcup T_q M^+ \sqcup T_q M^-,$$

where $T_q M^+$ consists of outwards pointing vectors (outward vectors for short) and $T_q M^-$ consists of inwards pointing vectors.

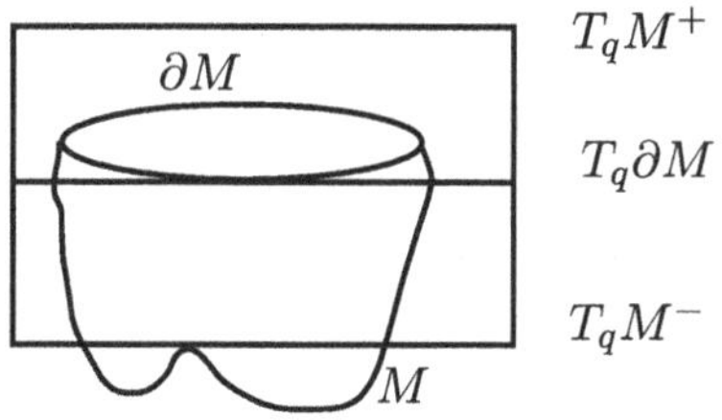

Let $f : (M, \partial M) \to \mathbb{R}$ be a smooth function. In this situation, a point $q \in \partial M$ is a correct critical point of f if q is a critical point of $f_{|\partial M} : \partial M \to \mathbb{R}$ and $Df(q)_{|T_q M}$ is not identically zero. Now let us assume that $(M, \partial M)$ is compact and that $f : (M, \partial M) \to \mathbb{R}$ is a correct Morse function. Let us denote by $p_1, \ldots, p_k$ the critical points of $f_{|M \setminus \partial M}$ and by $q_1, \ldots, q_l$ those of $f_{|\partial M}$. In this case, Theorem 2.3.14 becomes

$$\chi(M) = \sum_{i=1}^{k} (-1)^{\lambda(p_i)} + \sum_{j \mid \nabla_M f(q_j) \text{ inward}} (-1)^{\mu(q_j)},$$

where $\lambda(p_i)$ is the Morse index of f at p_i and $\mu(q_j)$ is the Morse index of $f_{|\partial M}$ at q_j.

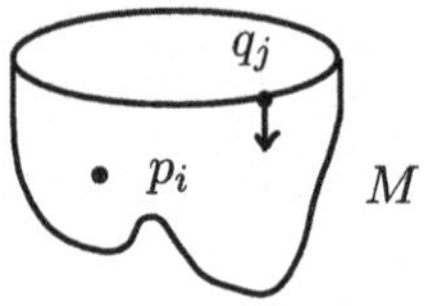

In the following chapters of this mini-course, we will often use relative versions of the previous theorems on Morse theory.

Theorem 2.3.15 *Let $M \subset \mathbb{R}^N$ be a smooth compact manifold and let $f : M \to \mathbb{R}$ be a Morse function with critical points $p_1, \ldots, p_k$. For any $\alpha \in \mathbb{R}$, we have:*

$$\chi(M \cap \{f \geq \alpha\}, M \cap \{f = \alpha\}) = \sum_{i \mid f(p_i) > \alpha} (-1)^{\lambda(p_i)},$$

where $\lambda(p_i)$ is the Morse index of p_i.

Proof See [22], p. 161, Theorem 3.4. □

Theorem 2.3.14 has a similar relative version.

3 The Eisenbud–Levine Formula, the Khimshiashvili Formula and Some Generalizations

3.1 The Eisenbud–Levine Formula

As seen in the second chapter, the Poincaré–Hopf index of a vector field plays an important role in the topology of manifolds. Here we present an algebraic formula for this index.

Let $f = (f_1, \ldots, f_n) : (\mathbb{R}^n, 0) \to (\mathbb{R}^n, 0)$ be a C^∞ map-germ (this is exactly the local expression of a vector field on a smooth manifold). We assume that

$$Q(f) = \frac{C^\infty(\mathbb{R}^n, 0)}{(f_1, \ldots, f_n)},$$

is a finite dimensional vector space over $\mathbb{R}$. Here $C^\infty(\mathbb{R}^n, 0)$ is the algebra of germs at $0 \in \mathbb{R}^n$ of C^∞ real valued functions and $(f_1, \ldots, f_n)$ is the ideal generated by the components $f_1, \ldots, f_n$ of f. We write $\dim_{\mathbb{R}} Q(f) < +\infty$. We denote by J_f the jacobian of the map-germ f. Namely, we have

$$J_f = \frac{\partial(f_1, \ldots, f_n)}{\partial(x_1, \ldots, x_n)}.$$

Theorem 3.1.1 (The Eisenbud–Levine formula) *Let $f : (\mathbb{R}^n, 0) \to (\mathbb{R}^n, 0)$ be a C^∞ map-germ such that* $\dim_{\mathbb{R}} Q(f) < +\infty$. *Then we have*

1. *0 is isolated in $f^{-1}(0)$,*
2. *$J_f \neq 0$ in $Q(f)$,*
3. *$\forall g \in Q(f)$, $gJ_f = g(0)J_f$ in $Q(f)$,*
4. *let $\varphi : Q(f) \to \mathbb{R}$ be a linear form such that $\varphi(J_f) > 0$ and let $\Phi : Q(f) \times Q(f) \to \mathbb{R}$ be the bilinear symmetric form defined by $\Phi(g, h) = \varphi(gh)$. Then Φ is non-degenerate and* signature $\Phi = \operatorname{Ind}(f, 0)$.

Proof See [1, 17] or [4]. For a first approach, see [16]. □

Example: Let f be the map-germ defined by

$$\begin{array}{rcccc} f & : & (\mathbb{R}^2, 0) & \to & (\mathbb{R}^2, 0) \\ & & (x, y) & \mapsto & (x^2 - y^2, 2xy). \end{array}$$

We have $Q(f) = \frac{C^\infty(\mathbb{R}^2,0)}{(x^2-y^2,2xy)}$. We see that $\dim_{\mathbb{R}} Q(f) = 4$ and that $\bar{1}, \bar{x}, \bar{y}$ and $\overline{x^2 + y^2}$ form a basis of $Q(f)$. It is clear that 0 is isolated in $f^{-1}(0)$. Let us compute J_f:

$$J_f(x, y) = \begin{vmatrix} 2x & -2y \\ 2y & 2x \end{vmatrix} = 4(x^2 + y^2).$$

Let $\varphi : Q(f) \to \mathbb{R}$ be the linear form given by

$$\varphi(\bar{1}) = \varphi(\bar{x}) = \varphi(\bar{y}) = 0 \text{ and } \varphi(\overline{x^2 + y^2}) = \frac{1}{4}.$$

Then $\varphi(J_f) = 1$. Let Φ be the linear symmetric form defined by $\Phi(P, Q) = \varphi(PQ)$. Let us compute its matrix in the basis $(\bar{1}, \bar{x}, \bar{y}, \overline{x^2 + y^2})$. We have

$$\Phi(\bar{1}, \bar{1}) = \varphi(\bar{1}) = 0,\ \Phi(\bar{1}, \bar{x}) = \Phi(\bar{x}, \bar{1}) = 0,\ \Phi(\bar{1}, \bar{y}) = \Phi(\bar{y}, \bar{1}) = 0,$$

$$\Phi(\bar{x}, \bar{x}) = \varphi(\bar{x}^2) = \varphi\left(\frac{1}{2}\overline{x^2 + y^2}\right) = \frac{1}{8},$$

$$\Phi(\bar{y}, \bar{y}) = \varphi(\bar{y}^2) = \varphi\left(\frac{1}{2}\overline{x^2 + y^2}\right) = \frac{1}{8},$$

$$\Phi(\bar{x}, \bar{y}) = \varphi(\overline{xy}) = \varphi(\bar{0}) = 0 = \Phi(\bar{y}, \bar{x}),$$

$$\Phi(\bar{1}, \overline{x^2 + y^2}) = \frac{1}{4}, \Phi(\bar{x}, \overline{x^2 + y^2}) = \varphi(\bar{x}^3 + \overline{xy^2}) = \varphi(\bar{0}) = 0,$$

$$\Phi(\bar{y}, \overline{x^2 + y^2}) = 0, \Phi(\overline{x^2 + y^2}, \overline{x^2 + y^2}) = \varphi\big(\overline{(x^2 + y^2)(x^2 + y^2)}\big) = 0.$$

So this matrix is

$$\begin{bmatrix} 0 & 0 & 0 & \frac{1}{4} \\ 0 & \frac{1}{8} & 0 & 0 \\ 0 & 0 & \frac{1}{8} & 0 \\ \frac{1}{4} & 0 & 0 & 0 \end{bmatrix}.$$

The eigenvalues are $\frac{1}{8}$ with multiplicity 2, $\frac{1}{4}$ with multiplicity 1 and $-\frac{1}{4}$ with multiplicity 1. So the signature of Φ is $3 - 1 = 2 = \text{Ind}(f, 0)$.

The Eisenbud–Levine formula gives an algebraic formula for the index of a vector field, hence an algebraic and "effective" way to compute a topological data. In the sequel, using technics introduced in the second chapter, we will present several formulas relating topological invariants to indices of vector fields. Thanks to the Eisenbud–Levine formula, these topological invariants become algebraically computable.

3.2 *The Khimshiashvili Formula*

Let $f : (\mathbb{R}^n, 0) \to (\mathbb{R}, 0)$ be an analytic-map germ with an isolated critical point at 0. The Khimshiashvili formula (see [24]) relates the Poincaré–Hopf index of the gradient vector field ∇f of f to the topology of a small regular level of f.

Theorem 3.2.1 *We have*

$$\chi\big(f^{-1}(\delta) \cap B_\epsilon^n\big) = 1 - \text{sign}(-\delta)^n \text{Ind}(\nabla f, 0), \tag{1}$$

where δ is a regular value of f, $0 < |\delta| \ll \epsilon \ll 1$, and

$$\chi(\{f \geq \delta\} \cap B_\epsilon^n) - \chi(\{f \leq \delta\} \cap B_\epsilon^n) = \text{sign}(-\delta)^{n+1} \text{Ind}(\nabla f, 0). \tag{2}$$

Proof Let U be a small open subset of $\mathbb{R}^n$ such that $0 \in U$, and f is defined in U. We perturb f into a Morse function $\tilde{f} : U \to \mathbb{R}$. Let $p_1, \ldots, p_k$ be the critical points of $\tilde{f}$, with respective indices $\lambda_1, \ldots, \lambda_k$. Let $\delta > 0$, by Morse theory we have

$$\chi\big(f^{-1}([-\delta,\delta]) \cap B^n_\epsilon\big) - \chi\big(f^{-1}(-\delta) \cap B^n_\epsilon\big) = \sum_{i=1}^{k} (-1)^{\lambda_i}.$$

Actually we can choose $\tilde{f}$ sufficiently close to f so that the p_i's lie in $f^{-1}([-\frac{\delta}{4}, \frac{\delta}{4}])$. Now, $f^{-1}([-\delta,\delta]) \cap B^n_\epsilon$ retracts to the central fibre $f^{-1}(0) \cap B^n_\epsilon$ and $f^{-1}(0) \cap B^n_\epsilon$ is the cone over $f^{-1}(0) \cap S^{n-1}_\epsilon$ (see [30]) so

$$\chi\big(f^{-1}([-\delta,\delta]) \cap B^n_\epsilon\big) = 1.$$

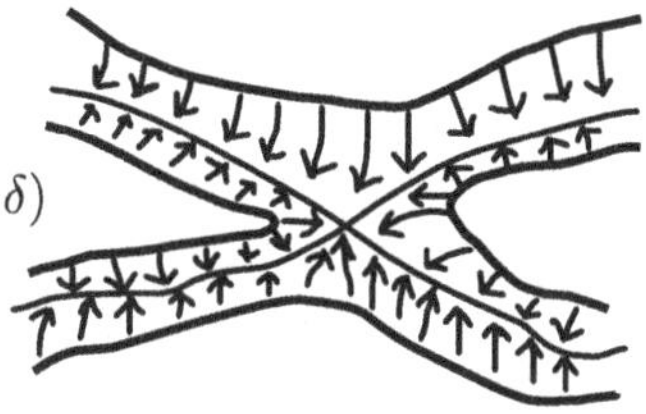

Moreover, we have

$$\sum_{i=1}^{k} (-1)^{\lambda_i} = \sum_{i=1}^{k} \text{sign det}[D(\nabla \tilde{f})(p_i)].$$

The sum on the right hand-side is the degree of the map $\frac{\nabla \tilde{f}}{|\nabla \tilde{f}|} : S^{n-1}_\epsilon \to S^{n-1}$ which is equal, by homotopy, to the degree of $\frac{\nabla f}{|\nabla f|} : S^{n-1}_\epsilon \to S^{n-1}$. By definition, this last degree is $\text{Ind}(\nabla f, 0)$. This gives the result for a negative regular value. For a positive regular value, we apply the result to $-f$ and use the relation $\text{Ind}(-\nabla f, 0) = (-1)^n \text{Ind}(\nabla f, 0)$. This proves formula (1). Formula (2) is proved with similar arguments. □

We will call $f^{-1}(\delta) \cap B^n_\epsilon$ the (positive or negative) real Milnor fibre. The following formulas are due to Arnol'd [2] and Wall [39].

Corollary 3.2.2 *With the same hypothesis on f, we have*

$$\chi(\{f \le 0\} \cap S^{n-1}_\epsilon) = 1 - \text{Ind}(\nabla f, 0),$$

$$\chi(\{f \ge 0\} \cap S^{n-1}_\epsilon) = 1 + (-1)^{n-1}\text{Ind}(\nabla f, 0).$$

If n is even, we have:

$$\chi(\{f = 0\} \cap S^{n-1}_\epsilon) = 2 - 2\,\text{Ind}(\nabla f, 0).$$

Proof By a deformation argument due to Milnor [30], $f(-\delta) \cap B_\epsilon^n$, $\delta > 0$, is homeomorphic to $\{f \leq -\delta\} \cap S_\epsilon^{n-1}$, which is homeomorphic to $\{f \leq 0\} \cap S_\epsilon^{n-1}$ if δ is very small.

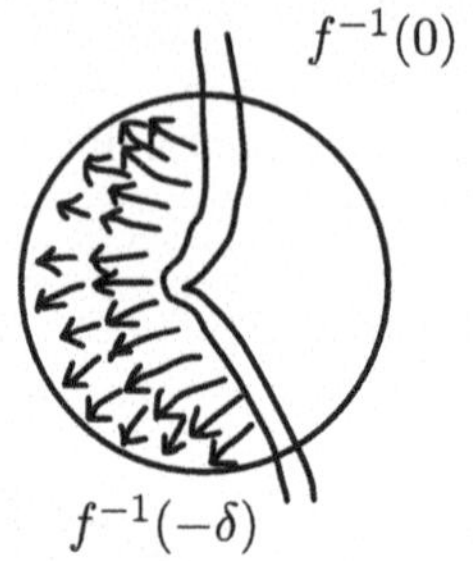

□

3.3 Non-isolated Critical Points

We would like to obtain similar results in the case when 0 is not an isolated critical point of $f : (\mathbb{R}^n, 0) \to (\mathbb{R}, 0)$. The following result is due to Szafraniec [33].

Let $\omega(x) = x_1^2 + \cdots + x_n^2$. We suppose that f is defined in an open set $U \subset \mathbb{R}^n$ and $0 \in U$. Let $\Sigma_{f,\omega}$ be the following polar set:

$$\Sigma_{f,\omega} = \{x \in U \mid \mathrm{rank}(\nabla\omega(x), \nabla f(x)) \leq 1\},$$

and let V be defined by

$$V = \left\{(x, \epsilon, y) \in U \times \mathbb{R} \times \mathbb{R} \mid \omega(x) = \epsilon^2, x \in \Sigma_{f,\omega}, y = f(x)\right\}.$$

Then V is an analytic subset of $\mathbb{R}^n \times \mathbb{R} \times \mathbb{R}$. Let $\pi : \mathbb{R}^n \times \mathbb{R} \times \mathbb{R} \to \mathbb{R} \times \mathbb{R}$ be the projection $\pi(x, \epsilon, y) = (\epsilon, y)$. Then $\pi_{|V} : V \to \mathbb{R} \times \mathbb{R}$ is proper in the neighborhood of the origin. Hence $\pi(V)$ is closed and semi-analytic in the neighborhood of the origin. Let us set $Y_1 = \mathbb{R} \times \{0\}$ and $Y_2 = \overline{\pi(V) \setminus Y_1}$. The set Y_2 is semi-analytic.

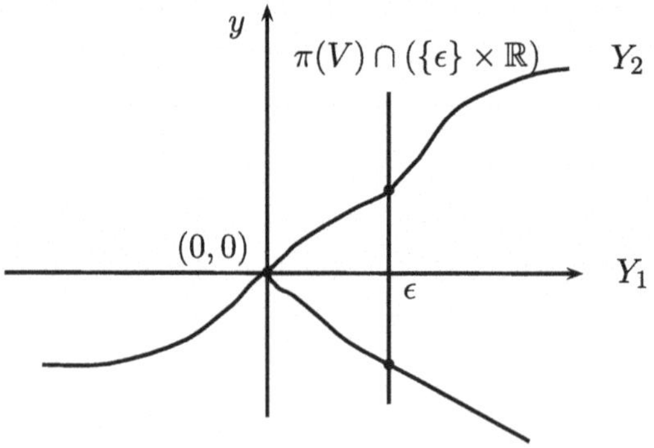

If $\epsilon \neq 0$ then

$$\pi(V) \cap (\{\epsilon\} \times \mathbb{R}) = \{\epsilon\} \times \left\{\text{critical values of } f_{|S_\epsilon^{n-1}}\right\}.$$

So $\pi(V) \cap (\{\epsilon\} \times \mathbb{R})$ is finite and $\dim \pi(V) = \dim Y_2 = 1$. Therefore 0 is an isolated point of $Y_1 \cap Y_2$. By the Łojasiewicz's inequality [26], there exist $C > 0$ and $\alpha > 0$ such that

$$d\left((\epsilon, y), Y_1\right) \geq Cd\left((\epsilon, y), \{0\}\right)^{2\alpha},$$

for $(\epsilon, y) \in Y_2$. Here $d((\epsilon, y), Z)$ is the distance from (ϵ, y) to the set Z. This implies that $|y| \geq C\epsilon^{2\alpha}$ for $(\epsilon, y) \in Y_2$ sufficiently close to $(0, 0)$ and that for $x \in \Sigma_{f,\omega} \setminus \{f = 0\}$ close to the origin,

$$|f(x)| \geq C\omega(x)^q, \qquad (*)$$

if $q \in \mathbb{N}$ and $q \geq \alpha$.

Proposition 3.3.1 ([33], p. 412, Lemma 1) *Let $f : (\mathbb{R}^n, 0) \to (\mathbb{R}, 0)$ be a real analytic function germ. Let $g : (\mathbb{R}^n, 0) \to (\mathbb{R}, 0)$ be defined by $g = f - c\omega^q$ with $c \in]0, C[$, $q \in \mathbb{N}$ and $q \geq \alpha$. Then g has an isolated critical point at the origin. Moreover, for $0 < \epsilon \ll 1$,*

$$\chi(\{f \leq 0\} \cap S_\epsilon^{n-1}) = 1 - \text{Ind}(\nabla g, 0).$$

Proof We work in an open set U, $0 \in U$, as above. Note that $\Sigma_{f,\omega} = \Sigma_{g,\omega}$ in the neighborhood of the origin. Furthermore we see that g does not vanish on $\Sigma_{g,\omega} \setminus \{0\}$ by the inequality $(*)$. Therefore if ϵ is small enough, 0 is a regular of $g_{|S_\epsilon^{n-1}}$. This implies that g has an isolated critical point at the origin because $\{\nabla g = 0\} \subset \{g = 0\}$ and $\{\nabla g = 0\} \subset \Sigma_{g,\omega}$ in a neighborhood of the origin.

Let us fix $\epsilon > 0$ sufficiently small. Let $N_f = \{x \in S_\epsilon^{n-1} \mid f(x) \leq 0\}$ and

$$N_g = \{x \in S_\epsilon^{n-1} \mid g(x) \leq 0\} = \{x \in S_\epsilon^{n-1} \mid f(x) \leq c\epsilon^q\}.$$

Then $N_f \subset N_g$. Furthermore, $f_{|S_\epsilon^{n-1}}$ has no critical point on $N_g \setminus N_f$ by the inequality $(*)$ and so N_f is a retract by deformation of N_g. But by Corollary 3.2.2,

$$\chi(N_g) = 1 - \text{Ind}(\nabla g, 0). \qquad \square$$

Of course, applying this argument to $-f$, we obtain a similar formula for $\chi(\{f \geq 0\} \cap S_\epsilon^{n-1})$ and, as a consequence of the Mayer–Vietoris sequence, a formula for $\chi(\{f = 0\} \cap S_\epsilon^{n-1})$.

When f is homogeneous, we can improve the previous result.

Proposition 3.3.2 ([6], p. 550, Theorem 5 or [35], p. 242, Lemma 3) *Let $f : \mathbb{R}^n \to \mathbb{R}$ be a homogeneous polynomial of degree d. Let $q = d + 1$ if d is odd or $q = d + 2$ if d is even. Then the function g defined by*

$$g(x) = f(x) - \frac{x_1^q + \cdots + x_n^q}{q},$$

has an algebraically isolated critical point at 0 *(i.e.* 0 *is an isolated zero of* $\nabla g_{\mathbb{C}}^{-1}(0)$*) and*

$$\chi\left(\{f \leq 0\} \cap S^{n-1}\right) = 1 - \mathrm{Ind}(\nabla g, 0).$$

Remark 3.3.3 Since 0 is an algebraically isolated critical point of g, we can use the Eisenbud–Levine formula to compute $\mathrm{Ind}(\nabla g, 0)$, which was not necessarily the case in Proposition 3.3.1.

Remark 3.3.4 In [35], Szafraniec studied weighted homogeneous polynomials. The above formula is a particular case of the results proved in [35].

3.4 Semi-analytic Versions

As an application of Morse theory for manifolds with corners, we extend the Khimshiashvili formula to a class of semi-analytic sets.

Let $f : (\mathbb{R}^n, 0) \to (\mathbb{R}, 0)$ be an analytic function-germ with an isolated critical point at 0. We will denote by M_δ the Milnor fibre $f^{-1}(\delta) \cap B_\epsilon^n$, by $\mathcal{L}$ the link of the singularity $f^{-1}(0) \cap S_\epsilon^{n-1}$ and by $\mathcal{A}_+$ (resp. $\mathcal{A}_-$) the set $\{x \in S_\epsilon^{n-1} \mid f(x) \geq 0\}$ (resp. $\{x \in S_\epsilon^{n-1} \mid f(x) \leq 0\}$). For any subset W of $\mathbb{R}^n$, for all $k \in \{1, \ldots, n\}$ and for all $\nu = (\nu_1, \ldots, \nu_k) \in \{0, 1\}^k$, we write

$$W(\nu) = W(\nu_1, \ldots, \nu_k) = W \cap \{(-1)^{\nu_1} x_1 \geq 0, \ldots, (-1)^{\nu_k} x_k \geq 0\},$$

and $|\nu| = \sum_{i=1}^k \nu_i$. For each $k \in \{1, \ldots, k\}$, let $H_k(f) : (\mathbb{R}^n, 0) \to (\mathbb{R}^n, 0)$ be defined by

$$H_k(f) = \left(x_1 \frac{\partial f}{\partial x_1}, \ldots, x_k \frac{\partial f}{\partial x_k}, \frac{\partial f}{\partial x_{k+1}}, \ldots, \frac{\partial f}{\partial x_n}\right).$$

The next results were proved in [12].

Theorem 3.4.1 *If* 0 *is isolated in* $H_k(f)^{-1}(0)$ *then*

$$\sum_{\nu \in \{0,1\}^k} (-1)^{|\nu|} \chi((M_\delta(\nu)) = -\mathrm{sign}(-\delta)^n \mathrm{Ind}(H_k(f), 0).$$

Proof For each $l \in \{0, \ldots, k\}$, we write I_l for a subset of $\{1, \ldots, k\}$ with l elements and $\overline{I_l}$ for its complement in $\{1, \ldots, k\}$. We put

$$F_{I_l} = \{x \mid x_i = 0 \text{ for each } i \in I_l\}.$$

By our assumption on $H_k(f)$, f admits an isolated critical point at 0 on each space $F_{I_l}, l \in \{0, \dots, k\}$. This implies that, on each of these spaces, $f^{-1}(0)$ intersects S_ϵ^{n-1} transversally for ϵ sufficiently small. By transversality, the same is true for a regular fibre $f^{-1}(\delta)$, $0 < |\delta| \ll \epsilon$. Moving f a little if necessary, we can assume that f is a Morse function on each manifold with corners $\{(-1)^{\nu_1} x_1 \geq 0, \dots, (-1)^{\nu_k} x_k \geq 0\}$, $(\nu_1, \dots, \nu_k) \in \{0, 1\}^k$. Let $\delta > 0$ be a small regular value of f and let us consider the set $\{P\}$ of correct critical points of f on these 2^k manifolds such that $|f(P)| < \delta$. Note that none of these points belongs to S_ϵ^{n-1} because of our above transversality argument and that $\{P\}$ is exactly the zero set of $H_k(f)$, after the small perturbation of f.

For each set I_l and for each k-tuple ν of $\{0, 1\}^k$, we define a subset $C_{I_l}^\nu$ of $\{P\}$ as follows:

$$C_{I_l}^\nu = \left\{ P \in F_{I_l} \mid (-1)^{\nu_i} \frac{\partial f}{\partial x_i}(P) > 0 \text{ if } i \in I_l \text{ and } (-1)^{\nu_j} x_j(P) > 0 \text{ if } j \in \overline{I_l} \right\}.$$

By Theorem 2.3.14, we have for a fixed ν

$$\chi\big(f^{-1}([-\delta, \delta]) \cap B_\epsilon^n \cap \{(-1)^{\nu_i} x_i \geq 0, i = 1, \dots, k\}, M_{-\delta}(\nu)\big) = \sum_{l=0}^{k} \sum_{I_l} \sum_{P \in C_{I_l}^\nu} (-1)^{\lambda(P)},$$

where $\lambda(P)$ is the Morse index of P. Since the semi-analytic set

$$f^{-1}([-\delta, \delta]) \cap B_\epsilon^n \cap \{(-1)^{\nu_i} x_i \geq 0, i = 1, \dots, k\}$$

is contractible, this gives

$$(-1)^{|\nu|} \chi(M_{-\delta}(\nu)) = (-1)^{|\nu|} - (-1)^{|\nu|} \sum_{l=0}^{k} \sum_{I_l} \sum_{P \in C_{I_l}^\nu} (-1)^{\lambda(P)}.$$

It remains to relate $(-1)^{|\nu|} \cdot (-1)^{\lambda(P)}$ to the Poincaré–Hopf of $H_k(f)$ at P. This index is the sign of the Jacobian determinant of the map $H_k(f)$ at P, which is easily seen to be

$$\left(\prod_{i \in I_l} \operatorname{sign} \frac{\partial f}{\partial x_i}(P) \prod_{j \in \overline{I_l}} \operatorname{sign} x_j(P) \right) \cdot (-1)^{\lambda(P)}.$$

The product of the signs in front of $(-1)^{\lambda(P)}$ is equal to $(-1)^{|\nu|}$ and we just have to sum over all the k-tuples ν to get the result for a negative regular value of f. The formula for a positive regular value is obtained replacing f by $-f$. □

The following corollary is obtained with the same deformation arguments and using the Mayer–Vietoris sequence as in Corollary 3.2.2.

Corollary 3.4.2 *If* 0 *is isolated in* $H_k(f)^{-1}(0)$, *then*

$$\sum_{\nu\in\{0,1\}^k} (-1)^{|\nu|}\chi(\mathcal{A}_-(\nu)) = -\mathrm{Ind}(H_k(f),0),$$

$$\sum_{\nu\in\{0,1\}^k} (-1)^{|\nu|}\chi(\mathcal{A}_+(\nu)) = (-1)^{n-1}\mathrm{Ind}(H_k(f),0).$$

If n is even, we have

$$\sum_{\nu\in\{0,1\}^k} (-1)^{|\nu|}\chi(\mathcal{L}(\nu)) = -2\ \mathrm{Ind}(H_k(f),0).$$

If n is odd, we have

$$\sum_{\nu\in\{0,1\}^k} (-1)^{|\nu|}\chi(\mathcal{L}(\nu)) = 0.$$

Let us apply this corollary to get a generalization of Proposition 3.3.1. We consider an analytic function-germ $f : (\mathbb{R}^n, 0) \to (\mathbb{R}, 0)$ and we do not assume that 0 is an isolated critical point of f. We suppose that f is defined in an open set $U \subset \mathbb{R}^n$ and $0 \in U$. For $i, j = 1, \ldots, n$, $i \neq j$, let

$$\Delta_{ij}(f) = \begin{vmatrix} \frac{\partial\omega}{\partial x_i} & \frac{\partial\omega}{\partial x_j} \\ \frac{\partial f}{\partial x_i} & \frac{\partial f}{\partial x_j} \end{vmatrix}.$$

For each set I_l, we define

$$\Sigma_{I_l}(f) = \left\{x \in F_{I_l} \cap U \mid \Delta_{ij}(f)(x) = 0 \text{ for all } (i,j) \in \overline{I_l}^2,\ i \neq j\right\}.$$

Note that for $l < n$, $\Sigma_{I_l}(f) \cap S_\epsilon^{n-1}$ is the set of critical points of $f_{|F_{I_l}\cap S_\epsilon^{n-1}}$. If $l = n$ then $I_n = \{1, \ldots, n\}$ and $\Sigma_{I_n}(f) = \{0\}$.

Applying the method of Szafraniec explained in Sect. 3.3 and because there is a finite number of I_l's, we see that there exist $C > 0$ and $\alpha > 0$ such that for $x \in \left(\cup_{l=1}^k \cup_{I_l} \Sigma_{I_l}(f)\right) \setminus \{f = 0\}$,

$$|f(x)| \geq C\omega(x)^q,$$

for $q \in \mathbb{N}$ and $q \geq \alpha$.

Proposition 3.4.3 *Let* $f : (\mathbb{R}^n, 0) \to (\mathbb{R}, 0)$ *be a real analytic function-germ. Let* $g : (\mathbb{R}^n, 0) \to (\mathbb{R}, 0)$ *be defined by* $g = f - c\omega^q$ *with* $c \in]0, C[$, $q \in \mathbb{N}$ *and* $q \geq \alpha$. *Then for* $k \in \{1, \ldots, n\}$, $H_k(g)$ *has an isolated zero at the origin. Moreover,*

$$\sum_{\nu \in \{0,1\}^k} (-1)^{|\nu|} \chi(\mathcal{A}_-(\nu)) = -\text{Ind}(H_k(g), 0).$$

Proof We see that $H_k(g)$ has an isolated zero at 0 if and only if for each set I_l, $l \in \{0, \ldots, k\}$, $g_{|F_{I_l}} : (F_{I_l}, 0) \to (\mathbb{R}, 0)$ has an isolated critical point at $0 \in F_{I_l}$. So we can conclude that $H_k(g)$ has an isolated zero as in Proposition 3.3.1. The rest of the proof is similar to Proposition 3.3.1. □

Replacing f by $-f$, we obtain a similar formula for

$$\sum_{\nu \in \{0,1\}^k} (-1)^{|\nu|} \chi(\mathcal{A}_+(\nu)),$$

and applying the Mayer–Vietoris sequence, a formula for

$$\sum_{\nu \in \{0,1\}^k} (-1)^{|\nu|} \chi(\mathcal{L}(\nu)).$$

When f is homogeneous, we can improve this result.

Proposition 3.4.4 *Let $f : \mathbb{R}^n \to \mathbb{R}$ be a homogeneous polynomial of degree d. Let $q = d + 1$ if d is odd and $q = d + 2$ if d is even. Then for $k \in \{1, \ldots, n\}$, the map $H_k(g)$ has an algebraically isolated zero at 0 where g is defined by*

$$g(x) = f(x) - \frac{x_1^q + \cdots + x_n^q}{q}.$$

Moreover, we have

$$\sum_{\nu \in \{0,1\}^k} (-1)^{|\nu|} \chi(\mathcal{A}_-(\nu)) = -\text{Ind}(H_k(g), 0).$$

As already mentioned, we can use the Eisenbud–Levine formula to compute $\text{Ind}(H_k(g), 0)$.

Remark 3.4.5 In [12], we studied the more general case of weighted homogeneous polynomials.

4 Degree Formulas and Signature Formulas for the Euler Characteristic of Algebraic Sets and Semi-algebraic Sets

Let $F = (F_1, \ldots, F_k) : \mathbb{R}^n \to \mathbb{R}^k$ be a polynomial mapping and let $W = F^{-1}(0)$. Let $G_1, \ldots, G_l : \mathbb{R}^n \to \mathbb{R}$ be polynomials. We would like to compute

$$\chi(W \cap \{G_1?_1 0, \ldots, G_l?_l 0\}),$$

where $?_j \in \{\leq, <, >, \geq\}$ for $j \in \{1, \ldots, l\}$, in terms of the polynomials $F_1, \ldots, F_k$ and $G_1, \ldots, G_l$. We will see that it is possible in some cases. We will start with the case of algebraic sets and then give results for some classes of semi-algebraic sets.

4.1 Case of 0-Dimensional Algebraic Sets

Let us consider the algebra $A_F = \frac{\mathbb{R}[x_1,\ldots,x_n]}{(F_1,\ldots,F_k)}$ and let us assume that

$$\dim_{\mathbb{R}} A_F < +\infty.$$

This implies that W is a finite number of points and that $\chi(W) = \#W$.

For each $f \in A_F$, let $L_f : A_F \to A_F$ be the linear endomorphism of multiplication by f, i.e. $\forall g \in A_F$, $L_f(g) = fg$. Let Q be the quadratic form A_F defined by $Q(f) = \text{Trace}(L_{f^2})$.

Theorem 4.1.1 ([3], p. 280, Proposition 4.2 or [31], p. 205, Theorem 2.1) *We have*

$$\text{signature } Q = \#W = \chi(W).$$

4.2 Case of a Compact Algebraic Set

Here we present a formula due to Bruce [6] and Szafraniec [33, 36] that expresses the Euler characteristic of a compact algebraic set of $\mathbb{R}^n$ as the signature of a quadratic form. It relies strongly on the formula for the Euler characteristic of the link of a homogeneous singularity (Proposition 3.3.2).

We recall that $W = F_1^{-1}(0) \cap \cdots \cap F_k^{-1}(0)$ and we assume that W is compact. Let d be the highest degree of the F_i's. For $i = 1, \ldots, k$, let G_i be the homogeneous polynomial function in the variables $x_0, x_1, \ldots, x_n$ of degree $d+1$ defined by

$$G_i(x_0, x_1, \ldots, x_n) = x_0^{d+1} F_i\left(\frac{x_1}{x_0}, \ldots, \frac{x_n}{x_0}\right).$$

Let G be the polynomial function given by $G = G_1^2 + \cdots + G_k^2$. Then G is homogeneous of degree $2d+2$. Let $q = 2d+4$ and let $\bar{G} : \mathbb{R}^{n+1} \to \mathbb{R}$ be defined by

$$\bar{G}(x_0, x_1, \ldots, x_n) = G(x_0, x_1, \ldots, x_n) - \frac{x_0^q + x_1^q + \cdots + x_n^q}{q}.$$

Theorem 4.2.1 ([6], p. 550, Proposition 7 or [33], p. 412, Theorem 1) *The polynomial function $\bar{G}$ has an algebraically isolated critical point at the origin and*

$$\chi(W) = \frac{1}{2}\left((-1)^n - \text{Ind}(\nabla\bar{G}, 0)\right).$$

Proof Since the G_i's are homogeneous, W is homeomorphic to the sets

$$\left\{x \in S^n \mid x_0 > 0, G_1(x) = \cdots = G_k(x) = 0\right\},$$

and

$$\left\{x \in S^n \mid x_0 < 0, G_1(x) = \cdots = G_k(x) = 0\right\}.$$

Since W is compact,

$$\chi\left(\left\{x \in S^n \mid G_1(x) = \cdots = G_k(x) = 0\right\}\right) = 2\chi(W) + \chi(S^{n-1}).$$

So we get that

$$\chi(W) = \frac{1}{2}\left(\chi\left(\left\{x \in S^n \mid G_1(x) = \cdots = G_k(x) = 0\right\}\right) - 1 - (-1)^{n-1}\right) = \frac{1}{2}\left(\chi\left(\left\{x \in S^n \mid G(x) \leq 0\right\}\right) - 1 + (-1)^n\right),$$

and, by Proposition 3.3.2,

$$\chi(W) = \frac{1}{2}\left(1 - \text{Ind}(\nabla\bar{G}, 0) - 1 + (-1)^n\right).$$

□

Remark 4.2.2 1. As observed by Zbigniew Szafraniec, if W is not compact, the formula is still valid replacing $\chi(W)$ by $\chi_c(W)$.
2. Since $\bar{G}$ has an algebraically isolated critical point at 0, $\text{Ind}(\nabla\bar{G}, 0)$ can be computed as the signature of a quadratic form on

$$\frac{\mathbb{R}\{x_0, x_1, \ldots, x_n\}}{\left(\frac{\partial\bar{G}}{\partial x_0}, \frac{\partial\bar{G}}{\partial x_1}, \ldots, \frac{\partial\bar{G}}{\partial x_n}\right)},$$

thanks to the Eisenbud–Levine formula.

4.3 Case of a Smooth Complete Intersection

The above formula works only for compact algebraic sets of $\mathbb{R}^n$. When W is not compact, Bruce and Szafraniec also obtained a signature formula for $\chi(W)$, but it is not effective since one can not compute explicitly one of the constant involved in the formula.

In this section, we present a method due to Szafraniec [34] in the case where W is a smooth complete intersection of codimension k. We recall that $F = (F_1, \ldots, F_k) : \mathbb{R}^n \to \mathbb{R}^k$ is a polynomial mapping and that $W = F^{-1}(0)$. Furthermore, we assume that $1 \leq k \leq n-1$ and that for $x \in W$, $\text{rank}[DF(x)] = k$. This implies that W is a smooth manifold of dimension $n-k$ (maybe empty). Therefore, we can use tools of differential topology to study the topology of W.

Let $(x; \lambda) = (x_1, \ldots, x_n; \lambda_1, \ldots, \lambda_k)$ be a coordinate system of $\mathbb{R}^{n+k}$. Let H be the following polynomial mapping:

$$\begin{array}{cccc} H : & \mathbb{R}^{n+k} & \to & \mathbb{R}^{n+k} \\ & (x_1, \ldots, x_n; \lambda_1, \ldots, \lambda_k) & \mapsto & (x + \sum_{i=1}^k \lambda_i \nabla F_i(x), F(x)). \end{array}$$

Theorem 4.3.1 ([34], p. 199, Theorem 2.2) *The set $H^{-1}(0)$ is compact. Let $R > 0$ be such that $H^{-1}(0) \subset B_R^{n+k}$ then*

$$\chi(W) = (-1)^k \text{Ind}(H, \infty),$$

where $\text{Ind}(H, \infty)$ *is the Brouwer degree of the map* $\frac{H}{|H|} : S_R^{n+k-1} \to S^{n+k-1}$.

Proof We just give a sketch of proof. The reader can refer back to Szafraniec's paper for the details. The fact that $H^{-1}(0)$ is compact is not difficult to prove. For simplicity, we assume that H has only non-degenerate roots. This means that for each $(p, \beta) \in H^{-1}(0)$, $\det[DH(p, \beta)] \neq 0$, where DH is the Jacobian matrix of H. Let us write

$$H^{-1}(0) = \{(p_1, \beta_1), \ldots, (p_r, \beta_r)\}.$$

Let $\rho(x) = \frac{1}{2}(x_1^2 + \cdots + x_n^2)$, then $\nabla \rho(x) = x$ for $x \in \mathbb{R}^n$. The points p_i are exactly the critical points of $\rho_{|W}$. Furthermore they are Morse critical points and, if s_i is the Morse index of $\rho_{|W}$ at p_i then, by a computation of determinants, we get

$$(-1)^{s_i} = (-1)^k \text{sign} \det[DH(p_i, \beta_i)].$$

We can apply Morse theory to $\rho_{|W}$ and we obtain that

$$\chi(W) = \sum_{i=1}^r (-1)^{s_i} = \sum_{i=1}^r (-1)^k \text{sign} \det[DH(p_i, \beta_i)] =$$

$$(-1)^k \left(\sum_{i=1}^r \text{sign} \det[DH(p_i, \beta_i)] \right).$$

But by the application after Theorem 2.1.7, we know that

$$\text{Ind}(H, \infty) = \sum_{i=1}^{r} \text{sign det}[DH(p_i, \beta_i)]. \qquad \square$$

The formula in the above theorem is completely explicit and if

$$\dim_{\mathbb{R}} \frac{\mathbb{R}[x_1, \ldots, x_n; \lambda_1, \ldots, \lambda_k]}{\langle H \rangle} < +\infty,$$

where $\langle H \rangle$ is the ideal generated by the components of H, then $\text{Ind}(H, \infty)$ can be computed as the signature of a quadratic form on this polynomial algebra (see [38]). However, we have to add k more variables and this makes the computations difficult in practice. In the sequel, we will briefly present a second method due to Szafraniec and his collaborators which is more algebraic and more effective, because it does not require any extra variable.

Let A be the following algebra:

$$A = \frac{\mathbb{R}[x_1, \ldots, x_n]}{\left(F_1, \ldots, F_k; \frac{\partial(\rho, F_1, \ldots, F_k)}{\partial(x_{i_1}, \ldots, x_{i_{k+1}})}\right)},$$

where $\frac{\partial(\rho, F_1, \ldots, F_k)}{\partial(x_{i_1}, \ldots, x_{i_{k+1}})}$ is the $(k+1) \times (k+1)$-minor of the Jacobian matrix $D(\rho, F_1, \ldots, F_k)$ with respect to the variables $x_{i_1}, \ldots, x_{i_{k+1}}$. We assume that A is a finite dimensional real vector space. The following theorem was proved in [9].

Theorem 4.3.2 ([9], p. 374, Theorem 1.4) *We have*

$$\chi(W) \equiv \dim_{\mathbb{R}} A \mod 2.$$

This result was later improved by Szafraniec in [37]. He managed to construct a linear functional φ on A and with this linear functional, he defined two bilinear symmetric forms Φ and Φ_M in the following way:

$$\forall (f, g) \in A \times A, \quad \Phi(f, g) = \varphi(fg) \text{ and } \Phi_M(f, g) = \varphi(Mfg),$$

where

$$M = \frac{\partial(F_1, \ldots, F_k)}{\partial(x_1, \ldots, x_k)}.$$

The following theorem relates $\chi(W)$ to the signatures of Φ and Φ_M.

Theorem 4.3.3 ([37], p. 357, Theorem 5.5) *If $n - k$ is odd then we have*

$$\chi(W) = (-1)^k \text{signature } \Phi.$$

If $n - k$ is even then we have

$$\chi(W) = \text{signature } \Phi_M.$$

4.4 Case of Complete Intersections with Isolated Singularities

In two papers [11, 13], we improved the above results of Szafraniec in some cases where W has isolated singularities.

In [11], we did not assume that W was smooth but we still assumed that the dimension of the above algebra A was finite. This implies that W can admit a finite number of singularities that we denote by $q_1, \ldots, q_r$.

Theorem 4.4.1 ([11], p. 132, Theorem 2.2) *We have*

$$\chi(W) + \sum_{j=1}^{r} \mu(F_{\mathbb{C}}, q_j) \equiv \dim_{\mathbb{R}} A \bmod 2,$$

where $F_{\mathbb{C}}$ is the complexification of F and $\mu(F_{\mathbb{C}}, q_j)$ is its Milnor number at q_j.

In the same paper, we were able to replace this mod 2 equality with an equality in $\mathbb{Z}$ in the case of curves ($k = n - 1$) ([11], Theorem 5.2, Corollary 5.3 and Theorem 5.4) and for odd dimensional hypersurfaces (n even, $k = 1$) ([11], Theorem 4.3 and Corollary 4.4).

Later in [13], we found a new method that works for all hypersurfaces, even or odd dimensional ones. Let us explain it now.

Let $F : \mathbb{R}^n \to \mathbb{R}$ be a polynomial function such that its set of critical points $\nabla F^{-1}(0)$ is finite and such that $F(0) > 0$. This implies that $W = F^{-1}(0)$ is a hypersurface with isolated singularities and that $0 \notin W$. Let $(x, \lambda) = (x_1, \ldots, x_n, \lambda)$ be a coordinate system of $\mathbb{R}^{n+1}$ and let us define four polynomial mappings H, K, L_1 and L_2 as follows:

$$H(x, \lambda) = (\lambda x + \nabla F, F), \ K(x, \lambda) = (\lambda x + \nabla F, \lambda F), \ L_1(x, \lambda) = (\nabla F, \lambda F - 1),$$

$$\text{and } L_2(x, \lambda) = (\nabla F, \lambda F^2 - 1).$$

The next theorem can be viewed as a global version of the Khimshiashvili formula (Theorem 3.2.1).

Theorem 4.4.2 ([13], p. 330, Theorem 5.10) *The sets* $H^{-1}(0)$, $K^{-1}(0)$, $L_1^{-1}(0)$ *and* $L_2^{-1}(0)$ *are compact. If n is even, then*

$$\chi(W) = \text{Ind}(H,\infty) + \text{Ind}(\nabla F,\infty) - \text{Ind}(L_2,\infty),$$

$$\chi(\{F \geq 0\}) - \chi(\{F \leq 0\}) = 1 - \text{Ind}(K,\infty) + \text{Ind}(L_1,\infty).$$

If n is odd, then

$$\chi(W) = \text{Ind}(K,\infty) - \text{Ind}(L_1,\infty),$$

$$\chi(\{F \geq 0\}) - \chi(\{F \leq 0\}) = 1 - \text{Ind}(H,\infty) - \text{Ind}(\nabla F,\infty) + \text{Ind}(L_2,\infty).$$

Proof We will just prove briefly the equality

$$\chi(W) = \text{Ind}(K,\infty) - \text{Ind}(L_1,\infty),$$

when n is odd. The fact that $K^{-1}(0)$ and $L_1^{-1}(0)$ are compact is easy to prove. Let us write

$$\nabla F^{-1}(0) = \{q_1,\dots,q_t\} = \{q_1,\dots,q_r\} \cup \{q_{r+1},\dots,q_t\},$$

where for $i \in \{1,\dots,r\}$, $q_i \in W$ and for $j \in \{r+1,\dots,t\}$, $q_j \notin W$.

Let us choose $R' \gg 1$ sufficiently big so that $\chi(W) = \chi(W \cap B^n_{R'})$ and $\nabla F^{-1}(0) \subset B^n_{R'}$. Let δ, $0 < |\delta| \ll 1$, be a small regular value of F. The following equality is proved in the same way as the Khimshiashvili formula (Theorem 3.2.1):

$$\chi(F^{-1}(\delta) \cap B^n_{R'}) = \chi(W \cap B^n_{R'}) + (\text{sign }\delta) \sum_{i=1}^{r} \text{Ind}(\nabla F, q_i).$$

Let K_δ be the following perturbation of K:

$$K_\delta(x,\lambda) = (\lambda x + \nabla F, \lambda(F-\delta)).$$

Let $R > 0$ be such that $K^{-1}(0) \subset B^{n+1}_R$. If R is sufficiently big and δ sufficiently small then the mapping $\frac{K}{|K|} : S^n_R \to S^n$ and $\frac{K_\delta}{|K_\delta|} : S^n_R \to S^n$ are homotopic. This implies that $\text{Ind}(K,\infty)$ is equal to the topological degree of the map $\frac{K_\delta}{|K_\delta|} : S^n_R \to S^n$.

Let $\rho(x) = \frac{1}{2}(x_1^2 + \cdots + x_n^2)$. Then it is not difficult to see that

$$K_\delta^{-1}(0) \cap B^{n+1}_R = \left\{(p,\lambda) \mid p \text{ critical point of } \rho_{|F^{-1}(\delta)\cap B^n_{R'}}\right\} \bigcup \sqcup_{j=1}^{t} \{(q_j,0)\}.$$

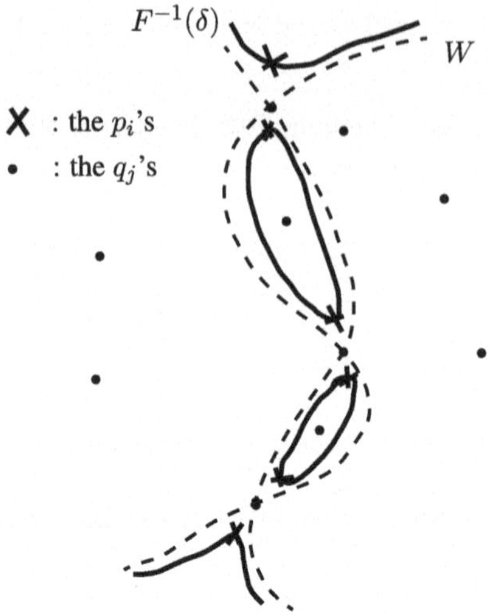

For simplicity, let us assume that $\rho_{|F^{-1}(\delta)\cap B^n_{R'}}$ is a Morse function. Then it admits a finite number of critical points $p_1, \ldots, p_m$, with respective indices $s_1, \ldots, s_m$, and

$$K_\delta^{-1}(0) \cap B_R^{n+1} = \{(p_1, \lambda_1), \ldots, (p_m, \lambda_m)\} \bigcup \sqcup_{j=1}^t \{(q_j, 0)\}.$$

A computation of determinants gives that

$$(-1)^{s_i} = \text{sign}\left[(\lambda_i)^{n-1}\text{det}[DK_\delta(p_i, \lambda_i)]\right],$$

for $i = 1, \ldots, m$. By Morse theory and since n is odd, we have

$$\chi(F^{-1}(\delta) \cap B^n_{R'}) = \sum_{i=1}^m (-1)^{s_i} = \sum_{i=1}^m \text{sign det}[DK_\delta(p_i, \lambda_i)].$$

It remains to study $\text{Ind}(K_\delta, (q_j, 0))$ for $j = 1, \ldots, t$. It is not difficult to see that

$$\text{Ind}(K_\delta, (q_j, 0)) = \text{sign}(F(q_j) - \delta)\text{Ind}(\nabla F, q_j).$$

Actually, this equality is easy to prove when q_j is a non-degenerate zero of ∇F. To establish it in the general case, we just use a small perturbation of F. Putting together all these results, we obtain that

$$\text{Ind}(K, \infty) = \chi(F^{-1}(\delta) \cap B^n_{R'}) + \sum_{j=1}^t \text{sign}(F(q_j) - \delta)\text{Ind}(\nabla F, q_j) =$$

$$\chi(F^{-1}(\delta) \cap B^n_{R'}) + \sum_{j=1}^{r} \text{sign}(F(q_j) - \delta)\text{Ind}(\nabla F, q_j) + \sum_{j=r+1}^{t} \text{sign}(F(q_j) - \delta)\text{Ind}(\nabla F, q_j).$$

For $j = 1, \ldots, r$, we have $\text{sign}(F(q_j) - \delta) = -\text{sign}(\delta)$ and for $j = r+1, \ldots, t$, $\text{sign}(F(q_j) - \delta) = \text{sign}(F(q_j))$ because δ is a small regular value of f. Therefore, we get

$$\text{Ind}(K, \infty) = \chi(F^{-1}(\delta) \cap B^n_{R'}) - \text{sign}(\delta)\sum_{j=1}^{r} \text{Ind}(\nabla F, q_j) +$$

$$\sum_{j=r+1}^{t} \text{sign}(F(q_j))\text{Ind}(\nabla F, q_j) = \chi(W) + \sum_{j=r+1}^{t} \text{sign}(F(q_j))\text{Ind}(\nabla F, q_j).$$

But the zero set of L_1 is exactly $\cup_{j=r+1}^{t}(q_j, \frac{1}{F(q_j)})$ and a straightforward calculation shows that

$$\text{Ind}\left(L_1, \left(q_j, \frac{1}{F(q_j)}\right)\right) = \text{sign}(F(q_j))\text{Ind}(\nabla F, q_j). \qquad \square$$

Let us end this section with some remarks.

Remark 4.4.3 In [14], we generalized Szafraniec's results (Theorem 4.3.1) to the case when W admits a compact singular set, i.e. the set

$$\{x \in W \mid \text{rank}[DF(x)] < k\},$$

is compact. However, although the formula is completely explicit, it is not effective in practice because we need to add $k + 1$ variables and we use a Łojasiewicz exponent of size $O(2kd)^n$, where d is the degree of the map F.

4.5 Semi-algebraic Versions

4.5.1 0-Dimensional Semi-algebraic Sets

We consider a polynomial mapping $F = (F_1, \ldots, F_k) : \mathbb{R}^n \to \mathbb{R}^k$ such that the algebra $A_F = \frac{\mathbb{R}[x_1,\ldots,x_n]}{(F_1,\ldots,F_k)}$ is finite dimensional, i.e. $\dim_{\mathbb{R}} A_F < +\infty$. Let $G : \mathbb{R}^n \to \mathbb{R}$ be another polynomial. Keeping the notations of Sect. 4.1, we define a quadratic form Q_G on A_F as follows:

$$\forall f \in A_F,\ Q_G(f) = \text{Trace}(L_{f^2 G}).$$

Theorem 4.5.1 ([3], p. 280, Proposition 4.2 or [31], p. 205, Theorem 2.1) *We have*

$$\begin{aligned}&\text{signature } Q_G\\ &\quad = \#W \cap \{G > 0\} - \#W \cap \{G < 0\}\\ &\qquad\qquad\qquad\qquad = \chi(W \cap \{G > 0\}) - \chi(W \cap \{G < 0\}).\end{aligned}$$

When we have several polynomials $G_1, \ldots, G_l$ then applying the above formula to the polynomials $G_1^{\alpha_1} \ldots G_l^{\alpha_l}$, where $\alpha_i \in \{0, 1, 2\}$ for $i \in \{1, \ldots, l\}$, it is possible to compute the cardinalities

$$\#W \cap \{G_{i_1} *_{i_1} 0, \ldots, G_{i_m} *_{i_m} 0\},$$

where $m \in \{1, \ldots, l\}$, $\{i_1, \ldots, i_m\} \subset \{1, \ldots, l\}$ and $*_{i_j} \in \{=, <, >\}$ for $j \in \{1, \ldots, m\}$. An algorithm for such a computation is explained in [32], Sect. 5.1, in the univariate case, but it also works in the multivariate case.

4.5.2 Compact Semi-algebraic Sets

We assume that $W = F_1^{-1}(0) \cap \cdots \cap F_k^{-1}(0)$ is compact. Keeping the notations of Sect. 3.4 and Proposition 3.4.4, we obtain a formula for

$$\sum_{\nu \in \{0,1\}^l} (-1)^{|\nu|} \chi(W(\nu)),$$

l even and $l \in \{1, \ldots, n\}$. The method is similar to the one applied by Bruce and Szafraniec for $\chi(W)$.

Let d be the highest degree of the F_i's. For $i = 1, \ldots, k$, let G_i be the homogeneous polynomial function in the variables $x_1, \ldots, x_n, x_{n+1}$ of degree $d + 1$ defined by

$$G_i(x_1, \ldots, x_n, x_{n+1}) = x_{n+1}^{d+1} F_i \left(\frac{x_1}{x_{n+1}}, \ldots, \frac{x_n}{x_{n+1}} \right).$$

Let G be the polynomial function given by $G = G_1^2 + \cdots + G_k^2$. Then G is homogeneous of degree $2d + 2$. Let $q = 2d + 4$ and let $\bar{G} : \mathbb{R}^{n+1} \to \mathbb{R}$ be defined by

$$\bar{G}(x_1, \ldots, x_{n+1}) = G(x_1, \ldots, x_{n+1}) - \frac{x_1^q + \cdots + x_{n+1}^q}{q}.$$

Theorem 4.5.2 *For l even in $\{1, \ldots, n\}$, the polynomial function $H_l(\bar{G})$ has an algebraically isolated zero at the origin and*

$$\sum_{\nu\in\{0,1\}^l} (-1)^{|\nu|}\chi(W(\nu)) = -\frac{1}{2}\mathrm{Ind}(H_l(\bar{G},0)).$$

Proof For each $\nu \in \{0,1\}^l$, let $\bar{\nu} \in \{0,1\}^l$ be defined by $\bar{\nu}_i = 0$ if $\nu_i = 1$ and $\bar{\nu}_i = 1$ if $\nu_i = 0$. It is clear that $|\nu| + |\bar{\nu}| = l$ and that $(-1)^{|\nu|} = (-1)^{|\bar{\nu}|}$. Since the G_i's are homogeneous, $W(\nu)$ is homeomorphic to

$$\left\{x \in S^n \mid G_1(x) = \cdots = G_k(x) = 0, x_{n+1} > 0\right\} \\ \cap \left\{(-1)^{\nu_1}x_1 \geq 0, \ldots, (-1)^{\nu_l}x_l \geq 0\right\},$$

and $W(\bar{\nu})$ is homeomorphic to

$$\left\{x \in S^n \mid G_1(x) = \cdots = G_k(x) = 0, x_{n+1} < 0\right\} \\ \cap \left\{(-1)^{\nu_1}x_1 \geq 0, \ldots, (-1)^{\nu_l}x_l \geq 0\right\}.$$

Let $\mathcal{L} = \{x \in S^n \mid G_1(x) = \cdots = G_k(x) = 0\}$. Since W is compact, we have

$$\chi(\mathcal{L}(\nu)) = \chi(W(\nu)) + \chi(W(\bar{\nu})) + \chi(S^{n-1}(\nu)).$$

Using the equality $(-1)^{|\nu|} = (-1)^{|\bar{\nu}|}$, we find that

$$2\sum_{\nu\in\{0,1\}^l} (-1)^{|\nu|}\chi(W(\nu)) = \sum_{\nu\in\{0,1\}^l} (-1)^{|\nu|}\chi(\mathcal{L}(\nu)) - \sum_{\nu\in\{0,1\}^l} (-1)^{|\nu|}\chi(S^{n-1}(\nu)).$$

Applying Corollary 3.4.2 to the function $\omega(x) = x_1^2 + \cdots + x_n^2$, we see that

$$\sum_{\nu\in\{0,1\}^l} (-1)^{|\nu|}\chi(S^{n-1}(\nu)) = 0,$$

because for $l \in \{1, \ldots, n\}$, $\mathrm{Ind}(H_l(\omega), 0) = 0$ since at least one of the components of $H_l(\omega)$ is non-negative. Proposition 3.4.4 enables us to conclude because

$$\mathcal{L}(\nu) = \{x \in S^n \mid G(x) \leq 0\} \cap \{(-1)^{\nu_i}x_i \geq 0, i = 1, \ldots, l\}.$$ □

Remark 4.5.3 In [12], Theorem 5.1, we proved a more general formula without assumption on the parity of l. However it is more complicated than the one presented above since it is based on a version of Proposition 3.4.4 for weighted homogeneous polynomials, inspired by the results of Szafraniec [35, 36] on the link of a weighted homogeneous singularity.

4.5.3 Non-compact Semi-algebraic Sets

We present an adaptation, due to Julie Lapébie (Ph.D. student of the author), to semi-algebraic sets of Szafraniec's result explained in Theorem 4.3.1.

We still consider a polynomial mapping $F = (F_1, \dots, F_k) : \mathbb{R}^n \to \mathbb{R}^k$, with $1 \leq k \leq n-1$, and we assume that for $x \in W = F^{-1}(0)$, $\text{rank}[DF(x)] = k$. Let $G = (G_1, \dots, G_l) : \mathbb{R}^n \to \mathbb{R}^l$ be another polynomial map such that $k + l \leq n$. We assume that the sets W and $G_i^{-1}(0)$ fulfill the following transversality condition:

$$\forall \{i_1, \dots, i_s\} \subset \{1, \dots, k\}, \forall x \in W \cap G_{i_1}^{-1}(0) \cap \dots \cap G_{i_s}^{-1}(0),$$

$$\text{rank}[DF(x), DG_{i_1}(x), \dots, DG_{i_s}(x)] = k + s.$$

Let $(x, \lambda, \mu) = (x_1, \dots, x_n; \lambda_1, \dots, \lambda_k; \mu_1, \dots, \mu_l)$ be a coordinate system of $\mathbb{R}^{n+k+l}$. Let L be the following mapping:

$$\begin{array}{rccc} L & : & \mathbb{R}^{n+k+l} & \to & \mathbb{R}^{n+k+l} \\ & & (x, \lambda, \mu) & \mapsto & \Big(x + \sum_{i=1}^k \lambda_i \nabla F_i(x) + \sum_{j=1}^l \mu_j \nabla G_j(x), \\ & & & & F(x), \mu_1 G_1(x), \dots, \mu_l G_l(x) \Big). \end{array}$$

For any subset Y of $\mathbb{R}^n$ and for all $\nu = (\nu_1, \dots, \nu_l) \in \{0, 1\}^l$, we set

$$\begin{aligned} Y_G(\nu) &= Y_G(\nu_1, \dots, \nu_l) \\ &= Y \cap \{x \in \mathbb{R}^n \mid (-1)^{\nu_1} G_1(x) \geq 0, \dots, (-1)^{\nu_l} G_l(x) \geq 0\}. \end{aligned}$$

In her Ph.D. thesis [25] Lapébie proved the following theorem.

Theorem 4.5.4 *The set $L^{-1}(0)$ is compact. Let $R > 0$ be such that $L^{-1}(0) \subset B_R^{n+k+l}$ then*

$$\sum_{\nu \in \{0,1\}^l} (-1)^{|\nu|} \chi(W_G(\nu)) = (-1)^k \text{Ind}(L, \infty),$$

where $\text{Ind}(L, \infty)$ *is the Brouwer degree of the map* $\frac{L}{|L|} : S_R^{n+k+l-1} \to S^{n+k+l-1}$.

Proof We give a sketch of proof in the case $l = 1$ and $k + 1 < n$. We write $G = G_1$ and then

$$L(x, \lambda, \mu) = \left(x + \sum_{i=1}^k \lambda_i \nabla F_i(x) + \mu \nabla G(x), F(x), \mu G(x) \right).$$

We have to show that

$$\chi(W \cap \{G \geq 0\}) - \chi(W \cap \{G \leq 0\}) = (-1)^k \text{Ind}(L, \infty).$$

The fact that $L^{-1}(0)$ is compact is not too difficult to prove. Let $\rho(x) = \frac{1}{2}(x_1^2 + \cdots + x_n^2)$ then $\nabla\rho(x) = x$ for all $x \in \mathbb{R}^n$. For simplicity, let us assume that $\rho_{|W\cap\{G\geq 0\}}$ and $\rho_{|W\cap\{G\leq 0\}}$ are Morse correct functions. Then $L^{-1}(0)$ splits into two subsets:

$$L^{-1}(0) = \{(p_1, \beta_1, 0), \ldots, (p_m, \beta_m, 0)\} \sqcup \{(q_1, \beta'_1, \gamma_1), \ldots, (q_s, \beta'_s, \gamma_s)\},$$

where $p_1, \ldots, p_m$ are the critical points of $\rho_{|W}$ and $q_1, \ldots, q_s$ are those of $\rho_{|W\cap\{G=0\}}$. Note that since $\rho_{|W\cap\{G\geq 0\}}$ and $\rho_{|W\cap\{G\leq 0\}}$ are correct, $G(p_i) \neq 0$ for $i = 1, \ldots, m$ and $\gamma_j \neq 0$ for $j = 1, \ldots, s$. Moreover the points $(p_1, \beta_1, 0), \ldots, (p_m, \beta_m, 0)$ are non-degenerate zeroes of L and a computation of determinants gives that for $i \in \{1, \ldots, m\}$,

$$\text{sign det } DL(p_i, \beta_i, 0) = (\text{sign } G(p_i))\,(-1)^{\sigma_i + k},$$

where σ_i is the Morse index of $\rho_{|W}$ at p_i. Similarly the points $(q_1, \beta'_1, \gamma_1), \ldots, (q_s, \beta'_s, \gamma_s)$ are non-degenerate zeroes of L and for $j \in \{1, \ldots, s\}$,

$$\text{sign det } DL(q_j, \beta'_j, \gamma_j) = \left(\text{sign } \gamma_j\right)(-1)^{\tau_j + k + 1},$$

where τ_j is the Morse index of $\rho_{|W\cap\{G=0\}}$ at q_j. By Morse theory for manifolds with boundary, we have

$$\chi(W \cap \{G \geq 0\}) = \sum_{i \mid g(p_i)>0} (-1)^{\sigma_i} + \sum_{j \mid \gamma_j<0} (-1)^{\tau_j},$$

$$\chi(W \cap \{G \leq 0\}) = \sum_{i \mid g(p_i)<0} (-1)^{\sigma_i} + \sum_{j \mid \gamma_j>0} (-1)^{\tau_j},$$

and so

$$\begin{aligned}\chi(W \cap \{G \geq 0\}) - {} & \chi(W \cap \{G \leq 0\}) \\ &= \sum_{i=1}^{m} (\text{sign } G(p_i))(-1)^{\sigma_i} - \sum_{j=1}^{s} (\text{sign } \gamma_j)(-1)^{\tau_j}.\end{aligned}$$

Therefore we get

$$\begin{aligned}&\chi(W \cap \{G \geq 0\}) - \chi(W \cap \{G \leq 0\}) \\ &= (-1)^k \left(\sum_{i=1}^{m} \text{sign det } DL(p_j, \beta_j, 0) + \sum_{j=1}^{s} \text{sign det } DL(q_j, \beta'_j, \gamma_j) \right) \\ &= (-1)^k \text{Ind}(L, \infty).\end{aligned}$$

The proof of the general case is similar and uses Theorem 2.3.14. The extra difficulties come from the notations and the fact that we have to study the critical points of ρ restricted to 2^l manifolds with corners, as in the proof of Theorem 3.4.1. □

Remark 4.5.5 1. In her Ph.D. thesis, Lapébie also proved generalizations of Theorem 4.4.2.
2. In Theorem 4.5.2 and 4.5.4, we gave formulas for weighted sums of Euler characteristics of some semi-algebraic sets, the weights been equal to $+1$ or -1, but not for the Euler characteristic of the semi-algebraic sets considered. In [12], we presented a method that enables us to compute these Euler characteristics from the formulas for weighted sums of Euler characteristics.

5 Topology of the Real Milnor Fibration

In this chapter, we review some results about the real Milnor fibration and the topology of the real Milnor fibre. We start with some results due to Massey [27] which generalize the ones of Milnor [30], Chap. 11.

5.1 Milnor's Conditions (a) and (b) and the Fibration Theorem

Let $F = (f_1, \ldots, f_k) : (\mathbb{R}^n, 0) \to (\mathbb{R}^k, 0)$ be an analytic map-germ, $1 \le k \le n-1$, $V = F^{-1}(0)$ and Σ_F be the set of critical points of F, i.e.

$$\Sigma_F = \left\{x \in \mathbb{R}^n \mid \text{rank}(\nabla f_1(x), \ldots, \nabla f_k(x)) < k\right\},$$

(of course F is not constant).

Let $\rho(x) = \frac{1}{2}(x_1^2 + \cdots + x_n^2)$ and let us denote by $\Sigma_{F,\rho}$ the set of critical points of the pair (F, ρ), i.e.

$$\Sigma_{F,\rho} = \left\{x \in \mathbb{R}^n \mid \text{rank}(\nabla f_1(x), \ldots, \nabla f_k(x), \nabla \rho(x)) < k+1\right\}.$$

It is clear that $\Sigma_F \subseteq \Sigma_{F,\rho}$.

The following two conditions were introduced by Massey [27].

Definition 5.1.1 ([27]) Let F and ρ be as above.

1. We say that F satisfies Milnor's condition (a) at the origin if $\Sigma_F \subset V$ in a neighborhood of the origin.
2. We say that F satisfies Milnor's condition (b) at the origin if 0 is isolated in $V \cap \overline{\Sigma_{F,\rho} \setminus V}$ in a neighborhood of the origin.

Examples: (1) Let $f : (\mathbb{C}^n, 0) \to (\mathbb{C}, 0)$ be a holomorphic function-germ. It is well known that $\Sigma_f \subset f^{-1}(0)$ in a neighborhood of the origin so it satisfies Milnor's conditions (a).

Furthermore, Hamm and Lê [21], p. 323, proved that the Thom a_f-condition is satisfied for a Whitney stratification of V. Hence, Milnor's condition (b) is also satisfied.

(2) Similarly if $f : (\mathbb{R}^n, 0) \to (\mathbb{R}, 0)$ is an analytic function-germ then it satisfies Milnor's conditions (a) and (b). Condition (b) follows from the fact that the a_f-condition is satisfied (see [5], Proposition 10).

(3) Let $F : (\mathbb{R}^n, 0) \to (\mathbb{R}^k, 0)$ be an analytic map-germ with an isolated singular point at origin. Then Milnor's conditions (a) and (b) hold. Milnor's condition (b) is satisfied because the zero locus V of F is transverse to all small spheres.

We say that $\epsilon > 0$ is a *Milnor radius for F at the origin* if

$$B_\epsilon^n \cap (\overline{\Sigma_F - V}) = \emptyset \text{ and } B_\epsilon^n \cap V \cap (\overline{\Sigma_{F,\rho} \setminus V}) \subseteq \{0\}.$$

Under Milnor's conditions (a) and (b), we see that for $0 < |\delta| \ll \epsilon \ll 1$, the fibre $F^{-1}(\delta)$ is smooth and transverse to S_ϵ^{n-1}.

Theorem 5.1.2 ([27], p. 284, Theorem 4.3) *Let $F = (f_1, \ldots, f_k) : (\mathbb{R}^n, 0) \to (\mathbb{R}^k, 0)$, $2 \leq k \leq n-1$, be an analytic map-germ satisfying Milnor's conditions (a) and (b) at the origin and let $\epsilon_0 > 0$ be a Milnor radius for F at origin. Then, for each $0 < \epsilon \leq \epsilon_0$, there exists $\delta \in \mathbb{R}^k$ with $0 < |\delta| \ll \epsilon$, such that*

$$\bar{F} : B_\epsilon^n \cap F^{-1}(\mathring{B}_\delta^k \setminus \{0\}) \to \mathring{B}_\delta^k \setminus \{0\}, \tag{5.1}$$

where $\bar{F}$ is the restriction of F to $B_\epsilon^n \cap F^{-1}(\mathring{B}_\delta^k \setminus \{0\})$, is the projection of a smooth locally trivial fibration.

Proof It relies on Ehresmann's fibration theorem for manifolds with boundary. It is proved in [27], Theorem 4.3 in the general case and in [30], Theorem 11.2 and Lemma 11.3 in the case of an isolated critical point. □

From now on, we will denote by M_F the fibre of this fibration.

Example: This example is inspired by examples in [8].

Let $F : (\mathbb{R}^3, 0) \to (\mathbb{R}^2, 0)$ be defined by $F(x, y, z) = (x^2 z + y^2, x)$. It is easy to see that $\Sigma_F = \{(0, 0, z) : z \in \mathbb{R}\} \subseteq V$ and so Milnor's condition (a) holds. However, for any $\delta > 0$ the fibres $F^{-1}(\delta, 0)$ and $F^{-1}(-\delta, 0)$ are not homeomorphic because $F^{-1}(-\delta, 0)$ is empty. Hence, Milnor's condition (b) does not hold.

5.2 *Topology of the Real Milnor Fibre*

In this section, we give a method to obtain a topological degree formula for the Euler characteristic of M_F. We still consider $F = (f_1, \ldots, f_k) : (\mathbb{R}^n, 0) \to (\mathbb{R}^k, 0)$,

$1 \leq k \leq n-1$, an analytic map-germ. Let $l \in \{1, \ldots, k\}$ and $I = \{i_1, \ldots, i_l\}$ be a subset of l pairwise distinct elements of $\{1, \ldots, k\}$. We denote by f_I the mapping $(f_{i_1}, \ldots, f_{i_l}) : (\mathbb{R}^n, 0) \to (\mathbb{R}^l, 0)$. Suppose that F satisfies Milnor's condition (a) at the origin. Then, we have

$$\Sigma_{f_I} \subset \Sigma_F \subset F^{-1}(0) \subset f_I^{-1}(0),$$

and so the map f_I also satisfies Milnor's condition (a) at the origin.

The following results are proved in [15], Sect. 4.

Lemma 5.2.1 *Assume that F satisfies Milnor's conditions* (a) *and* (b) *at the origin. Then, for* $l \in \{1, \ldots, k\}$ *and* $I = \{i_1, \ldots, i_l\} \subset \{1, \ldots, k\}$, *the map* $f_I : (\mathbb{R}^n, 0) \to (\mathbb{R}^l, 0)$ *satisfies Milnor's conditions* (a) *and* (b).

Corollary 5.2.2 *There exists* $\epsilon_0 > 0$ *such that, for all* $l \in \{1, \ldots, k\}$ *and* $I = \{i_1, \ldots, i_l\} \subset \{1, \ldots, k\}$, *the maps* $f_I : (\mathbb{R}^n, 0) \to (\mathbb{R}^l, 0)$ *have* ϵ_0 *as a Milnor radius.*

Now we give a result which was stated as a conjecture in Milnor [30], Chap. 11. It was first proved by King in his unpublished Ph.D.-Thesis in the isolated singularity case and by the author and Araújo dos Santos [15] in the general case. We note that Jacquemard [23] also proved a version of this Milnor conjecture.

Theorem 5.2.3 (Milnor's conjecture) *Let* $F = (f_1, \ldots, f_k) : (\mathbb{R}^n, 0) \to (\mathbb{R}^k, 0)$, $3 \leq k \leq n-1$, *be an analytic map-germ that satisfies Milnor's conditions (a) and (b) at the origin and let* ϕ *be the mapping* $(f_1, \ldots, f_{k-1}) : (\mathbb{R}^n, 0) \to (\mathbb{R}^{k-1}, 0)$. *Let* M_F *be the Milnor fibre of F and let* M_ϕ *be the Milnor fibre of* ϕ. *Then the fibre* M_ϕ *is homeomorphic to* $M_F \times [-1, 1]$.

Proof We just give an idea of the proof for an isolated singularity. We denote by g the function f_k. By an easy application of the Curve Selection Lemma, we can prove that for $\epsilon > 0$ sufficiently small, the critical points of $g_{|\phi^{-1}(0) \cap S_\epsilon^{n-1}}$ lie in $\{g \neq 0\}$ (see [15], Lemma 3.1) and are outwards-pointing in $\{g > 0\}$ and inwards-pointing in $\{g < 0\}$ ([15], Lemma 3.3). So let $\epsilon > 0$ be a Milnor radius for F and ϕ and let $\delta \in \mathbb{R}^{k-1}$ be a regular value of ϕ. We can assume that

1. $0 < |\delta| \ll \epsilon \ll 1$,
2. M_ϕ is homeomorphic to $\phi^{-1}(\delta) \cap B_\epsilon^n$,
3. M_F is homeomorphic to $\phi^{-1}(\delta) \cap g^{-1}(0) \cap B_\epsilon^n$ because $\Sigma_F \subset F^{-1}(0)$,
4. the critical points of g restricted to $\phi^{-1}(\delta) \cap B_\epsilon^n$ lie in $\{g \neq 0\}$, are outwards-pointing in $\{g > 0\}$ and inwards-pointing in $\{g < 0\}$.

Note that $g_{|\phi^{-1}(\delta) \cap \mathring{B}_\epsilon^n}$ has no critical points because $\Sigma_{\phi,g} \subset \phi^{-1}(0) \cap g^{-1}(0)$. By a version of Ehresmann's fibration theorem for manifolds with boundary (see for instance [15], Lemma 6.1), $\phi^{-1}(\delta) \cap B_\epsilon^n \cap \{g \geq 0\}$ is homeomorphic to $\phi^{-1}(\delta) \cap B_\epsilon^n \cap \{g = 0\} \times [0, 1]$. Similarly $\phi^{-1}(\delta) \cap B_\epsilon^n \cap \{g \leq 0\}$ is homeomorphic to $\phi^{-1}(\delta) \cap B_\epsilon^n \cap \{g = 0\} \times [-1, 0]$. Therefore $\phi^{-1}(\delta) \cap B_\epsilon^n$ is homeomorphic to $\phi^{-1}(\delta) \cap B_\epsilon^n \cap \{g = 0\} \times [-1, 1]$ because it is homeomorphic to the gluing of $\phi^{-1}(\delta) \cap B_\epsilon^n \cap \{g \geq 0\}$ and $\phi^{-1}(\delta) \cap B_\epsilon^n \cap \{g \leq 0\}$ along $\phi^{-1}(\delta) \cap B_\epsilon^n \cap \{g = 0\}$. □

Corollary 5.2.4 *Under the above conditions, we have* $\chi(M_F) = \chi(M_\phi)$.

Corollary 5.2.5 *Let* $l \in \{2, \ldots, k\}$ *and let* $I = \{i_1, \ldots, i_l\}$ *be a subset of* l *pairwise distinct elements of* $\{1, \ldots, k\}$. *Then we have* $\chi(M_{f_I}) = \chi(M_F)$.

It remains to consider the fibres of the function $f_j : (\mathbb{R}^n, 0) \to (\mathbb{R}, 0)$. Here such a function admits two Milnor fibres : $M^+_{f_{\{j\}}} = f_j^{-1}(\delta) \cap B^n_\epsilon$ and $M^-_{f_{\{j\}}} = f_j^{-1}(-\delta) \cap B^n_\epsilon$, where $0 < \delta \ll \epsilon \ll 1$.

Let us write for instance $f = f_1$ and $g = f_2$. Using the same argument as above, we see that M^+_f is homeomorphic to $M_{(f,g)} \times [-1, 1]$ and that M^-_f is also homeomorphic $M_{(f,g)} \times [-1, 1]$.

Corollary 5.2.6 *For every* $j \in \{1, \ldots, k\}$, *we have* $\chi(M^+_{f_{\{j\}}}) = \chi(M^-_{f_{\{j\}}}) = \chi(M_F)$.

In the case of an isolated singularity, we can easily deduce from this corollary and the Khimshiashvili formula a topological degree formula for $\chi(M_F)$.

Proposition 5.2.7 *Let* $F = (f_1, \ldots, f_k) : (\mathbb{R}^n, 0) \to (\mathbb{R}^k, 0)$, $1 \le k \le n-1$, *be an analytic map-germ with an isolated singularity at the origin. Then for* $i = 1, \ldots, k$, f_i *also has an isolated singularity at the origin. The following holds:*

(i) if n *is even, then* $\mathrm{Ind}(\nabla f_1, 0) = \cdots = \mathrm{Ind}(\nabla f_k, 0)$ *and*

$$\chi(M_F) = 1 - \mathrm{Ind}(\nabla f_1, 0),$$

(ii) if n *is odd, then* $\mathrm{Ind}(\nabla f_1, 0) = \cdots = \mathrm{Ind}(\nabla f_k, 0) = 0$ *and*

$$\chi(M_F) = 1.$$

Proof We know that for all $i \in \{1, \ldots, k\}$,

$$\chi(M_F) = \chi(M^+_{f_{\{i\}}}) = \chi(M^-_{f_{\{i\}}}).$$

If n is even, by the Khimshiashvili formula,

$$\chi(M^+_{f_{\{i\}}}) = \chi(M^-_{f_{\{i\}}}) = 1 - \mathrm{Ind}(\nabla f_i, 0).$$

If n is odd, we have

$$\chi(M^+_{f_{\{i\}}}) = 1 + \mathrm{Ind}(\nabla f_i, 0) = \chi(M^-_{f_{\{i\}}}) = 1 - \mathrm{Ind}(\nabla f_i, 0). \qquad \square$$

Now we would like to explain how to establish a similar formula in the case of a mapping F satisfying Milnor's conditions (a) and (b). The strategy is to relate $\chi(M_F)$ to the Euler characteristic of the link of a real analytic singularity and apply Szafraniec's result (Proposition 3.3.1).

Let $F = (F_1, \ldots, F_k) : (\mathbb{R}^n, 0) \to (\mathbb{R}^k, 0)$, $1 \le k \le n-1$, be an analytic map-germ satisfying Milnor's conditions (a) and (b) at the origin. Let us choose $l \in$

$\{2, \ldots, k\}$ and $I = \{i_1, \ldots, i_l\}$ a subset of l pairwise distinct elements of $\{1, \ldots, k\}$. We write $J = \{i_1, \ldots, i_{l-1}\}$ and $g = f_{i_l}$. We also denote by $\mathcal{L}_I$ (resp. $\mathcal{L}_J$) the link of the zero-set of f_I (resp. f_J). If $l = 1$ then we put $J = \emptyset$ and $f_J = 0$.

Proposition 5.2.8 *We have:*

$$\chi(\mathcal{L}_J) - \chi(\mathcal{L}_I) = (-1)^{n-l} 2\chi(M_F).$$

Proof Let us write $V_J = f_J^{-1}(0)$. By the deformation argument due to Milnor (see Corollary 3.2.2), $V_J \cap g^{-1}(\delta) \cap B_\epsilon^n$ is homeomorphic to $V_J \cap \{g \geq \delta\} \cap S_\epsilon^{n-1}$ and $V_J \cap g^{-1}(-\delta) \cap B_\epsilon^n$ is homeomorphic to $V_J \cap \{g \leq -\delta\} \cap S_\epsilon^{n-1}$ for $0 < \delta \ll \epsilon \ll 1$. By the Mayer–Vietoris sequence, we can write:

$$\chi(V_J \cap S_\epsilon^{n-1}) = \chi(V_J \cap S_\epsilon^{n-1} \cap \{g \geq \delta\}) + \chi(V_J \cap S_\epsilon^{n-1} \cap \{g \leq -\delta\}) +$$

$$\chi(V_J \cap S_\epsilon^{n-1} \cap \{-\delta \leq g \leq \delta\}) - \chi(V_J \cap S_\epsilon^{n-1} \cap \{g = \delta\}) - \chi(V_J \cap S_\epsilon^{n-1} \cap \{g = -\delta\}).$$

By the above remark and Corollaries 5.2.5 and 5.2.6, the first two terms of the right-hand side of this equality are equal to $\chi(M_F)$. The third term is equal to $\chi(\mathcal{L}_I)$ because by Durfee's result ([10], Proposition 1.6), $\mathcal{L}_I$ is a retract by deformation of $V_J \cap S_\epsilon^{n-1} \cap \{-\delta \leq g \leq \delta\}$. Furthermore, if $n - l$ is even then the last two Euler characteristics are equal to 0 because $V_J \cap S_\epsilon^{n-1} \cap \{g = \delta\}$ and $V_J \cap S_\epsilon^{n-1} \cap \{g = -\delta\}$ are odd-dimensional compact manifolds. If $n - l$ is odd, they are equal to $2\chi(M_F)$ because they are boundaries of odd-dimensional Milnor fibres of f_I. □

Corollary 5.2.9 *Let $j \in \{1, \ldots, k\}$. If n is even, then we have $\chi(\mathcal{L}_{\{j\}}) = 2\chi(M_F)$ and if n is odd, then we have $\chi(\mathcal{L}_{\{j\}}) = 2 - 2\chi(M_F)$.*

Proof We apply the previous proposition to the case $l = 1$. In this case, if n is even then $\chi(\mathcal{L}_J) = 0$ and if n is odd then $\chi(\mathcal{L}_J) = 2$. □

Corollary 5.2.10 *Let $l \in \{3, \ldots, k\}$ and let $I = \{i_1, \ldots, i_l\} \subset \{1, \ldots, k\}$. Let K be a subset of $l - 2$ pairwise distinct elements of I. Then we have $\chi(\mathcal{L}_K) = \chi(\mathcal{L}_I)$.*

Proof Let J be a subset of $l - 1$ pairwise distinct elements of I built from adding to K one element of $I \setminus K$. By the previous proposition, we see that $\chi(\mathcal{L}_J) - \chi(\mathcal{L}_I) = \chi(\mathcal{L}_J) - \chi(\mathcal{L}_K)$. □

So, in order to express the Euler characteristics of all the links $\mathcal{L}_I$, we just need to compute the Euler characteristic of a link $\mathcal{L}_I$ where $\#I = 2$. Let us set $I = \{1, 2\}$. By Proposition 5.2.8, we find that

$$\chi(\mathcal{L}_I) = \chi(\mathcal{L}_{\{1\}}) - (-1)^n 2\chi(M_F).$$

So we see that $\chi(\mathcal{L}_I) = 0$ if n is even and that $\chi(\mathcal{L}_I) = 2$ if n is odd. We can summarize all these results in the following theorem.

Theorem 5.2.11 *Let $l \in \{1, \ldots, k\}$ and let $I = \{i_1, \ldots, i_l\}$ be a subset of l pairwise distinct elements of $\{1, \ldots, k\}$. If n is even, then we have*

$$\chi(\mathcal{L}_I) = 2\chi(M_F) \text{ if } l \text{ is odd and } \chi(\mathcal{L}_I) = 0 \text{ if } l \text{ is even.}$$

If n is odd, then we have

$$\chi(\mathcal{L}_I) = 2 - 2\chi(M_F) \text{ if } l \text{ is odd and } \chi(\mathcal{L}_I) = 2 \text{ if } l \text{ is even.}$$

Using this last theorem and Szafraniec's formula (Proposition 3.3.1), we can state a Poincaré–Hopf type formula for non-isolated singularities.

Proposition 5.2.12 *Let $F = (f_1, \ldots, f_k) : (\mathbb{R}^n, 0) \to (\mathbb{R}^k, 0)$, $1 \leq k \leq n-1$, be an analytic map-germ that satisfies Milnor's conditions (a) and (b) at the origin. Then there exist $c > 0$ and an integer K such that for $g(x) = f_1^2(x) - c\rho(x)^K$, we have*

(i) if n is even, then $\chi(M_F) = \frac{1}{2}(1 - \mathrm{Ind}(\nabla g, 0))$,
(ii) if n is odd, then $\chi(M_F) = \frac{1}{2}(1 + \mathrm{Ind}(\nabla g, 0))$.

Of course if one of the components of F is homogeneous, we can use the Bruce-Szafraniec formula (Proposition 3.3.2) to get a signature for $\chi(M_F)$.

References

1. Arnold, V.I., Gusein-Zade, S.M., Varchenko, A.N.: Singularities of Differentiable Maps, vol. 1. Birkhauser, Boston (1988)
2. Arnol'd, V.I.: Index of a singular point of a vector field, the Petrovski-Oleinik inequality, and mixed Hodge structures. Funct. Anal. Appl. **12**, 1–14 (1978)
3. Becker, E., Woermann, T.: On the trace formula for quadratic forms. Recent Advances in Real Algebraic Geometry and Quadratic Forms (Berkeley, CA, 1990/1991; San Francisco, CA, 1991). Contemporary Mathematics, vol. 155, pp. 271–291. American Mathematical Society, Providence (1994)
4. Becker, E., Cardinal, J.P., Roy, M.F., Szafraniec, S.: Multivariate Bezoutians, Kronecker symbol and Eisenbud and Levine formula. Algorithms in Algebraic Geometry and Applications. Progress in Mathematics, vol. 143, pp. 79–104. Birkhauser, Boston (1996)
5. Bekka, K.: Regular stratification of subanalytic sets. Bull. Lond. Math. Soc. **25**(1), 7–16 (1993)
6. Bruce, J.W.: Euler characteristics of real varieties. Bull. Lond. Math. Soc. **22**, 547–552 (1990)
7. Cerf, J.: Topologie de certains espaces de plongements. Bull. Soc. Math. Fr. **89**, 227–380 (1961)
8. Cisneros-Molina, J., Seade, J., Snoussi, J.: Milnor fibrations and the concept of d-regularity for analytic map germs. Contemporary Mathematics, vol. 569, pp. 01–28. American Mathematical Society, Providence (2012)
9. Dudzinski, P., Lecki, A., Nowak-Przygodzki, P., Szafraniec, Z.: On the topological invariance of the Milnor number mod 2. Topology **32**, 573–576 (1993)
10. Durfee, A.: Neighborhoods of algebraic sets. Trans. Am. Math. Soc. **276**(2), 517–530 (1983)
11. Dutertre, N.: On affine complete intersection with isolated singularities. J. Pure Appl. Algebra **164**(1–2), 129–147 (2001)
12. Dutertre, N.: On the Euler-Poincaré characteristic of semi-analytic sets and semi-algebraic sets. Math. Proc. Camb. Philos. Soc. **135**(3), 527–538 (2003)

13. Dutertre, N.: On topological invariants associated with a polynomial with isolated critical points. Glasg. Math. J. **46**(2), 323–334 (2004)
14. Dutertre, N.: Degree formulas and signature formulas for the Euler characteristic of real algebraic sets. J. Math. Sci. (N. Y.) **195**(2), 131–138 (2013)
15. Dutertre, N., Araújo dos Santos, R.: Topology of real Milnor fibration for non-isolated singularities. Int. Math. Res. Not. **2016**(16), 4849–4866 (2016)
16. Eisenbud, D.: An algebraic approach to the topological degree of a smooth map. Bull. Am. Math. Soc. **84**(5), 751–764 (1978)
17. Eisenbud, D., Levine, H.I.: An algebraic formula for the degree of a C^∞ map-germ. Ann. Math. **106**, 19–44 (1977)
18. Golubitsky, M., Guillemin, V.: Stable Mappings and Their Singularities. Graduate Texts in Mathematics, vol. 14. Springer, New York (1973)
19. Goresky, M., Macpherson, R.: Stratified Morse Theory. Springer, Berlin (1988)
20. Guillemin, V., Pollack, A.: Differential Topology. Prentice-Hall Inc., Englewood Cliffs (1974)
21. Hamm, H., Lê, D.T.: Un théorème de Zariski du type de Lefschetz. Ann. Sci. École. Norm. Sup. (3) **6**, 317–355 (1973)
22. Hirsch, M.: Differential Topology. Graduate Texts in Mathematics, vol. 33. Springer, New York (1976)
23. Jacquemard, A.: On the fiber of the compound of a real analytic function by a projection. Bollettino dell'Unione Matematica Italiana, serie 8, **2-B**(2), 263–278 (1999)
24. Khimshiashvili, G.M.: On the local degree of a smooth map. Soobshch. Akad. Nauk Gruz. SSR **85**, 309–311 (1977)
25. Lapébie, J.: Sur la topologie des ensembles semi-algébriques: caractéristique d'Euler, degré topologique et indice radial. Thèse de doctorat, Aix-Marseille Université (2015)
26. Łojasiewicz, S.: Ensembles Semi-analytiques. Institut des Hautes Études Scientifiques, Bures-sur-Yvette (1965)
27. Massey, D.: Real analytic Milnor fibrations and a strong Łojasiewicz inequality. Real and Complex Singularities. London Mathematical Society Lecture Note Series, vol. 380, pp. 268–292. Cambridge University Press, Cambridge (2010)
28. Milnor, J.: Morse Theory. Annals of Mathematics Studies, vol. 51. Princeton University Press, Princeton (1963)
29. Milnor, J.: Topology from the Differentiable Viewpoint. The University Press of Virginia, Charlottesville (1965)
30. Milnor, J.: Singular Points of Complex Hypersurfaces. Annals of Mathematics Studies, vol. 61. Princeton University Press, Princeton (1968)
31. Pedersen, P., Roy, M.-F., Szpirglas, A.: Counting real zeros in the multivariate case. Computational Algebraic Geometry (Nice, 1992). Progress in Mathematics, vol. 109, pp. 203–224. Birkhauser, Boston (1993)
32. Roy, M.-F.: Basic algorithms in real algebraic geometry and their complexity: from Sturm's theorem to the existential theory of reals. Lectures in Real Geometry. De Gruyter Expositions in Mathematics, vol. 23, pp. 1–67. de Gruyter, Berlin (1996)
33. Szafraniec, Z.: On the Euler characteristic of analytic and algebraic sets. Topology **26**, 411–414 (1986)
34. Szafraniec, Z.: The Euler characteristic of algebraic complete intersections. J. reine angew Math. **397**, 194–201 (1989)
35. Szafraniec, Z.: Topological invariants of weighted homogeneous polynomials. Glasg. Math. J. **33**, 241–245 (1991)
36. Szafraniec, Z.: Topological invariants of real analytic sets. Habilitation thesis, Uniwersytet Gdański (1993)
37. Szafraniec, Z.: A formula for the Euler characteristic of a real algebraic manifold. Manuscr. Math. **85**, 345–360 (1994)
38. Szafraniec, Z.: Topological degree and quadratic forms. J. Pure Appl. Algebra **141**(3), 299–314 (1999)
39. Wall, C.T.C.: Topological invariance of the Milnor number mod 2. Topology **22**, 345–350 (1983)

Equisingularity and the Theory of Integral Closure

Terence Gaffney

Abstract This is an introduction to the study of the equisingularity of sets using the theory of the integral closure of ideals and modules as the main tool. It introduces the notion of the landscape of a singularity as the right setting for equisingularity problems.

Keywords Equisingularity · Multiplicity of ideals and modules · Integral closure of ideals · Integral closure of modules

Introduction

"Let me now take a new tack which promises a better wind. Instead of dealing with a pair of hypersurfaces, let us consider analytic families of hypersurfaces V_r, all having a singular point at the origin and depending on a set of parameters." O. Zariski, Presidential Address, Bulletin A.M.S. 77 No. 4 (1971), 481–491 [41].

Given a family of sets or maps, when are all the members the same? When are some of the members different? Equisingularity is the study of these questions. As Zariski noticed, it is easier to say when a member of family is different, than it is to say when two sets or two maps are the same. Often the change in a single invariant suffices to pick out the members which are out of step with the rest.

A basic question is what do we mean by "the same"? And how do we tell when a family of sets are the same using invariants of the members of the family? These questions are explored in these lectures.

As Zariski indicates earlier in his address, equisingularity had its roots in both differential topology and algebraic geometry, and both areas continue to contribute important ideas. The use of algebraic geometry naturally leads to the use of commutative algebra to count and to control.

T. Gaffney (✉)
Department of Mathematics, Northeastern University, 567 Lake Hall, Boston, MA 02115, USA
e-mail: t.gaffney@northeastern.edu

R. N. Araújo dos Santos et al. (eds.), *Singularities and Foliations. Geometry, Topology and Applications*, Springer Proceedings in Mathematics & Statistics 222, https://doi.org/10.1007/978-3-319-73639-6_3

In answering the question of what "the same" means a topologist might ask that the members of the family be homeomorphic; a differential topologist would ask that some of the infinitesimal structure, such as limiting tangent planes and secant lines be preserved as well, while an algebraic geometer might ask that the singularities have the same multiplicity.

In these lectures we work in the complex analytic case using the Whitney conditions or Verdier's W, known to be equivalent in the complex analytic case [38], to say when the members of a family are the same. These conditions imply all three of the above possible answers. The theory of integral closure of ideals and modules provides an algebraic description of these conditions from which we may abstract the invariants which control them in families.

Here is an overview of my current approach to equisingularity questions. Given a set X, decide on the landscape that the set is part of. This means deciding on the allowable families that include the set, and the generic elements that appear in allowable families. Each set should have a unique generic element that it deforms to, and some elements of the topology of this generic element should be important invariants of our set. Describing the connection between the infinitesimal geometry of X and the topology of the generic element related to X is part of understanding the landscape. Based on the allowable deformations, determine the corresponding first order infinitesimal deformations of X. These make up a module $N(X)$. The Jacobian module of X, $JM(X)$ is the module generated by the partial derivatives of a set of defining equations for X. These can be viewed also as the infinitesimally trivial deformations of X. For the case of sets, the invariants we need for checking condition W come from the pair $(JM(X), N(X))$ and $N(X)$ by itself. A change at the infinitesimal level of the family is always tied to a change in the topology of the generic related elements.

Those who have studied maps using stabilizations [31] will recognize many elements of the overview in that context.

This paper is divided into three lectures with an afterword. They are designed to help you reach the point where the overview makes sense. In the afterword we will look at the overview again, using determinantal singularities as an example.

The first lecture introduces the Whitney conditions and Verdier's condition W, and shows how Verdier's condition W can be described using analytic inequalities. In the second lecture, the theory of the integral closure of ideals and modules is introduced, allowing us to recast the analytic inequalities of the first lecture in algebraic terms. This lecture contains a new and shorter proof of the integral closure formulation of Whitney equisingularity, Proposition 2.34. The third lecture introduces the main source of our invariants–the multiplicity of ideals and modules. In applications these multiplicities are infinitesimal objects, being intersection numbers connected with conormal spaces. The polar variety of a module is defined, and in the applications, these are local objects on our families. Through the Multiplicity Polar Theorem 3.22, they are connected to our infinitesimal invariants. The third lecture continues by applying all of these ideas to the study of determinantal singularities, which are a reasonable next step in complexity beyond complete intersections.

For complete intersections our families are obtained by varying the equations directly; for determinantal singularities we cannot vary the equations freely, but we can vary the entries of the matrix defining the singularity freely. This is the connection with complete intersections. However, since determinantal singularities are the inverse images of generic determinantal singularities, the polar varieties of the generic determinantal singularities contribute to the invariants we need to describe Whitney equisingularity in this context. (Cf. Theorem 3.28.)

Since these lectures are meant to be a tool for students to enter the subject, there are many exercises scattered through the lectures. I encourage you to try all of them. There are also some readings which fill in gaps in the proofs or provide deeper understanding. I encourage you to try these as well.

A first reading which gives an overview of how the material in these lectures developed can be found on the conference web site, along with the abstract for the course. It is a PDF of the talk I gave at Aussois in June '15 to celebrate the 70th birthday of Bernard Teissier. Teissier has made all of his papers available on his web site, (webusers.imj-prg.fr/ bernard.teissier/articles-Teissier.html) and many of the suggested readings can be found there.

It is a pleasure to thank the organizers of the conference for giving me the chance to speak about these beautiful ideas, and to share some of my thoughts about them.

1 Equisingularity Conditions

We start with some notation to describe a family of sets. In the diagram:

$$\begin{array}{ccccc} X^d(0) & \subset & \mathcal{X}^{d+k} & \subset & Y \times \mathbb{C}^N \\ \downarrow{\scriptstyle p_Y} & & \downarrow{\scriptstyle \pi_Y} & & \\ 0 \in & & Y = \mathbb{C}^k & & \end{array}$$

the parameter space is Y, $X(0)$ denotes the fiber of the family over $\{0\}$, $\mathcal{X}^{d+k}$ denotes the total space of the family which is contained in $Y \times \mathbb{C}^N$. We usually assume $Y \subset \mathcal{X}^{d+k}$, and $\mathcal{X} = F^{-1}(0)$, $X(y) = {f_y}^{-1}(0)$, where $f_y(z) = F(y, z)$.

Given a family of map germs as above, we say the family is holomorphically trivial if there exists a holomorphic family of origin preserving bi-holomorphic germs r_y such that $r_y(X(0)) = X(y)$. If the map-germs are only homeomorphisms we say the family is C^0 trivial.

Every subject needs a good example to start with. Here is ours:

Example 1.1 Let $\mathcal{X}$ be the family of four moving lines in the plane with equation $F(x, y, z) = xz(z + x)(z - (1 + y)x) = 0$. Here y is the parameter, the x and z axis are fixed, as is the line $z + x = 0$ while the line $z - (1 + y)x = 0$ moves with y. Here is a picture of the total space of the family:

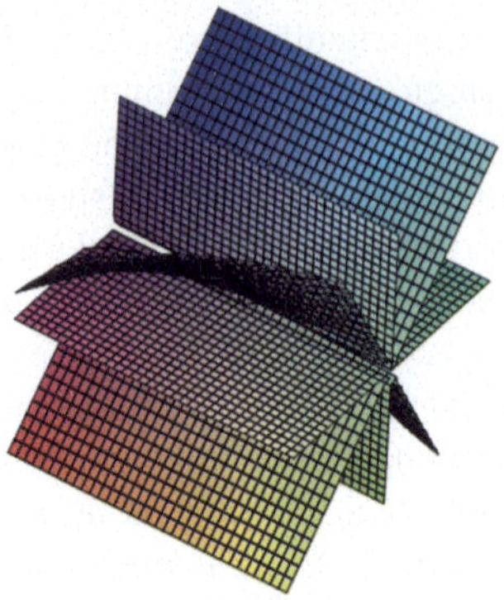

This family is not holomorphically trivial as the next exercise shows, but it should be equisingular for any reasonable definition of equisingularity.

Problem 1.2 *Show that the family of 4 lines is not homomorphically trivial by following the hints and proving them: If r_y is a trivialization of the family of sets, $Dr_y(0)$ must carry the tangent lines of $X(0)$ to $X(y)$. If a linear map preserves the lines defined by $x = 0$, $z = 0$, $z = -x$ then the linear map must be a multiple of the identity. Hence r_y can't map $z = x$ to $z = (1 + y)x$, $y \neq 0$.*

Thus, we need a notion of equisingularity that is less restrictive than holomorphic equivalence.

The Whitney conditions imply C^0 triviality but also imply the family is well-behaved at the infinitesimal level.

If X is an analytic set, X_0 the set of smooth points on X, Y a smooth subset of X, then the pair (X_0, Y) satisfies **Whitney's condition A** at $y \in Y$ if for all sequences $\{x_i\}$ of points of X_0,

$$\begin{matrix} \{x_i\} \to y \\ \{TX_{x_i}\} \to T \end{matrix} \Rightarrow T \supset TY_y$$

The pair (X_0, Y) satisfies **Whitney's condition B** at $y \in Y$ if for all sequences $\{x_i\}$ of points of X_0,

$$\begin{matrix} \{x_i\} \to y \\ \{TX_{x_i}\} \to T \\ sec(x_i, \pi_Y(x_i)) \to L \end{matrix} \Rightarrow T \supset L$$

Problem 1.3 *Show that the family of 4 lines satisfies the Whitney conditions. (Hint: The family consists of submanifolds meeting pairwise transversely.)*

Example 1.4 This is a famous example used in many singularities talks. $\mathcal{X}$ is defined by $F(x, y, z) = z^3 + x^2 - y^2z^2 = 0$. The members of the family $X(y)$ consist of node singularities where the loop is pulled smaller and smaller as y tends to zero, becoming a cusp at $y = 0$. Here is a picture:

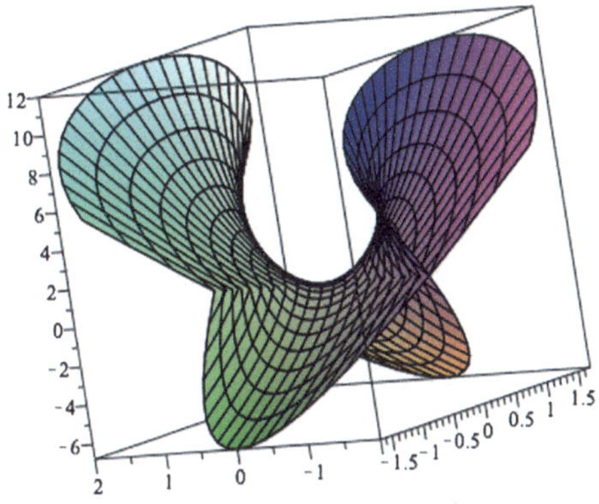

The singular locus is the y-axis. Whitney A holds because every limiting tangent plane contains the y-axis. But Whitney B fails. Notice that the parabola $z = y^2$ is in the surface, and letting $x_i = (0, t_i, t_i^2)$ and $y_i = (0, t_i, 0)$, t_i any sequence tending to 0, we see that the limiting secant line is the z-axis, while the limiting tangent plane along this curve is the xy-plane.

We see that the dimension of the limiting tangent planes at the origin is 1, while it is zero elsewhere on the y-axis. This kind of drastic change at the infinitesimal level is prevented by the Whitney conditions.

Reading You can read about the Whitney conditions in many places. Two references are the first chapter of [22], and Chap. III of [38]. The latter is more in the spirit of the way we are developing the subject, though harder. When you begin to study the polar varieties of a module in the third lecture, the lectures of Teissier [36] on the historical development of the polar variety of a space, and its connections to the Whitney conditions are very interesting. (Among other things, he explains why they are called "polar" varieties.)

Verdier's Condition W

The next condition, while equivalent to the Whitney conditions in the complex analytic case (proved by Teissier [38]) is easier to work with using algebra.

Condition W says that the distance between between the tangent space to X at a point x_i of X_0 and the tangent space to Y at y goes to zero as fast as the distance between x_i and Y. We first need to define what we mean by the distance between two linear spaces.

Suppose A, B are linear subspaces at the origin in $\mathbb{C}^N$, then define the distance from A to B as:

$$\operatorname{dist}(A, B) = \sup_{\substack{u \in B^{\perp} - \{0\} \\ v \in A - \{0\}}} \frac{\|(u, v)\|}{\|u\| \, \|v\|}.$$

In the applications B is the "big" space and A the "small" space. The inner product is the Hermitian inner product when we work over $\mathbb{C}$. The same formula also works over $\mathbb{R}$.

Example 1.5 For this example, we work with linear subspaces of $\mathbb{R}^3$. Let $A = x$-axis, B a plane with unit normal u_0, then the formula for the distance from A to B reduces to $\cos\theta$, where θ is the small angle between u_0 and the x-axis, in the plane they determine. So when the distance is 0, B contains the x-axis.

We recall Verdier's condition W.

Definition 1.6 Suppose $Y \subset \bar{X}$, where X, Y are strata in a stratification of an analytic space, and $\text{dist}(TY_0, TX_x) \leq C\text{dist}(x, Y)$ for all x close to Y. Then the pair (X, Y) satisfies **Verdier's condition** W at $0 \in Y$.

Problem 1.7 *Show that W fails for Teissier's example for X_0, Y where Y is the y-axis at the origin.*

As a first step to understanding the condition, we consider the case where X is a hypersurface in $\mathbb{C}^n$. We would like to re-write this condition in terms of F where F defines X. This will allow us to develop an algebraic formulation of the W condition.

Set-up: We use the basic set-up with $\mathcal{X}^{k+n}$ a family of hypersurfaces in $Y^k \times \mathbb{C}^{n+1}$.

Proposition 1.8 *Condition W holds for $(\mathcal{X}_0, Y)$ at $(0, 0)$ if and only if there exists U a neighborhood of $(0, 0)$ in $\mathcal{X}$ and $C > 0$ such that*

$$\|\frac{\partial F}{\partial y_l}(y, z)\| \leq C \sup_{i,j} \|z_i \frac{\partial F}{\partial z_j}(y, z)\|$$

for all $(y, z) \in U$ and for $1 \leq l \leq k$.

Proof In this set-up, Y is a k-plane, so we will set $A = Y$, and calculate the distance between Y and a tangent plane to $\mathcal{X}_0$ at (y, z) which is our B. At a smooth point of $\mathcal{X}^{k+n}$, we can use $\overline{DF(y, z)}/\|DF(y, z)\|$ for $u \in B^{\perp}$, and the standard basis for the vectors from A.

Then the distance formula says that condition W holds if and only if

$$\sup_{1 \leq l \leq k} \frac{\|\frac{\partial F}{\partial y_l}(y, z)\|}{\|DF(y, z)\|} \leq C''\text{dist}((y, z), Y) = C' \sup_{1 \leq i \leq n+1} \|z_i\|$$

This is equivalent to

$$\|\frac{\partial F}{\partial y_l}(y, z)\| \leq C \sup_{1 \leq i \leq n+1} \|z_i\| \sup_{1 \leq j \leq n+1} \|\frac{\partial F}{\partial z_j}(y, z)\|$$

From which the desired result follows. □

Denote the ideal generated by the partial derivatives of F with respect to the z variables by $J_z(F)$, and the ideal generated by z_j by m_Y. Then $z_i\frac{\partial F}{\partial z_j}$ are a set of generators for $m_Y J_z(F)$. The inequality above says that the partial derivatives of F with respect to y_l go to zero as fast as the ideal $m_Y J_z(F)$. We will examine the implications of this in the next section.

Reading After you read a little about the integral closure of ideals, reading pp. 589–605 [37] will give you a good background on the integral closure approach to Whitney equisingularity for hypersurfaces with isolated singularities.

2 The Theory of Integral Closure of Ideals and Modules

Many operations on ideals and submodules of a free module come from operations on rings. (For other examples of this, see [14, 15, 18].)

We illustrate this idea by reviewing the notions of the integral closure of a ring and the normalization of an analytic space.

Definition 2.1 Let A, B be commutative Noetherian rings with unit, $A \subset B$ a subring. Then $h \in B$ is integrally dependent on A if there exists a monic polynomial $f(T) = T^n + \sum_{i=0} f_i T^i$, $f_i \in A$ such that $f(h) = 0$. The integral closure of A in B consists of all elements of B integrally dependent on A.

Example 2.2 Let A be the ring of convergent power series in the germs t^2 and t^3, denoted $\mathbb{C}\{t^2, t^3\}$, $B = \mathbb{C}\{t\}$. Then if $f(T) = T^2 - t^2$ we have $f(t) = 0$, so t is integrally dependent on A. In fact, B is the integral closure of A in B.

Definition 2.3 Let A be the local ring of an analytic space X, x, B the ring of meromorphic functions on X at x; the space associated with the integral closure of A in B is the normalization of X.

Example 2.4 Let $A = \mathbb{C}\{t^2, t^3\}$, $B = \mathbb{C}\{t\}$. Then A is the local ring at the origin of the cusp $x^3 - y^2 = 0$, and since $t^3/t^2 = t$, the ring of meromorphic functions on X at the origin is $\mathbb{C}\{t\}$. So by the previous example the normalization of the cusp is a line.

In this context a ring A is normal if the integral closure of A in its quotient field is A. A space germ is normal if its local ring is normal. Normal spaces have nice properties–they are non-singular in codimension 1 and the Riemann removable singularities theorem is true for them. Given a space germ X, we always have a map π_{NX} from the normalization of X, denoted NX, to X which is finite and generically 1-1. NX and π_{NX} are unique up to holomorphic right equivalence. You can read proofs of these facts in [23] pp. 154–163, working backwards as necessary.

The following exercise is easy assuming the facts in the last paragraph.

Problem 2.5 *Show that the normalization of an irreducible curve germ X, x is $\mathbb{C}, 0$.*

If you know a little bit about singularities of maps, the next exercise is also easy.

Problem 2.6 *Suppose* $f : \mathbb{C}^n, 0 \to \mathbb{C}^p, 0$*,* $n < p$ *and* f *is a finitely determined map-germ. Show* $(\mathbb{C}^n, 0)$*,* f *is a normalization of the image of* f*.*

Basic Results from the Theory of Integral Closure for Ideals

The operation of integral closure of rings creates, as we shall see, an operation on ideals, the operation of forming the integral closure of I, which is an ideal, denoted $\overline{I}$. Assume I is an ideal in $\mathcal{O}_{X,x}$, $f \in \mathcal{O}_{X,x}$. In discussing the properties of integral closure, sometime we work on a small neighborhood of X. In this case, I refers to the coherent sheaf I generates on U.

List of Basic Properties f is integrally dependent on I if one of the following equivalent conditions obtain:

(i) There exists a positive integer k and elements a_j in I^j, so that f satisfies the relation $f^k + a_1 f^{k-1} + \cdots + a_{k-1} f + a_k = 0$ in $\mathcal{O}_{X,0}$.

(ii) There exists a neighborhood U of 0 in $\mathbb{C}^N$, a positive real number C, representatives of the space germ X, the function germ f, and generators $g_1, \ldots, g_m$ of I on U, which we identify with the corresponding germs, so that for all x in U we have: $\|f(x)\| \leq C \max\{\|g_1(x)\|, \ldots, \|g_m(x)\|\}$.

(iii) For all analytic path germs $\phi : (\mathbb{C}, 0) \to (X, 0)$ the pull–back $\phi^* f = f \circ \phi$ is contained in the ideal generated by $\phi^*(I)$ in the local ring of $\mathbb{C}$ at 0. If for all paths $\phi^* f$ is contained in $\phi^*(I) m_1$, then we say f is strictly dependent on I and write $f \in I^{\dagger}$.

Let NB denote the normalization of the blowup of X by I, $\bar{D}$ the pullback of the exceptional divisor of the blowup of X by I to NB by the normalization map. Then we have:

(iv) For any component C of the underlying set of $\bar{D}$, the order of vanishing of the pullback of f to NB along C is no smaller than the order of the divisor $\bar{D}$ along C. This implies that the pullback of f lies in the ideal sheaf generated by the pullback of I.

The set of all elements of $\mathcal{O}_{X,x}$ which are integrally dependent on I is the *integral closure of I* and is denoted $\overline{I}$.

Proposition 2.7 *If* I *is an ideal in* $\mathcal{O}_{X,x}$*, then so is* $\overline{I}$*.*

Proof We use property (iii). Let $\phi : (\mathbb{C}, 0) \to (X, 0)$ be any analytic curve, $g \in \mathcal{O}_{X,x}$, f_1, f_2 in $\overline{I}$. Then $(g f_1 + f_2) \circ \phi = (g \circ \phi)(f_1 \circ \phi) + (f_2 \circ \phi) \in \phi^*(I)$, since $\phi^*(I)$ is an ideal in $\mathcal{O}_1$. □

The proof of this for general rings is Corollary 1.3.1 of [35].

The first property is usually taken as the definition, and shows that integral dependence is an algebraic idea. This permits the extension of the concept to ideals in any ring. For the development of the idea of the integral closure of an ideal or module from the algebraic point of view see [35].

The second property is used to control equisingularity conditions. It already appeared in the discussion of Verdier's condition W in the hypersurface case earlier, and we will revisit it shortly.

The third property is convenient for computations, and often for proofs as the proof of the previous proposition shows. It is also helpful in understanding conditions involving limits. In the analytic setting, definitions that use sequences of points, such as the Whitney conditions, can be checked with curves, often leading to an interpretation of the condition in terms of the integral closure of an ideal or module. We will see an example of this in the study of limiting tangent hyperplanes in the next section.

The notion of strict dependence defined in the third property is used to describe properties like Whitney A, or Thom's A_f condition where integral dependence is insufficient–see the problem later on about Whitney A.

Given a curve $\phi(s)$, and a germ f, if $f \circ \phi$ is defined, it is equal to $cs^r \mod m_1^{r+1}$ for $c \neq 0$ for some r. We call r the order of f on ϕ and write $f_\phi = r$, and J_ϕ for the order of an ideal J on ϕ.

Because the exceptional divisor of the blow-up of the Jacobian ideal tracks limiting infinitesimal information, the fourth property is perhaps the most important. Since NB is normal, each component of the exceptional divisor is generically a smooth submanifold of a manifold, so the ideal vanishing on the component is locally principal. This means we can talk about the order of vanishing on each component. The order of the divisor $\bar{D}$ is just the order of vanishing along the component of the pull-back of I to NB. Concretely, pick a local generator u of the ideal of the component, and write the elements of I in terms of u. The smallest power of u that appears is the order of I along C.

The fourth property also shows how a closure operation on rings gives a closure operation on ideals– start with a ring and an ideal, enlarge the ring by a closure operation, look at the ideal generated in the new ring, then intersect with the original ring to define the closure operation on the ideal.

Reading For detailed proofs of the equivalences between these properties see [28] pp. 18–27. You can download this paper from Teissier's list of publications–it is #15. Try this after reading the proofs of the equivalences contained here.

In the next example, we practice using the first property.

Example 2.8 Let $A = \mathcal{O}_2$, $I = (x^n, y^n)$. Suppose $f = x^i y^j$, $i + j \geq n$. Consider the monic polynomial $h(T) = T^n - (x^n)^i (y^n)^j$. Since $(x^n)^i (y^n)^j$ is in $(I^i)(I^j) \subset I^{i+j} \subset I^n$, and $h(f) = 0$, then $f \in \overline{I}$.

Now we do a computation using the third property.

Example 2.9 Let $A = \mathcal{O}_2$, $I = (x^a, y^b)$. Given $m = x^i y^j$ define the weight of m to be $bi + aj$, given $f(x, y)$, define the weight of f to be the minimum weight of all monomials appearing in a power expansion of f. We will show that $\overline{I}$ consists of all f such that weight of $f \geq ab$.

First, we'll show weight of $m \geq ab$ implies $m \in \overline{I}$. It suffices to check this for curves $\phi(t) = (t^r, t^s)$ as higher order terms don't affect the order of I or m on the curve. Since $\overline{I}$ is an ideal, this will show that $f \in \overline{I}$.

We have $I_\phi = \min\{ra, sb\}$; assume $ra \leq sb$.

It is convenient to think of the monomial $x^i y^j$ as the point (i, j) in the xy-plane. Consider the parallel lines $rx + sy = c$. Then if m is any monomial on this line, $m_\phi = c$, and $m_\phi > c$ if m lies above this line. If the weight of $m \geq ab$ then m lies above or on the line connecting $(a, 0)$ and $(0, b)$, so it will lie above or on any line passing through $(a, 0)$, which lies below or on $(0, b)$. This implies that $m_\phi \geq ra$ and shows $m \in \overline{I}$.

Suppose the power expansion of f contains a monomial m which lies below the line connecting $(a, 0)$ and $(0, b)$. Then the convex hull of the monomials appearing in f has a vertex m' which lies below the line connecting $(a, 0)$ and $(0, b)$. We can find a line passing through this vertex which lies below $(a, 0)$ and $(0, b)$. Then for the curve ψ defined by this line,

$$f_\psi = m'_\psi < I_\psi$$

which shows that $f \notin \overline{I}$.

This kind of reasoning is very useful in studying properties of ideals which are well connected to their Newton polygons. In this example, the Newton polygon of I is all the points of $\mathbb{R}^2$ above or on the line connecting $(a, 0)$ and $(0, b)$ in the first quadrant. For more examples and details see [39], which is #46 on Teissier's publication list or [34].

Next, we use property 2 to characterize Verdier's W in the hypersurface case.
Set-up: We use the basic set-up with $\mathcal{X}^{k+n}$ a family of hypersurfaces in $Y^k \times \mathbb{C}^{n+1}$.

Proposition 2.10 *Condition W holds for $(\mathcal{X}_0, Y)$ at $(0, 0)$ if and only if $\frac{\partial F}{\partial y_l} \in \overline{m_Y J_z(F)}$ for $1 \leq l \leq k$.*

Proof By the last proposition of the first section we know that W holds if and only if

$$\|\frac{\partial F}{\partial y_l}(y, z)\| \leq C \sup_{i,j} \|z_i \frac{\partial F}{\partial z_j}(y, z)\|$$

But, by property 2 this is equivalent to $\frac{\partial F}{\partial y_l} \in \overline{m_Y J_z(F)}$ for $1 \leq l \leq k$. □

If we have a curve ϕ on $\mathcal{X}^{k+n}$, $\phi(0) = 0$, and the image of ϕ in $\mathcal{X}_0^{k+n}$ except at 0, and $J(F)_\phi = r$ then we can calculate the limiting tangent hyperplane to $\mathcal{X}^{k+n}$ along ϕ as

$$\underset{s \to 0}{limit}(1/s^r)(DF(\phi(s)))$$

If $\frac{\partial F}{\partial y_l} \in \overline{J_z(F)}$ for $1 \leq l \leq k$, then the limiting plane is never vertical, but it does not necessarily contain Y.

Problem 2.11 *Show that if* $\frac{\partial F}{\partial y_l}$ *for* $1 \le l \le k$ *is strictly dependent on* $J_z(F)$ *then every limit of tangent planes along every curve* ϕ *not in* $V(J_z(F))$ *contains* Y.

Problem 2.12 *Show that if* $\frac{\partial F}{\partial y_l}$ *for* $1 \le l \le k$ *is strictly dependent on* $J_z(F)$ *then WA holds.*

We will prove a few of the implications showing the equivalence of the basic properties.

Proposition 2.13 *Property 1 implies property 3*

Proof Let f satisfy the relation $f^k + a_1 f^{k-1} + \cdots + a_{k-1} f + a_k = 0$ in $\mathcal{O}_{X,0}$, and let $\phi : \mathbb{C}, 0 \to X, 0$. Choose $g \in I$ such that $g_\phi = I_\phi$. We may assume the image of ϕ does not lie in $V(I)$. Then

$$\frac{(f \circ \phi)^k}{(g \circ \phi)^k} + \frac{a_1 \circ \phi}{(g \circ \phi)} \frac{(f \circ \phi)^{k-1}}{(g \circ \phi)^{k-1}} + \cdots + \frac{a_{k-1} \circ \phi}{(g \circ \phi)^{k-1}} \frac{(f \circ \phi)}{(g \circ \phi)} + \frac{a_k \circ \phi}{(g \circ \phi)^k} = 0$$

and $\frac{a_i \circ \phi}{(g \circ \phi)^i}$ is holomorphic for all i. Since $\mathcal{O}_1$ is normal, it follows that $\frac{(f \circ \phi)}{(g \circ \phi)}$ is holomorphic, hence $f \circ \phi \in \phi^*(I)$.

Proposition 2.14 *Property 3 implies property 4*

Proof We will only prove this for the case where $V(I) = 0$.

Consider the components $\{C_i\}$ of $\bar{D}$. Since NB is normal and the C_i have codimension 1, we can pick out points c_i on each C_i and curves $\tilde{\phi}_i$, such that $\tilde{\phi}_i(0) = c_i$, and $\tilde{\phi}_i$ is transverse to C_i. We can choose c_i so that $\pi^*_{NB}(I)$ vanishes only on C_i in a neighborhood of c_i, and the same is true for $f \circ \pi_{NB}$. If u_i defines C_i at c_i, then we have $f \circ \pi_{NB} = h_i {u_i}^{f_i}, h_i$ a unit. The exponent f_i is the order of vanishing of f along C_i. Since $\tilde{\phi}_i$ is transverse to C_i at c_i, $u_i \circ \phi_i(t) = t$, so $f \circ \pi_{NB} \circ \phi_i(t) = h'_i(t) t^{f_i}$, h' a unit.

We can also find local generators of $\pi^*_{NB}(I)$ of form ${u_i}^{l_i}$ where l_i is the order of I along C_i. Now $\pi_{NB} \circ \tilde{\phi}_i$ is a map from $\mathbb{C}, 0 \to X, 0$, since $\pi_{NB}(C_i) = 0$, and hence $\pi_{NB}(c_i) = 0$. (This is the reason for restricting to this case.) Hence, if property 3 holds, $f_i \ge l_i$ for all i. If we work at any point of $\bar{D}$ since $\pi^*_{NB}(I)$ is principal, we can find $g \circ \pi_{NB}$ a local generator then $f \circ \pi_{NB} / g \circ \pi_{NB}$ is a meromorphic function which is well defined off a set of codimension 2. Since NB is normal, the function is analytic, so $f \circ \pi_{NB} \in \pi^*_{NB}(I)$. □

Proposition 2.15 *Property 4 implies property 2*

Proof Choose a compact neighborhood U of 0, and consider its inverse image in NB. The inverse image must be compact as well. So, since $f \circ \pi_{NB} \in \pi^*_{NB}(I)$, we can cover $\pi^{-1}_{NB}(U)$ with a finite number of sets and choose elements of I such that

$$\|f \circ \pi_{NB}(p')\| \le C \max\{\|g_1 \circ \pi_{NB}(p')\|, \ldots, \|g_m \circ \pi_{NB}(p')\|\}$$

holds on $\pi_{NB}^{-1}(U)$. Then it is clear that

$$\|f(\pi_{NB}(p'))\| \leq C \max\{\|g_1(\pi_{NB}(p'))\|, \ldots, \|g_m(\pi_{NB}(p'))\|\}.$$

Since π_{NB} surjects on U, this finishes the proof. □

There is a nice corollary of the method of proof used in the previous proposition and of property 2 which we now describe. Given a subset S of an analytic set X, $f\ X, S \to Y, y$ where $S = f^{-1}(y)$ denotes the germ of an analytic map along S. Given an ideal I in $\mathcal{O}_{Y,y}$, $f^*(I)$ denotes the ideal sheaf along S obtained by pulling back I by f.

Proposition 2.16 *Suppose $f\ X, S \to Y, y$ where $S = f^{-1}(y)$, f proper and surjective. Suppose I an ideal of $\mathcal{O}_{Y,y}$, $h \in \mathcal{O}_{Y,y}$. Then $h \in \overline{I}$ if and only if $h \circ f \in \overline{f^*(I)}$ along S.*

Proof Since f is proper, S is compact, and as in the last proof we can cover S with a collection of neighborhoods such that on the union the germ of a function along S is in $\overline{f^*(I)}$ if an only if it satisfies an analytic inequality of the type described by property 2. Since f is surjective, the inequalities push down/pullback to Y, y. □

Problem 2.17 *Use the finite map $f(x, y) = (x^b, y^a)$ to give another proof that $\overline{(x^a, y^b)}$ consists of all g such that weight of $g \geq ab$.*

We have Proposition 2.10 to describe W for hypersurfaces, but what about sets of higher codimension? We will see that the theory of integral closure of modules provides the tools we need to describe the higher codimension case.

The Theory of Integral Closure for Modules: Motivation

Verdier's condition W is based on the distance between the tangent space TX_x to X at smooth points x and the tangent space T to Y. Recall this distance is defined as

$$\text{dist}(T, TX_x) = \sup_{\substack{u \in TX_x^{\perp} - \{0\} \\ v \in T - \{0\}}} \frac{\|(u, v)\|}{\|u\|\,\|v\|}.$$

If $u \in TX_x^{\perp} - \{0\}$, then the set of points perpendicular to u consists of a hyperplane which contains TX_x. These hyperplanes are called *tangent hyperplanes*; denote a tangent hyperplane to X, x by H_x, and the collection of all tangent hyperplanes to X, x by $C(X)_x$. Then we can rephrase the distance formula as

$$\text{dist}(T, TX_x) = \sup_{H_x \in C(X)_x} \text{dist}(T, H_x)$$

If $X = F^{-1}(0)$ where $F\ \mathbb{C}^n \to \mathbb{C}^p$, then at a smooth point p of X, the projectivisation of the rowspace of the matrix of partial derivatives of F is $C(X)_p$. Since the tangent hyperplanes are what we need to control the distance between the tangent space of X, p and $TY, 0$, this suggests we should look at the module generated by the partial derivatives of F denoted $JM(X)$, just as we looked at $J(F)$ in the hypersurface case.

Basic Results from the Theory of Integral Closure for Modules

Notation: $M \subset N \subset F^p$, F^p a free $\mathcal{O}_{X,x}$ module of rank p, M, N submodules of F. If M is generated by g generators $\{m_i\}$, then let $[M]$ be the matrix of generators whose columns are the $\{m_i\}$.

We will develop properties for modules similar to those for ideals; however a convenient entry way into the theory is:

Definition 2.18 If $h \in F^p$ then h is integrally dependent on M, if for all curves ϕ, $h \circ \phi \in \phi^*(M)$. The integral closure of M denoted $\overline{M}$ consists of all h integrally dependent on M.

A good very basic reference on properties of integral closure of modules is [9, pp. 301–307]. The development of these ideas in the setting of modules over commutative rings can be found in [35] starting with the chapter "Integral Closure of Modules".

Problem 2.19 *$\overline{M}$ is a module, $\overline{\overline{M}} = \overline{M}$*

Example 2.20 Let $[M] = \begin{bmatrix} x & y & 0 \\ 0 & x & y \end{bmatrix}$, then $\overline{M} = m_2\mathcal{O}_2^2$.

It is clear that $\overline{M} \subset m_2\mathcal{O}_2^2$; we will show that $\begin{pmatrix} y \\ 0 \end{pmatrix} \in \overline{M}$.

Given a curve ϕ we can assume $y_\phi < x_\phi$ otherwise $\begin{pmatrix} y \circ \phi \\ 0 \end{pmatrix} \in \begin{pmatrix} x \circ \phi \\ 0 \end{pmatrix} \mathcal{O}_1$.

Then

$$\begin{pmatrix} y \\ 0 \end{pmatrix} \circ \phi = \begin{pmatrix} y \\ x \end{pmatrix} \circ \phi - x/y \circ \phi \begin{pmatrix} 0 \\ y \end{pmatrix} \circ \phi$$

where $x/y \circ \phi \in \mathcal{O}_1$.

Connection with the Theory of Integral Closure of Ideals I

Notation: Given an element $h \in F$ and a submodule M, then (h, M) denotes the submodule generated by h and the elements of M. Given a submodule N of F, $J_k(N)$ denotes the ideal generated by the set of k by k minors of a matrix whose columns are a set of generators of N. If M is an $\mathcal{O}_X$ module then the rank of M is k on a component V of X if $J_k(M) \neq (0)$ on V and k is the largest value for which this is true.

Theorem 2.21 (Jacobian principle) *Suppose the rank of (h, M) is k on each component of (X, x). Then $h \in \overline{M}$ if and only if $J_k(h, M) \subset \overline{J_k(M)}$*

Proof The complete proof appears in [9, p. 304]. The easy part is to show that $h \in \overline{M}$ implies $J_k(h, M) \subset \overline{J_k(M)}$.

We have

$$\phi^*(J_k(h, M)) = J_k(\phi^*(h, M)) = J_k(\phi^*(M) = \phi^*(J_k(M))$$

which implies the result.

The problem in the other direction is checking for curves which lie in the set of points where the rank is less than maximal, so that all the elements of $J_k(h, M)$ vanish, but h doesn't vanish. We approach this problem in two steps.

Assume first that the image of our curve ϕ does not lie entirely in $V(J_k(h, M))$.

Then, by hypothesis $\phi^*(J_k(h, M)) = \phi^*(J_k(M)) \neq 0$. So, there is a non-zero minor of the matrix of generators $[M]$, of M, $J(I, K)$ such that $J(I, K) \circ \phi$ is generator of $\phi^*(J_k(M))$. Here I is an index of the rows and K an index of the columns which comprise the $k \times k$ submatrix whose determinant is $J(I, K)$.

Consider $M_{I,K}$ the submodule of F^k defined using as matrix of generators the square submatrix of $[M]$ whose determinant is $J(I, K)$, and let h_I be the element obtained from h by using the entries indexed by I.

Applying Cramer's rule, we have that $h_I \circ \phi \in \phi^*(M_{I,K})$, where $h_I \circ \phi(t) = ([M_{I,K}] \circ \phi(t))\xi(t)$ for some column vector $\xi(t)$, given by composing the output of Cramer's rule with $\phi(t)$. Let $[M_K]$ be the submatrix of $[M]$ using the columns indexed by K. Consider $h_I \circ \phi(t) - ([M_K] \circ \phi(t))\xi(t)$. If this is zero, we have checked the condition for ϕ. If it is not zero, then $\phi^*(h, M)$ has rank greater than k which is a contradiction.

Now suppose the image of ϕ does lie entirely in $V(J_k(h, M))$, so $\phi^*(J_k(h, M)) = 0$.

Here the argument breaks into two parts again. We first assume X is smooth so that we can vary the curve freely, then we use the resolution of singularities to reduce to the smooth case.

Suppose $\phi^*(M) \neq \phi^*(h, M)$. Now, by the Artin–Rees theorem we know that there exists $\nu_0 > 0$, $\nu_0 \in \mathbb{Z}$ such that

$$m_1^l \mathcal{O}_1^p \cap \phi^*(h, M) = m_1^{l-\nu_0}(m_1^{\nu_0} \mathcal{O}_1^p \cap \phi^*(h, M)).$$

This implies, that in fact,

$$\phi^*(M) \neq \phi^*(h, M) \mod m_1^l \mathcal{O}_1^p$$

for any $l > \nu_0$. If not, then $h \circ \phi = g \mod \phi^*(M)$, with $g \in m_1^l \mathcal{O}_1^p$, and so

$$g \in m_1^l \mathcal{O}_1^p \cap \phi^*(h, M),$$

hence

$$g, h \circ \phi \in \phi^*(M) + m_1(m_1^{\nu_0} \mathcal{O}_1^p \cap \phi^*(h, M)).$$

Since $\phi^*(M) + m_1\phi^*(h, M) = \phi^*(h, M)$, Nakayma's lemma would imply the result.

Now choose $l > \nu_0$; since X is smooth, we can find a curve ϕ_1, by changing terms of the power series expansion ϕ of order $\geq l$, such that the image of ϕ_1 does not lie in $V(J_k(h, M))$.

This implies that

$$\phi_1^*(M) = \phi^*(M) \mod m_1^l \mathcal{O}_1^p$$

$$\phi_1^*(h, M) = \phi^*(h, M) \mod m_1^l \mathcal{O}_1^p$$

$$\phi_1^*(M) = \phi_1^*(h, M)$$

This gives a contradiction in this case.

If X is not smooth, then we can make a resolution, $\tilde{X}$, π, of singularities of X, lift ϕ to $\tilde{\phi}$ on $\tilde{X}$. Then $\phi^*(M) \neq \phi^*(h, M)$ if and only if $\tilde{\phi}^*\pi^*(M) \neq \tilde{\phi}^*\pi^*(h, M)$, then we can again vary $\tilde{\phi}^*$ as before. □

If $h \in \overline{M}$, this last proposition allows us to to do more than show $h \in M$ along curves.

Proposition 2.22 *Suppose $h \in \overline{M}$, then there exists an open cover $\{U_{I,K}\}$ of the complement of $V(J(M))$, such that on each $U_{I,K}$, $h = [M]\xi_{I,K}$, where the entries of $\xi_{I,K}$ are locally bounded on $U_{I,K}$.*

Proof The open cover $\{U_{I,K}\}$ is constructed by constructing an open cover $\{V_{I,K}\}$ of the fiber over the origin in $NB_{J(M)}(X)$ such that on each $V_{I,K}$, the pullback of $J(I, K)$ is a local generator of the pullback of $J(M)$. Then Cramer's rule applies, and the pullbacks of the $\xi_{I,K}$ are holomorphic, hence locally bounded on the images of the $V_{I,K}$ which are the $U_{I,K}$. □

As another application we can develop the analogue of property 2 for ideals.

Proposition 2.23 ([9], Proposition 1.11) *Suppose $h \in \mathcal{O}_{X,x}^p$, M a submodule of $\mathcal{O}_{X,x}^p$ of generic rank k on each component of X. Then $h \in \overline{M}$ if and only if for each choice of generators $\{s_i\}$ of M, there exists a constant $C > 0$ and a neighborhood U of x such that for all $\psi \in \Gamma(Hom(\mathbb{C}^p, \mathbb{C}))$,*

$$\|\psi(z) \cdot h(z)\| \leq C \sup_i \|\psi(z) \cdot s_i(z)\|$$

for all $z \in U$.

For each choice of ψ, the $\{\psi \cdot s_i(z)\}$ give a linear combination of the rows of $[M]$ at each point, while $\psi(z) \cdot h(z)$ is the analogous combination of the entries of h. So the inequality of the theorem relates the size of row vectors of $[M(x)]$ to corresponding combinations of the entries of h. The constant C and the neighborhood U depend on h and M but not on ψ.

Proof We will use the Jacobian principle to show that the inequality implies the integral closure inclusion, by using special ψ_i.

Let S_I be a $k \times (k-1)$ submatrix of $[M]$, going through all such submatrices as I varies, let h_I be a k-tuple gotten by dropping the same entries from h as rows from $[M]$ in forming S_I. Let $\psi_I(z)(h(z)) = \det[h_I(z), S_I(z)]$. Note that $\psi_I(z)s_i(z) = \det[s_i(z), S_I(z)]$, a generator of $J_k(M)$.

The inequality which we are assuming then shows that $J_k(h, M) \subset \overline{J_k(M)}$, which gives the result by the Jacobian principle.

A weaker version of the other direction is easy; if $h \in \overline{M}$, then for any curve ϕ, $(\psi(z) \cdot h(z)) \circ \phi \in \phi^*(\{\psi(z) \cdot s_i(z)\})$, hence $(\psi(z) \cdot h(z)) \in \overline{(\{\psi(z) \cdot s_i(z)\})}$. Then the result follows by property 2 for ideals. However, here the constant does depend on ψ.

Instead we argue like this. Let $\{s_i\}$ be a set of generators of M. Applying property 2 to the finite set of elements $\{g_i\}$ that make up the numerators of the entries of the $\xi_{I,K}$ in the last proposition, we have that there exists U and C such that if g_i is such a numerator, then

$$\|g_i(z)\| \leq C \sup \|J_{I,K}(z)\|.$$

We have that $J_{I,K}(z)h(z) = \sum g_i s_i$ for appropriate g_i. Then working first at $z \notin V(J(M))$

$$\|\psi(z) \cdot h(z)\| = \|\sum (g_i/J(I,K))(z)\psi(z) \cdot s_i(z)\| \leq CN \sup_i \|\psi(z) \cdot s_i(z)\|$$

where N is the number of terms in the sum. Since the inequality is between continuous functions and holds on an open dense subset of U it holds on U. □

Corollary 2.24 *Suppose $h \in \mathcal{O}^p_{X,x}$, M a submodule of $\mathcal{O}^p_{X,x}$ of generic rank k on each component of X. Then $h \in \overline{M}$ if and only if for each choice of generators $\{s_i\}$ of M, there exists a constant $C > 0$ and a neighborhood U of x such that for all $T \in \mathbb{C}^p$,*

$$\|T \cdot h(z)\| \leq C \sup_i \|T \cdot s_i(z)\|$$

for all $z \in U$.

Proof In one direction, take ψ to be constant; in the other we can replace T by ψ, using the fact that the constant C is independent of the choice of T. □

The corollary reflects the equivalence of $h \in \overline{M}$ and $\rho(h) \in \mathcal{M}$. (The notions of $\rho(h)$, $\mathcal{M}$ and the equivalence will be developed later.)

There is a useful variant of the last Proposition.

Proposition 2.25 ([17]) *For a section $h \in \mathcal{O}^p_X$ to be integrally dependent on M at 0, it is necessary that, for all maps $\phi : (\mathbb{C}, 0) \to (X, 0)$ and $\psi : (\mathbb{C}, 0) \to (Hom(\mathbb{C}^p, \mathbb{C}), \lambda)$ with $\lambda \neq 0$, the function $\psi(h \circ \phi)$ on $\mathbb{C}$ belong to the ideal $\psi(M \circ \phi)$.*

Conversely, it is sufficient that this condition obtain for every ϕ whose image meets any given dense Zariski open subset of X.

We will use these ideas to extend our criterion for condition W to equidimensional sets of any codimension, but first we develop the analogue of property 4 for modules.

Blowing Up Modules and Connection with Ideals II

We now develop the analogue of property 4 for modules. We will want a construction that works for pairs of submodules, not just a single submodule.

Given a submodule M of a free $\mathcal{O}_{X^d}$ module F of rank p, we can associate a subalgebra $\mathcal{R}(M)$ of the symmetric $\mathcal{O}_{X^d}$ algebra on p generators. This is known as the Rees algebra of M. If $(m_1, \ldots, m_p)$ is an element of M then $\sum m_i T_i$ is the corresponding element of $\mathcal{R}(M)$. Then $\mathrm{Projan}(\mathcal{R}(M))$, the projective analytic spectrum of $\mathcal{R}(M)$ is the closure of the projectivised row spaces of M at points where the rank of a matrix of generators of M is maximal. Denote the projection to X^d by c, or by c_M where there is ambiguity.

Example 2.26 If M is the Jacobian module of X, then $\mathrm{Projan}(\mathcal{R}(M))$ is $C(X)$, the projectivised conormal space of X.

If M is a submodule of N or h is a section of N, then h and M generate ideals on $\mathrm{Projan}\,\mathcal{R}(N)$; denote them by $\rho(h)$ and $\mathcal{M}$. If we can express h in terms of a set of generators $\{n_i\}$ of N as $\sum g_i n_i$, then in the chart in which $T_1 \neq 0$, we can express a generator of $\rho(h)$ by $\sum g_i T_i/T_1$.

Example 2.27 If M is the Jacobian module of X and $N = F^p$ then $V(\mathcal{M})$ consists of pairs (x, L) where $x \in X$ and $L \in \mathbb{P}Hom(\mathbb{C}^p, \mathbb{C})$, and $L \circ DF(x) = 0$. If H is the hyperplane which is the kernel of L, then the image of $DF(x)$ lies in H.

Using Proposition 2.23 it is easy to show that h is integrally dependent on M at the origin, if and only the ideal sheaf induced from h is integrally dependent as an ideal sheaf on $\mathcal{M}$ along $0 \times \mathbb{P}^{p-1}$. In other words, if and only if $\rho(h)$ is integrally dependent on $\mathcal{M}$. The combination $\psi(t)$, $\phi(t)$ amounts to giving path on $X \times \mathbb{P}^{p-1}$. This is the second connection between integral closure of ideals and modules.

Looking at a pair (M, N) allows us to "strip out" one copy of N from M, as the following example shows.

Example 2.28 Let $M = I = (x^2, xy, z) = J(z^2 - x^2y)$ and $N = J = (x, z)$. M is the Jacobian ideal of the Whitney umbrella, and N defines the singular locus of the umbrella. So, working on $\mathbb{C}^3$, $\mathrm{Projan}\,\mathcal{R}(N) = B_J(\mathbb{C}^3)$, which has ring $R = \mathbb{C}[T_1, T_2]/(zT_1 - xT_2)$, and where the map from $\mathcal{R}(N)$ to R is given by $x \to T_1, z \to T_2$. Writing the generators of I in terms of the generators of J as $x^2 = x \cdot x, xy = y \cdot x, z = z$ the map from $\mathcal{R}(I)$ to R has image (xT_1, yT_1, T_2) and this induces the ideal sheaf $\mathcal{I}$ on $\mathrm{Projan}\,\mathcal{R}(N)$. We see that this is supported only at the point $(0, [1, 0])$.

The next proposition and the ideas behind it, is very useful in the study of determinantal singularities. It is also a good example of stripping a copy of a module N from M.

Proposition 2.29 *Suppose $M \subset N \subset \mathcal{O}^p_{X,0}$ are $\mathcal{O}^p_X$ modules with matrix of generators $[M]$, $[N]$, and $[F]$ is a matrix such that $[M] = [N][F]$. Let $\mathcal{F}$ be the ideal sheaf induced on* Projan$(\mathcal{R}(N))$ *by the module F with matrix of generators $[F]$. Then $\overline{M} = \overline{N}$ if and only if $V(\mathcal{F})$ is empty.*

Proof We are going to apply Proposition 2.25, so we must show that for all maps $\phi\ (\mathbb{C}, 0) \to (X, 0)$ and $\psi\ (\mathbb{C}, 0) \to (Hom(\mathbb{C}^p, \mathbb{C}), \lambda)$, that the order in t of $\psi(t)[M] \circ \phi(t)$ and $\psi(t)[N] \circ \phi(t)$ are the same. We have

$$\psi(t)[M] \circ \phi(t) = \psi(t)[N][F] \circ \phi(t).$$

Suppose the order of $\psi(t)[N] \circ \phi(t)$ in t is k. Then we can lift ϕ, ψ to a curve on Projan$(\mathcal{R}(N))$ as follows. Define $\Phi : \mathbb{C}, 0 \to X \times \mathbb{P}^{g(N)-1}$, by $\Phi(t) = (\phi(t), [(1/t^k)(\psi(t)[N] \circ \phi(t)])$. We have $\Phi(0) = (0, \lim_{t\to 0}(1/t^k)(\psi(t)[N] \circ \phi(t))$, and the image of Φ for $t \neq 0$ clearly lies in Projan$(\mathcal{R}(N))$.

Given an element $f \in \mathcal{F}$, the value of f along Φ is $(\phi(t), [(1/t^k)(\psi(t)[N]\tilde{f} \circ \phi(t)])$, where $\tilde{f}$ is the element of F which induces f. Then $V(\mathcal{F})$ is empty if and only if the order of $\mathcal{F}$ along all Φ is zero. Since $[M] = [N][F]$ this is equivalent to the order of M and N being the same on (ψ, ϕ). □

Notice that if $M \subset N$ and $\mathcal{F}$ are as above then the inclusion of M in N always induces a map from Projan$(\mathcal{R}(N))\backslash V(\mathcal{F})$ to Projan$(\mathcal{R}(M))$. The map is given by taking (x, p) to $(x, \mathcal{F}(p))$, where $\mathcal{F}(p)$ is evaluation of the set of generators of $\mathcal{F}$ which come from the columns of $[F]$. The next corollary includes this setting in our discussion of reduction.

Corollary 2.30 *Suppose M and N as above, then the following are equivalent:*

1. *M is reduction of N.*
2. *$V(\mathcal{F})$ is empty.*
3. *The induced map is a finite map from* Projan$(\mathcal{R}(N))$ *to* Projan$(\mathcal{R}(M))$.

Proof (1) and (2) are equivalent by the previous proposition. The material in Sect. 2 of [26] shows that the induced map is finite if and only if $V(\mathcal{F})$ is empty. □

Here is a typical way that (3) is used.

Proposition 2.31 *Suppose $N \subset F$, F a free $\mathcal{O}_{X,x}$ module, and suppose the fiber of* Projan $\mathcal{R}(N)$ *over x has dimension k. Then N has a reduction M, where M is generated by $k + 1$ elements.*

Proof Let g be the number of generators of N, so we view Projan $\mathcal{R}(N)$ as a subset of $X \times \mathbb{P}^{g-1}$. For a generic choice of plane P in $\mathbb{P}^{g-1}$ of codimension $k + 1$, the intersection of P and the fiber of Projan $\mathcal{R}(N)$ over x is empty. We can choose coordinates on $\mathbb{P}^{g-1}$ so that the plane given by $T_1 = \cdots = T_{k+1} = 0$ is such a plane, T_i coordinates on $\mathbb{P}^{g-1}$. Choosing coordinates on $\mathbb{P}^{g-1}$ is equivalent to choosing generators on N. Let M be the submodule of N generated by the first $k + 1$ generators of

N after the new choice of generators. Then the projection onto the first $k+1$ coordinates of $\mathbb{P}^{g-1}$, when restricted to Projan $\mathcal{R}(N)$ gives a finite map to Projan $\mathcal{R}(M)$. Hence M is a reduction of N by (3). □

Corollary 2.32 *Suppose $N \subset F$, F a free $\mathcal{O}_{X,x}$ module, X^d equidimensional, N has generic rank e on each component of X, x, then N has a reduction with $d+e-1$ generators.*

Proof Since the generic rank of N is e, the generic fiber dimension of Projan $\mathcal{R}(N)$ is $e-1$, so the dimension of Projan $\mathcal{R}(N)$ is $d+e-1$. Then $d+e-2$ is the largest the dimension of the fiber of Projan $\mathcal{R}(N)$ over x can be, so N has a reduction with $(d+e-2)+1$ generators. □

Having defined the ideal sheaf $\mathcal{M}$, we blow up by it. The advantages of this we will see in the next section, as it gives a constructive/geometric way to calculate the multiplicity of a pair of modules. But for now, this gives the context for which property 4 in the ideal case holds. As an example of how the blow up comes up, if we are in the basic set-up, and $M = m_Y JM(\mathcal{X})$ then the blow up by $\mathcal{M}$ is the blowup of the conormal of $\mathcal{X}$ by the ideal defining the stratum Y. Teissier has shown [38] that condition W holds for the pair $(\mathcal{X}_0, Y)$ at the origin if and only if the exceptional divisor of this blow up is equidimensional over Y. We will see the proof of one direction of this in the next section as well.

To state our result some more notation is needed. Given M a submodule of $N \subset F^p$, $h \in N$, let $NB_{\mathcal{M}}(\text{Projan } \mathcal{R}(N)), \pi_{\mathcal{M}}$ be the normalized blow-up of Projan $\mathcal{R}(N)$ by $\mathcal{M}$ with projection $\pi_{\mathcal{M}}$ to Projan $\mathcal{R}(N)$.

Proposition 2.33 (Analogue of Property 4 for ideals) *In the above set-up $h \in \overline{M}$ if and only if $\pi^*_{\mathcal{M}}(\rho(h)) \in \pi^*_{\mathcal{M}}(\mathcal{M})$.*

Proof We give the proof for the case where N is free for simplicity. We apply Corollary 2.24, so $h \in \overline{M}$ if and only if for all $\phi\ (\mathbb{C}, 0) \to (X, 0)$ and $\psi\ (\mathbb{C}, 0) \to (Hom(\mathbb{C}^p, \mathbb{C}), \lambda)$, we have the function $\psi(h \circ \phi)$ on $\mathbb{C}$ belongs to the ideal $\psi(M \circ \phi)$. Giving the pair (ϕ, ψ) is equivalent to giving a path on $X \times \mathbb{P}^{p-1}$, the order of $\rho(h)$ on the path is the order of $\psi(h \circ \phi)$. So 2.23 is equivalent to : $h \in \overline{M}$ if and only if the ideal sheaf induced by $\rho(h)$ is in the integral closure of the ideal sheaf $\mathcal{M}$. In turn, by property 4 for ideals, this implies the result. □

As an application we can extend our criterion for condition W to equidimensional sets of any codimension.

Set-up: We use the basic set-up with $\mathcal{X}^{k+n}$ an equidimensional family of equidimensional sets, $\mathcal{X}^{k+n} \subset Y^k \times \mathbb{C}^N$, $JM(X) \subset \mathcal{O}^p$.

Proposition 2.34 *Condition W holds for $(\mathcal{X}_0, Y)$ at $(0,0)$ if and only if $\frac{\partial F}{\partial y_l} \in \overline{m_Y JM(F)}$ for $1 \le l \le k$.*

Proof We re-work the form of Verdier's condition W to fit our current framework. If we work at a smooth point x of X, then a conormal vector u of X at x can always be written as $S \cdot DF(x)$, where $S \in \mathbb{C}^p$; S is not unique unless $DF(x)$ has rank p. Conversely, any such S gives a conormal vector. It is clear also that W holds if the distance inequality holds for the standard basis for the tangent space T of Y. Then

$$\operatorname{dist}(T, TX_x) = \sup_{\substack{u \in TX_x^{\perp} - \{0\} \\ v \in T - \{0\}}} \frac{\|(u, v)\|}{\|u\| \, \|v\|}.$$

becomes

$$\operatorname{dist}(T, TX_x) = \sup_{\substack{S \in \mathbb{C}^p - \{0\} \\ 1 \leq i \leq k, \, S \cdot DF(x) \neq 0}} \frac{\|S \cdot \frac{\partial f}{\partial y_i}\|}{\|S \cdot DF(x)\|}$$

because $\|u\| = \|S \cdot DF(x)\|$, and $\|v\| = 1$.

So Verdier's condition W becomes:

$$\sup_{\substack{S \in \mathbb{C}^p \\ 1 \leq i \leq k}} \|S \cdot \frac{\partial f}{\partial y_i}\| \leq C\|z\| \, \|S \cdot DF(x)\| .$$

Since the functions are analytic and the inequality holds on a Z-open set of X, we can assume it holds on a neighborhood of the origin.

Now consider the integral closure condition, $\frac{\partial F}{\partial y_l} \in \overline{m_Y JM(F)}$ for $1 \leq l \leq k$. Using Corollary 2.4, we have $\frac{\partial F}{\partial y_l} \in \overline{m_Y JM(F)}$ for $1 \leq l \leq k$ if and only if

$$\sup_{\substack{S \in \mathbb{C}^p \\ 1 \leq i \leq k}} \|S \cdot \frac{\partial f}{\partial y_i}\| \leq C \sup_{1 \leq i \leq n} \|z_i S \cdot DF(x)\| .$$

But this is easily seen to be equivalent to the previous inequality. □

This last result shows that Verdier's condition W is exactly the geometric meaning of the ideal sheaf induced by the $\frac{\partial f}{\partial y_i}$ being in the integral closure of the ideal sheaf induced by $m_Y JM(X)$ on $X \times \mathbb{P}^{p-1}$.

In the next section we will see how to describe and control equisingularity conditions using multiplicity of ideals and modules.

3 Multiplicities, Integral Closure and the Multiplicity-Polar Theorem

The multiplicity of an ideal or module or pair of modules is one of the most important invariants we can associate to an m-primary module. It is intimately connected with integral closure. It has both a length theoretic definition and intersection theoretic definition. We give the definition in terms of length first, for ideals, and submodules of a free module. Denote the length of a module M by $l(M)$.

Theorem/Definition 3.1 *(Buchsbaum–Rim [1]) Suppose $M \subset F$, M, F both A-modules, F free of rank p, A a Noetherian local ring of dimension d, F/M of finite length, $\mathcal{F} = A[T_1, \ldots, T_p]$, $\mathcal{R}(M) \subset \mathcal{F}$, then*

$\lambda(n) = l(\mathcal{F}_n/\mathcal{M}_n)$ is eventually a polynomial $P(M, F)$ of degree d+p-1.

Writing the leading coefficient of $P(M, F)$ as $e(M)/(d + p - 1)!$, then we define $e(M)$ as the multiplicity of M.

It is possible to compute simple ideal examples by hand as we show:

Example 3.2 Let $M = I = (x^2, xy, y^2) \subset \mathcal{O}_2$. Then $e(M) = 4$.

We have $p = 1$, $F = \mathcal{O}_2$, and we work with $\mathcal{F} = \mathcal{O}_2[T_1]$. (Notice that Projan $\mathcal{F} = \mathbb{C}^2$.)

Now $\mathcal{M}_n = I^n T^n = m_2^{2n} T^n$, so

$$l(\mathcal{F}_n/\mathcal{M}_n) = l(\mathcal{O}_2/m^{2n}) = (2n)(2n + 1)/2 = 4n^2/2! + (l.o.t.)$$

So $e(M) = 4$.

Problem 3.3 *Let $M = I = (x^2, y^2) \subset \mathcal{O}_2$. Show $e(M) = 4$. (Hint: Try to show that the terms that are missing in this problem due to the missing xy term, grow only linearly with n, so the leading term of the polynomial is the same.)*

It is possible to do the very simplest module examples by hand easily as well.

Problem 3.4 *Let $M = m_2\mathcal{O}_2^2$. Show $e(M) = 3$.*

The next problem is harder–try to use the same strategy as in Problem 3.3.

Problem 3.5 *Let $[M] = \begin{bmatrix} x & y & 0 \\ 0 & x & y \end{bmatrix}$. Show $e(M) = 3$.*

If $\mathcal{O}_{X^d,x}$ is Cohen–Macaulay, and M has $d + p - 1$ generators where $M \subset F^p$, then there is a useful relation between M and its ideal of maximal minors and the multiplicity of both of them. The multiplicity of M is the colength of M, and is also the colength of the ideal of maximal minors, by some theorems of Buchsbaum and Rim [1], 2.4 p. 207, 4.3 and 4.5 p. 223. A proof of this theorem in the context of analytic geometry using the Multiplicity Polar theorem is given in [13]. Using this result, it is easy to do Problem 3.5.

Challenge Problem 3.6 *Buchsbaum and Rim showed $e(M) = l(F^p/M)$, if M has $d + p - 1$ generators, F a module over a Cohen–Macaulay ring. What is a generalization of this to $e(M, N)$? (If M and N are ideals there is something along these lines in [12] Theorem 2.3.)*

An important theorem both for computational and theoretical purposes was proved by Rees in the ideal case. A proof of a generalization to modules appears in [26].

Theorem 3.7 *Suppose $M \subset N$ are m primary submodules of F^p, and $\overline{M} = \overline{N}$. Then $e(M) = e(N)$. Suppose further that $\mathcal{O}_{X,x}$ is equidimensional, then $e(M) = e(N)$ implies $\overline{M} = \overline{N}$.*

Several generalizations of this result exist: Kleiman and Thorup [[26], (6.8)(b)] proved a similar result in which F^p is replaced by an arbitrary finitely generated module whose support is equidimensional; they also proved an additivity result in Theorem (6.7b)(i) of [26] for the three pairs of modules arising from three nested modules. Generalizations also exist where the multiplicity is not defined. Gaffney and Gassler did the case of ideals [16], and Gaffney for modules [10], while Ulrich and Valadoshti have an approach using the epsilon multiplicity.

For computational purposes, this is coupled with another result–given any $M \subset F^p$, M a module over a local ring of dimension d, there exists a submodule R of M with $d + p - 1$ generators such that $\overline{M} = \overline{R}$. Such an R is called a *reduction* of M.

So if $\mathcal{O}_{X^d,x}$ is Cohen–Macaulay, we can try to find a reduction R of M with the right number of generators $d + p - 1$, then calculate the length of F/R. (This length is also called the colength of R.) Here is a very simple example.

Problem 3.8 *Suppose I is any ideal in $m_2^n\mathcal{O}_2$ which contains x^n, y^n. Then $e(I) = n^2$.*

Now we want to give an intersection theoretic definition of the multiplicity. This definition applies to pairs of modules as well.

The next diagram shows the spaces that come into the definition.

$$\begin{array}{ccc} B_{\mathcal{M}}(\text{Projan}\,\mathcal{R}(N)) & \stackrel{\pi_N}{\rightarrow} & \text{Projan}\,\mathcal{R}(N) \\ \downarrow{\scriptstyle \pi_M} & & \downarrow{\scriptstyle \pi_{XN}} \\ \text{Projan}\,\mathcal{R}(M) & \xrightarrow{\pi_{XM}} & X \end{array}$$

On the blow up $B_{\mathcal{M}}(\text{Projan}\,\mathcal{R}(N))$ we have two tautological bundles. One is the pullback of the bundle on Projan $\mathcal{R}(N)$. The other comes from Projan $\mathcal{R}(M)$. Denote the corresponding Chern classes by c_M and c_N, and denote the exceptional divisor by $D_{M,N}$. Suppose the generic rank of N (and hence of M) is g.

Then the multiplicity of a pair of modules M, N is:

$$e(M, N) = \sum_{j=0}^{d+g-2} \int D_{M,N} \cdot c_M^{d+g-2-j} \cdot c_N^j.$$

Kleiman and Thorup show that this multiplicity is well defined at $x \in X$ as long as $\overline{M} = \overline{N}$ on a deleted neighborhood of x. This condition implies that $D_{M,N}$ lies in the fiber over x, hence is compact. Notice that when $N = F$ and M has finite colength in F then $e(M, N)$ is the Buchsbaum-Rim multiplicity $e(M, \mathcal{O}^p_{X,x})$.

Kleiman and Thorup also showed that $e(M, N)$ vanishes if and only if M and N have the same integral closure, provided the support of N is equidimensional. ([26], (6.3)(ii).)

Remark 3.9 We have seen that there is a map from Projan $\mathcal{R}(N)\backslash V(\mathcal{F}) \to$ Projan $\mathcal{R}(M)$. The diagram used in the definition of $e(M, N)$ can be used to make this more precise. Namely, the complement of $\pi_M D_{M,N}$ is the largest open subset V of Projan $\mathcal{R}(M)$ such that the map $\pi_M^{-1} V\backslash D_{M,N} \to V$ is finite. Plainly, π_N is an isomorphism over the complement U of $V(\mathcal{F})$, and $\pi_N^{-1}U$ contains $\pi_M^{-1}V$.

Let's re-calculate two examples using this definition.

Example 3.10 Let $M = I = (x^2, xy, y^2) \subset \mathcal{O}_2$. Then $e(M) = 4$.

Here $d = 2$, $p = g = 1$, Projan $\mathcal{R}(N) = \mathbb{C}^2$, Projan$(\mathcal{M}) = B_I(\mathbb{C}^2) = B_{\mathcal{M}}$ (Projan $\mathcal{R}(N)$), and Projan$(\mathcal{M}) \subset \mathbb{C}^2 \times \mathbb{P}^1$. So the only term we need to calculate is $\int D_{M,N} \cdot c_M$. We can calculate this term as follows: Intersect $B_I(\mathbb{C}^2)$ with $\mathbb{C}^2 \times H$, H a generic hyperplane in $\mathbb{P}^1$, which represents $c(M)$. Project this curve to $\mathbb{C}^2$, and calculate the order of I on the curve. Projecting the curve to $\mathbb{C}^2$ amounts to setting a generic combination of the generators to zero, and looking at the curve obtained, removing any components in $V(I)$. In this case a generic curve is $x^2 - ay^2 = 0$, $a \neq 0$. This consists of two branches ($x - y = 0$ and $x + y = 0$ if $a = 1$) and the colength of the ideal on each branch is 2 so the multiplicity is $2 + 2 = 4$.

Example 3.11 Let $[M] = \begin{bmatrix} x & y & 0 \\ 0 & x & y \end{bmatrix}$. Show $e(M) = 3$.

Here $d = 2$, $p = g = 2$, $N = \mathcal{O}_2^2$, Projan $\mathcal{R}(N) = \mathbb{C}^2 \times \mathbb{P}^1$, Projan $\mathcal{R}(M) \subset \mathbb{C}^2 \times \mathbb{P}^2$, dimension of $B_{\mathcal{M}}$(Projan $\mathcal{R}(N)$) is 3. So we need to calculate $\int D_{M,N} \cdot c_M^2$, $\int D_{M,N} \cdot c_M \cdot c_N$ (Notice that $c_N^2 = 0$, since we are working on Projan $\mathcal{R}(N) = \mathbb{C}^2 \times \mathbb{P}^1$.) Now we have two choices: as before we intersect a representative of each class with the blow-up then push down to X, then see what the multiplicity of M is on each curve. Or, we can push down to Projan $\mathcal{R}(N)$ and evaluate $\mathcal{M}$ on each curve. (For details of how this approach works, the reader should consult [11] Theorem 3.1 and the two examples which follow.)

Taking the second route, projecting the intersection of the blow-up with a hyperplane from $\mathbb{C}^2 \times \mathbb{P}^1$ and a hyperplane from $\mathbb{C}^2 \times \mathbb{P}^2$, is a curve on $\mathbb{C}^2 \times \mathbb{P}^1$, defined by a linear relation $T_1 = aT_2$, and by setting one of the elements of $\mathcal{M}$ restricted to this set to zero. The restriction of $\mathcal{M}$ to the locus $T_1 = aT_2$ is the ideal generated by the entries of the linear combination of the first row and a times the second row from the original matrix. A generic curve is given by setting $x + ay = 0$, and the multiplicity of $\mathcal{M}$ on this curve is 1. So, $\int D_{M,N} \cdot c_M \cdot c_N = 1$.

Projecting the intersection of the blow-up with two hyperplanes from $\mathbb{C}^2 \times \mathbb{P}^2$, amounts to setting two generic elements of $\mathcal{M}$ to zero and removing any components of $V(\mathcal{M})$. Setting $xT_1 + yT_2$ and $yT_1 + xT_2 = 0$ gives two curves. One curve is $x = y$, $T_1 = 1 = -T_2$ and the other curve is $x = -y$, $T_1 = 1 = T_2$.

The restriction of $\mathcal{M}$ to the first curve is x so the multiplicity is 1; as it is on the second curve as well, for a total of 3.

Notice that in the last example $3 = e(M) \neq e(J(M)) = 4$. (J(M) is the ideal of maximal order non-vanishing minors, and is (x^2, xy, y^2) in this case.) But,

Problem 3.12 *Suppose $M \subset N \subset F$ are m primary $\mathcal{O}_{X,x}$ modules, X, x equidimensional. Show that $e(M) = e(N)$ if and only if $e(J(M)) = e(J(N))$.*

There are examples though, where there is a family of ICIS singularities where $e(JM(X_y))$ is independent of y, but $e(J(JM(X_y)))$ is not. In the example due to Henry and Merle, the embedding dimension of the singularity changes at $y = 0$–the singularity goes from being codimension 2 to being codimension 1, because one of the defining equations is no longer singular off the origin. Is this the only way for the connection between the two invariants to break?

Challenge Problem 3.13 *Give a geometric characterization of when $e(JM(X_y))$ is independent of y, but $e(J(JM(X_y)))$ is not.*

This problem is connected with the difference between using the conormal modification to study equisingularity conditions and using the Nash modification, which is why it is interesting. In the ICIS case a difference in the value of the multiplicity between the generic point y and the origin implies there is a jump in the dimension of the fiber of the exceptional divisor over the origin. So if the value of $e(JM(X_y))$ is independent of y, but $e(J(JM(X_y)))$ is not, then the set of limiting tangent planes has a jump in dimension at the origin, but the set of limiting tangent hyperplanes does not.

Reading In Sect. 3 of [11] these ideas are developed further. It also contains the example due to Henry and Merle mentioned above.

There is an important case where it is easy to calculate the multiplicity of the pair. Suppose we are given $\mathcal{O}_X$ modules $M \subset N \subset F$, where F is free, X has dimension 1, and $e(M, N)$ is defined. We want a procedure to calculate $e(M, N)$. The first step is to find a normalization $\tilde{X}, n$ of the curve. Then we can use the following proposition.

Proposition 3.14 *Suppose X is a curve singularity, then $e(M, N) = e(n^*(M), n^*(N))$.*

Proof This is a corollary of Theorem 5.1 of [25]. □

We'll illustrate the rest of the procedure with an example taken from [7]. The procedure is also described in [25].

The curves we consider are the X_l, defined by the minors of

$$F_l = \begin{bmatrix} z & x \\ y & z \\ x^l & y \end{bmatrix}.$$

We assume $l - 1$ is not divisible by 3. With this assumption we have a normalization given by $(\mathbb{C}, n_l)$ where $n_l(t) = (t^3, t^{2l+1}, t^{l+2})$. The assumption on l means that the exponents on the first and last terms in the formula for n are relatively prime. The form of n is a reflection of the fact that X_l is weighted homogeneous with weights $(3, 2l + 1, l + 2)$.

In this example the module N is $F_l^*(JM(\Sigma^2))$ where Σ^2 is the linear maps of rank < 2, and we view F_l as map from $\mathbb{C}^3 \to Hom(\mathbb{C}^2, \mathbb{C}^3)$. Then $M = JM(X_l)$.

The next step is to find a minimal set of generators for $n_l^*(N)$ and $n_l^*(M)$. Pulling back the generators of $JM(\Sigma^2)$ using $F_l \circ n_l$, we get:

$$n_l^*(N) = \begin{bmatrix} t^{l+2} & -t^3 & 0 & -t^{2l+1} & t^{l+2} & 0 \\ 0 & t^{2l+1} & -t^{l+2} & 0 & -t^{3l} & t^{2l+1} \\ t^{2l+1} & 0 & -t^3 & -t^{3l} & 0 & t^{l+2} \end{bmatrix}.$$

As this matrix has generic rank 2, $n_l^*(N)$ can be generated freely by 2 generators since we are working over $\mathcal{O}_1$, so a matrix of generators R_N of $n_l^*(N)$ with a minimal number of columns is

$$R_N = \begin{bmatrix} -t^3 & 0 \\ t^{2l+1} & -t^{l+2} \\ 0 & -t^3 \end{bmatrix}.$$

A calculation shows that $n_l^*(JM(X))$ is generated by the columns of:

$$R_{JM} = \begin{bmatrix} -t^3 & 2t^{l+2} \\ 2t^{2l+1} & -t^{3l} \\ t^{l+2} & t^{2l+1} \end{bmatrix}.$$

Note that

$$R_{JM} = R_N \begin{bmatrix} 1 & -2t^{l-1} \\ -t^{l-1} & -t^{2l-2} \end{bmatrix}.$$

Denote the submodule of $\mathcal{O}_1^2$ whose matrix of generators is the 2×2 matrix in the last line by K. Since $n_l^*(N)$ is freely generated, it is isomorphic to $\mathcal{O}_1^2$. The isomorphism carries the pair $(n_l^*(JM(X)), n_l^*(N))$ to $(K, \mathcal{O}_1^2)$. Then $e(n_l^*(JM(X)), n_l^*(N)) = e(K, \mathcal{O}_1^2)$. Since $\mathcal{O}_1$ is Cohen–Macaulay, the multiplicity of the second pair is the colength of the determinant of the matrix of generators of K, which is $2l - 2$.

Polar Varieties of a Module

Intuitively, the polar varieties of a module measure the "curvature" of Projan $\mathcal{R}(M)$, and we have encountered them in the examples of the previous paragraph. As we shall see, the projection of $B_{\mathcal{M}}(\text{Projan } \mathcal{R}(N)) \cdot c_M^2$ to $\mathbb{C}^2$, studied in Example 3.11 is the polar curve of M.

The *polar variety of codimension* l of M in X, denoted $\Gamma_l(M)$, is constructed by intersecting Projan $\mathcal{R}(M)$ with $X \times H_{g+l-1}$ where H_{g+l-1} is a general plane of codimension $g + l - 1$, then projecting to X.

So, in the setting of Example 3.11, $g = 2$, and $g + l - 1 = 2 + 1 - 1 = 2$, and the projection of $B_{\mathcal{M}}(\text{Projan}\,\mathcal{R}(N)) \cdot c_M^2$ to Projan $\mathcal{R}(M)$ is the intersection of $\mathbb{C}^2 \times H_2$ with Projan $\mathcal{R}(M)$. Thus the projection of $B_{\mathcal{M}}(\text{Projan}\,\mathcal{R}(N)) \cdot c_M^2$ to $\mathbb{C}^2$ is $\Gamma_1(M)$.

The polar varieties of M can be constructed by working only on X. The plane H_{g+l-1} consists of all hyperplanes containing a fixed plane H_K of dimension $g+l-1$. By multiplying the matrix of generators of M by a basis of H_K we obtain a submodule of M denoted M_H.

Proposition 3.15 *In this set-up the polar variety of codimension l consists of the closure in X of the set of points where the rank of M_H is less than g, and the rank of M is g.*

Proof Since H_{g+l-1} is generic, the general point of Projan $\mathcal{R}(M) \cap X \times H_{g+l-1}$ lies over points where the rank of M is g. Choose coordinates so that a basis for H_K consists of the last $g+l-1$ elements of the standard basis of $\mathbb{C}^j$, j the number of generators of M. We can find v such that $v[M_H] = 0$ but $v[M] \neq 0$ if and only if we are at a point where the rank of $M_H < g$. The existence of v is equivalent to being able to find a combination of the rows of $[M]$, such that the last $g+l-1$ entries are 0. This row is a hyperplane which lies in H_{g+l-1}. □

Teissier [36, 38] defined the polar varieties of an analytic germ $(X^d, x) \subset \mathbb{C}^n$ of codimension l as follows: take a generic projection π of $X^d \to \mathbb{C}^{d-l+1}$, and take the closure of the critical points of the restriction of the projection to the smooth points of X. Using the last proposition, it is easy to see that these polar varieties are the polar varieties of the Jacobian module of X.

For, given $(X^d, x) \subset \mathbb{C}^n$, the generic rank g of the Jacobian module of X is $n-d$. The kernel of a generic projection to $\mathbb{C}^{d-l+1}$ has dimension $n-d+l-1 = g+l-1$. Let the fixed plane H_K in the previous proposition be the kernel of π. Then the rank of M_H is less than maximal at a smooth point of X if and only if the tangent space of X has larger than expected intersection with the kernel of π. Thus, a tangent hyperplane of X contains H_K at a smooth point of X if and only if x is a critical point for the restriction of the projection to X at x. Thus the two notions of polar variety coincide.

If M is an ideal and we are working on X, then M_H is a sheaf of ideals and the polar varieties are the closure of the set defined by this sheaf on the complement of $V(M)$.

Problem 3.16 *Given $M \subset N \subset \mathcal{O}^p_{X,x}$, M and N both $\mathcal{O}_X$ modules, M induces an ideal sheaf on* Projan $\mathcal{R}(N)$, *and we can define the polar varieties of this ideal sheaf. (To do this we must work on the fiber of* Projan $\mathcal{R}(N)$ *over x.) Show that the projection of the polar of dimension d defined in this way to X is $\Gamma^d(M)$.*

Thus, there are 4 different settings for studying the polar varieties. It is often useful in proofs to move between them.

There is a special case which will be important to us. The diagram below represents the smoothing of an isolated singularity.

$$\begin{array}{ccccc} X^d(0) \subset & \mathcal{X}^{d+1} & \subset Y \times \mathbb{C}^N & \supset \mathcal{X}(y) \\ \downarrow & & \downarrow{\scriptstyle p_Y} & \downarrow{\scriptstyle \pi_Y} \\ 0 \in & & Y = \mathbb{C} & \supset y \neq 0 \end{array}$$

Let $M = JM_z(\mathcal{X})$, Then $\Gamma_d(\mathcal{X})$ by the previous proposition is defined by selecting $N-1$ generic generators of $JM_z(\mathcal{X})$, and looking to see where they have less than maximal rank. Assume coordinates chosen so that the first $N-1$ columns of $[JM(\mathcal{X})]$ are generic. Then the points where the polar intersects $\mathcal{X}(y)$ are the critical points of z_N restricted to $\mathcal{X}(y)$. The number of such points is the number of sheets of $\Gamma_d(\mathcal{X})$ over Y is the multiplicity of $\Gamma_d(\mathcal{X})$ over Y at the origin. If the smoothing is unique up to diffeomorphism, then the invariant is denoted $m_d(X)$. It is clear that the number of critical points of a generic linear form on a smoothing of X is important to the topology of $\mathcal{X}(y)$, so this number is an important invariant of X.

By construction, the existence of a polar variety of M at $x \in X$ is tied to the dimension of the fiber of Projan $\mathcal{R}(M)$ over x.

Problem 3.17 *Suppose X^d, x equidimensional and M has the same generic rank g on each component of X at x. Show that $\Gamma_l(M, x)$ is non-empty if and only if the dimension of the fiber of* Projan $\mathcal{R}(M)$ *over x is greater than or equal to $l + g - 1$.*

There is a strong connection between polar varieties and integral closure thanks to an important result of Kleiman and Thorup [26, 27], which we next discuss. The following theorem ties the dimension of this fiber to integral closure conditions. Set-up: X the germ of a reduced analytic space of pure dimension d, F a free $\mathcal{O}_X$-module, $M \subset N \subset F$ two nested submodules with $M \neq N$, M and N are generically equal and free of rank e. Set $r := d + e - 1$. Set $C :=$ Projan$(\mathcal{R}(M))$ where $\mathcal{R}(M) \subset \mathcal{S}ym\mathcal{F}$ is the subalgebra induced by M in the symmetric algebra on F. Let $c : C \to X$ denote the structure map. Let W be the closed set in X where N is not integral over M, and set $E := c^{-1}W$.

Theorem 3.18 (Kleiman-Thorup, [26, 27]) *If N is not integral over M, then E has dimension $r - 1$, the maximum possible.*

Proof Since this theorem is so important to us, we give a concise version, due to Kleiman [24], of the proof that appears in [27].

Given an element $h \in N$ that's not integral over M, let H be the module generated by h and M. Now we use the notation of the diagram used in the definition of $e(M, N)$. We have $D_{M,H}$ is nonempty by Remark 3.9, so of dimension $r - 1$ where $r :=$ dimProjan $\mathcal{R}(M)$. But π_H embeds $D_{M,H}$ in Projan $\mathcal{R}(M)$ because H/M is cyclic. Moreover, Remark 3.9 implies that N is integral over M locally off $\pi_M D_{M,N}$; so H is too; so Remark 3.9 implies that $\pi_M D_{M,N}$ contains $\pi_M D_{H,N}$. Plainly, $\pi_M D_{H,N}$ lies in E. Thus dim $E = r - 1$. □

A recent proof in a more general setting appears in [33].

We give an example the usefulness of this Theorem by giving a simple proof of one direction of a theorem of Teissier describing Whitney equisingularity.

Set-up: Suppose $Y^k, 0 \subset X^{d+k}, 0$, Y^k smooth, $\underline{y}$ coordinates on Y, $I(Y) = m_Y$. Set $M = m_Y JM(X)$, $N = M + \mathbb{C}\{\frac{\partial f}{\partial \underline{y}}\}$, then $\text{Projan}(\mathcal{R}(M)) = B_{m_Y}(C(X))$, $M = N$ off Y.

Let E denote the exceptional divisor of $B_{m_Y}(C(X))$.

Theorem 3.19 (Teissier, [38]) *If the fibers of E, the exceptional divisor of $B_{m_Y}(C(X))$ over Y, have the same dimension, then the Whitney conditions hold along Y.*

Proof If the Whitney conditions fail along Y, they do so on a proper closed subset $S \subset Y$. Then S is the set where $\overline{M} \neq \overline{N}$ [9]. By the Kleiman-Thorup theorem there must be a component of E over S, so the fibers of E have larger dimension over points in S than over the generic point of Y. □

For the ICIS case we can use the machinery of multiplicities, together with the Kleiman-Thorup theorem to get criteria for a family of sets to be Whitney equisingular, in which the criteria depend only on the members of the family, not the total space. We describe how this developed.

The first theorem is a generalization of a result of Teissier, who used it in conjunction with hypersurfaces. This theorem is useful in showing that if invariants are independent of parameter then equisingularity conditions hold.

Theorem 3.20 (Principle of Specialization of Integral Dependence) *Assume that X is equidimensional, and that $y \mapsto e(y)$ is constant on Y^k. Let h be a section of a free $\mathcal{O}_X$ module E whose image in $E(y)$ is integrally dependent on the image of $M(y)$ for all y in a dense Zariski open subset of Y. Then h is integrally dependent on M.*

Proof Cf. Theorem 1.8 [17]. □

The proof of the PSID proceeds by showing that the constancy of the multiplicity means that M has a reduction M_R which is generated by $\dim(X(y)) + p - 1$ generators, which is the minimum possible if $e(M(y))$ is well defined for all y. To do this, first we find such an M_R whose restriction $M_R(0)$ to $X(0)$ is a reduction of M restricted to $X(0)$, so $e(M_R(0)) = e(M(0))$ by Theorem 3.7. Then the uppersemicontinuity of the multiplicity ([17], 1.1 p. 547), implies $e(M_R(0)) \geq e(M_R(y))$, while $M_R(y) \subset M(y)$ implies $e(M_R(y)) \geq e(M(y))$. This gives us the inequality:

$$e(M(0)) = e(M_R(0)) \geq e(M_R(y)) \geq e(M(y)) = e(M(0)).$$

Thus, by Theorem 3.7, $M_R(y)$ is a reduction of $M(y)$ for all y.

Now replace M by the submodule generated by M_R and g, where g may be h or any element of M not in M_R. A lemma ([17] 1.2, p. 548) shows that if the set of points where g fiberwise is not integrally dependent on M_R is a proper Zariski closed subset of Y, then the set of points where g is not integrally dependent on M_R is also a proper Zariski closed subset W of Y. This implies that M_R is a reduction of (M, h) off a Zariski closed set of Y as this is true fiberwise.

Now, the dimension of the fiber of $\text{Projan}(\mathcal{R}(M_R))$ over our base point $x_0 \in X$ is at most $\dim(X(y)) + p - 2$, which is one less than the number of generators. Now the inverse image of W in $\text{Projan}(\mathcal{R}(M_R))$ must have dimension at most $\dim(X(y)) + p - 2 + k - 1$. Then since

$$\begin{aligned}\dim(X(y)) + p - 2 + k - 1 &\leq (\dim(X(y)) + k) + (p - 1) - 2\\ &= (\dim(X) + p - 1) - 2,\end{aligned}$$

the Kleiman-Thorup theorem then shows that $\bar{M}_R = \bar{M}$, which gives the result.

In order to show that the equisingularity condition implied that the invariants were independent of y more ideas are necessary. These are discussed in the proof of the next theorem.

Theorem 3.21 *Suppose* $(X^{d+k}, 0) \subset (\mathbb{C}^{n+k}, 0)$ *is a complete intersection,* $X = F^{-1}(0)$*,* $F : \mathbb{C}^{n+k} \to \mathbb{C}^p$*,* Y *a smooth subset of* X*, coordinates chosen so that* $\mathbb{C}^k \times 0 = Y$*. Then the following are equivalent:*

(i) the pair $(X - Y, Y)$ *satisfies* W *at* 0*;*
(ii) The sets $X(y)$ *are complete intersections with isolated singularities and* $e(m_y JM(X_y))$ *is independent of* y *for all* $y \in Y$ *near* 0.

Proof For the proof that (ii) implies (i), the condition on $e(m_y JM(X_y))$ implies that the singularities do not split, so that $X - Y$ is smooth. Since the integral closure condition is a generic condition, the PSID applies.

For the proof that (i) implies (ii) the proof is more complicated. An expansion formula shows that $e(m_y JM(X_y))$ is a sum of multiplicities. Each multiplicity that appears is the sum of two Milnor numbers of plane sections of the ICIS $X(y)$. Since Whitney equisingularity of X implies the Whitney equisingularity of the plane sections of X, and the Milnor numbers of the sections are topological invariants, the multiplicities, and hence their sum is invariant as well. □

With this result you can see that the Whitney conditions imply in the ICIS case, that the fiber of $B_{m_Y}(C(X))$, the blow-up of the conormal modification along Y, is equidimensional over Y. For the Whitney conditions imply that the multiplicity of $m_Y JM(X(y))$ along Y is constant. Then the technique of proof used in the Principle of Specialization implies that we can pick $d + p - 1$ elements of $m_Y JM(X(0))$ which generate a reduction N first of $m_Y JM(X(0))$, then of $m_Y JM(X)$. This implies that there exists a finite map from $B_{m_Y}(C(X))$ to $\text{Projan}(\mathcal{R}(N))$. Now since $\text{Projan}(\mathcal{R}(N)) \subset X \times \mathbb{P}^{d+p-2}$, the fiber dimension of $B_{m_Y}(C(X))$ over $0 \in X$ is less than or equal to $d + p - 2 = n - 2$ which is the minimum possible.

For an ICIS X, we use the multiplicity of $m_Y JM(X)$ to control the Whitney equisingularity type. What do we use when $e(m_Y JM(X))$ is not defined? Since $e(m_Y JM(X))$ is defined only when $JM(X)$ has finite codimension in $\mathcal{O}^p_X$, it is only defined for ICIS.

Looking at the ideas relating $e(m_Y JM(X))$ to the Whitney conditions, though the connections are beautiful, the proofs that Whitney implies the constancy of the multiplicities seem unnecessarily round about. The Whitney conditions themselves are described by the behavior of the exceptional divisor of $B_{m_Y}(C(X))$. Is there a direct link between $e(m_Y JM(X))$ and the exceptional divisor, so that it would not be necessary to go through topology to show that Whitney implies the constancy of $e(m_Y JM(X))$?

To answer the first question, start with thinking about the pair of modules $(JM(X), \mathcal{O}^p_X)$. The module $JM(X)$ can be viewed as the module of infinitesimal, first order trivial deformations of X. (Trivial with respect to biholomorphic equivalences of $\mathbb{C}^n$.) The module $\mathcal{O}^p_X$ is then the module of all infinitesimal, first order deformations of X since we can deform the equations of X freely, and get a family of ICIS. It is known that if X has an isolated singularity, then again the codimension of $JM(X)$ inside the module $N(X)$ of all infinitesimal, first order deformations of X is finite. This suggests using $e(JM(X), N(X))$.

However, two problems surface. We want specialization of N from the total space of a family to the fibers. This is necessary if the results are to depend only on the fibers of the family and not on the total space. This will be true, provided any first order linear infinitesimal deformation of a space lifts to a deformation of the family. However this is clearly false, if the base space of the versal deformation space has components. If the base space of the verbal deformation space is smooth for example, then the specialization property is true.

Another problem enters because $N(X)$ may have curvature. Here we are making an analogy between N and $JM(X)$. Moonen [29] has shown that the multiplicities of the polar varieties of X, x are related to the curvature of X at x. (In the real case see also [4]) This curvature then is related to the limiting tangent hyperplanes of X at x. Since the polar varieties of N are related to limiting hyperplanes defined by row vectors of a matrix of generators of N, it is reasonable to call the phenomena picked up by polar multiplicities of N as the curvature of N. How this curvature enters into the invariants we want will be a main theme of the next section.

In the next section we give also an example which shows the multiplicity of the pair may be zero, but the curvature contribution of N gives a non-zero invariant.

Since the Whitney conditions are controlled by the dimension of the fiber of the exceptional divisor of $B_{m_Y}(C(\mathcal{X}))$, and the dimension of the fibers are detected by the presence of the polar varieties of the relative Jacobian module, it is reasonable to look for a connection between invariants associated with integral closure and those associated with polar varieties.

An approach for linking the behavior of the multiplicity of an ideal in a family to the degree of the exceptional divisor is given by Teissier in [38, p. 345]. We include an excerpt from this reference where this idea is mentioned.

Soient $f : (X,0) \to (\mathbb{D},0)$ un morphisme d'espaces réduits, I un idéal de $\mathcal{O}_X$ définissant un sous-espace $Y \subset X$ tel que $f|Y : Y \to \mathbb{D}$ soit fini, $p : X' \to X$ l'éclatement de Y, $D_{vert.}$ la réunion des composantes du diviseur exceptionnel D (non nécessairement réduites) dont l'image ensembliste par p est 0, $\deg D_{vert.} = \deg(\mathcal{O}_{D_{vert.}}(1))$. Pour tout représentant suffisamment petit du germe de f en 0, on a l'égalité

$$\deg D_{vert.} = e(I \cdot \mathcal{O}_{X(0)}) - e(I \cdot \mathcal{O}_{X(s)}) \quad (\text{pour } s \neq 0) \quad .$$

En particulier, on a "$e(I.\mathcal{O}_{X(s)})$ est indépendant de $s \in \mathbb{D}$" si et seulement si $\dim p^{-1}(0) = \dim X - 2$.

Here is how we can understand Teissier's formula. The fiber of the exceptional divisor over $0 \in X^{d+1}$ is a projective variety so it has a degree. When we intersect this variety with a linear space of complementary dimension, on the one hand, the number of points we get is the degree of the variety, on the other, because intersecting $B_I(X)$ with this linear space defines the polar curve of I, it is the number of points in the polar curve over a generic t value. Call this number $m_d(I, X)$. Now one way to define the polar curve is to pick d generic elements of I, chosen so that they define a reduction of $I(0)$ and are a reduction of I on the total space over $\mathbb{D} - 0$, and see where they are zero. Call this ideal J. By construction the points where they are zero outside of $V(I)$, will be a Z-open and dense set of the polar curve, and at points of $V(I)$, $\overline{I(y)} = \overline{J(y)}$ and so $e(I(y)) = e(J(y))$ at such points. Since J is generated by d elements, a lemma shows that $e(J(y))$ is independent of y. So

$$degD_{vert} = m_d(I, X) = e(J \cdot \mathcal{O}_{X(0)}) - e(J \cdot \mathcal{O}_{X(y)}) = e(I \cdot \mathcal{O}_{X(0)}) - e(I \cdot \mathcal{O}_{X(y)}).$$

If we extend this approach to pairs of modules we find that the polar variety of N enters as well as the polar variety for M.
Set-up: $M \subset N \subset F$, a free $\mathcal{O}_X$ module, X equidimensional, a family of sets over Y, with equidimensional fibers, Y smooth, $\overline{M} = \overline{N}$ off a set C of dimension k which is finite over Y.

Let $\Delta(e(M, N)) = e(M(0), N(0), \mathcal{O}_{X(0)}, 0) - e(M(y), N(y), \mathcal{O}_{X(y)}, (y, x))$ be the change in the multiplicity of the pair (M, N) as the parameter changes from y to 0.

Theorem 3.22 (Multiplicity Polar Theorem [6, 11]):

$$\Delta(e(M, N)) = mult_y \Gamma_d(M) - mult_y \Gamma_d(N)$$

Many applications of this theorem can be found in: [7, 11–13, 19].
To show its power we give a simple proof of Theorem 3.21 which links $e(JM(X_y))$ and the Whitney conditions. The proof that that (ii) implies (i) avoids the use of topology.

Proof (*of* 3.21): (i) implies (ii) The Whitney conditions imply that the fiber of $D \subset B_{m_Y}(C(X))$, the exceptional divisor is equidimensional over Y. Because the dimension of the fiber is small, there is no polar variety of codimension d for $m_Y JM(X)$. Since $\mathcal{O}_X^p$ has no polar varieties, the Multiplicity Polar Theorem implies that $e(mJM(X_y))$ is independent of y.

(ii) implies (i) The independence of $e(mJM(X_y))$ from y implies that there is no polar variety of codimension d for $m_Y JM(X)$, and hence the fiber of D over Y^k is equidimensional. At this point we apply the theorem of Kleiman-Thorup (3.18). We know that $JM_Y(X)$, the submodule generated by the partial derivatives taken with respect to coordinates on Y, is in the integral closure of $m_Y JM(X)$ at points in a Z-open subset of Y. Since the dimension of the set of points of $\text{Projan}(m_Y JM(X))$ over the set of points where the integral closure condition does not hold is at most $(k-1)+(d+g-2) < (d+k)+(g-1)-1$, it follows that $JM_Y(X)$ is in the integral closure of $m_Y JM(X)$ at all points of Y. □

In the next part, we examine an important class of singularities for which the module N of first order deformations does specialize as we desire.

Determinantal Singularities

We begin with F, a $(n+k, n)$ matrix, with entries in $\mathcal{O}_q$; we view F as a map from $\mathbb{C}^q \to Hom(\mathbb{C}^n, \mathbb{C}^{n+k})$. Let Σ^r denote elements of $Hom(\mathbb{C}^n, \mathbb{C}^{n+k})$ of rank less than r. Let I_r be the ideal in $\mathcal{O}_{n^2+nk}$ generated by the minors of size r of elements of $Hom(\mathbb{C}^n, \mathbb{C}^{n+k})$. It is easy to check that the codimension of Σ^r in $Hom(\mathbb{C}^n, \mathbb{C}^{n+k})$ is $(n-r+1)(n+k-r+1)$. The elements of $Hom(\mathbb{C}^n, \mathbb{C}^{n+k})$ of rank r, $0 \leq r \leq n$ give a stratification of $Hom(\mathbb{C}^n, \mathbb{C}^{n+k})$ which we call the rank stratification.

Assume F is transverse to the rank stratification of $Hom(\mathbb{C}^n, \mathbb{C}^{n+k})$ on $\mathbb{C}^q - 0$. Let $\Sigma^r(F) := V(F^*(I_r))$, then $F^*(I_r)$ is generated by the minors of size r of F. $\Sigma^r(F)$ is determinantal i.e. codim $\Sigma^r(F)$ =codim Σ^r. If $q < (n-r+2)(n+k-r+2)$ then $\Sigma^r(F)$ has a smoothing, because when we deform F so that it is transverse to the rank stratification there will be no points where the rank$< r-1$.

We fix the class of deformations and fix a unique smoothing by only considering deformations of $\Sigma^r(F)$ which come from deformations of the entries of F. As we shall see, the geometric meaning of the invariants we develop is tied to the topology of the smoothing.

We may freely vary the entries of F and deformations of the entries of F induce deformations of the generators of $F^*(I_r)$; first order deformations define the module $N(X_F)$. Generators of $N(X_F)$ are tuples of minors of F of size $r-1$. If F and r are understood we simply write $N(X)$.

Properties of $N(X)$

The operation of forming $N(X)$ has some nice properties.

- N is universal. If the entries of F are coordinates on $Hom(\mathbb{C}^n, \mathbb{C}^{n+k})$ denote $N(X)$ by N_U. Then for any M, $N(X_F) = F^* N_U$.
- N_U is stable; $N_U = JM(\Sigma^r)$. Coupled with universality this implies $N(X_F) = F^* JM(\Sigma^r)$, which explains why the generators of $N(X_F)$ are tuples of minors of F of size $r-1$. (We say that the first order linear infinitesimal deformations are

stable if they are trivial. Here the first order linear trivial infinitesimal deformations are deformations are $JM(\Sigma^r)$.)

- Stability implies the polar varieties of Σ^r are the polar varieties of N_U.
- Universality implies $\Gamma_i(N(X_F)) = F^*\Gamma_i(N_U)$.
- Together they imply if $\tilde{F}$ defines a smoothing $\tilde{X}$ of X_F^d, then

$$mult_{\mathbb{C}}\Gamma_d(N(\tilde{X}_{\tilde{F}})) = F(\mathbb{C}^q) \cdot \Gamma_d(\Sigma).$$

In general, the intersection number $F(\mathbb{C}^q) \cdot \Gamma_d(\Sigma)$ is defined as follows. Work on $\mathbb{C}^q \times Hom(\mathbb{C}^n, \mathbb{C}^{n+k})$ and consider the intersection of the graph of F with $\mathbb{C}^q \times \Gamma_d(\Sigma)$, where, since $\Gamma_d(\Sigma)$ is the polar variety of codimension d in $\Gamma_d(\Sigma)$, the graph of F and $\mathbb{C}^q \times \Gamma_d(\Sigma)$ have complementary dimension in $\mathbb{C}^q \times Hom(\mathbb{C}^n, \mathbb{C}^{n+k})$. If F is one to one, then the intersection number is that of the image of F with $\Gamma_d(\Sigma)$ in $Hom(\mathbb{C}^n, \mathbb{C}^{n+k})$.

If $r = n$ which is the case that I_r is the ideal of maximal minors, $F(\mathbb{C}^q) \cdot \Gamma_d(\Sigma)$ is computed in terms of the entries of F in [7]. We give a brief introduction to the formula in this paper in order to continue the study of curve singularities begun at the end of the section on multiplicities. This will also show why singularities defined by maximal minors are easier to study.

To study the polar varieties of $F^*JM(\Sigma^n)$, we need to understand $\text{Projan}(\mathcal{R}(F^*JM(\Sigma^n)))$. At a smooth point M of X_F, consider pairs (l_1, l_2) where $l_1 \in ker M^t$, $l_2 \in ker M$. Here $l_2 \in \mathbb{P}^{n-1}$ is unique, while the set of $l_1 \in \mathbb{P}^{n+k-1}$ has dimension k. Take the closure of this set in $X_F \times \mathbb{P}^{n+k-1} \times \mathbb{P}^{n-1}$. This is the M-transform of X, denoted X_M. In [7], this is shown to be isomorphic to $\text{Projan}(\mathcal{R}(F^*JM(\Sigma^n)))$. The isomorphism is defined by

$$\Phi(x, (T_1, \dots, T_{n+k}), (S_1, \dots, S_n)) = (x, T \cdot S),$$

where $T \cdot S$ is an element of $\mathbb{P}Hom(n, n+k)$.

If $\tilde{F}$ is defines a smoothing $\mathcal{X}$ of X^d, then we want to calculate the degree over the base $\mathbb{C}$ of the polar curve of $\tilde{F}^*JM(\Sigma^n)$, denoted $m_d(F^*JM(\Sigma^n))$. Ideally, we would want to find the equations of the polar variety of Σ^n of complementary dimension to q, pull them back to $\mathcal{X}$ and take degree. This seems difficult. Instead, we will define "mixed polars" for which we can find equations, and which will define Cohen–Macaulay germs. To define these we look again at the construction of the polar varieties of Σ^n and their pull backs–the polars of $\tilde{F}^*JM(\Sigma^n)$.

First, denote the fiber over the origin in $\mathcal{X}$ of $\text{Projan}\,\mathcal{R}(\tilde{F}^*JM(\Sigma^n))$ by E. The generic rank of $\tilde{F}^*JM(\Sigma^n)$ is the same as the generic rank of $JM_z(\mathcal{X})$ which is $k+1$, the codimension of the generic fiber of $\mathcal{X}$. Then the polar curve is gotten by intersecting $\text{Projan}\,\mathcal{R}(\tilde{F}^*JM(\Sigma^n))$ with $d+k$ hyperplanes and projecting to $\mathcal{X}$. The degree of the polar curve over $\mathbb{C}$ is just $E \cdot h^{d+k}$ in $\mathbb{P}Hom(n, n+k)$, where h is the hyperplane class of $\mathbb{P}Hom(n, n+k)$. Now we use the isomorphism between $\text{Projan}\,\mathcal{R}(\tilde{F}^*JM(\Sigma^n))$ and $\mathcal{X}_M$. Denote the hyperplane classes on $\mathcal{X} \times \mathbb{P}^{n-1}$ and

$\mathcal{X} \times \mathbb{P}^{n+k-1}$ by h_2 and h_1 respectively. As classes, the pullback of h to $\mathcal{X} \times \mathbb{P}^{n+k-1} \times \mathbb{P}^{n-1}$ by the Veronese V is $h_1 + h_2$. So,

$$m_d(F^* JM(\Sigma^n)) = \sum_{i=0}^{d+k} \binom{d+k}{i} h_1^i h_2^{d+k-i} \cdot E.$$

The simple description we have of $C(\Sigma^n)$ which permits the decomposition of the last formula seems to be unique to $r = n$. This decomposition is the key to being able to write $m_d(F^* JM(\Sigma^n))$ as the alternating sum of colengths of ideals defined using the entries of F.

Define $\Gamma_{i,j}(\tilde{F}^* JM(\Sigma^n))$ to be $\pi_{\mathcal{X}}(\mathcal{X}_M \cap h_1^i h_2^j)$. We call these the mixed polars of type (i, j) of $\tilde{F}^* JM(\Sigma^n)$. Denote the degree of this mixed polar over $\mathbb{C}$ by $h_1^i h_2^j$.Then

$$m_d(F^* JM(\Sigma^n)) = \sum_{i=0}^{d+k} \binom{d+k}{i} h_1^i h_2^{d+k-i}.$$

It is shown in [7] that the mixed polars are related to certain determinantal varieties, and that the $h_1^i h_2^j$ are the alternating sum of degrees of these determinantal varieties. These degrees are just the lengths of the rings gotten by modding out the local ring of the associated determinantal variety by the coordinate on $\mathbb{C}$. In turn, these are just the lengths of the pullbacks by F of the rings defining the corresponding varieties on Σ^n. So, these numbers depend only on the component functions of F.

Now we consider again the determinantal space curves X_l defined by $F_{X_l}^{-1}(\Sigma^2)$,

$$F_{X_l} = \begin{bmatrix} z & x \\ y & z \\ x^l & y \end{bmatrix}.$$

We have $n = 2, k = 1, d = 1$, so

$$m_1(F_{X_l}^* JM(\Sigma^n)) = h_1^2 + 2h_1 h_2 + h_2^2.$$

The h_2^2 term is zero, because we are working on $X_l \times \mathbb{P}^2 \times \mathbb{P}^1$, and the square of the hyperplane class on $\mathbb{P}^1$ is zero.

To calculate h_1^2, note that if we choose $(1, 0, 0)$ as the point of intersection of our two hyperplanes on $\mathbb{P}^2$, the ideal of $\Gamma_{2,0}$ for this choice on $Hom(2, 3)$, is $(a_{1,1}, a_{1,2}, a_{2,1}a_{3,2} - a_{2,2}a_{3,1})$, for these are the points of Σ^2 for which $(1, 0, 0)$ is in the kernel of M^t, $M \in Hom(2, 3)$. Pulling this ideal back by $F_{X_l}^*$ gives (x, z, y^2), which has colength 2, so $h_1^2 = 2$.

To compute $h_1 h_2$, choose $(0, 1)$ as the point on $\mathbb{P}^1$ defined by the hyperplane, and let $(0, 0, 1)$ be the hyperplane on $\mathbb{P}^2$. So, we are looking for M such that $(0, 1)$ is in the kernel of M and some line defined by $(a, b, 0)$ is in the kernel of M^t. The ideal that defines this set is $(a_{2,1}, a_{2,2}, a_{2,3})$. This is already determinantal, so our procedure simplifies in this case. We get $h_1 h_2$ is the colength of (x, y, z) which is 1, so $m_1(F^*_{X_l} JM(\Sigma^n)) = 2 + 2(1) + 0 = 4$ for all l.

Putting together our previous work, we see that if $l = 1$, then $e(JM(X_1), F_1^*(JM(\Sigma^2))) = 0$, but $e(JM(X_1), F_1^*(JM(\Sigma^2))) + m_1(F_1^*(JM(\Sigma^2))) = 3$. In fact, for isolated space curve singularities, the invariant $e(JM(X_F), F^*(JM(\Sigma^k))) + m_1(F_1^*(JM(\Sigma^k)))$ is never zero, since the polar of codimension 1 of Σ^r is non-empty for $r > 1$. (If $r = 1$, then F defines an ICIS, and $e(JM(X_F), F^*(JM(\Sigma^1))) = e(JM(X_F)) \neq 0$.)

It is important to understand when an invariant is zero. The next proposition gives a geometric criterion for when $e(JM(X_F), F^*(JM(Y)) = 0$, and also relates this invariant to the map F.

Proposition 3.23 *Suppose* $F\ \mathbb{C}^q, 0 \to \mathbb{C}^n, 0$, $(Y, 0) \subset \mathbb{C}^n$, *$Y$ reduced and X_F defined with reduced structure also. Then* $e(JM(X_F), F^*(JM(Y))) = 0$ *if and only if no limiting tangent hyperplane to Y along the image of F contains the image of* $DF(0)$.

Proof Let $G = 0$ define Y with reduced structure. By hypothesis, $G \circ F$ defines X with reduced structure also. This implies that $JM(X_F) \subset F^*(JM(Y))$, by the Chain rule. The condition that $e(JM(X_F), F^*(JM(Y))) = 0$ is equivalent to $\overline{JM(X_F)} = \overline{F^*(JM(Y))}$. By Proposition 2.29 this is exactly the condition that the ideal sheaf induced by $JM(X_F)$ on Projan $\mathcal{R}(F^*(JM(Y)))$ is irrelevant ie. does not vanish on the fiber of Projan $\mathcal{R}(F^*(JM(Y)))$ over $0 \in X_F$. Since Projan$(JM(Y))$ is $C(Y)$, the fiber of Projan $\mathcal{R}(F^*(JM(Y)))$ over 0 is just limiting tangent hyperplanes to Y along the image of F.

The set Projan $\mathcal{R}(F^*(JM(Y)))$ is a subset of $X_F \times \mathbb{P}^{n-1}$. By the Chain Rule we know $DG(F(x)) \circ DF(x) = D(G \circ F)(x)$. Now, $DF(x)$ induces an ideal sheaf on $X \times \mathbb{P}^{n-1}$, because F has n component functions. If we restrict this sheaf to Projan $\mathcal{R}(F^*(JM(Y)))$, we get the ideal sheaf induced by $JM(X)$ on Projan $\mathcal{R}(F^*(JM(Y)))$, because this ideal sheaf arises from writing the generators of $JM(X)$ in terms of the generators of $F^*(JM(Y))$, and this is exactly what the Chain Rule does for us. Denote this sheaf by $\mathcal{F}$.

The condition that this ideal sheaf vanish at a point $(x, H) \in$ Projan $\mathcal{R}(F^*(JM(Y)))$ is just that the linear form defining H when applied to each of the generators of $\mathcal{F}$ give zero. For, the value of the i-th generator, $\sum_1^n \frac{\partial F_j}{\partial z_i} T_j$ on $(x, (a_1, \ldots, a_n))$ is $\sum_1^n \frac{\partial F_j}{\partial z_i}(x) a_j$. Since the fiber over 0 is the limiting tangent hyperplanes to Y along F at the origin the result follows. □

We explore the case of three lines in $\mathbb{C}^3$ ($l = 1$) further. It is simpler to do this if we use the map

$$F = \begin{bmatrix} z & 0 \\ y & y \\ 0 & x \end{bmatrix}.$$

For this F, X_F is the coordinate axes.

Example 3.24 The fiber of Projan $\mathcal{R}(F^*JM(\Sigma^2))$ over 0 consists of three copies of $\mathbb{P}^1$, namely,

$$\begin{bmatrix} 0 & 0 \\ 0 & a \\ 0 & b \end{bmatrix}, \begin{bmatrix} a & 0 \\ b & 0 \\ 0 & 0 \end{bmatrix}, \begin{bmatrix} a & -a \\ 0 & 0 \\ b & -b \end{bmatrix}, (a, b) \in \mathbb{P}^1.$$

Further, the image of $DF(0)$ is not contained in any element of the fiber of Projan $\mathcal{R}(F^*JM(\Sigma^2))$ over 0.

To see why these assertions are true, note that the fiber of Projan $\mathcal{R}(F^*JM(\Sigma^2))$ is constant over the z axis for $z \neq 0$. This is because F and Σ^2 are homogeneous.

We have a general result which describes the fiber of $C(\Sigma^r)$ which we can apply here, which we now describe.

We know that the fiber to the normal bundle to the smooth manifold $\Sigma^{r+1} - \Sigma^r$ at $M \in \Sigma^{r+1} - \Sigma^r$, is $Hom(K(M), C(M))$ where $K(M)$ denotes the kernel of M and $C(M)$ denotes the cokernel, which we think of as the vectors in $\mathbb{C}^{n+k}$ which annihilate the image of M.

So up to some identifications, the fiber of $C(\Sigma^{r+1})$ at M is inside $\mathbb{P}Hom(K(M), C(M))$. Let $\Sigma_j(M)$ denote the elements of $Hom(K(M), C(M))$ of kernel rank j.

Let X_j denote the projective variety determined by $\overline{\Sigma}_j$. If $M \in Hom(\mathbb{C}^n, \mathbb{C}^{n+k})$, then we denote $\mathbb{P}(\overline{\Sigma}_r(M))$ by $X_r(M)$.

Theorem 3.25 (Conormal fiber Theorem) *Suppose M is in Σ_s, $s > r$. Then the fiber of the conormal of $C(\overline{\Sigma}_r)$ at M is $X_{s-r}(M)$.*

Proof See the Conormal Fiber Theorem at the end of Sect. 2 of [7]. □

In the case of singularities defined by maximal minors if we know the M-modification of X_F we can compute these fibers. For example, at points on the z axis of X_F, $z \neq 0$, we see that the fiber is $(0, a, b) \times (0, 1)$, because $(0, 1)$ is the kernel of $F(0, 0, 1)$, and $(0, a, b)$ is the kernel of $F^t(0, 0, 1)$. Then a point of the fiber maps to

$$\begin{bmatrix} 0 \cdot 0 & 0 \cdot 0 \\ 0 \cdot a & 1 \cdot a \\ 0 \cdot b & 1 \cdot b \end{bmatrix}$$

The condition that the image of $DF(0)$ is contained in a limiting tangent hyperplane implies that

$$\frac{\partial F}{\partial x} \cdot \begin{bmatrix} 0 & 0 \\ 0 & a \\ 0 & b \end{bmatrix} = 0, \ \frac{\partial F}{\partial y} \cdot \begin{bmatrix} 0 & 0 \\ 0 & a \\ 0 & b \end{bmatrix} = 0.$$

Expanding we get:

$$\begin{bmatrix} 0 & 0 \\ 0 & 0 \\ 0 & 1 \end{bmatrix} \cdot \begin{bmatrix} 0 & 0 \\ 0 & a \\ 0 & b \end{bmatrix} = \begin{bmatrix} 0 \cdot 0 & 0 \cdot 0 \\ 0 \cdot 0 & 0 \cdot a \\ 0 & 1 \cdot b \end{bmatrix} = 0, \begin{bmatrix} 0 & 0 \\ 1 & 1 \\ 0 & 0 \end{bmatrix} \cdot \begin{bmatrix} 0 & 0 \\ 0 & a \\ 0 & b \end{bmatrix} = \begin{bmatrix} 0 \cdot 0 & 0 \cdot 0 \\ 1 \cdot 0 & 1 \cdot a \\ 0 & 0 \cdot b \end{bmatrix} = 0.$$

This implies that $a = b = 0$; thus no element of the fiber which is a limit of tangent hyperplanes to Σ^2 along the image of the z axis in Σ^2 can contain the image of $DF(0)$.

Problem 3.26 *Prove the rest of the assertions of the last example.*

We can use this simple example to get some idea of the possible ways our invariants can change in a family. Given a family of singularities $\{X_t^d\}$, with parameter t, let $e(JM(X_t), F_t^*JM(\Sigma^r), t)$ denote the sum of $e(JM(X_t), F_t^*JM(\Sigma^r), x)$ over all $x \in X_t$; let

$$e_\Gamma(M, F_t^*JM(\Sigma^r), x) = e(JM(X_t), F_t^*JM(\Sigma^r), x) + m_d(F_t^*JM(\Sigma^r), x),$$

and define $e_\Gamma(M, F_t^*JM(\Sigma^r), t)$ in a way similar to $e(JM(X_t), F_t^*JM(\Sigma^r), t)$.

Example 3.27 Let $F_t = \begin{bmatrix} z & 0 \\ y-t & y+t \\ 0 & x \end{bmatrix}$. Let $X_t = X_{F_t}$, then X_t for $t \neq 0$ consists of three lines which intersect in two plane curve singularities–both ordinary nodes. Further $e(JM(X_t), F_t^*JM(\Sigma^2), t)$ is 0 for $t = 0$ and 4 for $t \neq 0$, hence is not upper semicontinuous. The invariant $e_\Gamma(M, F_t^*JM(\Sigma^2), t) = 4$, for all t.

The example shows that $e_\Gamma(M, F_t^*JM(\Sigma^2), t) = 4$ being independent of t does not prevent the singularity from splitting. If we assume the parameter space is embedded in X as $\mathbb{C} \times 0$, and ask that $e_\Gamma(M, F_t^*JM(\Sigma^2), (t, 0))$ is independent of t, then splitting cannot occur because $e_\Gamma(M, F_t^*JM(\Sigma^2), t)$ is upper semicontinuous, and $e_\Gamma(M, F_t^*JM(\Sigma^2), x)$ is always non-zero in the curve case if x is singular.

Equisingularity of Determinantal Varieties

In this section we bring together many elements of these lectures to prove a theorem on the Whitney equisingularity of families of determinantal singularities.

The key invariant is the generalization of the invariant $m_d(X^d)$ in the ICIS case. As in the definition of $m_d(F^*JM(\Sigma^n))$ we pick a smoothing $\tilde{F}$ of F. We can extend the sheaf $JM(X_F)$ over $X_{\tilde{F}}$ by considering the sheaf of modules generated by the partial derivatives of $\tilde{F}$ with respect to the variables of $\mathbb{C}^q$, the ambient space of X_F. Denote this by $JM_z(X_{\tilde{F}})$. Now assume $X = X_F = F^{-1}(\Sigma^r)$; for simplicity, assume X has a smoothing. Applying the MPT to this set-up (3.22), we know that

$$m_d(X) = e(JM(X_F), F^*(JM(\Sigma^r))) + F(\mathbb{C}^q) \cdot \Gamma_d(\Sigma^r) := e_\Gamma(JM(X_F), F^*(JM(\Sigma^r))).$$

In an analogous way we can define $m_d(mJM(X))$, and again we have as a corollary of the MPT,

$$m_d(mJM(X)) = e(mJM(X_F), F^*(JM(\Sigma^r))) + F(\mathbb{C}^q) \cdot \Gamma_d(\Sigma^r) :$$
$$= e_\Gamma(mJM(X_F), F^*(JM(\Sigma^r))).$$

(In picking the smoothing it is necessary to ensure that $\tilde{F}(t,0) \notin \Sigma^r$ for $t \neq 0$.) We use the notation e_Γ for the multiplicity of a pair corrected by the curvature of the larger module.

If we have a family of sets X_F defined by $F : \mathbb{C}^t \times \mathbb{C}^q \to Hom(\mathbb{C}^n, \mathbb{C}^{n+k})$, $Y = \mathbb{C}^t \times 0 \subset X_F$, we show that $m_d(mJM(X))$ controls the Whitney conditions for the open stratum of X_F along Y. The precise statement follows.

Theorem 3.28 *Suppose* $(X^{d+t}, 0) \subset (\mathbb{C}^{q+t}, 0)$, $X = F^{-1}(\Sigma^r)$, $F : \mathbb{C}^{q+t} \to Hom(\mathbb{C}^n, \mathbb{C}^{n+k})$, *$Y$ a smooth subset of X, coordinates chosen so that $\mathbb{C}^t \times 0 = Y$, F induced from a deformation of the presentation matrix of $X(0)$, X equidimensional with equidimensional fibers, of expected dimension, $X(y)$ has only isolated singularities for all y.*

(A) Suppose the singular set of X is Y. Suppose $e_\Gamma(m_y JM(X_y), F_y^* JM(\Sigma^r))$ is independent of y. Then the union of the singular points of $X(y)$ is Y, and the pair of strata $(X - Y, Y)$ satisfies condition W.

(B) Suppose the singular set of X is Y and the pair $(X - Y, Y)$ satisfies condition W. Then $e_\Gamma(m_y JM(F_y), F_y^* JM(\Sigma^r))$ is independent of y.

Proof First, we prove (A). We can embed the family in a restricted versal unfolding with smooth base $\tilde{Y}^l$. Consider the polar variety of $m_Y JM_z(F)$ of dimension l, and the degree of its projection to $\tilde{Y}^l$ along points of Y. The hypothesis on $e_\Gamma(m_y JM(X_y), F_y^* JM(\Sigma^r))$ implies by the multiplicity polar theorem that this degree is constant over Y. In turn this implies that the polar variety over Y does not split, hence the polar of the original deformation is empty. This implies that the fiber of the exceptional divisor of B_{m_Y} Projan$(JM_z(F))$ cannot be maximal, since there is no polar variety. By the theorem of Kleiman-Thorup on the dimension of this fiber, it then follows that $JM_Y(F) \subset \overline{m_Y JM_z(F)}$ which implies W.

This also implies that $\overline{JM(F)} \subset \overline{JM_z(F)}$. Hence the union of the singular points of F_y which is the cosupport of $\overline{JM_z(F)}$ is equal to the cosupport of $\overline{JM(F)}$ which is Y. Then the inclusion $JM_Y(F) \subset \overline{m_Y JM_z(F)}$ implies W for $(X - Y, Y)$. (Cf. [9].)

Now we prove (B). W implies $JM_Y(F) \subset \overline{m_Y JM_z(F)}$ which implies that $\overline{m_Y JM(F)} = \overline{m_Y JM_z(F)}$. We know by [38] that condition W implies that the fiber dimension of the exceptional divisor of $B_{m_Y}(C(X))$ over each point of Y is as small as possible. The integral closure condition $\overline{m_Y JM(F)} = \overline{m_Y JM_z(F)}$ implies that the same is true for B_{m_Y}(Projan $\mathcal{R}(JM_z(F))$). This implies that the polar of $m_Y JM_z(F)$ is empty, hence by the multiplicity polar formula the invariant $e_\Gamma(mJM(F_y), F_y^* JM(\Sigma^r))$ is independent of y. □

We also have a geometric description of our invariant based on the smoothing and the existence of a unique Milnor fiber.

Theorem 3.29 $e(JM(X_y), F_y^* JM(\Sigma^r)) + F(y)(\mathbb{C}^q) \cdot \Gamma_d(\Sigma^r) = (-1)^d \chi(X_{s,y}) + (-1)^{d-1} \chi((X \cap H)_{s,y})$, *$X_{s,y}$ a smoothing of $X(y)$.*

Proof (Cf. [11, p. 130], [32].) □

Example 3.30 Consider the family of curves X_l, defined by the minors of

$$F_{X_l} = \begin{bmatrix} z & x \\ y & z \\ x^l & y \end{bmatrix}.$$

Then by our previous work we have $e_\Gamma(JM(X_l), F_l^* JM(\Sigma^2)) = 2l + 2$, for $l = 1$ or $l - 1$ not divisible by 3, $l > 1$. Since $-\chi(X_{s,l}) + \chi((X \cap H)_{s,l}) = \mu(X_l) + m(X_l) - 1$, we have $\mu(X_l) = 2l$ recovering a result of Watanabe et al. [30].

Challenge Problems and Further Directions in Determinantal Singularities

- In the maximal minor case, the work of [7] gives a formula for the Euler characteristic of a smoothing of a nondeterminantal singularity. Can we say something about the Betti-numbers of a smoothing when there is more than 1? (Frühbis–Krüger and Zach have some results for three-folds. Cf. [5, 40].)
- What is the connection between the results of [7] on the Euler characteristic of a smoothing and Damon–Pike [3] in the (2,3) case?
- What is the relation in the curve case between between the results of [7] and those of Greuel and Buchweitz [2] and Rosenlicht differentials?
- For what determinantal singularities is the invariant $m_d(X^d) = 0$? Hopefully, we can classify them. In May 2015, work was done giving the dimensions in which they can appear, and a transversality condition that must be satisfied. In September of 2016 as part of a project with Ruas and Pedersen, normal forms for the space curve-maximal minor case were found.
- What additional invariants are needed to ensure the singular locus of a family does not split? In the ICIS case the independence from parameter of $m_d(X_y)$ ensures the singular locus is the parameter axis. Because some determinantal singularities have $m_d(X) = 0$, this is not true for families of determinantal singularities, even in the maximal minor $(2, 3)$ case.
- Is there a way to connect the terms that appear in the calculation of the multiplicity of the polar of $F^* JM(\Sigma^n)$ with the geometry of X_F in the $(n, n+k)$ case?
- What is a formula in terms of the entries of the presentation matrix for $F(\mathbb{C}^q) \cdot \Gamma_d(\Sigma^r)$, $1 < r < n$?
- What can we say about EIDS (Essentially Isolated Determinantal Singularities)? These include determinantal singularities which are isolated, but cannot be smoothed, because the dimension of the domain is too large, as well as determinantal singularities which are non-isolated, but which are well behaved away from the origin.) Some work on these has been done in [7, 20] and other papers mentioned in their bibliographies.
- Can we calculate the multiplicities of the polar varieties of $\Sigma^r \subset Hom(\mathbb{C}^n, \mathbb{C}^{n+k})$ at the origin of $Hom(\mathbb{C}^n, \mathbb{C}^{n+k})$? This is known for the cases $r = n, 2$ ([8]). This will give a lower bound on the size of the contribution of $F(\mathbb{C}^q) \cdot \Gamma_d(\Sigma^r)$ to $e_\Gamma(JM(X), F^* JM(\Sigma^r))$. Since the Σ^r are homogeneous, their ideals define projective varieties, and these multiplicities will be the degrees of the polar classes of the projective varieties.

- There are other invariants associated with X such as the index of differential forms and the Milnor number(?) of functions with isolated singularities. Compute these in terms of infinitesimal invariants similar to those of these lectures.(Cf. [19] for a framework for doing this.)

4 Afterword: Examples of the Point of View of the Introduction

We will talk about two examples of our point of view.

Hypersurfaces with isolated singularities are our first example. Suppose $X^n, 0$ has an isolated singularity at the origin, $X = f^{-1}(0)$.

Choose the landscape This is done by looking at the possible deformations of X. We see we can deform f freely, and still, for small deformations, get a hypersurface with at most isolated singularities. So, the landscape will be all hypersurfaces in $\mathbb{C}^{n+1}$ with at most isolated singularities. The generic element that X deforms to is its Milnor fiber.

Describe the connection between X and its generic element To do this, deform X to its Milnor fiber, using $F(y, z) = f(z) - y$. Then the ideal $J_z(F)$, when restricted to the graph, vanishes only at $(0, 0)$, so its polar curve is given by the vanishing of the first n partial derivatives, in generic coordinates. Applying the MPT, we get $e(J(f), \mathcal{O}_{X,0}) = mult_{\mathbb{C}}\Gamma_n(J_z(F))$.

In turn $mult_{\mathbb{C}}\Gamma_n(J_z(F))$ is the colength of the ideal $(f, \frac{\partial f}{\partial z_1}, \ldots, \frac{\partial f}{\partial z_n})$ in $\mathcal{O}_{n+1}$. This is $\mu(X) + \mu(X \cap H)$, H a generic hyperplane.

Determine the first order infinitesimal deformations Since $f \to f + tg$ where g is arbitrary, is a first order defomation, and the corresponding infinitesimal first order deformation is $f \to \frac{\partial f + tg}{\partial t} = g$, the first order infinitesimal deformations are just $\mathcal{O}_{X,0}$.

Our invariant for controlling Whitney equisingularity is $e(mJ(f), \mathcal{O}_{X,0})$.

If we have a family of hypersurfaces $\mathcal{X}$, then if $\mu(X) + \mu(X \cap H)$ changes, then so must $e(J(f), \mathcal{O}_{X,0})$, and the exceptional divisor of $B_{J_z(F)}(\mathcal{X})$ must pick up a vertical component and vice-versa. The change in the topology of the landscape is reflected in a dramatic change in the fibers of the exceptional divisor, which is the infinitesimal information.

For determinantal singularities the story is similar.

If we look at all possible deformations, then we have examples where the same singularity can be deformed in two different ways, even giving Whitney equisingular families in which the generic fiber has non-homeomorphic smoothings [7]. So, we restrict our deformations by using the same size presentation matrix. The entries of the matrix can be deformed freely.

Then, the landscape will be the determinantal singularities corresponding to a matrix of fixed size. The generic element associated to X will be smooth, given some dimension restriction; otherwise we can say what the stabilizations of the singularity are, and can begin to study those [20].

In the case of smoothable singularities, by use of the multiplicity polar theorem and some topology, we get Theorem 3.29 which gives the connection between the topology of smoothing and the algebraic invariants of the singularity, which are connected to its infinitesimal geometry. This is generalized in [20] to the EIDS case.

The first order infinitesimal deformations of X can be explicitly computed; deform an entry of the presentation matrix by t, calculate the minors of the order used to define X; taking derivative with respect to t then gives a map from the defining equations for X into tuples in $\mathcal{O}^g_{X,0}$, where g is the number of defining equations. These give the generators of $N(X)$. It is clear from this formulation that N is universal and specializes well in families. We can calculate $JM(\Sigma)$ explicitly–the partial with respect to the (i, j) entry of the matrix is just the corresponding generator of N. So Σ is stable. The geometric representation of $C(\Sigma^n)$ in terms of kernels of M and M^t gives the formula for computing $mult_{\mathbb{C}}\Gamma_d(N(\tilde{X}_{\tilde{M}}))$ using the presentation matrix, but leaves the formula in terms of the entries still to be determined in general.

Once again, a change at the infinitesimal level of the family is always tied to a change in topology of the generic related elements. Here, the infinitesimal level of a family $\mathcal{X}^{t+d} \subset \mathbb{C}^{t+q}$ is the relative conormal modification $C_Y(\mathcal{X})$ of $\mathcal{X}$, which is the limits of tangent hyperplanes in $\mathbb{C}^q$ to the fibers of $\mathcal{X}$ over $\mathbb{C}^t$. Assume the singular locus of the family is $\mathbb{C}^t \times 0$. By a change at the infinitesimal level, we mean that the dimension of the fiber of $C_Y(\mathcal{X})$ over the origin in $\mathcal{X}(0)$ jumps in dimension from the generic value of $q - d - 1$ to at least $q - 1$. This is equivalent to the polar variety of dimension t of the module $JM_z(\mathcal{X})$ at $(0, 0)$ being non-empty. In turn by the MPT, this implies that $m_d(\mathcal{X}(0)) > m_d(\mathcal{X}(y))$, y a generic value of $\mathbb{C}^t$. By Theorem 3.29, this implies that the topology of the smoothings of $\mathcal{X}(0)$ and $\mathcal{X}(y)$ are different.

Acknowledgements The author was partially supported by Conselho Nacional de Desenvolvimento Científico e Tecnológico - CNPq, Brazil, grant PVE 401565/2014-9. These lectures grew out of a short course of lectures given at the Universidade Federal Fluminense in January of 2015. It is pleasure to thank these institutions for their support. I also thank two individuals who had a strong impact on shaping my thinking on the subject–Steven Kleiman and Bernard Teissier. Their contributions to this work will be obvious to any serious reader. I also thank Steven Kleiman for several helpful comments which improved the exposition.

References

1. Buchsbaum, D.A., Rim, D.S.: A generalized Koszul complex. II. Depth and multiplicity. Trans. AMS **111**, 197–224 (1963)
2. Buchweitz, R., Greuel, G.: The Milnor number of complex curve singularities. Invent. Math. **58**, 241–281 (1980)
3. Damon, J., Pike, B.: Solvable groups, free divisors and nonisolated matrix singularities II: vanishing topology. Geom. Topol. **18**(2), 911–962 (2014)

4. Dutertre, N.: Stratified critical points on the real milnor fibre and integral-geometric formulas. In: Proceedings of Geometry and Topology of Singular Spaces, CIRM (2012) (J. Singul. **13**, 87–106 (2015)
5. Frühbis-Krüger, A., Zach, M.: On the vanishing topology of isolated Cohen–Macaulay codimension 2 singularities (2015)
6. Gaffney, T.: The multiplicity polar theorem. arXiv:math/0703650v1 [math.CV] (2007)
7. Gaffney, T., Rangachev, A.: Pairs of modules and determinantal isolated singularities. arXiv:1501.00201 (2017)
8. Gaffney, T.: Duality of polar multiplicities, manuscript (2016)
9. Gaffney, T.: Integral closure of modules and Whitney equisingularity. Invent. Math. **107**, 301–322 (1992)
10. Gaffney, T.: Generalized Buchsbaum–Rim multiplicities and a Theorem of Rees. Commun. Algebra **31**, 3811–3828 (2003)
11. Gaffney, T.: Polar methods, invariants of pairs of modules and equisingularity. In: Gaffney, T., Ruas, M. (eds.) Real and Complex Singularities (São Carlos, 2002). Contemporary Mathematics, vol. 354, pp. 113–136. American Mathematical Society, Providence (2004)
12. Gaffney, T.: The multiplicity of pairs of modules and hypersurface singularities. In: Brasselet, J.-P., Ruas, M.A.S. (eds.) Real and Complex Singularities (São Carlos VIII). Trends in Mathematics, pp. 143–168. Birkhäuser, Basel (2007)
13. Gaffney, T.: The Multiplicity polar theorem and isolated singularities. J. Algebr. Geom. **18**(3), 547–574 (2009)
14. Gaffney, T.: Bi-Lipschitz equivalence, integral closure and invariants. Real and Complex Singularities. London Mathematical Society Lecture Note Series, vol. 380, pp. 125–137. Cambridge University Press, Cambridge (2010)
15. Gaffney, T.: The genericity of the infinitesimal Lipschitz condition for hypersurfaces. J. Singul. **10**, 108–123 (2017)
16. Gaffney, T., Gassler, R.: Segre numbers and hypersurface singularities. J. Algebr. Geom. **8**, 695–736 (1999)
17. Gaffney, T., Kleiman, S.: Specialization of integral dependence for modules. Invent. Math. **137**, 541–574 (1999)
18. Gaffney, T., Vitulli, M.: Weak subintegral closure of ideals. Adv. Math. **226**(3), 2089–2117 (2011)
19. Gaffney, T., Grulha Jr., N.: The multiplicity polar theorem, collections of 1-forms and Chern numbers. J. Singul. **7**, 39–60 (2013)
20. Gaffney, T., Ruas, M.A.S.: Equisingularity and EIDS. arXiv:1602.00362 (2016)
21. Gaffney, T., Grulha Jr., N.G, Ruas, M.A.S.: The local Euler obstruction and topology of the stabilization of associated determinantal varieties. arXiv:1611.00749 (2016)
22. Gibson, G.C., Looijenga, E., du Plessis, A., Wirthmuller, K.: Topological Stability of Smooth Maps, SLN #555. Springer, New York (1976)
23. Gunning, R.C.: Lectures on Complex Analytic Varieties: The Local Parametrization Theorem. Princeton University Press, Princeton (1970)
24. Kleiman, S.: Personal correspondence (2017)
25. Kleiman, S.: Two formulas for the BR multiplicity. Ann. Univ. Ferrara **63**, 147–158 (2017). https://doi.org/10.1007/s11565-016-0250-2
26. Kleiman, S., Thorup, A.: A geometric theory of the Buchsbaum–Rim multiplicity. J. Algebra **167**, 168–231 (1994)
27. Kleiman, S., Thorup, A.: The exceptional fiber of a generalized conormal space. In: Siersma, D., Wall, C.T.C., Zakalyukin, V. (eds.) New Developments in Singularity Theory. Nato Science series, II Mathematics, Physics and Chemistry, vol. 21, pp. 401–404. Kluwer Academic Publishers, Dodrecht (2001)
28. Lejeune-Jalabert, M., Teissier, B.: Cloture intégrale des idéaux et èquisingularité, avec 7 compléments. Annales de la Faculté des Sciences de Toulouse **XVII**(4), 781–859 (2008)
29. Moonen, B.: The fundamental cycle of a coherent analytic sheaf and polar multiplicities as limits of curvature integrals, Mathematisches Institut der Universitat zu Krln, Im Weyertal, pp. 86–90, 5000 Krln 40, W. Germany (1985)

30. Nari, H., Numata, T., Watanabe, K.: Genus of numerical semigroups generated by three elements. arXiv:1104.3600 [math.GR] (2011)
31. Nuño-Ballesteros, J.J., Mond, D.: Singularities of Mappings. Book in draft. http://homepages.warwick.ac.uk/~masbm/LectureNotes/book.pdf (2016)
32. Nuño-Ballesteros, J.J., Orfice-Okamoto, B., Tomazella, J.N.: The vanishing Euler characteristic of an isolated determinantal singularity. Isr. J. Math. **197**(1), 475–495 (2013)
33. Rangachev, A.: Associated points and integral closure of modules. arXiv:1607.07979 [math.AG] (2016)
34. Saia, M.: The integral closure of ideals and the Newton filtration. J. Algebr. Geom. **5**(#1), 1–11 (1996)
35. Swanson, I., Huneke, C.: Integral Closure of Ideals, Rings, and Modules. London Mathematical Society Lecture Note series, vol. 336. Cambridge University Press, Cambridge (2006)
36. Teissier, B., Flores, A.: Local polar varieties in the geometric study of singularities. arXiv:1607.07979 [math.AG] (2016)
37. Teissier, B.: The hunting of invariants in the geometry of the discriminant. In: Holm, P. (ed.) Real and Complex Singularities, Oslo 1976, pp. 565–678. Sijthoff & Noordhoff, Alphen aan den Rijn (1977)
38. Teissier, B.: Multiplicités polaires, sections planes, et conditions de Whitney. In: Aroca, J.M., Buchweitz, R., Giusti M., Merle M. (eds.) Proceedings La Rábida, 1981. Lecture Notes in Mathematics, vol. 961, pp. 314–491. Springer, Berlin (1982)
39. Teissier, B.: Monômes, volumes et multiplicités. (French) [Monomials, volumes and multiplicities] Introduction à la théorie des singularités, II, vol. 37, pp. 127–141, Travaux en Cours, Hermann, Paris (1988)
40. Zach, M.: Vanishing cycles of smoothable isolated Cohen–Macaulay codimension 2 singularities of type 2. arXiv:1607.07527 (2016)
41. Zariski, O.: Presidential address. Bull. A.M.S. **77**(4), 481–491 (1971)

Part II
Surveys Papers on Advances in Foliations and Singularity Theory: Topology Geometry and Applications

A Brief Survey on Singularities of Geodesic Flows in Smooth Signature Changing Metrics on 2-Surfaces

N. G. Pavlova and A. O. Remizov

Abstract We present a survey on generic singularities of geodesic flows in smooth signature changing metrics (often called pseudo-Riemannian) in dimension 2. Generically, a pseudo-Riemannian metric on a 2-manifold S changes its signature (degenerates) along a curve S_0, which locally separates S into a Riemannian (R) and a Lorentzian (L) domain. The geodesic flow does not have singularities over R and L, and for any point $q \in R \cup L$ and every tangential direction $p \in \mathbb{RP}$ there exists a unique geodesic passing through the point q with the direction p. On the contrary, geodesics cannot pass through a point $q \in S_0$ in arbitrary tangential directions, but only in some admissible directions; the number of admissible directions is 1 or 2 or 3. We study this phenomenon and the local properties of geodesics near $q \in S_0$.

Keywords Pseudo-Riemannian metrics · Geodesics · Singular points · Normal forms

2010 Mathematics Subject classification 53C22 · 53B30 · 34C05

1 Introduction

Let S be a real smooth manifold, dim $S = n \geq 2$. By *metric* on S we mean a symmetrical covariant tensor field of the second order on the tangent bundle TS, not necessary positive defined. Moreover, metrics whose signature has different signs at

N. G. Pavlova
Department of Nonlinear Analysis and Optimization, RUDN University, Moscow, Russia
e-mail: natasharussia@mail.ru

A. O. Remizov (✉)
V.A. Trapeznikov Institute of Control Sciences of Russian Academy of Sciences, Moscow, Russia
e-mail: alexey-remizov@yandex.ru

A. O. Remizov
Laboratory of Dynamical Systems, National Research University Higher School of Economics, Moscow, Russia

R. N. Araújo dos Santos et al. (eds.), *Singularities and Foliations. Geometry, Topology and Applications*, Springer Proceedings in Mathematics & Statistics 222, https://doi.org/10.1007/978-3-319-73639-6_4

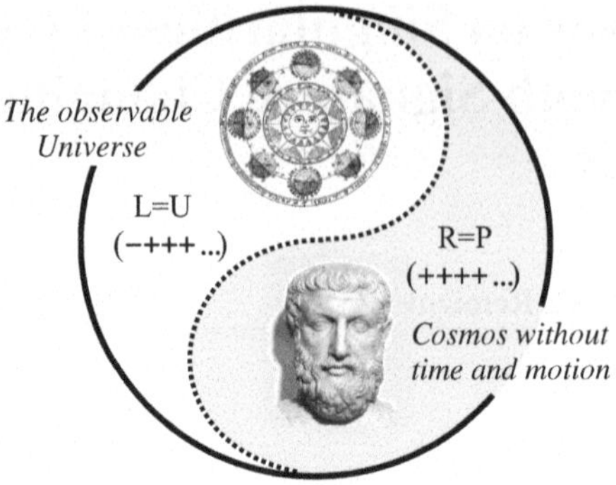

Fig. 1 Sakharov's cosmological model. Here the region L (which apparently includes the observable Universe) is denoted by U and the region R (cosmos without time and motion) is denoted by P (after Parmenides, a Greek philosopher theorized about space and time). The hypersurface S_0 is represented by the dotted line

different points of S, are of the special interest. For instance, in the quantum theory of gravitation and general relativity two types of signature changing metrics are considered:

- **Smooth**. The metric is degenerate on a hypersurface $S_0 \subset S$ that divides the *Riemannian* region $R \subset S$ with signature $(+ \cdots + +)$ from the *Lorentzian* region $L \subset S$ with signature $(+ \cdots + -)$. Example: $ds^2 = dx_1^2 + \cdots + dx_{n-1}^2 + x_n dx_n^2$.
- **Discontinuous**. The metric is smooth and non-degenerate everywhere except for a hypersurface $S_0 \subset S$ (which separates R and L defined as above), where it fails to be continuous. Example: $ds^2 = dx_1^2 + \cdots + dx_{n-1}^2 + \frac{1}{x_n} dx_n^2$.

In the paper [27], Russian physicist A.D. Sakharov conjectured there exist states of the physical continuum which include regions with different signatures of the metric; the observed Universe and an infinite number of other Universes arose as a result of quantum transitions with a change in the signature of the metric. This concept is exemplified by Fig. 1.

In his cosmological model, Sakharov used discontinuous metrics. However, some other authors consider models with smooth signature changing metrics; see e.g., [1, 18–20] and the references therein. From physical viewpoint, the difference between smooth and discontinuous signature changing metrics corresponds to different physical proposals, in particular, different solutions of the Einstein equation. *Euclidean-Lorentzian transitions* (junctions) between the domains R and L play an important role, both in the smooth and discontinuous models. The term *Euclidean* is used in sense of *Riemannian*, that is typical for physical literature, see e.g., [2]. Similarly, the term *Lorentzian* is referred to non-degenerate indefinite metrics.

In this paper, we discuss a purely mathematical problem connected with smooth signature changing metrics (further called *pseudo-Riemannian*): the local behavior of geodesics in a neighborhood of the points where the metric has a generic degeneracy. Such points are singular points of the geodesic flow, and the standard existence and uniqueness theorem for ordinary differential equations is not applicable. This leads to an interesting geometric phenomenon: geodesics cannot pass through a degenerate point in arbitrary tangential directions, but only in certain directions said to be *admissible*.

A study of this phenomenon for two-dimensional pseudo-Riemannian metrics is started in [13, 24–26]; similar results in three-dimensional case were announced in [22]. In these works, mainly the local properties of geodesics and geodesic flows were considered, some global properties of geodesics of pseudo-Riemannian metrics with differentiable groups of symmetries are investigated in [25]. This allows, in particular, to obtain the phase portraits of geodesics on surfaces of revolution (sphere, torus, etc.) embedded in three-dimensional Minkowski space.

Various other aspects of pseudo-Riemannian metrics (including the Gauss–Bonnet formula) are treated by many authors, see e.g., [12, 16, 17, 19–21, 28] and the references therein. However, there exist a number of unsolved problem connected with degeneracy of metrics. According to our knowledge, the problem of local geodesic equivalence of pseudo-Riemannian metrics at degenerate points is not studied yet, although it is well studied for Riemannian and Lorentzian metrics, see e.g., [7] (in this paper, the authors call *pseudo-Riemannian* what we call *Lorentzian*, i.e., non-degenerate indefinite metrics).

From now we always assume that $\dim S = 2$.

Similarly, just as Riemannian metrics naturally appear on surfaces embedded in Euclidean space, pseudo-Riemannian metrics can be generated in pseudo-Euclidean space. Let S be a smooth surface embedded in 3D Minkowski space (X, Y, Z) with the pseudo-Euclidean metric $dX^2 + dY^2 - dZ^2$. Then the pseudo-Euclidean metric in the ambient (X, Y, Z)-space induced a pseudo-Riemannian metric on S. For instance, let S be the standard Euclidean sphere

$$X^2 + Y^2 + Z^2 = 1.$$

The metric induced on the sphere S degenerates on two parallels $Z = \pm 1/\sqrt{2}$, which separate S into three regions, where the metric has constant signatures. The North $\big(Z > 1/\sqrt{2}\big)$ and the South $\big(Z < -1/\sqrt{2}\big)$ regions are Riemannian, while the equatorial region $|Z| < 1/\sqrt{2}$ is Lorentzian; see Fig. 2 (left). The condition of the point $q \in S$ belonging to R or S_0 or L depends on the mutual relationships between the tangent plane $T_q S$ and the *isotropic* (*light*) *cone*

$$dX^2 + dY^2 - dZ^2 = 0;$$

see Fig. 2 (right).

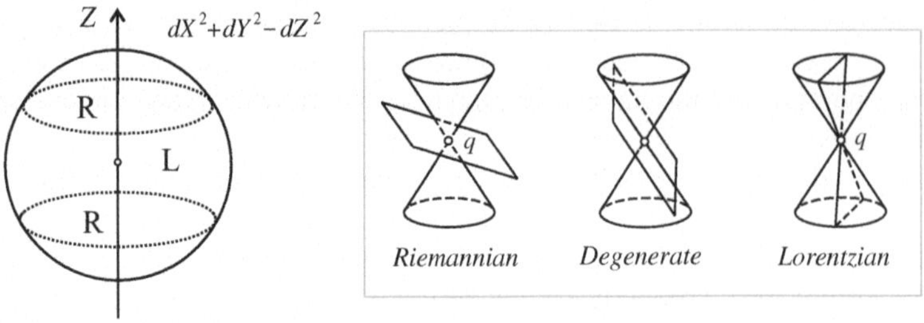

Fig. 2 On the left: pseudo-Riemannian metric on the sphere $X^2 + Y^2 + Z^2 = 1$ in 3D Minkowski space. Here S_0 consists of two parallels $Z = \pm 1/\sqrt{2}$ depicted as dotted lines. On the right: intersections of the light cone with the tangent plane $T_q S$, $q \in S$

2 Definition of Geodesics

Consider a two-dimensional manifold (surface) S with pseudo-Riemannian metric

$$ds^2 = a(x, y)\,dx^2 + 2b(x, y)\,dxdy + c(x, y)\,dy^2, \tag{1}$$

whose coefficients are smooth (i.e., C^∞). Geodesics in the metric (1) can be defined via variational principles similarly to the Riemannian case, with additional nuances.

For instance, the arc-length parametrization is not defined for the *isotropic lines* (or *lightlike lines* or *null curves*). Moreover, the Lagrangian of the length functional

$$J_l(\gamma) = \int_\gamma \sqrt{a\dot{x}^2 + 2b\dot{x}\dot{y} + c\dot{y}^2}\,dt \;\to\; \text{extr},$$

where the dot means differentiation by the parameter t, fails to be differentiable on the *isotropic surface* $\mathscr{F}$

$$a(x, y)\,dx^2 + 2b(x, y)\,dxdy + c(x, y)\,dy^2 = 0, \tag{2}$$

and the Euler–Lagrangian equation for the length functional is not defined on $\mathscr{F}$. Note that Eq. (2) defines the isotropic surface $\mathscr{F}$ in the complement of the zero section of TS or, equivalently, in the projectivized tangent bundle PTS.

Binary differential equation (2) defines a direction field on $\mathscr{F}$, whose integral curves correspond to isotropic lines in the metric (1). This equation plays an important role for understanding the behavior of geodesics, and we consider it in more detail below.

As already mentioned above, the Euler–Lagrangian equation for the length functional J_l does not allow to define extremals on $\mathscr{F}$. However, this problem does not arise if we define geodesics as extremals of the action functional

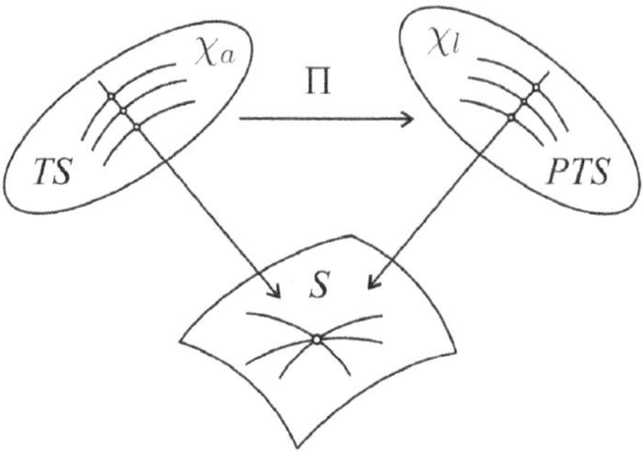

Fig. 3 The relationship between the fields χ_l and χ_a. The horizontal arrow means the projectivization $\Pi: TS \to PTS$. Two lower arrows are the standard projections $TS \to S$ and $PTS \to S$

$$J_a(\gamma) = \int_\gamma (a\dot{x}^2 + 2b\dot{x}\dot{y} + c\dot{y}^2)\, dt \ \to \ \text{extr.}$$

The corresponding Euler–Lagrange reads

$$\begin{cases} 2(a\ddot{x} + b\ddot{y}) = (c_x - 2b_y)\dot{y}^2 - 2a_y\dot{x}\dot{y} - a_x\dot{x}^2, \\ 2(b\ddot{x} + c\ddot{y}) = (a_y - 2b_x)\dot{x}^2 - 2c_x\dot{x}\dot{y} - c_y\dot{y}^2, \end{cases} \tag{3}$$

and the corresponding parametrization is called *natural* or *canonical*. Obviously, the definition of geodesics as auto-parallel curves in the Levi–Civita connection generated by the metric (1) leads to the same Eq. (3).

The natural parametrization is well defined for all types of geodesics, including isotropic. For non-isotropic geodesics it coincides with the arc-length (of course, here the length to be real or imaginary). The functionals J_l (length) and J_a (action) define the corresponding fields of extremals: χ_l on PTS away of $\mathscr{F}$ and χ_a on the complement of the zero section of TS (including $\mathscr{F}$). The relationship between the fields χ_l and χ_a is as follows (see also Fig. 3).

The natural projectivization $\Pi: TS \to PTS$ sends the field χ_a to a direction field on PTS, which is parallel to the vector field

$$\vec{V} = 2\Delta\left(\frac{\partial}{\partial x} + p\frac{\partial}{\partial y}\right) + M\frac{\partial}{\partial p}, \quad p = \frac{dy}{dx}, \tag{4}$$

where

$$\Delta(x, y) = ac - b^2, \quad M(x, y, p) = \sum_{i=0}^{3} \mu_i(x, y)p^i,$$

with the coefficients

$$\begin{aligned} \mu_0 &= a(a_y - 2b_x) + a_x b, \\ \mu_1 &= b(3a_y - 2b_x) + a_x c - 2ac_x, \\ \mu_2 &= b(2b_y - 3c_x) + 2a_y c - ac_y, \\ \mu_3 &= c(2b_y - c_x) - bc_y. \end{aligned} \tag{5}$$

The vector field $\vec{V}$ given by (4) is defined and smooth at all points of PTS including the isotropic surface $\mathscr{F}$. It is worth observing that the direction field χ_l is parallel to (4) at all points where χ_l is defined, i.e., at all points away from the surface $\mathscr{F}$. One can interpret the direction field given by (4) as a natural extension of χ_l to $\mathscr{F}$. This brings us to the following definition: the projections of integral curves of the field (4) from PTS to S distinguished from a point are *non-parametrized geodesics* in the pseudo-Riemannian metric (1).

Moreover, let $\vec{W}$ be the vector field on PTS (determined uniquely up to multiplication by a non-vanishing scalar factor) that corresponds to the length functional J_l. Since the length functional is invariant with respect to reparametrizations, one can put $t = x$ and take as $\vec{W}$ the vector field corresponding to the Euler–Lagrange equation with the Lagrangian $\sqrt{F}$, where $F(x, y, p) = a(x, y) + 2b(x, y)p + c(x, y)p^2$. A straightforward calculation (see [25]) shows that

$$\vec{W} = \frac{1}{2F^{\frac{3}{2}}}\vec{V} \quad \text{and} \quad \operatorname{div} \vec{W} = 0 \ \text{ at all points where } \ F \neq 0. \tag{6}$$

The field $\vec{W}$ is divergence-free, since it comes directly from an Euler–Lagrange equation, while $\vec{V}$ is not, since it is obtained via an additional procedure, the projectivization $\Pi\colon TS \to PTS$. The property (6) plays an important role, due to the following general fact:

Theorem 1 ([13]) *Let $\vec{V}(\xi)$, $\xi \in \mathbb{R}^n$, be a smooth vector field, $f(\xi)$ be a smooth scalar function such that the hypersurface $\mathscr{F} = \{\xi : f(\xi) = 0\}$ is regular, r be a positive real number. Suppose that the field $\vec{W}(\xi) = f^{-r}(\xi)\vec{V}(\xi)$ is divergence-free at all points where it is defined, i.e., at all points $\xi \notin \mathscr{F}$. Then $\mathscr{F}$ is an invariant hypersurface of the field $\vec{V}$. Moreover, let $\xi_* \in \mathscr{F}$ be a singular point of $\vec{V}$ and $\lambda_1, \ldots, \lambda_n$ be the eigenvalues of the linearization of $\vec{V}$ at ξ_*. Then $\lambda_1 + \cdots + \lambda_n = r\lambda_j$ for at least one j.*

By Theorem 1, we have the following assertions:

- The isotropic surface $\mathscr{F}$ is an invariant surface of the field (4) and all isotropic lines are geodesics (with identically zero length).[1]
- Geodesics do not change their type (timelike, spacelike, isotropic) away of degenerate points. This statement follows from the previous one.

[1]The first assertion is valid for any $\dim S \geq 2$, while the second assertion (about isotropic lines) is valid for $\dim S = 2$ only. Indeed, in the case $\dim S > 2$ there exist isotropic lines that are not geodesics; see the example in [25].

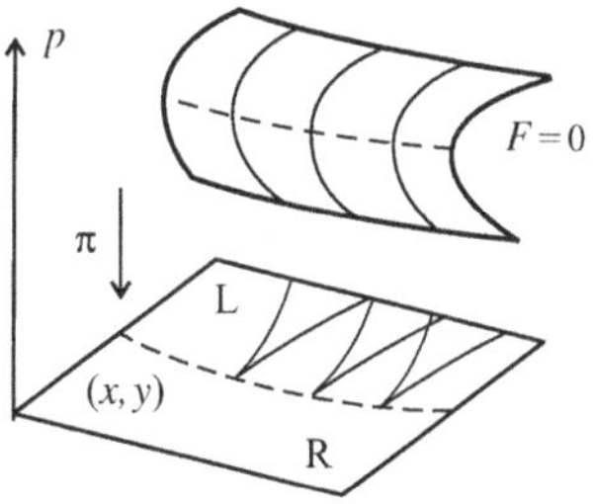

Fig. 4 The isotropic surface $\mathscr{F}$ in *PTS*, integral curves of the field X (top) and integral curves of the equation $F(x, y, p) = 0$ (down). The dashed lines represent the criminant (top) and the discriminant curve (down)

3 Equation of Isotropic Lines

Suppose that set

$$S_0 = \{q = (x, y) \in S \colon \Delta(x, y) = 0\}$$

is a regular curve. It is called the *degenerate* or *discriminant* curve of the metric (1), and points $q \in S_0$ are called *degenerate points* of the metric. Then the coefficients a, b, c do not vanish simultaneously, and the isotropic direction

$$p_0(q) = -\frac{a}{b}(q) = -\frac{b}{c}(q), \quad q \in S_0, \tag{7}$$

is defined and unique at every point $q \in S_0$.

The projectivization $\Pi \colon TS \to PTS$ transforms binary differential equation (2) into the implicit differential equation

$$F(x, y, p) = 0, \quad \text{where} \quad F = a(x, y) + 2b(x, y)p + c(x, y)p^2. \tag{8}$$

In the space PTS, the surface $\mathscr{F}$ composes a two-sheeted covering of the Lorentzian domain of S ($\Delta < 0$) with branching along the discriminant curve S_0. Over the Riemannian domain ($\Delta > 0$), the surface $\mathscr{F}$ does not pass. See Fig. 4.

A well-known geometrical approach to study implicit equation (8) consists of the lift the multivalued direction field on S to a single-valued direction field X on the surface $\mathscr{F}$.[2] The field X is an intersection of the contact planes $dy = pdx$ with the tangent planes to the surface $\mathscr{F}$, that is, X is defined by the vector field

$$\dot{x} = F_p, \quad \dot{y} = pF_p, \quad \dot{p} = -(F_x + pF_y), \tag{9}$$

whose integral curves become isotropic lines of the metric (1) after the projection $\pi : \mathscr{F} \to S$ along the p-direction. Further we shall call this direction *vertical* in the space PTS. The locus of the projection $\pi : \mathscr{F} \to S$ (given by the equations

[2]This approach is applicable to implicit differential equations $F(x, y, p) = 0$ with a smooth function F not necessarily quadratic in p. The idea goes back to H. Poincaré and A. Clebsch, see [23] for details.

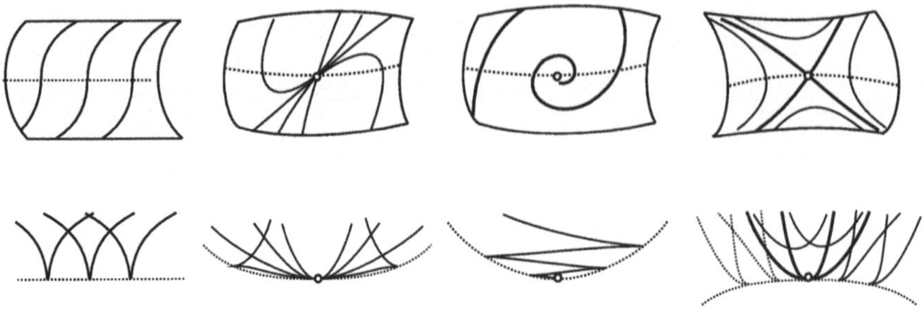

C D_n D_f D_s

Fig. 5 From the top to the bottom: integral curves of the field (9) on the isotropic surface $\mathscr{F}$ and isotropic lines, obtained by the projection $\pi : \mathscr{F} \to S$. The dashed lines represent the criminant (top) and the discriminant curve (down)

$F = F_p = 0$) is called the *criminant* of Eq. (8). It is not hard to see that the criminant consists of the points $(q, p_0(q))$, $q \in S_0$ (see formula (7)).

Since $\mathscr{F}$ is an invariant surface of the field (4) and both fields (4) and (9) are tangent to the contact planes $dy = pdx$, the restriction of (4) to the invariant surface $\mathscr{F}$ is parallel to (9). Moreover, the restriction of the field (4) to $\mathscr{F}$ is equal to the field (9) multiplied by a smooth scalar function vanishing along the criminant (see [13]). Generically, here there are two possible cases:

- The case C: the isotropic direction $p_0(q)$ is transversal to S_0. Then the field (9) at the point $(q, p_0(q))$, $q \in S_0$, is non-singular, and binary equation (2) has *Cibrario normal form* $dx^2 = y\,dy^2$. See Fig. 5 (left).
- The case D: the isotropic direction $p_0(q)$ is tangent to S_0. The field (9) at $(q, p_0(q))$, $q \in S_0$, has a non-degenerate singular point: saddle or node or focus (subcases D_s, D_n, D_f, respectively). Under certain additional conditions (formulated below), binary equation (2) has *Dara–Davydov normal form*

$$dy^2 = (y - \varepsilon x^2)\,dx^2, \tag{10}$$

where $\varepsilon < 0$ (if saddle) or $0 < \varepsilon < \frac{1}{16}$ (if node) or $\varepsilon > \frac{1}{16}$ (if focus). See Fig. 5.

The normal form $dx^2 = y\,dy^2$ is named after Italian mathematician Maria Cibrario who established it first in C^ω (real analytic) category when studying second-order linear partial differential equations of the mixed type [8]. Later on, a general (and rather simple) proof of the Cibrario normal form (in C^ω and C^∞ categories) was presented in the famous Arnold's book [5].

The normal form (10) was firstly conjectured by Brazilian mathematician Lak Dara [9] and then proved by A.A. Davydov [10] under the following genericity conditions. Let $\alpha_{1,2}$ be the eigenvalues of the linearization of the vector field (9) at the singular point considered. Then $\alpha_{1,2}$ are roots of the characteristic equation $\alpha^2 - \alpha + 4\varepsilon = 0$, and the excluded values $\varepsilon = 0$ and $\varepsilon = \frac{1}{16}$ correspond to a degenerate singular point (saddle-node or degenerate node, respectively). The additional conditions required for the normal form (10) are the following.

First, the ratio of $\alpha_{1,2}$ is different from ± 1, and the eigendirections are not tangent to the criminant. Second, the germ of the vector field (9) is C^∞-linearizable, i.e., it is C^∞-smoothly equivalent to its linear part. The C^∞-linearizability condition holds true, for instance, if between the eigenvalues $\alpha_{1,2}$ there are no resonant relations $\alpha_i = n_1\alpha_1 + n_2\alpha_2$ with integers $n_{1,2} \geq 0$, $n_1 + n_2 \geq 2$ (SternbergChen Theorem, see e.g., [4, 14]). The proof presented in [10] is done in C^∞ category, but is valid in C^ω as well (the requirement of C^∞-linearizability should be replaced with C^ω-linearizability), see also the recent paper [6].

4 Singular Points of the Geodesic Flow

In addition to the isotropic surface $\mathscr{F}$, the vector field $\vec{V}$ given by (4) has one more evident invariant surface – the *vertical* surface

$$\overline{S}_0 = \{(q, p)\ \ q = (x, y) \in S_0,\ \ p \in \mathbb{RP}\}.$$

The restriction of the field (4) to $\overline{S}_0$ is vertical at almost all points (except for the points where $M = 0$, and the field vanishes). Hence the surface $\overline{S}_0$ is filled with vertical integral curves of the filed (4) and its singular points.

Singular points of the field (4) are given by two equations:

$$\Delta(x, y) = 0 \quad \text{and} \quad M(x, y, p) = 0, \tag{11}$$

and consequently, they are not isolated, but form a curve (or curves) in PTS. Algebraically, this property can be expresses in the following form: all components of the vector field (4) belong to the ideal I (in the ring of smooth functions) generated by two of them, namely, $I = \langle \Delta, M \rangle$.

Remark 1 The fact that the *horizontal* generator $\Delta(x, y)$ of the field (4) does not depend on p and the *vertical* generator $M(x, y, p)$ is a cubic polynomial in p, plays a crucial role in a general geometrical context, e.g., in the framework of Cartan's theory of the projective connection [3, 5].

Let us list those of the properties of the field (4) that we are going to use:

- Singular points of the field (4) are given by Eq. (11) and form a curve (or several curves) in PTS.
- The spectrum of the linearization of the field (4) at every singular point contains one zero eigenvalue and two real eigenvalues $\lambda_{1,2}$, which vanish (simultaneously) at those points where the cubic polynomial $M(q, p)$ has a double root p. The latter condition is equivalent to the direction p is tangent to S_0 at the point q.

- For every point $q \in S_0$ and any $p \in \mathbb{RP}$ such that $M(q, p) \neq 0$ there exists a unique integral curve of the field (4) that passes through the point (q, p) – a vertical straight line, whose projection on S is not a geodesic. Consequently, the vertical surface $\overline{S}_0$ is an invariant surface of (4).
- Geodesics cannot enter a point $q \in S_0$ in arbitrary tangential direction, but only in *admissible* directions p that satisfy the condition $M(q, p) = 0$.
- The isotropic direction $p_0(q)$ given by formula (7) is admissible at every point $q \in S_0$, i.e., $M(q, p_0(q)) = 0$ for all $q \in S_0$.

Depending on the roots of the cubic polynomial M (see Fig. 6), we have four cases:

- C_1: the isotropic direction p_0 is a unique real root of M,
- C_2: M has a simple root p_0 and a double non-isotropic real root $p_1 = p_2$,
- C_3: M has three simple real roots: isotropic p_0 and non-isotropic p_1, p_2,
- D : the isotropic double root $p_0 = p_1$ and a simple non-isotropic root p_2.

If $\mathrm{Re}\,\lambda_{1,2} \neq 0$, the set W is the center manifold of the field, and the restriction of the field to W is identically zero. Hence in a neighborhood of every singular point where $\mathrm{Re}\,\lambda_{1,2} \neq 0$, the phase portrait of the field has a very simple topological structure. Indeed, the reduction principle [4, 15] asserts that the germ of the field is orbitally topologically equivalent to the direct product of the standard 2-dimensional node (if $\mathrm{Re}\,\lambda_{1,2}$ have the same sign) or saddle (if $\mathrm{Re}\,\lambda_{1,2}$ have different signs) and 1-dimensional zero vector field. However, the topological classification is not enough.

The paper [23] presents finite-smooth local normal forms of such fields, [26] contains a brief survey (Appendix A) on the smooth and C^ω classifications. These results allow to establish smooth local normal forms of the field (4) at all singular points (q, p_i), $q \in S_0$, where p_i is a simple real root of $M(q, p)$. This gives the description of geodesics that enter a degenerate point with all possible admissible directions for the cases C_1, C_3. To study geodesics with the isotropic admissible direction in the cases C_2 and D, one can use a blow-up procedure.

Choosing appropriate local coordinates, we shall further assume that in a neighborhood of the point $q \in S_0$, Eq. (2) has the form $dx^2 = y\,dy^2$ in the case C and (10) in the case D. Consequently, the discriminant curve S_0 is the axis $y = 0$ in the case C and the parabola $y - \varepsilon x^2 = 0$ in the case D. Since multiplication the metric by the factor -1 does not change the geodesic flow, without loss of generality, assume that $y > \varepsilon x^2$ and $y < \varepsilon x^2$ (including the case $\varepsilon = 0$) are Lorentzian and Riemannian domains, respectively.

From now on, we shall consider geodesics outgoing from a degenerate point $q \in S_0$ with the isotropic admissible direction $p_0(q)$ as semitrajectories starting from q. We distinguish geodesics outgoing into the Lorentzian (resp. Riemannian) domains using the superscript $+$ (resp. $-$). Let us clarify this with the following example.

Example 1 For the metric $ds^2 = dx^2 - ydy^2$, the discriminant curve $S_0 = \{y = 0\}$ divides the plane into the Lorentzian $(y > 0)$ and Riemannian $(y < 0)$ domains. Formula (5) yields $M(q, p) = p^2$, and we have the case C_2.

At every degenerate point $q \in S_0$ there exist two admissible directions: $p_1 = 0$ (non-isotropic, double root) and $p_0 = \infty$ (isotropic). To see that the direction $p_0 = \infty$ is admissible, it is convenient to interchange x and y. In the new coordinates $\bar{x} = y$, $\bar{y} = x$, $\bar{p} = 1/p$, the polynomial $M(q, \bar{p}) = -\bar{p}$ has the root $\bar{p} = 0$.

The corresponding field (4) has a unique integral curve $y = 0$ that pass through every point $q \in S_0$ with tangential direction $p_1 = 0$. Substituting $y = 0$ directly in (3), one can see that $y = 0$ is an extremal of the action functional and its natural parametrization is given by the equation $\ddot{x} = 0$. Moreover, given degenerate point $q \in S_0$ there exists a one-parameter family of geodesics outgoing from q with the tangential direction $p = \infty$. For instance, consider the family Γ_0 of geodesics γ_α, $\alpha \in \mathbb{R}$, outgoing from the origin; see Fig. 8 (right). They can be presented in the agreed upon way as

$$\gamma_\alpha = \begin{cases} \gamma_\alpha^+ : x = \alpha y^{\frac{3}{2}}, & y \geq 0, \\ \gamma_\alpha^- : x = \alpha(-y)^{\frac{3}{2}}, & y \leq 0. \end{cases} \tag{12}$$

4.1 The Case C

The linearization of the field (4) at every singular point (q, p_i), $q \in S_0$, $i = 0, 1, 2$, has the spectrum $(\lambda_1, \lambda_2, 0)$ with non-zero real eigenvalues $\lambda_{1,2}$. Moreover, at a singular point (q, p_0) corresponding to the isotropic admissible direction the resonant relation $\lambda_1 = 2\lambda_2$ holds. On the other hand, at a singular point (q, p_i), $i = 1, 2$, corresponding to non-isotropic admissible direction the resonant relation $\lambda_1 + \lambda_2 = 0$ holds.[3] Using the smooth classification of vector fields with non-isolated singular points (see e.g., [26], Appendix A), we have the following results.

The germ of the field (4) at any point (q, p_0), $q \in S_0$, has C^∞ orbital normal form

$$2\xi\frac{\partial}{\partial\xi} + \eta\frac{\partial}{\partial\eta} + 0\frac{\partial}{\partial\zeta} \tag{13}$$

with the first integrals $I_1 = \xi/\eta^2$ and $I_2 = \zeta$. The germ of the field (4) at any point (q, p_i), $q \in S_0$, $i = 1, 2$, has C^∞ orbital normal form

$$\xi\frac{\partial}{\partial\xi} - \eta\frac{\partial}{\partial\eta} + \xi\eta\frac{\partial}{\partial\zeta} \tag{14}$$

[3] The relation $\lambda_1 = 2\lambda_2$ is a corollary of $\lambda_1 + \lambda_2 + \lambda_3 = r\lambda_1$ with $r = \frac{3}{2}$ and $\lambda_3 = 0$, see Theorem 1 and formula (6). The relation $\lambda_1 + \lambda_2 = 0$ follows form the fact that the field $\vec{W}$ is divergence-free and the function F does not vanish in a neighborhood of (q, p_i), $i = 1, 2$.

with the first integral $I = \xi\eta$. One can see that to every singular point of the field (13) corresponds a one-parameter family of integral curves passing through this point, while to every singular point of the field (14) correspond only two integral curves. Projecting the integral curves down, we obtain the following results.

Theorem 2 ([24, 25]) *Suppose that C holds true. Then to the isotropic direction p_0 corresponds a one-parameter family Γ_0 of geodesics outgoing from the point q. There exist smooth local coordinates centered at q such that the discriminant curve S_0 coincides with the x-axis, the isotropic direction $p_0(q) = \infty$ and the geodesics $\gamma_\alpha^\pm \in \Gamma_0$ are semi-cubic parabolas*

$$x = \alpha\tau^3 X_\alpha^\pm(\tau), \quad y = \tau^2 Y_\alpha^\pm(\tau), \quad \alpha \geq 0, \tag{15}$$

where $X_\alpha^\pm$, $Y_\alpha^\pm$ are smooth functions, $X_\alpha^\pm(0) = 1$, $Y_\alpha^\pm(0) = \pm 1$.

Theorem 3 ([24, 25]) *Suppose that C_3 holds true. Then to each admissible direction p_i, $i = 1, 2$, corresponds a unique geodesic passing through the point q. Both these geodesics are smooth and timelike.*

In the left panel of Fig. 7 we present the invariant foliations of the field (4) in a neighborhood of the point (q, p_0), $q \in S_0$, that correspond to the first integrals $I_1 = \xi/\eta^2$ (left) and $I_2 = \zeta$ (right) of the normal form (13). Intersection of these foliations gives the family of integral curves of (4). The family Γ_0 of the geodesics (15) is obtained (by the projection $PTS \to S$) from the family of integral curves of the field (4) that pass through its singular point (q, p_0). The subfamily $\Gamma_0^+ \subset \Gamma_0$ of the geodesics (15) outgoing into the Lorentzian semiplane, contains timelike, spacelike, and isotropic geodesics.

In the right panel of Fig. 7 we present those of the leaves of the invariant foliation of the field (4) in a neighborhood of the point (q, p_i), $q \in S_0$, $i = 1, 2$, that pass through (q, p_i). This foliation corresponds to the first integral $I = \xi\eta$ in the normal form (14), and the leaves passing through (q, p_i) coincide with the planes $\xi = 0$ and $\eta = 0$, while none of the remaining leaves contains singular points of (4). One of these leaves coincides with the vertical surface $\overline{S}_0$ filled with vertical integral curves whose projection on S are points of S_0. Another invariant surface is filled with non-vertical integral curves, through every point (q, p_i), $q \in S_0$, there pass exactly one curve.

Example 2 To illustrate the above, return to Example 1. In the coordinates $\bar{x} = y$, $\bar{y} = x$, $\bar{p} = 1/p$, the equation of isotropic lines coincides with Cibrario normal form. After multiplication by -1, the corresponding vector field (4) reads

$$\vec{V} = 2\bar{x}\left(\frac{\partial}{\partial\bar{x}} + \bar{p}\frac{\partial}{\partial\bar{y}}\right) + \bar{p}\frac{\partial}{\partial\bar{p}}. \tag{16}$$

It is easy to check that the field (16) possesses the invariant foliation $\bar{x} = c\bar{p}^2$, which includes, in particular, the vertical surface $\overline{S}_0$ (for $c = 0$), the isotropic surface

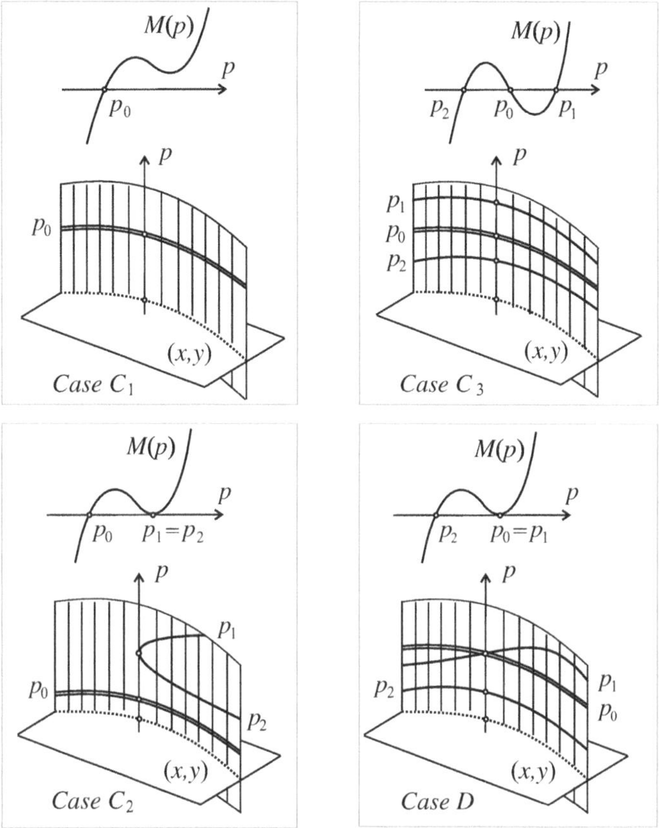

Fig. 6 Real roots of the cubic polynomial $M(p)$ and the set of singular point of the field (4). The double line presents $\{(q, p_0), q \in S_0\}$, the bold lines present $\{(q, p_i), q \in S_0\}$, $i = 1, 2$, the dotted line presents S_0

(for $c = 1$). This foliation is presented in the left side of the left panel of Fig. 7. The restriction of the field (16) to every invariant leaf $\bar{x} = c\bar{p}^2$ reads $2c\bar{p}^3 \frac{\partial}{\partial \bar{y}} + \bar{p} \frac{\partial}{\partial \bar{p}}$. Canceling the factor $\bar{p}$, we obtain the non-singular field $2c\bar{p}^2 \frac{\partial}{\partial \bar{y}} + \frac{\partial}{\partial \bar{p}}$, whose integral curves are presented in Fig. 5 (left). Fixing a degenerate point $q \in S_0$, in going through all invariant leaves $\bar{x} = c\bar{p}^2$ and projecting down, we obtain the family (12) of geodesics γ_α^+ (for $c > 0$) and γ_α^- (for $c < 0$) presented in Fig. 8 (right).[4] In Fig. 9, two more examples of geodesics outgoing from a degenerate point are presented: the family (12) in the case C_1 (on the left) and the family (12) together with two geodesics that have non-isotropic admissible directions in the case C_3 (on the right).

[4]The attentive reader may remark that this invariant foliation contains also the leaf $\bar{p} = 0$, which can be considered as the limiting case for $c \to \infty$. The restriction of (16) to this leaf is filled with integral curves parallel to the $\bar{x}$-axis. This gives the family of geodesics $x = \text{const}$, which are the limiting case of the semi-cubic parabolas (15): the two branches are glued together.

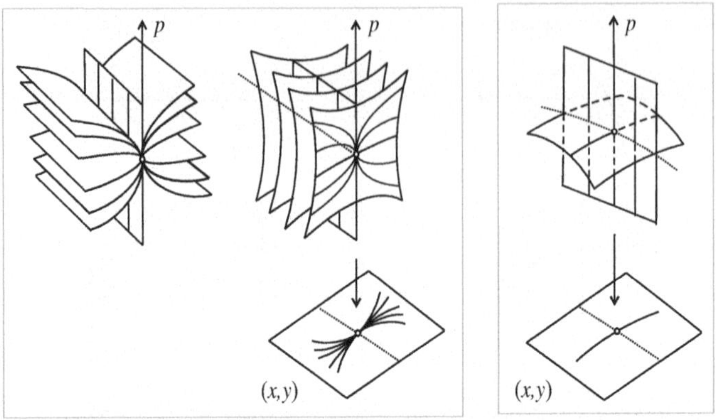

Fig. 7 The cases C_1, C_3. Left panel: two invariant foliations of the field (4) near the point (q, p_0), $q \in S_0$. Right panel: the invariant leaves of the field (4) passing through the point (q, p_i), $q \in S_0$, $i = 1, 2$. The dotted lines present the set of singular points of the field (4) and its projection, the discriminant curve

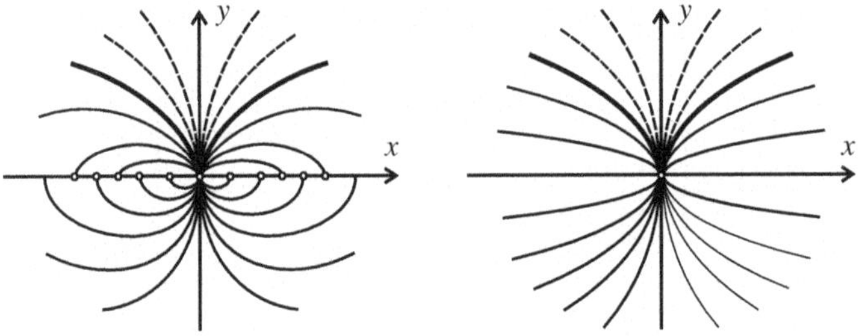

Fig. 8 Two examples of the family Γ_0 in Theorem 2. Geodesics in the metrics $e^y dx^2 - y dy^2$ (left) and $dx^2 - y dy^2$ (right) outgoing from $q = 0$. Timelike, spacelike, and isotropic geodesics are depicted as solid, dashed, and bold solid lines respectively

Remark 2 If the pseudo-Riemannian metric on the surface S is induced by the pseudo-Euclidean metric $dX^2 + dY^2 - dZ^2$ of the ambient space (see the example above), the difference between the cases C_1 and C_3 has a graphical interpretation. Namely, C_1 and C_3 correspond to positive and negative Gaussian curvature of the surface S calculated in the Euclidean metric $dX^2 + dY^2 + dZ^2$.

Theorem 4 *Suppose that C_2 holds true. Generically, the point q locally separates the curve S_0 in two parts, filled with C_1 and C_3 points, respectively, and there exist smooth local coordinates centered at q such that the metric has the form*

$$ds^2 = a(x, y)\, dx^2 + ye(x, y)\, dy^2, \quad a(0) \neq 0, \; e(0) \neq 0, \; a_y(0) = 0, \; a_{xy}(0) \neq 0. \tag{17}$$

Then to the double admissible direction $p_1 = p_2$ corresponds a unique geodesic passing through the point q, a semicubic parabola with branches outgoing from q

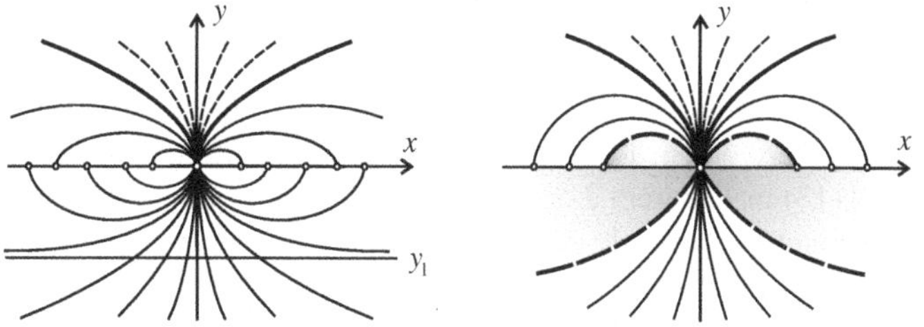

Fig. 9 Geodesics in the metrics $ds^2 = a(y)dx^2 - ydy^2$ with $a(y) = 1 + (y + y_1)^2$ outgoing from $q = 0$. The case C_1 ($y_1 < 0$, left) and the case C_3 ($y_1 > 0$, right). Timelike, spacelike, and isotropic geodesics are depicted as solid, dashed, and bold solid lines respectively. Geodesics passing through 0 with non-isotropic admissible directions (right) are depicted as long-dashed bold lines. The grey domains do not contain geodesics passing though 0

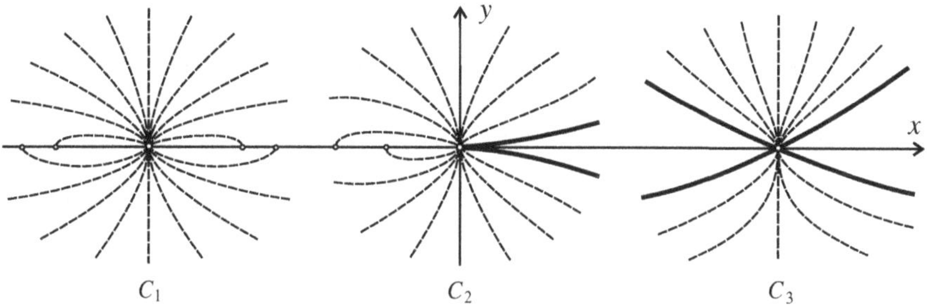

Fig. 10 Geodesics in the metrics (17) outgoing from three different degenerate points: C_1 ($x < 0$, left), C_2 ($x = 0$, center), and C_3 ($x > 0$, right). Here timelike, spacelike, and isotropic geodesics outgoing from $q \in S_0$ with the isotropic direction $p_0 = \infty$ are depicted as dashed lines, while the geodesics outgoing from $q \in S_0$ with non-isotropic admissible direction $p_1 = 0$ are depicted as bold lines

into the Lorentzian and Riemannian domains (depicted as long-dashed line in Fig. 10, center).

The proof is not published yet. In Example 1 considered above, we deal with a non-generic case C_2, since the condition $a_{xy}(0) \neq 0$ in (17) does not hold true. This leads to the geodesic $y = 0$ instead of a semicubic parabola mentioned in Theorem 4.

4.2 *The Case D*

The cubic polynomial M at $q \in S_0$ has the isotropic double root $p_0 = p_1$ and a simple non-isotropic root p_2. For the admissible direction p_2, the analogous assertion to Theorem 3 holds true: the germ of the field (4) at (q, p_2) has C^∞ normal form (14), and to the direction p_2 corresponds a unique smooth geodesic passing through the point q. However, the study of geodesics with the isotropic direction is more complicated.

A special feature of the case D is that the linear part of the germ (4) at (q, p_0), $q \in S_0$, has three zero eigenvalues. This prevents the possibility to obtain a normal form similar to (13) in Theorem 2 or similar to (14) in Theorem 3. Moreover, in this case even the reduction principle does not allow to establish the topological normal form of this filed, since the *center subspace*[5] of the germ (4) at (q, p_0) coincides with the whole tangent space, see [4]. However, using appropriate blowing up procedure, one can reduce the germ (4) at (q, p_0) to a smooth vector field with non-zero spectrum and study the obtained vector field using the standard methods.

Further we always assume that in the cases D_s and D_n the following genericity condition holds true: there are no non-trivial integer relations

$$n_1\alpha_1 + n_2\alpha_2 + n_3\alpha_3 = \alpha_j, \quad n_1 + n_2 + n_3 \geq 1, \quad n_i \in \mathbb{Z}_+, \quad j = 1, 2, 3,$$

where $\alpha_{1,2}$ are the eigenvalues of the linearization of the vector field (9) at (q, p_0) and $\alpha_3 = 2$. This condition implies the germ of a vector field obtained from (4) by the blowing up procedure is linearizable, as well as the germ of the field (9).

4.2.1 The Cases D_n and D_f

In a neighborhood of the considered point (q, p_0), $q \in S_0$, the field (4) above the Lorentzian domain has an invariant foliation $\{\mathscr{F}_\alpha\}$ presented in the left panel of Fig. 11. Here the invariant leaf $\mathscr{F}_0$ coincides with the isotropic surface $\mathscr{F}$. The invariant leaves above the Riemannian domain are not depicted, since they contain no integral curves that pass through (q, p_0).

The linear part of the restriction of the field (4) to every invariant leaf $\mathscr{F}_\alpha$ at its singular point (q, p_0) is equal to (9) multiplied by a smooth scalar function σ_α vanishing along the criminant. Therefore, the restriction of the field (4) to every invariant leaf $\mathscr{F}_\alpha$ has the local phase portrait of the same type: node or focus. See Fig. 11 (right panel). In going through all invariant leaves $\mathscr{F}_\alpha$ and projecting the integral curves down, we obtain the following result.

Theorem 5 ([26]) *Let the case D_n or D_f holds true. Then to the isotropic direction p_0 corresponds a two-parameter family Γ_0 of C^2-smooth geodesics γ_α^+ outgoing from q into the Lorentzian domain, while there are no geodesics outgoing from q into the Riemannian domain. Given α, the geodesics $\gamma_{\alpha,\beta}^+ \in \Gamma_0$ with fixed α and varying β are projections of the integral curves from the leaf $\mathscr{F}_\alpha$; see Fig. 11, center for D_n and right for D_f. The geodesics $\gamma_{\alpha,\beta}^+ \in \Gamma_0$ are timelike if $\alpha < 0$, spacelike if $\alpha > 0$ and isotropic if $\alpha = 0$.*

[5] The center subspace T_c of a vector filed $\vec{V}$ at its singular point 0 is spanned by the generalized eigenvectors of the linearization of $\vec{V}$ at 0 corresponding to the eigenvalues λ with $\mathrm{Re}\,\lambda = 0$.

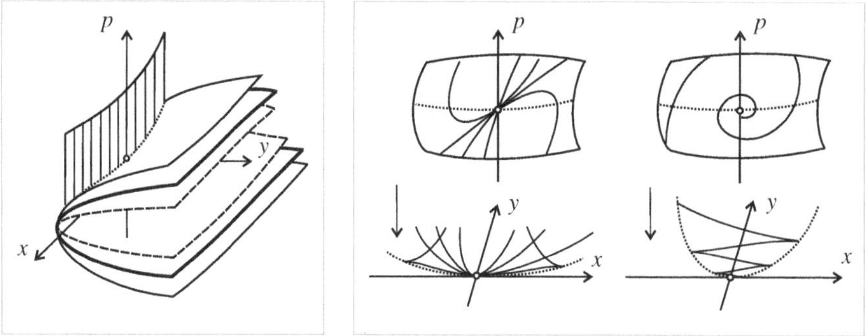

Fig. 11 The cases D_n and D_f. The left panel: the invariant foliation $\{\mathscr{F}_\alpha\}$ of the field (4) above the Lorentzian domain (the isotropic surface $\mathscr{F} = \mathscr{F}_0$ is depicted as bold). Here the leaves filled with timelike, spacelike and isotropic geodesics are depicted as solid, dashed and bold solid lines, respectively. The right panel: integral curves of the restriction of field (4) to $\mathscr{F}_\alpha$ and their projections (the case D_n on the left and D_f on the right). The criminant and the discriminant curve are depicted as dotted lines

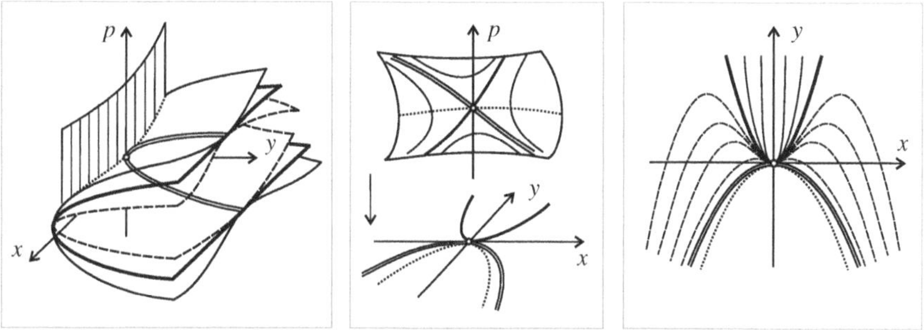

Fig. 12 The case D_s. On the left: invariant foliation $\{\mathscr{F}_\alpha\}$ of the field (4) above the Lorentzian domain (the isotropic surface $\mathscr{F} = \mathscr{F}_0$ is depicted as bold). Center: integral curves of the field (4) on an invariant leaf $\mathscr{F}_\alpha$. On the right: geodesics outgoing from the point $q \in S_0$. Timelike and spacelike geodesics are depicted as solid and dashed lines, respectively. The bold solid and the double solid lines present two isotropic geodesics. The criminant and the discriminant curve are depicted as dotted lines

4.2.2 The Case D_s

In a neighborhood of the considered point (q, p_0), $q \in S_0$, the field (4) above the Lorentzian domain has an invariant foliation $\{\mathscr{F}_\alpha\}$ presented in the left panel of Fig. 12. Here the invariant leaf $\mathscr{F}_0$ coincides with the isotropic surface $\mathscr{F}$. The invariant leaves above the Riemannian domain are not depicted, since they contain no integral curves that pass through (q, p_0).

The linear part of the restriction of the field (4) to every invariant leaf $\mathscr{F}_\alpha$ at its singular point (q, p_0) is equal to (9) multiplied by a smooth scalar function σ_α

Fig. 13 Computer generated geodesics (solid lines) of the metric $dy^2 + (\varepsilon x^2 - y)dx^2$ with $\varepsilon < 0$ (the case D_s), $0 < \varepsilon < \frac{1}{16}$ (the case D_n), and $\varepsilon > \frac{1}{16}$ (the case D_f). The parabola depicted as dotted line is the discriminant curve S_0

vanishing along the criminant. Therefore, the restriction of the field (4) to every invariant leaf $\mathscr{F}_\alpha$ has a saddle at (q, p_0). See Fig. 12 (right panel).

Theorem 6 ([26]) *Let the case D_s holds true. Then to the isotropic direction p_0 corresponds a one-parameter family Γ_0 of C^2-smooth geodesics outgoing from q into the Lorentzian domain, while there are no geodesics outgoing from q into the Riemannian domain. There exist smooth local coordinates centered at q such that S_0 is the parabola $y = \varepsilon x^2$ and the geodesics $\gamma_\alpha^+ \in \Gamma_0$ outgoing from q have the form*

$$y = \frac{\varepsilon_1}{2}x^2 + Y_\alpha(x), \quad Y_\alpha(x) = o(x^2), \quad \alpha \in \mathbb{R}, \tag{18}$$

together with one additional isotropic geodesic

$$y = \frac{\varepsilon_2}{2}x^2 + Y(x), \quad Y(x) = o(x^2), \tag{19}$$

where $\varepsilon_1\varepsilon_2 = \varepsilon$, $\varepsilon_1 + \varepsilon_2 = \frac{1}{2}$, $\varepsilon_1 > \frac{1}{2}$, $\varepsilon_2 < 0$. Geodesics (18) *are timelike if $\alpha < 0$, spacelike if $\alpha > 0$, isotropic if $\alpha = 0$; see Fig. 12, right.*

It is interesting to note that invariant foliations in the cases D_n, D_f and D_s have the different topological structures (compare the left panels of Figs. 11 and 12). In the cases D_n, D_f all invariant leaves intersect on the criminant only, while in the case D_s they intersect on the criminant (dotted line) and on the double line, whose projection is the isotropic geodesic (19). In Fig. 13, we present computer generated geodesics in the cases D_s, D_n, D_f.

4.3 Example: Clairaut Type

It is of interest to observe an important difference between the families Γ_0 in the cases C_1, C_3 and D. In the cases C_1, C_3, the family Γ_0 is symmetric with respect to S_0 in the following sense: it contains an infinite number of geodesics $\gamma_\alpha^+ \in \Gamma_0$ outgoing into the Lorentzian domain and an infinite number of geodesics $\gamma_\alpha^- \in \Gamma_0$ outgoing into the Riemannian domain. On the contrary, in the case D, the family Γ_0 is non-symmetric: it contains an infinite number of geodesics $\gamma_\alpha^+ \in \Gamma_0$ outgoing into the Lorentzian domain and no geodesics $\gamma_\alpha^- \in \Gamma_0$ outgoing into the Riemannian domain.

To understand this phenomenon better, consider the case when the isotropic direction p_0 is tangent to the curve S_0 *at all points* $q \in S_0$, for instance, the metric $dy^2 + (\varepsilon x^2 - y)dx^2$. The equation of geodesics in the metric $ds^2 = dy^2 - ydx^2$ can be studied using qualitative methods, see [25] (Sect. 3). The Lagrangian of the length functional $L = \sqrt{p^2 - y}$ does not depend on the variable x, hence the field (4) possesses the energy integral $H = L - pL_p$. After evident transformations, equation $H = \text{const}$ can be reduced to

$$p^2 = y - \alpha y^2, \quad \alpha \in \mathbb{R}, \tag{20}$$

which is a family of implicit differential equations *of Clairaut type* [11].

Every (unparameterized) geodesic in the metric $ds^2 = dy^2 - ydx^2$ is a solution of Eq. (20). Conversely, every solution of (20) is a geodesic except the horizontal lines $y \equiv \text{const}$, each of which is the envelop of the family of integral curves of (20) for a given α (see [25]). For instance, the value $\alpha = 0$ corresponds to the isotropic surface $p^2 = y$ (a parabolic cylinder) and gives, in particular, the isotropic geodesic $y = \frac{1}{4}x^2$ passing through the origin.

For determining non-isotropic geodesics, observe that every invariant surface (20) is a cylinder whose generatrices are parallel to the x-axis and the base is an ellipse (if $\alpha > 0$) or a hyperbola (if $\alpha < 0$). In the latter case, the hyperbolic cylinder $p^2 = y - \alpha y^2$ consists of two connected components: *positive* and *negative* lying in the domains $y \geq 0$ and $y \leq \alpha^{-1}$, respectively. Positive components of the hyperbolic cylinders ($\alpha < 0$) together with all other cylinders ($\alpha \geq 0$) form an invariant foliation over the Lorentzian domain $y > 0$. Negative components of the hyperbolic cylinders form an invariant foliation over the Riemannian domain $y < 0$; they do not intersect the plane $y = 0$, and consequently, do not contain integral curves whose projections to the (x, y)-plane are geodesics passing through the x-axis. See Fig. 14 (left).

Thus to every $\alpha \geq 0$ corresponds a geodesic $\gamma_\alpha^+ \in \Gamma_0$ which is timelike if $\alpha > 0$ or isotropic if $\alpha = 0$. To every $\alpha < 0$ corresponds a spacelike geodesic $\gamma_\alpha^+ \in \Gamma_0$, whose lift belongs to the positive component of the hyperbolic cylinder $p^2 = y - \alpha y^2$. In contrast to this, the negative component of the same cylinder is filled with integral curves of the field (4) whose projections on the (x, y)-plane are separated from the x-axis by the horizontal strip $\alpha^{-1} < y < 0$. Therefore, there are no geodesics outgoing into the Riemannian domain. See Fig. 14, right.

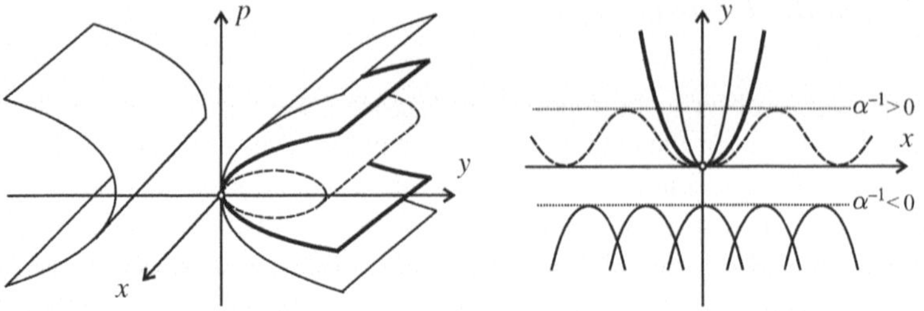

Fig. 14 The invariant foliation $p^2 = y - \alpha y^2$ in the (x, y, p)-space (left) and the corresponding geodesics (right). Timelike, spacelike and isotropic geodesics are solid, dashed and bold solid lines, respectively

Acknowledgements The publication was supported by the Russian Foundation for Basic Research (research projects 16-01-00766, 17-01-00849) and the Laboratory of Dynamical Systems NRU HSE.

References

1. Aguirre, E., Fernandez, V., Lafuente, J.: On the conformal geometry of transverse Riemann-Lorentz manifolds. J. Geom. Phys. **57**, 1541–1547 (2007)
2. Al'tshuler, B.L., Barvinski, A.O.: Quantum cosmology and physics of transitions with a change of spacetime signature. Uspekhi Fiz. Nauk **166**(5), 459–492 (1996); English transl. in Physics-Uspekhi **39**, 429
3. Aminova, A.V.: Projective transformations and symmetries of differential equations. Mat. Sb. **186**(12), 21–36 (1995)
4. Anosov, D.V., Arnold, V.I. (eds.): Dynamical systems I. Ordinary differential equations and smooth dynamical systems. Encyclopaedia of Mathematical Sciences 1. Springer, Berlin (1988)
5. Arnol'd, V.I.: Geometrical Methods in the Theory of Ordinary Differential Equations. Springer, New York (1988)
6. Bogaevsky, I.A.: Implicit ordinary differential equations: bifurcations and sharpening of equivalence. Izvestiya Math. **78**(6), 1063–1078 (2014)
7. Bolsinov, A.V., Matveev, V.S.: Local normal forms for geodesically equivalent pseudo-Riemannian metrics. Trans. Am. Math. Soc. **367**, 6719–6749 (2015)
8. Cibrario, M.: Sulla reduzione a forma delle equationi lineari alle derviate parziale di secondo ordine di tipo misto. Accademia di Scienze e Lettere, Instituto Lombardo Redicconti **65**, 889–906 (1932)
9. Dara, L.: Singularités générique des équations différentielles multiformes. Bol. Soc. Bras. Math. **6**(2), 95–128 (1975)
10. Davydov, A.A.: The normal form of a differential equation, that is not solved with respect to the derivative, in the neighborhood of its singular point. Funktsional. Anal. i Prilozhen. **19**, 1–10 (1985)
11. Davydov, A.A., Ishikawa, G., Izumiya, S., Sun, W.-Z.: Generic singularities of implicit systems of first order differential equations on the plane. Jpn. J. Math. **3**, 93–119 (2008)
12. Genin, D., Khesin, B., Tabachnikov, S.: Geodesics on an ellipsoid in Minkowski space. Enseign. Math. **53**, 307–331 (2007)
13. Ghezzi, R., Remizov, A.O.: On a class of vector fields with discontinuities of divide-by-zero type and its applications to geodesics in singular metrics. J. Dyn. Control Syst. **18**, 135–158 (2012)

14. Hartman, Ph: Ordinary Differential Equations. Birkhauser, Boston (1982)
15. Hirsch, M.W., Pugh, C.C., Shub, M.: Invariant Manifolds. Lecture Notes in Mathematics, vol. 583. Springer, Berlin (1977)
16. Khesin, B., Tabachnikov, S.: Pseudo-Riemannian geodesics and billiards. Adv. Math. **221**, 1364–1396 (2009)
17. Kossowski, M.: Pseudo-Riemannian metrics singularities and the extendability of parallel transport. Proc. Amer. Math. Soc. **99**, 147–154 (1987)
18. Kossowski, M., Kriele, M.: Smooth and discontinuous signature type change in general relativity. Class. Quantum Grav. **10**, 2363–2371 (1993)
19. Kossowski, M., Kriele, M.: Transverse, type changing, pseudo-Riemannian metrics and the extendability of geodesics. Proc. Roy. Soc. Lond. Ser. A Math. Phys. **444**(1921), 297–306 (1994)
20. Kossowski, M., Kriele, M.: The Einstein equation for signature type changing spacetimes. Proc. Roy. Soc. Lond. Ser. A Math. Phys. **446**(1926), 115–126 (1994)
21. Miernowski, T.: Formes normales d'une métrique mixte analytique réelle générique. Ann. Fac. Sci. Toulouse Math. **16**, 923–946 (2007)
22. Pavlova, N.G., Remizov, A.O.: Geodesics on hypersurfaces in the Minkowski space: singularities of signature change. Russian Math. Surveys **66**, 1201–1203 (2011)
23. Remizov, A.O.: Multidimensional Poincaré construction and singularities of lifted fields for implicit differential equations. J. Math. Sci. (N.Y.) **151**(6), 3561–3602 (2008)
24. Remizov, A.O.: Geodesics on 2-surfaces with pseudo-Riemannian metric: singularities of changes of signature. Mat. Sb. **200**(3), 75–94 (2009)
25. Remizov, A.O.: On the local and global properties of geodesics in pseudo-Riemannian metrics. Differ. Geom. Appl. **39**, 36–58 (2015)
26. Remizov, A.O., Tari, F.: Singularities of the geodesic flow on surfaces with pseudo-Riemannian metrics. Geom. Dedicata. **185**(1), 131–153 (2016)
27. Sakharov, A.D.: Cosmological transitions with changes in the signature of the metric. Zh. Eksper. Teor. Fiz. **87**(2) (8) 375–383 (1984). English transl. in Soviet Phys. JETP **60**(2), August 1984, 214–218
28. Steller, M.: A Gauss-Bonnet formula for metrics with varying signature. Z. Anal. Anwend. **25**, 143–162 (2006)

Orbital Formal Rigidity for Germs of Holomorphic and Real Analytic Vector Fields

Jessica Angélica Jaurez-Rosas

Abstract This survey paper is focused on discussing the main facts from the (orbital) formal rigidity phenomenon for germs of holomorphic and real analytic vector fields in the complex and real planes, exploring their similar and different properties.

Keywords Holomorphic vector fields · Real analytic vector fields · Orbital formal rigidity

1 Introduction

The (orbital) formal rigidity and (orbital) real-formal rigidity, as any rigidity phenomenon, take place when *a weaker equivalence implies a stronger equivalence*: the (orbital) formal rigidity phenomenon takes place when a *formal relation* between germs of holomorphic vector fields at $\widehat{0} \in \mathbb{C}^2$ implies that such germs coincide under a (an orbital) holomorphic change of coordinates (Sect. 2.1), while the (orbital) real-formal rigidity phenomenon takes place when a formal relation between germs of real analytic vector fields at $\widehat{0} \in \mathbb{R}^2$ implies that such germs coincide under a (an orbital) real analytic change of coordinates (Sect. 2.2).

Despite the fact that the complexification of a germ of real analytic vector field at $\widehat{0} \in \mathbb{R}^2$ allows us to consider it as a germ of holomorphic vector field at $\widehat{0} \in \mathbb{C}^2$, the (orbital) real-formal rigidity phenomenon is not an immediate consequence of the (orbital) formal rigidity phenomenon (Sects. 2.3, 3.1 and 4).

First of all, let us briefly discuss the (orbital) formal rigidity. Poincaré's linearization theorem precedes the study of this phenomenon: Poincaré's theorem states that, given a germ of holomorphic vector field with a singularity at the origin in $\mathbb{C}^2$, if its linear part has nonresonant spectrum belonging to the Poincaré domain (Definitions 3.1 and 3.2), then there exists a germ of biholomorphism at $\widehat{0} \in \mathbb{C}^2$ sending the germ of vector field to the vector field induced by its linear part (see Theorem 3.4).

J. A. Jaurez-Rosas (✉)
Departamento de Álgebra, Análisis Matemático, Geometría y Topología, Universidad de Valladolid, 7 Paseo de Belén, Campus Miguel Delibes, Valladolid, Spain
e-mail: jessica@im.unam.mx

R. N. Araújo dos Santos et al. (eds.), *Singularities and Foliations. Geometry, Topology and Applications*, Springer Proceedings in Mathematics & Statistics 222, https://doi.org/10.1007/978-3-319-73639-6_5

The search of understanding the cases which do not satisfy the genericity conditions of Poincaré's linearization theorem led to the study of the (orbital) formal rigidity phenomenon. This point of view has allowed to understand the invariants of the (orbital) analytic classification of germs of holomorphic vector fields, as mentioned in Sects. 3 and 4.

The well-known results due to Poincaré-Dulac and Brjuno (Theorems 3.4 and 3.7) allow us to conclude that the (orbital) formal rigidity takes place for germs of real analytic vector fields with generic nonzero linear part. But this phenomenon in general fails for germs with nonzero linear part which do not satisfy the genericity assumptions of these rigidity theorems (Sect. 3.2).

The case with zero linear part was recently studied by Laura Ortiz-Bobadilla, Ernesto Rosales-González and Serguei M. Voronin (Sect. 4). In particular, Voronin proved that the orbital formal rigidity phenomenon takes place for a generic class of germs of holomorphic vector fields with zero linear part (Theorem 4.1).

Now, we briefly refer the case of the (orbital) real-formal rigidity phenomenon. Theorems 3.5 and 3.8 are the real analytic versions of Poincaré-Dulac's and Brjuno's results. These theorems allow us to conclude that, for germs of real-analytic vector fields with generic nonzero linear part, the (orbital) real-formal rigidity takes place. The Euler vector field shows that this phenomenon fails when germs with nonzero linear part do not satisfy the generic assumptions of these results: the vector field

$$x^2 \frac{\partial}{\partial x} + (y - x)\frac{\partial}{\partial y}$$

has a unique *complex separatrix*, that is, an invariant irreducible analytic curve passing through the singularity, namely the line $\{x = 0\}$. This vector field is (orbitally) real-formally equivalent to a real analytic vector field with two different separatrices, namely its *(orbital) formal normal form* (see [14]). As a consequence, they are not (orbitally) real-analytic equivalent.

In [15] it was proved the real-analytic version of Voronin's rigidity theorem: under generic conditions the orbital real-formal rigidity takes place for germs of real analytic vector fields with zero linear part (see Theorem 4.1).

We briefly mention the structure of this survey paper. Section 2 is focused on describing the (orbital) formal rigidity phenomena for holomorphic and real analytic germs. The (orbital) formal rigidity problem for germs with nonzero linear part is discussed in Sect. 3.

Section 4 is focused on discussing the main results from the orbital rigidity phenomenon for holomorphic and real analytic germs with zero linear part. In this section is included a detailed outline of the proof of Voronin's rigidity theorem and its real-analytic version (Theorem 4.1), having as principal aim to discuss further similarities and differences between the holomorphic and the real-analytic cases.

2 Analytic Equivalence and Formal Rigidity

2.1 Holomorphic Vector Fields

Let $\mathcal{V}$ be the class of germs of holomorphic vector fields with an isolated singularity at the origin $\widehat{0}$ in the complex plane $\mathbb{C}^2$. Given $v, w \in \mathcal{V}$, we denote by $\mathscr{F}_v$ and $\mathscr{F}_w$ the *singular foliations induced by representatives of* v *and* w, that is, the partition of $(\mathbb{C}^2, \widehat{0})$ into the complex trajectories of v and w (see [14]).

We say that v and w are *orbitally analytically equivalent* (or their foliations $\mathscr{F}_v$ and $\mathscr{F}_w$ are analytically equivalent) if there exists a holomorphic change of coordinates mapping the leaves of $\mathscr{F}_v$ into the leaves of $\mathscr{F}_w$. Equivalently, there exist a germ of biholomorphism $\mathscr{H}$ at $\widehat{0} \in \mathbb{C}^2$ and a germ of holomorphic map $\mathscr{K}$ at $\widehat{0} \in \mathbb{C}^2$ which is nonzero at the origin, satisfying

$$w = \mathscr{K}\, \mathscr{H}_*(v) := \mathscr{K} \left[(\mathrm{D}\, \mathscr{H} \cdot v) \circ \mathscr{H}^{-1} \right]. \tag{1}$$

If the change of coordinates preserves the parametrization of the complex trajectories of v and w, then $\mathscr{K} \equiv 1$. In such case we shall say that v and w are *analytically equivalent.*

Differentiating both sides of the equality (1) we obtain a *formal relation* induced by the Taylor series of $\mathscr{H}$ and $\mathscr{K}$ between the Taylor series of v and w. In general, this relation is described as follows: there exist an invertible formal transformation $H \in \left(\mathbb{C}[[x, y]]\right)^2$ with zero constant term in each component and a formal series $K \in \mathbb{C}[[x, y]]$ with nonzero constant term satisfying

$$\widehat{w} = K\, H_*(\widehat{v}) := K \left[(\mathrm{D}\, H \cdot \widehat{v}) \circ H^{-1} \right], \tag{2}$$

where $\widehat{v}$ and $\widehat{w}$ are Taylor series of v and w, respectively. In this case, we shall say that v and w are *orbitally formally equivalent* (or their foliations $\mathscr{F}_v$ and $\mathscr{F}_w$ are formally equivalent). If $K \equiv 1$ we shall say that v and w are *formally equivalent.*

Remark 2.1 The (orbital) analytic equivalence and the (orbital) formal equivalence induce equivalence relations on the class $\mathcal{V}$.

As we have seen, the existence of a (an orbital) formal equivalence is a necessary condition for the existence of an (orbital) analytic equivalence between two germs of holomorphic vector fields. Since it is easier to prove the existence of a (an orbital) formal equivalence, it is important to know whether it is also a sufficient condition. This phenomenon will be called *(orbital) formal rigidity.*

2.2 Real Analytic Vector Fields

Let $\mathsf{v} = (\mathsf{v}_1, \mathsf{v}_2)$ and $\mathsf{w} = (\mathsf{w}_1, \mathsf{w}_2)$ be two germs of real analytic vector fields with *an algebraic isolated singularity at* $\widehat{0} \in \mathbb{R}^2$, that is, v_1 and v_2 (w_1 and w_2) are coprime in the ring of convergent complex series $\mathbb{C}\langle x, y\rangle$.

The real singular foliations induced by representatives of v and w (i.e., the partition of $(\mathbb{R}^2, \widehat{0})$ into their real trajectories) will be denoted by $\mathscr{F}_{\mathsf{v}}$ and $\mathscr{F}_{\mathsf{w}}$, respectively.

As above, we shall define equivalence relations on the class of germs of real analytic vector fields with an algebraic isolated singularity at $\widehat{0} \in \mathbb{R}^2$.

We say that v and w are *orbitally real-analytically equivalent* if there exist a germ of real analytic diffeomorphism mapping the leaves of the foliation $\mathscr{F}_{\mathsf{v}}$ into the leaves of $\mathscr{F}_{\mathsf{w}}$, that is, if v and w satisfy the relation (1), taking $\mathscr{H}$ as a germ of real analytic diffeomorphism at $\widehat{0} \in \mathbb{R}^2$ and $\mathscr{K}$ as a germ of real analytic map which is nonzero at the origin. If the germ $\mathscr{H}$ preserves the parametrization of the real trajectories of v and w, then $\mathscr{K} \equiv 1$, in which case v and w are *real-analytically equivalent.*

These geometric equivalent relations between v and w imply the following formal relations: v and w are *orbitally real-formally equivalent* if the Taylor series of v and w satisfy the equality (2), where $H \in \left(\mathbb{R}[[x, y]]\right)^2$ has zero constant term in each component and $K \in \mathbb{R}[[x, y]]$ has nonzero constant term. If $K \equiv 1$ we say that v and w are *real-formally equivalent.*

As in the complex case, whether the (orbital) real-formal equivalence is also a sufficient condition for the (orbital) real-analytic equivalence, we shall say that the *(orbital) real-formal rigidity phenomenon* takes place.

2.3 Formal Rigidity and Real-Formal Rigidity

In this subsection we shall relate the (orbital) formal rigidity with the (orbital) real-formal rigidity. As a result we may show their basic differences.

In what follows we refer to $(\mathbb{R}^2, \widehat{0})$ and $(\mathbb{C}^2, \widehat{0})$ as neighborhoods of the origin in $\mathbb{R}^2$ and $\mathbb{C}^2$, respectively.

Given v a real vector field defined on $(\mathbb{R}^2, \widehat{0})$ we shall denote by $\mathsf{v}^{\mathbb{C}}$ its *complexification*, that is, the holomorphic vector field defined on $(\mathbb{C}^2, \widehat{0})$ which results from the extension of the domain of v to an open neighborhood of the origin in $\mathbb{C}^2$.

The complexification process establishes an one-to-one correspondence between the class of germs of real analytic vector fields with an algebraic isolated singularity at $\widehat{0} \in \mathbb{R}^2$ and the class of germs of holomorphic vector fields with an isolated singularity at $\widehat{0} \in \mathbb{C}^2$ whose components are convergent series with real coefficients. The last class will be denoted by $\mathcal{V}^{\mathbb{R}}$.

Under such identification, the (orbital) real-analytic equivalence and the (orbital) real-formal equivalence induce equivalence relations on $\mathcal{V}^{\mathbb{R}}$, which will be called in the same way.

Remark 2.2 Two germs $\mathsf{v}, \mathsf{w} \in \mathcal{V}^{\mathbb{R}}$ are (orbital) real-analytic equivalent if and only if they are (orbital) analytic equivalent and $\mathcal{H}$ (the pair $(\mathcal{H}, \mathcal{K})$) realizing the equivalence as in (1) is the complexification of the germ of a real analytic diffeomorphism (pair). As a consequence $\mathcal{H}$ preserves the real plane $\mathbb{R}^2$.

On the other hand, $\mathsf{v}, \mathsf{w} \in \mathcal{V}^{\mathbb{R}}$ are (orbital) real-formal equivalent if and only if they are (orbital) formal equivalent and H (the pair (H, K)) realizing the equivalence as in (2) has real coefficients.

As in Sect. 2.2, we shall say that the *(orbital) real-formal rigidity phenomenon* takes place $\mathcal{V}^{\mathbb{R}}$ if the (orbital) real-analytic equivalence and the (orbital) real-formal equivalence coincide.

The orbital formal rigidity phenomenon does not imply the orbital real-formal rigidity phenomenon: if $\mathsf{v}^{\mathbb{C}}$ and $\mathsf{w}^{\mathbb{C}}$ are orbitally analytically equivalent and the germ of biholomorphism $\mathcal{H}$ realizes the analytic equivalence between their foliations, then does not necessarily preserve the real plane, as the following examples shows.

Example 2.3 We consider $\mathsf{v}^{\mathbb{C}} \in \mathcal{V}^{\mathbb{R}}$. Let $\phi_{\mathsf{v}}(t; (x, y))$ be the flow of a representative of v with initial condition $(x, y) = \phi_{\mathsf{v}}(0; (x, y))$. Its complexification, denoted by $\phi_{\mathsf{v}}^{\mathbb{C}}(t; (x, y))$, is defined on $\mathrm{D}_\epsilon \times \mathrm{B}_\delta$, where D_ϵ is an open disc of radius $\epsilon > 0$ centered at $0 \in \mathbb{C}$ and B_δ is an open disc of radius $\delta > 0$ centered at $\widehat{0} \in \mathbb{C}^2$.

It is defined below a holomorphic map $\mathsf{t}(x, y)$ such that the transformation $(x, y) \mapsto \phi_{\mathsf{v}}^{\mathbb{C}}(\mathsf{t}(x, y), (x, y))$ does not preserve the real plane $\mathbb{R}^2$. As a consequence, we obtain an orbital automorphism of $\mathsf{v}^{\mathbb{C}}$ which is not the complexification of a real analytic diffeomorphism.

Let $(x_0, y_0) \in \mathbb{R}^2$ be a nonsingular point of v with $x_0 \neq 0$ belonging to B_δ. Since (x_0, y_0) is nonsingular then there exists $\alpha \in \mathbb{C} \setminus \mathbb{R}$ as close to 0 as desired, such that $\phi_{\mathsf{v}}(\alpha, (x_0, y_0))$ does not belong to $\mathbb{R}^2$. Let $\alpha_0 \in \mathrm{D}_\epsilon$ such a number.

We consider $\mathsf{t}(x, y) := \beta_0 x$, where $\beta_0 := \alpha_0 / x_0$. The set

$$A := \mathsf{t}^{-1}(\mathrm{D}_\epsilon) \cap \mathrm{B}_\delta = \left\{ (x, y) \in \mathbb{C}^2 \mid \|(x, y)\| < \delta\,,\ |x| < \tfrac{\epsilon}{|\beta_0|} \right\}$$

satisfies the following properties:

- It is an open convex subset of $\mathbb{C}^2$, that is, the line segment connecting two elements of A is contained in A. As a consequence A is connected.
- It is invariant by complex conjugacy, that is, $(x, y) \in A$ if and only if $(\bar{x}, \bar{y}) \in A$.
- The origin $\widehat{0}$ and the point (x_0, y_0) belong to A.

As a consequence the holomorphic transformation $H(x, y) := \phi_{\mathsf{v}}^{\mathbb{C}}(\mathsf{t}(x, y); (x, y))$ is defined on A. Moreover, it is a biholomorphism around the origin in $\mathbb{C}^2$.

We shall show that H does not preserve the real plane $\mathbb{R}^2$. Otherwise, it must coincide with the holomorphic transformation $\widetilde{H}(x, y) := \overline{H(\bar{x}, \bar{y})}$ in an open neighborhood of the origin. Since A is invariant by complex conjugacy, both transformations are defined on A. Moreover, H and $\widetilde{H}$ must coincide in A, since A is a connected open neighborhood of $\widehat{0} \in \mathbb{C}^2$ (see Identity Theorem, [11]). But

$H(x_0, y_0) = \phi_{\mathrm{v}}^{\mathbb{C}}(\alpha_0; (x_0, y_0))$ does not belongs to $\mathbb{R}^2$ by the choice of α_0, leading to a contradiction.

3 Formal Rigidity for Elementary Singularities

In this subsection we shall discuss Poincaré-Dulac's and Brjuno's theorems together with their real analytic versions. These theorems state that the formal and real-formal rigidity phenomenon take place for generic germs of vector fields with nonzero linear part. Even though these results are valid for $\mathbb{C}^m$ and $\mathbb{R}^m$, we are only focused on $\mathbb{C}^2$ and $\mathbb{R}^2$. At the end of this subsection we shall consider *special cases* where the formal rigidity phenomenon does not take place.

3.1 Poincaré-Dulac's and Brjuno's Theorems

In what follows v, w will be germs of holomorphic vector fields with an isolated singularity at $\widehat{0} \in \mathbb{C}^2$ (that is $v, w \in \mathcal{V}$) and A will be the linear part of v at the singular point.

Definition 3.1 *((Non)resonant pairs and resonant vector monomials)* The pair $(\lambda_1, \lambda_2) \in \mathbb{C}^2$ is called *resonant*, or more precisely, *additive resonant* if there exists $(m_1, m_2) \in \mathbb{N}^2$ such that $m_1 + m_2 \geqslant 2$, satisfying

$$\lambda_i = m_1\lambda_1 + m_2\lambda_2 , \quad \text{for some} \quad i \in \{1, 2\} . \tag{3}$$

The equality (3) will be called *resonance* and the monomial vector field $x_1^{m_1} x_2^{m_2} \frac{\partial}{\partial x_i}$ will be *the resonant vector monomial corresponding to the resonance* (3). If there does not exist a pair of natural numbers satisfying the property (3), we shall say that (λ_1, λ_2) *is nonresonant.*

When the vector field v has nonzero linear part, we shall say that v *is resonant* if the spectrum of its linear part is resonant. Otherwise, v will be *nonresonant.*

If v has nonzero linear part A, it is possible to show that there exists a formal change of coordinates H such that $H_*(v) = A + \widetilde{v}$, where $\widetilde{v}$ is a formal vector field with order greater than 1 having only resonant vector monomials corresponding to the resonances of the spectrum of A: the nonresonant vector monomials can be eliminated recursively by means of suitable polynomial transformations. More specifically, the nonresonant vector monomial $x_1^{n_1} x_2^{n_2} \frac{\partial}{\partial x_i}$ is eliminated by a polynomial transformation having a coefficient with factor $(\lambda_i - n_1\lambda_1 - n_2\lambda_2)^{-1}$, where λ_1, λ_2 are the eigenvalues of A; the number $\lambda_i - n_1\lambda_1 - n_2\lambda_2$ will be called *small denominator.*

This formal vector field is called *the formal normal form of* v and it is unique with respect to such properties. In particular $\widetilde{v} \equiv 0$ whenever the spectrum of A is nonresonant, that is, if v is nonresonant then it is formally linearizable. Accurate proofs of the previous assertions can be found in [1, 14].

The following properties will be useful to describe when the (orbital) formal rigidity phenomenon takes place.

Definition 3.2 (*Poincaré and Siegel domains*) *The Poincaré domain* is the collection of ordered pairs $(\lambda_1, \lambda_2) \in \mathbb{C}^2$ such that $0 \in \mathbb{C}$ does not belong to the convex hull of λ_1, λ_2 in $\mathbb{C}$. The *Siegel domain* is the complement of the Poincaré domain in $\mathbb{C}^2$.

Remark 3.3 By the definition, (λ_1, λ_2) belongs to the Poincaré domain if and only if $\lambda_1, \lambda_2 \neq 0$ and its quotient is not a negative real number. Therefore this is a generic property.

It follows directly that (λ_1, λ_2) has a finite number of resonances when it belongs to the Poincaré domain. As a consequence v has a polynomial formal normal form whenever the spectrum of its linear part A belongs to the Poincaré domain.

If (λ_1, λ_2) belongs to the Siegel domain and $\lambda_1, \lambda_2 \neq 0$ has a negative irrational quotient, then the pair is nonresonant. Otherwise (λ_1, λ_2) has an infinite number of resonances.

Theorems 3.4 and 3.7 state that the (orbital) formal rigidity phenomenon takes place under generic conditions over the linear part of germs of nondegenerate holomorphic vector fields. The real analytic versions of such results (Theorems 3.5 and 3.8) are not their immediate corollaries, as will be seen from the outlines of their proofs appearing at the end of this subsection.

Theorem 3.4 (Poincaré-Dulac) *If the spectrum of A, the linear part of $v \in \mathcal{V}$ at the singular point, belongs to the Poincaré domain then v and w are (orbitally) formally equivalent if and only if they are (orbitally) analytically equivalent.*

Theorem 3.5 (Poincaré-Dulac, real analytic version) *If the spectrum of A belongs to the Poincaré domain and v and w are the complexification of germs of real analytic vector fields being (orbitally) formally equivalent then they are (orbitally) real-analytically equivalent.*

Accurate proofs of Theorems 3.4 and 3.5 can be found in [1, 5, 14, 26].

Definition 3.6 (*Brjuno's condition*) A nonresonant pair $(\lambda_1, \lambda_2) \in \mathbb{C}^2$ satisfies *the Brjuno's condition* if there exist $C, \epsilon > 0$ such that

$$\left|\lambda_i - (m_1\lambda_1 + m_2\lambda_2)\right|^{-1} \leqslant C \exp(|m|^{1-\epsilon})$$

for $m := (m_1, m_2) \in \mathbb{N}^2$, with $|m| := m_1 + m_2$ large enough.

In what follows we formulate Brjuno's theorem and its real analytic version, whose proofs can be found in [1, 3, 14].

Theorem 3.7 (Brjuno) *If the spectrum of A, being nonresonant, belongs to the Siegel domain and satisfies the Brjuno's condition, then v is analytically linearizable. As a consequence, if v and w are (orbitally) formally equivalent then they are (orbitally) analytically equivalent.*

An earlier result is Siegel's theorem which has the same conclusion as Brjuno's theorem assuming that the spectrum of A is *Diophantine*. Even though this condition is more restrictive than the Brjuno's condition, the pairs in $\mathbb{C}^2$ which are not Diophantine have Lebesgue measure zero (see [1]).

Theorem 3.8 (Brjuno, real analytic version) *If v and w are the complexifications of germs of real analytic vector fields being (orbitally) formally equivalent, under the conditions of Brjuno's theorem, they are (orbitally) real-analytically equivalent.*

Finally, we briefly outline the main ideas of the proofs of the above results. Under the conditions of Theorems 3.4 and 3.7 we can conclude that the formal normal form of v is a polynomial vector field. Moreover, the formal transformation H mapping v into its formal normal form is convergent, since in these cases the small denominators decay no faster than $\frac{1}{C}\exp(|m|^{\epsilon-1})$ and as a consequence, the growth of the coefficients of H is controlled.

If v and w are (orbitally) formally equivalent, then we can suppose that they have the same k-jet for an arbitrary finite order k after a polynomial change of coordinates (multiplied by a polynomial with nonzero constant term if the equivalence is orbital), which is induced by a suitable finite jet of the formal transformations realizing the equivalence. As a consequence, they have the same formal normal form (by the uniqueness of the formal normal form). Therefore w will be analytically equivalent to this formal normal form, and in this way one can conclude Theorems 3.4 and 3.7.

The real analytic versions of Poincaré-Dulac's and Brjuno's theorems (Theorems 3.5 and 3.8) are proved by the previous arguments, after proving that the formal normal form of the complexification of a real analytic vector field has real coefficients, in the same way as the (orbital) formal change of coordinates which conjugates them.

3.2 Cases Where Formal Rigidity Fails

What happens when the conditions of Poincaré-Dulac's and Brjuno's theorems are not satisfied? We shall briefly mention some cases where this conditions are violated and as a consequence, the formal rigidity phenomenon does not take place: *Cremer saddles, resonant complex saddles, and complex saddle-nodes.*

We consider the germ of a holomorphic vector field at $\widehat{0} \in \mathbb{C}^2$. It will be called *complex saddle* if its linear part has nonzero eigenvalues belonging to the Siegel domain. By Hadamard-Perron's theorem this germ has two separatrices. After an analytic change of coordinates we may assume that these separatrices are the axes, and multiplying this latter vector field by a nonzero constant we can assume that its linear part is $\frac{\partial}{\partial x} + \lambda \frac{\partial}{\partial y}$, where λ is the respective quotient of the eigenvalues.

In what follows the holonomy map over a positive small loop on the x-axis will be referred as *the holonomy map of the complex saddle*. Note that the linear coefficient of this holonomy map is $\exp(2\pi i\lambda) \in \mathbb{S}^1$.

There exist complex saddles, called *Cremer saddles*, which are formally linearizable but are not analytically linearizable. The analytic classification of such complex saddles is unknown. Below we shall briefly indicate the arguments to guarantee the existence of this type of complex saddles.

The only analytic invariant of the complex saddles with the same linear part is their holonomy map [8, 14]. On the other hand we have the following realization theorem whose proof can be found in [25] or [9].

Theorem 3.9 *For any conformal germ $f(z) = \exp(2\pi i\phi)z + O(z^2)$, $\phi \in \mathbb{R}$ and any $\lambda < 0$ such that $\lambda \equiv \phi \mod \mathbb{Z}$, there exists a vector field with linear part $\frac{\partial}{\partial x} + \lambda\frac{\partial}{\partial y}$ whose holonomy map coincides with f.*

From the above results and the following theorem due to Yoccoz (see [35, 36]), the existence of the complex saddles which are formally linearizable but are not analytically linearizable follows.

Theorem 3.10 (Yoccoz) *If the complex number $\mu = \exp(2\pi i l)$, with $l \in \mathbb{R}$, violates the Brjuno's condition, then there exists a holomorphic germ at $0 \in \mathbb{C}$ with linear part μz which is not analytically linearizable.*

Recall that a complex number $\mu = \exp(2\pi i l)$, with $l \in \mathbb{R}$, satisfies the *Brjuno's condition* if it is not a root of unity and there exist $\epsilon, C > 0$ such that for all $k \in \mathbb{N}$, $\mid \mu^k - 1 \mid^{-1} < C\exp(k^{1-\epsilon})$.

Now we shall mention the case of *resonant complex saddles* (eigenvalues have rational negative ratio). The (orbital) formal classification of these saddles depends on scalar parameters (see [12, 13]).

Given a formal equivalence class of resonant complex saddles, there is a functional modulus of analytic classification which is known as *Ecalle–Voronin's modulus* (this modulus was discovered independently by Ecalle and Voronin, [6, 13, 18, 31, 33]). The existence of such functional moduli implies that the formal rigidity phenomenon cannot take place.

This modulus is obtained from the analytic classification of the holonomy maps of the resonant complex saddles, passing through the analytic classification of *parabolic germs* (i.e., holomorphic germs at $(\mathbb{C}, 0)$ with linear part equal to 1). A complete exposition of these results can be found in [14].

Finally we shall briefly discuss the germs of holomorphic vector fields at $\widehat{0} \in \mathbb{C}^2$ whose linear part has one zero and one nonzero eigenvalue. It will be called *saddle-node*.

A saddle-node is orbitally formally equivalent to

$$\frac{x^{r+1}}{1+ax^r}\frac{\partial}{\partial x}-y\frac{\partial}{\partial y}$$

for some $a \in \mathbb{C}$ (the coordinate axes are separatrices of these vector fields). In general a saddle-node does not have a holomorphic separatrix related to the zero eigenvalue. J. Martinet and J.-P. Ramis obtained the moduli of analytic classification for every formal equivalence class of saddle-nodes. Such moduli are functional moduli (equivalent to Ecalle–Voronin's moduli) and it is known as *Martinet-Ramis's moduli* [7, 13, 17, 18, 34]. Hence the orbital formal classification has one-dimensional moduli ($a \in \mathbb{C}$), and the orbital analytic classification has functional moduli. Thus, the formal rigidity phenomenon cannot take place.

Exceptionally, orbital formal rigidity phenomenon takes place for germs of vector fields having nilpotent linear part, that is, the orbital formal and orbital analytic classification coincide [4, 9, 16, 19, 29, 30].

4 Rigidity for Singularities with Zero Linear Part

We shall denote by $\mathcal{V}_n$ the class of germs of holomorphic vector fields with an isolated singularity at $\widehat{0} \in \mathbb{C}^2$, zero $(n-1)$-jet, and nonzero n-jet, for $n \geqslant 2$. The class $\mathcal{V}_n^{\mathbb{R}}$ is the intersection $\mathcal{V}_n \cap \mathcal{V}^{\mathbb{R}}$, that is, is the class of germs of complexifications of real analytic vector fields belonging to $\mathcal{V}_n$.

In [32] Voronin proved that the orbital formal rigidity phenomenon takes place for generic germs of vector fields in the class $\mathcal{V}_n$ (Theorem 4.1). Later Ortiz, Rosales, and Voronin proved that the formal rigidity phenomenon also takes place for generic germs of vector fields in $\mathcal{V}_n$ [20]. After that, they proved that formal rigidity and orbital formal rigidity takes place for generic dicritical germs of vector fields in $\mathcal{V}_n$, obtaining in addition the minimal invariants for the strict orbital analytic classification of such germs [21]. Recently they obtained the minimal invariants for strict orbital analytic classification of generic nondicritical germs [23]. It is important to mention that in order to obtain the minimal invariants, they constructed orbital formal normal forms for both dicritical and nondicritical germs, which are in fact analytic normal forms for generic cases [22, 24].

The main goal of the rest of this section is to give a detailed outline of the proof of Theorem 4.1 which includes Voronin's rigidity theorem [32] and its real analytic version proved recently in [15]. The presented outline is an adjustment of the proof appearing in [15].

Theorem 4.1 (Voronin-Jaurez) *Under generic conditions, $v, w \in \mathcal{V}_n$ are orbitally formally equivalent if and only if they are orbitally analytically equivalent.*

Whenever v and w are the complexifications of real analytic germs, they are orbitally real-formally equivalent if and only if they are orbitally real-analytically equivalent.

In Sect. 4.1 are introduced the basic notions used in the rest of this section. The generic conditions of Theorem 4.1 are specified in Sect. 4.2. The outline of the proof is achieved in Sect. 4.3, while Sect. 4.4 collects the fundamental results used in the proof of Theorem 4.1.

4.1 Basic Notions and Definitions

4.1.1 Complex and Real Möbius Band

The complex Möbius band $\mathbb{M}_{\mathbb{C}}$ is the closure in $\mathbb{C}^2 \times \mathbb{CP}^1$ of the set constituted by the pairs $((x, y), (x; y))$, where $(x; y)$ is the element in $\mathbb{CP}^1$ generated by the point $(x, y) \in \mathbb{C}^2 \setminus \{\widehat{0}\}$. The boundary of this subset in $\mathbb{C}^2 \times \mathbb{CP}^1$ is $\mathbb{D}_{\mathbb{C}} := \{\widehat{0}\} \times \mathbb{CP}^1$ called *the complex exceptional divisor*.

The complex Möbius band $\mathbb{M}_{\mathbb{C}}$ has an analytic manifold structure induced by the following embeddings

$$\begin{array}{ccc} \mathbb{C}^2 & \xrightarrow{\Phi} & \mathbb{C}^2 \times \mathbb{CP}^1 \\ (x, u) & \longmapsto & ((x, xu), (1; u)) \end{array} \quad , \quad \begin{array}{ccc} \mathbb{C}^2 & \xrightarrow{\Psi} & \mathbb{C}^2 \times \mathbb{CP}^1 \\ (v, y) & \longmapsto & ((vy, y), (v; 1)) \end{array}$$

The Cartesian projection $\pi : \mathbb{M}_{\mathbb{C}} \longrightarrow \mathbb{C}^2$ is called *the (standard) monoidal map* or the *blow-down*. Its inverse map $\pi^{-1} : \mathbb{C}^2 \setminus \{\widehat{0}\} \longrightarrow \mathbb{M}_{\mathbb{C}} \setminus \mathbb{D}_{\mathbb{C}}$ is defined as $(x, y) \mapsto ((x, y), (x; y))$. This map is known as *the complex blow-up*.

The set $\mathbb{M}_{\mathbb{C}} \cap \left(\mathbb{R}^2 \times \mathbb{RP}^1\right)$ will be called *the real Möbius band* and it will be denoted by $\mathbb{M}_{\mathbb{R}}$. Its boundary in $\mathbb{R}^2 \times \mathbb{RP}^1$ is $\{\widehat{0}\} \times \mathbb{RP}^1 = \mathbb{D}_{\mathbb{C}} \cap \left(\mathbb{R}^2 \times \mathbb{RP}^1\right) =: \mathbb{D}_{\mathbb{R}}$ and it will be called *the real exceptional divisor*.

In what follows, the map Φ (Ψ) together with its domain will be called *the coordinate chart* $(x, u = y/x)$ (*the coordinate chart* $(v = x/y, y)$).

4.1.2 Vanishing Holonomy Group

Definition 4.2 Given $v \in \mathcal{V}_n$, the foliation induced by v on $(\mathbb{C}^2, \widehat{0})$ will be denoted by $\mathscr{F}_v$. *The blow-up of* $\mathscr{F}_v$ (or *the blow-up of* v), denoted by $\widetilde{\mathscr{F}}_v$, is the foliation obtained by extending $\pi^{-1}(\mathscr{F}_v)$ on the complex exceptional divisor $\mathbb{D}_{\mathbb{C}}$ (see [14]).

Let (P_n, Q_n) be the n-jet of $v \in \mathcal{V}_n$. We shall say that $v \in \mathcal{V}_n$ *is nondicritical* if the homogeneous polynomial $x Q_n - y P_n$ does not vanishes identically, or equivalently, if the complex exceptional divisor $\mathbb{D}_{\mathbb{C}}$ is a separatrix of the foliation $\widetilde{\mathscr{F}}_v$. Otherwise v will be called *dicritical*.

Let $v \in \mathcal{V}_n$ be a nondicritical germ. In this case the complex exceptional divisor $\mathbb{D}_\mathbb{C}$ is an invariant set of the foliation $\widetilde{\mathscr{F}}_v$ which has (not necessarily different) $n+1$ singular points $\mathfrak{d}_1, \dots, \mathfrak{d}_{n+1}$ in $\mathbb{D}_\mathbb{C}$. Let $\mathfrak{d}_0, \mathfrak{d}_0' \in \mathbb{D}_\mathbb{C} \setminus \{\mathfrak{d}_1, \dots, \mathfrak{d}_{n+1}\}$ nonsingular points and $\gamma\colon [0,1] \longrightarrow \mathbb{D}_\mathbb{C} \setminus \{\mathfrak{d}_1, \dots, \mathfrak{d}_{n+1}\}$ be a path beginning at $\mathfrak{d}_0$ and ending at $\mathfrak{d}_0'$.

We shall consider $T_{\mathfrak{d}_0}$ and $T'_{\mathfrak{d}_0'}$ two germs of complex curves which are contained in $\mathbb{M}_\mathbb{C}$ and intersect $\mathbb{D}_\mathbb{C}$ transversally at $\mathfrak{d}_0$ and $\mathfrak{d}_0'$, respectively (i.e., $T_{\mathfrak{d}_0}$ *and* $T'_{\mathfrak{d}_0'}$ *are cross-sections to* $\mathbb{D}_\mathbb{C}$).

Definition 4.3 (*Correspondence map over a path*) The map obtained by the lift of the path γ on each leaf sufficiently close to the complex exceptional divisor $\mathbb{D}_\mathbb{C}$ will be called *the correspondence map for the foliation* $\widetilde{\mathscr{F}}_v^\mathbb{C}$ *over the path* γ. We shall denote this map by $\Delta_\gamma^v\colon \big(T_{\mathfrak{d}_0}, \mathfrak{d}_0\big) \longrightarrow \big(T'_{\mathfrak{d}_0'}, \mathfrak{d}_0'\big)$.

A careful exposition of the previous concept and its resulting properties can be found in [14].

It is important to notice that the correspondence map Δ_γ^v is a germ of biholomorphism whose inverse map is $\Delta_{\gamma^{-1}}^v : \big(T'_{\mathfrak{d}_0'}, \mathfrak{d}_0'\big) \longrightarrow \big(T_{\mathfrak{d}_0}, \mathfrak{d}_0\big)$, where γ^{-1} is defined as $t \mapsto \gamma(1-t)$.

Given $\boldsymbol{\alpha} \in \Pi_1\big(\mathbb{D}_\mathbb{C} \setminus \{\mathfrak{d}_1, \dots, \mathfrak{d}_{n+1}\}, \mathfrak{d}_0\big)$ and two representatives $\sigma, \widetilde{\sigma} \in \alpha$, it can be verified that $\Delta_\sigma^v = \Delta_{\widetilde{\sigma}}^v : \big(T_{\mathfrak{d}_0}, \mathfrak{d}_0\big) \longrightarrow \big(T'_{\mathfrak{d}_0}, \mathfrak{d}_0\big)$. As a consequence we shall denote by $\Delta_{\boldsymbol{\alpha}}^v : \big(T_{\mathfrak{d}_0}, \mathfrak{d}_0\big) \longrightarrow \big(T'_{\mathfrak{d}_0}, \mathfrak{d}_0\big)$ the correspondence map over any representative of α. In particular, if $T_{\mathfrak{d}_0} = T'_{\mathfrak{d}_0}$, the map $\Delta_{\boldsymbol{\alpha}}^v$ will be called *the holonomy map for the foliation* $\widetilde{\mathscr{F}}_v^\mathbb{C}$ *over* α.

Definition 4.4 (*Vanishing holonomy group*) The group of germs

$$\mathcal{G}_v = \Big\{ \Delta_{\boldsymbol{\alpha}}^v : \big(T_{\mathfrak{d}_0}, \mathfrak{d}_0\big) \longrightarrow \big(T_{\mathfrak{d}_0}, \mathfrak{d}_0\big) \,\big|\, \boldsymbol{\alpha} \in \Pi_1\big(\mathbb{D}_\mathbb{C} \setminus \{\mathfrak{d}_1, \dots, \mathfrak{d}_{n+1}\}, \mathfrak{d}_0\big) \Big\}$$

is called the *vanishing holonomy group of* $\widetilde{\mathscr{F}}_v^\mathbb{C}$ *on* $\big(T_{\mathfrak{d}_0}, \mathfrak{d}_0\big)$.

This group modulo a simultaneous conjugacy will be referred to as *holonomy group of* $\widetilde{\mathscr{F}}_v^\mathbb{C}$, being independent of a cross-section or even a base point.

4.2 Generic Conditions: Classes Σ_n and $\Sigma_n^\mathbb{R}$

A germ $v \in \mathcal{V}_n$ belongs to the class Σ_n if it satisfies the following conditions:

i. The germ v is nondicritic. Furthermore, if (P_n, Q_n) is n-jet of v, the homogeneous polynomial $xQ_n - yP_n$ has $n+1$ simple linear factors.

As a consequence, the foliation $\widetilde{\mathscr{F}}_v$ has $n+1$ pairwise different singular points $\mathfrak{d}_1, \cdots, \mathfrak{d}_{n+1}$ on the complex exceptional divisor $\mathbb{D}_\mathbb{C}$. Given $1 \le j \le n+1$ we consider λ_j^1, λ_j^2 the eigenvalues of the linear part of the foliation $\widetilde{\mathscr{F}}_v$ at the singular point $\mathfrak{d}_j$, being λ_j^2 the eigenvalue related to the complex exceptional divisor. It can verified that λ_j^2 is nonzero. The ratio $\lambda_j := \lambda_j^1/\lambda_j^2$ is *the characteristic number (or Camacho-Sad index) of the foliation* $\widetilde{\mathscr{F}}_v$ *at the singular point* $\mathfrak{d}_j$.

ii. For all $1 \le j \le n+1$, $\lambda_j \in \mathbb{C} \setminus \left(\mathbb{Q}_+ \cup \{\widehat{0}\}\right)$.
iii. The vanishing holonomy group of the foliation $\widetilde{\mathscr{F}}_v$ is nonsolvable.

The class $\Sigma_n^{\mathbb{R}}$ is defined as the intersection $\mathcal{V}_n^{\mathbb{R}} \cap \Sigma_n$, that is, a germ $v \in \mathcal{V}_n$ belongs to $\Sigma_n^{\mathbb{R}}$ if and only if v is the complexification of a real analytic vector field and satisfies the properties i, ii, and iii.

The conditions i and iii are generic on the class $\mathcal{V}_n$ ($\mathcal{V}_n^{\mathbb{R}}$) in the algebraic sense: the n-jet of a germ in $\mathcal{V}_n$ ($\mathcal{V}_n^{\mathbb{R}}$) violating the condition i satisfies a finite number of (real) polynomial identities, and if a germ in $\mathcal{V}_n$ ($\mathcal{V}_n^{\mathbb{R}}$) does not meet the condition iii, its $(n+2)$-jet satisfies a finite number of real polynomial identities (see [28]). In the sense of Lebesgue measure, the condition ii is satisfied by most germs in $\mathcal{V}_n^{\mathbb{R}}$ ($\mathcal{V}_n^{\mathbb{R}}$) whose blow-ups have singularities with positive characteristic numbers.

Remark 4.5 By the conditions i and ii, the blow-up of a germ v in Σ_n is the desingularization of the foliation $\mathscr{F}_v$. In what follows we refer to $\widetilde{\mathscr{F}}_v$ as *the desingularization of* v.

4.3 Detailed Outline of the Proof of Theorem 4.1

In what follows, $v, w \in \Sigma_n$ are orbitally formally equivalent, or orbitally real-formally equivalent if v and w are the complexifications of real analytic vector fields. The pair (H, K) realizes the equivalence as it is expressed by Eq. (2). By Lemma 4.6 we may assume without loss of generality that H has linear part equal to the identity, K has constant term 1, and moreover, v and w have the same separatrices $S_1, \ldots, S_{n+1}$. As a consequence, the desingularizations $\widetilde{\mathscr{F}}_v$ and $\widetilde{\mathscr{F}}_w$ have the same singular points $\mathfrak{p}_1, \ldots, \mathfrak{p}_{n+1} \in \mathbb{D}_{\mathbb{C}}$ together with their respective characteristic numbers. These singular points belong to the chart (x, u) except for a linear change of variables in v and w, which preserves the real plane whenever $v, w \in \Sigma_n^{\mathbb{R}}$.

Let $\mathfrak{p}_0 \in \mathbb{D}_{\mathbb{C}}$ be a nonsingular point belonging to the chart (x, u). We consider a cross-section $\Gamma_{\mathfrak{p}_0}$ to the complex exceptional divisor $\mathbb{D}_{\mathbb{C}}$ at the point $\mathfrak{p}_0$. Given a loop γ with base point $\mathfrak{p}_0$ and image contained in $\mathbb{D}_{\mathbb{C}} \setminus \{\mathfrak{p}_1, \ldots, \mathfrak{p}_{n+1}\}$, we shall denote by $\Delta_\gamma^v, \Delta_\gamma^w \colon \left(\Gamma_{\mathfrak{p}_0}, \mathfrak{p}_0\right) \longrightarrow \left(\Gamma_{\mathfrak{p}_0}, \mathfrak{p}_0\right)$ the holonomy maps for the foliations $\widetilde{\mathscr{F}}_v$ and $\widetilde{\mathscr{F}}_w$ over γ, respectively.

Theorems 4.7 and 4.8 state that there exists $h \colon \left(\Gamma_{\mathfrak{p}_0}, \mathfrak{p}_0\right) \longrightarrow \left(\Gamma_{\mathfrak{p}_0}, \mathfrak{p}_0\right)$ a holomorphic map with linear part equal to the identity such that

$$\Delta_\gamma^w = h \circ \Delta_\gamma^v \circ h^{-1}, \tag{4}$$

for all loop γ with base point $\mathfrak{p}_0$ and image contained in $\mathbb{D}_{\mathbb{C}} \setminus \{\mathfrak{p}_1, \ldots, \mathfrak{p}_{n+1}\}$. Moreover, if v and w are the complexifications of real analytic vector fields, $\mathfrak{p}_0 \in \mathbb{D}_{\mathbb{R}}$ and the cross-section $\Gamma_{\mathfrak{p}_0}$ is the complexification of a real analytic curve then h is the complexification of a real analytic diffeomorphism defined around $\widehat{0} \in \mathbb{R}^2$.

The main idea of the rest of the proof is to extend h as a biholomorphism defined on some neighborhood of the complex exceptional divisor, in such a way that the biholomorphism sends the leaves of the foliation $\widetilde{\mathscr{F}}_v$ into the leaves of $\widetilde{\mathscr{F}}_w$. In this way, the blow-down of the biholomorphism obtained will be an orbital analytic equivalence between v and w. Furthermore, we look for the invariance of the real Möbius band by the biholomorphism, whenever v and w are the complexifications of real analytic vector fields.

Below it is presented an heuristic discussion about the extension of h on a neighborhood of $\mathbb{D}_{\mathbb{C}}$ but far from the singular points $\mathfrak{p}_1, \ldots, \mathfrak{p}_{n+1}$. This discussion will serve as a motivation for the notion of *auxiliary foliation* and, at the same time, it will allow us to show the main problems appeared in the rest of the proof.

4.3.1 Heuristic Discussion on the Extension of *h*

Suppose that we obtained a dicritical holomorphic foliation $\widetilde{\widetilde{\mathscr{F}}}$ in $(\mathbb{M}_{\mathbb{C}}, \mathbb{D}_{\mathbb{C}})$ without singularities or tangency points, such that the desingularization of the separatrices are invariant subsets of this foliation. Since its leaves cross the complex exceptional divisor $\mathbb{D}_{\mathbb{C}}$ transversally, then we can extend the map h given by (4) by means of analytic continuation of the leaves of $\widetilde{\mathscr{F}}_\nu$ and $\widetilde{\mathscr{F}}_\omega$ along any path free from singularities. In this way one can construct a biholomorphism $\mathcal{H}$ far from the singular points. More specifically, if $\mathfrak{p}_0 \in \mathbb{D}_{\mathbb{C}}$ is a nonsingular point as above, $\mathfrak{q}_0 \in \mathbb{D}_{\mathbb{C}}$ is another nonsingular point, and $\mathscr{L}_{\mathfrak{p}_0}, \mathscr{L}_{\mathfrak{q}_0}$ are the leaves of the dicritical foliation $\widetilde{\widetilde{\mathscr{F}}}$ passing through $\mathfrak{p}_0$ and $\mathfrak{q}_0$, respectively, then

$$\mathcal{H}|_{\mathscr{L}_{\mathfrak{q}_0}} := \Delta^{\omega}_{\tau} \circ h \circ \Delta^{\nu}_{\tau^{-1}}, \tag{5}$$

where τ is any path contained in $\mathbb{D}_{\mathbb{C}} \setminus \{\mathfrak{p}_1, \ldots, \mathfrak{p}_{n+1}\}$, beginning at $\mathfrak{p}_0$, and ending at $\mathfrak{q}_0$, and $\Delta^{v}_{\tau}, \Delta^{w}_{\tau} \colon \left(\mathscr{L}_{\mathfrak{p}_0}, \mathfrak{p}_0\right) \longrightarrow \left(\mathscr{L}_{\mathfrak{q}_0}, \mathfrak{q}_0\right)$ are the correspondence maps for the foliations $\widetilde{\mathscr{F}}_v$ and $\widetilde{\mathscr{F}}_w$ over τ, respectively. As a consequence of the conjugation (4), $\mathcal{H}$ is a well defined biholomorphism outside the desingularization of the separatrices.

In such a situation we would need to holomorphically extend $\mathcal{H}$ to a neighborhood of the desingularization of each separatrix. In this way we would conclude that the blow-down of this transformation onto $\mathbb{C}^2$ conjugates the foliation $\mathscr{F}_\nu$ with the foliation $\mathscr{F}_\omega$.

Moreover, we would like to prove that the biholomorphism $\mathcal{H}$ leaves invariant the real Möbius band, whenever v and w are the complexifications of real analytic vector fields and $\mathfrak{p}_0 \in \mathbb{D}_{\mathbb{R}}$, since in this way the blow-down of this biholomorphism leaves invariant the real plane $\mathbb{R}^2$.

Unfortunately, the existence of such a dicritical holomorphic foliation in $(\mathbb{M}_{\mathbb{C}}, \mathbb{D}_{\mathbb{C}})$ is equivalent to the existence of a holomorphic change of coordinates which simultaneously rectifies the separatrices (that is, it transforms all separatrices into straight lines), but in general this is impossible (see [10]). However, the assumptions on the dicritical foliation can be considerably relaxed, as explained below.

4.3.2 Extension of h Using an Auxiliary Foliation

Lemma 4.9 states that given $\mathfrak{q}_1, \ldots, \mathfrak{q}_{n-2} \in \mathbb{D}_{\mathbb{C}}$ pairwise different nonsingular points belonging to the chart (x, u), there exists a dicritical holomorphic vector field X defined around $\widehat{0} \in \mathbb{C}^2$ whose blow-up $\widetilde{\mathscr{F}}_X$, called *auxiliary foliation*, satisfies the following properties: the desingularization of the separatrices $S_1, \ldots, S_{n+1}$ are invariant subsets of $\widetilde{\mathscr{F}}_X$, its leaves intersect $\mathbb{D}_{\mathbb{C}}$ at all its points transversally, except for $\mathfrak{q}_1, \ldots, \mathfrak{q}_{n-2}$, where the leaves have simple tangencies with $\mathbb{D}_{\mathbb{C}}$. As a consequence, $\widetilde{\mathscr{F}}_X$ does not have any singularity on $(\mathbb{M}_{\mathbb{C}}, \mathbb{D}_{\mathbb{C}})$.

The auxiliary foliation provides a way to extend the transversal conjugation h outside the locus of tangency between the foliation $\widetilde{\mathscr{F}}_v$ and the auxiliary foliation $\widetilde{\mathscr{F}}_X$, as in Eq. (5). This locus consists of the desingularization of the separatrices $S_1, \ldots, S_{n+1}$ and $n-2$ smooth analytic curves $T_1, \ldots, T_{n-2}$ crossing the exceptional divisor $\mathbb{D}_{\mathbb{C}}$ transversally at the points $\mathfrak{q}_1, \ldots, \mathfrak{q}_{n-2}$, respectively (Lemma 4.10). The last curves are called *polar curves*.

It is important to mention that, whenever v and w are the complexifications of real analytic vector fields there are other special properties: if $\mathfrak{q}_1, \ldots, \mathfrak{q}_{n-2} \in \mathbb{D}_{\mathbb{R}}$ there exists an auxiliary foliation induced by the complexification of a real analytic vector field and in this case, the polar curves $T_1, \ldots, T_{n-2}$ are the complexification of real analytic curves.

In what follows we shall denote by $\mathcal{H}$ the extension of h far from the separatrices and the polar curves, as in (5). It should be mentioned that the extension of $\mathcal{H}$ will send the leaves of the auxiliary foliation $\widetilde{\mathscr{F}}_X$ into themselves.

4.3.3 Preparation for the Extension of $\mathcal{H}$

Let $\mathfrak{p}_0 \in \mathbb{D}_{\mathbb{C}}$ be a fixed nonsingular point such that the leaf $\mathscr{L}_{\mathfrak{p}_0}$ of the auxiliary foliation passing through $\mathfrak{p}_0$ is transversal to the exceptional divisor $\mathbb{D}_{\mathbb{C}}$. Given $j = 1, \ldots, n-2$ we choose a path γ_j contained in $\mathbb{D}_{\mathbb{C}} \setminus \{\mathfrak{p}_1, \ldots, \mathfrak{q}_{n+1}\}$, beginning at $\mathfrak{q}_j$, and ending at $\mathfrak{q}_0$. We denote by $\Delta^v_{\gamma_j}, \Delta^w_{\gamma_j} : (T_j, \mathfrak{q}_j) \longrightarrow (\mathscr{L}_{\mathfrak{p}_0}, \mathfrak{p}_0)$ the correspondence maps over γ_j for the foliations $\widetilde{\mathscr{F}}_v$ and $\widetilde{\mathscr{F}}_w$, respectively.

In [32] it is proved that, except for a holomorphic change of variables in w whose linear part is the identity, the locus of tangency points of the foliations $(\widetilde{\mathscr{F}}_w, \widetilde{\mathscr{F}}_X)$ coincides with the locus of tangency points of the pair $(\widetilde{\mathscr{F}}_v, \widetilde{\mathscr{F}}_X)$ (that is, it is the union of the separatrices $S_1, \ldots, S_{n+1}$ and the polar curves $T_1, \ldots, T_{n-2}$). Moreover, the biholomorphism $h\colon (\mathscr{L}_{\mathfrak{p}_0}, \mathfrak{p}_0) \longrightarrow (\mathscr{L}_{\mathfrak{p}_0}, \mathfrak{p}_0)$ conjugating the holonomy maps as in (4) is the identity map, and the correspondence maps $\Delta^v_{\gamma_j}$ and $\Delta^w_{\gamma_j}$ coincide for all $j = 1, \ldots, n-2$. In [15] it is proved that, whenever v and w are the complexifications of real analytic vector fields and $\mathfrak{p}_0$ belongs to $\mathbb{D}_{\mathbb{R}}$, the holomorphic change of variables is the complexification of a real analytic diffeomorphism.

In this way, for all $\mathfrak{q}_0 \in \mathbb{D}_{\mathbb{C}}$ nonsingular point such that the leaf $\mathscr{L}_{\mathfrak{q}_0}$ of the auxiliary foliation passing through $\mathfrak{q}_0$ is transversal to the complex exceptional divisor $\mathbb{M}_{\mathbb{C}}$, the restriction of $\mathcal{H}$ on $\mathscr{L}_{\mathfrak{q}_0}$ is

$$\mathcal{H}\big|_{\mathscr{L}_{\mathfrak{q}_0}} = \Delta^{w}_{\tau} \circ \Delta^{v}_{\tau^{-1}}\,, \tag{6}$$

where τ is a path contained in $\mathbb{D}_{\mathbb{C}} \setminus \{\mathfrak{p}_1, \ldots, \mathfrak{p}_{n+1}\}$, beginning at $\mathfrak{p}_0$, and ending at $\mathfrak{q}_0$, and $\Delta^{w}_{\tau}, \Delta^{v}_{\tau} \colon (\mathscr{L}_{\mathfrak{p}_0}, \mathfrak{p}_0) \longrightarrow (\mathscr{L}_{\mathfrak{q}_0}, \mathfrak{q}_0)$ are the correspondence maps for the foliations $\widetilde{\mathscr{F}}_v$ and $\widetilde{\mathscr{F}}_w$ over τ, respectively.

4.3.4 Extension of $\mathcal{H}$ Around Polar Curves and Separatrices

Now we shall briefly discuss the key arguments used to extend $\mathcal{H}$ around the polar curves and the separatrices.

By (6), the transformation $\mathcal{H}$ is the identity on $\mathbb{D}_{\mathbb{C}}$, except for finitely many points. As a consequence, its extension should send every leaf of the auxiliary foliation $\widetilde{\mathscr{F}}_X$ and every polar curve T_j onto itself. Moreover, $\mathcal{H}|_{T_j}$ should be the identity map, since $\Delta^{w}_{\gamma_j} \circ \Delta^{v}_{\gamma_j^{-1}}$ is the identity map.

To extend $\mathcal{H}$ around the polar curve T_j are considered local models of foliations $(\mathscr{F}, \mathscr{F}_g)$, where $\mathscr{F}$ is the trivial foliation $\{x = c\}_{c\in(\mathbb{C},0)}$ and $\mathscr{F}_g$ is the foliation $\{x + g(u) = c\}_{c\in(\mathbb{C},0)}$, with g a nonconstant holomorphic map at $0 \in \mathbb{C}$ satisfying $g(0) = g'(0) = 0$. Every parametrization of the polar curve T_j determines uniquely the local models and the holomorphic change of coordinates defined around $\mathfrak{q}_j \in \mathbb{D}_{\mathbb{C}}$ sending the pairs of foliations $(\widetilde{\mathscr{F}}_v, \widetilde{\mathscr{F}}_X)$ and $(\widetilde{\mathscr{F}}_w, \widetilde{\mathscr{F}}_X)$ to the respective models. In [15, 32] it is proved that, given a parametrization of the polar curve T_j, the models coincide and the biholomorphism $\mathcal{H}$ is extended as the identity transformation with respect to the fixed model.

Now we shall discuss the extension of $\mathcal{H}$ around the separatrices. The characteristic number corresponding to the singular point $\mathfrak{p}_i$ is denoted by λ_i, for all $i = 1, \ldots, n + 1$.

Since the complex saddles have the property of saturation of short cross-sections and, as a consequence, the holonomy maps are the moduli for the orbital analytic classification of germs of complex saddles (see [8] or [14]), whenever λ_i is a negative number, the transformation $\mathcal{H}$ is holomorphically extended at $\mathfrak{p}_i$.

In the case of λ_i belongs to $\mathbb{C} \setminus (\mathbb{R}_- \cup \mathbb{Q}_+ \cup \{0\})$, the pairs of foliations $(\widetilde{\mathscr{F}}_v, \widetilde{\mathscr{F}}_X)$ and $(\widetilde{\mathscr{F}}_w, \widetilde{\mathscr{F}}_X)$ are identified with the local model $(\mathscr{F}_i, \{u = c\}_{c\in(\mathbb{C},0)})$ where $\mathscr{F}_i$ is the foliation induced by the linear vector field $\lambda_i \times \frac{\partial}{\partial x} + u\frac{\partial}{\partial u}$. These identifications, induced by local changes of coordinates, are consequences of the Poincaré's linearization theorem. With respect to the described model, $\mathcal{H}$ is extended around $\mathfrak{p}_i$ by the transformation $(x, u) \mapsto (\alpha_i\, x, u)$ for some $\alpha_i \in \mathbb{C} \setminus \{0\}$ [15, 32].

4.3.5 Invariance of the Real Analytic Möbius Band $\mathbb{M}_{\mathbb{R}}$

In what follows, v and w will be the complexifications of real analytic vector fields. It will be proved that in this case the biholomorphism $\mathcal{H}$ constructed above, leaves invariant the real Möbius band $\mathbb{M}_{\mathbb{R}}$. For that purpose we shall recall some key properties used in the construction.

The auxiliary foliation $\widetilde{\mathscr{F}}_X$ of v is the blow-up of the complexification of a real analytic vector field, and the polar curves $T_1, \ldots, T_{n-2}$ of the pair $(\widetilde{\mathscr{F}}_v, \widetilde{\mathscr{F}}_X)$ are the complexification of $n-2$ smooth real analytic curves, which cross the exceptional divisor $\mathbb{D}_{\mathbb{C}}$ transversally at the tangency points $\mathfrak{q}_1, \ldots, \mathfrak{q}_{n-2} \in \mathbb{D}_{\mathbb{R}}$ (Lemmas 4.9 and 4.10).

Let $\mathfrak{p}_0 \in \mathbb{D}_{\mathbb{R}}$ be a nonsingular point such that the leaf $\mathscr{L}_{\mathfrak{p}_0}$ of the auxiliary foliation cross the exceptional divisor $\mathbb{D}_{\mathbb{C}}$ transversally. In [15] it is proved that, except for a change of variables in w which is the complexification of a real analytic diffeomorphism with linear part equal to the identity, we may assume without loss of generality that w and v have the same separatrices, the pairs $(\widetilde{\mathscr{F}}_v, \widetilde{\mathscr{F}}_X)$ and $(\widetilde{\mathscr{F}}_w, \widetilde{\mathscr{F}}_X)$ have the same locus of tangency points, and the conjugation between the holonomy groups of v and w on $(\mathscr{L}_{\mathfrak{p}_0}, \mathfrak{p}_0)$ is the identity (that is, for all loop γ with base point $\mathfrak{p}_0$ and image contained in $\mathbb{D}_{\mathbb{C}} \setminus \{\mathfrak{p}_1, \ldots, \mathfrak{p}_{n+1}\}$, the holonomy maps on $(\mathscr{L}_{\mathfrak{p}_0}, \mathfrak{p}_0)$ for the foliations $\widetilde{\mathscr{F}}_v$ and $\widetilde{\mathscr{F}}_w$ over γ, coincide).

The biholomorphism $\mathcal{H}$ conjugating $\widetilde{\mathscr{F}}_v$ with $\widetilde{\mathscr{F}}_w$ is defined on a neighborhood of the complex exceptional divisor $\mathbb{D}_{\mathbb{C}}$ in the complex Möbius band $\mathbb{M}_{\mathbb{C}}$ and preserves the auxiliary foliation $\widetilde{\mathscr{F}}_X$. Far from the singular points $\mathfrak{p}_1, \ldots, \mathfrak{p}_{n+1}$ and the tangency points $\mathfrak{q}_1, \ldots, \mathfrak{q}_{n-2}$, the biholomorphism is defined in the following way: given a nonsingular point $\mathfrak{q}_0 \in \mathbb{D}_{\mathbb{C}}$ such that the leaf of the auxiliary foliation $\mathscr{L}_{\mathfrak{q}_0}$ passing through $\mathfrak{q}_0$ crosses the exceptional divisor $\mathbb{D}_{\mathbb{C}}$ transversally, then the restriction of $\mathcal{H}$ on the leaf $\mathscr{L}_{\mathfrak{q}_0}$ is defined as

$$\mathcal{H}\big|_{\mathscr{L}_{\mathfrak{q}_0}} := \Delta^{w}_{\gamma_{\mathfrak{q}_0}} \circ \Delta^{v}_{\gamma^{-1}_{\mathfrak{q}_0}},$$

where $\gamma_{\mathfrak{q}_0}$ is a path contained in $\mathbb{D}_{\mathbb{C}} \setminus \{\mathfrak{p}_1, \ldots, \mathfrak{p}_{n+1}\}$, beginning at $\mathfrak{p}_0$, and ending at $\mathfrak{q}_0$, and $\Delta^{v}_{\gamma_{\mathfrak{q}_0}}, \Delta^{w}_{\gamma_{\mathfrak{q}_0}} : \left(\mathscr{L}_{\mathfrak{p}_0}, \mathfrak{p}_0\right) \longrightarrow \left(\mathscr{L}_{\mathfrak{q}_0}, \mathfrak{q}_0\right)$ are the correspondence maps over $\gamma_{\mathfrak{q}_0}$ for the foliations $\widetilde{\mathscr{F}}_v$ and $\widetilde{\mathscr{F}}_w$, respectively.

To prove the biholomorphism $\mathcal{H}$ leaves invariant the real Möbius band, we shall stress that it is enough to prove the existence of a neighborhood of $\mathfrak{p}_0$ where the biholomorphism leaves invariant the real Möbius band $\mathbb{M}_{\mathbb{R}}$, that is, $\mathcal{H}(x_0, u_0) = \overline{\mathcal{H}(\overline{x}_0, \overline{u}_0)}$ for all (x_0, y_0) in a neighborhood of $(0, p_0)$, where $u(\mathfrak{p}_0) = p_0$.

Without loss of generality we may assume that $\mathcal{H}$ is defined on $U \subseteq \mathbb{C}^2$ a connected open neighborhood of the axis $\{x = 0\}$ which is invariant by complex conjugation. Since the map $(x, u) \mapsto \overline{\mathcal{H}(\bar{x}, \bar{u})}$ is a well defined holomorphic map in U, then the equality $\mathcal{H}(x_0, u_0) = \overline{\mathcal{H}(\bar{x}_0, \bar{u}_0)}$ for (x_0, u_0) in a neighborhood of $(0, p_0)$ implies that the same happens all over U (see Identity Theorem, [11]). Hence, for $(a, b) \in \mathbb{R}^2 \cap U$, $\mathcal{H}(a, b) = \overline{\mathcal{H}\left(\overline{a}, \overline{b}\right)} = \overline{\mathcal{H}(a, b)}$. Therefore for any $\mathfrak{q} \in \mathbb{D}_{\mathbb{R}}$ which belongs to the coordinate chart (x, u), $\mathscr{L}_{\mathfrak{q}} \cap \mathbb{M}_{\mathbb{R}}$ is invariant by $\mathcal{H}$. In the coordinate chart $(v = x/y, y)$ we denote by $\mathscr{L}_{\infty}$ the leaf of the auxiliary foliation passing through $(v, y) = (0, 0)$. For $\mathfrak{p} \in \mathscr{L}_{\infty} \cap \mathbb{M}_{\mathbb{R}}$ we take a sequence of points $(\widetilde{\mathfrak{p}}_i)_{i\in\mathbb{N}}$ tending to $\mathfrak{p}$, such that $\widetilde{\mathfrak{p}}_i \in \mathscr{L}_{\widetilde{\mathfrak{q}}_i} \cap \mathbb{M}_{\mathbb{R}}$ for $\widetilde{\mathfrak{q}}_i \in \mathbb{D}_{\mathbb{R}}$ belonging to the chart (x, u). The continuity of $\mathcal{H}$ implies that $\left(\mathcal{H}(\widetilde{\mathfrak{p}}_i)\right)_{i\in\mathbb{N}} = \mathcal{H}(\mathfrak{p})$. Since in (v, y) this is a Cauchy sequence in $\mathbb{R}^2$, then $\mathcal{H}(\mathfrak{p}) \in \mathbb{R}^2$. Thus $\mathscr{L}_{\infty} \cap \mathbb{M}_{\mathbb{R}}$ is invariant by $\mathcal{H}$.

It remains to prove the existence of a neighborhood of $\mathfrak{p}_0$ in $\mathbb{M}_{\mathbb{C}}$ in which $\mathcal{H}$ leaves invariant the real Möbius band $\mathbb{M}_{\mathbb{R}}$. We consider $\mathfrak{q} \in \mathbb{D}_{\mathbb{R}}$ belonging to (x, u), with $u(\mathfrak{q}) = q$, such that the path $\delta_{\mathfrak{q}}$ defined as $t \mapsto (0, tq + (1-t)p_0)$ with respect to the coordinate chart (x, u), does not pass through any singular point. Given that v and w are the complexifications of real analytic vector fields, the restrictions of the correspondence maps $\Delta^{v}_{\delta_{\mathfrak{q}}}$ and $\Delta^{w}_{\delta_{\mathfrak{q}}}$ on $\mathscr{L}_{\mathfrak{p}_0} \cap \mathbb{M}_{\mathbb{R}}$ are the continuation of solutions of the real analytic vector fields describing the desingularizations of v and w. As a consequence, $\Delta^{v}_{\delta_{\mathfrak{q}}}$ and $\Delta^{w}_{\delta_{\mathfrak{q}}}$ take the intersection $\mathscr{L}_{\mathfrak{p}_0} \cap \mathbb{M}_{\mathbb{R}}$ into $\mathscr{L}_{\mathfrak{q}} \cap \mathbb{M}_{\mathbb{R}}$, and so, $\mathcal{H}\big|_{\mathscr{L}_{\mathfrak{q}}} = \Delta^{\omega}_{\delta_{\mathfrak{q}}} \circ \Delta^{\nu}_{\delta^{-1}_{\mathfrak{q}}}$ leaves invariant $\mathscr{L}_{\mathfrak{q}} \cap \mathbb{M}_{\mathbb{R}}$. Since $\mathfrak{p}_0$ is a nonsingular point, there is a sufficiently small connected neighborhood of $\mathfrak{p}_0$ in $\mathbb{M}_{\mathbb{C}}$ such that any point in the intersection of this neighborhood with $\mathbb{M}_{\mathbb{R}}$ satisfies the previous condition. Therefore in this neighborhood $\mathcal{H}$ leaves invariant the real Möbius band $\mathbb{M}_{\mathbb{R}}$.

4.4 Fundamental Results Used in the Proof of Theorem 4.1

4.4.1 Equivalence of Separatrices

Let $v, w \in \Sigma_n$ be orbitally formally equivalent holomorphic vector fields. Furthermore, if $v, w \in \Sigma^{\mathbb{R}}_n$ then they are orbitally real-formally equivalent.

We denote the singular points of $\widetilde{\mathscr{F}}_v$ in the complex exceptional divisor $\mathbb{D}_{\mathbb{C}}$ by $\mathfrak{p}_1, \ldots, \mathfrak{p}_{n+1}$. These singularities belong to the chart $(x, u = y/x)$ except for a linear change of variables in v, which preserves the real plane whenever v is the complexification of a real analytic vector field.

Lemma 4.6 *There exists w_1 an orbital analytic equivalent vector field to w with the same separatrices. Moreover, the invertible formal transformation and the scalar formal series realizing the equivalence between w_1 and v have linear part equal to the identity and constant term* 1, *respectively. If v and w are the complexifications of real analytic vector fields, then w_1 is orbitally real-analytical equivalent to w.*

Sketch of the proof of Lemma 4.6. We consider $H \in (\mathbb{C}[[x, y]])^2$ the invertible formal transformation and the formal series $K \in \mathbb{C}[[x, y]]$ with nonzero constant term satisfying

$$w = K\left[(\mathrm{D}H \cdot v) \circ H^{-1}\right].$$

Given $m := n + 2$, we denote H_m the m-jet of H and K_{m-1} the $(m-1)$-jet of $1/K \circ H_m^{-1}$. The holomorphic vector field

$$w_0 := K_{m-1} \left[(\mathrm{D}\, H_m \cdot w) \circ H_m^{-1} \right]$$

belongs to the class Σ_n and has the same $2n+2$ as v. As a consequence, if the curves $\{y = \phi_i(x)\}$ and $y = \{y = \psi_i(x)\}$ are the separatrices of v and w_0 whose desingularization intersect $\mathbb{D}_{\mathbb{C}}$ at the singular point $\mathfrak{p}_i$, then the difference $\phi_i(x) - \psi_i(x)$ has order at 0 greater or equal to $n+2$.

Thus the transformation $G(x, y) = (x, y + g(x, y))$ defined around $\widehat{0} \in \mathbb{C}^2$, where $g(0, y) = 0$ and

$$g(x, y) = \sum_{i=1}^{n+1} \frac{(\phi_i(x) - \psi_i(x)) \prod_{j \neq i} (y - \psi_j(x))}{\prod_{j \neq i} (\psi_i(x) - \psi_j(x))}, \qquad \text{when} \quad x \neq 0,$$

is an invertible holomorphic function with linear part equal to the identity. Then we define w_1 as $(\mathrm{D}\, G \cdot w_0) \circ G^{-1}$.

If $v, w \in \Sigma_n^{\mathbb{R}}$ are orbitally real-analytically equivalent, then $w_0 \in \Sigma_n^{\mathbb{R}}$ is orbital real-analytic equivalent to w and the Taylor series of G has real coefficients since for each summand there is another term conjugated to it. Therefore, w_1 belongs to $\Sigma_n^{\mathbb{R}}$. ■

As a consequence of Lemma 4.6, we can suppose without loss of generality that v and w have the same separatrices, and their desingularizations $\widetilde{\mathscr{F}}_v$ and $\widetilde{\mathscr{F}}_w$ have the same singular points $\mathfrak{p}_1, \ldots, \mathfrak{p}_{n+1} \in \mathbb{D}_{\mathbb{C}}$ and their respective characteristic numbers.

4.4.2 Formal and Analytic Conjugation of Holonomy Groups

Let $\mathfrak{p}_0 \in \mathbb{D}_{\mathbb{C}}$ be a nonsingular point belonging to the coordinate chart (x, u) and $\Gamma_0 = \{u = p_0 := u(\mathfrak{p}_0)\}$ be the straight line passing through $\mathfrak{p}_0$. Given a loop γ with base point $\mathfrak{p}_0$ whose image is contained in $\mathbb{D}_{\mathbb{C}} \setminus \{\mathfrak{p}_1, \ldots, \mathfrak{p}_{n+1}\}$, we shall denote by $\Delta_\gamma^v, \Delta_\gamma^w \colon (\Gamma_0, \mathfrak{p}_0) \longrightarrow (\Gamma_0, \mathfrak{p}_0)$ the holonomy maps for the foliations $\widetilde{\mathscr{F}}_v$ y $\widetilde{\mathscr{F}}_w$ over γ, respectively.

Theorem 4.7 *There exists a formal transformation $\rho \in \mathbb{C}[[x]]$ with zero constant term and linear part x such that for all loop γ with base point $\mathfrak{p}_0$ and image contained in $\mathbb{D}_{\mathbb{C}} \setminus \{\mathfrak{p}_1, \ldots, \mathfrak{p}_{n+1}\}$,*

$$\rho \circ \Delta_\gamma^v \circ \rho^{-1} = \Delta_\gamma^w . \tag{7}$$

The invertible formal transformation ρ belongs to $\mathbb{R}[[x]]$ whenever v and w are the complexification of real analytic vector fields and $\mathfrak{p}_0 \in \mathbb{D}_{\mathbb{R}}$.

As a consequence of the following formal rigidity theorem for nonsolvable finitely generated groups of germs of holomorphic self-maps on $(\mathbb{C}, 0)$ due to D. Cerveau, R. Moussu [4] and J. P. Ramis [27], ρ is convergent, that is, ρ is the germ of a holomorphic self-map. Moreover, it is the complexification of the germ of an invertible real analytic diffeomorphism at $(\mathbb{R}, 0)$ provided that v and w are the complexification of real analytic vector fields and $\mathfrak{p}_0 \in \mathbb{D}_{\mathbb{R}}$.

Theorem 4.8 *Let G and H be nonsolvable finitely generated groups of germs of holomorphic self-maps on $(\mathbb{C}, 0)$. Suppose that there exists an invertible formal transformation with zero constant term $h_0 \in C[[z]]$ which conjugates G with H, that is*

$$H = h_0 G h_0^{-1} := \left\{h_0 \circ g \circ h_0^{-1} \mid g \in G\right\} .$$

Then h_0 is a convergent series.

In what follows we shall construct ρ stated in Theorem 4.7 for the purpose of showing that it belongs to $\mathbb{R}[[x]]$ whenever $v, w \in \Sigma_n^{\mathbb{R}}$ and $\mathfrak{p}_0 \in \mathbb{D}_{\mathbb{R}}$.

We consider the invertible formal transformation with complex coefficients and linear part equal to the identity $H(x, y) = (H_1(x, y), H_2(x, y))$ which conjugates v with w orbitally. As a consequence, $\widetilde{H}$, the blow-up of H, conjugates the foliation $\widetilde{\mathscr{F}}_v$ with the foliation $\widetilde{\mathscr{F}}_w$. In the coordinate chart $(x, u = y/x)$ the transformation $\widetilde{H}$ is defined as follows.

$$\widetilde{H}(x, u) = \left(\widetilde{H}_1, \widetilde{H}_2\right)(x, u) := \left(H_1(x, xu), \tfrac{H_2(x,xu)}{H_1(x,xu)} .\right)$$

We define the formal curve $\widehat{\Gamma}_0 := \left\{(x, u) = \left(\widetilde{H}_1(x, p_0), \widetilde{H}_2(x, p_0)\right)\right\}$ which is a formal cross-section to the exceptional divisor $\mathbb{D}_{\mathbb{C}}$ at the point $\mathfrak{p}_0$.

Let $\widetilde{v}$ and $\widetilde{w} = (\widetilde{w}_1, \widetilde{w}_2)$ be the vector fields in (x, u) which induce the foliations $\widetilde{\mathscr{F}}_v$ and $\widetilde{\mathscr{F}}_w$, respectively. We define the nonautonomous differential equation

$$\frac{dx}{du} = \frac{\widetilde{w}_1(x, u)}{\widetilde{w}_2(x, u)} = \sum_{k \geqslant 1} S_k(u) x^k , \qquad (8)$$

where S_k is a rational function in u with poles at $u(\mathfrak{p}_1), \ldots, u(\mathfrak{p}_{n+1})$. Then the extended phase portrait of the differential equation (8) coincides with the foliation $\widetilde{\mathscr{F}}_w$ in a neighborhood of the axis $\{x = 0\}$ without the separatrices of $\widetilde{\mathscr{F}}_w$.

Let $U, V \subseteq \mathbb{C}$ be open discs centered at p_0 and 0, respectively, such that for all $\widetilde{u} \in U, x \in V$ the complex trajectory $u \mapsto \mathrm{X}(u, \widetilde{u}, x)$ is the solution of Eq. (8) with initial condition $x = \mathrm{X}(\widetilde{u}, \widetilde{u}, x)$. Then the flow map X is defined on $U^2 \times V$.

We define the formal series $\rho(x) := \mathrm{X}\left(p_0, \widetilde{H}_2(x, p_0), \widetilde{H}_1(x, p_0)\right)$ which has linear part equal to the identity and as a consequence, it is an invertible formal transformation. It is proved that such transformation conjugates the holonomy group of v on $(\Gamma_0, \mathfrak{p}_0)$ with the holonomy group of w on $(\Gamma_0, \mathfrak{p}_0)$, as follows:

$$\rho \circ \Delta_\gamma^v \circ \rho^{-1} = \Delta_\gamma^w ,$$

for all loop γ with based point $\mathfrak{p}_0$ whose image is contained in $\mathbb{D}_C \setminus \{\mathfrak{p}_1, \ldots, \mathfrak{p}_{n+1}\}$.

If $v, w \in \Sigma_n^{\mathbb{R}}$ and $\mathfrak{p}_0 \in \mathbb{D}_{\mathbb{R}}$, $\widetilde{H}$ has real coefficients and X is the complexification of a real analytic map. As a result, ρ belongs to $\mathbb{R}[[x]]$.

4.4.3 Auxiliary Foliation and Polar Curves

This part is devoted to give a sketch of the proof of the existence of an auxiliary foliation. At the end the notion of polar curve is introduced.

We consider $\vartheta \in \Sigma_n$ and its separatrices $C_1, \ldots, C_{n+1}$. The singular points of the desingularization $\widetilde{\mathcal{F}}_\vartheta$ belong to the chart (x, u) except for a linear change of variables in ϑ, which preserves the real plane whenever ϑ is the complexification of a real analytic vector field. Let $\widetilde{\mathfrak{q}}_1, \ldots, \widetilde{\mathfrak{q}}_{n-2} \in \mathbb{D}_\mathbb{C}$ be $n-2$ pairwise different nonsingular points belonging to the chart (x, u).

Lemma 4.9 *There exists a dicritical holomorphic vector field Y at $(\mathbb{C}^2, \widehat{0})$ whose blow-up $\widetilde{\mathcal{F}}_Y$ is a holomorphic foliation on $(\mathbb{M}_\mathbb{C}, \mathbb{D}_\mathbb{C})$ satisfying the following conditions:*

1. *The desingularization of the separatrices $C_1, \ldots, C_{n+1}$ are invariant subsets of $\widetilde{\mathcal{F}}_Y$.*
2. *For all $\mathfrak{q} \in \mathbb{D}_\mathbb{C} \setminus \{\widetilde{\mathfrak{q}}_1, \ldots, \widetilde{\mathfrak{q}}_{n-2}\}$, the leaf of the foliation $\widetilde{\mathcal{F}}_Y$ passing through $\mathfrak{q}$ intersects the exceptional divisor $\mathbb{D}_\mathbb{C}$ transversally.*
3. *Around the point $\widetilde{\mathfrak{q}}_j$ the foliation $\widetilde{\mathcal{F}}_Y$ is described by the ordinary differential equation*

$$\frac{dx}{du} = f_j(u) + g_j(x, u) ,$$

where f_j, g_j are holomorphic analytic functions defined around $u(\widetilde{\mathfrak{q}}_j) = \widetilde{q}_j$ and $(0, \widetilde{q}_j)$, respectively. Moreover, $g_j(x, u) = O(x)$ and $\operatorname{ord}_{\widetilde{\mathfrak{q}}_j}(f_j) = 1$, that is, the leaf of the foliation $\widetilde{\mathcal{F}}_Y$ passing through $\widetilde{\mathfrak{q}}_j$ has a simple tangency with $\mathbb{D}_\mathbb{C}$.
4. *Whenever the separatrices are invariant by complex conjugation (as in the case $\vartheta \in \Sigma_n^{\mathbb{R}}$) and $\widetilde{\mathfrak{q}}_1, \ldots, \widetilde{\mathfrak{q}}_{n-2}$ belong to the real exceptional divisor $\mathbb{D}_\mathbb{R}$, it is possible to choose Y being the complexification of a dicritical real analytic vector field. In this case for all $1 \leqslant j \leqslant n-2$, f_j, g_j are the complexifications of real analytic maps.*

In Lemma 4.9 we say that *the separatrices are invariant by complex conjugation* in the following sense. We consider m smooth analytic curves $\mathcal{L}_i = \{y = \psi_i(x)\}$ at $(\mathbb{C}^2, \widehat{0})$ satisfying $\psi_j'(0) \neq \psi_k'(0)$ if $j \neq k$. The curves $\mathcal{L}_1, \ldots, \mathcal{L}_m$ are said to be *invariant by complex conjugation* if given $1 \leqslant i \leqslant m$ such that $\psi_i(x) = \sum_{r \geqslant 1} a_r x^r$ there exists $1 \leqslant j \leqslant m$ satisfying $\psi_j(x) = \sum_{r \geqslant 1} \bar{a}_r x^r$, where $\bar{a}$ is the complex conjugate of $a \in \mathbb{C}$.

From the properties 2 and 3 of Lemma 4.9 it is concluded that the foliation $\widetilde{\mathscr{F}}_Y$, called *auxiliary foliation of the vector field* ϑ, does not have any singular point on $(\mathbb{M}_{\mathbb{C}}, \mathbb{D}_{\mathbb{C}})$.

Sketch of the Proof of Lemma 4.9. The separatrices $C_i = \{y = \widetilde{\phi}_i(x)\}$ are invariant by a holomorphic vector field $\widetilde{P}\frac{\partial}{\partial x} + \widetilde{Q}\frac{\partial}{\partial y}$ if and only if there exists a convergent series $C \in \mathbb{C}\langle x, y\rangle$ such that

$$\widetilde{P}R_x + \widetilde{Q}R_y = RC\,, \tag{9}$$

being $R(x, y) = \prod_{i=1}^{n+1}\left(y - \widetilde{\phi}_i(x)\right) = r_{n+1}(x, y) + r_{n+2}(x, y) + \cdots$, where r_i is an homogeneous polynomial of degree i with complex coefficients. In what follows, we shall construct holomorphic vector fields satisfying Eq. (9) and some additional conditions.

Considering $h(x, y) := \prod_{j=1}^{n-2}\left(y - \widetilde{q}_j x\right)$, by substituing $C := (n+1)h$, $\widetilde{P} := xh + \widehat{P}$ and $\widetilde{Q} := yh + \widehat{Q}$ into Eq. (9), it is obtained

$$\widehat{P}R_x + \widehat{Q}R_y = h\left((n+1)R - xR_x - yR_y\right)\,. \tag{10}$$

Since the right-hand side of Eq. (10) has order greater or equal to $2n$, it belongs to the gradient ideal of R, $\mathcal{I}_R = \{AR_x + BR_y \mid A, B \in \mathbb{C}\langle x, y\rangle\}$, as it is proved in [2, 32]. Then it is possible to find $\widehat{P}$ and $\widehat{Q}$ satisfying the equality (10) having order greater or equal to n at the origin $\widehat{0}$, since $r_{n+1,x}$ and $r_{n+1,y}$ are relative primes in $\mathbb{C}\langle x, y\rangle$.

Considering $\widetilde{P} = xh + \widehat{P}$, $\widetilde{Q} = yh + \widehat{Q}$, $C = (n+1)h$ as before and $c \in \mathbb{C}$, the holomorphic vector field $\widetilde{P}_c\frac{\partial}{\partial_, x} + \widetilde{Q}_c\frac{\partial}{\partial_, y}$ satisfy Eq. (9), being

$$\widetilde{P}_c := \widetilde{P} - cR_y\,, \quad \widetilde{Q}_c := \widetilde{Q} + cR_x\,.$$

We denote $\widehat{\mathfrak{p}}_n$ and $\widehat{\mathfrak{q}}_n$ the homogeneous components of degree n of $\widehat{P}$ and $\widehat{Q}$, respectively. Since $\widetilde{\mathfrak{q}}_1, \ldots, \widetilde{\mathfrak{q}}_{n-2}$ are nonsingular points, the property $\widehat{\mathfrak{q}}_n(1, \widetilde{q}_j) - \widetilde{q}_j\widehat{\mathfrak{p}}_n(1, \widetilde{q}_j) + c(n+1)r_{n+1}(1, \widetilde{q}_j) \neq 0$ for all $1 \leqslant j \leqslant n-2$, is satisfied for all $c \in \mathbb{C}$ except for a finitely many complex number. Choosing one such $c \in \mathbb{C}$, it is verified that the holomorphic vector field $\widetilde{P}_c\frac{\partial}{\partial x} + \widetilde{Q}_c\frac{\partial}{\partial y}$ satisfies all required conditions of Lemma 4.9.

Whenever the separatrices are invariant by complex conjugation and the points $\widetilde{\mathfrak{q}}_1, \ldots, \widetilde{\mathfrak{q}}_{n-2} \in \mathbb{D}_{\mathbb{R}}$, then R and h are the complexifications of real analytic maps. As a consequence, the right-hand side of the equality (10) belongs to the gradient ideal of R with respecto to $\mathbb{R}\langle x, y\rangle$, $\widetilde{\mathcal{I}}_R = \left\{AR_x + BR_y \mid A, B \in \mathbb{R}\langle x, y\rangle\right\}$, as it is proved in [15].

Then it is possible to find $\widehat{P}, \widehat{Q} \in \mathbb{R}\langle x, y\rangle$ satisfying the equality (10). Choosing $c \in \mathbb{R}$ such that the number $\widehat{\mathfrak{q}}_n(1, \widetilde{q}_j) - \widetilde{q}_j\widehat{\mathfrak{p}}_n(1, \widetilde{q}_j) + c(n+1)r_{n+1}(1, \widetilde{q}_j)$ is nonzero for all $1 \leqslant j \leqslant n-2$, we take $\widetilde{P}_c$, $\widetilde{Q}_c$ defined as above. The vector field $\widetilde{P}_c\frac{\partial}{\partial x} + \widetilde{Q}_c\frac{\partial}{\partial y}$ is the complexification of a real analytic vector field which satisfies the required properties of Lemma 4.9. ■

The following result describes the locus $\widetilde{T}$ on $(\mathbb{M}_{\mathbb{C}}, \mathbb{D}_{\mathbb{C}})$ where the leaves of the foliation $\widetilde{\mathscr{F}}_{\vartheta}$ and its auxiliary foliation $\widetilde{\mathscr{F}}_{Y}$ are tangent. This result is an immediate consequence of the holomorphic and the real analytic implicit function theorems.

Lemma 4.10 *The set $\widetilde{T}$ consists of the desingularization of the separatrices of ϑ, $C_1, \ldots, C_{n+1}$, and $n-2$ analytic curves $\widetilde{T}_1, \ldots, \widetilde{T}_{n-2}$ which cross the exceptional divisor $\mathbb{D}_{\mathbb{C}}$ transversally at the points $\widetilde{\mathfrak{q}}_1, \ldots, \widetilde{\mathfrak{q}}_{n-2}$, respectively.*

Moreover, if v is the complexification of a real analytic vector field and the points $\widetilde{\mathfrak{q}}_1, \ldots, \widetilde{\mathfrak{q}}_{n-2}$ belong to $\mathbb{D}_{\mathbb{R}}$ then the curves $\widetilde{T}_1, \ldots, \widetilde{T}_{n-2}$ are the complexifications of real analytic curves.

Remark 4.11 The set $\widetilde{T}$ is *the locus of tangency points of the pair* $(\widetilde{\mathscr{F}}_{\vartheta}, \widetilde{\mathscr{F}}_{Y})$, and the curves $\widetilde{T}_1, \ldots, \widetilde{T}_{n-2}$ are called *the polar curves of the pair* $(\widetilde{\mathscr{F}}_{\vartheta}, \widetilde{\mathscr{F}}_{Y})$.

References

1. Arnold, V.I.: Geometrical Methods in the Theory of Ordinary Differential Equations. Springer, New York (1983)
2. Arnold, V.I., Gusein-Zade, S.M., Varchenko, A.N.: Singularities of Differentiable Maps, Vol. I. The Classification of Critical Points, Caustics and Wave Fronts. Birkhäuser Boston Inc, Boston (1985)
3. Brjuno, A.D.: Analytic form of differential equations. I, II, (Russian) Trudy Moskov. Mat. Obshch. **25**, 119–262 (1971); **26**, 199–239 (1972). English translation: Trans. Moscow Math. Soc. **25** (1971); **26** (1972)
4. Cerveau, D., Moussu, R.: Groups d'automorphismes de $(C, 0)$ et équations différentielles $y\,dy + \ldots = 0$. Bull. Soc. Math. Fr. **116**, 459–488 (1988)
5. Dulac, H.: Recherches sur les points singuliers des équations differ-entielles. Journal de l'École Polytechnique **2**(9), 1–125 (1904)
6. Écalle, J.: Théorie itérative, Introduction à la théorie des invariants holomorphes. J. Math. Pures et Appl. **54**, 183–258 (1975)
7. Écalle, J.: Les fonctions résurgentes, Tome I, II, Orsay: Publications Mathématiques d'Orsay 81, 5. Université de Paris-Sud, Département de Mathématique (1981)
8. Elizarov, P.M., Ilyashenko, Yu S.: Remarks on the orbital analytic classification of germs of vector fields. Math. USSR Sbornik **49**(1), 111–124 (1984)
9. Elizarov, P.M., Ilyashenko, Yu S., Shcherbakov, A.A., Voronin, S.M.: Finitely generated groups of germs of one-dimensional conformal mappings, and invariants for complex singular points of analytic foliations of the complex plane. Nonlinear Stokes Phenomena. Advances in Soviet Mathematics, vol. 14, pp. 57–105. American Mathematical Society, Providence (1993)
10. Granger, J.M.: Sur un Espace de Modules de germe de Courbe Plane, Bull. Sc. Math., 2e série, 3–16 (1979)
11. Gunning, R.C.: Introduction to Holomorphic Functions of Several Variables, vol. II, Local Theory, Wadsworth and Brooks/Cole, Advance Books and Software, California (1990)
12. Ilyashenko, Yu S.: Dulac's memoir "On limit cycles" and related questions of the local theory of differential equations. (Russian) Uspekhi Mat. Nauk **40**(6), 41–78 (1985)
13. Ilyashenko, Yu S.: Nonlinear Stokes phenomena. Nonlinear Stokes Phenomena. Advances in Soviet Mathematics, vol. 14, pp. 1–55. American Mathematical Society, Providence, RI (1993)
14. Ilyashenko, YuS, Yakovenko, SYu.: Lectures on Analytic Differential Equations. Graduate Studies in Mathematics, vol. 86. American Mathematical Society, Providence (2008)

15. Jaurez-Rosas, J.A.: Real-formal orbital rigidity for germs of real analytic vector fields on the real plane. J. Dyn. Control Syst. **2**(22) (2016). https://doi.org/10.1007/s10883-016-9314-y
16. Loray, F.: Rèduction formalle des singularities cuspidales de champs de vecteurs analytiques. J. Differ. Equ. **158**, 152–173 (1999)
17. Martinet, J., Ramis, J.-P.: Problèmes de modules pour des équations différentielles non linéaires du premier ordre, Inst. Hautes Études Sci. Publ. Math. vol. 5, pp. 63–164 (1982)
18. Martinet, J., Ramis, J.-P.: Classification analytique des équations différentielles non linéaires résonnantes du premier ordre, Ann. Sci. École Norm. Sup. (4) **16** (1983); 4, 571–621 (1984)
19. Mattei, J.-F.: Modules de feuilletages holomorphes singuliers I. Équisingularité. Invent. Math. (French) **23**(2), 297–325 (1991)
20. Ortiz-Bobadilla, L., Rosales-González, E., Voronin, S.M.: Rigidity theorem for degenerate singular points of germs of holomorphic vector fields in the complex plane. J. Dyn. Control Syst. **7**(4), 553–599 (2001)
21. Ortiz-Bobadilla, L., Rosales-González, E., Voronin, S.M.: Rigidity theorems for generic holomorphic germs of dicritic foliations and vector fields in $(\mathbb{C}^2, 0)$. Mosc. Math. J. **5**(1), 171–206 (2005)
22. Ortiz-Bobadilla, L., Rosales-González, E., Voronin, S.M.: Analytic normal forms of germs of holomorphic dicritic foliations. Mosc. Math. J. **8**(3), 521–545 (2008)
23. Ortiz-Bobadilla, L., Rosales-González, E., Voronin, S.M.: Thom's problem for degenerate singular points of holomorphic foliations in the plane. Mosc. Math. J. **12**(4), 825–862 (2012)
24. Ortiz-Bobadilla, L., Rosales-González, E., Voronin, S.M.: Formal and analytic normal forms of germs of holomorphic nondicritic foliations. J. Singul. **9**, 168–192 (2014)
25. Pérez-Marco, R.: Total convergence or general divergence in small divisors. Commun. Math. Phys. **223**(3), 451–464 (2001)
26. Poincaré, H.: Note sur les propriétés des fonctions définies par des équations differéntielles. Journal de l'École Polytechnique **45**, 13–26 (1878)
27. Ramis, J.P.: Confluence et resurgence. J. Fac. Sci. Univ. Tokyo Sect. IA Math. **36**, 706–716 (1989)
28. Shcherbakov, A.A.: Topological and analytical conjugacy of noncommutative groups of germs of conformal mappings. Trudy Sem. Petrovsk. **10**, 170–196 (1984)
29. Strozyna, E., Zoladek, H.: The analytic and formal normal forms for the nilpotent singularity. J. Differ. Equ. **179**, 479–537 (2002)
30. Takens, F.: Normal forms for certain singularities of vector fields, Colloque International sur l'Analyse et la Topologie Différentielle, Ann. Inst. Fourier (Grenoble) **23**(2), 163–195 (1973)
31. Voronin, S.M.: Analytic classification of germs of conformal mappings $(\mathbb{C}, 0) \to (\mathbb{C}, 0)$ with identity linear part. Funct. Anal. Appl. **15**(1), 1–17 (1981)
32. Voronin, S.M.: Orbital analytic equivalence of degenerate singular points of holomorphic vector fields on the complex plane, Tr. Mat. Inst. Steklova **213**, 30–49 (1997); Differ. Uravn. s Veshchestv. i Kompleks. Vrem., 35–55 (Russian)
33. Voronin, S.M., Grintchy, A.A.: An analytic classifications of saddle resonant singular points of holomorphic vector fields in the complex plane. J. Dyn. Control Syst. **2**(1), 21–53 (1996)
34. Voronin, S.M., Meshcheryakova, Y.I.: Analytic classification of generic degenerate elementary singular points of gems of holomorphic vector fields on the complex plane, (Russian) Izv. Vyssh. Uchebn. Zaved. Mat. **46**(1), 13–16 (2002). Translation in Russian Math. (Iz. VUZ). **46**(1), 11–14 (2002)
35. Yoccoz, J.-C.: Linéarisation des germes de diffeómorphismes holomorphes de $(\mathbb{C}, 0)$. C. R. Acad. Sci. Paris Sér. I Math. **306**(1), 55–58 (1988)
36. Yoccoz, J.-C.: Théorème de Siegel, nombres de Bruno et polynômes quadratiques. Astérisque **231**, 3–88 (1995)

On Singular Holomorphic Foliations with Projective Transverse Structure

Bruno Scárdua

Abstract In this paper we study holomorphic foliations with singularities having a homogeneous transverse structure of projective model (i.e., $\mathbb{P}SL(2, \mathbb{C})$ model). Our basic situation is the case of a foliation with singularities $\mathcal{F}$ on a complex analytic space M of dimension two and the structure exists in the complement of some analytic subset $S \subset M$ of codimension one. The main case occurs, as we shall see, when the analytic set is invariant by the foliation. We address both, the local and the global cases. This means two basic situations: (i) M is a projective surface (like $M = \mathbb{C}P(2)$ or $\overline{\mathbb{C}} \times \overline{\mathbb{C}}$) and (ii) $M = (\mathbb{C}^2, 0)$ which means the case of germs of foliations at the origin $0 \in \mathbb{C}^2$, having an isolated singularity at the origin. Our focus is the *extension* of the structure in a suitable sense. After performing a characterization of the existence of the structure in terms of suitable triples of differential forms, we consider the problem of extension of such structures to the analytic invariant set for germs of foliations and for foliations in complex projective spaces. Basic examples of this situation are given by logarithmic foliations and Riccati foliations. We also study the holonomy of such invariant sets, as a consequence of a strict link between this holonomy and the monodromy of a projective structure. These holonomy groups are proved to be solvable. Our final aim is the classification of such object under some mild conditions on the singularities they exhibit. In this work we perform this classification in the case where the singularities of the foliation are supposed to be non-dicritical and non-degenerate (more precisely, generalized curves). This case, we will see, corresponds to the transversely affine case and therefore to the class of logarithmic foliations. The more general case, which has to do with Riccati foliations, is dealt with by some extension results we prove and evoking results from Loray-Touzet-Vitorio.

Keywords Holomorphic foliation · Projective transverse structure · Holonomy group · Riccati foliation

B. Scárdua (✉)
Instituto de Matemática, Universidade Federal do Rio de Janeiro, CP. 68530, Rio de Janeiro-RJ 21945-970, Brazil
e-mail: scardua@im.ufrj.br

R. N. Araújo dos Santos et al. (eds.), *Singularities and Foliations. Geometry, Topology and Applications*, Springer Proceedings in Mathematics & Statistics 222, https://doi.org/10.1007/978-3-319-73639-6_6

1 Foliations and Transverse Structure

The *Riccati differential equation*

$$p(x)dy - (a(x)y^2 + b(x)y + c(x))dx = 0$$

where $(x, y) \in \mathbb{C}^2$ and p, a, b, c are complex polynomials is well-known to be a basic model for complex foliations, on projective surfaces, with projective transverse structure outside an invariant algebraic curve. Similarly the *Bernoulli equation*

$$p(x)dy - (a(x)y^{k+1} + b(x)y)dx = 0$$

stands for a model with an affine structure outside of an algebraic invariant curve [8, 27]. In this work we develop the study and classification of transversely projective holomorphic foliations. More precisely, we study codimension one holomorphic foliations with singularities, under the hypothesis of the existence of a projective transverse structure off some analytic codimension one subset.

Recall that a foliation (holomorphic of codimension one, with singularities) is *transversely projective* if the corresponding non-singular foliation is given by an atlas of local submersions with projective relations, i.e., two such submersions $y: U \to \overline{\mathbb{C}}$ and $\tilde{y}: \tilde{U} \to \overline{\mathbb{C}}$ are related by $\tilde{y} = \frac{ay+b}{cy+d}$ for some $a, b, c, d \in \mathbb{C}$ locally constant and satisfying $ad - bc = 1$. This is a particular case of foliation having a homogeneous transverse structure (cf. [4]) and in the holomorphic framework it is natural to consider the case where the foliation exhibits singularities and the transverse structure is defined in the complement of some analytic subset of codimension one [27]. This situation has two main examples given by the class of *logarithmic foliations*, i.e., foliations defined by simple poles closed meromorphic one-forms; and by the class of *Riccati foliations*, i.e., foliations induced by Riccati differential equations.

1.1 Holomorphic Foliations

The basic concepts of differentiable manifolds (as tangent space, tangent bundle, etc.) can be introduced in the complex holomorphic setting. This is also the case of the concept of foliation:

Definition 1 (*holomorphic foliation*) A *holomorphic foliation* $\mathcal{F}$ of (complex) dimension k on a complex manifold M is given by a *holomorphic atlas* $\{\varphi_j : U_j \subset M \to V_j \subset \mathbb{C}^n\}_{j \in J}$ with the *compatibility property*: Given any intersection $U_i \cap U_j \neq \emptyset$ the change of coordinates $\varphi_j \circ \varphi_i^{-1}$ preserves the horizontal fibration on $\mathbb{C}^n \simeq \mathbb{C}^k \times \mathbb{C}^{n-k}$.

Examples of such foliations are, like in the "real" case, given by non-singular holomorphic vector-fields, holomorphic submersions, holomorphic fibrations and locally free holomorphic complex Lie group actions on complex manifolds.

Remark 1 (i) As in the "real" case, the study of holomorphic foliations may be very useful in the classification theory of complex manifolds.
(ii) In a certain sense, the "holomorphic case" is closer to the "algebraic case" than the case of real foliations.

1.2 Holomorphic Foliations with Singularities

One of the most common compactifications of the complex affine space $\mathbb{C}^n$ is the complex projective space $\mathbb{C}P(n)$. It is well-known that any foliation (holomorphic) of codimension $k \geq 1$ on $\mathbb{C}P(n)$ must have some *singularity* (in other words, $\mathbb{C}P(n)$, for $n \geq 2$, exhibits no holomorphic foliation in the sense we have considered up to now, cf. [2].) Thus one may consider such objects: *singular (holomorphic) foliations* as part of the zoology. Let us illustrate this concept through some examples:

Example 1 (*Polynomial vector fields on* $\mathbb{C}^2$) Given affine coordinates $(x, y) \in \mathbb{C}^2$, let $X = P(x, y)(\partial/\partial x) + Q(x, y)(\partial/\partial y) = (P, Q)$ be a polynomial vector field (with isolated singularities) on $\mathbb{C}^2$. We have an ordinary differential equation:

$$\begin{cases} \dot{x} = P(x, y) \\ \dot{y} = Q(x, y) \end{cases}$$

The local solutions are given by Picard's Theorem (the existence and uniqueness theorem of ordinary differential equations):

$$\varphi(z) = (x(z), y(z))$$

$$\frac{d\varphi}{dz} = \dot{\varphi}(z) = X(\varphi(z))$$

Gluing the images of these unique local solutions, we can introduce the *orbits* of X on $\mathbb{C}^2$. The orbits are immersed Riemann surfaces on $\mathbb{C}^2$, which are locally given by the solutions of X.

Now we may be interested in what occurs these orbits in "a neighborhood of the infinity". We may for instance compactify $\mathbb{C}^2$ as the projective plane $\mathbb{C}P(2) = \mathbb{C}^2 \cup L_\infty$, $L_\infty \cong \mathbb{C}P(1)$.

1. What happens to X in a neighborhood of L_∞?
2. Is it still possible to consider its orbits around L_∞?

We may rewrite X as the coordinate system $(u, v) = (1/x, y/x)$: $X(u, v) = \frac{1}{u^m} Y(u, v)$, $m \in \mathbb{N} \cup \backslash 0$ where Y is a polynomial vector field, also with isolated singularities. The exterior product of X and Y is zero in common domain $U : X \wedge Y = 0$. Thus, orbits of Y (or X) are orbits of X (or Y), respectively in U. Then the orbits of X *extend* to the (u, v)-plane as the corresponding orbits of Y along L_∞. In this same way, we may consider the extension of the orbits to the $(r, s) = (x/y, 1/y)$ coordinate system. These extensions are called *leaves* of a foliation induced by X on $\mathbb{C}P(2)$. We obtain this way: A decomposition of $\mathbb{C}P(2)$ into immersed complex curves which are locally arrayed, as the orbits (solutions) of a complex vector field. This is a holomorphic foliation $\mathcal{F}$ with singularities of dimension one on $\mathbb{C}P(2)$.

Remark 2 (*singularities are defined by differential forms*) Assume that we have a holomorphic non-singular foliation $\mathcal{F}_0$ on $U \setminus \{0\}$, $0 \in \mathbb{C}^2$, $U \cap sing(\mathcal{F}) = \backslash 0$. Choose local coordinates (x, y) centered at 0 and define a meromorphic function $f : U \setminus \{0\} \to \overline{\mathbb{C}}$, $p \in U \setminus \{0\}$, as $f(p) =$ the inclination of the tangent to the leaf L_p of $\mathcal{F}_0$. By Hartogs' Extension Theorem [18, 34] f extends to a meromorphic function $f : U \to \overline{\mathbb{C}}$. We may write $f(x, y) = \frac{a(x,y)}{b(x,y)}$, $a, b \in \mathcal{O}(U)$ and define

$$\frac{dy}{dx} = f(x, y) = \frac{b(x, y)}{a(x, y)},$$

that is,

$$\begin{Bmatrix} \dot{x} = a(x, y) \\ \dot{y} = b(x, y). \end{Bmatrix}$$

Therefore, $\mathcal{F}$ is defined by a holomorphic 1-form $\omega = a(x, y)\, dy - b(x, y)\, dx$ in U.

The above remark also motivates the following definition:

Definition 2 (*holomorphic foliation with singularities*) Let M be a complex manifold. A *singular holomorphic foliation of codimension one* $\mathcal{F}$ on M is given by an open cover $M = \bigcup_{j \in J} U_j$ and holomorphic integrable 1-forms $\omega_j \in \bigwedge^1(U_j)$ such that if $U_j \cap U_j \neq \emptyset$, then $\omega_i = g_{ij}\omega_j$ in $U_i \cap U_j$, for some $g_{ij} \in \mathcal{O}^*(U_i \cap U_j)$. We put $sing(\mathcal{F}) \cap U_j = \{p \in U_j;\ \omega_j(p) = 0\}$ to obtain $sing(\mathcal{F}) \subset M$, a well-defined analytic subset of M, called *singular set* of $\mathcal{F}$. The open subset $M \setminus sing(\mathcal{F}) \subset M$ is foliated by a holomorphic codimension one (non-singular) foliation $\mathcal{F}_0$. By definition the *leaves* of $\mathcal{F}$ are the leaves of $\mathcal{F}_0$.

Remark 3 We may always assume that $sing(\mathcal{F}) \subset M$ has codimension ≥ 2. If $(f_j = 0)$ is an equation of codimension one component of $sing(\mathcal{F}) \cap U_j$, then we get $\omega_j = f_j^n \bar{\omega}_j$ where $\bar{\omega}_j$ is a holomorphic 1-form and $sing(\bar{\omega}_j)$ does not contain $(f_j = 0)$.

Remark 4 (*Convention*) Let M be a complex manifold. From now on, in the absence of a specific mention, by *foliation* on M we shall mean a codimension one holomorphic foliation with singularities. We shall also assume that the singular set $sing(\mathcal{F}) \subset M$ has codimension ≥ 2. In particular, if M has dimension two then $sing(\mathcal{F})$ is a discrete set of points of M.

Example 2 Let $f : M \to \overline{\mathbb{C}}$ be a meromorphic function on the complex manifold M. Then $\omega = df$ defines a holomorphic foliation of codimension one with singularities on M. The leaves are the connected components of the levels $\{f = c\}$, $c \in \overline{\mathbb{C}}$.

Example 3 Let G be a complex Lie group and $\varphi : G \times M \to M$ a holomorphic action of G on M. The action is foliated if all its orbits have a same fixed dimension. In this case there exists a holomorphic non-singular foliation $\mathcal{F}$ on M, whose leaves are orbits of φ. However, usually, actions are not foliated, though they may define singular holomorphic foliations. For instance, an action φ of $G = (\mathbb{C}, +)$ on M, $\varphi : \mathbb{C} \times M \to M$ is a holomorphic flows. We have a holomorphic complete vector field $X = \frac{\partial \phi}{\partial t}|_{t=0}$ on M. The singular set of X may be assumed to be of codimension ≥ 2 and we obtain a holomorphic singular foliation of dimension one $\mathcal{F}$ on M whose leaves are orbits of X, or equivalently, of φ.

Problem 1 Study and classify actions of complex Lie groups G on a given compact complex M, in terms of the corresponding foliation.

The general problem above may be therefore regarded under the stand-point of singular holomorphic foliations theory.

Example 4 (*Darboux foliations*) Let M be a complex manifold and let $f_j : M \to \overline{\mathbb{C}}$ be meromorphic functions and $\lambda_j \in \mathbb{C}^*$ complex numbers, $j = 1, \ldots, r$. The meromorphic integrable 1-form $\omega = (\prod\limits_{j=1}^{r} f_j) \sum\limits_{i=1}^{r} \lambda_i \frac{df_i}{f_i}$ defines a *Darboux foliation* $\mathcal{F} = \mathcal{F}(\omega)$ on M. The foliation $\mathcal{F}$ has $f = \prod\limits_{j=1}^{r} f_j^{\lambda_j}$ as a *logarithmic* first integral.

Example 5 (*Riccati foliations*) A *Riccati Foliation* on $\overline{\mathbb{C}} \times \overline{\mathbb{C}}$ is given in some affine chart $(x, y) \in \mathbb{C} \times \mathbb{C}$ by a polynomial one-form $\omega = p(x)dy - (y^2c(x) - yb(x) - a(x))dx$. This will be thoroughly studied in the next section.

The concept of holonomy in the singular case Let now $\mathcal{F}$ be a holomorphic foliation (with isolated singularities) on a complex manifold M. Given a leaf L_0 of $\mathcal{F}$ we choose any base point $p \in L_0 \subset M \setminus sing(\mathcal{F})$ and a transverse disc $\Sigma_p \subset M$ to $\mathcal{F}$ centered at p. Denote by $Diff(\Sigma_p, p)$ the group of germs of complex diffeomorphisms of Σ_p with a fixed point at p. The *holonomy group* of the leaf L_0 with respect to the disc Σ_p and to the base point p is the image of the representation $Hol\colon \pi_1(L_0, p) \to Diff(\Sigma_p, p)$ obtained by lifting closed paths in L_0 with base point p, to paths in the leaves of $\mathcal{F}$, starting at points $z \in \Sigma_p$, by means of a transverse fibration to $\mathcal{F}$ containing the disc Σ_p [6, 17]. Given a point $z \in \Sigma_p$ we denote the leaf through z by L_z. Given a closed path $\gamma \in \pi_1(L_0, p)$ we denote by $\tilde{\gamma}_z$ its lift to the leaf L_z and starting (the lifted path) at the point z. Then the image of the corresponding holonomy map is $h_{[\gamma]}(z) = \tilde{\gamma}_z(1)$, i.e., the final point of the lifted path $\tilde{\gamma}_z$. This defines a diffeomorphism germ map $h_{[\gamma]}\colon (\Sigma_p, p) \to (\Sigma_p, p)$ and also a group homomorphism $Hol\colon \pi_1(L_0, p) \to Diff(\Sigma_p, p)$. The image $Hol(\mathcal{F}, L_0, \Sigma_p, p) \subset Diff(\Sigma_p, p)$ of such homomorphism is called the *holonomy group* of the leaf L_0 with respect to Σ_p and p. By considering any parametrization $z\colon (\Sigma_p, p) \to (D, 0)$ we may identify (in a non-canonical way) the holonomy group with a subgroup of $Diff(\mathbb{C}, 0)$. It is clear from the construction that the maps in the holonomy group preserves the leaves of the foliation.

Separatrices and local holonomies Fix now a germ $\mathcal{F}$ of holomorphic foliation with a singularity at the origin $0 \in \mathbb{C}^2$. Choose a representative $\mathcal{F}(U)$ for $\mathcal{F}$, defined in an open neighborhood U of the origin. A leaf of $\mathcal{F}(U)$ accumulating only at 0 is closed off 0, thus by Remmert–Stein extension theorem [19] it is contained in an irreducible analytic curve through 0. Such a curve is called a local *separatrix* of $\mathcal{F}$ through 0. A separatrix is therefore the union of a leaf of $\mathcal{F}|_U$ which is closed off the singular point, and the singular point $0 \in \mathbb{C}^2$. By Newton–Puiseux parametrization theorem, if U is small enough, there is an analytic injective map $f: D \to U$ from the unit disk $D \subset \mathbb{C}$ into the separatrix, mapping the origin to $0 \in \mathbb{C}^2$, and non-singular outside the origin $0 \in D$. Therefore the leaf contained in a separatrix, locally has the topology of a punctured disk. In particular, given a separatrix Γ we may choose a loop $\gamma \in \Gamma \setminus \{0\}$ generating the (local) fundamental group $\pi_1(\Gamma \setminus \{0\})$. The corresponding holonomy map h_γ is defined in terms of a germ of complex diffeomorphism at the origin of a local disc Σ transverse to $\mathcal{F}$ and centered at a non-singular point $q \in \Gamma \setminus \{0\}$. This map is well-defined up to conjugacy by germs of holomorphic diffeomorphisms, and is generically referred to as *local holonomy* of the separatrix Γ with respect to the singularity $0 \in \mathbb{C}^2$.

1.3 Irreducible Singularities, Separatrices and Reduction of Singularities

Let $\omega = a(x, y)dx + b(x, y)dy$ be a holomorphic one-form defined in a neighborhood $0 \in U \in \mathbb{C}^2$. We say that $0 \in \mathbb{C}^2$ is a *singular* point of ω if $a(0, 0) = b(0, 0) = 0$, and a *non-singular* point otherwise. We say that $0 \in \mathbb{C}^2$ is an *irreducible* singular point of ω if the eigenvalues λ_1, λ_2 of the linear part of the corresponding dual vector field $X = -b(x, y)\frac{\partial}{\partial x} + a(x, y)\frac{\partial}{\partial y}$ at $0 \in \mathbb{C}^2$ satisfy one of the following conditions:

(1) $\lambda_1.\lambda_2 \neq 0$ and $\lambda_1/\lambda_2 \notin \mathbb{Q}_+$
(2) either $\lambda_1 \neq 0$ and $\lambda_2 = 0$, or *vice-versa.*

In case (1) there are two invariant curves tangent to the eigenvectors corresponding to λ_1 and λ_2. In case (2) there is an invariant curve tangent at $0 \in \mathbb{C}^2$ to the eigenspace corresponding to λ_1. These curves are called *separatrices* of the foliation.

Suppose that $0 \in \mathbb{C}^2$ is either a non-singular point or an irreducible singularity of a foliation $\mathcal{F}$. Then in suitable local coordinates (x, y) in a neighborhood $0 \in U \in \mathbb{C}^2$ of the origin, we have the following local normal forms for the one-forms defining this foliation [7]:

(Reg) $dy = 0$, whenever $0 \in \mathbb{C}^2$ is a non-singular point of $\mathcal{F}$.
and whenever $0 \in \mathbb{C}^2$ is an irreducible singularity of $\tilde{\mathcal{F}}$, then either
(Irr.1) $xdy - \lambda ydx + \omega_2(x, y) = 0$ where $\lambda \in \mathbb{C}\backslash\mathbb{Q}_+$, $\omega_2(x, y)$ is a holomorphic one-form with a zero of order ≥ 2 at $(0, 0)$. This is called *non-degenerate singularity*. Such a singularity is *resonant* if $\lambda \in \mathbb{Q}_-$ and *hyperbolic* if $\lambda \notin \mathbb{R}$, or

(Irr.2) $y^{t+1}dx - [x(1+\lambda y^t) + A(x, y)]dy = 0$, where $\lambda \in \mathbb{C}$, $t \in \mathbb{N} = \{1, 2, 3, \dots\}$ and $A(x, y)$ is a holomorphic function with a zero of order $\geq t + 2$ at $(0, 0)$. This is called *saddle-node singularity*. The *strong manifold* or *strong separatrix* of the saddle-node is given by $\{y = 0\}$. If the singularity admits another separatrix then it is necessarily smooth and transverse to the strong manifold, it can be taken as the other coordinate axis and will be called *central* manifold of the saddle-node. This class of irreducible singularity is thoroughly studied in [22].

Therefore, for a suitable choice of the coordinates, we have $\{y = 0\} \subset sep(\mathcal{F}, U) \subset \{xy = 0\}$, where $sep(\mathcal{F}, U)$ denotes the union of separatrices of $\mathcal{F}$ through $0 \in \mathbb{C}^2$.

An irreducible singularity $xdy - \lambda ydx + \dots = 0$ is in the *Poincaré domain* if $\lambda \notin \mathbb{R}_-$ and it is in the *Siegel domain* otherwise. For singularities in the Poincaré domain, the non-resonance condition ($\lambda \notin \mathbb{Q}$) actually implies, by Poincaré linearization theorem, that the singularity is analytically linearizable (cf. [16]). For singularities in the Siegel domain, this question is quite more delicate [23]).

Given a foliation $\mathcal{F}$ of dimension one on a complex surface M with finite singular set $sing(\mathcal{F})$, the Theorem of reduction of singularities of Seidenberg reads as follows:

Theorem 1 ([31]) *There is a proper holomorphic map $\pi : \tilde{M} \to M$ which is a finite composition of quadratic blowing-up's at the singular points of $\mathcal{F}$ in M such that the pull-back foliation $\tilde{\mathcal{F}} := \pi^*\mathcal{F}$ of $\mathcal{F}$ by π satisfies:*

(a) *$sing(\tilde{\mathcal{F}}) \subset \pi^{-1}(sing(\mathcal{F}))$, and*
(b) *Any singularity $\tilde{p} \in sing(\tilde{\mathcal{F}})$ is irreducible.*

Indeed, we can say more:

We call $\tilde{\mathcal{F}}$ the *desingularization* or *reduction of singularities* of $\mathcal{F}$. Moreover, the *exceptional divisor* $E = \pi^{-1}(sing(\mathcal{F})) \subset \tilde{M}$ of the reduction π can be written as $E = \bigcup_{j=1}^m \mathbb{P}_j$, where each $\mathbb{P}_j$ is diffeomorphic to an embedded projective line $\mathbb{C}P(1)$ introduced as a divisor of the successive blowing-up's. The $\mathbb{P}_j$ are called *components* of the divisor E. A singularity $q \in sing(\mathcal{F})$ is *non-dicritical* if $\pi^{-1}(q)$ is invariant by $\tilde{\mathcal{F}}$. Any two components $\mathbb{P}_i$ and $\mathbb{P}_j$, $i \neq j$, intersect (transversely) at most one point, which is called a *corner*. Moreover, there are no triple intersection points. Any non-invariant component of the exceptional divisor is transverse to the lifted foliation $\tilde{\mathcal{F}}$ at every point. Given any analytic curve $\Lambda \subset M$ we denote by $\tilde{\Lambda} := \overline{\pi^{-1}(\Lambda \setminus sing(\mathcal{F}))} \subset \tilde{M}$ the *strict transform of* Λ.

As seen above, a *separatrix* of $\mathcal{F}$ at $0 \in \mathbb{C}^2$ is the germ at $0 \in \mathbb{C}^2$ of an irreducible analytic curve, containing the singular point, which is invariant by $\mathcal{F}$. By the reduction of singularities (Theorem 1) we conclude that a separatrix Γ of $\mathcal{F}$ is the projection $\Gamma = \pi(\tilde{\Gamma})$ of a curve $\tilde{\Gamma}$ invariant by $\tilde{\mathcal{F}}$ and transverse to the exceptional divisor $\pi^{-1}(0)$. A singularity is called *dicritical* if it exhibits infinitely many separatrices. We shall say that a separatrix Γ is a *dicritical separatrix* if $\tilde{\Gamma}$ meets the exceptional divisor only at non-singular points. Equivalently, $\Gamma = \pi(\tilde{\Gamma})$ is non-dicritical if $\tilde{\Gamma}$ is a separatrix of some singularity of $\tilde{\mathcal{F}}$. A non-dicritical separatrix is geometrically characterized by the fact that it is *isolated* in the set of separatrices. Indeed, notice that a neighborhood of some projective line in a finite sequence of blowing-ups starting at the origin corresponds to what we call *sector* with vertex at the origin. Thus, from the Resolution theorem (Theorem 1) a dicritical separatrix is always one which is contained in the interior of a "sector of separatrices". Given a representative for the germ $\mathcal{F}$ in a neighborhood U of the singularity, we shall denote by $\mathcal{N}D(sep(\mathcal{F}, U)) \subset U$ the analytic set which is the union of the non-dicritical separatrices of $\mathcal{F}$ in U.

Definition 3 (*generalized curve* - [10] *p. 144*) A germ of a foliation singularity at the origin $0 \in \mathbb{C}^2$ is a *generalized curve* if (i) it is non-dicritical and (ii) it exhibits no saddle-node in its reduction by blow-ups.

Generalized curves play an important role in the zoology of the singularities of holomorphic foliations. They are those whose desingularization/reduction of singularities is like the one of a holomorphic function $f: \mathbb{C}^2, 0 \to \mathbb{C}, 0$ [10]. In this work we will consider a slightly more general concept which is the following:

Definition 4 (*(non-resonant) extended generalized curve*) A germ of a foliation singularity at the origin $0 \in \mathbb{C}^2$ will be called an *extended generalized curve* if the singularity exhibits no saddle-node in its reduction by blow-ups. This includes the case of dicritical singularities. An extended generalized curve singularity is called *non-resonant* if each connected component of the invariant part of exceptional divisor contains some non-resonant singularity.

2 Foliations with Projective Transverse Structure

2.1 Transversely Homogeneous Foliations

A (transversely) holomorphic foliation $\mathcal{F}$ on a smooth manifold M has a *holomorphic homogeneous transverse strucutre* if there are a complex Lie group G, a connected closed subgroup $H < G$ such that $\mathcal{F}$ admits an atlas of submersions $y_j: U_j \subset M \to G/H$ satisfying $y_i = g_{ij} \circ y_j$ for some locally constant map $g_{ij}: U_i \cap U_j \to G$ for each $U_i \cap U_j \neq \emptyset$. In other words, the transversely holomorphic atlas of submersions for $\mathcal{F}$ has transiction maps given by left translations on G and submersions taking values on the homogeneous space G/H. We shall say that $\mathcal{F}$ is transversely homogeneous *of model* G/H. Some important properties of transversely homogeneous holomorphic foliations are listed below:

1. Any transversely homogeneous holomorphic foliation is a transversely holomorphic foliation with a holomorphic homogeneous transverse structure.
2. Given a foliation $\mathcal{F}$ on M as in (1) with model G/H then any real submanifold $M \subset M$ transverse to $\mathcal{F}$ is equipped with a transversely holomorphic foliation $\mathcal{F}_1 = \mathcal{F}|_M$ with holomorphic homogeneous transverse structure of model G/H.
3. Let $F = G/H$ be an homogeneous space of a complex Lie group G ($H \lhd G$ is a closed Lie subgroup). Any homomorphism representation $\varphi: \pi_1(N) \to Aut(F)$

gives rise to a transversely holomorphic foliation $\mathcal{F}_\varphi$ on $(\widetilde{N} \times F)/\varphi = M_\varphi$ which is holomorphically transversely homogeneous of model G/H.

4. For the case $G = \mathbb{P}SL(2, \mathbb{C})$ and $H \subset G$ is the affine group $H = Aff(\mathbb{C})$ (isotropy group of the point at infinity $\infty \in \mathbb{C}P^1$), we have that the quotient $G/H \simeq \mathbb{C}P^1$ is the Riemann sphere and the foliations with this transverse model are called *transversely projective*.

More precisely we have, for the non-singular case:

Definition 5 (*transversely projective foliation: non-singular*) A codimension one non-singular holomorphic foliation $\mathcal{F}$ on a manifold M is called *transversely projective* if there is an open cover $\bigcup\limits_{j\in J} U_j = M$ such that in each U_j the foliation is given by a submersion $f_j : U_j \to \overline{\mathbb{C}}$ and if $U_i \cap U_j \neq \emptyset$ then we have $f_i = f_{ij} \circ f_j$ in $U_i \cap U_j$ where $f_{ij} : U_i \cap U_j \to \mathbb{P}SL(2, \mathbb{C})$ is locally constant. Thus, on each intersection $U_i \cap U_j \neq \emptyset$, we have $f_i = \frac{a_{ij} f_j + b_{ij}}{c_{ij} f_j + d_{ij}}$ for some locally constant functions $a_{ij}, b_{ij}, c_{ij}, d_{ij}$ with $a_{ij}d_{ij} - b_{ij}c_{ij} = 1$. The data $\mathcal{P} = \{U_j, f_j, f_{ij}, j \in J\}$ is called a *projective transverse structure* for $\mathcal{F}$.

Basic references for transversely affine and transversely projective foliations (in the non-singular case) are found in [17].

(5) Based on the Rieman-Koebe uniformization theorem we have:

Proposition 1 ([27] Theorem 6.1 p. 203).) *Let $\mathcal{F}$ be a transversely homogeneous holomorphic foliation of codimension one on M^n. Then $\mathcal{F}$ is transversely projective foliation on M^n.*

Proof We know that G/H is a simply-connected complex manifold of dimension one. By the Riemann-Koebe Uniformization theorem we have a conformal equivalence $G/H \equiv \overline{\mathbb{C}}$, $\mathbb{C}$ or D the unitary disc. This implies that either $G \subset Aut(\overline{\mathbb{C}}) = \mathbb{P}SL(2, \mathbb{C})$, $G \subset Aut(\mathbb{C}) = Aff(\mathbb{C})$ or $G \subset Aut(D) \cong \mathbb{P}SL(2, \mathbb{R})$. The proposition follows.

Let $\mathcal{F}$ be a codimension ℓ foliation on a manifold M. If $\mathcal{F}$ admits a Lie group transverse structure of model G, or a *G-transverse structure* for short, then we shall call $\mathcal{F}$ a *G-foliation* or, simply, *Lie foliation*. The characterization of G-foliations in terms of differential forms is given below. Let $\{\omega_1, \ldots, \omega_\ell\}$ be a basis of the Lie algebra of G. Then we have $d\omega_k = \sum\limits_{i<j} c_{ij}^k \omega_i \wedge \omega_j$ for a family constants $\{c_{ij}^k\}$ called the *structure constants* of the Lie algebra in the given basis.

Theorem 2 (Darboux-Lie, [17]) *Let G be a complex Lie group of dimension ℓ. Let $\{\omega_1, \ldots, \omega_\ell\}$ be a basis of the Lie algebra of G with structure constants $\{c_{ij}^k\}$. Suppose that a complex manifold V^m of dimension $m \geq \ell$ admits a system of one-forms $\Omega_1, \ldots, \Omega_\ell$ in M such that:*

(i) *$\{\Omega_1, \ldots, \Omega_\ell\}$ is a rank ℓ integrable system which defines $\mathcal{F}$.*

(ii) $d\Omega_k = \sum\limits_{i<j} c_{ij}^k \Omega_i \wedge \Omega_j$.

Then:

(iii) *For each point $p \in M$ there is a neighborhood $p \in U_p \subseteq M$ equipped with a submersion $f_p \colon U_p \to G$ which defines $\mathcal{F}$ in U_p such that $f_p^*(\omega_j) = \Omega_j$ in U_p, for all $j \in \{1, \dots, q\}$.*

(iv) *If $U_p \cap U_q \neq \emptyset$ then in the intersection we have $f_q = L_{g_{pq}}(f_p)$ for some locally constant left translation $L_{g_{pq}}$ in G.*

(v) *If M is simply-connected we can take $U_p = M$.*

2.2 Transversely Projective Foliations with Singularities

Let M be a complex manifold. As already stated, if no specific mention is made, by *foliation* on M we shall mean a codimension one holomorphic foliation with singularities and $\dim_{\mathbb{C}} M \geq 2$.

Definition 6 (*transversely projective: singular*) A foliation $\mathcal{F}$ on M is called *transversely projective* if the underlying "non-singular" foliation $\mathcal{F}_0 =: \mathcal{F}\big|_{M \setminus sing(\mathcal{F})}$ is transversely projective. This means that there is an open cover $\bigcup\limits_{j \in J} U_j = M \setminus sing(\mathcal{F})$ such that in each U_j the foliation is given by a submersion $f_j \colon U_j \to \overline{\mathbb{C}}$ and if $U_i \cap U_j \neq \emptyset$ then we have $f_i = f_{ij} \circ f_j$ in $U_i \cap U_j$ where $f_{ij} \colon U_i \cap U_j \to \mathbb{P}SL(2, \mathbb{C})$ is locally constant. Thus, on each intersection $U_i \cap U_j \neq \emptyset$, we have $f_i = \frac{a_{ij} f_j + b_{ij}}{c_{ij} f_j + d_{ij}}$ for some locally constant functions $a_{ij}, b_{ij}, c_{ij}, d_{ij}$ with $a_{ij} d_{ij} - b_{ij} c_{ij} = 1$.

As observed in [27] *the singularities of a foliation admitting a projective transverse structure are all of type $df = 0$ for some local meromorphic function* (indeed, if $\Delta \subset \mathbb{C}^n$ is a polydisc centered at the origin then $\Delta \setminus \{0\}$ is simply-connected for $n \geq 2$). In this work we will be considering foliations which are transversely projective in the complement of *codimension one invariant divisors*. Such divisors may, a priori, exhibit singularities which do not admit meromorphic first integrals.

2.3 Riccati Foliations

Example 6 (*Riccati Foliations*) The Riccati differential equation

$$p(x)dy - (a(x)y^2 + b(x)y + c(x))dx = 0$$

where $(x, y) \in \mathbb{C}^2$ and p, a, b, c are complex polynomials has been proved to be an important model for complex foliations, on projective surfaces. In the particular case when $c \equiv 0$, it as an important example of a foliation with affine transverse structure outside an algebraic invariant set [8, 27].

Fix affine coordinates $(x, y) \in \mathbb{C}^2$ and consider a polynomial one-form $\Omega = p(x)dy - \big(a(x)y^2 + b(x)y + c(x)\big)dx$ on $\mathbb{C}^2$. Then Ω defines a *Riccati foliation* $\mathcal{R}$ on $\overline{\mathbb{C}} \times \overline{\mathbb{C}}$ as follows: if we change coordinates via $u = \frac{1}{x}$, $v = \frac{1}{y}$ then we obtain $\Omega(x, v) = p(x)dv + \big(a(x) + b(x)v + c(x)v^2\big)dx$. Similarly for $\Omega(u, y) = u^{-n}[\tilde{p}(u)\,dy - \big(\tilde{a}(u)y^2 + \tilde{b}(u)y + \tilde{c}(u)\big)du]$ and $\Omega(u, v) = u^{-n}[\tilde{p}(u)\,dv - \big(\tilde{a}(u) + \tilde{b}(u)v + \tilde{c}(u)v^2\big)du]$. The similarity of these four expressions shows that Ω defines a holomorphic foliation $\mathcal{R}$ with isolated singularities on $\overline{\mathbb{C}} \times \overline{\mathbb{C}}$ and having a geometry as follows (see Fig. 1):

(i) $\mathcal{R}$ is transverse to the fibers $\{a\} \times \overline{\mathbb{C}}$ except for invariant fibers which are given in $\mathbb{C}^2$ by $\{p(x) = 0\}$.

(ii) If $S = \bigcup\limits_{j=1}^{r} \{a_j\} \times \overline{\mathbb{C}}$ is the set of invariant fibers then $\mathcal{R}$ is transversely projective in $(\overline{\mathbb{C}} \times \overline{\mathbb{C}})\backslash S$. Indeed, $\mathcal{R}|_{(\overline{\mathbb{C}}\times\overline{\mathbb{C}})\backslash S}$ is conjugate to the suspension of a representation $\varphi\colon \pi_1(\overline{\mathbb{C}}\backslash \bigcup\limits_{j=1}^{r}\{a_j\}) \to \mathbb{P}SL(2, \mathbb{C})$.

(iii) For a generic choice of the coefficients $a(x), b(x), c(x), p(x) \in \mathbb{C}[x]$ the singularities of $\mathcal{R}$ on $\overline{\mathbb{C}} \times \overline{\mathbb{C}}$ are hyperbolic, S is the only algebraic invariant set and therefore for each singularity $q \in sing(\mathcal{R}) \subset S$ there is a local separatrix of $\mathcal{R}$ transverse to S passing through q.

Now we consider the canonical way of passing from $\overline{\mathbb{C}} \times \overline{\mathbb{C}}$ to $\mathbb{C}P(2)$ by a map $\sigma : \overline{\mathbb{C}} \times \overline{\mathbb{C}} \to \mathbb{C}P(2)$ obtained as a sequence of birational maps as follows: first blow-up a point, for example the origin, of $\mathbb{C}^2 \subset \overline{\mathbb{C}} \times \overline{\mathbb{C}}$ then blow-down two suitable projective lines of self-intersection equals -1 as indicated in Fig. 1. Following this process step by step we conclude that the foliation $\mathcal{F} = \sigma_*(\mathcal{R}) = (\sigma^{-1})^*(\mathcal{R})$ induced by $\mathcal{R}$ on $\mathbb{C}P(2)$ has the following characteristics:

(i') $\mathcal{F}$ is transversely projective in $\mathbb{C}P(2)\backslash S$ where $S \subset \mathbb{C}P(2)$ is the union of a finite number of projective lines of the form $\bigcup\limits_{j=1}^{r} \overline{\{x = a_j\}} \subset \mathbb{C}P(2)$ in a suitable affine chart $(x, y) \in \mathbb{C}^2 \subset \mathbb{C}P(2)$.

(ii') For a generic choice of the coefficients of Ω, the singularities of $\mathcal{F}$ in S are hyperbolic except for one single dicritical singularity $q_\infty : (x = \infty, y = 0) \in \mathbb{C}P(2)$ which after one blow-up gives a non-singular foliation transverse to the projective line except for a single tangency point. This singularity will be called a *radial type singularity.* The foliation $\mathcal{F}$ also has two other nonhyperbolic singularities, belonging to the line at infinity $L_\infty = \mathbb{C}P(2) \setminus \mathbb{C}^2$, which is invariant, one linearizable with holomorphic first integral and the other dicritical of radial type, admitting a meromorphic first integral. Also, in general, $S \cup L_\infty$ is the only algebraic invariant set and $sing(\mathcal{F}) \subset S \cup L_\infty$.

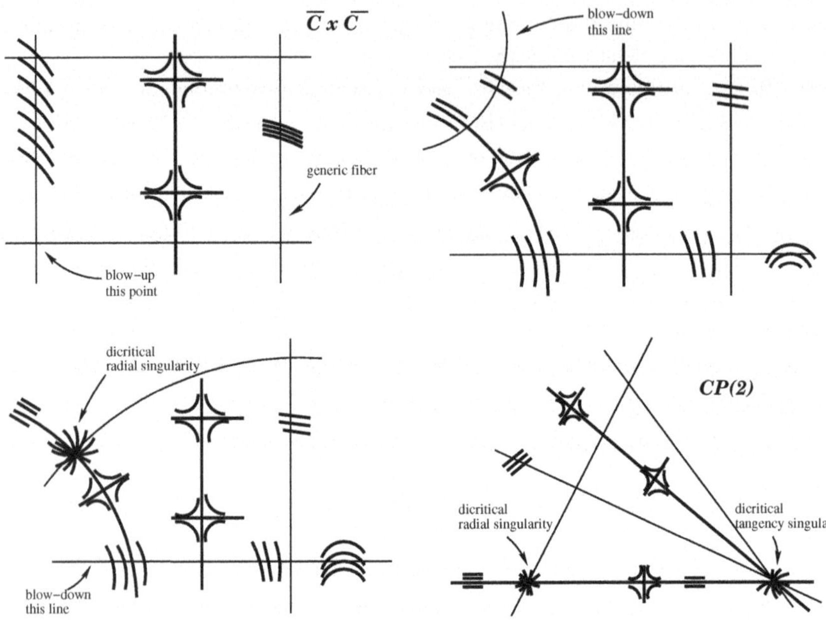

Fig. 1 A Riccati foliation from $\overline{\mathbb{C}} \times \overline{\mathbb{C}}$ to $\mathbb{C}P(2)$

(iii') Finally, we stress that on $\mathbb{C}P(2)$ the foliation $\mathcal{F}$ is transversely projective in a neighborhood of $L_\infty \setminus (L_\infty \cap sing(\mathcal{F}))$.

In this work we shall focus on the problem of extension of the structure to the analytic set, as well as on the consequences of this extension. The very basic result relating transversely homogeneous foliations and suitable systems of differential forms is the classic Darboux-Lie theorem [4, 17, 27].

Example 7 (*pull-backs*) Let $\mathcal{F}$ be a transversely projective foliation on M. Let $\pi : N \to M$ be a holomorphic map transverse to $\mathcal{F}$, then the pull-back foliation $\pi^*(\mathcal{F})$ is transversely projective in N. This can be used to construct examples of foliations on projective manifolds, which are transversely projective outside of some algebraic invariant curve. Take for instance a rational map $\pi : M \to \overline{\mathbb{C}} \times \overline{\mathbb{C}}$ where M is a non-singular projective manifold. Given a Riccati foliation $\mathcal{R}$ on $\overline{\mathbb{C}} \times \overline{\mathbb{C}}$ the pull-back $\mathcal{F} := \pi^*(\mathcal{R})$ is then a foliation on M which is transversely projective outside of some algebraic $C \subset M$ of codimension ≥ 1. As we will see, we can assume that C is invariant by $\mathcal{F}$, otherwise the projective structure extends to some component of C.

Example 8 (*suspensions of subgroups of* $\mathbb{P}SL(2, \mathbb{C})$) A well known way of constructing transversely homogeneous foliations on fibered spaces, having a prescribed holonomy group is the *suspension* of a foliation by a group of biholomorphisms. This construction is briefly described below: Let $G \subset Diff(N)$ be a finitely generated group of biholomorphisms of a complex manifold N. We can regard G as the image of a representation $h\colon \pi_1(M) \to Diff(N)$ of the fundamental group of a complex (connected) manifold M. Considering the universal holomorphic covering of $M, \pi : \widetilde{M} \to M$ we have a natural free action $\pi_1 : \pi_1(M) \times \widetilde{M} \to \widetilde{M}$, i.e., $\pi_1(M) \subset Diff(\widetilde{M})$ in a natural way. Using this we define an action $H : \pi_1(M) \times \widetilde{M} \times N \to \widetilde{M} \times N$ in the natural way: $H = (\pi_1, h)$. The quotient manifold $\frac{\widetilde{M} \times N}{H} = M_h$ is called the *suspension manifold* of the representation h. The group G appears as the *global holonomy* of a natural foliation $\mathcal{F}_h$ on M_h (see [17]), this foliation is called *suspension foliation* of G. When G is (isomorphic to) a finitely generated subgroup of $\mathbb{P}SL(2, \mathbb{C})$ the suspension foliation is transversely projective in M_h.

2.4 Development of a Transversely Projective Foliation

We recall the notion of development of a transversely projective foliation, first mentioned in the Introduction, already adapting it to our current framework. Let $\mathcal{G}$ be a (non-singular) holomorphic foliation on a complex manifold N. Suppose that $\mathcal{G}$ is transversely projective in N. There is a Galoisian (i.e., a transitive) covering $\pi : P \to N$ where π is holomorphic, a homomorphism $h\colon \pi_1(N) \to \mathbb{P}SL(2, \mathbb{C})$ and a holomorphic submersion $\Phi : P \to \mathbb{C}P^1$ such that:

(i) Φ is *h-equivariant*. This means that for any homotopy class $[\gamma] \in \pi_1(N)$, we have

$$h([\gamma])(\Phi(x)) = \Phi(\widetilde{[\gamma]}(x)), \ \forall x \in M \backslash S$$

where by $\widetilde{[\gamma]}\colon P \to P$ we denote the covering map induced by $[\gamma]$ in the Galoisian covering $p\colon P \to N$.

(ii) $\pi^*(\mathcal{G}\big|_N)$ is the foliation defined by the submersion Φ.

In the above construction of the development, we may take P as the universal covering $\pi : \widetilde{N} \to N$ of N. We shall refer to the submersion $\Theta\colon \widetilde{N} \to \mathbb{C}P^1$ as a *multiform first integral* of $\mathcal{G}$ given by the projective structure in N. Given a homotopy class $[\gamma] \in \pi_1(M \backslash S)$, the corresponding *monodromy map* is the image $h([\gamma]) \subset \mathbb{P}SL(2, \mathbb{C})$.

Definition 7 The *global monodromy* of the foliation, with respect to this development, is the image $\mathrm{Mon}(\mathcal{G}) = h(\pi_1(N)) \subset \mathbb{P}SL(2, \mathbb{C})$.

Remark 5 Some remarks about the above construction are: The construction of the development in [17] requires the foliation to be non-singular. Assume now that $\mathcal{F}$ is a foliation with singular set of codimension ≥ 2 on a complex manifold M.

Then $N = M \setminus sing(\mathcal{F})$ is a complex manifold and $\mathcal{G} := \mathcal{F}\big|_N$ is non-singular. By definition $\mathcal{F}$ is transversely projective if and only if $\mathcal{G}$ is transversely projective. Moreover, since $sing(\mathcal{F}) \subset M$ has real codimension ≥ 4, we conclude that there is a natural isomorphism $\pi_1(N) \cong \pi_1(M)$. In particular, we can assume in the above construction that $M = N$, i.e., the notion of development above introduced can be introduced for foliations with singularities. Finally, thanks to Hartogs' extension theorem [18], any holomorphic map from $M \backslash sing(\mathcal{F})$ to $\mathbb{C}P^1$ extends uniquely to a holomorphic map from M to $\mathbb{C}P^1$.

2.5 Holonomy Groups of Transversely Projective Foliations

In what follows we consider the following situation. Let $\mathcal{F}$ be a holomorphic foliation on a complex *surface* M, $\Lambda \subset M$ a closed analytic invariant subset of pure dimension one (a curve) and assume that $\mathcal{F}$ is transversely projective in $M \backslash \Lambda$. We will follow original ideas from [26] in the same vein as in [28].

Monodromy: Using the notion of development we can introduce the notion of *monodromy* of the projective transverse structure of $\mathcal{F}\big|_{M\backslash\Lambda}$ as follows:

Fix a base point $m_0 \in M\backslash\Lambda$ and a local determination f_{m_0} of the submersion Φ in a small ball B_{m_0} centered at m_0 (we have the following commutative diagram)

$$\begin{array}{ccccc} P & \supset & p^{-1}(B_{m_0}) & \Phi\big|_{p^{-1}(B_{m_0})} & \\ p\downarrow & & p\big|_{p^{-1}(B_{m_0})}\downarrow & \searrow & \\ M\backslash\Lambda & \supset & B_{m_0} & \xrightarrow{f_{m_0}} & \mathbb{C}P(1) \end{array}$$

Notice that $p^{-1}(B_{m_0}) = \bigcup_{\alpha\in\mathcal{A}} U_\alpha$, $p\big|_{U_\alpha} : U_\alpha \to B_{m_0}$ is a biholomorphism for each $\alpha \in \mathcal{A}$.

By construction, the total space of the covering $p \colon P \to M\backslash\Lambda$ is obtained by analytic continuation of f_{m_0} along all the elements in $\pi_1(M\backslash\Lambda, m_0)$.

The fiber $p^{-1}(m_0)$ is the set of all local determinations f_{m_0} at m_0. We can, by the general theory of transitive covering spaces, identify the group $Aut(P, p)$ of deck transformations of $p \colon P \to M\backslash\Lambda$ to the quotient $\pi_1(M\backslash\Lambda; m_0)\big/ p_\#\pi_1(P; f_{m_0})$. This is the *monodromy group* of $\mathcal{F}\big|_{M\backslash\Lambda}$ which will be denoted by $\mathrm{Mon}(\mathcal{F}, \Lambda)$.

The *monodromy map* is the natural projection

$$\rho \colon \pi_1(M\backslash\Lambda; m_0) \longrightarrow \pi_1(M\backslash\Lambda; m_0)\big/ p_\#\pi_1(P; f_{m_0}) =: \mathrm{Mon}(\mathcal{F}, \Lambda)$$

Our first remark is the following:

Lemma 1 *The monodromy group* $Mon(\mathcal{F}, \Lambda)$ *is naturally isomorphic to a subgroup of* $\mathbb{P}SL(2, \mathbb{C})$.

Proof This is clear since $\mathcal{F}\big|_{M\setminus\Lambda}$ is transversely projective on $M\setminus\Lambda$.

Holonomy In what follows we consider the following situation. Let $\mathcal{F}$ be a holomorphic foliation on a complex surface M, $\Lambda \subset M$ a closed analytic invariant subset of pure dimension one (a curve) and assume that $\mathcal{F}$ is transversely projective in $M\setminus\Lambda$. Let $S \subset \Lambda$ be an irreducible component of Λ. We suppose that each singular point in S is irreducible and exhibits at most one separatrix transverse to S.

Here we keep on following arguments originally in [26] and mimed in [28]. We proceed to study the holonomy of each irreducible component of M. It is enough to assume that M is the union of a smooth compact curve S and local analytic separatrices $sep(\mathcal{F}, S)$ of $\mathcal{F}$ transverse to S; $M = S \cup sep(\mathcal{F}, S)$, all of them smooth invariant and without triple points. We suppose that $sing(\mathcal{F}) \cap S \neq \emptyset$, each singular point in S is irreducible and, if it admits two separatrices then one is transverse to S). In this case we can consider a C^∞ retraction $r\colon W \to S$ from some tubular neighborhood W of S on M onto S such that, $\forall m \in S$ the fiber $r^{-1}(m)$ is either a disc transverse to $\mathcal{F}$ or a local branch of $sep(\mathcal{F}, S)$ at $m \in sing(\mathcal{F})$. We set $V = W\setminus(M \cap W)$ to obtain a C^∞ fibration $r\big|_V\colon V \to S\setminus sing(\mathcal{F})$ by punctured discs over $S\setminus sing(\mathcal{F})$. Since $\pi_2(S\setminus sing(\mathcal{F})) = 0$ the homotopy exact sequence of the above fibration gives the exact sequence

$$0 \longrightarrow \mathbb{Z} \longrightarrow \pi_1(V, \tilde{m}_0) \stackrel{\tau}{\longrightarrow} \pi_1(S\setminus sing(\mathcal{F}); m_0) \longrightarrow 0$$

where $\tilde{m}_0 \in V$ is a base point and $m_0 \in S\setminus sing(\mathcal{F})$ is its projection and $\tau = (r\big|_V)_\#$.

Now we consider the restriction of the covering space P to V; indeed for our purposes we may assume that $W = M$ and $V = M\setminus\Lambda$ so that we are just considering the space P itself. Let ρ be the monodromy map

$$\rho\colon \pi_1(V; \tilde{m}_0) \longrightarrow \pi_1(V; \tilde{m}_0)\big/ p_\#(\pi_1(p^{-1}(V);\, f_{\tilde{m}_0})) =: \mathrm{Mon}(\mathcal{F}, V)$$

Denote by $\mathrm{Mon}(\mathcal{F}, S)$ the quotient of $\mathrm{Mon}(\mathcal{F}, V)$ by the (normal) subgroup $Ker(\tau) \cong \mathbb{Z}$. Then there is a unique morphism $[\rho]$ such that the diagram commutes:

$$\begin{array}{ccccc} 0 \longrightarrow \mathbb{Z} \rightarrow & \pi_1(V; \tilde{m}_0) & \longrightarrow & \pi_1(S\setminus sing(\mathcal{F}); m_0) \rightarrow 0 \\ \searrow & \rho \downarrow & & [\rho] \downarrow \\ & \mathrm{Mon}(\mathcal{F}, V) & \longrightarrow & \mathrm{Mon}(\mathcal{F}, S) \rightarrow 0 \end{array}$$

The morphism $[\rho]$ is the monodromy of $\mathcal{F}\big|_V$ seen as follows:
given any element $[\gamma] \in \pi_1(S\setminus sing(\mathcal{F}); m_0)$ the monodromy $[\rho]([\gamma])$ is the analytic continuation of the local first integral f_{m_0} along γ and its holonomy lifting. This gives:

Lemma 2 *There exists a surjective group homomorphism* $\alpha\colon Hol(\mathcal{F}, S) \longrightarrow Mon(\mathcal{F}, S)$ *such that the diagram commutes*

$$\begin{array}{ccccc} & & \pi_1(S\backslash sing(\mathcal{F})) & & \\ Hol & \swarrow & & \searrow & [\rho] \\ Hol(\mathcal{F}.S) & & \xrightarrow{\alpha} & & Mon(\mathcal{F}; S) \end{array}$$

where $Hol\colon \pi_1(S\backslash sing(\mathcal{F})) \longrightarrow Hol(\mathcal{F}; S)$ *is the holonomy morphism of the leaf* $S\backslash sing(\mathcal{F})$ *of* $\mathcal{F}$, *and* $[\rho]\colon \pi_1(S\backslash sing(\mathcal{F})) \longrightarrow Mon(\mathcal{F}; S)$ *is as above.*

The kernel of α is the subgroup $Ker(\alpha) < Hol(\mathcal{F}; S)$ of those diffeomorphisms keeping fixed any element $\ell(z)$ of the fiber of $r\big|_V\colon V \to S\backslash sing(\mathcal{F})$ over $m_o \in S\backslash sing(\mathcal{F})$. The *invariance group* of ℓ, $Inv(\ell, z)$, defined as follows $Inv(\ell, z) = \{h \in Diff(\mathbb{C}, 0); \ell \circ h \equiv \ell\}$, where $Diff(\mathbb{C}, 0)$ denotes the group of germs of complex diffeomorphisms fixing the origin $0 \in \mathbb{C}$. Therefore $Ker(\alpha)$ is a subgroup of the invariance group $Inv(\ell, z)$, in the sense that if $p_\ell\colon V_\ell \to D^*$ is the covering space of the punctured disc $D^* = D\backslash\{0\}$ associated to ℓ then $\ell \circ h \equiv \ell$ *means that* $\forall\, m \in D^*, \forall\, \ell_m \in p_\ell^{-1}(m), \exists\, \ell_{h(m)} \in p_\ell^{-1}(h(m)),\ \ \ell_{h(m)} \circ h = \ell_m$.

In particular, to any element $h \in Inv(\ell, z)$ there is associated a pair $(\tilde{h}, h)$ where $\tilde{h}$ is the lifting of h to the covering space V_ℓ defined by $\tilde{h}\colon \ell_m \mapsto \ell_{h(m)}$.
Another lemma we need is:

Lemma 3 *Let* $0 \to G \to H \to K \to 0$ *be an exact sequence of groups. Then* H *is solvable if, and only if,* G *and* K *are solvable.*

From the above discussion we have an exact sequence

$$0 \longrightarrow Ker(\alpha) \longrightarrow Hol(\mathcal{F}, S) \xrightarrow{\alpha} \mathrm{Mon}(\mathcal{F}, S) \longrightarrow 0$$

We claim that $Inv(\ell, z)$ is solvable. Indeed, suppose the contrary. By Nakai's Density Lemma [25] the orbits of a non-solvable subgroup of $Diff(\mathbb{C}, 0)$ are locally dense in a neighborhood Γ of the origin. Let therefore $m \in \Gamma$ be a point and $\Gamma_m \subset \Gamma \setminus \{0\}$ be a small sector with vertex at the origin, such that the orbit of m in Γ_m is dense in Γ_m. Denote by ℓ_{Γ_m} a local determination of ℓ in Γ_m. Then ℓ_{Γ_m} is constant along each orbit of $Inv(\ell, z)$ in Γ_m and the orbit of m is dense in Γ_m so that ℓ_{Γ_m} is constant in Γ_m. By analytic continuation ℓ and the first integral Φ are constant yielding a contradiction. Thus the group $Inv(\ell, z)$ is solvable and therefore embeds in $\mathbb{P}SL(2, \mathbb{C})$. Hence $Hol(\mathcal{F}, S)\big/Ker(\alpha) \simeq \mathrm{Mon}(\mathcal{F}, S)$ embeds in $\mathbb{P}SL(2, \mathbb{C})$ but $Hol(\mathcal{F}, S)$ embeds in $Diff(\mathbb{C}, 0)$, as well as $Ker(\alpha)$ embeds in $Inv(\ell)$ which is a subgroup of $Diff(\mathbb{C}, 0)$ and therefore $Hol(\mathcal{F}, S)\big/Ker(\alpha)$ is isomorphic to a subgroup of $\mathbb{P}SL(2, \mathbb{C})$ with a fixed point. This implies that indeed, $Hol(\mathcal{F}, S)\big/Ker(\alpha)$ is solvable and conjugate to a subgroup of $Aff(\mathbb{C}, 0)$. Therefore $\mathrm{Mon}(\mathcal{F}, S)$ is solvable and by Lemma 3 the holonomy group $Hol(\mathcal{F}, S)$ is solvable.

Summarizing the above discussion we have:

Theorem 3 *Let $\mathcal{F}$ be a holomorphic foliation on a complex surface M, $\Lambda \subset M$ a closed analytic invariant curve and assume that $\mathcal{F}$ is transversely projective in $M \backslash \Lambda$. Let $S \subset \Lambda$ be an irreducible component of Λ. We suppose that each singular point in S is irreducible and exhibits a single separatrix transverse to S. Then the holonomy group $Hol(\mathcal{F}, S)$ of the leaf $S \backslash (sing(\mathcal{F}) \cap S)$ of $\mathcal{F}$ is a solvable group.*

2.6 Transversely Affine Foliations

A particular case of transversely projective foliations is described below. As above, we consider a codimension-one holomorphic foliation $\mathcal{F}$ on a complex manifold M^n, $n \geq 2$, with singular set $sing(\mathcal{F}) \subset M$ of codimension ≥ 2. We say that $\mathcal{F}$ is *transversely affine* in an open subset $U \subset M$ if there exists an open cover $\{U_\alpha\}_{\alpha \in A}$ of $U \backslash sing(\mathcal{F})$ such that there are holomorphic submersions $y_\alpha : U_\alpha \to \mathbb{C}$ such that $\mathcal{F}\big|_{U_\alpha}$ is given by $dy_\alpha = 0$, and for each $U_\alpha \cap U_\beta \neq \emptyset$ we have $y_\beta = a_{\alpha\beta}\, y_\alpha + b_{\alpha\beta}$ for some affine map $(z \mapsto a_{\alpha\beta}\, z + b_{\alpha\beta})$. Transversely affine foliations have been studied by several authors, in the real case [17, 32] and in the holomorphic case [3, 14, 27]. Examples of such complex foliations are *logarithmic foliations* and *Bernoulli foliations* as well as rational pull-backs of such foliations [8, 27]. For all of these, the foliation is transversely affine outside of some algebraic invariant curve $S \subset \mathbb{C}P(2)$. In [27] we find that a foliation $\mathcal{F}$ on $M = \mathbb{C}P(2)$ which is transversely affine outside some algebraic invariant curve $S \subset \mathbb{C}P(2)$ is a logarithmic foliation under some mild conditions on $sing(\mathcal{F}) \cap S$. Relaxing slightly the hypothesis on $sing(\mathcal{F}) \cap S$ we may prove that $\mathcal{F}$ admits a Liouvillian first integral as follows: Let ω be a polynomial one-form which defines $\mathcal{F}$ in some affine space $\mathbb{C}^2 \subset \mathbb{C}P(2)$, then ω admits a one-form η which is rational, with simple poles and such that $d\omega = \eta \wedge \omega$. We call the form η a *generalized integrating factor* for ω. The Liouvillian first integral for $\mathcal{F}$ is $F = \int \frac{\omega}{e^{\int \eta}}$ [8, 33]. Using [8] one may therefore conclude that, under some suitable hyperbolicity hypotheses, either $\mathcal{F}$ is given by a closed rational one-form on $\mathbb{C}P(2)$, or it is a rational pull-back of a Bernoulli foliation as follows $\mathcal{R}$: $p(x)dy - (y^2a(x) + yb(x))dx = 0$.

We separate the following useful definition:

Definition 8 (*generalized integrating factor*) Let Ω be a meromorphic one-form on a complex manifold M. A meromorphic one-form η in M is called a meromorphic *generalized integrating factor* for Ω if we have: (1) $d\Omega = \eta \wedge \Omega$ and (2) $d\eta = 0$. If this is the case then Ω is integrable and defines a foliation $\mathcal{F}$ (holomorphic, of codimension one, with singularities) on M. We shall say that η is a generalized integrating factor *for the foliation $\mathcal{F}$*.

3 Projective Structures and Differential Forms

3.1 Projective Triples

The very basic result relating transversely homogeneous foliations and suitable systems of differential forms is the classical Darboux-Lie theorem [4, 17, 27]. In the case of projective transverse structure this can be stated as:

Proposition 2 ([27], Proposition 1.1 p. 190) *Assume that $\mathcal{F}$ is given by an integrable holomorphic one-form Ω on M and suppose that there exists a holomorphic one-form η on M such that (Proj.1) $d\Omega = \eta \wedge \Omega$. Then $\mathcal{F}$ is transversely projective on M if and only if there exists a holomorphic one-form ξ on M such that (Proj.2) $d\eta = \Omega \wedge \xi$ and (Proj.3) $d\xi = \xi \wedge \eta$.*

The proof is found below.

3.2 Examples

Example 9 Let α be a closed meromorphic one-form on M and let $f : M \to \overline{\mathbb{C}}$ be a meromorphic function. Define (Ω, η, ξ) by: $\Omega = df - f^2\alpha$, $\eta = 2f\alpha$ and $\xi = 2\alpha$. Then (Ω, η, ξ) is a projective triple and therefore Ω defines a holomorphic foliation on M, transversely projective in the complement of the analytic invariant codimension one set $S \subset M$, $S = (\alpha)_\infty \cup (f)_\infty$. The same conclusion holds for $\Omega_\lambda = \Omega + \lambda\alpha$, where $\lambda \in \mathbb{C}$. The foliation $\mathcal{F}(\Omega_\lambda)$ is also transversely affine in some smaller open set of the form $M \backslash S'$ where $S' \supset S$, $S' = S \cup (f^2 - \lambda = 0)$. (In fact $\frac{\Omega_\lambda}{f^2-\lambda} = \frac{df}{f^2-\lambda} - \alpha$ is closed and holomorphic in $M \backslash S'$).

Example 10 Let $h : M \to \mathbb{C}^*$ be holomorphic such that $d\xi = -\frac{dh}{2h} \wedge \xi$ where ξ is holomorphic (we can write this condition as $d(\sqrt{h}.\xi) = 0$). Let F be any holomorphic function and write (for $\lambda \in \mathbb{C}$) $\Omega = F \cdot \left(\frac{dF}{F} - \frac{1}{2}\frac{dh}{h}\right) - \left(\frac{F^2}{2} - \frac{\lambda}{2}h\right).\xi$, $\eta = \frac{1}{2}\frac{dh}{h} + F \cdot \xi$. The triple (Ω, η, ξ) satisfies the conditions of Proposition 2 and then $\mathcal{F} = \mathcal{F}(\Omega)$ is a transversely projective foliation on M.

3.3 Proof of Proposition 2

Let us now give a proof for Proposition 2. We start with a remark about its need.

Remark 6 Proposition 2 is stated (for the real non-singular case) with an idea of its proof, in [17] (see Proposition 3.20, pp. 262). However, it seems that the suggested proof uses some triviality hypothesis on principal fiber-bundles of structural group

$Aff(\mathbb{C})$, over the manifold M (see [17] Proposition 3.6 pp. 249–250). In our case this is replaced by the existence of the form η in the statement. On the other hand, since some of its elements will be useful later, we supply a proof for Proposition 2.

We will use the two following lemmas whose proofs are straightforward consequences of Darboux-Lie theorem, Theorem 2, therefore left to the reader:

Lemma 4 *Let* $x, y, \widetilde{x}, \widetilde{y}\colon U \subset \mathbb{C}^n \to \overline{\mathbb{C}}$ *be meromorphic functions satisfying:*

(i) $ydx - xdy = \widetilde{y}d\widetilde{x} - \widetilde{x}d\widetilde{y}$;

(ii) $\frac{\widetilde{x}}{\widetilde{y}} = \frac{ax+by}{cx+dy}$, $\begin{pmatrix} a & b \\ c & d \end{pmatrix} \in \mathbb{P}SL(2, \mathbb{C})$.

Then $\widetilde{x} = \varepsilon.(ax + by)$ *and* $\widetilde{y} = \varepsilon.(cx + dy)$ *for some* $\varepsilon \in \mathbb{C}$, $\varepsilon^2 = 1$.

Lemma 5 *Let* $x, y, \widetilde{x}, \widetilde{y}\colon U \subset \mathbb{C}^n \to \overline{\mathbb{C}}$ *be meromorphic functions satisfying* $\widetilde{x} = ax + by$, $\widetilde{y} = cx + dy$ *for some* $\begin{pmatrix} a & b \\ c & d \end{pmatrix} \in \mathbb{P}SL(2, \mathbb{C})$. *Then* $xdy - ydx = \widetilde{x}d\widetilde{y} - \widetilde{y}d\widetilde{x}$.

Proof (*Proof of Proposition* 2) Suppose $\mathcal{F}$ is transversely projective in M^n, say, $\{f_i\colon U_i \to \mathbb{C}\}$ is a projective transverse structure for $\mathcal{F}$ in $M\backslash sing(\mathcal{F})$. In each U_i we have $\Omega = -g_i\, df_i$ for some holomorphic $g_i \in \mathcal{O}(U_i)^*$. In each $U_i \cap U_j \neq \phi$ we have: $g_i\, df_i = g_j\, df_j$ and (1) $f_i = \frac{a_{ij} f_j + b_{ij}}{c_{ij} f_j + d_{ij}}$ as in Definition 6. Since $d\Omega = d(-g_i\, df_i) = \frac{dg_i}{g_i} \wedge \Omega$ we have $\eta = \frac{dg_i}{g_i} - h_i \Omega$ for some holomorphic h_i in U_i. We define $x_i, y_i, u_i, v_i\colon U_i \to \mathbb{C}$ in the following way: (2) $y_i^2 = g_i$, $\frac{x_i}{y_i} = f_i$, $h_i = \frac{2v_i}{y_i}$ and$x_i v_i - y_i u_i = 1$. Thus we have: $\Omega = x_i\, dy_i - y_i\, dx_i$ and (3) $\eta = 2(v_i\, dx_i - u_i\, dy_i)$. This motivates us to define local models (see [17] Sect. 3.18 pp. 261): $\xi_i = 2(v_i\, du_i - u_i\, dv_i)$ in U_i. It is easy to check that we have $d\xi_i = \xi_i \wedge \eta$, $d\eta = \Omega \wedge \xi_i$ in U_i. We can assume that dx_i and dy_i are independent for all $i \in I$. In fact $dx_i \wedge dy_i = 0 \Rightarrow d\Omega\big|_{U_i} = 2\, dx_i \wedge dy_i = 0 \Rightarrow d\Omega = 0$ in M (we can assume M to be connected) $\Rightarrow$ we have $0 = d\Omega = \eta \wedge \Omega$ so that $\eta = h\Omega$ for some holomorphic function $h\colon M \to \mathbb{C} \Rightarrow$ we can choose $\xi = \frac{h^2 \Omega}{2} + h\eta + dh$ which satisfies the relations $d\eta = \Omega \wedge \xi$ and $d\xi = \xi \wedge \eta$.

Claim (1) *We have* $\xi_i = \xi_j$ *in each* $U_i \cap U_j \neq \phi$ *and therefore the* ξ_i*'s can be glued into a holomorphic one-form* ξ *in* $M\backslash sing(\mathcal{F})$ *satisfying the conditions of the statement.*

Proof From (1) and (2) we obtain $\frac{x_i}{y_i} = \frac{a_{ij} x_j + b_{ij} y_j}{c_{ij} x_j + d_{ij} y_j}$. Therefore according to Lemma 4 we have (4) $x_i = \varepsilon.(a_{ij} x_j + b_{ij} x_j)$, $y_i = \varepsilon.(c_{ij} x_j + d_{ij} y_j)\, \varepsilon^2 = 1$. Using (3) and (4) we obtain: $(a_{ij} v_i - c_{ij} u_i) dx_j + (b_{ij} v_i - d_{ij} u_i) dy_j = \varepsilon.(v_j\, dx_j - u_j\, dy_j)$ and therefore: (5) $v_j = \epsilon(a_{ij}\, v_i - c_{ij}\, u_i)$, $u_j = \epsilon(-b_{ij}\, v_i + d_{ij}\, u_j)$. It follows form (5) and Lemma 5 that $v_i\, du_i - u_i\, dv_i = v_j\, du_j - u_j\, dv_j$ which proves the claim.

Claim (2) *We have* $\xi = \xi_i = h_i^2 \frac{\Omega}{2} + h_i \eta + dh_i$ *in each* U_i.

Proof We have $h_i^2\Omega = \frac{4v_i^2}{y_i^2}(x_i\,dy_i - y_i\,dx_i)$, $h_i\eta = \frac{4v_i}{y_i}(v_i\,dx_i - u_i\,dy_i)$, $dh_i = 2d\left(\frac{v_i}{y_i}\right)$. Hence $\frac{h_i^2\Omega}{4} + \frac{h_i\eta}{2} + \frac{dh_i}{2} = \frac{v_i^2}{y_i}dx_i - \frac{v_i}{y_i^2}(x_iv_i - 1)dy_i + \frac{dv_i}{y_i}$.

On the other hand a straightforward calculation shows that $\frac{\xi_i}{2} = v_i\,du_i - u_i\,dv_i = \frac{v_i^2}{y_i}dx_i - \frac{v_i}{y_i}(x_iv_i - 1)dy_i + \frac{dv_i}{y_i}$. And thus Claim 2 is proved.

Since $codim\,sing(\mathcal{F}) \geq 2$ it follows that ξ extends holomorphically to M. This proves the first part. Now we assume that (Ω, η, ξ) is *holomorphic* as in the statement of the proposition:

Claim (3) *Given any $p \in M\backslash sing(\mathcal{F})$ there exist holomorphic $x, y, u, v\colon U \to \mathbb{C}$ defined in an open neighborhood $U \ni p$ such that: $\Omega = xdy - ydx$, $\eta = 2(vdx - udy)$ and $\xi = 2(vdu - udv)$.*

Proof This claim is a consequence of Darboux's Theorem (see [17] pp. 230), but we can give an alternative proof as follows: We write locally $\Omega = -gdf = xdy - ydx$ and $\eta = \frac{dg}{g} - h\Omega = 2(vdx - udy)$ as in the proof of the first part. Using Claim 2 above and the last part of Proposition 3 below we obtain locally $\xi = \frac{h^2\Omega}{2} + h\eta + dh + \ell.\Omega$; for some holomorphic function ℓ satisfying $\frac{d\ell}{-2\ell} \wedge \Omega = d\Omega$. This last equality implies that $d(\sqrt{\ell}.\Omega) = 0$ and then $\ell = \frac{r(f)}{g^2}$ for some holomorphic function $r(z)$. Now we look for holomorphic functions $\widetilde{f}$, $\widetilde{g}$ and $\widetilde{h}$ satisfying: $\Omega = -\widetilde{g}d\widetilde{f}$, $\eta = \frac{d\widetilde{g}}{\widetilde{g}} - \widetilde{h}\Omega$ and $\xi = \frac{\widetilde{h}^2\Omega}{2} + \widetilde{h}\eta + d\widetilde{h}$. We try $\widetilde{f} = U(f)$ for some holomorphic non-vanishing $U(z)$. Using $\Omega = gdf = -\widetilde{g}d\widetilde{f}$ we get $\widetilde{g} = \frac{g}{U'(f)}$. Using $\eta = \frac{dg}{g} - d\Omega = \frac{d\widetilde{g}}{\widetilde{g}} - \widetilde{h}\Omega$ we get $\widetilde{h} = h - \frac{U''}{gU'}$. Using $\xi = \frac{h^2\Omega}{2} + h\eta + dh + \ell\Omega = \frac{\widetilde{h}^2\Omega}{2} + \widetilde{h}\eta + d\widetilde{h}$ we get $d\left(\frac{U''(f)}{U'(f)}\right) = r(f)df$.

Therefore it is possible to write Ω, η and ξ as in the statement of the claim: define $x = \widetilde{f}y$, $y = \sqrt{\widetilde{g}}$, $v = \frac{\widetilde{h}y}{2}$ and $u = \frac{xv-1}{y}$ as in the first part of the proof. This proves Claim 3.

Using Claim 3 we prove that $\mathcal{F}$ is transversely projective in $M\backslash sing(\mathcal{F})$, that is in M. The last part of Proposition 2 can be proved using the relation stated above between the projective structure and the local trivializations for Ω, η and ξ. For instance we prove the following.

Claim (4) *The triples (Ω, η, ξ) and $(f\Omega, \eta + \frac{df}{f}, \frac{1}{f}\xi)$ define the same projective structure for $\mathcal{F}$, for any holomorphic $f\colon M \to \mathbb{C}^*$.*

Proof Using the notation of the first part we define $\hat{x}_i = \sqrt{f}.x_i$, $\hat{y}_i = \sqrt{f}.y_i$, $\hat{u}_i = \frac{1}{\sqrt{f}}.u_i$ and $\hat{v}_i = \frac{1}{\sqrt{f}}.v_i$. Then: $f\Omega = \hat{x}_i\,d\hat{y}_i - \hat{y}_i\,d\hat{x}_i$, $\eta + \frac{df}{f} = 2(\hat{v}_i\,d\hat{x}_i - \hat{u}_i\,d\hat{y}_i)$ and $\frac{1}{f}\xi = 2(\hat{v}_i\,d\hat{u}_i - \hat{u}_i\,d\hat{v}_i)$. Furthermore we have $\frac{\hat{x}_i}{\hat{y}_i} = \frac{x_i}{y_i} = \frac{a_{ij}x_j + b_{ij}y_j}{c_{ij}x_j + d_{ij}y_j} = \frac{a_{ij}\hat{x}_j + b_{ij}\hat{y}_j}{c_{ij}\hat{x}_j + d_{ij}\hat{y}_j}$, and this proves the claim and finishes the holomorphic part of the proof.

Now we only have to observe that if (Ω, η) is a pair of meromorphic one-forms and if $\mathcal{F}$ is transversely projective in M, then the same steps of the first part of the proof apply to construct a meromorphic one-form ξ satisfying the relations of the statement.

Let $\mathcal{F}$ be a codimension one holomorphic foliation with singular set $sing(\mathcal{F})$ of codimension ≥ 2 on a complex manifold M. As mentioned in the Introduction, the existence of a projective transverse structure for $\mathcal{F}$ is equivalent to the existence of suitable triples of differential forms (cf. Proposition 2, see also [27] Sect. 3, page 193):
This motivates the following definition:

Definition 9 (*projective triple*) Given holomorphic one-forms (respectively, meromorphic one-forms) Ω, η and ξ on M we shall say that (Ω, η, ξ) is a *holomorphic projective triple* (respectively, a *meromorphic projective triple*) if they satisfy relations *(Proj.1)*, *(Proj.2)* and *(Proj.3)* above. The foliation $\mathcal{F}^\perp$ defined by the one-form ξ is called *transverse foliation* corresponding to the projective triple. If η is not identically zero then $\mathcal{F}^\perp$ is really a foliation on M which is transverse to $\mathcal{F}$ outside of a proper analytic subset.

The following definition will play a fundamental role in the last section of this work.

Definition 10 (*moderate growth (transversely projective foliations)*) A foliation $\mathcal{F}$ on M will be called *transversely projective of moderate growth* if it admits a meromorphic projective triple (Ω, η, ξ) defined in M. This means that $\mathcal{F}$ is transversely projective in some the complementar of some analytic subset $\Lambda \subset M$ of codimension one.

The termonilogy *foliation with moderate growth* has already been introduced in [35]. With the above definitions, and the notation of Proposition 2, this last says that $\mathcal{F}$ is transversely projective on M if and only if the holomorphic pair (Ω, η) may be completed to a holomorphic projective triple. Moreover, a foliation $\mathcal{F}$ which is transversely projective of moderate growth exhibits a projective transverse structure $\mathcal{P}$ in the complement of some codimension divisor $D \subset M$ (D contained in the polar set of the projective triple). One question then is whether the projective transverse structure $\mathcal{P}$ extends to the divisor D. The other question, apparently simpler, is whether the foliation $\mathcal{F}$ is actually projective of moderate growth. According to [27] we may perform modifications in a projective triple as follows:

Proposition 3 ([27]) *Let M be a connected complex manifold.*

(i) *Given a meromorphic projective triple (Ω, η, ξ) and meromorphic functions g, h on M we can define a new meromorphic projective triple as follows:*
(Mod.1) $\Omega' = g\,\Omega$
(Mod.2) $\eta' = \eta + \frac{dg}{g} + h\,\Omega$
(Mod.3) $\xi' = \frac{1}{g}\left(\xi - dh - h\eta - \frac{h^2}{2}\,\Omega\right)$

(ii) *Two holomorphic projective triples (Ω, η, ξ) and (Ω', η', ξ') define the same projective transverse structure for a given foliation $\mathcal{F}$ if and only if we have (Mod.1), (Mod.2) and (Mod.3) for some holomorphic functions g, h with g non-vanishing.*

(iii) *Let* (Ω, η, ξ) *and* (Ω, η, ξ') *be meromorphic projective triples. Then* $\xi' = \xi + F\,\Omega$ *for some meromorphic function* F *in* M *with* $d\,\Omega = -\frac{1}{2}\frac{dF}{F} \wedge \Omega$.

This last proposition implies that suitable meromorphic projective triples also define projective transverse structures. We can rewrite condition (iii) on F as $d(\sqrt{F}\,\Omega) = 0$. This implies that if the projective triples (Ω, η, ξ) and (Ω, η, ξ') are not identical then the foliation defined by Ω is transversely affine outside the codimension one analytical invariant subset $S = \{F = 0\} \cup \{F = \infty\}$ [27].

This approach is useful because of the following proposition:

Proposition 4 ([27] Theorem 4.1 p. 197) *Let* $\mathcal{F}$ *be a foliation on* M *where* M *is either an open polydisc* $M \subset \mathbb{C}^m$ *or a projective manifold over* $\mathbb{C}$ *of dimension* $m \geq 2$. *Assume that* $\mathcal{F}$ *admits a meromorphic projective triple* (Ω, η, ξ) *defined in* M. *If* ξ *admits a meromorphic first integral in* U *then* $\mathcal{F}$ *is a meromorphic pull-back of a Riccati foliation.*

Proof By hypothesis, ξ defines a foliation which admits a meromorphic first integral. Since we are either on a projective manifold or in a polydisc centered at the origin, we can write $\xi = g\,dR$ for some meromorphic functions g and R (these functions are rational in the case of a projective surface). Then we may replace the meromorphic triple (Ω, η, ξ) by (Ω', η', ξ') where $\Omega' = g\Omega$, $\eta' = \eta + \frac{dg}{g}$ and $\xi' = \frac{1}{g}\xi = dR$. The relations $d\Omega' = \eta' \wedge \xi'$, $d\eta' = \Omega' \wedge \xi'$, $d\xi' = \xi \wedge \eta'$ imply that $\eta' = HdR$ for some meromorphic function H. Now we define $\omega := \frac{H^2}{2}\xi' - H\eta' + dH = \frac{1}{2}H^2 dR + dH$ one-form such that $d\omega = -HdH \wedge dR$. On the other hand $\eta' \wedge \omega = HdR \wedge dH = -HdH \wedge dR$. Thus $d\omega = \eta' \wedge \omega$. We also have $d\eta' = dH \wedge dR = (-\frac{1}{2}H^2 dR + dH) \wedge dR = \omega \wedge \xi'$. The meromorphic triple (ω, η', ξ') satisfies the projective relations $d\omega = \eta' \wedge \omega$, $d\eta' = \omega \wedge \xi'$, $d\xi' = \xi' \wedge \eta'$ and therefore by Proposition 3 (iii) we conclude that $\Omega' = \omega + F.\xi'$ for some meromorphic function F such that $d\xi' = \xi' \wedge \frac{1}{2}\frac{dF}{F}$ · This implies $dF \wedge dR \equiv 0$. By the classical Stein Factorization theorem we may assume from the beginning that R has connected fibers and therefore $dF \wedge dR \equiv 0$ implies $F = \varphi(R)$ for some one-variable meromorphic function $\varphi(z) \in \mathbb{C}(z)$. *In the case where* M *is a projective manifold all the meromorphic objects are rational and therefore* $\varphi(z)$ *is also a rational function.* We obtain therefore $\Omega' = -\frac{1}{2}H^2 dR + dH + \varphi(R)dR == dH - (\frac{1}{2}H^2 - \varphi(R))dR$. If we define a meromorphic map $\sigma : M \to \overline{\mathbb{C}} \times \overline{\mathbb{C}}$ by $\sigma(p) = \big(R(p), H(p)\big)$ then clearly $\Omega' = \sigma^*(dy - (\frac{1}{2}y^2 - \varphi(x))dx)$ and therefore $\mathcal{F}$ is the pull-back $\mathcal{F} = \sigma^*(\mathcal{R})$ of the Riccati foliation $\mathcal{R}$ given on $\overline{\mathbb{C}} \times \overline{\mathbb{C}}$ by the meromorphic (rational if M is a projective manifold) one-form $\Omega_\varphi := dy - (\frac{1}{2}y^2 - \varphi(x))dx$.

Definition 11 A meromorphic projective triple (Ω', η', ξ') is *geometric* if it can be written locally as in *(Mod.1)*, *(Mod.2)* and *(Mod.3)* for some (locally defined) holomorphic projective triple (Ω, η, ξ) and some (locally defined) meromorphic functions g, h.

As an immediate consequence we obtain:

Proposition 5 *A geometric projective triple (Ω', η', ξ') defines a transversely projective foliation $\mathcal{F}$ given by Ω' on M.*

Example 11 (*Riccati Foliations - revisited*) Fix affine coordinates $(x, y) \in \mathbb{C}^2$ and consider a polynomial one-form $\Omega = p(x)dy - (y^2 c(x) - yb(x) - a(x))dx$. Then Ω defines a *Riccati foliation* $\mathcal{R}$ on $\overline{\mathbb{C}} \times \overline{\mathbb{C}}$ as seen in Example 6 above. Now we study the Lie Algebra associated to this example. Put $\eta = 2\frac{dy}{y} + \frac{p'+b}{p}\,dx + \frac{2a}{yp}\,dx$ and $\xi = \frac{-2a}{y^2p^2}\,dx$. Then (Ω, η, ξ) satisfies the projective relations stated in Proposition 2. This shows that $\mathcal{F}$ is transversely projective in $\overline{\mathbb{C}} \times \overline{\mathbb{C}}$ minus the algebraic subset $\overline{\{x \in \mathbb{C} \mid p(x) = 0\} \times \overline{\mathbb{C}} \cup \overline{\mathbb{C}} \times \{y = 0\}}$. But since in the case $a(x) \not\equiv 0$, only the subset $S = \{p(x) = 0\} \times \overline{\mathbb{C}}$ is $\mathcal{F}$ invariant it follows that the transverse projective structure extends to $\overline{\mathbb{C}} \times \overline{\mathbb{C}} \backslash S$. Indeed according to Proposition 3 if we define $g = \frac{-1}{p(x)y}$ then $\eta' = \eta + 2g\Omega = \frac{p'-b+2yc}{p}\,dx$ and $\xi' = \xi - 2dg - 2g\eta - 2g^2\Omega = \frac{2c}{p^2}\,dx$; define a triple (Ω, η', ξ') holomorphic in $(\overline{\mathbb{C}} \times \overline{\mathbb{C}}) \setminus S$ which gives a projective structure for $\mathcal{F}$ in this affine set. This projective structure coincides with the one given in $(\overline{\mathbb{C}} \times \overline{\mathbb{C}}) \setminus (S \cup \overline{\mathbb{C}} \times \{y = 0\})$ by (Ω, η, ξ). The one-form η is closed if and only if $a \equiv 0$. Therefore $\mathcal{F}$ is transversely affine in $\overline{\mathbb{C}} \times \overline{\mathbb{C}} \backslash (S \cup \overline{\mathbb{C}} \times \overline{\{y = 0\}})$ if the projective line $\{y = 0\}$ is invariant. The forms (Ω, η', ξ') define a rational projective triple and *the projective transverse structure of the foliation $\mathcal{F}^\perp$ defined by ξ extends from $\mathbb{C}^2 \backslash S$ to $\overline{\mathbb{C}} \times \overline{\mathbb{C}}$*. Indeed, $\mathcal{F}_\xi$ admits a rational first integral. We will see this is a general fact, under suitable hypothesis on the singularities of the foliation $\mathcal{F}$ on $\overline{\mathbb{C}} \times \overline{\mathbb{C}}$, admitting a projective transverse structure in the complementary of an algebraic one dimensional invariant subset $S \subset \overline{\mathbb{C}} \times \overline{\mathbb{C}}$.

Remark 7 (*Ricatti versus logarithmic*) In general, Ricatti foliations are not given by closed one-forms, hence are not logarithmic foliations.

3.4 Germs of Foliations and Foliations on Projective Surfaces

Let $\mathcal{F}$ be a holomorphic foliation aaa of codimension one on $\mathbb{C}P^2$ having singular set $sing(\mathcal{F}) \subset \mathbb{C}P^2$. As it is well-known we can assume that $sing(\mathcal{F})$ is of codimension ≥ 2 and $\mathcal{F}$ is given in any affine space $\mathbb{C}^2 \subset \mathbb{C}P^2$ with coordinates (x, y), by a polynomial one-form $\Omega(x, y) = A(x, y)dx + B(x, y)dy$ with $sing(\mathcal{F}) \cap \mathbb{C}^2 = sing(\Omega)$. In particular $sing(\mathcal{F}) \subset \mathbb{C}P^2$ is a nonempty finite set of points. Given any algebraic subset $S \subset \mathbb{C}P^2$ of dimension one we can therefore always obtain a meromorphic (rational) one-form Ω on $\mathbb{C}P^2$ such that Ω defines $\mathcal{F}$, $(\Omega)_\infty$ is non-invariant and in general position (indeed, we can assume that $(\Omega)_\infty$ is any projective line in $\mathbb{C}P(2)$). Also if we take $\eta_0 = \frac{B_x}{B}\,dx + \frac{A_y}{A}\,dy$ then we obtain a rational one-form such that $d\Omega = \eta_0 \wedge \Omega$ and with polar set given by $(\eta_0)_\infty = \{(x, y) \in \mathbb{C}^2 : A(x, y) = 0\} \cup \{(x, y) \in \mathbb{C}^2 : B(x, y) = 0\} \cup (\Omega)_\infty$. In particular, $(\eta_0)_\infty \cap \mathbb{C}^2$ has order one and the "residue" of η_0 along any component T of $(\Omega)_\infty$ equals $-k$ where k is the order of T as a set of poles of Ω. Any rational one-form η such that $d\Omega = \eta \wedge \Omega$ writes $\eta = \eta_0 + h\Omega$ for some rational function h. We obtain in this way one-forms η with appropriately located set of poles, with respect to $\mathcal{F}$, and applying Propositions 2 and 3 we obtain:

Proposition 6 (foliations on projective spaces) *Let $\mathcal{F}$ be a holomorphic foliation on $\mathbb{C}P(2)$. Assume that $\mathcal{F}$ is transversely projective in $\mathbb{C}P(2)\backslash S$ for some algebraic subset S of dimension one. Then $\mathcal{F}$ has a projective triple (Ω, η, ξ) on $\mathbb{C}P(2) \setminus S$ where Ω and η are rational one-forms and ξ is meromorphic on $\mathbb{C}P(2)\backslash S$. In particular ξ defines a transverse foliation $\mathcal{F}^\perp$ to $\mathcal{F}$ on $\mathbb{C}P(2)\backslash S$ having a projective transverse structure.*

The same holds, with a very similar proof, for germs of foliations at the origin $0 \in \mathbb{C}^2$ where the curve S is replaced by a finite set of local branches of separatrices of the foliation through the singularity. More precisely:

Proposition 7 (germs of foliations) *Let $\mathcal{F}$ be a germ of a holomorphic foliation with a singularity at the origin $0 \in \mathbb{C}^2$. Assume that $\mathcal{F}$ is transversely projective in the complement of an analytic subset $S \subset sep(\mathcal{F}, 0)$ of the set of separatrices through the origin. Then, for a sufficiently small bidisc $0 \in M \subset \mathbb{C}^2$ the germ $\mathcal{F}$ has a projective triple (Ω, η, ξ) where Ω is a holomorphic one-form in M, the form η is meromorphic in M and ξ is meromorphic in $M\backslash S$.*

Remark 8 (*Generalizations for algebraic projective manifolds*) Let us consider M^2 a non-singular algebraic projective surface. Let $\mathcal{F}$ be a foliation on M^2. Since we can define in a natural way, polynomial and rational functions on M^2 we can define in a natural way *algebraic leaves of $\mathcal{F}$*. Let $S \subset M$ be an *algebraic curve*, i.e., a pure codimension one analytic subset. The condition that $M\backslash S$ is affine is equivalent to say that it is a Stein manifold. This does not hold in general, very much depending on the curve $S \subset M$. Any meromorphic function on a projective surface is a rational function. A foliation $\mathcal{F}$ on M is therefore given by a rational one-form Ω on M admitting a rational one-form η such that $d\Omega = \eta \wedge \Omega$. We have then natural versions of Propositions 2, 3 and 6 to this situation.

4 Extension of Projective Triples

In this section we address the following basic problem. Let $\mathcal{F}$ be a foliation on a complex manifold M with a projective transverse structure in $M\backslash S$ for some codimension one analytic subset $S \subset M$. *Under which conditions does the projective structure $\mathcal{P}$ extends to S?*. A more appropriate question may be as follows: suppose that the projective transverse structure $\mathcal{P}$ on $M \setminus S$ is given by a projective triple

(Ω, η, ξ) with (as it is natural to assume), Ω and η meromorphic in M. *Under which conditions does the one-form ξ admits a meromorphic extension to S?* We shall focus on two main cases.

(1) The *local case*, where M is a neighborhood of the origin $0 \in \mathbb{C}^2$. In this case we regard $\mathcal{F}$ as a germ of a foliation at the origin $0 \in \mathbb{C}^2$ and consider S as a subset of its set of separatrices.
(2) The *projective case*, where M is a projective surface. In this case the objects are rational once they are meromorphic in M and $S \subset M$ is an algebraic curve.

4.1 Algebraic Leaves and Local Separatrices

Given a foliation $\mathcal{F}$ on a projective surface M, by an *algebraic leaf* of $\mathcal{F}$ we mean a leaf L of the foliation which is contained in an algebraic curve in M. Thanks to the Identity Principle and to Remmert–Stein extension theorem, a leaf L of $\mathcal{F}$ is algebraic if and only if it accumulates only at singular points of $\mathcal{F}$. In this case the algebraic curve consists of the leaf and such accumulation points. The following remark will be useful:

Lemma 6 ([30] Lemma 7.5 (iii)) *Let $\mathcal{F}$ and $\mathcal{F}_1$ be distinct foliations on a projective surface M. If a leaf L of $\mathcal{F}$ is also a leaf of $\mathcal{F}_1$ then this leaf is algebraic.*

Proof We choose affine coordinates $(x, y) \in M$ and polynomial equations for $\mathcal{F}$ and $\mathcal{F}_1$ in these coordinates, say: $\mathcal{F}$ is given by $\frac{dy}{dx} = \frac{P(x,y)}{Q(x,y)}$ and $\mathcal{F}_1$ by $\frac{dy}{dx} = \frac{P_1(x,y)}{Q_1(x,y)}$ where P, Q and P_1, Q_1 are relatively prime polynomials. Suppose $(x(z), y(z))$, $z \in V \subset \mathbb{C}$ is a common solution of the foliations $\mathcal{F}$ and $\mathcal{F}_1$ on M. Then we have

$$\frac{P(x(z), y(z)}{Q(x(z), y(z))} = \frac{dy/dz}{dx/dz} = \frac{P_1(x(z), y(z))}{Q_1(x(z), y(z))}$$

so that $(PQ_1 - P_1Q)(x(z), y(z)) = 0$. By hypothesis $PQ_1 - P_1Q \not\equiv 0$ so that L satisfies the non-trivial algebraic equation $PQ_1 - P_1Q = 0$. It follows that L is algebraic.

The following statement is about transversely projective foliations with moderate growth (cf. Definition 10). It is a compilation of some results above and a preparatory step for the final conclusion:

Theorem 4 *Let $\mathcal{F}$ be a foliation on a projective surface M, with a projective transverse structure outside of an algebraic subset $S \subset M$ of dimension one. Let (Ω, η, ξ) be a <u>rational</u> projective triple defining the projective transverse structure outside of the curve S. We have the following possibilities:*

1. *S contains all the non-dicritical separatrices of $\mathcal{F}$ in S.*
2. *There is some singularity $p \in sing(\mathcal{F}) \cap S$ and a (non-dicritical) separatrix Γ of $\mathcal{F}$ through p, which is not contained in S. In this case we have the following possibilities:*
   - *(a) The leaf containing Γ is not algebraic and $\mathcal{F}^\perp$-invariant. In this case $\mathcal{F}^\perp$ coincides with $\mathcal{F}$, η is closed and $\mathcal{F}$ admits a rational generalized integrating factor.*
   - *(b) The leaf containing Γ is not algebraic and is not $\mathcal{F}^\perp$ invariant.*
   - *(c) The leaf containing Γ is algebraic. In this case $\mathcal{F}$ is transversely affine in $M \setminus (S \cup A)$ for some algebraic invariant curve $A \subset M$ not contained in S.*

Proof We perform the resolution of singularities for $\mathcal{F}$ in S and obtain a projective surface $\tilde{M}$ and a resolution morphism $\sigma\colon \tilde{M} \to M$, a divisor $E = \sigma^{-1}(S) = D \cup \tilde{S}$, where D is the exceptional divisor and $\tilde{S}$ is the strict transform of S, equipped with a pull-back foliation $\tilde{\mathcal{F}} = \sigma^*(\mathcal{F})$ with irreducible singularities in E. The foliation $\tilde{\mathcal{F}}$ is transversely projective in $\tilde{M} \setminus E$. By Lemma 10 the projective transverse structure of $\tilde{\mathcal{F}}$ extends to the non-invariant part of D so that, for our purposes we may assume that D is $\tilde{\mathcal{F}}$-invariant, *though not necessarily connected.* If S contains all the non-dicritical separatrices of $\mathcal{F}$ in S then we are in case (1).

Thus, from now on we suppose that there is a singular point $\tilde{q} \in \tilde{S} \cap sing(\tilde{\mathcal{F}})$ such that $\tilde{\mathcal{F}}$ exhibits some local separatrix $\tilde{\Gamma}$ through $\tilde{q}$ which is not contained in E. Denote by $\tilde{\mathcal{F}}^\perp = \sigma^*(\mathcal{F}^\perp)$ the inverse image of $\mathcal{F}^\perp = \mathcal{F}_\xi$ on $\tilde{M}$. Assume that the leaf $\tilde{A}$ of $\tilde{\mathcal{F}}$ containing $\tilde{\Gamma}$ is not algebraic. In this case its projection $A = \sigma(\tilde{A})$ onto M is not algebraic. We have two possibilities. If $\tilde{\Gamma}$ is $\mathcal{F}_\xi$-invariant then by, Lemma 6, $\mathcal{F}$ coincides with $\mathcal{F}^\perp$ and we are in case (2)(a) in the statement. The second possibility is that $\tilde{\Gamma}$ is not $\tilde{\mathcal{F}}$-invariant. This corresponds to case (2)(b) in the statement.

Assume now that $\tilde{A} \supset \tilde{\Gamma}$ is an algebraic leaf of $\tilde{\mathcal{F}}$ not contained in E. This algebraic leaf projects onto an algebraic leaf A of $\mathcal{F}$, not contained in S. The projective transverse structure of $\mathcal{F}$ has A as a set of fixed points and therefore $\mathcal{F}$ is transversely affine in $M \setminus (S \cup A)$ what corresponds to case (2) (c) in the statement.

Thought the above statement already gives some information, it remains to study the last case, 2(c) above. We must explore the consequences of the existence of a non-dicritical separatrix which is not contained in the curve S, in the final description of the foliation. This is done in what follows. In few words, for the case of extended generalized curves, this allows to extend the projective triple, more precisely, the one-form ξ extends to the irreducible component of S that contains this separatrix.

4.2 Extension of Projective Triples (irreducible Case)

Our main extension result for projective triples is so far the following:

Theorem 5 *Let $\mathcal{F}$ be a holomorphic foliation (respectively a germ of a holomorphic foliation) on a projective surface U (respectively at the origin of $\mathbb{C}^2$). Assume that $\mathcal{F}$ is transversely projective in $U \setminus S$ where $S \subset U$ is an algebraic invariant curve in the projective surface (respectively a finite union of local branches of non-dicritical separatrices of $\mathcal{F}$ through the origin and U is a bidisc centered at the origin $0 \in \mathbb{C}^2$, where $\mathcal{F}$ has a representative). Suppose that the singularities of $\mathcal{F}$ in S are non-resonant extended generalized curves. Then $\mathcal{F}$ admits a meromorphic projective triple (respectively a germ of a meromorphic projective triple) (Ω, η, ξ) defined in U (respectively at the origin), which defines the projective transverse structure in $U \setminus S$ (respectively in the complement of S).*

In this section we pave the way to the proof of Theorem 5.

We recall the following fundamental result from [35]:

Theorem 6 (Touzet, [35] Theorem II.3.1 p. 821) *A non-degenerate non-resonant singularity $xdy - \lambda ydx + \Omega_2(x, y) = 0$, $\lambda \in \mathbb{C} \setminus \mathbb{Q}_+$, is analytically linearizable if and only if the corresponding foliation $\mathcal{F}$ is transversely projective in $U \setminus sep(\mathcal{F}, U)$ for some neighborhood U of the singularity.*

One other tool is discussed below. Let $\mathcal{F}$ be a germ of an *irreducible* singularity at the origin $0 \in \mathbb{C}^2$, assumed to be of resonant type or of saddle-node type. According to [35], Theorem II.4.2, *the foliation admits a meromorphic projective triple in a neighborhood U of the singularity if and only if in a neighborhood of the singularity $\mathcal{F}$ is the pull-back of a Riccati foliation on $\overline{\mathbb{C}} \times \overline{\mathbb{C}}$ by a meromorphic map*. The proof of this theorem is based in the study and classification of the Martinet-Ramis cocycles of the singularity expressed in terms of some classifying holonomy map of a separatrix of the singularity. For a resonant singularity any of the two separatrices has a classifying holonomy and for a saddle-node it is necessary to consider the strong manifold holonomy map. Thus we conclude that the proof given in [35] actually shows that:

Lemma 7 ([35], Theorem II.4.2) *Let $\mathcal{F}$ be a germ of an* irreducible *singularity at the origin $0 \in \mathbb{C}^2$, assumed to be of resonant type or of saddle-node type. The germ $\mathcal{F}$ is the pull-back of a Riccati foliation on $\overline{\mathbb{C}} \times \overline{\mathbb{C}}$ by a meromorphic map if and only if there exists a meromorphic projective triple (Ω', η', ξ') in a neighborhood U_0 of a separatrix $S \subset sep(\mathcal{F}, U)$ provided that S is the strong separatrix if the origin is a saddle-node.*

As a further motivation for our approach we mention two results which are proved in [9]. Such results imply the existence of a globally defined projective triple, parting from a *geometric* projective triple, in a situation similar to the one we are dealing with:

Proposition 8 ([9]) *Let $\mathcal{F}$ be a holomorphic foliation in a neighborhood V of the origin $0 \in \mathbb{C}^2$ given by the holomorphic one-form Ω admitting a meromorphic one-form η in V with $d\Omega = \eta \wedge \Omega$. Suppose that $\mathcal{F}$ has an irreducible non-degenerate singularity at the origin and is transversely projective in $U \setminus sep(\mathcal{F}, U)$ for some* *neighborhood* $U \subset V$ *of the origin where* $\mathcal{F}$ *has an expression in irreducible normal form. Let be given a holomorphic one-form* ξ *defined in* $U \setminus sep(\mathcal{F}, U)$ *such that* (Ω, η, ξ) *is a geometric projective triple in* $U \setminus sep(\mathcal{F}, U)$*. Then* ξ *extends as a meromorphic one-form to* U*. provided that, if the singularity is resonant,* ξ *extends as a meromorphic one-form to* $S^* = S - \{0\}$*, for some separatrix* $S \subset sep(\mathcal{F}, U)$*.*

This proposition or the Globalization theorem in [9] give for the *non-dicritical* case:

Proposition 9 (extension conditions) *Let* $\mathcal{F}$ *be a holomorphic foliation defined in a neighborhood* V *of* $0 \in \mathbb{C}^2$ *with an isolated non-dicritical singularity at the origin. Suppose that* $\mathcal{F}$ *is transversely projective in* $U \setminus sep(\mathcal{F}, U)$ *for some neighborhood* $U \subset V$ *of the origin where* $\mathcal{F}$ *is given by a holomorphic one-form* Ω *admitting a meromorphic one-form* η *such that* $d\Omega = \eta \wedge \Omega$ *in* U*. Let* ξ *be a meromorphic one-form defined in* $U \setminus sep(\mathcal{F}, U)$ *such that* (Ω, η, ξ) *is a geometric projective triple. Let* $\pi : \tilde{U} \to U$ *be the reduction morphism of the singularity and denote by* $(\tilde{\Omega}, \tilde{\eta}, \tilde{\xi})$ *the pull-back by* π *of the triple* (Ω, η, ξ)*. Then the one-form* ξ *extends to* U *provided that:*

- *(Ext.1) At any non-resonant irreducible singularity of the foliation, the form* $\tilde{\xi}$ *admits a meromorphic extension (from a neighborhood of the singularity minus its separatrices) to a neighborhood of the singularity.*
- *(Ext.2) At any resonant irreducible singularity of the foliation, the one-form* $\tilde{\xi}$ *admits a meromorphic extension (from a neighborhood of an annulus contained in one of the separatrices and around the singularity) to a neighborhood of the singularity.*

The (extension) conditions of the proposition above are satisfied in our current situation, as we will see below (cf. Proposition 10).

We shall reprove and extend these results by considering meromorphic triples, but which are not assumed to be geometric projective triples along the separatrices.

Remark 9 The above additional assumption (that ξ can be chosen holomorphic off the set of separatrices) is not restrictive. Indeed, in the sequel (in the paper), the foliation is assumed to be transversely projective off the set of local separatrices. Since Ω is defined meromorphic in a neighborhood of the singularity, we can assume that it is holomorphic otherwise we replace it conveniently (see also Lemma 11). Thus, if we write the one-form $\Omega = A(x, y)dx + B(x, y)dy$ with A, B holomorphic with an isolated common zero at the origin, then we can choose $\eta = \frac{B_x}{B}dx + \frac{A_y}{A}dy$. The polar set of η is contained in the curves $\{A = 0\}$ and $\{B = 0\}$. So we can assume in the case of a non-degenerate non-resonant singularity that the poles of η are contained in the separatrices, which are the coordinate axes in suitable coordinates. Under this hypothesis, the hypothesis of existence of a projective transverse structure off the separatrices gives a holomorphic one-form ξ in the complement of the separatrices, such that Ω, η, ξ is a (holomorphic) geometric projective triple off the axes.

Next we show that the (extension) conditions in Proposition 9 are satisfied and that we can apply some of these techniques also in the dicritical case. In order to do this we remake the basic steps with the necessary changes. The starting point is the non-resonant case considered below:

Proposition 10 (non-resonant case) *Suppose that the origin is a nondegenerate non-resonant singularity. Assume that $\mathcal{F}$ is transversely projective on $U \setminus sep(\mathcal{F}, U)$. Let η be a meromorphic one-form on U and ξ be a meromorphic one-form on $U \setminus sep(\mathcal{F}, U)$ such that on $U \setminus sep(\mathcal{F}, U)$ the one-forms Ω, η, ξ define a projective triple. Then ξ extends as a meromorphic one-form to U.*

Before going into the proof we state a lemma:

Lemma 8 (non-resonant case) *Let ℓ be a meromorphic function in $U^* = U \setminus \{xy = 0\}$ such that $d\Omega = -\frac{1}{2}\frac{d\ell}{\ell} \wedge \Omega$ where $\Omega = g(xdy - \lambda ydx)$ for some holomorphic non-vanishing function g in U and $\lambda \in \mathbb{C} \setminus \mathbb{Q}$. Then $\ell = \tilde{c}.(gxy)^{-2}$ for some constant $\tilde{c} \in \mathbb{C}$.*

Proof Fix a complex number $a \in \mathbb{C}$ and introduce the one-form $\eta_0 = \frac{d(xyg)}{xyg} + a(\frac{dy}{y} - \lambda\frac{dx}{x})$ in U. Since $\frac{\Omega}{gxy} = \frac{dy}{y} - \lambda\frac{dx}{x}$ is closed it follows that $d\Omega = \eta_0 \wedge \Omega$. Thus the one-form $\Theta := -\frac{1}{2}\frac{d\ell}{\ell} - \eta_0$ is closed meromorphic in U^* and satisfies $\Theta \wedge \Omega = d\Omega - d\Omega = 0$. This implies that $\Theta \wedge (\frac{dy}{y} - \lambda\frac{dx}{x}) = 0$ in U^* and therefore we have $\Theta = h.(\frac{dy}{y} - \lambda\frac{dx}{x})$ for some meromorphic function h in U^*. Taking exterior derivatives we conclude that $dh \wedge (\frac{dy}{y} - \lambda\frac{dx}{x}) = 0$ in U^* and therefore h is a meromorphic first integral for Ω in U^*. Since $\lambda \notin \mathbb{Q}$ we must have $h = c$, a constant: indeed, write $h = \sum\limits_{i,j \in \mathbb{Z}} h_{ij}x^i y^j$ in Laurent series in a small bidisc around the origin. Then from $dh \wedge (\frac{dy}{y} - \lambda\frac{dx}{x}) = 0$ we obtain $(i + \lambda j)h_{ij} = 0$, $\forall (i, j) \in \mathbb{Z} \times \mathbb{Z}$ and since $\lambda \notin \mathbb{Q}$ this implies that $\lambda_{ij} = 0$, $\forall (i, j) \neq (0, 0)$.

This already shows that the one-form Θ always extends as a meromorphic one-form with simple poles to U and therefore the function ℓ extends as a meromorphic function to U. The residue of Θ along the axis $\{y = 0\}$ is given by $Res_{\{y=0\}}\Theta = -Res_{\{y=0\}}\frac{1}{2}\frac{d\ell}{\ell} - Res_{\{y=0\}}\eta_0 = -\frac{1}{2}k - (1 + a)$ where $k \in \mathbb{N}$ is the order of $\{y = 0\}$ as a set of zeroes of ℓ or minus the order of $\{y = 0\}$ as a set of poles of ℓ. Thus by a suitable choice of a we can assume that $Res_{\{y=0\}}\Theta = 0$ and therefore by the expression $\Theta = c(\frac{dy}{y} - \lambda\frac{dx}{x})$ we conclude that, for such a choice of a, we have $0 = \Theta = -\frac{1}{2}\frac{d\ell}{\ell} - \eta_0$ and thus $-\frac{1}{2}\frac{d\ell}{\ell} = \frac{dx}{x} + \frac{dy}{y} + \frac{dg}{g} + a(\frac{dy}{y} - \lambda\frac{dx}{x})$ and therefore, comparing residues along the axes $\{y = 0\}$ and $\{x = 0\}$ we obtain that $1 + a \in \mathbb{Q}$ and $1 - a\lambda \in \mathbb{Q}$. Since $\lambda \notin \mathbb{Q}$ the only possibility is $a = 0$. This proves that indeed $-\frac{1}{2}\frac{d\ell}{\ell} = \frac{dx}{x} + \frac{dy}{y} + \frac{dg}{g}$ in U and integrating this last expression we obtain $\ell = \tilde{c}(gxy)^{-2}$ for some constant $\tilde{c} \in \mathbb{C}$. This proves the lemma.

Now we can prove Proposition 10.

Proof (*Proof of Proposition* 10) By hypothesis the foliation is given in suitable local coordinates around the origin by $xdy - \lambda ydx + \Omega_2(x, y) = 0$ where $\lambda \in \mathbb{C}\backslash\mathbb{Q}$, $\Omega_2(x, y)$ is a holomorphic one-form of order ≥ 2 at $0 \in \mathbb{C}^2$.

Claim *The singularity is analytically linearizable.*

Indeed, if $\lambda \notin \mathbb{R}_-$ then the singularity is in the Poincaré domain with no resonance and by Poincaré-Linearization Theorem the singularity is analytically linearizable. Assume now that $\lambda \in \mathbb{R}_-\backslash\mathbb{Q}_-$. In this case the singularity is in the Siegel domain and, a priori, it is not clear that the singularity is linearizable. Nevertheless, by hypothesis $\mathcal{F}$ is transversely projective in $U^* = U \setminus sep(\mathcal{F}, U)$ and by Theorem 6 the singularity $p \in sing(\mathcal{F})$ is analytically linearizable. This proves the claim.

Therefore we can suppose that $\Omega\big|_U = g(xdy - \lambda ydx)$ for some holomorphic non-vanishing function g in U. We define $\eta_0 = \frac{dg}{g} + \frac{dx}{x} + \frac{dy}{y}$ in U. Then η_0 is meromorphic and satisfies $d\Omega = \eta_0 \wedge \Omega$ so that $\eta = \eta_0 + h\Omega$ for some meromorphic function h in U. We also take $\xi_0 = 0$ so that $d\eta_0 = 0 = \Omega \wedge \xi_0$ and $d\xi_0 = 0 = \xi_0 \wedge \eta$. The triple (Ω, η_0, ξ_0) is a meromorphic projective triple in U so that according to Proposition 3 we can define a meromorphic projective triple (Ω, η, ξ_1) in U by setting $\xi_1 = \xi_0 - dh - h\eta_0 - \frac{h^2}{2}\Omega = -dh - h\eta_0 - \frac{h^2}{2}\Omega$. Then we have by Proposition 3 (iii) that $\xi = \xi_1 + \ell.\Omega$ for some meromorphic function ℓ in U^* such that $d\Omega = -\frac{1}{2}\frac{d\ell}{\ell} \wedge \Omega$.

By Lemma 8 above we have $\ell = \tilde{c}.(gxy)^{-2}$ in U^* and therefore ξ extends to U as $\xi = \xi_1 + \tilde{c}.(gxy)^{-2}$ in U^*. This proves the lemma.

Now we deal with the second extension condition (Ext. 2) in Proposition 9. The first step is:

Lemma 9 (resonant case) *Let $\mathcal{F}$ be a germ of a holomorphic foliation with a resonant irreducible (non-degenerate) singularity at the origin $0 \in \mathbb{C}^2$ and let $0 \in U \subset \mathbb{C}^2$ be a bidisc centered at the origin where $\mathcal{F}$ is defined by a holomorphic one-form Ω. Denote by $sep(\mathcal{F}, U)$ the set of local separatrices of $\mathcal{F}$ through the origin in U. Let ℓ be a meromorphic function in $U \setminus sep(\mathcal{F}, U)$ such that $d\,\Omega = -\frac{1}{2}\frac{d\ell}{\ell} \wedge \Omega$. Then ℓ extends as a meromorphic function to U provided that ℓ admits a meromorphic extension to (a neighborhood of) $S^* = S \setminus \{0\}$ for some separatrix $S \subset sep(\mathcal{F}, U)$. Indeed, we have the following possibilities for $\mathcal{F}$ in suitable coordinates in a neighborhood of the origin:*

- (i) *$\mathcal{F}$ is analytically linearizable, i.e., analytically conjugate to the form $xdy - \lambda ydx = 0$ for some $\lambda \in \mathbb{C} \setminus \{0\}$.*
- (ii) *$\mathcal{F}$ is a non-linearizable resonance analytically conjugate to the normal form: $\Omega_{n,m} = ny\,dx + mx(1 + \frac{\sqrt{-1}}{2\pi}x^n y^m)dy = 0$ where $n, m \in \mathbb{N}$.*

In all cases S is given by $\{y = 0\}$ and the function ℓ extends as meromorphic function to a neighborhood of the origin.

Proof We define the one-form $\eta = -\frac{1}{2}\frac{d\ell}{\ell}$. Then η is a closed meromorphic one-form in $U \setminus [sep(\mathcal{F}, U) \setminus S]$ such that $d\Omega = \eta \wedge \Omega$, moreover the polar set of η is contained in S and has order at most one. If η is holomorphic in $U \setminus [sep(\mathcal{F}, U) \setminus S]$

then the foliation $\mathcal{F}$ is transversely affine in $U \setminus [sep(\mathcal{F}, U) \setminus S]$ and therefore the holonomy map of the leaf $L_0 = S \setminus \{0\}$ is linearizable. Since the origin is irreducible and S is not a central manifold the conjugacy class of this holonomy map classifies the foliation up to analytic conjugation. Thus the singularity is itself linearizable. Assume now that $(\eta)_\infty \neq \emptyset$. In this case we have the residue of η along S given by $Res_S \eta = -\frac{1}{2}k$ where k is either the order of S as zero of ℓ or minus the order of S as pole of ℓ. We have two possibilities:

(a) If $-\frac{1}{2}k \notin \{2, 3, \ldots\}$ then by [27], Lemma 3.1, the holonomy map of the leaf L_0 is analytically linearizable and the same holds for the singularity.

(b) If $-\frac{1}{2}k = t + 1 \geq 2$ for some $t \in \mathbb{N}$ then by [27], Lemma 3.1, the holonomy map of L_0 is conjugate to a map of the form $h(z) = \frac{\alpha z}{(1+\beta z^t)^{\frac{1}{t}}}$, i.e., this is a finite ramified covering of an homography. Suppose that the singularity is nondegenerate say $\Omega = xdy - \lambda ydx + \ldots$. If $\lambda \notin \mathbb{Q}$ then the map $h(z)$ is analytically linearizable and therefore, again, the singularity is linearizable. Suppose now that the map h is not analytically linearizable. Then we must have $\lambda = -\frac{n}{m}$ for some $n, m \in \mathbb{N}$, $< n, m >= 1$ and the holonomy h is analytically conjugate to the corresponding holonomy of the germ of singularity $\Omega_{n,m} = ny\,dx + mx(1 + \frac{\sqrt{-1}}{2\pi} x^n y^m)dy$; such a singularity is called a *non-linearizable resonant saddle*. As it is well-known, in the Siegel domain and in particular in the class of resonant singularities, the analytical classification of the holonomy implies the analytical classification of the singularity. More precisely, by [23, 24] we may assume that $\mathcal{F}$ is of the form $\Omega_{n,m} = 0$ in the variables $(x, y) \in U$. So far we have proved that the following are the possibilities for the singularities:

(1) The singularity is analytically linearizable, this is the case if it is not a resonance.
(2) The singularity is analytically conjugated to $\Omega_{n,m}$ if it is resonant and not analytically linearizable.

Let us now use these two models in order to conclude the extension of ℓ to U.
Case 1. In the linearizable case we can write $S : \{y = 0\}$ and $\Omega = g(xdy - \lambda ydx)$ for some holomorphic non-vanishing function g in U. If we introduce $\eta_0 = \frac{d(gxy)}{gxy}$ then we have $d\Omega = \eta_0 \wedge \Omega$ and therefore $(\eta - \eta_0) \wedge \Omega = 0$ so that $(\eta - \eta_0) \wedge (\frac{dy}{y} - \lambda\frac{dx}{x}) = 0$ and then $\eta = \eta_0 + F.(\frac{dy}{y} - \lambda\frac{dx}{x}) = 0$ for some meromorphic function F in $U_0 := U \setminus [sep(\mathcal{F}, U) \setminus S]$. Since η and η_0 are closed we conclude that $d(F.(\frac{dy}{y} - \lambda\frac{dx}{x})) = 0$ in U_0. Write now $F = \sum\limits_{i,j \in \mathbb{Z}} F_{ij} x^i y^j$ in Laurent series in a small bidisc around the origin. We obtain from the last equation that $(i + \lambda j)F_{ij} = 0$, $\forall i, j \in \mathbb{Z}$. If $\lambda \notin \mathbb{Q}$ this implies that $F = F_{00}$ is constant. Assume now that $\lambda = -\frac{n}{m} \in \mathbb{Q}_-$. Then we have $\Omega \wedge d(x^n y^m) = 0$ and also $F = \varphi(x^n y^m)$ for some function $\varphi(z) = \sum\limits_{t \in \mathbb{Z}} \varphi_t z^t$ defined in a punctured disc around the origin. Nevertheless, the function F is meromorphic along the axis $\{y = 0\}$ and therefore φ admits a meromorphic extension to the origin $0 \in \mathbb{C}$ and thus F extends as a meromorphic function $F = \varphi(x^n y^m)$ to a neighborhood of the origin.
Case 2. In the non-linearizable (resonant) case we can write $S : \{y = 0\}$ and $\Omega = g\,\Omega_{n,m} = g(ny\,dx + mx(1 + \frac{\sqrt{-1}}{2\pi} x^n y^m)dy)$ for some holomorphic non-vanishing

function g on U. Define $\eta_0 = \frac{d(gx^{n+1}y^{m+1})}{gx^{n+1}y^{m+1}}$. As above we conclude that $\eta = \eta_0 + F.(n\frac{dx}{x^{n+1}y^m} + m\frac{dy}{x^n y^{m+1}} + \frac{m\sqrt{-1}}{2\pi}\frac{dy}{y})$ for some meromorphic function F in U_0 such that $dF \wedge (n\frac{dx}{x^{n+1}y^m} + m\frac{dy}{x^n y^{m+1}} + \frac{m\sqrt{-1}}{2\pi}\frac{dy}{y}) = 0$. In other words, F is a meromorphic first integral in U_0 for the foliation $\mathcal{F}$. This implies that F is constant. In order to see this it is enough to use Laurent series as above. Alternatively one can argue as follows. If F is not constant then the holonomy map h of the leaf $L_0 \subset S$ leaves invariant a nonconstant meromorphic map (the restriction of the first integral F to a small transverse disc to S). This implies that h is a map with finite orbits and indeed h is periodic. Nevertheless this is never the case of the holonomy map of the separatrix $\{y = 0\}$ of the foliation $\Omega_{n,m}$. Thus the only possibility is that F is constant.

Summarizing the above discussion, we have proved that in all cases $\eta = \eta_0 + F.\omega$ for some meromorphic function F in U and some meromorphic closed one-form ω in U. Moreover, F is constant except in the resonant case. This shows that $\eta = -\frac{1}{2}\frac{d\ell}{\ell}$ admits a meromorphic extension to U and therefore also ℓ admits a extends meromorphic extension to U. The lemma is proved.

The remaining step for the irreducible resonant case is the following:

Proposition 11 (resonant case) *Let $\mathcal{F}$ be a germ of a holomorphic foliation with a resonant (irreducible) singularity at the origin $0 \in \mathbb{C}^2$ and let $0 \in U \subset \mathbb{C}^2$ be a bidisc centered at the origin where $\mathcal{F}$ is defined by a holomorphic one-form Ω. Fix a separatrix $S \subset sep(\mathcal{F}, U)$. Let η be a meromorphic one-form in U and ξ be a meromorphic one-form in $(U \setminus sep(\mathcal{F}, U)) \cup S$ such that in $U \setminus sep(\mathcal{F}, U)$ the one-forms Ω, η, ξ define a projective triple. Then ξ extends as a meromorphic one-form to U.*

Proof By hypothesis we are in the resonant case, i.e., $\Omega = g(xdy - \lambda ydx + \ldots)$ with $\lambda = -\frac{n}{m} \in \mathbb{Q}_-$. Suppose first that the singularity is not analytically linearizable. As we have seen in Lemma 7, $\mathcal{F}$ is the pull-back of a Riccati foliation on $\overline{\mathbb{C}} \times \overline{\mathbb{C}}$ by some meromorphic map $\sigma : U \to \overline{\mathbb{C}} \times \overline{\mathbb{C}}$ provided that there is a meromorphic projective triple (Ω', η', ξ') in a neighborhood W of a separatrix $S \subset sep(\mathcal{F}, U)$. From our hypothesis such a projective triple is given by the restrictions of Ω and η to $U \setminus [sep(\mathcal{F}, U) \setminus S]$ and by the one-form ξ. Thus we conclude that $\mathcal{F}$ is a meromorphic pull-back of a Riccati foliation and in particular there is a one-form ξ' defined in a neighborhood $\tilde{U}$ of the origin such that (Ω, η, ξ') is a projective triple in this neighborhood. This implies that $\xi = \xi' + \ell.\Omega$ in $\tilde{U}$ for some meromorphic function ℓ in $\tilde{U}$ such that $d\Omega = -\frac{1}{2}\frac{d\ell}{\ell}$ in $\tilde{U}$. Now we have two possibilities. Either $\xi = \xi'$ in $\tilde{U}$ or $\ell \not\equiv 0$. In the first case ξ extends meromorphically to U as $\xi = \xi'$. In the second case we apply Lemma 9 above in order to conclude that the singularity is analytically normalizable and ℓ extends as a meromorphic function to U. Suppose now that the singularity is resonant analytically linearizable and $\mathcal{F}$ is given in U by $\Omega = g\big(xdy + \frac{n}{m}\,ydx\big)$ where $n, m \in \mathbb{N}$ and g is a meromorphic function in U. In this case as above we define $\eta_0 = \frac{dg}{g} + \frac{dx}{x} + \frac{dy}{y}$, write $\eta = \eta_0 + h\Omega$ and define $\xi_0 = 0$, $\xi_1 = \xi_0 - dh - h\eta_0 - \frac{h^2}{2}\,\Omega = -dh - h\eta_0 - \frac{h^2}{2}\,\Omega$. Now we have $\xi = \xi_1 + \ell\Omega$

for some meromorphic function ℓ in U^*. In this case we have from $d\ell = -\frac{1}{2}\frac{d\ell}{\ell} \wedge \Omega$ that $\ell(gxy)^2 = [\varphi(x^n y^m)]^2$ for some meromorphic function $\varphi(z)$ defined in a punctured neighborhood of the origin $0 \in \mathbb{C}$. In particular we conclude that since ξ extends meromorphically to some separatrix $\{x = 0\}$ or $\{y = 0\}$ then it extends meromorphically to U.

Thus ξ extends as a meromorphic one-form to U in all cases proving the desired result.

Remark 10 Propositions 6, 10 and 11 already prove Theorem 5 in the case of a germ of a foliation with an irreducible singularity at the origin.

4.3 Extension to Non-invariant Divisors

Since we are considering the possibility of existence of non-invariant components in the exceptional divisor, we shall be able to extend the projective triple to such components. This is done by means of the following lemma regarding the non-invariant case:

Lemma 10 (non-invariant divisor, [9]) *Let be given a holomorphic foliation $\mathcal{F}$ on a complex surface M. Suppose that $\mathcal{F}$ is given by a meromorphic integrable one-form Ω which admits a meromorphic one-form η on M such that $d\Omega = \eta \wedge \Omega$. If $\mathcal{F}$ is transversely projective in $M \backslash S$ for some non-invariant irreducible analytic subset $S \subset M$ of codimension one then $\mathcal{F}$ is transversely projective in M. Indeed, the projective transverse structure for $\mathcal{F}$ in $M \backslash S$ extends to M as a projective transverse structure for $\mathcal{F}$. In particular, if ξ is a meromorphic one-form in $M \setminus U$ such that (Ω, η, ξ) is a projective triple on $M \backslash S$, then ξ admits a meromorphic extension to S.*

Proof Our argumentation is local, i.e., we consider a small neighborhood U of a generic point $q \in S$ where $\mathcal{F}$ is transverse to S. Thus, since S is not invariant by $\mathcal{F}$, performing changes as $\Omega' = g_1 \Omega$ and $\eta' = \eta + \frac{dg_1}{g_1}$ we can assume that Ω and η have poles in general position with respect to S in U. The existence of a projective transverse structure for $\mathcal{F}$ in $M \backslash S$ then gives a meromorphic one-form ξ in $M \setminus S$ such (Ω, η, ξ) is a geometric projective triple in $M \setminus S$. For U small enough we can assume that for suitable local coordinates $(x, y) \in U$ we have $S \cap U = \{x = 0\}$ and also

$$\Omega = g dy, \eta = \frac{dg}{g} + h dy$$

for some holomorphic function $g, h \colon U \to \mathbb{C}$ with $1/g$ also holomorphic in U. Then we have

$$\xi = -\frac{1}{g}[dh + \frac{h^2}{2} dy]$$

where

$$d(\sqrt{\ell} g dy) = 0$$

Thus, $\sqrt{\ell} g = \varphi(y)$ for some meromorphic function $\varphi(y)$ defined for $x \neq 0$ and therefore for $x = 0$. This shows that ξ extends to U as a *holomorphic one-form* and then the projective structure extends to U. This shows that the transverse structure extends to S.

4.4 Extended Generalized Curves

Let us consider the general case, where we allow singularities which are not irreducible, but belong to the the class of (non-resonant) *generalized curve*. For this type of singularity we have the following extension result:

Proposition 12 (extension - generalized curve) *Let $\mathcal{F}$ be a germ of a non-resonant (and non-dicritical) generalized curve at the origin $0 \in \mathbb{C}^2$. Suppose that $\mathcal{F}$ is transversely projective in $U \setminus sep(\mathcal{F}, U)$, for some bidisc U centered at the origin, and let (Ω, η, ξ) be a meromorphic projective triple in $U \setminus sep(\mathcal{F}, U)$ with Ω holomorphic in U, η meromorphic in U and ξ meromorphic in $U \setminus sep(\mathcal{F}, U)$. Then the one-form ξ extends to U as a meromorphic one-form.*

Proof Let $\pi : \tilde{U} \to U$ be the reduction morphism of the singularity and denote by $(\tilde{\Omega}, \tilde{\eta}, \tilde{\xi})$ the pull-back by π of the triple (Ω, η, ξ). Because the singularity is non-dicritical, the exceptional divisor $E = \pi^{-1}(0) = \bigcup_{j=1}^{r} \mathbb{P}_j$ in the reduction process is connected and invariant. By the non-resonance hypothesis, this divisor contains some non-resonant singularity say $p_0 \in \mathbb{P}_{j_0}$. Thanks to Hartogs' extension theorem, the one-form $\tilde{\xi}$ also extends to the irreducible component $\mathbb{P}_{j_0}$ minus the singularities of the lifted foliation. Now according to Propositions 10 and 11, the form $\tilde{\xi}$ also extends to all the components $\mathbb{P}_j$ intersecting $\mathbb{P}_{j_0}$. The same argument and the connectedness of E show that the projective triple $(\tilde{\Omega}, \tilde{\eta}, \tilde{\xi})$ extends to a neighborhood of the exceptional divisor.

We will prove a more general case in what follows. In a natural extension of the arguments in the proof of Proposition 12 we obtain the following result which is Theorem 5 in the local case.

Proposition 13 (extension - extended generalized curve) *Let $\mathcal{F}$ be a germ of a holomorphic foliation at the origin $0 \in \mathbb{C}^2$. Suppose that for some small bidisc U centered at the origin, the representative of $\mathcal{F}$ is transversely projective in $U \setminus S$ where $S \subset \mathcal{ND}(sep(\mathcal{F}, U)) \subset sep(\mathcal{F}, U)$ is a (finite) union of local branches, all of them corresponding to non-dicritical separatrices. Assume that the singularity $0 \in S$ is a non-resonant extended generalized curve. Then $\mathcal{F}$ admits in U a meromorphic projective triple. Indeed, let (Ω, η, ξ) be a meromorphic triple in $U \setminus sep(\mathcal{F}, U)$ with*

Ω holomorphic in U, η meromorphic in U and ξ meromorphic in $U \setminus sep(\mathcal{F}, U)$. Then the one-form ξ extends to U as a meromorphic one-form.

Proof (Proof of Theorem 5) The proof is similar to the one given for the case of generalized curve (Proposition 12). The existence of a meromorphic projective triple (Ω, η, ξ) with Ω, η meromorphic in U and ξ in $U \setminus S$ is granted by Proposition 6. Notice that by hypothesis each branch S_j in S is a non-dicritical separatrix and therefore it meets the exceptional divisor in some singular point $\tilde{p}_j \in sing(\tilde{\mathcal{F}})$ of an invariant component $\mathbb{P}(S_j)$. We have $S_j \cap \mathbb{P}(S_j) = \{\tilde{p}_j\}$, where we still denote by S_j the strict transform of S_j.

By the non-resonance hypothesis the component $\mathbb{P}(S_j)$ belongs to a connected component $E(S_j)$ of the invariant part of E, which contains some non-resonant singularity $\tilde{q}_j \in E(S_j) \cap sing(\tilde{\mathcal{F}})$. Therefore, by the same arguments in the proof of Proposition 11 we conclude that third form of the pull-back projective triple $(\tilde{\Omega}, \tilde{\eta}, \tilde{\xi})$ extends as a meromorphic one-form to this component each connected component of the invariant part of the exceptional divisor E obtained in the reduction of the singularity. The extension of $\tilde{\xi}$ to the non-invariant components of the exceptional divisor is granted by Lemma 10. If a connected component E_i of the invariant part of E does not contain a singularity belonging to a branch of S, still it contains some non-resonant singularity and the extension to E_i is assured as above. Thus Theorem 5 is proved in the local situation. The global case, i.e., the case of foliations on projective surfaces, is proved in the same way.

For the case of projective surfaces we promptly have:

Theorem 7 *Let $\mathcal{F}$ be a holomorphic foliation by curves on a projective manifold M. Assume that $\mathcal{F}$ is transversely projective in $M \backslash S$ where $S \subset M$ is an algebraic curve. Suppose that the singularities of $\mathcal{F}$ in S are non-resonant extended generalized curves. Then $\mathcal{F}$ admits a rational projective triple (Ω, η, ξ), which defines the projective structure for $\mathcal{F}$ in $M \setminus S$.*

4.5 Extension of Projective Structures $\mathcal{P}$

In this section we investigate the extension not only of meromorphic projective triples but, of projective transverse structures (generically denoted by $\mathcal{P}$) to a codimension one divisor. According to Lemma 10 we may assume that the divisor is invariant by the foliation.

Proposition 14 (extension through a point) *Let (Ω, η, ξ) be a meromorphic projective triple on a complex surface M^2, and $S \subset M$ an irreducible analytic subset of dimension one. Suppose that the triple defines a projective transverse structure $\mathcal{P}$ outside S. If there is a point $q \in S$ and a neighborhood $q \in U \subset M$ to which the projective structure $\mathcal{P}$ extends, then this projective structure extends to M.*

Proof According to the preceding lemma, we may assume that S is $\mathcal{F}$-invariant. We consider the local case where the foliation $\mathcal{F}$ is given by a holomorphic one-form Ω in an open subset $W \subset \mathbb{C}^n$ with isolated zeros and admitting a meromorphic one-form η on W satisfying $d\Omega = \eta \wedge \Omega$. We can assume that Ω and η have poles in general position with respect to S.

For $U \subset W$ small enough we can find a holomorphic submersion $y\colon U \to \mathbb{C}$ and meromorphic functions $g,\ h$ in U such that

$$\Omega = g dy, \eta = \frac{dg}{g} + h dy,\ \xi = -\frac{1}{g}\big[dh + \frac{h^2}{2} dy\big] + \ell g dy$$

where

$$d(\sqrt{\ell} g dy) = 0.$$

Thus, $\sqrt{\ell} g = \varphi(y)$ for some meromorphic function $\varphi(z)$ and therefore $\ell = \frac{\varphi^2(y)}{g^2}$. Hence we have

$$\Omega = g dy, \eta = \frac{dg}{g} + h dy,\ \xi = -\frac{1}{g}\big[dh + \frac{h^2}{2} dy\big] + \frac{\varphi^2(y)}{g} dy$$

We investigate under which conditions we can write

$$\Omega = \tilde{g} d\tilde{y}, \eta = \frac{d\tilde{g}}{\tilde{g}} + \tilde{h} d\tilde{y},\ \xi = -\frac{1}{\tilde{g}}\big[d\tilde{h} + \frac{\tilde{h}^2}{2} d\tilde{y}\big]$$

for some suitable meromorphic functions $\tilde{g}, \tilde{h}, \tilde{y}$.

Imposing the above equations we obtain

$$\begin{cases} g dy = \tilde{g} d\tilde{y} \\ \frac{dg}{g} + h dy = \frac{d\tilde{g}}{\tilde{g}} + \tilde{h} d\tilde{y} \\ -\frac{1}{g}\big[dh + \frac{h^2}{2} dy\big] + \frac{\varphi^2(y)}{g} dy = -\frac{1}{\tilde{g}}\big[d\tilde{h} + \frac{\tilde{h}^2}{2} d\tilde{y}\big] \end{cases} \tag{1}$$

We shall refer to equations in (1) as *main equations*. From $g dy = \tilde{g} d\tilde{y}$ we obtain $g = r(y)\tilde{g}$ for some meromorphic function $r(y)$. This implies $d\tilde{y} = r(y) dy$ and then $\frac{dg}{g} + h dy = \frac{d\tilde{g}}{\tilde{g}} + \frac{r'(y)}{r(y)} dy + h dy$ so that replacing in the second main equation we obtain $\frac{d\tilde{g}}{\tilde{g}} + \tilde{h} d\tilde{y} = \frac{d\tilde{g}}{\tilde{g}} + \frac{r'(y)}{r(y)} dy + h dy$ and then $\frac{r'(y)}{r(y)} dy + h dy = \tilde{h} d\tilde{y} = \tilde{h} r(y) dy$. This last equation rewrites

$$\frac{r'(y)}{r(y)} + h = \tilde{h} r(y) \tag{2}$$

and the final form

$$\tilde{h} = \frac{1}{r(y)}\big[\frac{r'(y)}{r(y)} + h\big] \tag{3}$$

Let us turn our attention to the third main equation. From this we obtain

$$\frac{1}{g}\left[dh+\left(\frac{h^2}{2}-\varphi^2(y)\right)dy\right]=\frac{1}{\tilde{g}}\left[d\tilde{h}+\frac{\tilde{h}^2}{2}d\tilde{y}\right]$$

Then

$$\frac{\tilde{g}}{g}\left[dh+\left(\frac{h^2}{2}-\varphi^2(y)\right)dy\right]=d\tilde{h}+\frac{\tilde{h}^2}{2}d\tilde{y}$$

$$\frac{1}{r(y)}\left[dh+\left(\frac{h^2}{2}-\varphi^2(y)\right)dy\right]=d\tilde{h}+\frac{\tilde{h}^2}{2}d\tilde{y}$$

$$\frac{1}{r(y)}\left[dh+\left(\frac{h^2}{2}-\varphi^2(y)\right)dy\right]=d\tilde{h}+\frac{\tilde{h}^2}{2}r(y)dy$$

$$dh+\left(\frac{h^2}{2}-\varphi^2(y)\right)dy=r(y)\left[d\left(\frac{1}{r(y)}\left(\frac{r'(y)}{r(y)}+h\right)\right)+\frac{1}{2r(y)^2}\left(\frac{r'(y)}{r(y)}+h\right)^2r(y)dy\right]$$

$$dh+\left(\frac{h^2}{2}-\varphi^2(y)\right)dy=r(y)\left[d\left(\frac{1}{r(y)}\left(\frac{r'(y)}{r(y)}+h\right)\right)+\frac{1}{2}\frac{1}{r(y)}\left(\frac{r'(y)}{r(y)}+h\right)^2dy\right]$$

$$dh+\left(\frac{h^2}{2}-\varphi^2(y)\right)dy=\frac{1}{2}\left(\frac{r'(y)+h}{r(y)}\right)^2dy-\frac{r'(y)}{r(y)}\left(\frac{r'(y)}{r(y)}+h\right)dy+d\left(\frac{r'(y)}{r(y)}+h\right)$$

This last equation is equivalent to

$$-\varphi^2(y)=-\frac{1}{2}\left(\frac{r'(y)}{r(y)}\right)^2+\left(\frac{r'(y)}{r(y)}\right)' \tag{4}$$

Let us put

$$s(y):=\frac{r'(y)}{r(y)}$$

Then equation (4) rewrites

$$s'-\frac{1}{2}s^2=-\varphi^2 \tag{5}$$

So, the original question is reduced to find conditions under which the equation above has a holomorphic solution. This is the case, for instance if φ is holomorphic. Now we need to return to equation $\frac{r'(y)}{r(y)}=s(y)$ and study its solutions. It is clear from integration that there is a holomorphic solution, which must be given by $r(y)=$

$e^{\int s(y)dy}$, if and only if the given data $s(y)$ is either holomorphic or meromorphic with a simple pole and integral positive residue at $y = 0$.

First case. If $s(y)$ has a simple pole at $y = 0$. We may assume for simplicity that $s(y) = a/y$ for some $a \in \mathbb{C}^*$. In this case from the differential equation $s' - s^2/2 = -\varphi^2$ we obtain $\varphi = \frac{\sqrt{2a-a^2}}{y}$. Integrating $r(y) = e^{\int s(y)dy}$ we obtain $r(y) = y^a$. Since $r(y) = g/\tilde{g}$ we have that $r(y)$ is holomorphic without zeros. In particular we cannot have $a \neq 0$, contradiction.

Second case. If $s(y)$ has a pole of order $m + 1 \geq 2$ at $y = 0$. In this case we can assume that $s(y) = a/y^{m+1}$ for some $m \geq 1$ and integration gives $r(y) = e^{-\frac{a}{my^m}}$ which is not meromorphic at the origin, contradiction.

Third case. If $s(y)$ is holomorphic at $y = 0$. In this case we write $s(y) = ay^m$ for some $m \geq 0$. We obtain $r(y) = e^{\frac{a}{m+1}y^{m+1}}$ which is holomorphic and non-vanishing.

Let us now finish the proof. Because the projective structure extends to U the equation (1) has a holomorphic solution and this implies that $\varphi(y)$ is holomorphic according to the above considerations. As a consequence the one-form ξ is also holomorphic in U and therefore admits a holomorphic extension to $S \setminus [(\Omega)_\infty \cup (\eta)_\infty]$. Hence, the projective structure extends to $S \setminus [(\Omega)_\infty \cup (\eta)_\infty]$ and then to S.

The next lemma shows that once we have fixed the forms Ω and η associated to a transverse projective structure, then we may replace the third form ξ without changing the invariant set S to which we wish to extend the structure.

Lemma 11 *Let (Ω, η, ξ) be a meromorphic projective triple in a complex surface M. Assume that the triple defines a projective transverse structure for $\mathcal{F}$ in $M \backslash S$ for some invariant codimension one analytic subset $S \subset M$. Let ξ' be a meromorphic one-form in M such that (Ω, η, ξ') is also a projective triple. Then S is ξ-invariant if and only if it is ξ'-invariant.*

Proof We fix a local coordinate system $(x, y) \in U$ centered at a point $p \in M$ such that $\mathcal{F}$ is given in these coordinates by $\Omega = gdy$ and S by $\{y = 0\}$. We may write $\xi' = \xi + \ell\Omega$ where $d(\sqrt{\ell}\Omega) = 0$. Then we have $\ell = \frac{\varphi^2(y)}{g}$ for some meromorphic function $\varphi(z)$. Assume by contradiction that S is not ξ-invariant but S is ξ'-invariant. We may assume that the polar set of ξ has no irreducible component contained in S and therefore $\varphi(y)$ and g have no poles on $\{y = 0\}$. Write $\xi' = Adx + Bdy$ with holomorphic coefficients $A(x, y)$, $B(x, y)$. Since S is ξ'-invariant we have $A(x, y) = y A_1(x, y)$ for some holomorphic function $A_1(x, y)$. Then from $\xi' = \xi + \ell\Omega$ we get $\xi = yA_1(x, y)dx + (B(x, y) - \frac{\varphi^2(y)}{g})dy$. Since A_1 and $B(x, y) - \frac{\varphi^2(y)}{g}$ have no poles in $\{y = 0\}$ we conclude from the above expression that S is ξ-invariant, contradiction.

5 Classification of Transversely Projective Foliations

5.1 Classification of Transversely Projective Foliations: Non-dicritical Case

We consider now an application of the above study to the classification of foliations with projective transverse structure. Nevertheless, *because of the non-dicriticalness hypothesis on the singularities, we will still be dealing with the affine case* (see Remark 11 (i)). The (dicritical) projective *non-affine* case will be dealt with later on in this work. We point out that the non-dicriticity hypothesis excludes the "pure" transversely projective case, i.e., the case where the structure is not transversely affine in some other "affine" subset. We prove:

Theorem 8 *Let $\mathcal{F}$ be a germ of a (non-dicritical) holomorphic foliation at the origin $0 \in \mathbb{C}^2$. Suppose that:*

(i) *$\mathcal{F}$ is a germ of a non-resonant generalized curve and can be reduced with a single blow-up.*
(ii) *$\mathcal{F}$ is transversely projective outside of the set $sep(\mathcal{F}, 0)$ of local separatrices of $\mathcal{F}$ through 0.*

Then $\mathcal{F}$ admits a generalized integrating factor. In particular, $\mathcal{F}$ is transversely affine in some neighborhood of the origin minus its set of local separatrices $sep(\mathcal{F}, 0)$.

As for the global case we have:

Theorem 9 *Let $\mathcal{F}$ be a foliation on a compact projective surface M. Assume that $\mathcal{F}$ is transversely projective in the complement of an algebraic invariant curve $S \subset M$. Suppose that for some smooth irreducible component $S_0 \subset S$ we have:*

(i) *The singularities of $\mathcal{F}$ in S_0 are irreducible and non-degenerate, one of which is non-resonant.*
(ii) *$M \setminus S_0$ is a Stein manifold.*

Then $\mathcal{F}$ admits a rational generalized integrating factor. In particular $\mathcal{F}$ is transversely affine in an open subset $M \setminus C$ for some algebraic curve $C \subset M$.

We point-out that, since the singularities in S_0 are irreducible non-degenerate, usually the non-resonance hypothesis appearing in Theorem 8 is automatic. Indeed, for instance for the case of the projective plane $M = \mathbb{C}P(2)$ this is a consequence of the Index theorem [7] and of the special geometry of $\mathbb{C}P(2)$. Actually, we can state:

Theorem 10 *Let $\mathcal{F}$ be a foliation on the projective plane $\mathbb{C}P(2)$, which is transversely projective in the complement of an algebraic curve $S \subset \mathbb{C}P(2)$. Suppose that for some smooth irreducible component $S_0 \subset S$ the singularities of $\mathcal{F}$ in S_0 are irreducible and non-degenerate. Then $\mathcal{F}$ admits a rational generalized integrating factor. In particular $\mathcal{F}$ is transversely affine in an open subset $M \setminus C$ for some algebraic curve $S \subset C \subset \mathbb{C}P(2)$.*

As we see from Examples 6 and 11 the general Riccati case appears when we allow the curve S to have some dicritical singularities.

Let us pave the way to the proof of Theorems 8 and 9. Let $G \subset Diff(\mathbb{C}, 0)$ be a solvable subgroup of germs of complex diffeomorphisms fixing the origin $0 \in \mathbb{C}$. We recall that [13] if the group of commutators $[G, G]$ is not cyclic (in particular G is solvable not abelian) then G is analytically conjugate to a subgroup of $\mathbb{H}_k = \left\{z \mapsto \frac{az}{\sqrt[k]{1+bz^k}}; a \in \mathbb{C} \setminus \{0\}, b \in \mathbb{C}\right\}$ for some $k \in \mathbb{N}$. This is the case if G (is solvable and) contains some non-resonant element. Using this we can state the following well-known technical result.

Lemma 12 *Let $G < Diff(\mathbb{C}, 0)$ be a solvable subgroup of germs of holomorphic diffeomorphisms fixing the origin $0 \in \mathbb{C}$ containing some non-resonant element $f \in G$ of the form $f(z) = e^{2\pi i\lambda} z + \dots$ with $\lambda \in \mathbb{C}\backslash\mathbb{Q}$. We have the following possibilities:*

(i) *G is abelian. In this case G admits a formal meromorphic invariant one-form.*
(ii) *If G is not abelian then f is analytically linearizable in a coordinate that also embeds G into some $\mathbb{H}_k$.*

Proof (i) is in [13]. Indeed, it is well-known that G admits a formal invariant holomorphic vector field say $\hat{M}$ with an isolated singularity at the origin $0 \in \mathbb{C}$. Such a vector field can be written $\hat{M}(z) = \frac{y^{k+1}}{1+\lambda y^k}\frac{\partial}{\partial y}$, for some $k \in \mathbb{N}$ and some $\lambda \in \mathbb{C}$. Moreover, according to [13] (see also [5, 28, 29]), because this group contains some non-resonant element, this vector field is indeed analytic. Now we take the corresponding dual one-form $\hat{\omega} = \frac{\lambda y^k+1}{y^{k+1}} dy$. Since $\hat{M}$ is invariant by the maps in G the same holds for $\hat{\omega}$. This proves (i).

Now we prove (ii). Since G contains a non-resonant element we can, as already observed above, choose a holomorphic coordinate $z \in (\mathbb{C}, 0)$ which embeds G as a subgroup of the group $\mathbb{H}_k$ for some $k \in \mathbb{N}$. Given then a non-resonant map $f \in G$ we can write $f(z) = \frac{e^{2\pi i\lambda} z}{\sqrt[k]{1+bz^k}}$ for some $k \in \mathbb{N}$, $b \in \mathbb{C}$. Since $\lambda \in \mathbb{C}\backslash\mathbb{Q}$ the homography $H(z) = \frac{e^{2\pi i\lambda} z}{1+bz}$ is conjugate by another homography to its linear part $z \mapsto e^{2\pi i\lambda} z$ and therefore f is analytically linearizable in a coordinate that also embeds G into $\mathbb{H}_k$.

Proof (*Proof of Theorem* 8) Let $\mathcal{F}$ be defined in an open bidisc $0 \in U \subset \mathbb{C}^2$ by the holomorphic one-form Ω. Put $\tilde{\mathcal{F}} = \pi^*(\mathcal{F})$ in $\tilde{U} = \pi^{-1}(U)$ where $\pi : \tilde{\mathbb{C}}^2_0 \to \mathbb{C}^2$ is the blow-up of $\mathbb{C}^2$ at $0 \in \mathbb{C}^2$. Let also $\tilde{\Omega} = \pi^*(\Omega)$ be the lift of Ω to $\tilde{U}$. The exceptional divisor $S = \pi^{-1}(0)$ is a compact invariant curve (a projective line). Each singularity of $\tilde{\mathcal{F}}$ in S is irreducible and exhibits a separatrix transverse to S. This set of separatrices (of $\tilde{\mathcal{F}}$ transverse to S) is $sep(\tilde{\mathcal{F}}, S) = \overline{\pi^{-1}(sep(\mathcal{F}, 0)\backslash\{0\})} = \pi^{-1}(sep(\mathcal{F}, 0))\backslash S$ in $\tilde{U}$. Now, because of (ii) the pull-back foliation $\tilde{\mathcal{F}}$ is transversely projective in $\tilde{U}\backslash\tilde{M}$ where $\tilde{M} = S \cup sep(\tilde{\mathcal{F}}, S)$. According to Theorem 3 this implies that the holonomy group $Hol(\tilde{\mathcal{F}}, S)$ of the leaf $S\backslash sing(\tilde{\mathcal{F}})$ of $\tilde{\mathcal{F}}$ is solvable. We have two cases to consider:

Case 1. The group $Hol(\tilde{\mathcal{F}}, S)$ is abelian.

Because this holonomy group is analytically conjugate to an abelian subgroup of $Diff(\mathbb{C}, 0)$, it follows from Lemma 12 (i) that there exists a meromorphic integrating factor $\tilde{h}$ for $\tilde{\Omega}$, defined over the open curve $S_0 = S \backslash \text{sing}(\tilde{\mathcal{F}})$. By this we mean a meromorphic function $\tilde{h}$ defined in a neighborhood of S_0 such that the form $\frac{1}{\tilde{h}}\tilde{\Omega}$ is closed. Moreover, according to [5, 28, 29], because of the hypothesis on the singularities in S, this integrating factor extends as a meromorphic integrating factor for $\tilde{\Omega}$ in a neighborhood of S. Therefore, the foliation $\tilde{\mathcal{F}}$ is defined by a closed meromorphic one-form $\tilde{\omega} = \frac{1}{\tilde{h}}\tilde{\Omega}$ in a neighborhood of S.

Case 2. The holonomy group $Hol(\tilde{\mathcal{F}}, S)$ is solvable but not abelian. By the non-resonance hypothesis this group contains some element of the form $f(z) = e^{2\pi i\lambda} z + \dots$ with $\lambda \in \mathbb{C}\backslash\mathbb{Q}$. By Lemma 12 this map f is analytically linearizable and in the same analytic coordinate that embeds the holonomy group in the group

$$\mathbb{H}_k = \left\{\varphi(z) = \frac{az}{\sqrt[k]{1+bz^k}}, a \neq 0\right\} \quad \text{for some} \quad k \in \mathbb{N}.$$

According to Sect. 5 in [27] (see also [8, 28]), this implies that the foliation is transversely affine in the complement of its set of separatrices, admits a so called *closed logarithmic derivative* which is a *closed* meromorphic one-form $\tilde{\eta}_0$, with simple poles defined in a small neighborhood of the origin. The form $\tilde{\eta}_0$ satisfies

$$d\tilde{\Omega} = \tilde{\eta}_0 \wedge \tilde{\Omega}.$$

Now we can "project" the one-form $\tilde{\eta}_0$ via the blow-up map $\pi : \tilde{\mathbb{C}}_0^2 \to \mathbb{C}^2$ onto a one-form η_0 defined in a punctured neighborhood of the origin. This one-form satisfies $\tilde{\eta}_0 = \pi^*(\eta_0)$ and, by classical Hartogs' extension theorem [18] it extends (to the origin) as a meromorphic one-form in a neighborhood of the origin. It is clear that η_0 is closed and satisfies $d\Omega = \eta_0 \wedge \Omega$. This proves Theorem 8.

In the same line of reasoning we can prove Theorem 10:

Proof (*Proof of Theorem* 10) We know that $\mathcal{F}$ is given by a rational one-form Ω on $\mathbb{C}P(2)$. We shall prove that Ω admits a rational generalized integrating factor η on $\mathbb{C}P(2)$. This is partially done as in the proof of Theorem 8. Nevertheless, in order to mimic the proof of Theorem 8 we must prove:

Claim *Some singularity in S_0 is non-resonant.*

Proof (*Proof of Claim* 5.1) Recall that an irreducible non-degenerate singularity can be written in the form $xdy - \lambda ydx + h.o.t. = 0$, where $\lambda \in \mathbb{C} \setminus \mathbb{Q}_+$ and $\{xy = 0\}$ is the set of local separatrices. If we fix the separatrix $\{y = 0\}$ then the Index of the singularity with respect to this separatrix is given by λ. With respect to the other separatrix the index is $1/\lambda$. By the Index theorem [7] the sum of all indexes of singularities in S_0 with respect to the local branches of S is equal to a (natural) positive number, the self-intersection number of S_0 in the projective plane $\mathbb{C}P(2)$. This implies that not all indexes are rational negative. Since by definition the index of an irreducible singularity is never a positive rational number, this implies that some singularity has a non-rational index. This singularity is clearly non-resonant.

By the above claim, the holonomy group of (the leaf contained in) S_0 contains some non-resonant germ. From the proof of Theorem 8 there is a meromorphic generalized integrating factor η defined in some neighborhood V of S_0 in $\mathbb{C}P(2)$. Since $\mathbb{C}P(2) \setminus S_0$ is a Stein surface [34], by a theorem of Levi (see [11, 34]), we can conclude that the one-form η extends as a meromorphic one-form to $\mathbb{C}P(2)$ (see [10] for similar extension arguments). Finally, the extended one-form η must be rational because we are on a projective manifold. As in [27] the existence of η implies the final part of the statement.

Proof (*Proof of Theorem* 9) As for the proof of Theorem 9 very few remains to say. Indeed, the proof of Theorem 10 gives all the steps. The hypotheses (i) and (ii) are then necessary since we cannot prove a version of Claim 5.1 in this case.

Remark 11 (1) Theorems 8 and 10 above show that in order to capture the generic foliations in the class of Riccati foliations it is necessary to allow dicritical singularities or curves containing all of its separatrices.

(2) Theorem 8 completes an example given in [35] of a germ $\mathcal{F}$ satisfying (i) and (ii) but which is not a meromorphic pull-back of a Riccati foliation on an algebraic surface. Indeed, the construction given in [35] exhibits $\mathcal{F}$ having as projective holonomy group G, i.e., the holonomy group $G = Hol(\tilde{\mathcal{F}}, D)$, where D is the exceptional divisor of the blow-up, a non-abelian solvable group conjugate to a subgroup of $\mathbb{H}_1 = \left\{z \mapsto \frac{\lambda z}{1+\mu z}\right\}$.

(3) In [35] it is also given an example of a foliation $\mathcal{H}$ on a rational surface Y such that $\mathcal{H}$ is transversely projective on $Y \backslash M$ for some algebraic curve $M \subset Y$ and such that $\mathcal{H}$ is *not* birationally equivalent to a Riccati foliation on $\overline{\mathbb{C}} \times \overline{\mathbb{C}}$.

5.2 *Logarithmic Foliations, Separatrices and Invariant Curves*

Let us recall that a *logarithmic foliation* on a projective surface M is one given by a closed rational one-form β with simple poles. If $M = \overline{\mathbb{C}} \times \overline{\mathbb{C}}$ or $\mathbb{C}P(2)$ then a logarithmic foliation is given by a rational one-form β as follows: $\beta = \sum\limits_{j=1}^{r} \lambda_j \frac{df_j}{f_j}$, where the f_j are rational functions on V and $\lambda_j \in \mathbb{C} \setminus \{0\}$.

In [20] the author gives the following nice characterization of logarithmic foliations:

Theorem 11 (cf. [20], Theorem A) *Let $\mathcal{F}$ be a holomorphic foliation on a compact algebraic surface M and let S be a compact curve invariant by $\mathcal{F}$. Assume that one of the following conditions holds:*

(i) *$Pic(M)$ is isomorphic to $\mathbb{Z}$.*
(ii) *$Pic(M)$ is torsion free, $H^1(M, \mathbb{C}) = 0$, $S^2 > 0$ and* $\sum\limits_{p \in sing(\mathcal{F}) - S} BB_p(\mathcal{F}) \geq 0$.

Also assume that every local separatrix of $\mathcal{F}$ through any $p \subset sing(\mathcal{F}) \cap S$ is a local branch of S and that every singularity of $\mathcal{F}$ in S is a generalized curve. Then $\mathcal{F}$ is a logarithmic foliation.

Here, by $BB_p(\mathcal{F})$ we mean the Baum-Bott index associated to the Chern number c_1^2 of the normal sheaf of the foliation [1]. Also, $Pic(M) = H^1(M, \mathcal{O}_M^*)$ is the Picard group of M, while S^2 denotes the self-intersection number of S. We point-out that $Pic(M) = \mathbb{Z}$ for the case of projective spaces $M = \mathbb{C}P(m), m \geq 2$. Regarding condition (ii), the part $H^1(M, \mathbb{C}) = 0$, $S^2 > 0$ is verified for the case of projective spaces.

As a particular case we have:

Corollary 1 *Let $\mathcal{F}$ be a holomorphic foliation on $\mathbb{C}P(2)$ and let $S \subset \mathbb{C}P(2)$ be an invariant algebraic curve by $\mathcal{F}$. Assume that: (i) every local separatrix of $\mathcal{F}$ through any $p \subset sing(\mathcal{F}) \cap S$ is a local branch of S and that (ii) every singularity of $\mathcal{F}$ in S is a (non-dicritical) generalized curve. Then $\mathcal{F}$ is a logarithmic foliation.*

As for the last inequality in Theorem 11 (ii) we have:
The condition $\sum\limits_{p \in sing(\mathcal{F})-S} BB_p(\mathcal{F}) \geq 0$ holds if each singularity of $\mathcal{F}$ in $M \setminus S$ is linearly of Morse type (*i.e.* $\mathcal{F}$ is locally given by the holomorphic one-form $d(xy) + h.o.t.$). This condition also holds when $\mathcal{F}$ a has local holomorphic first integral around each point of M which is not in S. In particular we have:

Lemma 13 *Let $\mathcal{F}$ be a holomorphic foliation on a compact algebraic surface M and let S be an invariant compact curve by $\mathcal{F}$. If $\mathcal{F}$ is transversely projective in $M \backslash S$ and the singularities in $M \backslash S$ are all non-dicritical then* $\sum\limits_{p \in sing(\mathcal{F})-S} BB_p(\mathcal{F}) \geq O$.

Proof Since $\mathcal{F}$ is transversely projective in $M \setminus S$, any singularity $p \in sing(\mathcal{F}) \cap M \backslash S$ admits a meromorphic first integral. Because this singularity is non-dicritical, there is a holomorphic first integral. The conclusion follows from what we remarked above.

From this lemma and Theorem 11 and also Theorem 5 we promptly obtain:

Corollary 2 *Let $\mathcal{F}$ be a holomorphic foliation on a compact algebraic surface M and let S be an invariant compact curve by $\mathcal{F}$. Suppose that $\mathcal{F}$ is transversely projective in $M \backslash S$ and that every singularity of $\mathcal{F}$ in S is a generalized curve. Assume that one of the following conditions hold:*

(i) *$Pic(M)$ is isomorphic to $\mathbb{Z}$.*
(ii) *$Pic(M)$ is torsion free, $H^1(M, \mathbb{C}) = 0$, $S^2 > 0$ and the singularities off S are non-dicritical.*

There are two possibilities:

(a) *Every local separatrix of $\mathcal{F}$ through any $p \subset sing(\mathcal{F}) \cap S$ is a local branch of S. In this case $\mathcal{F}$ is a logarithmic foliation.*

(b) *There is a singular point $p \in S$ exhibiting a separatrix Γ_p not contained in S. In this case $\mathcal{F}$ admits a rational projective triple (Ω, η, ξ), defined on M.*

Let $\mathcal{F}$ be a holomorphic foliation on $\mathbb{C}P(2)$ of degree m, then $\sum_{p \in sing(\mathcal{F})} BB_p(\mathcal{F}) = (m+2)^2 \geq 4$. The author proves the following extension of the second part of Theorem 1 in [12] to compact complex surfaces (cf. [20] Proposition 3.1):

Proposition 15 ([20] Proposition 3.1) *Let $\mathcal{F}$ be a holomorphic foliation on a compact algebraic surface M with $H^1(M, \mathbb{C}) = 0$ and $Pic(M) = \mathbb{Z}$. Let S be an invariant compact curve with only nodal type singularities.*
If $\sum_{p \in sing(\mathcal{F}) - S} BB_p(\mathcal{F}) < S^2$, then $\mathcal{F}$ is logarithmic.

By taking a look at the proof given in [20] we conclude that the conclusion of Theorem 11 holds for a foliation $\mathcal{F}$ on the complex projective plane $\mathbb{C}P(2)$ having an invariant algebraic curve S such that each singularity of $\mathcal{F}$ in S is an extended generalized curve (cf. Definition 4) and if S contains each non-dicritical separatrix of each singularity of $\mathcal{F}$ in S.

Corollary 3 ([20], Corollary 3.1) *Let $\mathcal{F}$ be a holomorphic foliation on a compact algebraic surface M with $H^1(M, \mathbb{C}) = 0$ and $Pic(M) = \mathbb{Z}$. Let $S \subset M$ be an invariant compact curve with only nodal type singularities. If $sing(\mathcal{F}) \cap S = sing(S)$ and the singularities of $\mathcal{F}$ in S are non-degenerated, then $\mathcal{F}$ is a logarithmic foliation.*

5.2.1 Logarithmic Case and Moderate Growth

Theorem 12 *Let $\mathcal{F}$ be a foliation on a projective surface M such that $Pic(M)$ is isomorphic to $\mathbb{Z}$. Assume that $\mathcal{F}$ is transversely projective in $M \backslash S$ for some algebraic curve $S \subset M$ and that the singularities of $\mathcal{F}$ in S are (non-dicritical) non-resonant generalized curves. Then $\mathcal{F}$ is a logarithmic foliation or it is transversely projective of moderate growth.*

Proof (*Proof of Theorem* 12) We will follow the notation in the proof of Theorem 4. Because the singularities of $\mathcal{F}$ are non-dicritical, the resolution divisor $E = D \cup \tilde{S}$ is invariant by $\tilde{\mathcal{F}}$. Moreover, each connected component S^j of S originates a connected component of the resolution divisor. Therefore, for sake of simplicity of notation, let us assume that the singularities of $\mathcal{F}$ are already irreducible in M, i.e., *S and E exhibit the same number of connected components.* If we denote by $\sigma : \widetilde{M} \to M$ the resolution morphism for the singularities of $\mathcal{F}$ in S, then $\sigma|_{\widetilde{M \setminus S}} : M \setminus S \to M \setminus S$ is a diffeomorphism, in particular the fundamental groups $\pi_1(\widetilde{M \backslash S})$ and $\pi_1(M \backslash S)$ are isomorphic.

We have the following possibilities:

1. S contains all the separatrices of $\mathcal{F}$ in S.
2. There is a singularity of $\mathcal{F}$, say $q \in S$, exhibiting a separatrix Γ which is not contained in S.

In case (1), since the singularities are assumed to be generalized curves we may apply Theorem 11 and conclude that $\mathcal{F}$ is a logarithmic foliation.

Assume that we are in case (2). Then by Theorem 5 and other results in Sect. 4.5 we conclude that the projective structure in $M \setminus S$ defines a projective triple that extends to S. We have therefore a rational projective triple for $\mathcal{F}$ in M, i.e., $\mathcal{F}$ is transversely projective of moderate growth.

Clearly a logarithmic foliation is of moderate growth. Therefore we obtain:

Corollary 4 *Let $\mathcal{F}$ be a foliation on $M = \mathbb{C}P(2)$. Assume that $\mathcal{F}$ is transversely projective in $\mathbb{C}P(2) \setminus S$ for some algebraic curve $S \subset \mathbb{C}P(2)$ and that the singularities of $\mathcal{F}$ in S are (non-dicritical) non-resonant generalized curves. Then $\mathcal{F}$ is transversely projective of moderate growth.*

5.3 Classification of Projective Foliations: Moderate Growth On Projective Manifolds

In [21] we find the following definition of transversely projective foliation on a smooth projective manifold. *Let M be a smooth projective manifold over $\mathbb{C}$.A (holomorphic singular) codimension one foliation $\mathcal{F}$ on M.The foliation is said to be transversely projective if given a non zero rational 1-form ω defining $\mathcal{F}$ (and therefore satisfying the Frobenius integrability condition $\omega \wedge d\omega = 0$) we have that there are rational 1-forms α and β on M such that the sl_2-connection on the rank 2 trivial vector bundle defined by $\Delta = d + \begin{pmatrix} \alpha & \beta \\ \omega & -\alpha \end{pmatrix}$ is flat.*

Let us compare the above definition with the one we have been using so far in this work. Indeed, compared to Definitions 6 and 6 there is a difference, quite easy to explain. In the above definition, we already assume that the foliation admits a rational projective triple, i.e., a projective triple meromorphic defined everywhere in the manifold M. This is not necessarily the case if we just start with a foliation which is (according to our definition Definition 6) transversely projective in $M \setminus S$ for some algebraic curve $S \subset M$. Nevertheless, often we cannot extend the projective transverse structure to the curve S (for instance, in the case of Riccati foliations or logarithmic foliations). Thus what is considered in [21] are what we have called *transversely projective foliations with moderate growth* (cf. Definition 9). projective structure in $M \setminus S$.

The authors also introduce the following notion:

Definition 12 ([21]) A *Riccati foliation* over a projective manifold M consists of a pair $(\pi : P \to M; H) = (P; H)$ where $\pi : P \to M$ is a locally trivial $\mathbb{P}^1$ fiber bundle in the Zariski topology, this means that P is the projectivization of the total space of a rank two vector bundle E, and H is a codimension one foliation on P which is transverse to a general fiber of π. In the case of a clear context, the $\mathbb{P}^1$-bundle P is omitted from the notation. Then H is called a *Riccati foliation.* The foliation H is defined by the projectivization of horizontal sections of a (non unique) at meromorphic connection r on E. The connection r is uniquely determined by H and its trace on $\det(E)$. We say that the Riccati foliation H is *non-singular* if it lifts to a meromorphic connection r with at worst non-singular singularities (see [15]), and irnon-singular if not. It is said that a Riccati foliation $(P; H)$ over M *factors* through a projective manifold M' if there exists a Riccati foliation $(\pi' \colon P' \to M', H')$ over M', and rational maps $\phi \colon M \to M'$ and $\Phi \colon P \to P'$, such that $\pi' \circ \Phi = \phi \circ \pi$, and Φ has degree one when restricted to a general fiber of P, and $H = \Phi^* H'$.

Using the notion above, alternatively, in [21] the authors state that a foliation $\mathcal{F}$ on M is transversely projective if there exists a triple $\mathcal{P} = (P; H; \sigma)$ satisfying

1. $(P; H)$ is a Riccati foliation over M; and
2. $\sigma : M \to P$ is a rational section generically transverse to H such that $\mathcal{F} = \sigma^* H$.

After making the conversion between the notions of transversely projective foliation in [21] and the one we consider in our work, we can state the main classification result of [21] as follows:

Theorem 13 (cf. [21], Theorem D) *Let $\mathcal{F}$ be a codimension one transversely projective foliation of moderate growth on a projective manifold M. Then at least one of the following assertions holds true.*

1. *There exists a generically finite Galois morphism $f : Y \to M$ such that $f^*\mathcal{F}$ is defined by a closed rational one-form.*
2. *There exists a rational map $f : M \to S$ to a ruled surface S, and a Riccati foliation $\mathcal{R}$ on S such that $\mathcal{F} = f^* R$.*
3. *The transverse projective structure for $\mathcal{F}$ has at worst non-singular singularities, and the monodromy representation of $\mathcal{F}$ factors through one of the tautological representations of a polydisk Shimura modular orbifold $\mathcal{H}$.*

Combining this result and Theorem 7 we promptly obtain:

Theorem 14 *Let $\mathcal{F}$ be a holomorphic foliation by curves on a projective manifold M. Assume that $\mathcal{F}$ is transversely projective in $M \backslash S$ where $S \subset M$ is an algebraic curve. Suppose that the singularities of $\mathcal{F}$ in S are non-resonant extended generalized curves. Then at least one of the following assertions holds true.*

1. *There exists a generically finite Galois morphism $f : Y \to M$ such that $f^*\mathcal{F}$ is defined by a closed rational one-form.*
2. *There exists a rational map $f : M \to S$ to a ruled surface S, and a Riccati foliation $\mathcal{R}$ on S such that $\mathcal{F} = f^* R$.*

3. *The transverse projective structure for $\mathcal{F}$ has at worst non-singular singularities, and the monodromy representation of $\mathcal{F}$ factors through one of the tautological representations of a polydisk Shimura modular orbifold $\mathcal{H}$.*

References

1. Baum, P., Bott, R.: On the zeros of meromorphic vector fields, Essais en l'honeur de G. De Rham, 29–47 (1970)
2. Baum, P., Bott, R.: Singularities of holomorphic foliations. J. Differ. Geom. **7**, 279–342 (1972)
3. Berthier, M., Touzet, F.: Sur l'intégration des équations différentielles holomorphes réduites en dimension deux. Bol. Soc. Bras. Mat. **30**(3), 247–286 (1999)
4. Blumenthal, R.: Transversely homogeneous foliations. Ann. Inst. Fourier t. **29**(4), 143–158 (1979)
5. Camacho, C., Scárdua, B.: Foliations on complex projective spaces with algebraic limit sets. Géométrie complexe et systèmes dynamiques (Orsay, Astérisque **261**, 57–88 (2000); Soc. Math. France, Paris (2000)
6. Camacho, C., Lins Neto, A.: Geometric Theory of Foliations. Birkhauser Inc, Boston (1985)
7. Camacho, C., Sad, P.: Invariant varieties through singularities of holomorphic vector fields. Ann. Math. **115**, 579–595 (1982)
8. Camacho, C., Scárdua, B.: Holomorphic foliations with Liouvillian first integrals. Ergod. Theory Dyn. Syst. **21**, 717–756 (2001)
9. Camacho, C., Scárdua, B.: Extension theorems for analytic objects associated to foliations. Bulletin des Sciences Mathmatiques (Paris. 1885) **136**, 54–71 (2012). https://doi.org/10.1016/j.bulsci.2011.04.005
10. Camacho, C., Lins Neto, A., Sad, P.: Topological invariants and equidesingularization for holomorphic vector fields. J. of Differ. Geom. **20**(1), 143–174 (1984)
11. Camacho, C., Lins Neto, A., Sad, P.: Foliations with algebraic limit sets. Ann. Math. **136**, 429–446 (1992)
12. Cerveau, D., Lins Neto, A.: Holomorphic foliations in CP^2 having an invariant algebraic curve. Ann. Inst. Fourier, Grenoble **41**(4), 883–903 (1991)
13. Cerveau, D., Moussu, R.: Groupes d'automorphismes de $(\mathbb{C}, 0)$ et équations différentielles $ydy + \cdots = 0$. Bull. Soc. Math. Fr. **116**, 459–488 (1988)
14. Cerveau, D., Sad, P.: Liouvillian integration and Bernoulli foliations. Trans. Am. Math. Soc. **350**(8), 3065–3081 (1998)
15. Deligne, P.: Lecture Notes in Mathematics. Équations différentielles à points singuliers réguliers, vol. 163. Springer, Berlin (1970)
16. Dulac, H.: Solutions d'un système de équations différentiale dans le voisinage des valeus singulières. Bull. Soc. Math. Fr. **40**, 324–383 (1912)
17. Godbillon, C.: Progress in Mathematics. Feuilletages. Études géométriques. With a preface by G. Reeb, vol. 98. Birkhäuser, Basel (1991)
18. Gunning, R.C.: Introduction to holomorphic functions of several variables. Function Theory. Wadsworth & Brooks/Cole Advanced Books & Software, vol. I. Pacific Grove, CA (1990)
19. Gunning, R.C.: Introduction to holomorphic functions of several variables. Local Theory. The Wadsworth & Brooks/Cole Mathematics Series. Wadsworth & Brooks/Cole Advanced Books & Software, vol. II. Monterey, CA (1990)
20. Licanic, S.: Logarithmic foliations on compact algebraic surfaces. Bol. Soc. Brasil. Mat. (N.S.) **31**(1), 113–125 (2000)
21. Loray, F., Touzet, F., Vitrio, J.: Representations of quasiprojective groups, flat connections and transversely projective foliations (2015). arXiv:1402.1382 [math.AG]
22. Martinet, J., Ramis, J.-P.: Problème de modules pour des équations différentielles non lineaires du premier ordre. Publ. Math. Inst. Hautes Études Scientifiques **55**, 63–124 (1982)

23. Martinet, J., Ramis, J.-P.: Classification analytique des équations différentielles nonlineaires resonnants du premier ordre. Ann. Sc. Ec. Norm. Sup. **16**, 571–621 (1983)
24. Mattei, J.F., Moussu, R.: Holonomie et intégrales premières. Ann. Sci. École Norm. Sup. **13**(4), 469–523 (1980)
25. Nakai, I.: Separatrices for nonsolvable dynamics on $\mathbb{C}, 0$. Ann. Inst. Fourier (Grenoble) **44**, 569–599 (1994)
26. Nicolau, M., Paul, E.: A geometric proof of a Galois differential theory theorem. In: Mozo Fernandez, J. (ed.) Ecuaciones diferenciales y singularidades, Universidad de Valladolid, pp. 277–290 (1997)
27. Scárdua, B.: Transversely affine and transversely projective foliations. Ann. Sc. École Norm. Sup. 4e sériet **30**, 169–204 (1997)
28. Scárdua, B.: Integration of complex differential equations. J. Dyn. Control Syst. **1**(5), 1–50 (1999)
29. Scárdua, B.: Complex vector fields having orbits with bounded geometry. Tohoku Math. J. (2) **54**(3), 367–392 (2002)
30. Scárdua, B.: Differential algebra and Liouvillian first integrals of foliations. J. Pure Appl. Algebra **215**(5), 764–788 (2011)
31. Seidenberg, A.: Reduction of singularities of the differential equation $Ady = Bdx$. Am. J. Math. **90**, 248–269 (1968)
32. Seke, B.: Sur les structures transversalement affines des feuilletages de codimension un. Ann. Inst. Fourier, Grenoble **30**(1), 1–29 (1980)
33. Singer, M.F.: Liouvillian first integrals of differential equations. Trans. Am. Math. Soc. **333**(2), 673–688 (1992)
34. Siu, Y.: Techniques of Extension of Analytic Objects. Marcel Dekker, N.Y. (1974)
35. Touzet, F.: Sur les feuilletages holomorphes transversalement projectifs. Ann. Inst. Fourier **53**(3), 815–846 (2003)

Disentanglements of Corank 2 Map-Germs: Two Examples

David Mond

Abstract We compute the homology of the multiple point spaces of stable perturbations of two germs $(\mathbb{C}^n, 0) \to (\mathbb{C}^{n+1}, 0)$ of corank 2, using a variety of techniques based on the image computing spectral sequence ICSS. We provide a reasonably detailed introduction to the ICSS, including some low-dimensional examples of its use. The paper is partly expository.

Keywords Disentanglement · Multiple-point spaces

1991 Mathematics Subject Classification 14B05 · 32S30 · 32S25

1 Introduction

In studying a singularity of mapping from n-space to $(n + 1)$-space, a rôle analogous to that of Milnor fibre is played by a stable perturbation of the singularity, and in particular by its image. The image of a map acquires non-trivial homology through the identification of points of the domain, and these identifications are encoded in the multiple point spaces of the map. For germs of corank 1, these multiple point spaces are well understood. For germs of corank > 1 the situation is radically different.

In this paper we study the multiple point spaces of stable perturbations of two map-germs of corank 2 from n-space to $(n + 1)$-space. In one case $n = 3$ and in the other $n = 5$. Previous work of Marar, Nuño-Ballesteros and Peñafort, in [16, 17] has explored the case where $n = 2$. Increasing the dimension introduces new difficulties. Confronting these will require a range of new techniques.

D. Mond (✉)
Mathematics Institute, University of Warwick, Coventry CV4 7AL, UK
e-mail: d.m.q.mond@warwick.ac.uk

R. N. Araújo dos Santos et al. (eds.), *Singularities and Foliations. Geometry, Topology and Applications*, Springer Proceedings in Mathematics & Statistics 222, https://doi.org/10.1007/978-3-319-73639-6_7

Our work here is a preliminary exploration. Following the invitation of the editors to provide an accessible account, we have expanded the preliminary material on multiple-point spaces, disentanglements, and the image computing spectral sequence ICSS, our principle technical tool, and included, in Sect. 1.6, some examples of calculation using the ICSS.

The first of the two corank 2 map-germs we look at is the germ of lowest codimension in Sharland's list [24] of weighted homogeneous corank 2 map-germs $(\mathbb{C}^3, 0) \to (\mathbb{C}^4, 0)$:

$$f_0 : (\mathbb{C}^3, 0) \to (\mathbb{C}^4, 0), \quad f_0(x, y, z) = (x, y^2 + xz + x^2y, yz, z^2 + y^3). \tag{1.1}$$

This has $\mathcal{A}_e$-codimension 18.

The second is the lowest dimensional example of corank 2 map germ with $\mathcal{A}_e$-codimension 1,

$$f_0 : (\mathbb{C}^5, 0) \to (\mathbb{C}^6, 0), \quad f_0(x, y, a, b, c) = (x^2 + ax + by, xy, y^2 + cx + ay, a, b, c). \tag{1.2}$$

For each of these, we calculate a number of (topological) homology groups with rational coefficients, related to its disentanglement. By "disentanglement" we do not mean just the stable perturbation

$$f_t : U_t \twoheadrightarrow X_t \subset \mathbb{C}^{n+1}$$

of the germ f_0 (where U_t is a contractible neighbourhood of 0 in $\mathbb{C}^n$), as the term has been used by de Jong and van Straten in [4] and by Houston in [10] and subsequent papers. A richer picture is obtained by considering the "semi-simplicial resolution"

(1.3)

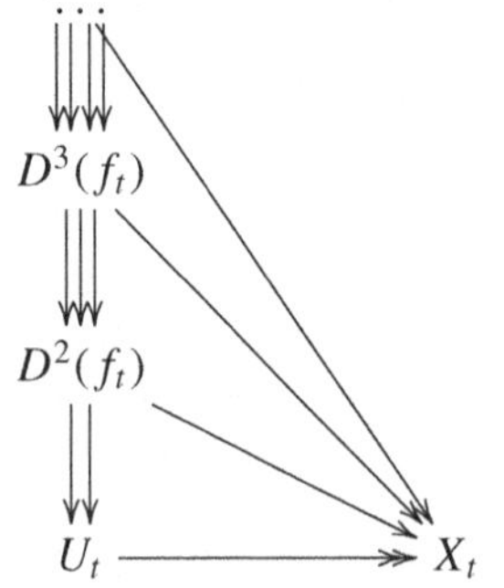

Here, for each integer $2 \le k \le n$, $D^k(f_t)$ is the closure, in U_t^k, of the set of k-tuples of pairwise distinct points ("strict" k-tuple points) sharing the same image, and the k distinct arrows $\pi_j^k : D^k(f_t) \to D^{k-1}(f_t)$, $1 \le j \le k$, are the restriction of the projections $U_t^k \to U_t^{k-1}$ obtained by forgetting the j'th factor in the product U_t^k.

For $k > n$, an $\mathcal{A}$-finite mono-germ $f : (\mathbb{C}^n, 0) \to (\mathbb{C}^{n+1}, 0)$ will have no strict k-tuple points, since the dimension of $D^k(f_0)$ at a strict k-tuple point is $n - k + 1$ (see Sect. 1.1 below). In this case $D^k(f_0)$ is defined by a slightly different procedure: we pick a stable unfolding $F : (\mathbb{C}^n \times \mathbb{C}^d, 0) \to (\mathbb{C}^{n+1} \times \mathbb{C}^d, 0)$ of f_0, define $D^k(F)$ as above, and then take $D^k(f_0)$ as the fibre over $0 \in \mathbb{C}^d$ of $D^k(F)$. We note that it is an easy consequence of the Mather–Gaffney criterion for $\mathcal{A}$-finiteness that if we apply this second procedure when $k \leq n$, we get the same space $D^k(f_0)$ as defined above.

The disentanglement, in this wider sense, contains complete information about the way that points of U_t are identified by f_t. The image X_t has the homotopy type of a wedge of n-spheres [25] whose number, the "image Milnor number" of f, $\mu_I(f)$, is the key geometric invariant of an $\mathcal{A}$-finite germ $(\mathbb{C}^n, 0) \to (\mathbb{C}^{n+1}, 0)$. Since the homology of X_t arises through the identifications induced by f_t, it is better described by the information attached to the diagram (1.3). This will become clearer in what follows.

Note that the π^k_j for fixed k and different j are left-right equivalent to one another thanks to the symmetric group actions on D^k and D^{k-1}, permuting the copies of U_t. In what follows we will consider only π^k_k, which we will refer to simply as π^k. We will denote the image of π^k in D^{k-1} by D^k_{k-1}, and, more generally, for $\ell < k$, we denote the image of $\pi^{\ell+1} \circ \cdots \circ \pi^k$ in D^ℓ by D^k_ℓ.

Remark 1.1 (1) Any finite map-germ $f : (\mathbb{C}^n, S) \to (\mathbb{C}^{n+1}, 0)$ is an embedding outside $D^2_1(f)$, which is the "non-embedding locus" of f. More generally each map $\pi^k : D^k(f) \to D^{k-1}(f)$ is an embedding outside $D^{k+1}_k(f)$, and each map π^{k+1} parameterises the non-embedding locus of its successor π^k. Thus the tower (1.3) shows a strong analogy with a free resolution of a module. If f is stable then $D^k(f)$, if not empty, is $n - k + 1$-dimensional. It follows that the length of this resolution is at most n.

(2) For maps $M^n \to N^{n+1}$ with $n < 6$ there is no stable singularity of corank 2. Every $\mathcal{A}$-finite germ is stable outside 0, so if $n < 6$, any singularity outside 0 of an $\mathcal{A}$-finite germ $f_0 : (\mathbb{C}^n, 0) \to (\mathbb{C}^{n+1}, 0)$, must be of corank 1. For stable germs of corank 1, all non-empty multiple point spaces are smooth [14]. It follows that for any $\mathcal{A}$-finite germ $f_0 : (\mathbb{C}^n, 0) \to (\mathbb{C}^{n+1}, 0)$ with $n < 6$, $D^k(f_0)$ has (at most) isolated singularity. It also follows that a stable perturbation f_t has no singularities of corank > 1. Therefore all of the non-empty multiple point spaces $D^k(f_t)$ are smooth – indeed, are smoothings of the isolated singularities $D^k(f_0)$. For any map f, $D^\ell(\pi^k)$ can be identified with $D^{k+\ell-1}(f)$, by the obvious map

$$\Big((x_1, \ldots, x_{k-1}, x_k^{(1)}), (x_1, \ldots, x_{k-1}, x_k^{(2)}), \ldots, (x_1, \ldots, x_{k-1}, x_k^{(\ell)})\Big) \longleftrightarrow \Big(x_1, \ldots, x_k^{(1)}, x_k^{(2)}, \ldots, x_k^{(\ell)}\Big) \tag{1.4}$$

– the left hand side here shows a point of $D^\ell(\pi^k)$, and the right hand side shows the corresponding point of $D^{k+\ell-1}(f)$. This observation is the basis of the "method of iteration" developed by Kleiman in [11]. From the smoothness of the $D^{k+j}(f_t)$ therefore follows smoothness of the multiple-point spaces of the projections π^k :

$D^k(f_t) \to D^{k-1}(f_t)$. The singularities of π^k are all of corank 1; this can be seen quite easily by writing f in linearly adapted coordinates, but see also [1]. By the characterisation of the stability of corank 1 map-germs by the smoothness of their multiple point spaces [14], it follows that provided $n < 6$, if f_t is stable then *all of the projections* π^k *are stable maps.*

1.1 Multiple Points

If $f_0 : (\mathbb{C}^n, 0) \to (\mathbb{C}^{n+1}, 0)$ is $\mathcal{A}$-finite then the set of strict k-tuple points is dense in $D^k(f_0)$, unless $D^k(f_0)$ consists only of the point $(0, \ldots, 0)$. The subset where f is an immersion at each of the x_i is still dense in $D^k(f_0)$. If $(x_1, \ldots, x_k)$ is such a k-tuple point, with $f_0(x_i) = y$ for $i = 1, \ldots, k$, then by the Mather–Gaffney criterion for $\mathcal{A}$-finiteness, the images of the germs $f_0 : (\mathbb{C}^n, x_i) \to (\mathbb{C}^{n+1}, y)$ meet in general position. It follows that their intersection has dimension $n + 1 - k$. This is therefore the dimension of $D^k(f_0)$, provided $k \leq n + 1$, and, similarly, of $D^k(f_t)$. If $k > n + 1$ then because f_t is stable, $D^k(f_t) = \emptyset$.

1.2 Alternating Homology

The developments in this section are due principally (but in some cases implicitly) to Goryunov in [6].
Notation For any topological space V, $C_k(V)$ is the free abelian group of singular k-chains in X, and $C_\bullet(V)$ is the singular chain complex. For a continuous map $\varphi : V \to W$, we denote by $\varphi_\#$ the map $C_j(V) \to C_j(W)$ induced by φ, and reserve the term φ_* for the corresponding map on homology.

Suppose $f : X \to Y$ is surjective. Recall the action of S_k on $D^k(f)$, permuting the copies of X. Define

$$C_j^{\text{Alt}}(D^k(f)) = \{c \in C_j(D^k(f)) : \sigma_\#(c) = \text{sign}(\sigma)c \text{ for all } \sigma \in S_k\}.$$

This gives a subcomplex, as $\partial(C_j^{\text{Alt}}) \subset C_{j-1}^{\text{Alt}}$, so we have *alternating homology*

$$H_j^{\text{Alt}}(D^k(f)).$$

Now observe also that $\pi_\#^k : C_j^{\text{Alt}}(D^k(f)) \subset C_j^{\text{Alt}}(D^{k-1}(f))$. To see this, let $\sigma \in S_{k-1}$, and define $\tilde{\sigma} \in S_k$ by setting $\tilde{\sigma}(i) = \sigma(i)$ for $1 \leq i \leq k - 1$ and $\tilde{\sigma}(k) = k$. Then $\text{sign}(\tilde{\sigma}) = \text{sign}(\sigma)$, and so if $c \in C_j^{\text{Alt}}(D^k(f))$,

$$\sigma_\#(\pi_\#^k(c)) = \pi_\#^k(\tilde{\sigma}_\#(c)) = \pi_\#^k(\text{sign}(\tilde{\sigma})c) = \text{sign}(\sigma)\pi_\#^k(c).$$

In fact we have a double complex: on $C_j^{\mathrm{Alt}}(D^k(f))$, $\pi_\#^{k-1} \circ \pi_\#^k = 0$; for

$$\pi_\#^{k-1} \circ \pi_\#^k = \pi_\#^{k-1} \circ \pi_\#^k \circ (k, k-1)_\#,$$

and on alternating chains $(k, k-1)_\#$ is multiplication by -1. By the same argument, $f_\# \circ \pi_\#^2 = 0$. Thus, denoting X by $D^1(f)$, Y by $D^{-1}(f)$, and f by π^1, we have

Proposition 1.2 *$(C_j^{Alt}(D^\bullet(f)), \pi^\bullet)$ is a complex, and $(C_\bullet^{Alt}(D^\bullet(f)), \partial, (-1)^\bullet \pi_\#^\bullet)$ is a double complex.* □

The relevance to the homology of the image can be seen from two short calculations. In each, "c_j^k" *always denotes an alternating chain, when $k \geq 2$.*

Example 1 let $c_j^2 \in Z_j^{\mathrm{Alt}}(D^2(f))$.

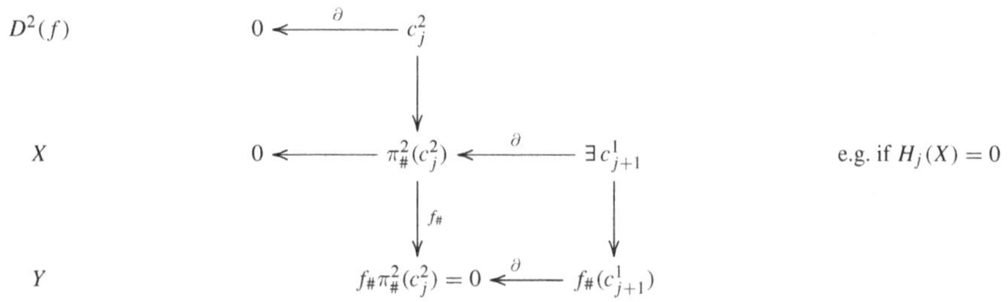

Because $f_\# \circ \pi_\#^2 = 0$ on alternating chains, $f_\#(c_{j+1}^1)$ is a cycle in Y. So from an alternating j-cycle c_j^2 in $D^2(f)$, we get a $j+1$ cycle on Y – provided $\pi_\#^2(c_j^2)$ is a boundary in X, i.e. provided $\pi_*^2[c_j^2] = 0$ in $H_j(X)$.

Example 2 let $c_j^3 \in Z_j^{\mathrm{Alt}}(D^3(f))$.

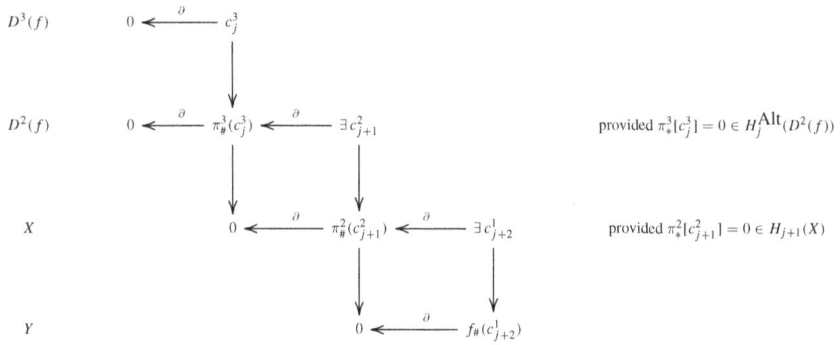

Here, a j-dimensional homology class in $D^3(f)$ leads to a $j+2$-dimensional class in Y, provided certain homology classes vanish.

Note that in both cases, if c_j^k is the cycle we begin with, then

- if $c_j^k = \pi_\#^{k+1}(c_j^{k+1})$ for some $c_j^{k+1} \in C_j^{\mathrm{Alt}}(D^{k+1}(f))$ then $\pi_\#^k(c_j^k) = 0$, and
- if $c_j^k = \partial c_{j+1}^k$ for some $c_{j+1}^k \in C_{j+1}^{\mathrm{Alt}}(D^k(f))$, then we can take $c_{j+1}^{k-1} = \pi_\#^k(c_{j+1}^k)$ so the homology class we get in $H_{j+1}^{\mathrm{Alt}}(D^{k-2}(f))$ is zero.

So we are really interested in

$$\frac{\ker \pi_*^k : H_j^{\mathrm{Alt}}(D^k(f)) \to H_j^{\mathrm{Alt}}(D^{k-1}(f))}{\operatorname{im} \pi_*^{k+1} : H_j^{\mathrm{Alt}}(D^{k+1}(f)) \to H_j^{\mathrm{Alt}}(D^k(f))}.$$

1.3 The Image-Computing Spectral Sequence

Lurking behind the two calculations we have just gone through is the *Image-computing spectral sequence*, ICSS. This was introduced in [7] and further developed in [6]. It calculates the homology of the image X_t in terms of the alternating homology $H_*^{\mathrm{Alt}}(D^k(f_t))$ of the multiple point spaces $D^k(f_t)$. The version introduced in [7] worked with the subspace of $H_*(D^k(f); \mathbb{Q})$ on which S_k acts by its sign representation:

$$\mathrm{Alt}\, H_j(D^k(f); \mathbb{Q}) = \{[c] \in H_j(D^k(f); \mathbb{Q}) : \sigma_*([c]) = \mathrm{sign}(\sigma)[c] \text{ for all } \sigma \in S_k\}.$$

If we take the complex of alternating chains described in the last paragraph and replace integer coefficients by rational coefficients, then the two versions coincide:

$$\mathrm{Alt}\, H_j(D^k(f); \mathbb{Q}) = H_j^{\mathrm{Alt}}(D^k(f); \mathbb{Q}).$$

The ICSS has $E_{p,q}^1 = H_q^{\mathrm{Alt}}(D^{p+1}(f_t))$ and converges to $H_{p+q}(X_t)$. The differential on the E^1 page, $d^1 : E_{p,q}^1 \to E_{p-1,q}^1$ is the simplicial differential $\pi_*^{p+1} : H_q^{\mathrm{Alt}}(D^{p+1}(f_t)) \to H_q^{\mathrm{Alt}}(D^p(f_t))$. In [7], a great deal hinges on the fact that for a stable perturbation f_t of an $\mathcal{A}$-finite germ f_0 of corank 1, the $D^k(f_t)$ are Milnor fibres of the isolated complete intersection singularities $D^k(f_0)$ (see [14]), and therefore their vanishing homology is confined to middle dimension. Since (over $\mathbb{Q}$) $H_i^{\mathrm{Alt}}(D^k(f_t)) \subset H_i(D^k(f_t))$, the vanishing alternating homology of $D^k(f_t)$ is also confined to middle dimension. From this it follows, in the case of a stable perturbation of a mono-germ, that the ICSS collapses at E^1: for all $r \geq 1$, $E_{p,q}^r = E_{p,q}^1$. The fact that the spectral sequence converges to $H_{p+q}(X_t)$ therefore means that, for map-germs $(\mathbb{C}^n, 0) \to (\mathbb{C}^{n+1}, 0)$, as $\mathbb{Q}$-vector space,

For a germ $(\mathbb{C}^n, 0) \to (\mathbb{C}^{n+c}, 0)$ with $c > 1$, the corresponding formula is

$$\tilde{H}_j(X_t) \simeq H_{n-(k-1)c}^{\mathrm{Alt}}(D^k(f_t)) \quad \text{if } j = n - (k-1)(c-1) \text{ with } 2 \leq k \leq \frac{n}{c} + 1,$$
$$= 0 \qquad \text{otherwise.}$$

$$H_n(X_t) \simeq H_{n-1}^{\mathrm{Alt}}(D^2(f_t)) \oplus H_{n-2}^{\mathrm{Alt}}(D^3(f_t)) \oplus \cdots \oplus H_0^{\mathrm{Alt}}(D^{n+1}(f_t)). \tag{1.5}$$

The argument for collapse is as follows: for each space $D^{p+1}(f_t)$ there is at most one non-zero alternating homology group, $H_{n-p}^{\mathrm{Alt}}(D^{p+1}(f_t))$, and therefore either the source or the target of every differential at E^1 is equal to 0. Thus $E^2_{p,q} = E^1_{p,q}$. The higher differentials $d^r : E^r_{p,q} \to E^r_{p-r,q+r-1}$ all vanish for exactly the same reason: for each one, either its source or its target is zero.

Notice that this is exactly what is needed to justify the assumptions we made in our two calculations in the previous paragraph. Whenever $D^k(f_t)$ has non-trivial alternating homology in dimension j, then $D^{k-1}(f)$ does not.

The situation for stable perturbations of multi-germs is slightly more complicated, as can be seen with the example of Reidemeister moves II and III in Sect. 1.6 below. Here $D^k(f_t)$ may have more than one connected component, and hence have vanishing alternating homology in dimension 0 as well as in middle dimension. As the calculations with Reidemeister moves II and III show, the differentials $\pi^k_* : H_0^{\mathrm{Alt}}(D^{p+1}(f_t)) \to H_0^{\mathrm{Alt}}(D^p(f_t))$ may not all be zero.

From (1.5) it follows that for a stable perturbation of a mono-germ

$$(\mathbb{C}^n, 0) \to (\mathbb{C}^{n+1}, 0)$$

$$\mu_I(f) = \sum_{k=2}^{n+1} \operatorname{rank} H_{n-k+1}^{\mathrm{Alt}}(D^k(f_t)). \tag{1.6}$$

In [10, Theorem 4.6], Kevin Houston showed the remarkable fact that that if f_t is a stable perturbation of an $\mathcal{A}$-finite mono-germ f_0 of *any* corank, then the *alternating* homology of $D^k(f_t)$ is once again confined to middle dimension, even though the ordinary homology of $D^k(f_0)$ may not be.[1] From Houston's theorem its follows that (1.5) and (1.6) hold for stable perturbations of mono-germs of any corank.

In both of our examples of corank 2 mono-germs, the multiplicity of f_0,

$$\dim_{\mathbb{C}} \frac{\mathcal{O}_{\mathbb{C}^n,0}}{f_0^*\mathfrak{m}_{\mathbb{C}^{n+1},0}\,\mathcal{O}_{\mathbb{C}^n,0}},$$

is equal to 3, so f_t has no quadruple or higher multiple points, and (1.6) reduces to

$$\mu_I(f_0) = \operatorname{rank} H_{n-1}^{\mathrm{Alt}}(D^2(f_t)) + \operatorname{rank} H_{n-2}^{\mathrm{Alt}}(D^3(f_t)). \tag{1.7}$$

If $f_0 : (\mathbb{C}^n, 0) \to (\mathbb{C}^{n+1}, 0)$ is a germ with $\mu_I(f_0) = 1$, then (1.6) implies that the vanishing homology of the image comes from just one of the multiple point spaces. It is an interesting project to determine, for each such f_0, which one this is. It is possible to show that *the answer depends only on the isomorphism class of the local*

[1] In fact for the stable perturbation f_t of the germ $(\mathbb{C}^5, 0) \to (\mathbb{C}^6, 0)$ described below, both $D^2(f_t)$ and $D^3(f_t)$ have non-trivial homology below middle dimension.

algebra of f_0. It is far from clear to me how to determine the answer from the local algebra. Nevertheless, examples support the following conjecture:

Conjecture 1.3 *If* $(n, n+1)$ *are in Mather's nice dimensions (i.e.* $n < 15$*) and if* $f_0 : (\mathbb{C}^n, 0) \to (\mathbb{C}^{n+1}, 0)$ *has* $\mu_I = 1$ *then the vanishing homology in the image of a stable perturbation* f_t *comes from* $D^k(f_t)$*, where k is the dimension of the local algebra of* f_0 *(and is the highest integer for which* $D^k(f_0) \neq \emptyset$*).*

This is proved for germs of corank 1 in [2, Sect. 4].

1.4 Symmetric Group Actions on the Homology of the Multiple Point Spaces

From here on, and in the rest of the paper, we will consider only germs of maps from n-space to $n+1$-space, we will consider only homology with rational coefficients, and by $H_i(D^k(f_t))$ we will mean always $H_i(D^k(f_t); \mathbb{Q})$.

As we have seen, each multiple-point space $D^k(f_t)$ is acted upon by the symmetric group S_k, permuting the factors of U_t^k. The resulting representation of S_k on $H_*(D^k(f_t); \mathbb{Q})$ splits as a direct sum of isotypal components, whose ranks are the principle numerical invariants of the disentanglement. We have

$$H_i(D^2(f_t)) \simeq H_i^T(D^2(f_t)) \oplus H_i^{\mathrm{Alt}}(D^2(f_t)),$$

where the two summands are the subspaces of $H_i(D^2(f_t))$ on which S_2 acts trivially, and by its sign representation, respectively, and

$$H_i(D^3(f_t)) = H_i^T(D^3(f_t)) \oplus H_i^{\mathrm{Alt}}(D^3(f_t)) \oplus H_i^{\rho}(D^3(f_t)),$$

where now the summands correspond to the trivial, sign and irreducible degree 2 representation of S_3.

Let $M_k(f_0)$ and $M_k(f_t)$ denote the set of *target* k-tuple points of f and f_0 respectively – points with at least k preimages, counting multiplicity. By e.g. [20], the germ $(M_k(f_0), 0)$ is defined by the $(k-1)$'st Fitting ideal of the $\mathcal{O}_{\mathbb{C}^{n+1},0}$-module $f_{0*}(\mathcal{O}_{\mathbb{C}^n})_0$, that is, the ideal generated by the $(m-k+1) \times (m-k+1)$ minors of the $m \times m$ matrix of a presentation of $f_{0*}(\mathcal{O}_{\mathbb{C}^n})_0$.

Lemma 1.4 *Let* $f_0 : (\mathbb{C}^n, 0) \to (\mathbb{C}^{n+1}, 0)$ *have multiplicity k and isolated instability, and suppose that* $M_k(f_0)$ *is non-singular. Let* f_t *be a stable perturbation of* f_0*. Then* $H_i^T(D^k(f_t)) = 0$ *for all* $i > 0$.

Proof Because the multiplicity of f_0 is k, f_t has no $(k+1)$-tuple points, and it follows that $M_k(f_t) \simeq D^k(f_t)/S_k$, and therefore $H_i(M_k(f_t)) \simeq H_i^T(D^k(f_t))$. Because $M_k(f_0)$ is smooth, $M_k(f_t)$ is contractible, and the result follows. □

Lemma 1.4, with $k = 3$, applies to both of the germs we consider. Smoothness of $M_3(f_t)$ can be seen in each case by considering a presentation of $f_{0*}(\mathcal{O}_{\mathbb{C}^n})_0$.

Suppose f has corank > 1. We have no closed formula for generators of the ideal defining $D^k(f)$ for $k \geq 3$, but for $D^2(f)$ there is a formula, introduced in [18], for germs of any corank. The ideal $(f \times f)^*(I_{\Delta_{n+1}})$ obtained by pulling back the ideal defining the diagonal in $\mathbb{C}^{n+1} \times \mathbb{C}^{n+1}$ vanishes on $D^2(f)$, but also on the diagonal in $\Delta_n \subset \mathbb{C}^n \times \mathbb{C}^n$. To remove Δ_n and leave only the points in the closure of the set of strict double points, we proceed as follows. The ideal $(f \times f)^*(I_{\Delta_{n+1}})$, generated by $f_i(x^{(1)}) - f_i(x^{(2)})$, for $i = 1, \ldots, n+1$, is contained in I_{Δ_n}, which is generated by $x_j^{(1)} - x_j^{(2)}$, $j = 1, \ldots, n$. Thus for $i = 1, \ldots, n+1$ there are functions $\alpha_{ij}(x^{(1)}, x^{(2)})$ such that

$$f_i(x^{(1)}) - f_i(x^{(2)}) = \sum_{j=1}^{n} \alpha_{ij}(x^{(1)}, x^{(2)}) \left(x_j^{(1)} - x_j^{(2)}\right).$$

The $(n+1) \times n$ matrix $\alpha = (\alpha_{ij})$ restricts to the jacobian matrix of f on Δ_n. We take

$$I_2(f) = (f \times f)^*(I_{\Delta_{n+1}}) + \min_n(\alpha).$$

Lemma 1.5 *Let $f_0 : (\mathbb{C}^n, 0) \to (\mathbb{C}^{n+1}, 0)$ be $\mathcal{A}$-finite and not an immersion. Then $D^2(f_0)$, as defined by $I_2(f_0)$, is Cohen–Macaulay of dimension $n-1$, and normal.*
□

The proof of Cohen–Macaulayness has been part of the folklore for some time, but has recently been written up carefully by Nuño-Ballesteros and Peñafort in [21]. When $n = 3$, $D^2(f_0)$ is therefore a normal surface singularity, and so by the Greuel–Steenbrink theorem, [9, Theorem 1], $H_1(D^2(f_t)) = 0$.

1.5 *Calculating $\mu_I(f)$*

Let $f_0 : (\mathbb{C}^n, 0) \to (\mathbb{C}^{n+1}, 0)$ have finite codimension and let

$$F : \left(\mathbb{C}^n \times \mathbb{C}^d, (0,0)\right) \to \left(\mathbb{C}^{n+1} \times \mathbb{C}^d, (0,0)\right), \quad F(x,u) = (f_u(x), u)$$

be a versal deformation. If G is a reduced equation for the image of F then for $u \in \mathbb{C}^d$, $g_u := G(\_, u)$ is a reduced equation for the image of f_u. By a theorem of Siersma [25], the image of g_u has the homotopy type of a wedge of n-spheres, whose number is equal to the number of critical points of g_u (counting multiplicity) which move off the zero level as u leaves 0. Note that the number of n-spheres is, by definition, the image Milnor number $\mu_I(f_0)$. We can therefore calculate $\mu_I(f_0)$ as follows: define the relative jacobian ideal J_G^{rel} by

$$J_G^{\mathrm{rel}} = \left(\frac{\partial G}{\partial y_1}, \ldots, \frac{\partial G}{\partial y_{n+1}}\right)$$

where $y_1, \ldots, y_{n+1}$ are coordinates on $(\mathbb{C}^{n+1}, 0)$. The relevant critical points of the functions g_t together make up the residual components of $V(J_G^{\mathrm{rel}})$ after removal of its components lying in $\{G = 0\}$. This residual set can be found as the zero-locus of the saturation $(J_G^{\mathrm{rel}} : G^\infty)$, defined as

$$\bigcup_{k\in\mathbb{N}} \{h \in \mathcal{O}_{\mathbb{C}^{n+1}\times\mathbb{C}^d,(0,0)} : hG^k \in J_G^{\mathrm{rel}}\}.$$

We denote the zero locus of $(J_G^{\mathrm{rel}} : G^\infty)$ by Σ. Thus the image Milnor number $\mu_I(f_0)$ is the degree of the projection $(\Sigma, 0) \to (\mathbb{C}^d, 0)$. This degree can be calculated as the intersection number $\big(\Sigma, \mathbb{C}^{n+1} \times \{0\}\big)_{(0,0)}$. If Σ is Cohen–Macaulay then

$$\mu_I(f_0) = \big(\Sigma, \mathbb{C}^{n+1} \times \{0\}\big)_{(0,0)} = \dim_{\mathbb{C}} \left(\frac{\mathcal{O}_{\mathbb{C}^{n+1}\times\mathbb{C}^d,(0,0)}}{(J_G^{\mathrm{rel}} : G^\infty) + (u_1, \ldots, u_d)}\right) \tag{1.8}$$

where $u_1, \ldots, u_d$ are coordinates on $(\mathbb{C}^d, 0)$.

In both of the examples considered here, this is the case, and it is a straightforward *Macaulay2* [8] calculation to follow this procedure (including to check the Cohen–Macaulayness of Σ) and find $\mu_I(f_0)$: it is 18 for the germ $(\mathbb{C}^3, 0) \to (\mathbb{C}^4, 0)$, and 1 for the germ $(\mathbb{C}^5, 0) \to (\mathbb{C}^6, 0)$.

If Σ is not Cohen–Macaulay, the intersection number can be calculated using Serre's *formule clef*, [23], which we use to calculate a related intersection number in Sect. 3.2 below.

Remark 1.6 The method outlined here gives no hint to any relation between $\mu_I(f_0)$ and the $\mathcal{A}_e$-codimension of f_0. It is conjectured that provided $(n, n+1)$ are nice dimensions, the standard "Milnor–Tjurina" relation holds, namely

$$\mathcal{A}_e\text{-codim } f_0 \leq \mu_I(f_0) \tag{1.9}$$

with equality if f_0 is weighted homogeneous. In [19] another slightly more complicated method for calculating μ_I is explained, with a similar case-by-case justification – verification of the Cohen Macaulayness of a certain relative T^1 module, $T^{1\,\mathrm{rel}}_{\mathcal{K}_{h,e}} i$, and consequent conservation of multiplicity. The virtue of this second method is that the relation (1.9) is an immediate consequence, whenever Cohen–Macaulayness of the relative T^1 can be shown, since $T^1_{\mathcal{A}_e} f_0$ is a quotient of $T^1_{\mathcal{K}_{h,e}} i_0$.

1.6 Examples

Example I: the ICSS for a stable perturbation of $f(x, y) = (x, y^3, xy + y^5)$. Here we apply the calculations described in Sect. 1.2 to a stable perturbation of the germ of the title of this subsection, of type H_2. For any map-germ $f : (\mathbb{C}^2, 0) \to (\mathbb{C}^3, 0)$ of the form $f(x, y) = (x, f_2(x, y), f_3(x, y))$, $D^2(f_t)$ is defined in (x, y_1, y_2)-space by the equations (see [14])

$$\frac{\begin{vmatrix} 1 & f_i(x, y_1) \\ 1 & f_i(x, y_2) \end{vmatrix}}{\begin{vmatrix} 1 & y_1 \\ 1 & y_2 \end{vmatrix}} \quad i = 2, 3 \tag{1.10}$$

and $D^3(f)$ is defined in (x, y_1, y_2, y_3)-space by the equations

$$\frac{\begin{vmatrix} 1 & f_i(x, y_1) & y_1^2 \\ 1 & f_i(x, y_2) & y_2^2 \\ 1 & f_i(x, y_3) & y_3^2 \end{vmatrix}}{\begin{vmatrix} 1 & y_1 & y_1^2 \\ 1 & y_2 & y_2^2 \\ 1 & y_3 & y_3^2 \end{vmatrix}}, \quad \frac{\begin{vmatrix} 1 & y_1 & f_i(x, y_1) \\ 1 & y_2 & f_i(x, y_2) \\ 1 & y_3 & f_i(x, y_3) \end{vmatrix}}{\begin{vmatrix} 1 & y_1 & y_1^2 \\ 1 & y_2 & y_2^2 \\ 1 & y_3 & y_3^2 \end{vmatrix}} \quad i = 2, 3. \tag{1.11}$$

In this case these give

$$y_1^2 + y_1 y_2 + y_2^2, \quad x + y_1^4 + y_1^3 y_2 + y_1^2 y_2^2 + y_1 y_2^3 + y_2^4$$

for $D^2(f)$ and

$$P_2(y_1, y_2, y_3), \quad y_1 + y_2 + y_3, \quad x + P_4(y_1, y_2, y_3), \quad P_3(y_1, y_2, y_3)$$

for $D^3(f)$, where each P_j is a symmetric polynomial of degree j. Thus $D^2(f)$ is an A_1 curve singularity and $D^3(f)$ is a non-reduced point of multiplicity 6. If f_t is a stable perturbation then $D^2(f_t)$ is a Milnor fibre of the A_1 singularity, homotopy equivalent to a circle, and $D^3(f_t)$ consists of 6 points forming a single S_3-orbit. By judicious choice of parameter values u and v in the miniversal deformation $f_{u,v}(x, y) = (x, y^3 + uy, xy + y^5 + vy^2)$ (see [15]), one can arrange that the *real* picture of $D^2(f_t)$ and $D^3(f_t)$, and their projections $D_1^3(f_t)$, $D_1^2(f_t)$, are as shown in the following diagram.

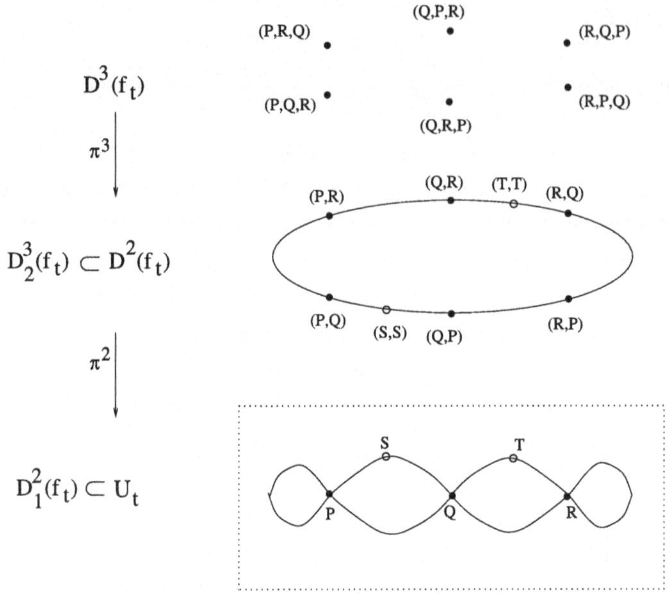

Here S and T in U_t are the non-immersive points of f_t. At each, the germ of f_t is equivalent to the parametrisation of the Whitney umbrella, $(x, y) \mapsto (x, y^2, xy)$, since this is the only stable non-immersive germ in this dimension range. The non-strict double points (S, S) and (T, T) are the fixed points of the involution $(1, 2)$ on $D^2(f_t)$, which, in our picture, is induced by a reflection in the straight line joining them.

As a single faithful S_3-orbit, $D^3(f_t)$ carries an alternating cycle,

$$c_0 = (P, Q, R) - (P, R, Q) + (R, P, Q) - (R, Q, P) + (Q, R, P) - (Q, P, R).$$

The projection of this cycle to $D^2(f_t)$, $\pi^3_\#(c_0)$, is an alternating boundary, as in Example 2 of Sect. 1.2: for instance

$$\pi^3_\#(c_0) = (P, Q) - (P, R) + (R, P) - (R, Q) + (Q, R) - (Q, P) = \partial(c_1)$$

where c_1 is the alternating 1-chain

$$[(T, T)(Q, R)] - [(T, T)(R, Q)] + [(Q, P)(R, P)] - [(P, Q)(P, R)]$$

(here, for any two non-antipodal points $A, B \in D^2(f_t)$, $[A, B]$ denotes the singular 1-simplex parametrising the shorter arc from A to B). The projection of c_1 to U_t is a 1-cycle in U_t, and is the boundary of a 2-chain c_2 with support equal to the union of the first and third bounded regions of the complement of $D^2_1(f_t)$, counting from left to right. And by the argument above, $f_{t\#}(c_2)$ is a cycle in the image X_t, indeed one of the two generators of $H_2(X_t)$. Another generator comes from the alternating 1-cycle c_1' on $D^2(f_t)$ consisting of the anticlockwise arc $[(S, S)(T, T)]$ minus the clockwise arc $[(S, S)(T, T)]$. I encourage the reader to find a 2-chain c_2' on U_t such that $\partial c_2' = \pi_{\#}^2(c_1')$.

Example II: the Reidemeister moves. The Reidemeister moves of knot theory are versal deformations of the three $\mathcal{A}_e$-codimension 1 singularities of mappings from the line to the plane. It is instructive to look at their disentanglements (in the sense described above), and at the resulting ICSS. The codimension 1 germs are shown in the middle column of the table below, and the right hand column shows a 1-parameter versal deformation, which, fixing $t \neq 0$, gives a stable perturbation.

I	$f_0 : x \mapsto (x^2, x^3)$	$f_t : x \mapsto (x^2, x^3 - tx)$
II	$f_0 : \begin{cases} x \mapsto (x, \ x^2) \\ y \mapsto (y, -y^2) \end{cases}$	$f_t : \begin{cases} x \mapsto (x, x^2 - t) \\ y \mapsto (y, -y^2) \end{cases}$
III	$f_0 : \begin{cases} x \mapsto (x, \ x) \\ y \mapsto (y, \ 0) \\ z \mapsto (z, -z) \end{cases}$	$f_t : \begin{cases} x \mapsto (x, \ x) \\ y \mapsto (y, \ t) \\ z \mapsto (z, -z) \end{cases}$

(1.12)

For all three cases, the non-trivial modules in the E^1 page of the ICSS for f_t are contained in the single column

$$\begin{array}{c} 0 \\ \downarrow \\ H_0^{\mathrm{Alt}}(D^3(f_t)) \\ \downarrow \pi_*^3 \\ H_0^{\mathrm{Alt}}(D^2(f_t)) \\ \downarrow \pi_*^2 \\ H_0(U_t) \\ \downarrow \\ 0 \end{array} \tag{1.13}$$

Reidemeister I. Take $F : (x, t) \mapsto (t, f_t(x))$ as stable unfolding. Since in order that $F(t_1, x_1) = F(t_2, x_2)$, we must have $t_1 = t_2$, we can embed $D^2(F)$ in $\mathbb{C}^3$ with coordinates t, x_1, x_2. There, following the recipe preceding Lemma 1.5 above, we find that $D^2(F)$ is defined by the equations

$$\frac{x_1^2 - x_2^2}{x_1 - x_2} = \frac{x_1^3 - tx_1 - (x_2^3 - tx_2)}{x_1 - x_2} = 0. \tag{1.14}$$

Simplifying, this gives

$$x_1 + x_2 = 0 \quad x_1^2 + x_1x_2 + x_2^2 = t. \tag{1.15}$$

Thus $D^2(f_0)$ is a 0-dimensional A_1 singularity. Setting $t > 0$ for a good real picture, and denoting $\sqrt{t}$ by P and $-\sqrt{t}$ by Q, $D^2(f_t)$ is its Milnor fibre, the point-pair $\{(P, Q), (Q, P)\}$. Then for $t \neq 0$, $H_0^{\text{Alt}}(D^2(f_t)) \simeq \mathbb{Q}$, generated by the class of $[(P, Q)] - [(Q, P)]$. Note that $H_0^{\text{Alt}}(D^2(f_0)) = 0$, since when $t = 0$, $P = Q$. For both f_0 and f_t, $D^3 = \emptyset$.

In the E^1 page (1.13), $H_0^{\text{Alt}}(D^3(f_t)) = 0$. We have $\pi_*^2 = 0$, for $\pi_*^2\big([(P, Q)] - [(Q, P)]\big) = [P] - [Q]$, and U_t is connected, so that $[P] = [Q]$. Hence the spectral sequence collapses at E^1, and for $t \neq 0$

$$H_0(X_t) = H_0(U_t) = \mathbb{Q}, \quad H_1(X_t) = H_0^{\text{Alt}}(D^2(f_t)) = \mathbb{Q}.$$

Reidemeister II. Here both branches of the bi-germ f_0 are immersions, so all multiple points are strict. Denote by 0_x and 0_y the origins of the coordinate systems with coordinates x and y respectively. The domain of the stable perturbation f_t is a disjoint union $U_t = U_{x,t} \cup U_{y,t}$, where $U_{x,t}$ is a contractible neighbourhood of 0_x and $U_{y,t}$ is a contractible neighbourhood of 0_y. Thus $H_0(U_t) \simeq \mathbb{Q}^2$. There are no triple points, and $D^2(f_t)$ consists of

$$\{(x, y) \in (\mathbb{C}, 0_x) \times (\mathbb{C}, 0_y) : x = y, x^2 - t = -y^2\} \tag{1.16}$$

together with its image under the involution $(1, 2)$ sending (x, y) to (y, x). When $t = 0$ this is a pair of 0-dimensional A_1 singularities, interchanged by $(1, 2)$. To describe $D^2(f_t)$ for $t \neq 0$, denote the points in $(\mathbb{C}, 0_x)$ with x coordinates $\sqrt{t/2}$ and $-\sqrt{t/2}$ by P_x and Q_x respectively, and the points in $(\mathbb{C}, 0_y)$ with y coordinates $\sqrt{t/2}$ and $-\sqrt{t/2}$ by P_y and Q_y. Then for $t \neq 0$,

$$D^2(f_t) = \{(P_x, P_y), (P_y, P_x), (Q_x, Q_y), (Q_y, Q_x)\}, \tag{1.17}$$

with the involution $(1, 2)$ interchanging the first and second points, and the third and fourth. For $t = 0$, this collapses just to

$$D^2(f_0) = \{(0_x, 0_y), (0_y, 0_x)\}.$$

Thus for $t \neq 0$, $H_0^{\text{Alt}}(D^2(f_t))$ is two-dimensional, with basis $[(P_x, P_y)] - [(P_y, P_x)]$, $[(Q_x, Q_y)] - [(Q_y, Q_x)]$, and for $t = 0$, $H_0^{\text{Alt}}(D^2(f_0))$ has basis $[(0_x, 0_y)] - [(0_y, 0_x)]$. With respect to the basis of $H_0^{\text{Alt}}(D^2(f_t))$ described above, and the basis $[P_x], [P_y]$ for $H_0(U_t)$, π_*^2 has matrix $\begin{pmatrix} 1 & 1 \\ -1 & -1 \end{pmatrix}$ when $t \neq 0$, and thus has 1-dimensional kernel and cokernel. The spectral sequence collapses at E^2, and

$$H_1(X_t) = E^2_{1,0} = \ker \pi^2_* \simeq \mathbb{Q}, \quad H_0(X_t) = E^2_{0,0} = \operatorname{Coker} \pi^2_* \simeq \mathbb{Q}.$$

Reidemeister III. We use the same conventions as for Reidemeister II. Let P_x and P_y denote the points in $U_{x,t}$ and $U_{y,t}$ with x and y coordinate t, and let Q_y and Q_z denote the points in $U_{y,t}$ and $U_{z,t}$ with y and z coordinate $-t$. Note that when $t = 0$, then $P_x = 0_x$, etc. Then

$$\begin{aligned} D^2(f_t) \cap \left(U_{x,t} \times U_{y,t}\right) &= \{(P_x, P_y)\} \\ D^2(f_t) \cap \left(U_{x,t} \times U_{z,t}\right) &= \{(0_x, 0_y)\} \\ D^2(f_t) \cap \left(U_{y,t} \times U_{z,t}\right) &= \{(Q_x, Q_z)\} \end{aligned} \tag{1.18}$$

and

$$D^3(f_0) \bigcap U_{x,t} \times U_{y,t} \times U_{z,t} = \{(0_x, 0_y, 0_z)\}.$$

Thus

$$\begin{aligned} &H_0^{\mathrm{Alt}}(D^3(f_0)) \simeq \mathbb{Q} \quad H_0^{\mathrm{Alt}}(D^3(f_t)) = 0 \\ &H_0^{\mathrm{Alt}}(D^2(f_0)) \simeq \mathbb{Q}^3 \; H_0^{\mathrm{Alt}}(D^2(f_t)) \simeq \mathbb{Q}^3 \\ &H_0(U_0) \simeq \mathbb{Q}^3 \qquad\quad H_0(U_t) \simeq \mathbb{Q}^3 \end{aligned} \tag{1.19}$$

with bases shown in the following table.

Module	Basis
$H_0^{\mathrm{Alt}}(D^3(f_0))$	$[(0_x, 0_y, 0_z)] - [(0_x, 0_z, 0_y)] + [(0_z, 0_x, 0_y)] - [(0_z, 0_y, 0_x)]$ $+[(0_y, 0_z, 0_x)] - [(0_y, 0_x, 0_z)]$
$H_0^{\mathrm{Alt}}(D^2(f_t))$	$[(P_x, P_y)] - [(P_y, P_x)],\ -[(0_x, 0_z)] + [(0_z, 0_x)],$ $[(Q_y, Q_z)] - [(Q_z, Q_y)]$
$H_0(U_t)$	$[0_x] = [P_x],\ [P_y] = [Q_y],\ [Q_z] = [0_z]$

With respect to these bases, the differentials π^k_* have the following matrices (with the first only for $t = 0$):

$$\pi^3_* = \begin{pmatrix} 1 \\ 1 \\ 1 \end{pmatrix} \qquad \pi^2_* = \begin{pmatrix} 1 & -1 & 0 \\ -1 & 0 & 1 \\ 0 & 1 & -1 \end{pmatrix}$$

In the spectral sequence for f_0, the image of π^3_* kills the kernel of π^2_*. When $t \neq 0$, D^3 vanishes, along with its homology, while $H_0^{\mathrm{Alt}}(D^2(f_t))$ remains unchanged.

The spectral sequence collapses at E^2, and

$$H_1(X_t) = E^2_{1,0} = \ker \pi^2_* \simeq \mathbb{Q}, \quad H_0(X_t) = E^2_{0,0} = \operatorname{Coker} \pi^2_* \simeq \mathbb{Q}.$$

2 New Examples: Disentanglements of Two Germs of Corank 2

2.1 *Summary of Results*

2.1.1 A Germ of Corank 2 from 3-Space to 4-Space

Let

$$f_0(x, y, z) = (x, y^2 + xz + x^2y, yz, z^2 + y^3).$$

The rows of the following table show relations between the ranks of the isotypal subspaces of the homology groups of $D^2(f_t)$ and $D^3(f_t)$ and of the homology groups of their projections to U_t, $D^2_1(f_t)$ and $D^3_1(f_t)$, and VD_∞, the number of Whitney umbrellas on $D^2_1(f_t)$, which plays a crucial role in our calculation. The left hand column shows where in the paper the calculation is made. Blank spaces indicate zeros.

datum	$H_2^T(D^2)$	$H_2^{Alt}(D^2)$	$H_2(D_1^2)$	$H_1^T(D^3)$	$H_1^{Alt}(D^3)$	$H_1^P(D^3)$	$H_1(D_1^3)$	VD_∞	
(1.7)		1			1				$= 18$
$\delta(D_1^3)$ in §4.2							1		$= 8$
$\delta(M_3)$ in §4.2				1					$= 0$
(3.4)				1	1	1	-2	-1	$= -1$
(3.6)								1	$= 10$
(3.7)				-1	1			-1	$= -1$
(3.9)			1						$= 27$
(3.10) – (3.13)	1	1	-1		1	$\frac{1}{2}$			$= 0$

(2.1)

The rank of the matrix of coefficients is 8, so we are able to compute all of the invariants. The following table shows their values.

$H_2^T(D^2)$	$H_2^{Alt}(D^2)$	$H_2(D_1^2)$	$H_1^T(D^3)$	$H_1^{Alt}(D^3)$	$H_1^P(D^3)$	$H_1(D_1^3)$	VD_∞
1	9	27	0	9	16	8	10

(2.2)

2.1.2 A Germ of Corank 2 from 5-Space to 6-Space

Let

$$f_0(x, y, a, b, c) = (x^2 + ax + by, xy, y^2 + cx + ay, a, b, c).$$

We are able to show

(a) $H_3^{\mathrm{Alt}}(D^3(f_t)) \simeq \mathbb{Q}$ and $H_4^{\mathrm{Alt}}(D^2(f_t)) = 0$, so the vanishing homology of the image comes from the triple points.
(b) $H_1(D^3(f_t)) = 0$, $H_2(D^3(f_t)) = H_2^{\rho}(D^3(f_t)) \simeq \mathbb{Q}^2$, and $H_3(D^3(f_t)) = H_3^{\mathrm{Alt}}(D^3(f_t)) \simeq \mathbb{Q}$.
(c) $H_1(D^2(f_t)) = 0$, $H_2(D^2(f_t)) = H_2^{T}(D^2(f_t)) \simeq \mathbb{Q}$.
(d) $\dim_{\mathbb{Q}} H_4(D^2(f_t)) = \dim_{\mathbb{Q}} H_3(D^2(f_t)) \leq 1$. Both groups are S_2-invariant, by Houston's theorem [10, Theorem 4.6]

Statements (a) and (b) are shown in Sect. 4.3, and (c) and (d) are shown in Sect. 4.4.

This is the first example I know of a stable perturbation of a map-germ $f_0 : (\mathbb{C}^n, 0) \to (\mathbb{C}^{n+1}, 0)$ for which the vanishing homology of the multiple point spaces is not confined to middle dimension, though of course many such examples are to be expected when f_0 has corank > 1.

3 Calculations for the Germ $(\mathbb{C}^3, 0) \to (\mathbb{C}^4, 0)$

3.1 Triple Points

Since no closed formula for a set of generators for the ideal defining $D^3(f)$ in $(\mathbb{C}^3)^3$ is known, we do not have direct access to any of the invariants of $D^3(f_t)$. However, we are able to build up a complete picture of the representation of S_3 on its homology, and in particular to calculate the dimension of $H_1^{\mathrm{Alt}}(D^3(f_t))$, by working our way up from its image under projection to U_t, $D_1^3(f_t)$.

Lemma 3.1 *$D_1^3(f_t)$ is a smoothing of $D_1^3(f)$*

Proof We have to show both that $D_1^3(f_t)$ is smooth, and that it is the fibre of a flat deformation of $D_1^3(f)$. The first statement is a consequence of the classification of stable map-germs. Up to $\mathcal{A}$-equivalence, the only stable germs of maps $\mathbb{C}^3 \to \mathbb{C}^4$ are

(a) a trivial unfolding of the parameterisation of the Whitney umbrella:

$$p_1(u, v, w) = (u, v, w^2, vw);$$

(b) a bi-germ whose two branches are a germ of type (a) and an immersion, meeting in general position in $\mathbb{C}^4$;
(c) a multi-germ consisting of k immersions meeting in general position, for $k = 1, 2, 3, 4$ (we denote these by (c1), ..., (c4)).

Since f_t is stable, every one if its germs is one of these types, and one can easily check that for each of them, except for (c4), the triple point locus D_1^3, where non-empty, is smooth. In the mapping f_t there are no points of type (c4), so $D_1^3 f_t)$ is smooth.

For the second statement, let $F : (\mathbb{C}^3 \times S, (0,0)) \to (\mathbb{C}^4 \times S, (0,0))$ be a stable unfolding of f over a smooth base S. Then $D^3(F)$ has dimension $1 + \dim S$. By the principle of iteration, $D^3_1(F) = M_2(\pi^2 : D^2(F) \to \mathbb{C}^3 \times S)$ (where M_2 means the set of double points in the target). Now $D^2(F)$ is Cohen–Macaulay, and π^2 is finite and generically 1-1, so $M_2(\pi^2)$ is also Cohen Macaulay [20]. Flatness of the projection $D^3_1(F) \to S$ now follows from the fact that the dimension of its fibre, $D^3_1(f_t)$, is equal to $\dim D^3_1(F) - \dim S$. □

It follows from the lemma that rank $H_1(D^3_1(f_t)) = \mu(D^3_1(f), 0)$. We find μ by using Milnor's formula $\mu = 2\delta - r + 1$ [13], where δ is the δ-invariant of a curve-germ and r the number of its branches. We find $D^3_1(f)$ as the zero locus of the ideal $f^*(\mathrm{Fitt}_2)$, where $\mathrm{Fitt}_2 := \mathrm{Fitt}_2(f_* \mathcal{O}_{\mathbb{C}^3,0}))$ is the second Fitting ideal of $\mathcal{O}_{\mathbb{C}^3,0}$ considered as $\mathcal{O}_{\mathbb{C}^4,0}$-module via f^*. *Macaulay2* [8] gives the following presentation of $f_*(\mathcal{O}_{\mathbb{C}^3})$:

$$\begin{pmatrix} -X^2U^2 - 2XUV + V^2 - UW & X^4 + U^2 + X^3V & X^3U + 2X^2V + XW \\ X^4 + U^2 + X^3V & -X^6 - 2X^2U - XV - W & -X^5 - XU + V \\ X^3U + 2X^2V + XW & -X^5 - XU + V & -X^4 - U \end{pmatrix} \quad (3.1)$$

from which we see that

$$\mathrm{Fitt}_2 = (X^4 + U, V, X^2U + W) \quad (3.2)$$

and

$$f^*\mathrm{Fitt}_2 = (x^4 + x^2y + y^2 + xz, yz, x^3z + z^2). \quad (3.3)$$

Primary decomposition of the ideal (3.3) shows that the curve $D^3_1(f)$ has three smooth components:

$$C_1 = V(y, x^3 + z), \quad C_2 = V(z, y - \xi x^2) \quad C_3 = V(z, y - \xi^2 x^2)$$

where $\xi = e^{2i\pi/3}$, with parameterisations

$$\gamma_1(t) = (t, 0, -t^3), \quad \gamma_2(u) = (u, \xi u^2, 0), \quad \gamma_3(v) = (v, \xi^2 v^2, 0).$$

Denoting by $\mathcal{O}_{\tilde{\Sigma}} := \mathbb{C}\{t\} \oplus \mathbb{C}\{u\} \oplus \mathbb{C}\{v\}$ the ring of the normalisation of Σ, we find that

$$n^*(\mathcal{O}_{\Sigma,0}) = (t^3) \oplus (u^3) \oplus (v^3) + \mathrm{Sp}\{(1,1,1), (t,u,v), (t^2,u^2,v^2), (0, \xi u^2, \xi^2 v^2)\} \subset \mathcal{O}_{\tilde{\Sigma}}.$$

Hence

$$\delta(D^3_1(f)) = \dim_{\mathbb{C}} \frac{\mathcal{O}_{\tilde{\Sigma}}, 0}{n^* \mathcal{O}_{\Sigma,0}} = 5,$$

and

$$\text{rank } H_1(D^3_1(f_t)) = \mu(D^3_1(f)) = 2\delta - 3 + 1 = 8.$$

The projection $D^3(f_t) \to D^3_1(f_t)$ is a double cover, with points (a, b, c) and (a, c, b) sharing the same image, and no points of higher multiplicity, as f has no quadruple points. The cover is simply branched at triple points of the form (a, b, b); there are no triple points of the form (a, a, a), since if there were, then f_t would have multiplicity ≥ 3 at a, and as we see in the list of stable germs in the proof of Lemma 3.1, none has multiplicity > 2. Thus

$$\chi(D^3(f_t)) = 2\chi(D^3_1(f_t)) - \#\text{branch points} = -14 - \#\text{branch points}. \qquad (3.4)$$

To complete the calculation of the Euler characteristic of $D^3(f_t)$, we have to compute the number of branch points. This seems not to be straightforward. Though the branch points are points of intersection of $D^3_1(f_t)$ and the non-immersive locus $R(f_t)$, both of these are curves so their intersection in U_t is not a proper intersection. Both curves lie in the surface $D^2_1(f_t)$, where the intersection is proper, but $D^3_1(f_t)$ is the singular locus of $D^2_1(f_t)$ and so again calculation of the intersection number is difficult. Instead we use the fact that the branch points are Whitney umbrella points of $D^2_1(f_t)$, which we explain in the next section, and count them using a theorem of Theo de Jong in [3].

3.2 Double Points

Lemma 3.2 $(a, b, b) \in D^3(f_t)$ *if and only if* $(a, b) \in D^2(f_t)$ *is a Whitney umbrella point of the projection* $\pi^2 : D^2(f_t) \to U_t$.

Proof This is, once again, the principle of iteration. The map

$$(a, b, c) \mapsto ((a, b), (a, c))$$

identifies $D^3(f_t)$ with $D^2(\pi^2 : D^2(f_t) \to U_t)$. A point of the form (a, b, b) becomes a fixed point of the involution $((a, b), (a, c)) \mapsto ((a, c), (a, b))$, and thus a non-immersive point of π^2. By Remark 1.1, this must be a Whitney umbrella point. □

From Lemma 3.2 we see that to find the dimension of $H_1^{\mathrm{Alt}}(D^3(f_t))$ we must count the number of Whitney umbrellas on $D^2_1(f_t)$. Let $W(D^2_1(f_t))$ denote the set of all such points. They appear when f_t has a bi-germ of type (b) in the list in the proof of Lemma 3.1: the Whitney umbrella appears on $D^2_1(f_t)$ at the source point of the immersive member of the bi-germ. If $R(f_t)$ is the set of non-immersive points of f_t, then $W(D^2_1(f_t)) = D^3_1(f_t) \cap R(f_t)$, so one might hope to calculate the number of points in $W(D^2_1(f_t))$ as an intersection number. But as remarked above, the intersection is improper: both $D^3_1(f_t)$ and $R(f_t)$ are curves. We are forced to look further afield, and use a theorem of Theo de Jong [3]. The *virtual number of* D_∞ *points* on a germ of singular surface $(S, x_0) \subset \mathbb{C}^3$, with 1-dimensional singular locus Σ, and with reduced equation h, is defined as follows. Let $\theta(h)$ be the restriction to Σ of the germs of vector fields on $(\mathbb{C}^3, x_0)$ tangent to all level sets of h.

Then $\theta(h) \subset \theta_\Sigma$. Let $\tilde{\Sigma}$ be the normalisation of Σ. Since vector fields lift uniquely to the normalisation we can consider the quotient $\theta_{\tilde{\Sigma},\tilde{x}_0}/\theta(h)$. De Jong defines

$$\mathrm{VD}_\infty(S) = \dim_{\mathbb{C}}\left(\frac{\theta_{\tilde{\Sigma},\tilde{x}_0}}{\theta(h)}\right) - 3\delta(\Sigma) \qquad (3.5)$$

and shows ([3, Theorem 2.5]) that $VD_\infty(S)$ is conserved in a flat deformation of S which induces a flat deformation of Σ.

Let us apply this to the case where S is the surface $D_1^2(f)$ for a finitely determined map-germ $f : (\mathbb{C}^3, 0) \to (\mathbb{C}^4, 0)$. In this case $\Sigma = D_1^3(f)$. A deformation of f over a smooth base S induces a flat deformation of $D_1^2(f)$, since this is a hypersurface. We have already seen that $D_1^3(f)$ deforms flat over S. Thus we may apply de Jong's theorem. The special points on $D_1^3(f_t)$, where $D_1^2(f_t)$ is not a normal crossing of two sheets, are of two types: Whitney umbrella points and triple points. We have already seen how Whitney umbrella points arise; triple points correspond to quadruple points of f_t, in which four pieces of $\mathbb{C}^3$ are mapped immersively and in general position. We denote the number of these by Q.

Corollary 3.3 $VD_\infty(D_1^2(f)) = |Fix(1,2)| - 8Q$

Proof Each Whitney umbrella point contributes 1 to $\mathrm{VD}_\infty(D_1^2)$. Each quadruple point gives rise to four triple points on $D_1^2(f)$. Each triple point contributes -2 to $\mathrm{VD}_\infty(D_1^2)$) [3, Example 2.3.3]. So

$$VD_\infty(D_1^2(f)) = \#\text{Whitney umbrellas} - 2\#\text{triple points} = |\mathrm{Fix}(1,2)| - 8Q.$$

□

Now we return to the map germ f of Sharland that is the focus of our interest.

To compute $VD_\infty(D_1^2)$ we need to find lifts to the normalisation $\tilde{\Sigma}$ of $D_1^3(f)$ of the vector fields annihilating the equation h of $D_1^2(f)$. A *Macaulay* calculation finds that modulo the defining ideal of $D_1^3(f)$, these vector fields are generated by

$$\chi_1 = (x^3y^2 + 2xy^3 - 3xz^2)\frac{\partial}{\partial x} + (2x^2y^3 + 4y^4)\frac{\partial}{\partial y} - 9z^3\frac{\partial}{\partial z}$$

$$\chi_2 = (y^4 + x^2z^2)\frac{\partial}{\partial x} - (2x^3y^3 - 2xy^4)\frac{\partial}{\partial y} + 3xz^3\frac{\partial}{\partial z}$$

These lift to

$$\tilde{\chi}_1 = \left(-3t^7\frac{\partial}{\partial t}, (2+\xi^2)u^7\frac{\partial}{\partial u}, (2+\xi)v^7\frac{\partial}{\partial v}\right), \quad \tilde{\chi}_2 = \left(t^8\frac{\partial}{\partial t}, \xi u^8\frac{\partial}{\partial u}, \xi^2 v^8\frac{\partial}{\partial v}\right)$$

in $\theta_{\tilde{\Sigma}} = \mathbb{C}\{t\}\partial_t \oplus \mathbb{C}\{u\}\partial_u \oplus \mathbb{C}\{v\}\partial_v$. The $\mathcal{O}_{\mathbb{C}^3}$-submodule of $\theta_{\tilde{\Sigma}}$ they generate is equal to

$$(t^{10})\partial_t \oplus (u^{10})\partial_u \oplus (v^{10})\partial_v + \mathrm{Sp}_{\mathbb{C}}\{\tilde{\chi}_1,\ x\tilde{\chi}_1,\ x^2\tilde{\chi}_1,\ \tilde{\chi}_2,\ x\tilde{\chi}_2\}.$$

Hence $\dim_{\mathbb{C}}(\theta_{\tilde{\Sigma}}/\theta(h)) = 25$ so that

$$VD_\infty(D_1^2) = 25 - 3 \times 5 = 10. \tag{3.6}$$

We have proved

Lemma 3.4 *The involution* $(2, 3)$ *has* 10 *fixed points on* $D^3(f_t)$. □

Corollary 3.5 $dim_{\mathbb{Q}} H_1(D^3(f_t); \mathbb{Q}) = 25$.

Proof By the lemma and (3.4), $\chi(D^3(f_t)) = -24$. □

Proposition 3.6 $dim_{\mathbb{Q}} H_1^{Alt}(D^3(f_t)) = 9$, $dim_{\mathbb{Q}} H_1^{\rho}(D^3(f_t)) = 16$, *and* $dim_{\mathbb{Q}} H_2^{Alt}(D^2(f_t)) = 9$.

Proof We use the Lefschetz fixed point theorem:

$$10 = \#\text{fixed points of } (2,3) = \sum_{k \geq 0} (-1)^k \left(\text{trace}(2,3)_* : H_k(D^3(f_t)) \to H_k(D^3(f_t))\right)$$

$$= 1 - \text{trace}(2,3)_* : H_1(D^3(f_t) \to H_1(D^3(f_t)) = 1 + \dim_{\mathbb{Q}} H_1^{\text{Alt}}(D^3(f_t)) - \dim_{\mathbb{Q}} H_1^{T}(D^3(f_t)). \tag{3.7}$$

The last equation here follows from the fact that the trace of $(2, 3)$ on the irreducible sign representation of S_3, on the trivial representation and on the irreducible 2-dimensional representation is -1, 1 and 0 respectively. It is straightforward to check that each fixed point of $(2, 3)$ is non-degenerate and therefore has Leftschetz number 1. Since $H_1^T(D^3(f_t)) = 0$, we obtain the first equality in the statement of the corollary. The second equality now follows by Corollary 3.5 and the third by the fact that $18 = \mu_I(f) = \dim_{\mathbb{Q}} H_2^{\text{Alt}}(D^2(f_t)) + \dim_{\mathbb{Q}} H_1^{\text{Alt}}(D^3(f_t))$. □

Now we compute $H_2(D^2(f_t))$. Although we have a formula for the ideal defining $D^2(f)$, we have no method of deriving from it a formula for the rank of the homology of $D^2(f_t)$. So once again we proceed indirectly, by calculating the homology of the image of its projection to U_t, $D_1^2(f_t)$.

Lemma 3.7 $D_1^2(f_t)$ *has the homotopy type of a wedge of 27 2-spheres.*

Proof We use the technique explained in Sect. 1.5, based on Siersma's theorem [25] that the rank of the vanishing homology of $D_1^2(f)$ is equal to the number of critical points of a reduced defining equation of $D_1^2(f)$ which move off the zero level as t moves off 0. The unfolding

$$F(t_1, t_2, t_3, x, y, z) = (t_1, t_2, t_3, x, y^2 + xz + x^2y, yz + t_1y + t_2z, z^2 + y^3 + t_3y)$$

is stable, by Mather's algorithm for the construction of stable germs as unfoldings of germs of rank 0, and $D_1^2(f_t)$ is the fibre of $D_1^2(F)$ over $t \in \mathbb{C}^3$. Let G be an equation for $D^2_1(F)$, let g_t be its restriction to $\{t\} \times \mathbb{C}^3$, and let J^{rel}_G be the relative jacobian ideal $(\partial G/\partial X, \partial G/\partial Y, \partial G/\partial Z)$. As in Sect. 1.5, we compute the number of critical points of a reduced defining equation of $D^2_1(f)$ which move off the zero level as t moves off 0, as the intersection multiplicity

$$\left(V(J^{\mathrm{rel}}_G : G^\infty) \cdot (\{0\} \times \mathbb{C}^3)\right)_{(0,0)}. \tag{3.8}$$

In fact calculation shows that in this case $(J^{\mathrm{rel}}_G : G^\infty)$ is equal to the transporter $(J^{\mathrm{rel}}_G : G)$. However, unlike the situation discussed in Sect. 1.5, $V(J^{\mathrm{rel}}_G : G^\infty)$ is not Cohen–Macaulay; it has projective dimension 5 as $\mathcal{O}_{\mathbb{C}^6,0}$-module, while its codimension is 3. To compute the intersection multiplicity, we have to use Serre's *formule clef*, from [23]. Denote $(J^{\mathrm{rel}}_G : G^\infty)$ by Q; then

$$(V(Q), \{0\} \times \mathbb{C}^3)_0 = \sum_j (-1)^j \dim_{\mathbb{C}} \mathrm{Tor}^{\mathcal{O}}_j \left(\frac{\mathcal{O}}{Q}, \frac{\mathcal{O}}{(t_1, t_2, t_3)}\right)$$

where $\mathcal{O} = \mathcal{O}_{\mathbb{C}^6,0}$. Since t_1, t_2, t_3 is a regular sequence there are at most three non-vanishing Tor modules, for $j = 0, 1, 2$. A *Macaulay* calculation shows that they have dimension 29, 3, 1 respectively, so that

$$\dim_{\mathbb{Q}} H_2(D^2_1(f_t)) = (V(Q), \{0\} \times \mathbb{C}^3)_{(0,0)} = 29 - 3 + 1 = 27. \tag{3.9}$$

□

It is striking that in this case $V(Q)$ is not Cohen–Macaulay. In all of the examples I know, where one uses the procedure of Sect. 1.5 to calculate μ_I, and G is the defining equation of the image of the stable unfolding F, the corresponding space $V(J^{\mathrm{rel}}_G : G^\infty)$ is Cohen Macaulay.

To relate the homology of $D^2_1(f_t)$ to the homology of $D^2(f_t)$, we use the image computing spectral sequence: $D^2_1(f_t)$ is the image of the projection $\pi^2 : D^2(f_t) \to U_t$. Taking account of the facts that f_t has no quadruple points, so that π^2 has no triple points, and that $H_1(D^2(f_t)) = 0$, the E^1 term is reduced to

$$\begin{array}{ccc} 0 = H_0^{\mathrm{Alt}}(D^2(\pi^2)) & H_1^{\mathrm{Alt}}(D^2(\pi^2)) & 0 \\ \downarrow & \downarrow d_1 & \\ H_0(D^2(f_t)) & 0 & H_2(D^2(f_t)) \end{array} \tag{3.10}$$

and the spectral sequence collapses here. So

$$27 = \dim_{\mathbb{C}} H_2(D^2_1(f_t)) = \dim_{\mathbb{C}} H_2(D^2(f_t)) + \dim_{\mathbb{C}} H_1^{\mathrm{Alt}}(D^2(\pi^2)). \tag{3.11}$$

Recall from Remark 1.1 the isomorphism $i : D^3(f_t) \to D^2(\pi^2 : D^2(f_t) \to U_t)$, given by $(a, b, c) \mapsto ((a, b), (a, c))$. The involution on $D^2(\pi^2)$ lifts to the transposition $(2, 3)$ on $D^3(f_t)$. Thus under the induced isomorphism of first homology, $H_1^{\mathrm{Alt}}(D^2(\pi^2))$ corresponds to the -1 eigenspace of $(2, 3)_*$ on $H_1(D^3(f_t))$. On each copy of the 2-dimensional irreducible representation ρ, and on each copy of the sign representation, $(2, 3)$ has 1-dimensional -1 eigenspace. Thus, using Proposition 3.6 for the second equality,

$$\dim_{\mathbb{Q}} H_1^{\mathrm{Alt}}(D^2(\pi^2)) = \dim_{\mathbb{Q}} H_1^{\mathrm{Alt}}(D^3(f_t)) + \frac{1}{2}\dim_{\mathbb{Q}} H_1^{\rho}(D^3(f_t)) = 17, \tag{3.12}$$

and, by (3.11),

$$\dim_{\mathbb{Q}} H_2(D^2(f_t)) = 10. \tag{3.13}$$

4 Calculations for the Germ $(\mathbb{C}^5, 0) \to (\mathbb{C}^6, 0)$

The germ

$$f_0(x, y, a, b, c) = (x^2 + ax + by, xy, y^2 + cx + ay, a, b, c)$$

has $\mu_I = \mathcal{A}_e$-codimension $= 1$, and versal unfolding

$$F(x, y, a, b, c, t) = (x^2 + ax + by, xy, y^2 + cx + (a + t)y, a, b, c, t).$$

Let $U_t \xrightarrow{f_t} X_t$ be a stable perturbation of f_0, with contractible domain $U_t \subset \mathbb{C}^5$. By (1.7),

$$1 = \operatorname{rank} H_5(X_t) = \operatorname{rank} H_4^{\mathrm{Alt}}(D^2(f_t)) + \operatorname{rank} H_3^{\mathrm{Alt}}(D^3(f_t)).$$

As in the previous section, we approach $D^3(f_t)$ via its projection to U_t, $D_1^3(f_t)$. As before, $D_1^3(f_t)$ is defined by the pull-back of the second Fitting ideal of $\mathcal{O}_{\mathbb{C}^5,0}$ considered as $\mathcal{O}_{\mathbb{C}^6,0}$-module. The $\mathcal{O}_{\mathbb{C}^6\times\mathbb{C},(0,0)}$-module $F_*\left(\mathcal{O}_{\mathbb{C}^5\times\mathbb{C},(0,0)}\right)$ has presentation

$$\begin{pmatrix} Y^2 - XZ - abZ - bcY + atY & aY + cX + tY & aY + bZ \\ aY + cX + tY & -Z - ac & Y - bc \\ aY + bZ & Y - bc & -X - ab - bt \end{pmatrix}. \tag{4.1}$$

so

$$\mathrm{Fitt}_2 = (Z + ac, Y - bc, X + (a + t)b)$$

and

$$
\begin{aligned}
F^*\mathrm{Fitt}_2 &= (y^2 + y(a+t) + xc + ac, xy - bc, x^2 + xa + yb + ab + bt) \\
&= \min_2 \begin{pmatrix} -y & -c \\ x+a & -y-a-t \\ b & x \end{pmatrix}
\end{aligned}
$$

The corresponding ideal for $t = 0$ defines the 3-fold singularity $D^3_1(f_0)$. A *Macaulay2* [8] calculation shows that the T^1 of $D^3_1(f_0)$) has dimension 1. Therefore $D^3_1(f_0)$ is isomorphic to the unique non-ICIS codimension 2 Cohen Macaulay 3-fold singularity with $\tau = 1$, which one can find in the table on p. 22 of [5]. This table also lists the Betti numbers of a smoothing, from which we obtain

$$
h_0(D^3_1(f_t)) = 1, \quad h_1(D^3_1(f_t)) = 0, \quad h_2(D^3_1(f_t)) = 1, \quad h_3(D^3_1(f_t)) = 0. \quad (4.2)
$$

In particular, $\chi(D^3_1(f_t)) = 2$.

Now $D^3(f_t)$ and $D^3_1(f_t)$ are smoothings of $D^3(f_0)$ and $D^3_1(f_0)$. Let $\pi = \pi^2 \circ \pi^3$ be the projection from $D^3(f_t)$ to U_t, $\pi(P, Q, R) = P$. Then $D^3(f_t)$ is a branched double cover of $D^3_1(f_t)$: for a generic point $P \in D^3_1(f_t)$, which shares its f_t-image with Q and R, $\pi^{-1}(P) = \{(P, Q, R), (P, R, Q)\}$. Because there are no quadruple points, the branching is of two types:

- over a point P where f_t has a stable singularity of type $\Sigma^{1,1,0}$, $\pi^{-1}(P) = \{(P, P, P)\}$. The set of all such points P is denoted $\Sigma^{1,1} f_t$. It lies in the closure of the set of branch points of the second kind:
- if $f_t(P) = f_t(Q)$ with f_t an immersion at P and of type $\Sigma^{1,0}$ at Q, then $\pi^{-1}(P) = \{(P, Q, Q)\}$, so (P, Q, Q) is a branch point.

We denote the set of all such points P by $D^3_{1,0}(f_t)$. Note that (Q, P, Q) and (Q, Q, P) are not branch points.

Thus

$$
\chi(D^3(f_t)) = 2\chi(D^3_1(f_t)) - \chi(D^3_{1,0}(f_t)) = 4 - \chi(D^3_{1,0}(f_t)). \quad (4.3)
$$

4.1 Equations for $\Sigma^{1,1} f$

The ramification ideal R_f, generated by the 5×5 minors of the jacobian matrix J of f_0 defines the non-immersive locus Σf of f_0. Then $\Sigma^{1,1}(f)$ is defined by the ideal of maximal minors of the matrix obtained by concatenating J with the jacobian matrix of a set of generators of R_f (see e.g. [12]). By removing from this ideal an $\mathfrak{m}$-primary component we obtain the ideal

$$
S := (3y + a, 3x + a, ac - 3bc, ab - 3bc, a^2 - 9bc),
$$

easily recognised as defining a curve isomorphic to the germ of the union of the three coordinate axes in $(\mathbb{C}^3, 0)$. This has $\delta = 2$ and therefore $\mu = 2\delta - r + 1 = 2$. It is not quite evident that this is deformed flat in a deformation of f_0, but nevertheless this is the case. The corresponding locus for the 1-parameter versal deformation F of f_0 has an $\mathfrak{m}$-primary component, whose removal leaves a 2-dimensional Cohen–Macaulay component which restricts to $\Sigma^{1,1} f_0$.

4.2 Equations for $D^3_{1,0}(f_0)$

By the description above, $D^3_{1,0}(f_0)$ is the "shadow component" of $f_0^{-1}(f(\Sigma f_0))$, that is, the closure of $f_0^{-1}(f_0(\Sigma f_0)) \setminus \Sigma f_0$. To find equations for it, we first look for equations for the support of $f_{0*}(\mathcal{O}_S/\mathcal{R}_{f_0})$. Let I_0 be the radical of the zero'th Fitting ideal of $f_{0*}(\mathcal{O}_S/\mathcal{R}_{f_0})$, let $I_1 = f_0^*(I_0)$, and let I_2 be the saturation $I_1 : \mathcal{R}_{f_0}^\infty$, in this case equal to $I_1 : \mathcal{R}_{f_0}^2$. After some effort one finds that I_2 is the ideal of maximal minors of the 2×4 matrix

$$\begin{pmatrix} a & b & x & y \\ -3y + a & x + a & -y - a & 3y - a + 4c \end{pmatrix}.$$

This is isomorphic to the cone over the rational normal curve of degree 4 (Pinkham's example). In the versal deformation F, the same construction leads to the ideal of maximal minors of the 2×4 matrix

$$\begin{pmatrix} a & b & x & y + t \\ -3y + a + t & x + a & -y - a - t & 3y - a + 4c - t \end{pmatrix}$$

One checks that this defines a smoothing of $D^3_{1,0}(f_0)$, over the Artin component of the base space (since it is given by the minors of a 2×4 matrix). So the only non-zero reduced Betti number is $\beta_2 = 1$ (see e.g. [22] p. 173). In particular

$$\chi(D^3_{1,0}(f_t)) = 2. \tag{4.4}$$

4.3 Homology of $D^3(f_t)$

By (1.7) and Lemma 1.4,

$$H_i(D^3(f_t)) = H_i^{\mathrm{Alt}} \oplus H_i^{\rho}. \tag{4.5}$$

Denote by h_i^{Alt} and h_i^{ρ} the ranks of these summands.

Because there are no quadruple points, $D^3_1(f_t)$ is the quotient of $D^3(f_t)$ by the $\mathbb{Z}_2$-action generated by the transposition $(2,3)(P,Q,R)=(P,R,Q)$. So $H_i(D^3_1(f_t))$ is the part of $H_i(D^3(f_t))$ invariant under $(2,3)_*$. Since $H_i^T(D^3(f_t))=0$ for $i>0$, the $(2,3)_*$ -invariant part of $H_i(D^3(f_t))$ is the $(2,3)_*$-invariant part of $H_i^\rho(D^3(f_t))$, and thus isomorphic to the sum of copies of the subspace of ρ invariant under $(2,3)$. The $(2,3)$-invariant subspace of ρ is 1-dimensional. Thus,

$$h_i(D^3_1(f_t))=\frac{1}{2}h_i^\rho(D^3(f_t)) \tag{4.6}$$

for $i>1$. Hence, by (4.2),

$$h_1^\rho(D^3(f_t))=0,\quad h_2^\rho(D^3(f_t))=2,\quad h_3^\rho(D^3(f_t))=0. \tag{4.7}$$

On the other hand, as $D^3(f_t)$ is a branched cover of degree 2 of $D^3_1(f_t)$, branched along $D^3_{1,0}(f_t)$, it follows that

$$\chi(D^3(f_t))=2\chi(D^3_1(f_t))-\chi(D^3_{1,0}(f_t))=2.$$

Putting this together with (4.6), we have

$$2=\chi(D^3(f_t))=1-\left(h_1^\rho+h_1^{\mathrm{Alt}}\right)+\left(h_2^\rho+h_2^{\mathrm{Alt}}\right)-\left(h_3^\rho+h_3^{\mathrm{Alt}}\right)=1-h_1^{\mathrm{Alt}}+h_2^{\mathrm{Alt}}+2-h_3^{\mathrm{Alt}}.$$

so

$$-1=-h_1^{\mathrm{Alt}}+h_2^{\mathrm{Alt}}-h_3^{\mathrm{Alt}}. \tag{4.8}$$

By [10, Theorem 4.6], the alternating homology of the multiple point spaces of a stable perturbation of a finitely determined map-germ is concentrated in middle dimension. As $D^3(f_t)$ is a 3-fold, this means $h_i^{\mathrm{Alt}}(D^3(f_t))=0$ for $i\neq 3$ and so from (4.8), $h_3^{\mathrm{Alt}}(D^3_1(f_t))=1$. Since here $\mu_I=1$, we conclude from (1.7) that $h_4^{\mathrm{Alt}}(D^2(f_t))=0$. Also, from (4.7) and (4.5), we conclude that $H_1(D^3(f_t))=0$, $\dim_{\mathbb{Q}}H_2(D^3(f_t))=2$ and $\dim_{\mathbb{Q}}H_3(D^3(f_t))=1$.

Remark 4.1 An application of the extended version of the Lefschetz Fixed Point Theorem gives the same conclusion: the fixed set of the involution $(2,3)$ on $D^3(f_t)$ is homeomorphic to the branch locus $D^3_{1,0}(f_t))$, and so by the extended version of the Lefschetz Fixed Point Theorem,

$$2=\chi(\mathrm{Fix}(2,3))=\sum_i(-1)^i\left(\mathrm{Tr}(2,3)_*:H_i(D^3(f_t))\to H_i(D^3(f_t))\right).$$

Because $\chi_\rho(2,3)=0$ and $H_i^T(D^3(f_t))=0$ for $i>0$, this alternating sum is equal to $1+h_1^{\mathrm{Alt}}-h_2^{\mathrm{Alt}}+h_3^{\mathrm{Alt}}$. This gives us the same information as (4.8).

4.4 Double Points

The double locus of F is defined by $F^*(\mathrm{Fitt}_2(F_*(\mathcal{O}_S))$, which is a principle ideal generated by the composite with F of the determinant of the lower right 2×2 submatrix of (4.1). We will call this composite G. As in Lemma 3.7, $D^2_1(f_t)$ has the homotopy type of a wedge of 4-spheres, whose number is the intersection number of $V(J^{\mathrm{rel}}_G : G^\infty)$ with $\mathbb{C}^5 \times \{0\}$. A *Macaulay2* [8] calculation gives

$$(J^{\mathrm{rel}}_G : G^\infty) = (J^{\mathrm{rel}}_G : G) = (3c - t, 3b + t, 2a + t, 6y + t, 6x - t). \tag{4.9}$$

The zero-locus of this ideal is a smooth curve of degree 1 over the t-axis, so $D^2_1(f_t)$ is homotopy-equivalent to a single 4-sphere, by Siersma's theorem [25].

The map of pairs $(D^2(f_t), D^3_2(f_t)) \xrightarrow{\pi^2_1} (D^2_1(f_t), D^3_1(f_t))$ induces a morphism between the long exact sequences of reduced homology of the pairs. Because $\pi^2_1 : D^2 \to D^2_1$ is an isomorphism outside D^3_2, the morphisms $H_i(D^2, D^3_2) \to H_i(D^2_1, D^3_1)$ are isomorphisms for all i. From the segment

$$\begin{array}{ccccccccc}
\cdots \longrightarrow & H_2(D^2, D^3_2) & \longrightarrow & H_1(D^3_2) & \longrightarrow & H_1(D^2) & \longrightarrow & H_1(D^2, D^3_2) & \longrightarrow 0 \\
& \| & & \downarrow & & \downarrow & & \| & \\
\cdots \longrightarrow & H_2(D^2_1, D^3_1) & \longrightarrow & H_1(D^3_1) & \longrightarrow & H_1(D^2_1) & \longrightarrow & H_1(D^2_1, D^3_1) & \longrightarrow 0
\end{array}$$

and the fact that $H_1(D^2_1) = 0 = H_1(D^3_2)$ (the latter equality because $D^3_2 \simeq D^3$, as there are no quadruple points) we see that $H_1(D^2(f_t)) = 0$. Because $H_2(D^2_1, D^3_1)$ is sandwiched between 0's in the lower sequence, continuing to the left we have

$$\begin{array}{ccccccccccccc}
& & \mathbb{Q} & & & & & & \mathbb{Q}^2 & & & & \\
& & \| & & & & & & \| & & & & \\
H_4(D^2, D^3_2) & \longrightarrow & H_3(D^3_2) & \longrightarrow & H_3(D^2) & \longrightarrow & H_3(D^2, D^3_2) & \longrightarrow & H_2(D^3_2) & \longrightarrow & H_2(D^2) & \longrightarrow & 0 \\
\| & & \downarrow & & \downarrow & & \| & & \downarrow & & \downarrow & & \\
H_4(D^2_1, D^3_1) & \longrightarrow & H_3(D^3_1) & \longrightarrow & H_3(D^2_1) & \longrightarrow & H_3(D^2_1, D^3_1) & \longrightarrow & H_2(D^3_1) & \longrightarrow & H_2(D^2_1) & \longrightarrow & 0 \\
\| & & \| & & \| & & \| & & \| & & \| & & \\
\mathbb{Q} & & 0 & & 0 & & \mathbb{Q} & & \mathbb{Q} & & 0 & &
\end{array}$$

We deduce successively

- $H_3(D^2, D^3_2) \simeq \mathbb{Q}$ and $H_3(D^2, D^3_2) \to H_2(D^3_2)$ is injective, and therefore
- $H_2(D^2) \simeq \mathbb{Q}$.
- $H_3(D^3_2) \to H_3(D^2)$ is surjective

The left end of the upper sequence is thus

$$0 \longrightarrow H_4(D^2) \longrightarrow H_4(D^2, D^3_2) \longrightarrow H_3(D^3_2) \longrightarrow H_3(D^2) \longrightarrow 0$$

with the two inner modules both isomorphic to $\mathbb{Q}$.

4.5 Homology of $M_2(f_t)$

By comparing the homology of $D^2(f_t)$ and $M_2(f_t)$ (which we will shortly determine), we might hope to gain some information about the homology of $D^2(f_t)$. All of the homology groups of $M_2(f_t)$ vanish. This can be seen with the help of the morphism f_{t*} from the long exact sequence of the pair $(U_t, D^2_1(f_t))$ to the long exact sequence of the pair $(X_t, M_2(f_t))$. Because f_t is an isomorphism outside $D^2_1(f_t)$, the morphisms of relative homology groups

$$f_{t*} : H_i(U_t, D^2_1(f_t)) \to H_i(X_t, M_2(f_t))$$

are all isomorphisms. From the top row of the diagram

$$\begin{array}{ccccccccc} 0 & \longrightarrow & 0 = H_5(U_t) & \longrightarrow & H_5(U_t, D^2_1) & \longrightarrow & H_4(D^2_1) & \longrightarrow & H_4(U_t) = 0 \\ & & \downarrow & & \downarrow & & \downarrow & & \downarrow \\ 0 & \longrightarrow & H_5(X_t) & \longrightarrow & H_5(X_t, M_2) & \longrightarrow & H_4(M_2) & \longrightarrow & H_4(X_t) = 0 \end{array}$$

we see that $H_5(U_t, D^2_1) \simeq \mathbb{Q}$. Hence $H_5(X_t, M_2) \simeq \mathbb{Q}$ also, and then from the bottom row it follows that $H_4(M_2) = 0$. A similar argument shows that $H_i(M_2) = 0$ for $0 < i < 4$.

In fact Houston shows in [10] by a rather more sophisticated argument that for a stable perturbation f_t of an $\mathcal{A}$-finite germ f_t, all of the $M_k(f_t)$ have the homotopy type of wedges of spheres in middle dimension. In this case the number of spheres in the wedge homotopy-equivalent to $M_2(f_t)$ is 0.

4.6 Relation Between D^2 and M_2

There is a surjective map $f_t^{(2)} : D^2(f_t) \to M_2(f_t)$, $f_t^{(2)}(P, Q) = f_t(P)$. The multiple point spaces of $f_t^{(2)}$ are related to those of f_t, but are not identical. Consider the following maps:

$$\alpha : D^2(f) \to D^2(f_t^{(2)}), \quad (P, Q) \mapsto \big((P, Q), (Q, P)\big)$$
$$\beta : D^2(f_t^{(2)}) \to D^2(f_t), \quad \big((P, Q), (R, S)\big) \mapsto (P, R).$$

Denote by $(1, 2)$ the usual involution on D^2. The diagrams

$$\begin{array}{ccc} D^2(f_t) & \xrightarrow{(1,2)} & D^2(f_t) \\ \alpha\downarrow & & \downarrow\alpha \\ D^2(f_t^{(2)}) & \xrightarrow{(1,2)} & D^2(f_t^{(2)}) \end{array} \qquad \begin{array}{ccc} D^2(f) & \xrightarrow{(1,2)} & D^2(f) \\ \beta\uparrow & & \uparrow\beta \\ D^2(f_t^{(2)}) & \xrightarrow{(1,2)} & D^2(f_t^{(2)}) \end{array}$$

both commute, and $\beta \circ \alpha$ is the identity on $D^2(f)$. It follows that α and β induce morphisms

$$H_i^{\mathrm{Alt}}(D^2(f_t)) \underset{\beta_*}{\overset{\alpha_*}{\rightleftarrows}} H_i^{\mathrm{Alt}}(D^2(f_t^{(2)}))$$

and $\beta_* \circ \alpha_*$ is the identity.

However α is not surjective and β is not injective. Suppose that $(P, Q, R) \in D^3(f_t)$ with P, Q, R pairwise distinct. Then

$$\big((P, Q), (Q, P)\big), \big((P, Q), (Q, R)\big), \big((P, R), (Q, P)\big), \big((P, R), (Q, R)\big)$$

all lie in $D^2(f_t^{(2)})$ and all are mapped by β to (P, Q). And, of these, only $\big((P, Q), (Q, P)\big)$ is in the image of α.

We draw no further conclusion from this, but ask whether further consideration of the multiple point spaces of the map $f_t^{(2)}$ and indeed of $f_t^{(k)}$ for higher k may provide useful information.

Nevertheless, from the vanishing of $H_1(M_2(f_t))$, and the image-computing spectral sequence, we obtain a second argument that $H_1(D^2(f_t)) = 0$.

Acknowledgements The calculations in Sect. 3 were begun in collaboration with Isaac Bird, as part of his final year MMath project at Warwick. I am grateful to him for agreement to use them here, and for his enthusiasm on the project, which contributed a great deal to its further development. I also thank Mike Stillman for help finding the 2×4 matrices of Sect. 4.2, and Juan José Nuño for pointing out the results of Greuel and Steenbrink in [9].

References

1. Altıntaş, A., Mond, D.: Free resolutions for multiple point spaces. Geom. Dedicata **162**, 177–190 (2013). MR 3009540
2. Cooper, T., Mond, D., Wik Atique, R.: Vanishing topology of codimension 1 multi-germs over $\mathbb{R}$ and $\mathbb{C}$. Compos. Math. **131**(2), 121–160 (2002). MR 1898432 (2004c:32052)
3. de Jong, T.: The virtual number of D_∞ points. I. Topology **29**(2), 175–184 (1990). MR 1056268 (91f:32043)
4. de Jong, T., van Straten, D.: Disentanglements. Singularity Theory and Its Applications, Part I (Coventry, 1988/1989). Lecture Notes in Mathematics, vol. 1462, pp. 199–211. Springer, Berlin (1991). MR 1129033 (93a:14003)

5. Fruehbis-Krueger, A., Zach, M.: On the vanishing topology of isolated Cohen-Macaulay codimension 2 singularities (2015). arXiv:1501.01915
6. Goryunov, V.V.: Semi-simplicial resolutions and homology of images and discriminants of mappings. Proc. Lond. Math. Soc. (3) **70**(2), 363–385 (1995). MR 1309234 (95j:32050)
7. Goryunov, V.V., Mond, D.: Vanishing cohomology of singularities of mappings. Compos. Math. **89**(1), 45–80 (1993). MR 1248891 (94k:32058)
8. Grayson, D.R., Stillman, M.E.: Macaulay2, a software system for research in algebraic geometry. http://www.math.uiuc.edu/Macaulay2/
9. Greuel, G.-M., Steenbrink, J.: On the topology of smoothable singularities. Singularities, Part 1 (Arcata, California, 1981). Proceedings of Symposia in Pure Mathematics, vol. 40, pp. 535–545. American Mathematical Society, Providence (1983). MR 713090
10. Houston, K.: Local topology of images of finite complex analytic maps. Topology **36**(5), 1077–1121 (1997). MR 1445555 (98g:32064)
11. Kleiman, S.L.: Multiple-point formulas. I. Iteration. Acta Math. **147**(1–2), 13–49 (1981). MR 631086 (83j:14006)
12. Mather, J.N.: On Thom-Boardman singularities. Dynamical Systems (Proceedings of a Symposium Held at the University of Bahia, Salvador, 1971), pp. 233–248. Academic Press, New York (1973). MR 0353359 (50 #5843)
13. Milnor, J.: Singular Points of Complex Hypersurfaces. Annals of Mathematics Studies, vol. 61. Princeton University Press, Princeton (1968)
14. Marar, W.L., Mond, D.: Multiple point schemes for corank 1 maps. J. Lond. Math. Soc. (2) **39**(3), 553–567 (1989). MR 1002466 (91c:58010)
15. Marar, W.L., Mond, D.: Real map-germs with good perturbations. Topology **35**(1), 157–165 (1996). MR 1367279
16. Marar, W.L., Nuño-Ballesteros, J.J.: A note on finite determinacy for corank 2 map germs from surfaces to 3-space. Math. Proc. Camb. Philos. Soc. **145**(1), 153–163 (2008). MR 2431646
17. Marar, W.L., Nuño-Ballesteros, J.J., Peñafort-Sanchis, G.: Double point curves for corank 2 map germs from $\mathbb{C}^2$ to $\mathbb{C}^3$. Topol. Appl. **159**(2), 526–536 (2012). MR 2868913
18. Mond, D.: Some remarks on the geometry and classification of germs of maps from surfaces to 3-space. Topology **26**(3), 361–383 (1987). MR 899055
19. Mond, D.: Some open problems in the theory of singularities of mappings. J. Singul. **12**, 141–155 (2015). MR 3317146
20. Mond, D., Pellikaan, R.: Fitting ideals and multiple points of analytic mappings. Algebraic Geometry and Complex Analysis (Pátzcuaro, 1987). Lecture Notes in Mathematics, vol. 1414, pp. 107–161. Springer, Berlin (1989). MR 1042359 (91e:32035)
21. Nuño-Ballesteros, J.J., Peñafort-Sanchis, G.: On multiple point schemes (2015). arXiv:1509.04990
22. Némethi, A.: Invariants of normal surface singularities. Real and Complex Singularities. Contemporary Mathematics, vol. 354, pp. 161–208. American Mathematical Society, Providence (2004). MR 2087811
23. Serre, J.-P.: Algèbre locale. Multiplicité. Lecture Notes in Mathematics, vol. 11. Springer, Berlin (1957)
24. Sharland, A.A.: Examples of finitely determined map-germs of corank 2 from n-space to $(n+1)$-space. Int. J. Math. **25**(5), 1450044, 17 (2014). MR 3215220
25. Siersma, D.: Vanishing cycles and special fibres. Singularity Theory and Its Applications, Part I (Coventry, 1988/1989). Lecture Notes in Mathematics, vol. 1462, pp. 292–301. Springer, Berlin (1991). MR 1129039 (92j:32129)

Singular Fibers of Stable Maps of Manifold Pairs and Their Applications

Osamu Saeki and Takahiro Yamamoto

Abstract Let (M, N) be a manifold pair, where M is a closed 3-dimensional manifold and N is a closed 2-dimensional submanifold of M. In this paper, we first classify singular fibers of C^∞ stable maps of (M, N) into surfaces. Then, we compute the cohomology groups of the associated universal complex of singular fibers, and obtain certain cobordism invariants for Morse functions on manifold pairs (M', N'), where M' is a closed surface and N' is a closed 1-dimensional submanifold of M'. We also give the 2-colored versions of all these results, when the submanifold separates the ambient manifold into two parts.

Keywords Manifold pair · Stable map · Singular fiber · Cobordism · 2-coloring

2000 Mathematics Subject Classification Primary 57R45 · Secondary 57R35 57R90 · 58K15 · 58K65

1 Introduction

Let M be a C^∞ manifold and N a closed C^∞ submanifold of M. In this paper, we call such a pair (M, N) a *manifold pair*. Its *dimension* is defined to be the pair $(\dim M, \dim N)$ of dimension of the ambient manifold and that of the submanifold. For a C^∞ map $f: M \to Q$ into another manifold Q, we often write it as $f: (M, N) \to Q$ when we need to specify the submanifold N. A C^∞ map $f: (M, N) \to Q$ is said to be C^∞ *stable* (or C^0 *stable*) if there exists an open neighborhood $N(f)$ of f in the mapping space $C^\infty(M, Q)$, endowed with the

O. Saeki (✉)
Institute of Mathematics for Industry, Kyushu University, Motooka 744, Nishi-ku, Fukuoka 819-0395, Japan
e-mail: saeki@imi.kyushu-u.ac.jp

T. Yamamoto
Department of Mathematics, Tokyo Gakugei University, 4-1-1 Nukuikita-machi, Koganei-shi, Tokyo 184-8501, Japan
e-mail: yamush@u-gakugei.ac.jp

R. N. Araújo dos Santos et al. (eds.), *Singularities and Foliations. Geometry, Topology and Applications*, Springer Proceedings in Mathematics & Statistics 222, https://doi.org/10.1007/978-3-319-73639-6_8

Whitney C^∞ topology, such that for every $g \in N(f)$ we have the commutative diagram

$$\begin{array}{ccc} (M,N) & \xrightarrow{\Phi} & (M,N) \\ f\big\downarrow & & \big\downarrow g \\ Q & \xrightarrow{\varphi} & Q \end{array} \tag{1.1}$$

for some diffeomorphisms (resp. homeomorphisms) Φ and φ with $\Phi(N) = N$. In the following, when we just use the terminology "stable map", it will mean "C^∞ stable map".

For a C^∞ map $f: (M, N) \to Q$ and a point $q \in Q$, the pre-image pair

$$(f^{-1}(q), f^{-1}(q) \cap N)$$

is called the *level set* of f over q. Furthermore, we call the map germ along the level set

$$f: ((M, N), (f^{-1}(q), f^{-1}(q) \cap N)) \to (Q, q)$$

the *fiber* over q, adopting the terminology introduced in [4]. If a point $q \in Q$ is a regular value of both f and $f|_N$, then we call the fiber (or the level set) over q a *regular fiber* (resp. a *regular level set*); otherwise, a *singular fiber* (resp. a *singular level set*).

Equivalence relations among fibers are defined as follows. Let $f_i: (M_i, N_i) \to Q_i$, $i = 0, 1$, be C^∞ maps. For $q_i \in Q_i$, $i = 0, 1$, we say that the fibers over q_0 and q_1 are C^∞ *equivalent* (or C^0 *equivalent*) if for some open neighborhoods U_i of q_i in Q_i, there exist diffeomorphisms (resp. homeomorphisms) $\Phi: (f_0^{-1}(U_0), f_0^{-1}(U_0) \cap N_0) \to (f_1^{-1}(U_1), f_1^{-1}(U_1) \cap N_1)$ and $\varphi: U_0 \to U_1$ with $\varphi(q_0) = q_1$ which make the following diagram commutative:

$$\begin{array}{ccc} ((f_0^{-1}(U_0), f_0^{-1}(U_0) \cap N_0), (f_0^{-1}(q_0), f_0^{-1}(q_0) \cap N_0)) & \xrightarrow{f_0} & (U_0, q_0) \\ \Phi\big\downarrow & & \big\downarrow\varphi \\ ((f_1^{-1}(U_1), f_1^{-1}(U_1) \cap N_1), (f_1^{-1}(q_1), f_1^{-1}(q_1) \cap N_1)) & \xrightarrow{f_1} & (U_1, q_1). \end{array}$$

Now, suppose that N is a codimension one submanifold of M and that N separates M into two parts R and B: i.e.,

$$R \cup B = M \setminus N, \ \overline{R} \cap \overline{B} = \partial R = \partial B = N,$$

where $\overline{R}$ and $\overline{B}$ denote the closures of R and B in M, respectively, and we set $\partial R = \overline{R} \setminus R$ and $\partial B = \overline{B} \setminus B$. In such a case, we say that (M, N) is a 2-*colored manifold pair*.[1] A C^∞ map $f: (M, N) \to Q$ of a 2-colored manifold pair is said to be C^∞ *stable* (resp. C^0 *stable*) if in the above definition of a stable map of a manifold pair, there exist diffeomorphisms (resp. homeomorphisms) Φ and φ as in (1.1) such that $\Phi(B) = B$ and $\Phi(R) = R$. We can define the equivalence relations for fibers similarly, which we call C^∞ *color equivalence* and C^0 *color equivalence*, respectively.

The notion of singular fibers of C^∞ maps between manifolds was first introduced in [4], where classifications of singular fibers of stable maps $M \to Q$ with $(\dim M, \dim Q) = (2, 1), (3, 2)$ and $(4, 3)$ were obtained. Later, singular fibers of stable maps of manifolds with or without boundary were studied in [4–10, 14–16], especially in connection with cobordisms. The first author [4] established the theory of universal complex of singular fibers of C^∞ maps as an analogy of the Vassiliev complex for map germs [3, 12]. This can be used for getting certain cobordism invariants of singular maps. For example, the first author [4] obtained cobordism invariants for stable Morse functions on closed surfaces, and the second author [16] studied the universal complex of singular fibers of two-colored C^∞ maps, computing its cohomology groups, where in [16], two-colorings are considered for the target. In these theories, for a certain set of singular fibers τ, cohomology classes of the universal complex of singular fibers of τ-maps provide τ-cobordism invariants for τ-maps.

In this paper, we study singular fibers of proper C^∞ stable maps of $(3, 2)$-dimensional manifold pairs into surfaces. By using such fibers, we obtain cobordism invariants for stable Morse functions on $(2, 1)$-dimensional manifold pairs. As far as the authors know, this is the very first study of singular fibers for generic maps of manifold pairs.

The paper is organized as follows. In Sect. 2, we classify fibers of proper C^∞ stable maps of $(3, 2)$-dimensional manifold pairs into surfaces. For this we use several known results on the classification of stable singularities of maps on manifold pairs together with the techniques developed by the first author in [4]. The equivalence relation that we consider in this paper is the following weak one: near the singular points, the diffeomorphisms (or the homeomorphisms) of the sources should preserve the submanifolds, and when the source manifold pairs are 2-colored, they should preserve the 2-colorings as well. However, we ignore the submanifolds away from the singular points (for details, see Definitions 2.6 and 2.14). We will see that the classification results are essentially different from those for singular fibers of stable maps of 3-dimensional manifolds with boundary into surfaces. This is because a fiber has "boundary points" if it intersects the boundary, while in the case of maps of manifold pairs, a fiber is not cut even if it intersects the submanifold. We will also see that the 2-colored version is somewhat different from the non-colored version as we take the 2-colorings into consideration.

[1] It is known that for a given manifold pair (M, N) with $\dim N = \dim M - 1$, it can be given a structure of a 2-colored manifold pair if and only if the Poincaré dual to the homology class represented by N vanishes in $H^1(M; \mathbb{Z}_2)$.

In Sect. 3, we obtain several co-existence formulae of singular fibers for C^∞ stable maps of (3, 2)-dimensional manifold pairs into surfaces. These formulae can be obtained by analyzing the adjacencies of the singular fibers.

In Sect. 4, we construct the universal complexes of singular fibers for proper C^∞ stable maps of (2-colored) manifold pairs. The constructions are similar to those given in [9, Sect. 4] for maps of manifolds with boundary. The equivalence relation for fibers that we consider takes into account the parity of the number of regular components. This is necessary for getting non-trivial cohomology classes. Then, we compute their cohomology groups of dimensions 0 and 1.

In Sect. 5, we use the cohomology classes of the universal complexes obtained in Sect. 4 to get cobordism invariants for singular maps. More explicitly, we obtain certain cobordism invariants for stable Morse functions on (2, 1)-dimensional (2-colored) manifold pairs. We will see that, unfortunately, these invariants are not fine enough to give complete invariants in most of the cases. However, in the case of stable Morse functions on 2-colored Morse functions on (2, 1)-dimensional manifold pairs, we get a complete invariant.

Finally in Appendices A and B, we will determine the cobordism groups of (2, 1)-dimensional manifold pairs and that of 2-colored ones, using standard and elementary techniques. These are included in this paper, since as far as the authors know, there are no such explicit results in the literature except for Wall's sophisticated result [13], and we need to calculate them for measuring the efficiency of the cobordism invariants obtained by using singular fibers.

Throughout the paper, all manifolds and maps between them are smooth of class C^∞ unless otherwise stated. For a smooth map $f\colon M \to Q$ between manifolds, we denote by $S(f)$ the set of points in M where the differential of f does not have maximal rank $\min\{\dim M, \dim Q\}$. For a space X, id_X denotes the identity map of X. For a (co)cycle c, we denote by $[c]$ the (co)homology class represented by c.

2 Classification of Singular Fibers

In this section, we classify singular fibers of proper C^∞ stable maps of manifold pairs (M, N) with $(\dim M, \dim N) = (3, 2)$ into surfaces Q, where a map $f\colon (M, N) \to Q$ is *proper* if $f^{-1}(K)$ is a compact subset of M for every compact subset K of Q.

2.1 Stable Maps of Manifold Pairs

For a proper C^∞ map $(M, N) \to Q$ as above, let us consider the associated map germ at a point in N. For the classification of such map germs, we use those source diffeomorphism germs of M which preserve the submanifold N, and this leads to a geometrically defined subgroup of $\mathcal{A}$ in the sense of Damon [1]. Consequently, the infinitesimal method can be used; for example, a map germ as above is C^∞

stable if and only if it is infinitesimally stable (see, for example, [1, Theorem 9.3]). Note that to a map germ as above are associated two map germs of 3-dimensional manifolds with boundary into surfaces, since N locally divides the source open set (of a representative of the map germ) into two parts. Then, the infinitesimal method implies that if a given map germ of a manifold pair is C^∞ stable, then the two associated map germs defined on manifolds with boundary are also C^∞ stable.

Therefore, we can use the known characterization of proper C^∞ stable maps of 3-dimensional manifolds with boundary into surfaces (for example, see [9]), which can be deduced by using results obtained in [2, 11], in order to get a similar characterization for maps of manifold pairs.

Note that an arbitrarily given pair of map germs of manifolds with boundary may not be associated with a map germ of manifold pairs: they should match along the boundary. On the other hand, when we work with maps of manifold pairs, source diffeomorphisms which interchange the two parts of the complement of N are also allowed so that the list of stable singularities may get smaller.

With these remarks in mind, we can get the following characterization of C^∞ stable maps of manifold pairs.

Proposition 2.1 *A proper C^∞ map $f\colon (M, N) \to Q$ with* $(\dim M, \dim N; \dim Q) = (3, 2; 2)$ *is C^∞ stable if and only if it satisfies the following conditions.*

(1) *(Local conditions) In the following, for $p \in N$, we use local coordinates (x, y, z) around p such that N corresponds to the set $\{z = 0\}$.*

(1a) *For $p \in M \setminus N$, the germ of f at p is right-left equivalent to one of the following:*

$$(x, y, z) \mapsto \begin{cases} (x, y), & p : \textit{regular point}, \\ (x, y^2 + z^2), & p : \textit{definite fold point}, \\ (x, y^2 - z^2), & p : \textit{indefinite fold point}, \\ (x, y^3 + xy - z^2), & p : \textit{cusp point}. \end{cases}$$

(1b) *For $p \in N \setminus S(f)$, the germ of f at p is right-left equivalent to one of the following:*

$$(x, y, z) \mapsto \begin{cases} (x, y), & p : \textit{regular point of } f|_N, \\ (x, y^2 + z), & p : \textit{relative fold point}, \\ (x, y^3 + xy + z), & p : \textit{relative cusp point}. \end{cases}$$

(1c) *For $p \in N \cap S(f)$, the germ of f at p is right-left equivalent to one of the following:*

$$(x, y, z) \mapsto \begin{cases} (x, y^2 + xz + z^2), & p : \textit{definite } B_2 \textit{ point}, \\ (x, y^2 + xz - z^2), & p : \textit{indefinite } B_2 \textit{ point}. \end{cases}$$

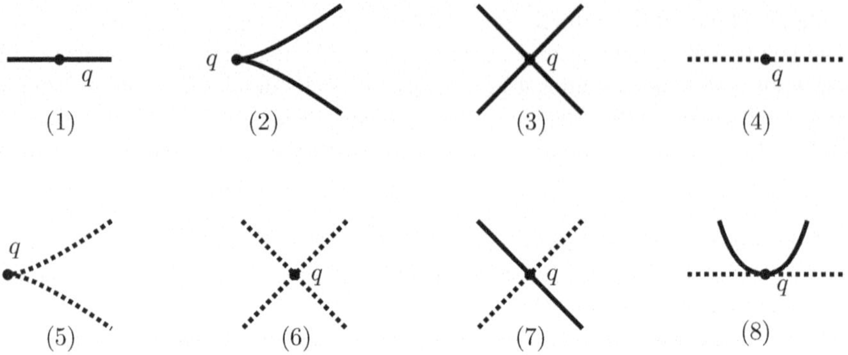

Fig. 1 The images of multi-germs of $f|_{S(f)\cup S(f|_N)}$

(2) *(Global conditions) For each* $q \in f(S(f)) \cup f(S(f|_N))$, *the multi-germ*

$$(f|_{S(f)\cup S(f|_N)}, f^{-1}(q) \cap (S(f) \cup S(f|_N)))$$

is right-left equivalent to one of the eight multi-germs whose images are depicted in Fig. 1, where the ordinary curves correspond to $f(S(f))$ *and the dotted curves to* $f(S(f|_N))$: (1) *and* (4) *represent immersion mono-germs* $(\mathbb{R}, 0) \ni t \mapsto (t, 0) \in (\mathbb{R}^2, 0)$ *which correspond to a single fold point and a single relative fold point respectively,* (3), (6) *and* (7) *represent normal crossings of two immersion germs, each of which corresponds to a fold point or a relative fold point,* (2) *and* (5) *represent cusp mono-germs* $(\mathbb{R}, 0) \ni t \mapsto (t^2, t^3) \in (\mathbb{R}^2, 0)$ *which correspond to a cusp point and a relative cusp point respectively, and* (8) *represents the restriction of the mono-germ* (1c), *corresponding to a single point in* $N \cap S(f)$, *to the singular point set.*

Note that if a C^∞ map $f: (M, N) \to Q$ is C^∞ stable, then so is $f|_N: N \to Q$.

Let M_i be smooth manifolds and $A_i \subset M_i$ be subsets, $i = 0, 1$. A continuous map $g: A_0 \to A_1$ is said to be *smooth* if for every $q \in A_0$, there exists a smooth map $\widetilde{g}: V \to M_1$ defined on a neighborhood V of $q \in M_0$ such that $\widetilde{g}|_{V\cap A_0} = g|_{V\cap A_0}$. Furthermore, a smooth map $g: A_0 \to A_1$ is called a *diffeomorphism* if it is a homeomorphism and its inverse is also smooth. When there exists a diffeomorphism between A_0 and A_1, we say that they are *diffeomorphic*.

By Proposition 2.1, we have the following local descriptions for singular level sets. In the statement, it may not be necessary to distinguish some of the cases if only the topology of level sets is concerned; nevertheless, we divide the cases as below in order to introduce symbols which take corresponding map germs into account.

Lemma 2.2 *Let* $f: (M, N) \to Q$ *be a proper* C^∞ *stable map of a manifold pair into* Q *with* $(\dim M, \dim N; \dim Q) = (3, 2; 2)$. *Then, every point* $p \in S(f) \cup S(f|_N)$ *has one of the following neighborhoods in its corresponding singular level set (see Figs. 2 and 3 for references):*

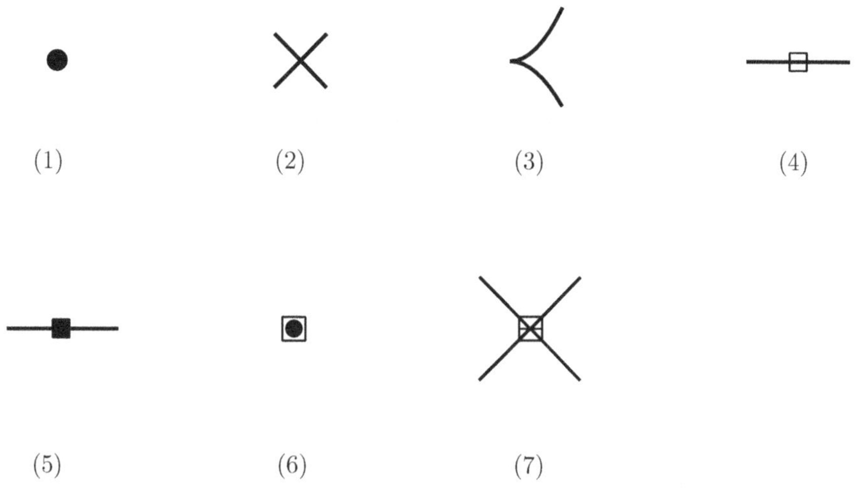

Fig. 2 Neighborhoods of singular points in their singular level sets

(1) *isolated point diffeomorphic to* $\{(y, z) \in \mathbb{R}^2 \mid y^2 + z^2 = 0\}$, *if* $p \in M \setminus N$ *is a definite fold point,*
(2) *union of two transverse arcs diffeomorphic to* $\{(y, z) \in \mathbb{R}^2 \mid y^2 - z^2 = 0\}$, *if* $p \in M \setminus N$ *is an indefinite fold point,*
(3) *cuspidal arc diffeomorphic to* $\{(y, z) \in \mathbb{R}^2 \mid y^3 - z^2 = 0\}$, *if* $p \in M \setminus N$ *is a cusp point,*
(4) *arc diffeomorphic to* $\{(y, z) \in \mathbb{R}^2 \mid y^2 + z = 0\}$, *if* $p \in N$ *is a relative fold point,*
(5) *arc diffeomorphic to* $\{(y, z) \in \mathbb{R}^2 \mid y^3 + z = 0\}$, *if* $p \in N$ *is a relative cusp point,*
(6) *isolated point diffeomorphic to* $\{(y, z) \in \mathbb{R}^2 \mid y^2 + z^2 = 0\}$, *if* $p \in N \cap S(f)$ *is a definite* B_2 *point,*
(7) *union of two transverse arcs diffeomorphic to* $\{(y, z) \in \mathbb{R}^2 \mid y^2 - z^2 = 0\}$, *if* $p \in N \cap S(f)$ *is an indefinite* B_2 *point.*

Remark 2.3 Note that in Fig. 2, a black dot (1) and a black dot surrounded by a square (6) both represent an isolated point; however, we use distinct symbols in order to distinguish their corresponding map germs. For an arc with square (4) and an arc with black square (5), we use distinct symbols by a similar reason, and for (2) and (7) as well. In the figures, squares represent points on the submanifold N; more precisely, they are points in $S(f|_N)$.

In Fig. 3, singular level sets that intersect the submanifold $N \subset M$ are depicted, where the surfaces appearing in the figures correspond to the submanifold $x = 0$, the intersection of $x = 0$ with N is depicted by thick curves, and the associated map germs restricted to $x = 0$ correspond to the respective height functions. In fact, for cases (4) and (5), the relevant map germs are right-left equivalent to the suspensions of the function germs in a sense similar to that in [9, Definition 4.2]. As to cases (6) and

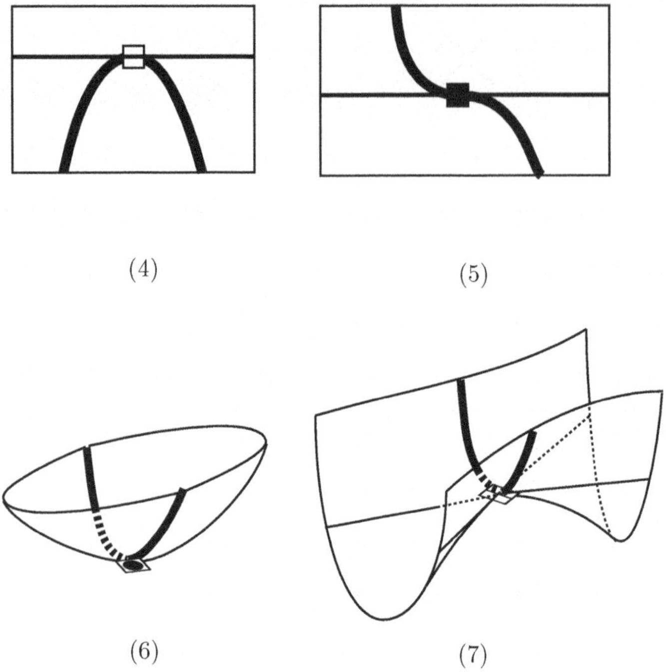

Fig. 3 Singular level sets touching the submanifold N

(7), the relevant map germs are obtained by using certain "generic deformations" of the function germs. As in Fig. 2(7), we use a square together with a small line segment inside. This line segment is chosen in such a way that it is tangent to the curve $N \cap \{x = 0\}$ as depicted in Fig. 3(7).

For the local nearby level sets, we have the following, which can be proved by direct calculations using the corresponding normal forms.

Lemma 2.4 *Let $f\colon (M, N) \to Q$ be a proper C^∞ stable map of a manifold pair into Q with* $(\dim M, \dim N; \dim Q) = (3, 2; 2)$. *For $p \in S(f) \cup S(f|_N)$ such that $f^{-1}(f(p)) \cap (S(f) \cup S(f|_N)) = \{p\}$, the level set near p is as depicted in Fig. 4:*

(1) *$p \in M \setminus N$ is a definite fold point,*
(2) *$p \in M \setminus N$ is an indefinite fold point,*
(3) *$p \in M \setminus N$ is a cusp point,*
(4) *$p \in N$ is a relative fold point,*
(5) *$p \in N$ is a relative cusp point,*
(6) *$p \in S(f) \cap N$ is a definite B_2 point,*
(7) *$p \in S(f) \cap N$ is an indefinite B_2 point,*

where each of the 0*-dimensional objects and the thin* 1*-dimensional objects represents a portion of the level set over the corresponding point in the target, each of the thick curves represents $f(S(f))$, and each of the dotted curves represents*

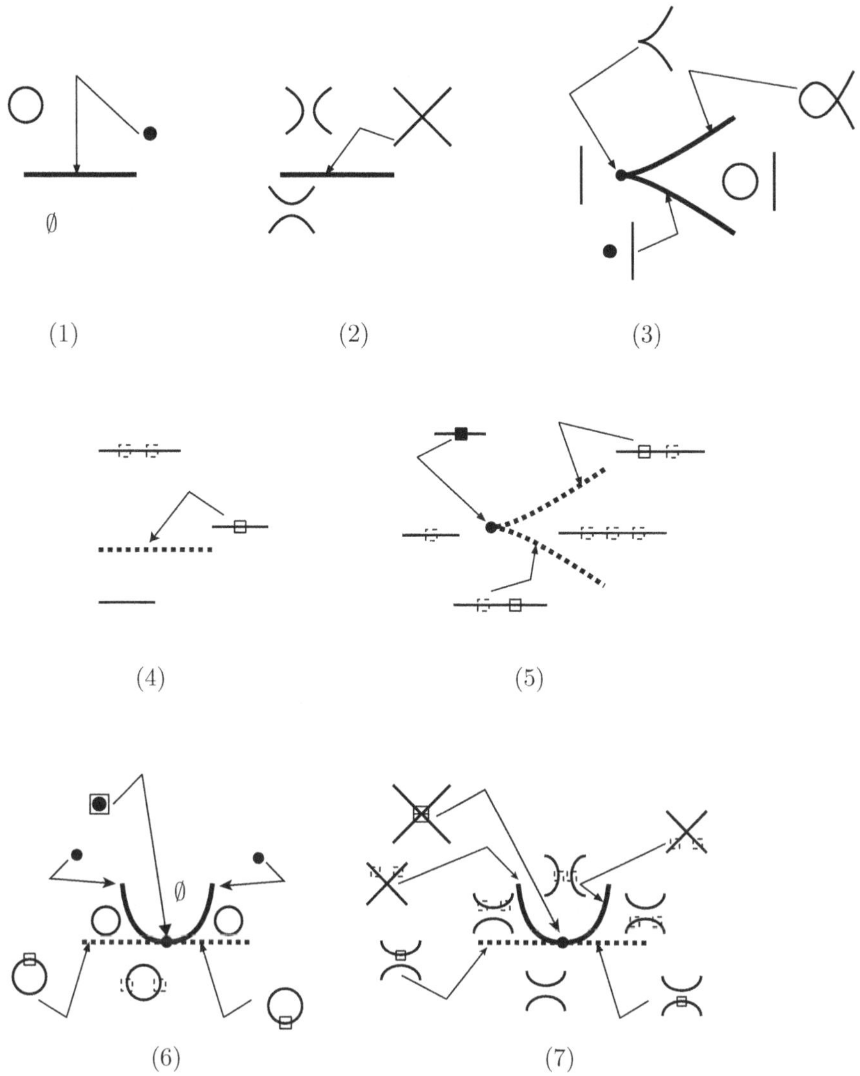

Fig. 4 Local degenerations of level sets

$f(S(f|_N))$ near $f(p)$. Furthermore, the dotted squares represent (transverse) intersections with N.

Definition 2.5 Suppose that we are given a finite number of fibers of smooth maps, where all the dimensions of the sources are the same and those of the targets are also the same. Then, their *disjoint union* is the fiber corresponding to the single map defined on the disjoint union of the sources, where the target spaces are all identified to a single small open disk. This definition clearly depends on such identifications; however, in the following, we can choose "generic identifications" in such a way that the resulting fiber is C^∞ stable, and the result is unique up to C^∞ equivalence as long as the above identifications are generic.

In order to get a classification result, let us introduce the following notion.

Definition 2.6 Let $f_i\colon (M_i, N_i) \to Q_i$, $i = 0, 1$, be C^∞ maps of manifold pairs. For $q_i \in Q_i$, $i = 0, 1$, we say that the fibers over q_0 and q_1 are *weakly C^∞ equivalent* (or *weakly C^0 equivalent*) if for some neighborhoods N'_i of $f_i^{-1}(q_i) \cap (S(f_i) \cup S(f_i|_{N_i}))$ in N_i and for some small open neighborhoods U_i of q_i in Q_i, there exist diffeomorphisms (resp. homeomorphisms) $\Phi\colon (f_0^{-1}(U_0), f_0^{-1}(U_0) \cap N'_0) \to (f_1^{-1}(U_1), f_1^{-1}(U_1) \cap N'_1)$, and $\varphi\colon U_0 \to U_1$ with $\varphi(q_0) = q_1$ which make the following diagram commutative:

$$\begin{array}{ccc}
((f_0^{-1}(U_0), f_0^{-1}(U_0) \cap N'_0), (f_0^{-1}(q_0), f_0^{-1}(q_0) \cap N'_0)) & \xrightarrow{f_0} & (U_0, q_0) \\
\Phi\downarrow & & \downarrow\varphi \\
((f_1^{-1}(U_1), f_1^{-1}(U_1) \cap N'_1), (f_1^{-1}(q_1), f_1^{-1}(q_1) \cap N'_1)) & \xrightarrow{f_1} & (U_1, q_1).
\end{array}$$

In other words, considering weak C^∞ equivalence classes instead of the usual ones corresponds to ignoring transverse intersections of the central level sets with the submanifold N_i away from the neighborhoods of singular points.

In the following, a *trivial circle bundle* refers to the projection $(S^1, P) \times B \to B$ to the second factor, where (S^1, P) is a manifold pair, P is a finite set of points in S^1, and B is a manifold.

Now, by using the method developed in [4], we get the following list of singular fibers. We omit the proof here.

Proposition 2.7 *Let us consider proper C^∞ stable maps $f\colon (M, N) \to Q$ with* $(\dim M, \dim N; \dim Q) = (3, 2; 2)$.

(i) *Each diagram in Fig. 5 uniquely determines a weak C^∞ equivalence class of fibers of a map f in such a way that the diagram represents the corresponding central level set modulo transverse intersections with N, under the convention described in Remark 2.3, where dotted squares are ignored.*

(ii) *Every fiber of f is weakly C^∞ equivalent to the disjoint union of one of the fibers in the following list and a finite number of copies of a fiber of a trivial circle bundle:*

(1) *fibers as depicted in Fig. 5, i.e.* $\widetilde{\mathrm{p0}}^0$, $\widetilde{\mathrm{pI}}^\mu$ *with* $1 \le \mu \le 4$, $\widetilde{\mathrm{pII}}^\mu$ *with* $5 \le \mu \le 12$, $\widetilde{\mathrm{pII}}^a$, $\widetilde{\mathrm{pII}}^b$, $\widetilde{\mathrm{pII}}^c$, $\widetilde{\mathrm{pII}}^d$, $\widetilde{\mathrm{pII}}^e$ *and* $\widetilde{\mathrm{pII}}^f$,
(2) *fibers* $\widetilde{\mathrm{pII}}^{\mu,\nu}$ *with* $1 \le \mu \le \nu \le 4$, *where* $\widetilde{\mathrm{pII}}^{\mu,\nu}$ *means the disjoint union of* $\widetilde{\mathrm{pI}}^\mu$ *and* $\widetilde{\mathrm{pI}}^\nu$.

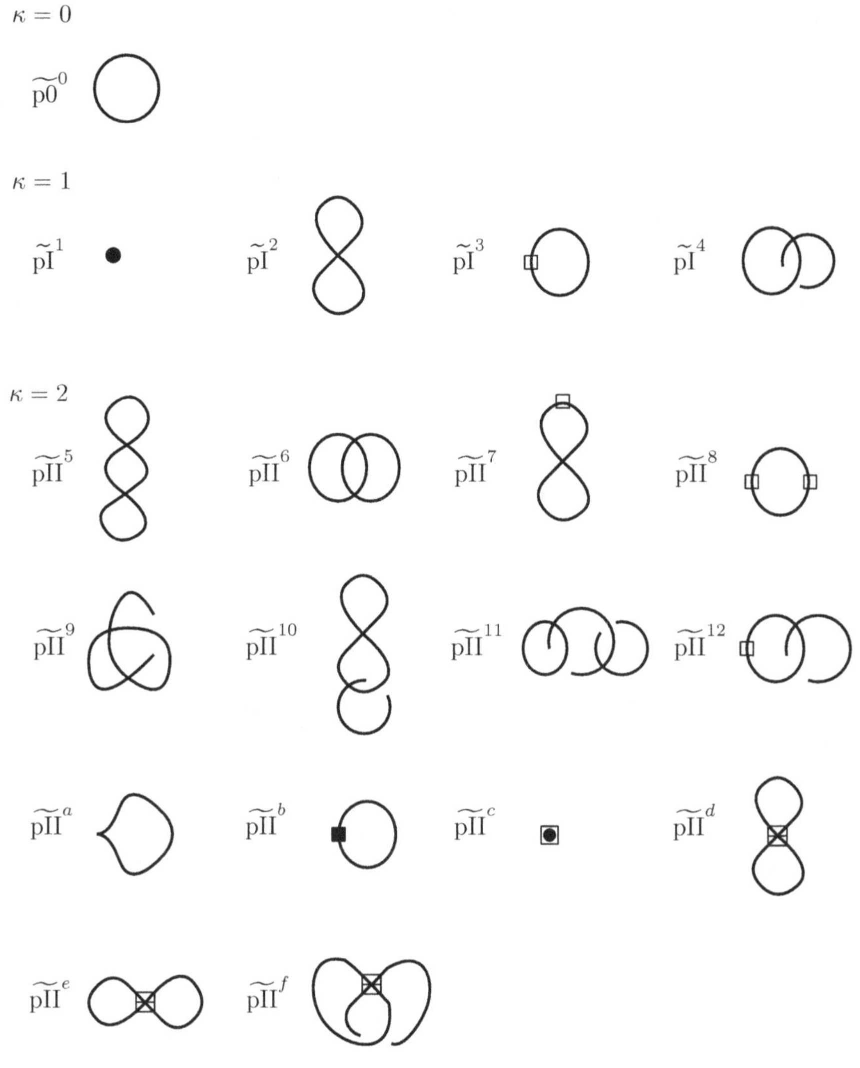

Fig. 5 List of the fibers of proper C^∞ stable maps $(M, N) \to Q$ with $(\dim M, \dim N; \dim Q) = (3, 2; 2)$ up to weak C^∞ equivalence

More precisely, two such fibers containing no singular points of the restriction to the submanifold N are weakly C^∞ equivalent if and only if their corresponding level sets are diffeomorphic. Therefore, in the figures, the associated level sets are depicted together with the information on the corresponding local map germs which are depicted in accordance with Lemmas 2.2 and 2.4.

In Fig. 5, κ denotes the *codimension* of the set of points in the target Q whose corresponding fibers are weakly C^∞ equivalent to the relevant one (see [4] for details).

Furthermore, the symbols $\widetilde{\mathrm{p0}}^*$, $\widetilde{\mathrm{pI}}^*$, and $\widetilde{\mathrm{pII}}^*$ mean the names of the corresponding (weak C^∞ equivalence classes of) fibers. Note that we have named the fibers so that each fiber with connected central level set has its own number or letter, and a fiber with disconnected central level set has the name consisting of the numbers of its "connected components".

We can prove Proposition 2.7 by using the relative version of Ehresmann's fibration theorem together with Proposition 2.1. See [4, Proof of Theorem 3.5] for details.

As to the usual C^∞ equivalence class, we see easily that two weakly C^∞ equivalent fibers are C^∞ equivalent if and only if there exists a homeomorphism between their central level sets which respects C^∞ equivalence classes of singular map germs of f and $f|_N$ at the points in $S(f) \cup S(f|_N)$ and which respects the transverse intersections with the submanifold N.

Then, we immediately obtain the following corollary. (For details, see [4, Proof of Corollary 3.9].)

Corollary 2.8 *Let $f_i\colon (M_i, N_i) \to Q_i$, $i = 0, 1$, be proper C^∞ stable maps with* $(\dim M_i, \dim N_i; \dim Q_i) = (3, 2; 2)$. *For $q_i \in Q_i$, $i = 0, 1$, the fibers over q_0 and q_1 are C^∞ equivalent if and only if they are C^0 equivalent.*

Remark 2.9 If the source 3-dimensional manifold is orientable, then the singular fibers of types $\widetilde{\mathrm{pI}}^4$, $\widetilde{\mathrm{pII}}^9$, $\widetilde{\mathrm{pII}}^{10}$, $\widetilde{\mathrm{pII}}^{11}$, $\widetilde{\mathrm{pII}}^{12}$, and $\widetilde{\mathrm{pII}}^f$ never appear.

Remark 2.10 Let (V, W) be a manifold pair with $(\dim V, \dim W) = (2, 1)$, and Q be the real line $\mathbb{R}$ or the circle S^1. A proper C^∞ function $f\colon (V, W) \to Q$ is C^∞ stable (i.e. it is a *stable Morse function*) if and only if it satisfies the following conditions.

(1) (Local conditions) In the following, for $p \in W$, we use local coordinates (x, y) around p such that W corresponds to the set $\{y = 0\}$.

(1a) For $p \in V \setminus W$, the germ of f at p is right-left equivalent to one of the following:

$$(x, y) \mapsto \begin{cases} x, & p: \text{regular point}, \\ x^2 \pm y^2, & p: \text{fold point or non-degenerate critical point}. \end{cases}$$

(1b) For $p \in W$, the germ of f at p is right-left equivalent to one of the following:

$$(x, y) \mapsto \begin{cases} x, & p: \text{regular point of } f|_W, \\ x^2 + y, & p: \text{relative fold point or} \\ & \quad \text{non-degenerate critical point of } f|_W. \end{cases}$$

(2) (Global conditions) $f(p_1) \neq f(p_2)$ if $p_1 \neq p_2 \in S(f) \cup S(f|_W)$.

The list of weak C^∞ equivalence classes of singular fibers of proper stable Morse functions $(V, W) \to Q$ can be obtained in a similar fashion. The result corresponds to those appearing in Fig. 5 with $\kappa = 0, 1$. In fact, it is not difficult to show that the suspensions of the fibers of such functions in a sense similar to that in [9, Definition 4.2] coincide with those appearing in the figure. However, in the following, by abuse of notation, we use the symbols in Fig. 5 with $\kappa = 0, 1$ for the fibers of stable Morse functions as well.

2.2 *Stable Maps of 2-Colored Manifold Pairs*

In this subsection, M is a 3-dimensional manifold and N is a closed 2-dimensional submanifold of M such that $M \setminus N$ consists of two disjoint open submanifolds R and B with $\partial R(=\overline{R} \setminus R)$ and $\partial B(=\overline{B} \setminus B)$ both coinciding with N. In other words, (M, N) is a 2-colored manifold pair. In this case, we call (R, B) the *coloring pair*. For a C^∞ map $(M, N) \to Q$ into a surface Q, we have the following characterization of C^∞ stable maps as in Proposition 2.1.

Proposition 2.11 *Let (M, N) be a 2-colored manifold pair with coloring pair (R, B) and with* $(\dim M, \dim N) = (3, 2)$*, and Q be a surface. A proper C^∞ map $f\colon (M, N) \to Q$ is C^∞ stable if and only if it satisfies the following conditions.*

(1) *(Local conditions) In the following, for $p \in N$, we use local coordinates (x, y, z) around p such that N, R and B correspond to the sets $\{z = 0\}$, $\{z > 0\}$ and $\{z < 0\}$, respectively.*

(1a) *For $p \in M \setminus N$, the germ of f at p is right-left equivalent to one of the following:*

$$(x, y, z) \mapsto \begin{cases} (x, y), & p: \textit{regular point}, \\ (x, y^2 + z^2), & p: \textit{definite fold point}, \\ (x, y^2 - z^2), & p: \textit{indefinite fold point}, \\ (x, y^3 + xy - z^2), & p: \textit{cusp point}. \end{cases}$$

(1b) *For $p \in N \setminus S(f)$, the germ of f at p is right-left equivalent to one of the following:*

$$(x, y, z) \mapsto \begin{cases} (x, y), & p: \textit{regular point of } f|_N, \\ (x, y^2 + z), & p: \textit{relative } B\textit{-fold point}, \\ (x, y^2 - z), & p: \textit{relative } R\textit{-fold point}, \\ (x, y^3 + xy + z), & p: \textit{relative cusp point}. \end{cases}$$

(1c) *For $p \in N \cap S(f)$, the germ of f at p is right-left equivalent to the map germ*

$$(x, y, z) \mapsto \begin{cases} (x, y^2 + xz + z^2), & p: \textit{definite } B_2 \textit{ point,} \\ (x, y^2 + xz - z^2), & p: \textit{indefinite } B_2 \textit{ point.} \end{cases}$$

(2) *(Global conditions) For each* $q \in f(S(f)) \cup f(S(f|_N))$, *the multi-germ*

$$(f|_{S(f) \cup S(f|_N)}, f^{-1}(q) \cap (S(f) \cup S(f|_N)))$$

is right-left equivalent to one of the eight multi-germs whose images are depicted in Fig. 1.

Please note that in Proposition 2.11 (1b), we have two cases for relative fold points, which was not present in Proposition 2.1. This distinction is necessary for distinguishing the region R from B.

Lemma 2.12 *Let* $f: (M, N) \to Q$ *be a proper* C^∞ *stable map of a 2-colored manifold pair with coloring pair* (R, B) *into* Q *with* $(\dim M, \dim N; \dim Q) = (3, 2; 2)$. *Then, every point* $p \in S(f) \cup S(f|_N)$ *has one of the following neighborhoods in its corresponding singular level set (see Figs. 6 and 7 for references):*

(1) *isolated point diffeomorphic to* $\{(y, z) \in \mathbb{R}^2 \mid y^2 + z^2 = 0\}$, *if* $p \in M \setminus N = R \cup B$ *is a definite fold point,*
(2) *union of two transverse arcs diffeomorphic to* $\{(y, z) \in \mathbb{R}^2 \mid y^2 - z^2 = 0\}$, *if* $p \in M \setminus N = R \cup B$ *is an indefinite fold point,*
(3) *cuspidal arc diffeomorphic to* $\{(y, z) \in \mathbb{R}^2 \mid y^3 - z^2 = 0\}$, *if* $p \in M \setminus N = R \cup B$ *is a cusp point,*
(4) *arc diffeomorphic to* $\{(y, z) \in \mathbb{R}^2 \mid y^2 + z = 0\}$ *or* $\{(y, z) \in \mathbb{R}^2 \mid y^2 - z = 0\}$, *if* $p \in N$ *is a relative fold point,*
(5) *arc diffeomorphic to* $\{(y, z) \in \mathbb{R}^2 \mid y^3 + z = 0\}$, *if* $p \in N$ *is a relative cusp point,*
(6) *isolated point diffeomorphic to* $\{(y, z) \in \mathbb{R}^2 \mid y^2 + z^2 = 0, \}$, *if* $p \in N \cap S(f)$ *is a definite* B_2 *point,*
(7) *union of two transverse arcs diffeomorphic to* $\{(y, z) \in \mathbb{R}^2 \mid y^2 - z^2 = 0\}$, *if* $p \in N \cap S(f)$ *is an indefinite* B_2 *point.*

In the figures, each black dot or black curve corresponds to the part in R, while each blue[2] *dot or curve corresponds to the part in B.*

In Figs. 6 and 7, we use the same conventions as in Figs. 2 and 3. However, as we need to distinguish the region R from B, we use colors and put shadows for the blue part.

For the local nearby level sets, we have the following, as before.

Lemma 2.13 *Let* $f: (M, N) \to Q$ *be a proper* C^∞ *stable map of a 2-colored manifold pair with coloring pair* (R, B) *into* Q *with* $(\dim M, \dim N; \dim Q) = (3, 2; 2)$. *For* $p \in S(f) \cup S(f|_N)$ *such that* $f^{-1}(f(p)) \cap (S(f) \cup S(f|_N)) = \{p\}$, *the level set near* p *as depicted in Fig. 8:*

---

[2]In the monochrome printing version, "blue" should be replaced by "gray".

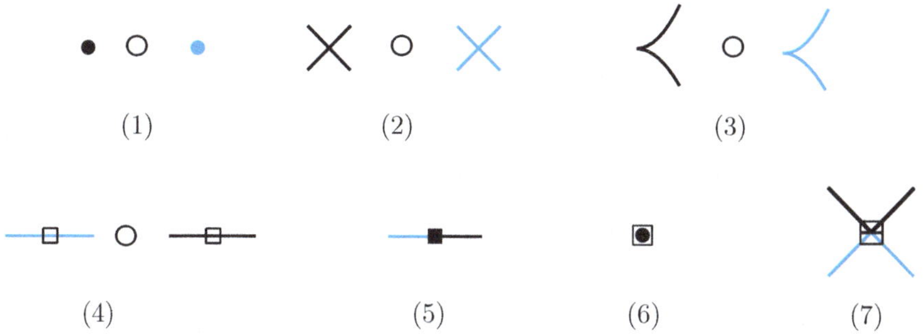

Fig. 6 Neighborhoods of singular points in their singular level sets

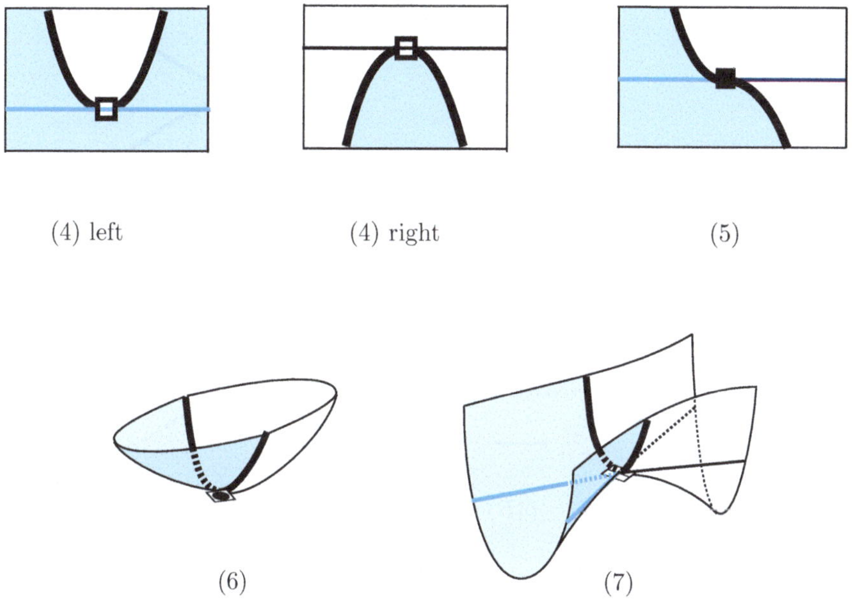

Fig. 7 Singular level sets touching the submanifold N: shadowed regions correspond to B

(1A) $p \in R$ *is a definite fold point,*
(1B) $p \in B$ *is a definite fold point,*
(2A) $p \in R$ *is an indefinite fold point,*
(2B) $p \in B$ *is an indefinite fold point,*
(3A) $p \in R$ *is a cusp point,*
(3B) $p \in B$ *is a cusp point,*
(4A) $p \in N$ *is a relative* B*-fold point,*
(4B) $p \in N$ *is a relative* R*-fold point,*
(5) $p \in N$ *is a relative cusp point,*

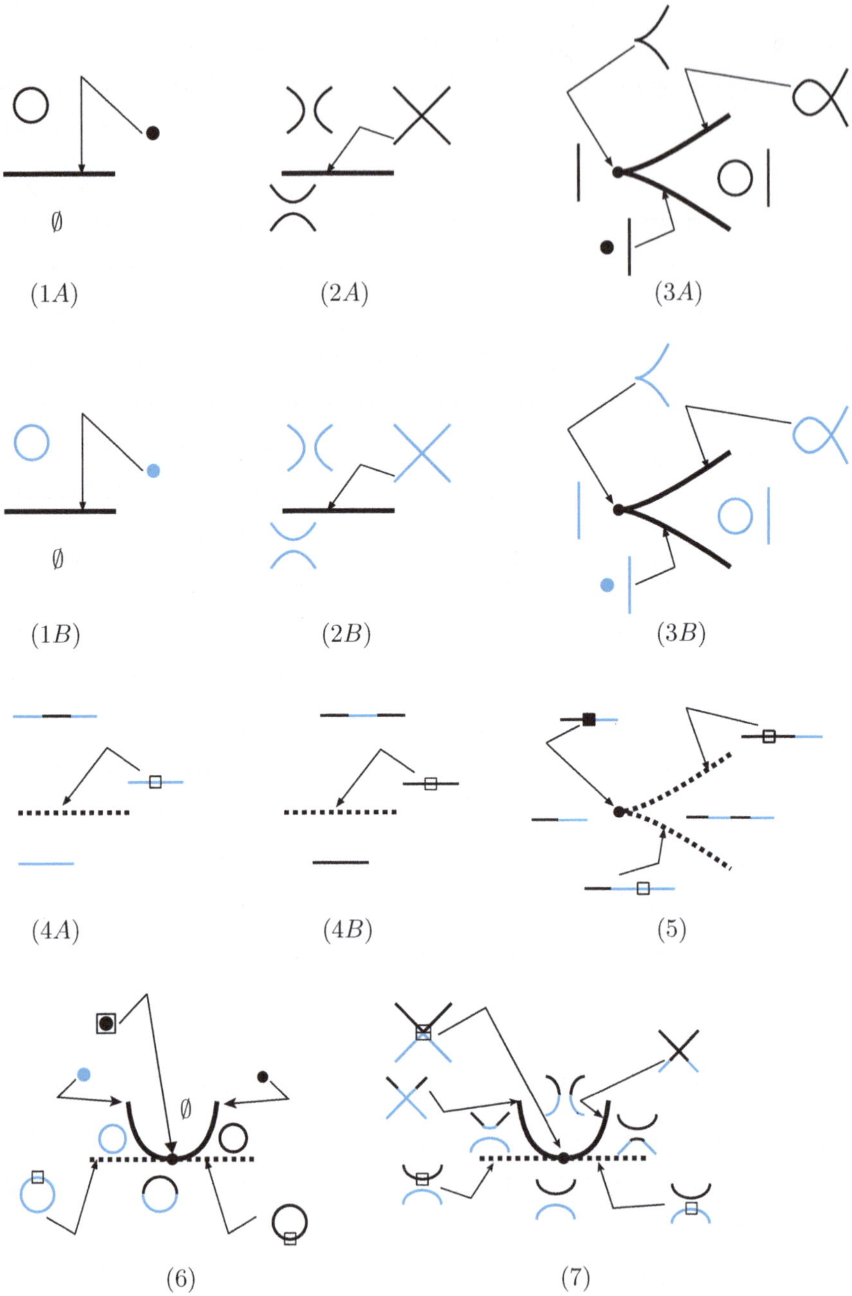

Fig. 8 Local degenerations of level sets

(6) *$p \in S(f) \cap N$ is a definite B_2 point,*
(7) *$p \in S(f) \cap N$ is an indefinite B_2 point,*

where we adopt the convention as in Figs. 4 and 6.

In order to get a classification result, let us introduce the following notion, similar to Definition 2.6.

Definition 2.14 Let $f_i\colon (M_i, N_i) \to Q_i$ be C^∞ maps of 2-colored manifold pairs with coloring pair (R_i, B_i), $i = 0, 1$. For $q_i \in Q_i$, $i = 0, 1$, we say that the fibers over q_0 and q_1 are *weakly C^∞ color equivalent* (or *weakly C^0 color equivalent*) if for some neighborhoods M_i' of $f_i^{-1}(q_i) \cap (S(f_i) \cup S(f_i|_{N_i}))$ in M_i and for some small open neighborhoods U_i of q_i in Q_i, there exist diffeomorphisms (resp. homeomorphisms) $\Phi\colon (f_0^{-1}(U_0), f_0^{-1}(U_0) \cap N_0') \to (f_1^{-1}(U_1), f_1^{-1}(U_1) \cap N_1')$ with $\Phi(M_0' \cap R_0) = M_1' \cap R_1$ and $\Phi(M_0' \cap B_0) = M_1' \cap B_1$, and $\varphi\colon U_0 \to U_1$ with $\varphi(q_0) = q_1$ which make the following diagram commutative:

$$\begin{array}{ccc}
((f_0^{-1}(U_0), f_0^{-1}(U_0) \cap N_0'), (f_0^{-1}(q_0), f_0^{-1}(q_0) \cap N_0')) & \xrightarrow{f_0} & (U_0, q_0) \\
\Phi \downarrow & & \downarrow \varphi \\
((f_1^{-1}(U_1), f_1^{-1}(U_1) \cap N_1'), (f_1^{-1}(q_1), f_1^{-1}(q_1) \cap N_1')) & \xrightarrow{f_1} & (U_1, q_1),
\end{array}$$

where $N_i' = N_i \cap M_i'$, $i = 0, 1$. In other words, considering weak C^∞ color equivalence classes instead of the usual C^∞ color equivalence classes corresponds to ignoring transverse intersections of the central level sets with the submanifold N_i away from the neighborhoods of singular points.

Then, we get the following list of singular fibers. We omit the proof here.

Proposition 2.15 *Let us consider proper C^∞ stable maps $f\colon (M, N) \to Q$ of 2-colored manifold pairs (M, N) with coloring pairs (R, B) into surfaces Q with* $(\dim M, \dim N; \dim Q) = (3, 2; 2)$.

(i) *Each diagram in Figs. 9 and 10 uniquely determines a weak C^∞ color equivalence class of fibers of a map f in such a way that the diagram represents the corresponding central level set modulo transverse intersections with N, under the convention described in Remark 2.3, where dotted squares are ignored.*

(ii) *Every fiber of f is weakly C^∞ color equivalent to the disjoint union of one of the fibers in the following list and a finite number of copies of a fiber of a trivial circle bundle:*

(1) *fibers as depicted in Fig. 9, i.e.* $\widetilde{\mathrm{c0}}^0$, $\widetilde{\mathrm{cI}}^{\mu A}$ *and* $\widetilde{\mathrm{cI}}^{\mu B}$ *with* $1 \leq \mu \leq 4$,
(2) *fibers* $\widetilde{\mathrm{cII}}^{\mu A, \nu A}$, $\widetilde{\mathrm{cII}}^{\mu A, \nu B}$ *and* $\widetilde{\mathrm{cII}}^{\mu B, \nu B}$ *with* $1 \leq \mu \leq \nu \leq 4$, *where* $\widetilde{\mathrm{cII}}^{\mu *, \nu \star}$ *means the disjoint union of* $\widetilde{\mathrm{cI}}^{\mu *}$ *and* $\widetilde{\mathrm{cI}}^{\nu \star}$ *for* $(*, \star) = (A, A), (A, B)$ *or* (B, B),
(3) *fibers as depicted in Fig. 10, i.e.* $\widetilde{\mathrm{cII}}^{\mu *}$, $\widetilde{\mathrm{cII}}^{b}$, $\widetilde{\mathrm{cII}}^{c}$, $\widetilde{\mathrm{cII}}^{d}$, $\widetilde{\mathrm{cII}}^{e}$ *and* $\widetilde{\mathrm{cII}}^{f}$, *where* $5 \leq \mu \leq 12$ *or* $\mu = a$ *and* $* = A, B, C$ *for* $\mu = 5, 6, 8, 9, 11$, $* = A, B, C, D$ *for* $\mu = 7, 10, 12$, *and* $* = A, B$ *for* $\mu = a$.

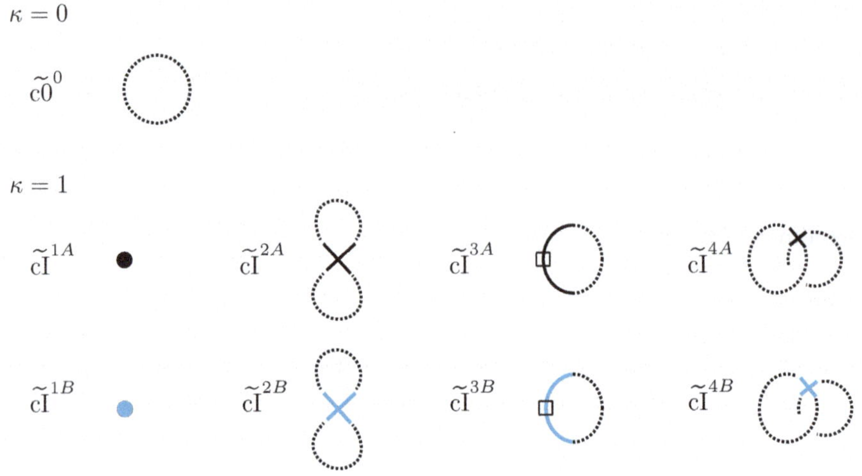

Fig. 9 List of the fibers of proper C^∞ stable maps $(M, N) \to Q$ of (3, 2)-dimensional 2-colored manifold pairs into surfaces up to weak C^∞ color equivalence, for $\kappa = 0, 1$

(Note that in the figures, dotted curves represent nonsingular curves whose colors or transverse intersections with N are ignored.) More precisely, two such fibers containing no singular points of the restriction to the submanifold N are weakly C^∞ color equivalent if and only if their corresponding level sets are diffeomorphic. Therefore, in the figures, the associated level sets are depicted together with the information on the corresponding local map germs which are depicted in accordance with Lemmas 2.12 and 2.13.

Then, we immediately obtain the following corollary as before. (For details, see [4, Proof of Corollary 3.9].)

Corollary 2.16 *Let $f_i \colon (M_i, N_i) \to Q_i$, $i = 0, 1$, be proper C^∞ stable maps of 2-colored manifold pairs with* $(\dim M_i, \dim N_i; \dim Q_i) = (3, 2; 2)$. *For $q_i \in Q_i$, $i = 0, 1$, the fibers over q_0 and q_1 are C^∞ equivalent if and only if they are C^0 equivalent.*

Remark 2.17 If the source 3-dimensional manifold is orientable, then the singular fibers of types $\widetilde{\mathrm{cI}}^{4A}$, $\widetilde{\mathrm{cI}}^{4B}$, $\widetilde{\mathrm{cII}}^{9A}$, $\widetilde{\mathrm{cII}}^{9B}$, $\widetilde{\mathrm{cII}}^{9C}$, $\widetilde{\mathrm{cII}}^{10A}$, $\widetilde{\mathrm{cII}}^{10B}$, $\widetilde{\mathrm{cII}}^{10C}$, $\widetilde{\mathrm{cII}}^{10D}$, $\widetilde{\mathrm{cII}}^{11A}$, $\widetilde{\mathrm{cII}}^{11B}$, $\widetilde{\mathrm{cII}}^{11C}$, $\widetilde{\mathrm{cII}}^{12A}$, $\widetilde{\mathrm{cII}}^{12B}$, $\widetilde{\mathrm{cII}}^{12C}$, $\widetilde{\mathrm{cII}}^{12D}$, and $\widetilde{\mathrm{cII}}^{f}$ never appear.

Remark 2.18 Let (V, W) be a 2-colored manifold pair with coloring pair (R, B) with $(\dim V, \dim W) = (2, 1)$, and Q be the real line $\mathbb{R}$ or the circle S^1. A proper C^∞ function $f \colon (V, W) \to Q$ is C^∞ stable (i.e. it is a *stable Morse function*) if and only if it satisfies the following conditions.

(1) (Local conditions) In the following, for $p \in W$, we use local coordinates (x, y) around p such that W, R and B correspond to the sets $\{y = 0\}$, $\{y > 0\}$ and $\{y < 0\}$, respectively.

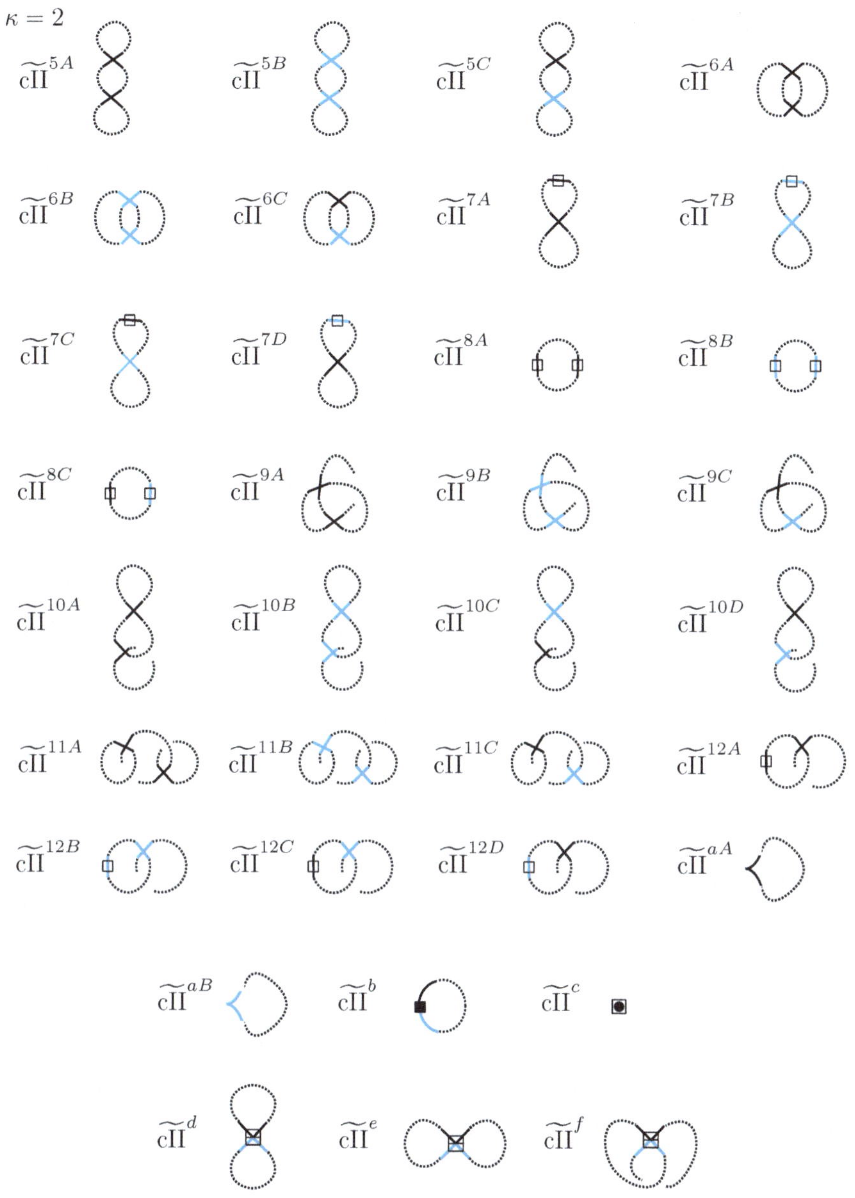

Fig. 10 List of the fibers of proper C^∞ stable maps $(M, N) \to Q$ of (3, 2)-dimensional 2-colored manifold pairs into surfaces up to weak C^∞ color equivalence, for $\kappa = 2$

(1a) For $p \in V \setminus W$, the germ of f at p is right-left equivalent to one of the following:

$$(x, y) \mapsto \begin{cases} x, & p : \text{regular point}, \\ x^2 \pm y^2, & p : \text{fold point or non-degenerate critical point}. \end{cases}$$

(1b) For $p \in W$, the germ of f at p is right-left equivalent to one of the following:

$$(x, y) \mapsto \begin{cases} x, & p : \text{regular point of } f|_W, \\ x^2 \pm y, & p : \text{relative fold point or} \\ & \quad \text{non-degenerate critical point of } f|_W. \end{cases}$$

(2) (Global conditions) $f(p_1) \neq f(p_2)$ if $p_1 \neq p_2 \in S(f) \cup S(f|_W)$.

The list of weak C^∞ color equivalence classes of singular fibers of proper stable Morse functions $(V, W) \to Q$ can be obtained in a similar fashion. The result corresponds to those appearing in Fig. 9 with $\kappa = 0, 1$. In fact, it is not difficult to show that the suspensions of the fibers of such functions in a sense similar to that in [9, Definition 4.2] coincide with those appearing in the figure. However, in the following, by abuse of notation, we use the symbols in Fig. 9 with $\kappa = 0, 1$ for the fibers of stable Morse functions as well.

Remark 2.19 In [16], the second author studied singular fibers of certain two-colored maps, where the coloring in that paper refers to that of the target.

3 Co-existence of Singular Fibers

3.1 Stable Maps of Manifold Pairs

Let $f\colon (M, N) \to Q$ be a C^∞ stable map of a manifold pair into Q with $(\dim M, \dim N; \dim Q) = (3, 2; 2)$. We suppose that M is closed. Let $\widetilde{\mathcal{F}}$ be a weak C^∞ equivalence class of singular fibers of codimension ≥ 1. Define $\widetilde{\mathcal{F}}(f)$ to be the set of points $q \in Q$ such that the fiber over q is weakly C^∞ equivalent to the disjoint union of $\widetilde{\mathcal{F}}$ and some copies of a fiber of a trivial circle bundle. Furthermore, define $\widetilde{\mathcal{F}}_o(f)$ (or $\widetilde{\mathcal{F}}_e(f)$) to be the subset of $\widetilde{\mathcal{F}}(f)$ which consists of the points $q \in Q$ such that the number of regular fibers, namely the total number of $\widetilde{\mathrm{p0}}^0$ components in the fiber, is odd (resp. even). For codimension zero fibers, by convention, we denote by $\widetilde{\mathrm{p0}}^0_o(f)$ (or $\widetilde{\mathrm{p0}}^0_e(f)$) the set of points $q \in Q$ over which lies a regular fiber consisting of an odd (resp. even) number of components.

If $\widetilde{\mathcal{F}}$ is of codimension 1, then the closure of $\widetilde{\mathcal{F}}_o(f)$ (or $\widetilde{\mathcal{F}}_e(f)$) is a finite graph embedded in Q. Its vertices correspond to points over which lies a singular fiber of codimension 2.

The handshake lemma of the classical graph theory implies the following formulae. In the following, for a finite set S, $|S|$ denotes its cardinality.

Proposition 3.1 *Let $f\colon (M, N) \to Q$ be a C^∞ stable map of a manifold pair into Q with* $(\dim M, \dim N; \dim Q) = (3, 2; 2)$. *We suppose that M is closed. Then, the following numbers are always even:*

(1) $|\widetilde{\mathrm{pII}}^{1,2}(f)| + |\widetilde{\mathrm{pII}}^{a}_{o}(f)|$,
(2) $|\widetilde{\mathrm{pII}}^{1,2}(f)| + |\widetilde{\mathrm{pII}}^{a}_{e}(f)|$,
(3) $|\widetilde{\mathrm{pII}}^{1,3}(f)| + |\widetilde{\mathrm{pII}}^{2,3}(f)| + |\widetilde{\mathrm{pII}}^{7}(f)|$,
(4) $|\widetilde{\mathrm{pII}}^{1,4}(f)| + |\widetilde{\mathrm{pII}}^{2,4}(f)| + |\widetilde{\mathrm{pII}}^{10}(f)|$.

By eliminating the terms of the forms $\mathcal{F}_o(f)$ and $\mathcal{F}_e(f)$, we obtain the following.

Corollary 3.2 *Let $f\colon (M, N) \to Q$ be a C^∞ stable map of a manifold pair into Q with* $(\dim M, \dim N; \dim Q) = (3, 2; 2)$. *We suppose that M is closed. Then, the following numbers are always even:*

(1) $|\widetilde{\mathrm{pII}}^{a}(f)|$,
(2) $|\widetilde{\mathrm{pII}}^{1,3}(f)| + |\widetilde{\mathrm{pII}}^{2,3}(f)| + |\widetilde{\mathrm{pII}}^{7}(f)|$,
(3) $|\widetilde{\mathrm{pII}}^{1,4}(f)| + |\widetilde{\mathrm{pII}}^{2,4}(f)| + |\widetilde{\mathrm{pII}}^{10}(f)|$.

Remark 3.3 Let (V, W) be a manifold pair with $(\dim V, \dim W) = (2, 1)$, and Q be the real line $\mathbb{R}$ or the circle S^1. Suppose that V is closed. By using the same method, we obtain similar co-existence results for singular fibers of a C^∞ stable Morse function $f\colon (V, W) \to Q$:

$$|\widetilde{\mathrm{pI}}^{1}(f)| + |\widetilde{\mathrm{pI}}^{2}(f)| \equiv 0 \mod 2,$$

where we are using the notation for the relevant fibers in the sense of Remark 2.10.

3.2 Stable Maps of 2-Colored Manifold Pairs

Let $f\colon (M, N) \to Q$ be a C^∞ stable map of a 2-colored manifold pair with coloring pair (R, B) into Q with $(\dim M, \dim N; \dim Q) = (3, 2; 2)$. We suppose that M is closed. By using notational conventions and arguments similar to those in the previous section, we get the following.

Proposition 3.4 *Let $f\colon (M, N) \to Q$ be a C^∞ stable map of a 2-colored manifold pair with coloring pair (R, B) into Q with* $(\dim M, \dim N; \dim Q) = (3, 2; 2)$. *We suppose that M is closed. Then, the following numbers are always even:*

(1) $|\widetilde{\mathrm{cII}}^{1A,1B}(f)| + |\widetilde{\mathrm{cII}}^{1A,2A}(f)| + |\widetilde{\mathrm{cII}}^{1A,2B}(f)| + |\widetilde{\mathrm{cII}}^{aA}_{e}(f)| + |\widetilde{\mathrm{cII}}^{c}_{o}(f)|$,
(2) $|\widetilde{\mathrm{cII}}^{1A,1B}(f)| + |\widetilde{\mathrm{cII}}^{1A,2A}(f)| + |\widetilde{\mathrm{cII}}^{1A,2B}(f)| + |\widetilde{\mathrm{cII}}^{aA}_{o}(f)| + |\widetilde{\mathrm{cII}}^{c}_{e}(f)|$,

(3) $|\widetilde{\mathrm{cII}}^{1A,1B}(f)| + |\widetilde{\mathrm{cII}}^{1B,2A}(f)| + |\widetilde{\mathrm{cII}}^{1B,2B}(f)| + |\widetilde{\mathrm{cII}}_e^{aB}(f)| + |\widetilde{\mathrm{cII}}_o^{c}(f)|$,

(4) $|\widetilde{\mathrm{cII}}^{1A,1B}(f)| + |\widetilde{\mathrm{cII}}^{1B,2A}(f)| + |\widetilde{\mathrm{cII}}^{1B,2B}(f)| + |\widetilde{\mathrm{cII}}_o^{aB}(f)| + |\widetilde{\mathrm{cII}}_e^{c}(f)|$,

(5) $|\widetilde{\mathrm{cII}}^{1A,2A}(f)| + |\widetilde{\mathrm{cII}}^{1B,2A}(f)| + |\widetilde{\mathrm{cII}}^{2A,2B}(f)| + |\widetilde{\mathrm{cII}}^{5C}(f)| + |\widetilde{\mathrm{cII}}_o^{9C}(f)|$
$+|\widetilde{\mathrm{cII}}_o^{aA}(f)| + |\widetilde{\mathrm{cII}}_o^{d}(f)| + |\widetilde{\mathrm{cII}}_o^{e}(f)|$,

(6) $|\widetilde{\mathrm{cII}}^{1A,2A}(f)| + |\widetilde{\mathrm{cII}}^{1B,2A}(f)| + |\widetilde{\mathrm{cII}}^{2A,2B}(f)| + |\widetilde{\mathrm{cII}}^{5C}(f)| + |\widetilde{\mathrm{cII}}_e^{9C}(f)|$
$+|\widetilde{\mathrm{cII}}_e^{aA}(f)| + |\widetilde{\mathrm{cII}}_e^{d}(f)| + |\widetilde{\mathrm{cII}}_e^{e}(f)|$,

(7) $|\widetilde{\mathrm{cII}}^{1A,2B}(f)| + |\widetilde{\mathrm{cII}}^{1B,2B}(f)| + |\widetilde{\mathrm{cII}}^{2A,2B}(f)| + |\widetilde{\mathrm{cII}}^{5C}(f)| + |\widetilde{\mathrm{cII}}_o^{9C}(f)|$
$+|\widetilde{\mathrm{cII}}_o^{aB}(f)| + |\widetilde{\mathrm{cII}}_o^{d}(f)| + |\widetilde{\mathrm{cII}}_o^{e}(f)|$,

(8) $|\widetilde{\mathrm{cII}}^{1A,2B}(f)| + |\widetilde{\mathrm{cII}}^{1B,2B}(f)| + |\widetilde{\mathrm{cII}}^{2A,2B}(f)| + |\widetilde{\mathrm{cII}}^{5C}(f)| + |\widetilde{\mathrm{cII}}_e^{9C}(f)|$
$+|\widetilde{\mathrm{cII}}_e^{aB}(f)| + |\widetilde{\mathrm{cII}}_e^{d}(f)| + |\widetilde{\mathrm{cII}}_e^{e}(f)|$,

(9) $|\widetilde{\mathrm{cII}}^{1A,3A}(f)| + |\widetilde{\mathrm{cII}}^{1B,3A}(f)| + |\widetilde{\mathrm{cII}}^{2A,3A}(f)| + |\widetilde{\mathrm{cII}}^{2B,3A}(f)| + |\widetilde{\mathrm{cII}}^{7A}(f)|$
$+|\widetilde{\mathrm{cII}}^{7C}(f)| + |\widetilde{\mathrm{cII}}_o^{b}(f)| + |\widetilde{\mathrm{cII}}_o^{c}(f)| + |\widetilde{\mathrm{cII}}_e^{d}(f)| + |\widetilde{\mathrm{cII}}_o^{e}(f)| + |\widetilde{\mathrm{cII}}_o^{f}(f)|$,

(10) $|\widetilde{\mathrm{cII}}^{1A,3A}(f)| + |\widetilde{\mathrm{cII}}^{1B,3A}(f)| + |\widetilde{\mathrm{cII}}^{2A,3A}(f)| + |\widetilde{\mathrm{cII}}^{2B,3A}(f)| + |\widetilde{\mathrm{cII}}^{7A}(f)|$
$+|\widetilde{\mathrm{cII}}^{7C}(f)| + |\widetilde{\mathrm{cII}}_e^{b}(f)| + |\widetilde{\mathrm{cII}}_e^{c}(f)| + |\widetilde{\mathrm{cII}}_o^{d}(f)| + |\widetilde{\mathrm{cII}}_e^{e}(f)| + |\widetilde{\mathrm{cII}}_e^{f}(f)|$,

(11) $|\widetilde{\mathrm{cII}}^{1A,3B}(f)| + |\widetilde{\mathrm{cII}}^{1B,3B}(f)| + |\widetilde{\mathrm{cII}}^{2A,3B}(f)| + |\widetilde{\mathrm{cII}}^{2B,3B}(f)| + |\widetilde{\mathrm{cII}}^{7B}(f)|$
$+|\widetilde{\mathrm{cII}}^{7D}(f)| + |\widetilde{\mathrm{cII}}_o^{b}(f)| + |\widetilde{\mathrm{cII}}_o^{c}(f)| + |\widetilde{\mathrm{cII}}_e^{d}(f)| + |\widetilde{\mathrm{cII}}_o^{e}(f)| + |\widetilde{\mathrm{cII}}_o^{f}(f)|$,

(12) $|\widetilde{\mathrm{cII}}^{1A,3B}(f)| + |\widetilde{\mathrm{cII}}^{1B,3B}(f)| + |\widetilde{\mathrm{cII}}^{2A,3B}(f)| + |\widetilde{\mathrm{cII}}^{2B,3B}(f)| + |\widetilde{\mathrm{cII}}^{7B}(f)|$
$+|\widetilde{\mathrm{cII}}^{7D}(f)| + |\widetilde{\mathrm{cII}}_e^{b}(f)| + |\widetilde{\mathrm{cII}}_e^{c}(f)| + |\widetilde{\mathrm{cII}}_o^{d}(f)| + |\widetilde{\mathrm{cII}}_e^{e}(f)| + |\widetilde{\mathrm{cII}}_e^{f}(f)|$,

(13) $|\widetilde{\mathrm{cII}}^{1A,4A}(f)| + |\widetilde{\mathrm{cII}}^{1B,4A}(f)| + |\widetilde{\mathrm{cII}}^{2A,4A}(f)| + |\widetilde{\mathrm{cII}}^{2B,4A}(f)| + |\widetilde{\mathrm{cII}}_o^{9C}(f)|$
$+|\widetilde{\mathrm{cII}}^{10A}(f)| + |\widetilde{\mathrm{cII}}^{10C}(f)| + |\widetilde{\mathrm{cII}}_o^{f}(f)|$,

(14) $|\widetilde{\mathrm{cII}}^{1A,4A}(f)| + |\widetilde{\mathrm{cII}}^{1B,4A}(f)| + |\widetilde{\mathrm{cII}}^{2A,4A}(f)| + |\widetilde{\mathrm{cII}}^{2B,4A}(f)| + |\widetilde{\mathrm{cII}}_e^{9C}(f)|$
$+|\widetilde{\mathrm{cII}}^{10A}(f)| + |\widetilde{\mathrm{cII}}^{10C}(f)| + |\widetilde{\mathrm{cII}}_e^{f}(f)|$,

(15) $|\widetilde{\mathrm{cII}}^{1A,4B}(f)| + |\widetilde{\mathrm{cII}}^{1B,4B}(f)| + |\widetilde{\mathrm{cII}}^{2A,4B}(f)| + |\widetilde{\mathrm{cII}}^{2B,4B}(f)| + |\widetilde{\mathrm{cII}}_o^{9C}(f)|$
$+|\widetilde{\mathrm{cII}}^{10B}(f)| + |\widetilde{\mathrm{cII}}^{10D}(f)| + |\widetilde{\mathrm{cII}}_o^{f}(f)|$,

(16) $|\widetilde{\mathrm{cII}}^{1A,4B}(f)| + |\widetilde{\mathrm{cII}}^{1B,4B}(f)| + |\widetilde{\mathrm{cII}}^{2A,4B}(f)| + |\widetilde{\mathrm{cII}}^{2B,4B}(f)| + |\widetilde{\mathrm{cII}}_e^{9C}(f)|$
$+|\widetilde{\mathrm{cII}}^{10B}(f)| + |\widetilde{\mathrm{cII}}^{10D}(f)| + |\widetilde{\mathrm{cII}}_e^{f}(f)|$.

By eliminating the terms of the forms $\mathcal{F}_o(f)$ and $\mathcal{F}_e(f)$, we obtain the following.

Corollary 3.5 *Let* $f\colon (M, N) \to Q$ *be a* C^∞ *stable map of a* 2*-colored manifold pair with coloring pair* (R, B) *into* Q *with* $(\dim M, \dim N; \dim Q) = (3, 2; 2)$*. We suppose that* M *is closed. Then, the following numbers are always even:*

(1) $|\widetilde{\mathrm{cII}}^{aA}(f)| + |\widetilde{\mathrm{cII}}^{c}(f)|$,

(2) $|\widetilde{\mathrm{cII}}^{aB}(f)| + |\widetilde{\mathrm{cII}}^{c}(f)|$,

(3) $|\widetilde{\mathrm{cII}}^{9C}(f)| + |\widetilde{\mathrm{cII}}^{aA}(f)| + |\widetilde{\mathrm{cII}}^{d}(f)| + |\widetilde{\mathrm{cII}}^{e}(f)|$,

(4) $|\widetilde{\mathrm{cII}}^{9C}(f)| + |\widetilde{\mathrm{cII}}^{aB}(f)| + |\widetilde{\mathrm{cII}}^{d}(f)| + |\widetilde{\mathrm{cII}}^{e}(f)|$,

(5) $|\widetilde{\mathrm{cII}}^{b}(f)| + |\widetilde{\mathrm{cII}}^{c}(f)| + |\widetilde{\mathrm{cII}}^{d}(f)| + |\widetilde{\mathrm{cII}}^{e}(f)| + |\widetilde{\mathrm{cII}}^{f}(f)|$,

(6) $|\widetilde{\mathrm{cII}}^{9C}(f)| + |\widetilde{\mathrm{cII}}^{f}(f)|$.

Remark 3.6 We see easily that the numbers in Corollary 3.5 are all even if and only if the following numbers are all even:

(1) $|\widetilde{\mathrm{cII}}^b(f)|$,
(2) $|\widetilde{\mathrm{cII}}^c(f)| + |\widetilde{\mathrm{cII}}^d(f)| + |\widetilde{\mathrm{cII}}^e(f)| + |\widetilde{\mathrm{cII}}^f(f)|$,
(3) $|\widetilde{\mathrm{cII}}^{aA}(f)| + |\widetilde{\mathrm{cII}}^{aB}(f)|$,
(4) $|\widetilde{\mathrm{cII}}^{aA}(f)| + |\widetilde{\mathrm{cII}}^c(f)|$,
(5) $|\widetilde{\mathrm{cII}}^{9C}(f)| + |\widetilde{\mathrm{cII}}^f(f)|$.

Remark 3.7 Let (V, W) be a manifold pair with $(\dim V, \dim W) = (2, 1)$, and Q be the real line $\mathbb{R}$ or the circle S^1. Suppose that V is closed. By using the same method, we obtain a similar co-existence result for singular fibers of a C^∞ stable Morse function $f\colon (V, W) \to Q$:

$$|\widetilde{\mathrm{cI}}^{1A}(f)| + |\widetilde{\mathrm{cI}}^{1B}(f)| + |\widetilde{\mathrm{cI}}^{2A}(f)| + |\widetilde{\mathrm{cI}}^{2B}(f)| \equiv 0 \mod 2,$$

where we are using the notation for the relevant fibers in the sense of Remark 2.18.

4 Universal Complex of Singular Fibers

In this section, we consider certain universal complexes of singular fibers for maps of manifold pairs, as we have discussed for maps of manifolds with boundary in [9, Sect. 4]. As all the notions and constructions are almost parallel to those given in [9], we omit most of the details here and describe only essential materials. Their applications to cobordisms of singular maps will be discussed in Sect. 5. We work with $\mathbb{Z}_2$-coefficients.

4.1 Stable Maps of Manifold Pairs

For a positive integer n, let

$$P\mathcal{S}_{\mathrm{pr}}(n, n-1; n-1)$$

be the set of fibers for proper C^0 stable Thom maps of manifold pairs (M, N) into Q with $(\dim M, \dim N; \dim Q) = (n, n-1; n-1)$. We put

$$P\mathcal{S}_{\mathrm{pr}} = \bigcup_{n=1}^{\infty} P\mathcal{S}_{\mathrm{pr}}(n, n-1; n-1).$$

Furthermore, let $\rho_{n,n-1;n-1}(2)$ be the *weak* C^0 *equivalence relation modulo two regular fibers* for fibers in $P\mathcal{S}_{\mathrm{pr}}(n, n-1; n-1)$: i.e., two fibers in $P\mathcal{S}_{\mathrm{pr}}(n, n-1; n-1)$ are $\rho_{n,n-1;n-1}(2)$-equivalent if they become weakly C^0 equivalent after we add some regular fibers to each of them with the numbers of added components having the same parity.

We denote by $\rho(2)$ the equivalence relation on $P\mathcal{S}_{\mathrm{pr}}$ induced by $\rho_{n,n-1;n-1}(2)$. Note that the set $P\mathcal{S}_{\mathrm{pr}}$ and the equivalence relation $\rho(2)$ satisfy conditions (a)–(e) in [9, Sect. 4].

Remark 4.1 For the construction of the universal complex of singular fibers, we use C^0 stable Thom maps for some technical reasons. However, for low dimensional cases (for example, the cases with $n = 2, 3$ that we are interested in), a map is a C^0 stable Thom map if and only if it is C^∞ stable. Therefore, we can use the classification results of fibers obtained in Sect. 2.

For a weak C^0 equivalence class $\widetilde{\mathcal{F}}$ of singular fibers, denote by $\widetilde{\mathcal{F}}_o$ (or $\widetilde{\mathcal{F}}_e$) the equivalence class with respect to $\rho_{n,n-1;n-1}(2)$ which consists of singular fibers of type $\widetilde{\mathcal{F}}$ with an odd number (resp. even number) of regular fiber components. For $n = 2, 3$, we denote by $\widetilde{\mathrm{p0}}^0_o$ (resp. $\widetilde{\mathrm{p0}}^0_e$) the equivalence class with respect to $\rho_{n,n-1;n-1}(2)$ which consists of regular fibers with an odd (resp. even) number of components.

Then, we can construct the universal complex $\mathcal{C}(P\mathcal{S}_{\mathrm{pr}}(3, 2; 2), \rho_{3,2;2}(2))$ of singular fibers (for details, see [4] or [9, Sect. 4]). Furthermore, we see that the coboundary homomorphism is given by the following formulae, which are obtained with the help of Lemma 2.4:

$$\begin{aligned}
\delta_0(\widetilde{\mathrm{p0}}^0_o) &= \delta_0(\widetilde{\mathrm{p0}}^0_e) = \widetilde{\mathrm{pI}}^1 + \widetilde{\mathrm{pI}}^2, \\
\delta_1(\widetilde{\mathrm{pI}}^1_o) &= \delta_1(\widetilde{\mathrm{pI}}^2_e) = \widetilde{\mathrm{pII}}^{1,2} + \widetilde{\mathrm{pII}}^a_e, \\
\delta_1(\widetilde{\mathrm{pI}}^1_e) &= \delta_1(\widetilde{\mathrm{pI}}^2_o) = \widetilde{\mathrm{pII}}^{1,2} + \widetilde{\mathrm{pII}}^a_o, \\
\delta_1(\widetilde{\mathrm{pI}}^3_o) &= \delta_1(\widetilde{\mathrm{pI}}^3_e) = \widetilde{\mathrm{pII}}^{1,3} + \widetilde{\mathrm{pII}}^{2,3} + \widetilde{\mathrm{pII}}^7, \\
\delta_1(\widetilde{\mathrm{pI}}^4_o) &= \delta_1(\widetilde{\mathrm{pI}}^4_e) = \widetilde{\mathrm{pII}}^{1,4} + \widetilde{\mathrm{pII}}^{2,4} + \widetilde{\mathrm{pII}}^{10},
\end{aligned}$$

where $\widetilde{\mathcal{F}}$ denotes $\widetilde{\mathcal{F}}_o + \widetilde{\mathcal{F}}_e$.

Then, by a straightforward calculation, we obtain the following.

Proposition 4.2 *The cohomology groups of*

$$\mathcal{C}(P\mathcal{S}_{\mathrm{pr}}(3, 2; 2), \rho_{3,2;2}(2))$$

of dimensions 0 *and* 1 *are described as follows:*

(1) $H^0(P\mathcal{S}_{\mathrm{pr}}(3, 2; 2), \rho_{3,2;2}(2)) \cong \mathbb{Z}_2$, *generated by* $[\widetilde{\mathrm{p0}}^0_o + \widetilde{\mathrm{p0}}^0_e]$,
(2) $H^1(P\mathcal{S}_{\mathrm{pr}}(3, 2; 2), \rho_{3,2;2}(2)) \cong \mathbb{Z}_2 \oplus \mathbb{Z}_2 \oplus \mathbb{Z}_2$, *generated by*

$$\alpha = [\widetilde{\mathrm{pI}}^4], \quad \beta = [\widetilde{\mathrm{pI}}^3], \quad \gamma = [\widetilde{\mathrm{pI}}^1_o + \widetilde{\mathrm{pI}}^2_e] = [\widetilde{\mathrm{pI}}^1_e + \widetilde{\mathrm{pI}}^2_o].$$

Note that the ranks of $C^{\kappa}(P\mathcal{S}_{\rm pr}(3,2;2), \rho_{3,2;2}(2))$, $\kappa = 0, 1, 2$, are equal to 2, 8 and 48, respectively.

4.2 Admissible Stable Maps of Manifold Pairs

Let us now consider a certain restricted class of stable maps. For a positive integer n, let

$$P\mathcal{AS}_{\rm pr}(n, n-1; n-1)$$

be the set of fibers for proper admissible C^0 stable Thom maps of manifold pairs (M, N) into Q with $(\dim M, \dim N; \dim Q) = (n, n-1; n-1)$. Here, a map $f\colon (M, N) \to Q$ is *admissible* if it is a submersion on a neighborhood of N. In particular, when $n = 3$, a stable map $f\colon (M, N) \to Q$ is admissible if and only if it has no definite or indefinite B_2 points. Note also that stable Morse functions on manifold pairs (M, N) with $(\dim M, \dim N) = (2, 1)$ and their suspensions are always admissible.

We set

$$P\mathcal{AS}_{\rm pr} = \bigcup_{n=1}^{\infty} P\mathcal{AS}_{\rm pr}(n, n-1; n-1).$$

Note that the above set together with the equivalence relation induced by $\rho(2)$, which we still denote by $\rho(2)$ by abuse of notation, satisfy conditions (a)–(e) mentioned in [9, Sect. 4]. Then, we can construct the universal complex

$$\mathcal{C}(P\mathcal{AS}_{\rm pr}(3, 2; 2), \rho_{3,2;2}(2))$$

of singular fibers for admissible C^∞ stable maps $(M, N) \to Q$ with

$$(\dim M, \dim N; \dim Q) = (3, 2; 2).$$

Observe that in the coboundary formulae for $\mathcal{C}(P\mathcal{S}_{\rm pr}(3, 2; 2), \rho_{3,2;2}(2))$ given just before Proposition 4.2, there appears no singular fiber containing a B_2 point. Therefore, we have the following.

Proposition 4.3 *The cohomology groups of*

$$\mathcal{C}(P\mathcal{AS}_{\rm pr}(3, 2; 2), \rho_{3,2;2}(2))$$

of dimensions 0 *and* 1 *are naturally isomorphic to those of the universal complex* $\mathcal{C}(P\mathcal{S}_{\rm pr}(3, 2; 2), \rho_{3,2;2}(2))$.

Note that the ranks of $C^{\kappa}(P\mathcal{AS}_{\rm pr}(3, 2; 2), \rho_{3,2;2}(2))$, $i = 0, 1, 2$, are equal to 2, 8 and 40, respectively.

4.3 Stable Maps of 2-Colored Manifold Pairs

For a positive integer n, let

$$P^c\mathcal{S}_{\mathrm{pr}}(n, n-1; n-1)$$

be the set of fibers for proper C^0 stable Thom maps $(M, N) \to Q$ of 2-colored manifold pairs with $(\dim M, \dim N; \dim Q) = (n, n-1; n-1)$. Recall that $M \setminus N$ is divided into two regions R and B, which are fixed and ordered. For example, if we interchange R and B, then it is considered to be a different 2-colored manifold pair.

We put

$$P^c\mathcal{S}_{\mathrm{pr}} = \bigcup_{n=1}^{\infty} P^c\mathcal{S}_{\mathrm{pr}}(n, n-1; n-1).$$

Furthermore, let $\rho^c_{n,n-1;n-1}(2)$ be the *weak C^0 color equivalence relation modulo two regular fibers* for fibers in $P^c\mathcal{S}_{\mathrm{pr}}(n, n-1; n-1)$: i.e., two fibers in $P^c\mathcal{S}_{\mathrm{pr}}(n, n-1; n-1)$ are $\rho^c_{n,n-1;n-1}(2)$-equivalent if they become weakly C^0 color equivalent after we add some regular fibers to each of them with the numbers of added components having the same parity.

We denote by $\rho^c(2)$ the equivalence relation on $P^c\mathcal{S}_{\mathrm{pr}}$ induced by $\rho^c_{n,n-1;n-1}(2)$, $n \geq 1$. Note that the set $P^c\mathcal{S}_{\mathrm{pr}}$ and the equivalence relation $\rho^c(2)$ satisfy conditions (a)–(e) described in [9, Sect. 4].

For a weak C^0 color equivalence class $\widetilde{\mathcal{F}}$ of singular fibers, denote by $\widetilde{\mathcal{F}}_o$ (or $\widetilde{\mathcal{F}}_e$) the equivalence class with respect to $\rho^c_{n,n-1;n-1}(2)$ which consists of singular fibers of type $\widetilde{\mathcal{F}}$ with an odd number (resp. even number) of regular fiber components. For $n = 2, 3$, we denote by $\widetilde{\mathrm{c0}}^0_o$ (resp. $\widetilde{\mathrm{c0}}^0_e$) the equivalence class with respect to $\rho^c_{n,n-1;n-1}(2)$ which consists of regular fibers with an odd (resp. even) number of components.

Then, we can construct the universal complex $\mathcal{C}(P^c\mathcal{S}_{\mathrm{pr}}(3, 2; 2), \rho^c_{3,2;2}(2))$ of singular fibers for maps of $(3, 2)$-dimensional 2-colored manifold pairs into 2-dimensional manifolds with respect to the relation $\rho^c_{3,2;2}(2)$. Its coboundary homomorphism is given by the following formulae, which are obtained with the help of Lemma 2.13:

$$\begin{aligned}
\delta_0(\widetilde{\mathrm{c0}}^0_o) &= \delta_0(\widetilde{\mathrm{c0}}^0_e) = \widetilde{\mathrm{cI}}^{1A} + \widetilde{\mathrm{cI}}^{1B} + \widetilde{\mathrm{cI}}^{2A} + \widetilde{\mathrm{cI}}^{2B},\\
\delta_1(\widetilde{\mathrm{cI}}^{1A}_o) &= \widetilde{\mathrm{cII}}^{1A,1B} + \widetilde{\mathrm{cII}}^{1A,2A} + \widetilde{\mathrm{cII}}^{1A,2B} + \widetilde{\mathrm{cII}}^{aA}_e + \widetilde{\mathrm{cII}}^{c}_o,\\
\delta_1(\widetilde{\mathrm{cI}}^{1A}_e) &= \widetilde{\mathrm{cII}}^{1A,1B} + \widetilde{\mathrm{cII}}^{1A,2A} + \widetilde{\mathrm{cII}}^{1A,2B} + \widetilde{\mathrm{cII}}^{aA}_o + \widetilde{\mathrm{cII}}^{c}_e,\\
\delta_1(\widetilde{\mathrm{cI}}^{1B}_o) &= \widetilde{\mathrm{cII}}^{1A,1B} + \widetilde{\mathrm{cII}}^{1B,2A} + \widetilde{\mathrm{cII}}^{1B,2B} + \widetilde{\mathrm{cII}}^{aB}_e + \widetilde{\mathrm{cII}}^{c}_o,\\
\delta_1(\widetilde{\mathrm{cI}}^{1B}_e) &= \widetilde{\mathrm{cII}}^{1A,1B} + \widetilde{\mathrm{cII}}^{1B,2A} + \widetilde{\mathrm{cII}}^{1B,2B} + \widetilde{\mathrm{cII}}^{aB}_o + \widetilde{\mathrm{cII}}^{c}_e,\\
\delta_1(\widetilde{\mathrm{cI}}^{2A}_o) &= \widetilde{\mathrm{cII}}^{1A,2A} + \widetilde{\mathrm{cII}}^{1B,2A} + \widetilde{\mathrm{cII}}^{2A,2B} + \widetilde{\mathrm{cII}}^{5C} + \widetilde{\mathrm{cII}}^{9C}_o + \widetilde{\mathrm{cII}}^{aA}_o\\
&\quad + \widetilde{\mathrm{cII}}^{d}_o + \widetilde{\mathrm{cII}}^{e}_o,
\end{aligned}$$

$$
\begin{aligned}
\delta_1(\widetilde{\mathrm{cI}}^{2A}_e) &= \widetilde{\mathrm{cII}}^{1A,2A} + \widetilde{\mathrm{cII}}^{1B,2A} + \widetilde{\mathrm{cII}}^{2A,2B} + \widetilde{\mathrm{cII}}^{5C} + \widetilde{\mathrm{cII}}^{9C}_e + \widetilde{\mathrm{cII}}^{aA}_e \\
&\quad + \widetilde{\mathrm{cII}}^{d}_e + \widetilde{\mathrm{cII}}^{e}_e, \\
\delta_1(\widetilde{\mathrm{cI}}^{2B}_o) &= \widetilde{\mathrm{cII}}^{1A,2B} + \widetilde{\mathrm{cII}}^{1B,2B} + \widetilde{\mathrm{cII}}^{2A,2B} + \widetilde{\mathrm{cII}}^{5C} + \widetilde{\mathrm{cII}}^{9C}_o + \widetilde{\mathrm{cII}}^{aB}_o \\
&\quad + \widetilde{\mathrm{cII}}^{d}_o + \widetilde{\mathrm{cII}}^{e}_o, \\
\delta_1(\widetilde{\mathrm{cI}}^{2B}_e) &= \widetilde{\mathrm{cII}}^{1A,2B} + \widetilde{\mathrm{cII}}^{1B,2B} + \widetilde{\mathrm{cII}}^{2A,2B} + \widetilde{\mathrm{cII}}^{5C} + \widetilde{\mathrm{cII}}^{9C}_e + \widetilde{\mathrm{cII}}^{aB}_e \\
&\quad + \widetilde{\mathrm{cII}}^{d}_e + \widetilde{\mathrm{cII}}^{e}_e, \\
\delta_1(\widetilde{\mathrm{cI}}^{3A}_o) &= \widetilde{\mathrm{cII}}^{1A,3A} + \widetilde{\mathrm{cII}}^{1B,3A} + \widetilde{\mathrm{cII}}^{2A,3A} + \widetilde{\mathrm{cII}}^{2B,3A} + \widetilde{\mathrm{cII}}^{7A} + \widetilde{\mathrm{cII}}^{7C} \\
&\quad + \widetilde{\mathrm{cII}}^{b}_o + \widetilde{\mathrm{cII}}^{c}_o + \widetilde{\mathrm{cII}}^{d}_e + \widetilde{\mathrm{cII}}^{e}_o + \widetilde{\mathrm{cII}}^{f}_o, \\
\delta_1(\widetilde{\mathrm{cI}}^{3A}_e) &= \widetilde{\mathrm{cII}}^{1A,3A} + \widetilde{\mathrm{cII}}^{1B,3A} + \widetilde{\mathrm{cII}}^{2A,3A} + \widetilde{\mathrm{cII}}^{2B,3A} + \widetilde{\mathrm{cII}}^{7A} + \widetilde{\mathrm{cII}}^{7C} \\
&\quad + \widetilde{\mathrm{cII}}^{b}_e + \widetilde{\mathrm{cII}}^{c}_e + \widetilde{\mathrm{cII}}^{d}_o + \widetilde{\mathrm{cII}}^{e}_e + \widetilde{\mathrm{cII}}^{f}_e, \\
\delta_1(\widetilde{\mathrm{cI}}^{3B}_o) &= \widetilde{\mathrm{cII}}^{1A,3B} + \widetilde{\mathrm{cII}}^{1B,3B} + \widetilde{\mathrm{cII}}^{2A,3B} + \widetilde{\mathrm{cII}}^{2B,3B} + \widetilde{\mathrm{cII}}^{7B} + \widetilde{\mathrm{cII}}^{7D} \\
&\quad + \widetilde{\mathrm{cII}}^{b}_o + \widetilde{\mathrm{cII}}^{c}_o + \widetilde{\mathrm{cII}}^{d}_e + \widetilde{\mathrm{cII}}^{e}_o + \widetilde{\mathrm{cII}}^{f}_o, \\
\delta_1(\widetilde{\mathrm{cI}}^{3B}_e) &= \widetilde{\mathrm{cII}}^{1A,3B} + \widetilde{\mathrm{cII}}^{1B,3B} + \widetilde{\mathrm{cII}}^{2A,3B} + \widetilde{\mathrm{cII}}^{2B,3B} + \widetilde{\mathrm{cII}}^{7B} + \widetilde{\mathrm{cII}}^{7D} \\
&\quad + \widetilde{\mathrm{cII}}^{b}_e + \widetilde{\mathrm{cII}}^{c}_e + \widetilde{\mathrm{cII}}^{d}_o + \widetilde{\mathrm{cII}}^{e}_e + \widetilde{\mathrm{cII}}^{f}_e, \\
\delta_1(\widetilde{\mathrm{cI}}^{4A}_o) &= \widetilde{\mathrm{cII}}^{1A,4A} + \widetilde{\mathrm{cII}}^{1B,4A} + \widetilde{\mathrm{cII}}^{2A,4A} + \widetilde{\mathrm{cII}}^{2B,4A} + \widetilde{\mathrm{cII}}^{9C}_o + \widetilde{\mathrm{cII}}^{10A} \\
&\quad + \widetilde{\mathrm{cII}}^{10C} + \widetilde{\mathrm{cII}}^{f}_o, \\
\delta_1(\widetilde{\mathrm{cI}}^{4A}_e) &= \widetilde{\mathrm{cII}}^{1A,4A} + \widetilde{\mathrm{cII}}^{1B,4A} + \widetilde{\mathrm{cII}}^{2A,4A} + \widetilde{\mathrm{cII}}^{2B,4A} + \widetilde{\mathrm{cII}}^{9C}_e + \widetilde{\mathrm{cII}}^{10A} \\
&\quad + \widetilde{\mathrm{cII}}^{10C} + \widetilde{\mathrm{cII}}^{f}_e, \\
\delta_1(\widetilde{\mathrm{cI}}^{4B}_o) &= \widetilde{\mathrm{cII}}^{1A,4B} + \widetilde{\mathrm{cII}}^{1B,4B} + \widetilde{\mathrm{cII}}^{2A,4B} + \widetilde{\mathrm{cII}}^{2B,4B} + \widetilde{\mathrm{cII}}^{9C}_o + \widetilde{\mathrm{cII}}^{10B} \\
&\quad + \widetilde{\mathrm{cII}}^{10D} + \widetilde{\mathrm{cII}}^{f}_o, \\
\delta_1(\widetilde{\mathrm{cI}}^{4B}_e) &= \widetilde{\mathrm{cII}}^{1A,4B} + \widetilde{\mathrm{cII}}^{1B,4B} + \widetilde{\mathrm{cII}}^{2A,4B} + \widetilde{\mathrm{cII}}^{2B,4B} + \widetilde{\mathrm{cII}}^{9C}_e + \widetilde{\mathrm{cII}}^{10B} \\
&\quad + \widetilde{\mathrm{cII}}^{10D} + \widetilde{\mathrm{cII}}^{f}_e,
\end{aligned}
$$

where $\widetilde{\mathcal{F}}$ denotes $\widetilde{\mathcal{F}}_o + \widetilde{\mathcal{F}}_e$.

Then, by a straightforward calculation, we obtain the following.

Proposition 4.4 *The cohomology groups of*

$$\mathcal{C}(P^c\mathcal{S}_{\mathrm{pr}}(3, 2; 2), \rho^c_{3,2;2}(2))$$

of dimensions 0 *and* 1 *are described as follows:*

(1) $H^0(P^c\mathcal{S}_{\mathrm{pr}}(3, 2; 2), \rho^c_{3,2;2}(2)) \cong \mathbb{Z}_2$, *generated by* $[\widetilde{\mathrm{c0}}^0_o + \widetilde{\mathrm{c0}}^0_e]$,

(2) $H^1(P^c\mathcal{S}_{\mathrm{pr}}(3, 2; 2), \rho^c_{3,2;2}(2)) \cong \mathbb{Z}_2 \oplus \mathbb{Z}_2 \oplus \mathbb{Z}_2$, *generated by*

$$\alpha = [\widetilde{\mathrm{cI}}^{4A} + \widetilde{\mathrm{cI}}^{4B}], \qquad \beta = [\widetilde{\mathrm{cI}}^{3A} + \widetilde{\mathrm{cI}}^{3B}],$$

$$\gamma = [\widetilde{\mathrm{cI}}_o^{1A} + \widetilde{\mathrm{cI}}_o^{1B} + \widetilde{\mathrm{cI}}_e^{2A} + \widetilde{\mathrm{cI}}_e^{2B}] = [\widetilde{\mathrm{cI}}_e^{1A} + \widetilde{\mathrm{cI}}_e^{1B} + \widetilde{\mathrm{cI}}_o^{2A} + \widetilde{\mathrm{cI}}_o^{2B}].$$

Note that the ranks of $C^\kappa(P^c\mathcal{S}_{\mathrm{pr}}(3,2;2), \rho^c_{3,2;2}(2))$, $\kappa = 0, 1, 2$, are equal to 2, 16 and 142, respectively.

4.4 Admissible Stable Maps of 2-Colored Manifold Pairs

For a positive integer n, let

$$P^c\mathcal{AS}_{\mathrm{pr}}(n, n-1; n-1)$$

be the set of fibers for proper admissible C^0 stable Thom maps of 2-colored manifold pairs (M, N) into Q with $(\dim M, \dim N; \dim Q) = (n, n-1; n-1)$.

We set

$$P^c\mathcal{AS}_{\mathrm{pr}} = \bigcup_{n=1}^{\infty} P^c\mathcal{AS}_{\mathrm{pr}}(n, n-1; n-1).$$

Note that the above set together with the equivalence relation induced by $\rho^c(2)$, which we still denote by $\rho^c(2)$ by abuse of notation, satisfy conditions (a)–(e) described in [9, Sect. 4]. Then we can construct the corresponding universal complex

$$\mathcal{C}(P^c\mathcal{AS}_{\mathrm{pr}}(3,2;2), \rho^c_{3,2;2}(2)).$$

Then, a straightforward calculation shows the following.

Proposition 4.5 *The cohomology groups of*

$$\mathcal{C}(P^c\mathcal{AS}_{\mathrm{pr}}(3,2;2), \rho^c_{3,2;2}(2))$$

of dimensions 0 *and* 1 *are described as follows:*

(1) $H^0(P^c\mathcal{AS}_{\mathrm{pr}}(3,2;2), \rho^c_{3,2;2}(2)) \cong \mathbb{Z}_2$, *generated by* $[\widetilde{\mathrm{c0}}_o^0 + \widetilde{\mathrm{c0}}_e^0]$,
(2) $H^1(P^c\mathcal{AS}_{\mathrm{pr}}(3,2;2), \rho^c_{3,2;2}(2)) \cong \mathbb{Z}_2 \oplus \mathbb{Z}_2 \oplus \mathbb{Z}_2 \oplus \mathbb{Z}_2$, *generated by*

$$\begin{aligned}
\alpha_1 &= [\widetilde{\mathrm{cI}}^{4A} + \widetilde{\mathrm{cI}}^{1A} + \widetilde{\mathrm{cI}}^{2A}] = [\widetilde{\mathrm{cI}}^{4A} + \widetilde{\mathrm{cI}}^{1B} + \widetilde{\mathrm{cI}}^{2B}],\\
\alpha_2 &= [\widetilde{\mathrm{cI}}^{4B} + \widetilde{\mathrm{cI}}^{1A} + \widetilde{\mathrm{cI}}^{2A}] = [\widetilde{\mathrm{cI}}^{4B} + \widetilde{\mathrm{cI}}^{1B} + \widetilde{\mathrm{cI}}^{2B}],\\
\beta &= [\widetilde{\mathrm{cI}}^{3A} + \widetilde{\mathrm{cI}}^{3B}],\\
\gamma &= [\widetilde{\mathrm{cI}}_o^{1A} + \widetilde{\mathrm{cI}}_o^{1B} + \widetilde{\mathrm{cI}}_e^{2A} + \widetilde{\mathrm{cI}}_e^{2B}] = [\widetilde{\mathrm{cI}}_e^{1A} + \widetilde{\mathrm{cI}}_e^{1B} + \widetilde{\mathrm{cI}}_o^{2A} + \widetilde{\mathrm{cI}}_o^{2B}].
\end{aligned}$$

Note that the ranks of $C^\kappa(P^c\mathcal{AS}_{\mathrm{pr}}(3,2;2), \rho^c_{3,2;2}(2))$, $\kappa = 0, 1, 2$, are equal to 2, 16 and 138, respectively.

5 Cobordisms of Singular Maps

In this section, we first review the concept of cobordism of singular maps, and then we see how cobordism invariants are induced from the cohomology classes of the universal complexes of singular fibers that have been obtained in the previous section.

Definition 5.1 (1) Let (M_i, N_i), $i = 0, 1$, be manifold pairs of the same dimension $(n, n-1)$, where M_0 and M_1 are closed, and let Q be a manifold of dimension $n-1$. Two C^∞ stable maps $f_i \colon (M_i, N_i) \to Q$, $i = 0, 1$, are *cobordant* if there exist a compact $(n+1)$-dimensional manifold X with boundary, a proper[3] closed submanifold Y of dimension n of X, and a C^∞ map $F \colon (X, Y) \to Q \times [0, 1]$ that satisfy the following conditions:

(i) ∂X is the disjoint union $M_0 \sqcup M_1$,
(ii) ∂Y is the disjoint union $N_0 \sqcup N_1$,
(iii) $F|_{M_0 \times [0,\varepsilon)} = f_0 \times \mathrm{id}_{[0,\varepsilon)} \colon (M_0, N_0) \times [0, \varepsilon) \to Q \times [0, \varepsilon)$ and $F|_{M_1 \times (1-\varepsilon, 1]} = f_1 \times \mathrm{id}_{(1-\varepsilon,1]} \colon (M_1, N_1) \times (1-\varepsilon, 1] \to Q \times (1-\varepsilon, 1]$, where $(M_0, N_0) \times [0, \varepsilon)$ and $(M_1, N_1) \times (1-\varepsilon, 1]$ denote collar neighborhoods of (M_0, N_0) and (M_1, N_1) in (X, Y), respectively,
(iv) $F^{-1}(Q \times \{i\}) = (M_i, N_i)$, $i = 0, 1$, and the restriction

$$F|_{X \setminus (M_0 \cup M_1)} \colon (X \setminus (M_0 \cup M_1), Y \setminus (N_0 \cup N_1)) \to Q \times (0, 1)$$

is a proper C^∞ stable map.

In this case, we call the map F a *cobordism* between f_0 and f_1.

(2) Suppose that the maps f_0 and f_1 as in (1) are admissible. Then, we say that f_0 and f_1 are *admissibly cobordant* if f_0 and f_1 are cobordant in the sense of (1) and the cobordism F can be chosen so that it is admissible, i.e. it is a submersion on a neighborhood of the proper closed submanifold Y.

(3) Suppose that the manifold pairs (M_i, N_i) are 2-colored with coloring pair (R_i, B_i), $i = 0, 1$. Then the two maps f_0 and f_1 as in (1) are *color cobordant* if f_0 and f_1 are cobordant in the sense of (1) and the manifold pair (X, Y) can be chosen to be 2-colored: i.e. for disjoint open submanifolds R and B of $X \setminus Y$, we have

$$R \cup B = X \setminus Y, \quad \overline{R} \cap \overline{B} = \overline{R} \setminus R = \overline{B} \setminus B = Y,$$

where $\overline{R}$ and $\overline{B}$ denote the closures of R and B in X, respectively. Furthermore, we impose the condition that

$$R \cap M_i = R_i, \quad B \cap M_i = B_i, \; i = 0, 1.$$

(4) Suppose that the manifold pairs (M_i, N_i) are 2-colored with coloring pair (R_i, B_i), $i = 0, 1$, and that the maps f_0 and f_1 as in (3) are admissible. Then f_0 and

[3] A submanifold Y of a manifold X with boundary is *proper* if $Y \cap \partial X = \partial Y$ and Y intersects ∂X transversely.

f_1 are *admissibly color cobordant* if they are color cobordant in the sense of (3) and the cobordism F can be chosen so that it is admissible, i.e. it is a submersion on a neighborhood of the proper closed submanifold Y.

Remark 5.2 It is not difficult to show that when $Q = \mathbb{R}^{n-1}$ is fixed, the set of all (admissible) cobordism classes of (admissible) C^∞ stable maps of closed $(n, n-1)$-dimensional (2-colored) manifold pairs into Q forms an abelian group under the disjoint union as the addition operation. We call this group the $(n, n-1; n-1)$-*dimensional* $P\mathcal{S}_{\mathrm{pr}}$*-cobordism group of* C^∞ *stable maps of manifold pairs* for Definition 5.1 (1), the $(n, n-1; n-1)$*-dimensional* $P\mathcal{AS}_{\mathrm{pr}}$*-cobordism group of admissible* C^∞ *stable maps of manifold pairs* for Definition 5.1 (2), the $(n, n-1; n-1)$*-dimensional* $P^c\mathcal{S}_{\mathrm{pr}}$*-cobordism group of* C^∞ *stable maps of* 2*-colored manifold pairs* for Definition 5.1 (3), and the $(n, n-1; n-1)$*-dimensional* $P^c\mathcal{AS}_{\mathrm{pr}}$*-cobordism group of admissible* C^∞ *stable maps of* 2*-colored manifold pairs* for Definition 5.1 (4).

As is explained in [9, Sect. 4], cohomology classes of the universal complex of singular fibers of a certain class of smooth maps induce cobordism invariants for the same class of maps. Let us first begin by the case of stable maps of manifold pairs.

Let

$$s_{\kappa *}\colon H^\kappa(P\mathcal{S}_{\mathrm{pr}}(3, 2; 2), \rho_{3,2;2}(2)) \to H^\kappa(P\mathcal{S}_{\mathrm{pr}}(2, 1; 1), \rho_{2,1;1}(2))$$

be the homomorphism induced by suspension s_κ. Note that $s_\kappa(\mathcal{F}_*) = \mathcal{F}_*$ for $\kappa = 0, 1$ and for each $\rho_{3,2;2}(2)$-class $\mathcal{F}_*$ of codimension κ (see Remark 2.10). Note also that $s_\kappa = 0$ for $\kappa = 2$.

Let us consider the cohomology classes α, β and γ in $H^1(P\mathcal{S}_{\mathrm{pr}}(3, 2; 2), \rho_{3,2;2}(2))$ obtained in Proposition 4.2. As has been explained in [9, Sect. 4], each element of the image of $s_{\kappa *}$ induces a cobordism invariant for stable Morse functions of (2, 1)-dimensional manifold pairs into 1-dimensional manifolds. By using the same method as in [4, Lemma 14.1], we can show that $s_{1*}\gamma$ induces a trivial $P\mathcal{S}_{\mathrm{pr}}$-cobordism invariant. Furthermore, $s_{1*}\beta$ also induces a trivial invariant, since on a circle submanifold, the number of critical points is always even.

Remark 5.3 There exists a stable Morse function $f\colon (\mathbb{R}P^2, W) \to \mathbb{R}$, for an arbitrary (possibly empty) closed 1-dimensional submanifold W of $\mathbb{R}P^2$, which has exactly one singular fiber of type $\widetilde{\mathrm{pI}}^4$. Thus, $s_{1*}\alpha$ induces a non-trivial $P\mathcal{S}_{\mathrm{pr}}$-cobordism invariant among stable Morse functions on manifold pairs (M, N) with $(\dim M, \dim N) = (2, 1)$. Note that the singular fiber of type $\widetilde{\mathrm{pI}}^4$ does not reflect the position of the 1-dimensional submanifold W in $\mathbb{R}P^2$. Recall that Saeki [4] showed that, for stable Morse functions on closed surfaces, the modulo two number of singular fibers of type $\widetilde{\mathrm{pI}}^4$ is a cobordism invariant.[4]

As we will see in Appendix A "Cobordism Group of Manifold Pairs of Dimension (2, 1)", the $P\mathcal{S}_{\mathrm{pr}}$-cobordism group of C^∞ stable Morse functions on (2, 1)-dimensional manifold pairs to $\mathbb{R}$ is isomorphic to $\mathbb{Z}_2 \oplus \mathbb{Z}_2$. The singular fiber of type

[4]The singular fiber of type $\widetilde{\mathrm{pI}}^4$ is denoted by $\widetilde{\mathrm{I}}^2$ in [4].

$\widetilde{\mathrm{pI}}^4$ detects the first $\mathbb{Z}_2$-factor. The second $\mathbb{Z}_2$-factor involves the number of components of the 1-dimensional submanifold whose normal bundles are non-trivial. It seems that such information cannot be detected by singular fibers.

For admissible cobordisms for admissible stable maps, as a consequence of Proposition 4.3, we obtain just one non-trivial $P\mathcal{AS}_{\mathrm{pr}}$-cobordism invariant for stable Morse functions on (2, 1)-dimensional manifold pairs, as we have seen in Remark 5.3. This should be compared with a result given in [9, 10], where it is shown that a cohomology class gives a complete invariant for the admissible cobordism group of C^∞ stable Morse functions on surfaces with boundary.

Let us now consider the case of stable maps of 2-colored manifold pairs. Let

$$s_{\kappa *}\colon\ H^\kappa(P^c\mathcal{S}_{\mathrm{pr}}(3, 2; 2), \rho^c_{3,2;2}(2)) \to H^\kappa(P^c\mathcal{S}_{\mathrm{pr}}(2, 1; 1), \rho^c_{2,1;1}(2))$$

be the homomorphism induced by suspension s_κ. Note that $s_\kappa(\mathcal{F}_*) = \mathcal{F}_*$ for $\kappa = 0, 1$ and for each $\rho^c_{3,2;2}(2)$-class $\mathcal{F}_*$ of codimension κ (see Remark 2.18). Note also that $s_\kappa = 0$ for $\kappa = 2$.

Let us consider the cohomology classes α, β and γ in $H^1(P^c\mathcal{S}_{\mathrm{pr}}(3, 2; 2), \rho^c_{3,2;2}(2))$ obtained in Proposition 4.4. As before, each element of the image of $s_{\kappa *}$ induces a cobordism invariant for stable Morse functions of (2, 1)-dimensional 2-colored manifold pairs into 1-dimensional manifolds. By using the same method as before, we can show that $s_{1*}\beta$ and $s_{1*}\gamma$ again induce trivial $P^c\mathcal{S}_{\mathrm{pr}}$-cobordism invariants.

Remark 5.4 The $P^c\mathcal{S}_{\mathrm{pr}}$-cobordism invariant induced by $s_{1*}\alpha$ is non-trivial. This can be seen as in Remark 5.3, where we choose W as a circle bounding a disk in $\mathbb{R}P^2$ in such a way that $(\mathbb{R}P^2, W)$ is a 2-colored manifold pair.

As we will see in Appendix B "Cobordism Group of 2-Colored Manifold Pairs of Dimension (2, 1)", the $P^c\mathcal{S}_{\mathrm{pr}}$-cobordism group of stable Morse functions on (2, 1)-dimensional 2-colored manifold pairs to $\mathbb{R}$ is isomorphic to $\mathbb{Z}_2$. This implies that the cobordism invariant induced by $s_{1*}\alpha$ is complete in this case.

Let us now consider admissible cobordisms for admissible stable maps of 2-colored manifold pairs. Let

$$s_{\kappa *}\colon\ H^\kappa(P^c\mathcal{AS}_{\mathrm{pr}}(3, 2; 2), \rho^c_{3,2;2}(2)) \to H^\kappa(P^c\mathcal{AS}_{\mathrm{pr}}(2, 1; 1), \rho^c_{2,1;1}(2))$$

be the homomorphism induced by suspension s_κ. Let us consider the cohomology classes $\alpha_1, \alpha_2, \beta$ and γ in $H^1(P^c\mathcal{AS}_{\mathrm{pr}}(3, 2; 2), \rho^c_{3,2;2}(2))$ obtained in Proposition 4.5. Then, we obtain the following cobordism invariants.

Proposition 5.5 *Let Q be the real line $\mathbb{R}$ or the circle S^1.*

(1) *The cohomology classes $s_{1*}\alpha_1$ and $s_{1*}\alpha_2$ induce $\mathbb{Z}_2$-linearly independent $P^c\mathcal{AS}_{\mathrm{pr}}$-cobordism invariants for stable Morse functions $f\colon (M, N) \to Q$ of (2, 1)-dimensional 2-colored manifold pairs (M, N).*
(2) *The cohomology classes $s_{1*}\beta$ and $s_{1*}\gamma$ induce trivial $P^c\mathcal{AS}_{\mathrm{pr}}$-cobordism invariants.*

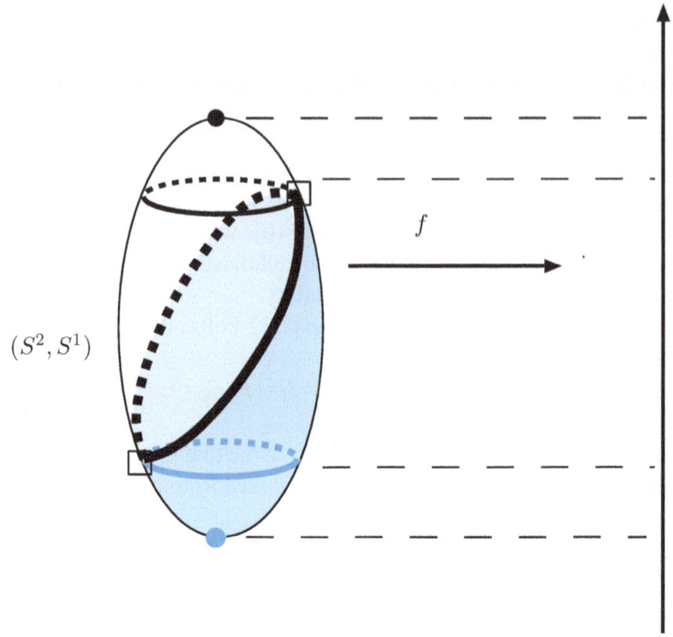

Fig. 11 Stable Morse function on a (2, 1)-dimensional 2-colored manifold pair

Proof (1) For the stable map $f\colon (S^2, S^1) \to \mathbb{R}$ given by the height function as depicted in Fig. 11, we have $s_{1*}\alpha_1(f) = 1$ and $s_{1*}\alpha_2(f) = 1$. Therefore, the $P^c\mathcal{AS}_{\mathrm{pr}}$-cobordism invariants $s_{1*}\alpha_1$ and $s_{1*}\alpha_2$ are both non-trivial.

Furthermore, we can construct stable Morse functions $f_1, f_2\colon (\mathbb{R}P^2, S^1) \to \mathbb{R}$ on a 2-colored manifold pair $(\mathbb{R}P^2, S^1)$ such that $(s_{1*}\alpha_1(f_1), s_{1*}\alpha_2(f_1)) = (1, 0)$ and $(s_{1*}\alpha_1(f_2), s_{1*}\alpha_2(f_2)) = (0, 1)$. Just consider the standard Morse function $\mathbb{R}P^2 \to \mathbb{R}$ with three critical points and take the submanifold S^1 as a small circle encircling the minimum point, which is a bit tilted. The other one is obtained by interchanging the colors. Therefore, the $P^c\mathcal{AS}_{\mathrm{pr}}$-cobordism invariants $s_{1*}\alpha_1$ and $s_{1*}\alpha_2$ are linearly independent over $\mathbb{Z}_2$.

(2) This can be proved by the same argument as before. □

Remark 5.6 For a stable Morse function $f\colon (M, N) \to Q$ of a (2, 1)-dimensional 2-colored manifold pair (M, N) with coloring pair (R, B), we have two stable maps $f_R = f|_{\overline{R}}\colon \overline{R} \to Q$ and $f_B = f|_{\overline{B}}\colon \overline{B} \to Q$ on surfaces with boundary. It is easy to show that these induce homomorphisms of the (2, 1)-dimensional $P^c\mathcal{AS}_{\mathrm{pr}}$-cobordism group to the cobordism group of admissible stable maps of surfaces with boundary. (Recall that the latter group is known to be isomorphic to $\mathbb{Z}_2$. See [10].) We can observe that the above linearly independent invariants are obtained by composing these homomorphisms with the admissible cobordism invariant obtained in [9].

We do not know if the above invariants induce an isomorphism between the admissible cobordism group of stable Morse functions on (2, 1)-dimensional 2-colored manifold pairs and $\mathbb{Z}_2 \oplus \mathbb{Z}_2$.

Acknowledgements The authors would like to thank the anonymous referee for some comments which improved the presentation of the paper. The first author has been supported in part by JSPS KAKENHI Grant Numbers JP23244008, JP23654028, JP25540041, JP15K13438, JP16H03936, JP16K13754, JP17H01090, JP17H06128. The second author has been supported in part by JSPS KAKENHI Grant Numbers JP23654028, JP15H03615, JP15K04880, JP15K13438, JP17H01090, JP17H06128.

A Appendix. Cobordism Group of Manifold Pairs of Dimension (2, 1)

Let $\mathfrak{N}_{n,n-1}$ be the (usual) cobordism group of $(n, n-1)$-dimensional manifold pairs (for example, see [13]). Let us first prove the following.

Proposition A.1 *The* (2, 1)*-dimensional cobordism group* $\mathfrak{N}_{2,1}$ *of manifold pairs is naturally isomorphic to the* (2, 1; 1)*-dimensional* $P\mathcal{S}_{\mathrm{pr}}$*-cobordism group of* C^∞ *stable maps of manifold pairs.*

In fact, the above proposition holds for every n such that $(n, n-1)$ is in the nice range.

Proof of Proposition A.1 For a stable Morse function $f\colon (M, N) \to \mathbb{R}$ on a (2, 1)-dimensional manifold pair, we associate the manifold pair (M, N). This defines a well-defined homomorphism between the relevant cobordism groups. It is surjective, since every (2, 1)-dimensional manifold pair admits a stable Morse function. It is also injective, since every (3, 2)-dimensional manifold pair giving a cobordism between the source manifold pairs of stable Morse functions f_0 and f_1 admits a C^∞ stable map into $\mathbb{R} \times [0, 1]$ extending f_0 and f_1. This completes the proof. □

Now, the main purpose of this section is to show the following elementary fact.

Proposition A.2 *The* (2, 1)*-dimensional cobordism group* $\mathfrak{N}_{2,1}$ *of manifold pairs is isomorphic to* $\mathbb{Z}_2 \oplus \mathbb{Z}_2$.

Thus, the (2, 1; 1)-dimensional $P\mathcal{S}_{\mathrm{pr}}$-cobordism group of C^∞ stable maps of manifold pairs is isomorphic to $\mathbb{Z}_2 \oplus \mathbb{Z}_2$.

In fact, Proposition A.2 is a direct consequence of a result of Wall [13]. Here, we give an elementary proof.

Proof of Proposition A.2 For a (2, 1)-dimensional manifold pair (M, N), we denote by $\varphi(M, N)$ the usual cobordism class of M in the 2-dimensional cobordism group $\mathfrak{N}_2 = \mathbb{Z}_2$. Furthermore, let $\psi(M, N)$ be the number modulo two of the components of N whose normal bundles in M are non-trivial.

Suppose that (M, N) and (M', N') are cobordant. Then, clearly M and M' are cobordant. Thus, φ induces a well-defined homomorphism $\mathfrak{N}_{2,1} \to \mathfrak{N}_2 = \mathbb{Z}_2$, which we still denote by φ. On the other hand, let (X, Y) be a cobordism between (M, N) and (M', N'). Since the disjoint union $N \cup N'$ bounds a surface Y in X, it represents zero in $H_1(X; \mathbb{Z}_2)$. On the other hand, for each component c of $N \cup N'$, its intersection number with $[Y, \partial Y] \in H_2(X, \partial X; \mathbb{Z}_2)$ in X is non-trivial if and only if its normal bundle in $M \cup M'$ is non-trivial. Since the intersection number of $N \cup N'$ with $[Y, \partial Y]$ must vanish, we see that $\psi(M, N) + \psi(M', N')$ must be zero modulo two. This means that ψ induces a well-defined homomorphism $\mathfrak{N}_{2,1} \to \mathbb{Z}_2$, which we still denote by ψ.

Suppose that (M, N) and (M', N') satisfy both $\varphi(M, N) = \varphi(M', N')$ and $\psi(M, N) = \psi(M', N')$. Let c and c' be a pair of components of $N \cup N'$ whose normal bundles are non-trivial. Let B be the Möbius band and b the center circle. We attach $B \times [-1, 1]$ to the disjoint union $(M \times [0, 1]) \cup (M' \times [0, 1])$ in such a way that $B \times \{-1, 1\}$ is attached to the tubular neighborhood of $(c \times \{1\}) \cup (c' \times \{1\})$ in $(M \times \{1\}) \cup (M' \times \{1\})$ and that $b \times \{-1, 1\}$ is attached to $(c \times \{1\}) \cup (c' \times \{1\})$ in $(M \times \{1\}) \cup (M' \times \{1\})$. Since $\psi(M, N) = \psi(M', N')$, we can repeat this procedure until there remains no component of $N \cup N'$ with non-trivial normal bundle. For each of the other components, we attach $D^2 \times [-1, 1]$ in a similar way. Note that the union of all $b \times [-1, 1]$ and $D^2 \times \{0\}$ together with appropriate collars in $(M \times [0, 1]) \cup (M' \times [0, 1])$ gives a null-cobordism of $N \cup N' = (N \times \{0\}) \cup (N' \times \{0\})$. Finally, by using our assumption that $\varphi(M, N) = \varphi(M', N')$, we can further attach a compact 3-manifold so that we obtain a null-cobordism for $(M, N) \cup (M', N') = ((M, N) \times \{0\}) \cup ((M', N') \times \{0\})$.

Thus, the homomorphism $\varphi \oplus \psi \colon \mathfrak{N}_{2,1} \to \mathbb{Z}_2 \oplus \mathbb{Z}_2$ is a monomorphism.

On the other hand, this is an epimorphism. For example, $(\mathbb{R}P^2, \emptyset)$ is sent to $(1, 0)$ and $(\mathbb{R}P^2, \ell)$ is sent to $(1, 1)$, where ℓ is a circle in $\mathbb{R}P^2$ with non-trivial normal bundle.

This completes the proof. □

Determining the group structure of the $(2, 1)$-dimensional $P\mathcal{AS}_{\text{pr}}$-cobordism group remains an open problem.

B Appendix. Cobordism Group of 2-Colored Manifold Pairs of Dimension (2, 1)

Let $\mathfrak{N}^c_{n,n-1}$ be the cobordism group of $(n, n-1)$-dimensional 2-colored manifold pairs. More precisely, let (M_i, N_i) be $(n, n-1)$-dimensional 2-colored manifold pairs with coloring pairs (R_i, B_i), where M_i are closed, $i = 0, 1$. Then, they are *color cobordant* if there exists an $(n+1, n)$-dimensional 2-colored manifold pair (X, Y) with coloring pair (R, B), where X is a compact manifold with boundary $\partial X = M_0 \sqcup M_1$, Y is a proper closed submanifold of Y with $\partial Y = N_0 \sqcup N_1$, $R \cap M_i = R_i$ and $B \cap M_1 = R_i$, $i = 0, 1$. It is easy to show that the set of all such color cobordism classes forms an abelian group under the disjoint union as the addition operation. This is the cobordism group of $(n, n-1)$-dimensional 2-colored manifold pairs.

As in Proposition A.1, we have the following.

Proposition B.1 *The* (2, 1)*-dimensional cobordism group* $\mathfrak{N}^c_{2,1}$ *of* 2*-colored manifold pairs is naturally isomorphic to the* (2, 1)*-dimensional* $P^c\mathcal{S}_{\mathrm{pr}}$*-cobordism group of stable Morse functions on* 2*-colored manifold pairs.*

In fact, the above proposition holds for every n such that $(n, n-1)$ is in the nice range.

The purpose of this section is to show the following elementary fact.

Proposition B.2 *The* (2, 1)*-dimensional cobordism group* $\mathfrak{N}^c_{2,1}$ *of* 2*-colored manifold pairs is isomorphic to* $\mathbb{Z}_2$.

Thus, the (2, 1)-dimensional $P^c\mathcal{S}_{\mathrm{pr}}$-cobordism group is isomorphic to $\mathbb{Z}_2$.

Proof of Proposition B.2 For a (2, 1)-dimensional 2-colored manifold pair (M, N) with coloring pair (R, B), define $\varphi^c(M, N) = M$. This induces a well-defined homomorphism $\mathfrak{N}^c_{2,1} \to \mathfrak{N}_2$, which we still denote by φ^c.

This homomorphism is clearly surjective.

Let (M, N) be a (2, 1)-dimensional 2-colored manifold pair with coloring pair (R, B) such that M is closed and bounds a compact 3-dimensional manifold. Let Y be a copy of the compact 3-dimensional manifold $\overline{R}$. Then, we can attach $Y \times [-1, 1]$ to $(M, N) \times [0, 1]$ along $(N \times [-1, 1]) \times \{1\}$, where $N \times [-1, 1]$ is a small tubular neighborhood of N in M and $Y \times \{0\}$ is attached to $N \times \{1\}$. Then, the resulting compact 3-manifold has boundary diffeomorphic to the disjoint union of the double of $\overline{R}$ and M. By assumption, both of these closed surfaces bound compact 3-dimensional manifolds. Then, we obtain a (3, 2)-dimensional 2-colored manifold pair whose boundary corresponds to $(M, N) \times \{0\}$. Thus, the homomorphism φ^c is injective.

This completes the proof. □

Determining the group structure of the (2, 1)-dimensional $P^c\mathcal{AS}_{\mathrm{pr}}$-cobordism group remains an open problem.

References

1. Damon, J.: The unfolding and determinacy theorems for subgroups of $\mathcal{A}$ and $\mathcal{K}$. Mem. Amer. Math. Soc. **50**(306) (1984), x+88 pp
2. Martins, L.F., Nabarro, A.C.: Projections of hypersurfaces in $\mathbb{R}^4$ with boundary to planes. Glasg. Math. J. **56**, 149–167 (2014). https://doi.org/10.1017/S001708951300013X
3. Ohmoto, T.: Vassiliev complex for contact classes of real smooth map-germs. Rep. Fac. Sci. Kagoshima Univ. Math. Phys. Chem. **27**, 1–12 (1994)
4. Saeki, O.: Topology of singular fibers of differentiable maps. Lect. Notes in Math. vol. 1854. Springer, Berlin (2004)
5. Saeki, O.: Cobordism of Morse functions on surfaces, the universal complex of singular fibers and their application to map germs. Algebr. Geom. Topol. **6**, 539–572 (2006)
6. Saeki, O., Yamamoto, T.: Singular fibers of stable maps and signatures of 4-manifolds. Geom. Topol. **10**, 359–399 (2006)
7. Saeki, O., Yamamoto, T.: Singular fibers and characteristic classes. Topol. Appl. **155**, 112–120 (2007)

8. Saeki, O., Yamamoto, T.: Co-orientable singular fibers of stable maps of 3-manifolds with boundary into surfaces. Sūrikaisekikenkyūsho Kōkyūroku, No. 1948, 137–152 (2015)
9. Saeki, O., Yamamoto, T.: Singular fibers of stable maps of 3-manifolds with boundary into surfaces and their applications. Algebr. Geom. Topol. **16**, 1379–1402 (2016)
10. Saeki, O., Yamamoto, T.: Cobordism group of Morse functions on surfaces with boundary. In: Real and Complex Singularities. Contemporary Mathematics, Proceedings of the XIII International Workshop on Real and Complex Singularities, São Carlos, 2014, vol. 675, pp. 279–297 (2016)
11. Shibata, N.: On non-singular stable maps of 3-manifolds with boundary into the plane. Hiroshima Math. J. **39**, 415–435 (2000)
12. Vassilyev, V.A.: Lagrange and Legendre characteristic classes, Translated from the Russian, Advanced Studies in Contemporary Mathematics, vol. 3. Gordon and Breach Science Publishers, New York (1988)
13. Wall, C.T.C.: Cobordism of pairs. Comment. Math. Helv. **35**, 136–145 (1961)
14. Yamamoto, T.: Classification of singular fibres of stable maps of 4-manifolds into 3-manifolds and its applications. J. Math. Soc. Jpn. **58**, 721–742 (2006)
15. Yamamoto, T.: Euler number formulas in terms of singular fibers of stable maps, in "Real and complex singularities". World Sci. Publ., Hackensack, NJ, pp. 427–457 (2007)
16. Yamamoto, T.: Singular fibers of two-colored maps and cobordism invariants. Pac. J. Math. **234**, 379–398 (2008)

A Global View on Generic Geometry

María del Carmen Romero Fuster

Abstract We describe how the study of the singularities of height and distance squared functions on submanifolds of Euclidean space, combined with adequate topological and geometrical tools, shows to be useful to obtain global geometrical properties. We illustrate this with several results concerning closed curves and surfaces immersed in $\mathbb{R}^n$ for $n = 3, 4, 5$.

Keywords Stratifications · Height functions · Distance squared functions
Curvature locus · Vertices · Semiumbilics · Inflection points · Convexity
2-regular immersions.

MS classification 58K05 · 58C27 · 53C42 · 57R30

1 Introduction

Our aim in this paper is to show the usefulness of Singularity Theory techniques in the global study of the Geometry of submanifolds. Some basic principles underneath this fact are the following:

(i) The local geometry of a submanifold is usually specified in terms of the geometrical properties of adequate models that are invariant under the action of the transformation group associated to the considered geometry. Such models must be chosen as those that better approach the submanifold at each point. In order to determine them we can analyze the singularities of appropriate families of functions and mappings on the considered submanifold. An important property to be considered is that the parameter spaces of such families may be stratified according to the singularity types of the different functions attached to them. Such stratifications are obtained as the pull-back of convenient stratifications of

M. C. Romero Fuster (✉)
Departament de Matemàtiques, Universitat de València,
46100 Burjassot (València), Spain
e-mail: carmen.romero@uv.es

R. N. Araújo dos Santos et al. (eds.), *Singularities and Foliations. Geometry, Topology and Applications*, Springer Proceedings in Mathematics & Statistics 222, https://doi.org/10.1007/978-3-319-73639-6_9

jet spaces of smooth functions (or mappings) on the submanifold. Then convenient multi-transversality conditions imposed by the genericity requirements in each case, allows us to use the known information on the incidence relations among the strata in order to obtain a complete picture of the local behavior of relevant geometrical subsets which are characterized through appropriate singular phenomena.

(ii) The contact directions associated to the families of contact functions determine foliations whose critical points have a relevant interpretation from the geometrical viewpoint. This allows us to apply well known topological techniques, such as the Poincaré–Hopf formula, in order to ensure the existence of critical points, or even to obtain lower bounds for their number.

Our descriptions in this paper will be essentially based in the analysis of the singularities of distance squared and height functions on the submanifolds in combination with appropriate geometrical and topological techniques. We shall see here how these tools have proven to be efficient for obtaining global geometrical results on closed submanifolds in Euclidean space. Clearly, the same principles can be transported to other ambient spaces, provided we have convenient families of contact functions attached to the geometrical properties that we want to analyze. We also remark the possibility of investigating general (non necessarily generic) properties as a limit of those of a sequence of generic objects.

Finally, we warn the reader that this work, rather than an exhaustive description of the known results in the area, intends to be an illustration of how the typical methods of the Singularity Theory become an important tool in the study of the Global Geometry of Submanifolds.

2 Contacts and Singularities

Let M_i, N_i $(i = 1, 2)$ be submanifolds of $\mathbb{R}^n$ with $\dim M_1 = \dim M_2$ and $\dim N_1 = \dim N_2$. We say that the *contact of* M_1 and N_1 at y_1 is of the same type as the *contact of* M_2 and N_2 at y_2 if there is a diffeomorphism germ $\Phi : (\mathbb{R}^n, y_1) \to (\mathbb{R}^n, y_2)$ such that $\Phi(X_1) = X_2$ and $\Phi(Y_1) = Y_2$. In this case we write $K(M_1, N_1; y_1) = K(M_2, N_2; y_2)$. This is clearly a local concept and we can replace in this definition $\mathbb{R}^n$ by any manifold. Montaldi [61] gave the following characterization of the notion of contact by using the terminology of singularity theory:

Theorem 2.1 *Let M_i, N_i $(i = 1, 2)$ be submanifolds of $\mathbb{R}^n$ with* $\dim M_1 = \dim M_2$ *and* $\dim N_1 = \dim N_2$. *Let $f_i : (M_i, x_i) \to (\mathbb{R}^n, y_i)$ be immersion germs and $g_i : (\mathbb{R}^n, y_i) \to (\mathbb{R}^r, 0)$ be submersion germs with $(N_i, y_i) = (g_i^{-1}(0), y_i)$. Then $K(M_1, N_1; y_1) = K(M_2, N_2; y_2)$ if and only if $g_1 \circ f_1$ and $g_2 \circ f_2$ are $\mathcal{K}$-equivalent.*

Therefore, given two submanifolds M and N of $\mathbb{R}^n$ with a common point p, an immersion germ $f : (M, x) \to (\mathbb{R}^n, p)$ and a submersion germ $g : (\mathbb{R}^n, p) \to (\mathbb{R}^r, 0)$, such that $N = g^{-1}(0)$, the contact of $M \equiv f(M)$ and N at p is completely determined by the $\mathcal{K}$-singularity type of the germ $(g \circ f, x)$ (see [29] for details on $\mathcal{K}$-equivalence). When N is a hypersurface, we have $r = 1$, and the function germ $(g \circ f, x)$ has a degenerate singularity if and only if its Hessian, $\mathcal{H}(g \circ f)(x)$, is a degenerate quadratic form. In such case, the tangent directions lying in the kernel of this quadratic form are called *contact directions* for M and N at p.

In order to study the local behavior, we can consider that the submanifold M is given by the image on an embedding $f : \mathbb{R}^m \to \mathbb{R}^n, n > m$. The methods that we shall describe in this work are mainly based in the analysis of two relevant families of functions on M whose behavior describes the geometrical properties attached to its contacts with hyperplanes and hyperspheres.
(1) The family of *Height functions* on a manifold M, (locally) given as the image of an embedding $f : U \to \mathbb{R}^n$ is (locally) given by

$$\begin{array}{rcl} H(f) : U \times S^{n-1} & \longrightarrow & \mathbb{R} \\ (u, v) & \longmapsto & \langle f(u), v \rangle = f_v(u). \end{array}$$

The singularities of the height functions describe the contacts of M with the hyperplanes of $\mathbb{R}^n$. A height function f_v has a singularity at $x = f(u) \in M$ if and only if the direction v is normal to M at x. The singularity type of f_v at u determines the contact of M with the (tangent) hyperplane which is orthogonal to v and passes through $x = f(u)$. The analysis of these singularities provides information on the "flat geometry"of M.

Suppose that M is a hypersurface in $\mathbb{R}^n$, so we can view it locally as the image of an embedding $f : U \to \mathbb{R}^n$, where U is an open subset of $\mathbb{R}^{n-1}$. Then the Gauss map on M is given by

$$\mathcal{N}(u) = \frac{\frac{\partial f}{\partial u_1} \times \cdots \times \frac{\partial f}{\partial u_{n-1}}}{\left|\left|\frac{\partial f}{\partial u_1} \times \cdots \times \frac{\partial f}{\partial u_{n-1}}\right|\right|}(u).$$

In appropriate local coordinates, we can identify the second fundamental form II_x of M at a point $x = f(u) \in M$ with the Hessian quadratic form of the height function $f_v : U \to \mathbb{R}$ at $u \in U$, with $v = \mathcal{N}(x)$. Now, the shape operator, $S_x = -D\mathcal{N}(x)$, at x satisfies

$$II_x(X) = -\langle S_x(X), X \rangle.$$

So the matrix of $D\mathcal{N}(x)$ is given by

$$-\operatorname{Hess}(f_v) = -\left[\frac{\partial^2 f_v}{\partial u_i, \partial u_j}\right](x),$$

where $v = \mathcal{N}(x)$. We thus have that a point $x = f(u)$ is a singular point of $\mathcal{N}$ if and only if u is a degenerate singularity of f_v, where $v = \mathcal{N}(x)$. That is, the contact direction associated to f_v at $x = f(u)$ is an asymptotic direction of M at x.

For submanifolds immersed in arbitrary codimension, we shall say that a normal direction v is a *binormal* at $x = f(u)$ provided u is a degenerate singularity of the height function f_v and the tangent directions lying in the kernel of the Hessian of f_v will be called *asymptotic* directions. This clearly generalizes the case of curves $\alpha : \mathbb{R} \to \mathbb{R}^3$, for which the binormal is the unique direction leading to a degenerate height function on the curve at each point. The generic behavior of height functions in connection with the geometry of a surface immersed in $\mathbb{R}^n$, $n > 3$ has been investigated in [53, 54, 56]. The properties associated to these contacts correspond to what we call the *Flat Geometry* of the surface. The existence of binormal and asymptotic directions has been studied in [53] for the case of generic surfaces in 4-space, and in [54] for the generic submanifolds of codimension 2 in Euclidean space. We observe that in the last case the binormal and the degenerate directions coincide.

The asymptotic directions can be characterized in terms of normal sections of M as follows: Let v be a degenerate direction at a point $x = f(u)$ of M such that $corank(Hess(f_v)(u)) = 1$, and let θ be a tangent vector in the kernel of the quadratic form $Hess(f_v)(u)$. We denote by γ_θ the normal section of the surface M in the tangent direction θ. That is, γ_θ is a curve in the $(k+1)$-space $V_\theta = \langle\theta\rangle \oplus N_xM$, obtained as the intersection of this $(k+1)$-space with M. Then we have,

Proposition 2.2 ([53]) *Let $x = f(u) \in M$ and $v \in N_xM$ a degenerate direction for M at x. Let θ be a tangent direction lying in $Ker(Hess(f_v)(u))$. Then θ is an asymptotic direction corresponding to the binormal v if and only if v is the binormal direction at x for the curve γ_θ in the $(k+1)$-space V_θ.*

The binormal and asymptotic directions on generic surfaces in $\mathbb{R}^5$ were first introduced in [56], where it was shown that there exist at least one and at most five at each point of such surfaces. The number of these directions is determined by the number of real roots of certain polynomials and jumps by two when crossing the discriminant set, which consists of closed regular curves made of points at which the considered polynomials admit multiple roots. The generic behavior of the asymptotic lines near the critical points and the discriminant is described in [87].

(b) The *Distance squared functions family* over an m-submanifold M, given as the image $f(U)$ of an embedding $f : U \to \mathbb{R}^n$, with U an open subset of $\mathbb{R}^m$, is (locally) defined by

$$\begin{aligned} d(f) : U \times \mathbb{R}^n &\longrightarrow \mathbb{R} \\ (u, a) &\longmapsto d_a(u) = \|f(u) - a\|^2. \end{aligned}$$

The singularities of this family describe the contacts of M with the hyperspheres of $\mathbb{R}^n$ centered at the point $a \in \mathbb{R}^n$. We have that $x = f(u) \in M$ is a singular point of a function d_a if and only if the vector $a - x$ lies in the normal subspace N_xM of M at x, that is $a = f(u) + \lambda N(u)$, where $N(u)$ is the normal vector to M at $x = f(u)$. In the case of a hypersurface, we have that the singularity of d_a at u is degenerate if and only if $\lambda = \frac{1}{\kappa_i(x)}$, where $\kappa_i(x)$ is one of the principal curvatures of M at the point $x = f(u)$, in other words, a is a curvature center of M at x. The singularity type of d_a at u determines the contact of M with the hypersphere with center a passing through the point $x = f(u)$. In this sense, we can say that the singularities of distance squared functions on a submanifold M describe the properties belonging to the "round geometry" of M. So the bifurcation set of d coincides with the *focal set of* M,

$$\mathcal{F} = \{a \in \mathbb{R}^n : \exists u \in U \text{ such that } d_a \text{ has a degenerate singularity at } u\}.$$

This is classically known to be the image of the singular set of the exponential map $exp_M : NM \to \mathbb{R}^4$. The points of $\mathcal{F}$ are the *focal centers* of M. A pioneer work in this direction is [75].

Observe that a point $x = f(u) \in M$ is a degenerate singularity for a distance-squared function d_a if and only if the rank of the Hessian matrix $Hess(d_a)(u)$ is not maximum and those directions lying in the kernel of $Hess(d_a)(u)$ are said to be a *spherical contact direction* of M at $x = f(u)$. In fact, these are the direction along which M has a closer contact with the focal hypersphere at x with center at $a \in N_x M$ and radius $r = \|x - a\|\ x$. It is classically known that given a focal center a at a point x of M, the contact direction $X \in T_x M$ of M with the focal hypersphere $S(a, r)$, is an eigenvector of the shape operator associated to the normal direction $x - p$ at x and thus a principal direction of M at x.

The centers of the focal hyperspheres of M which have contact of type A_k, $k \geq 3$ with the surface are called *(k-order) ribs* and the corresponding contact directions on M with this hyperspheres are the *strong principal directions*. The integral lines of these direction fields are called *strong principal lines*. A point $x \in M$ which is a singular point of type A_k, with $k \geq 4$ for some distance-squared function d_a is said to be a *k-order ridge point*. On a generically immersed surface, the k-order ridge points, $k \geq 4$ form regular curves known as *ridges* at which the 5-order ridge points are isolated. Moreover, the corresponding 4-order ribs form curves in the focal set having the 5-order ribs as isolated singularities.

In the case of a generically immersed surface M in $\mathbb{R}^4$, the focal set is a stratified subset of dimension 3, whose ribs are the union of the strata of dimension ≤ 2. It was shown in [59] and [62] that there are at most 5 and at least 1 rib points on the normal space at any point of the surface. Centered at each one of these points, we have a 3-sphere whose contact with the surface is given by a singularity of type A_k, $k \geq 3$. The corresponding contact direction are called *strong principal direction* at $m \in M$. By integration of these direction fields on the surface, we get the *strong principal lines*. We thus have up to 5 strong principal curves passing through each point. These foliations may have singularities along a discriminant set that separates regions with different number of strong principal directions. Then it can be shown that *a ridge point of M is a higher order ridge if and only if it is a vertex of one of the strong principal lines considered as curves in 4-dimensional Euclidean space* [91].

The corank 2 singularities of distance squared functions on surfaces in 3-space correspond to the umbilic points [76] and are generically isolated points on such surfaces. For higher codimensions we have a more sophisticated picture. For instance, the corank 2 singularities on generically immersed surfaces in $\mathbb{R}^4$ form regular simple closed curves made of *semiumbilic* points, such points are characterized by the fact that the curvature ellipse degenerates to a segment (see [59] or [46]). Moreover, they coincide with the set of critical points of the different principal configurations associated to normal fields on the surface [92]. For surfaces generically immersed in $\mathbb{R}^4$, the semiumbilic points are singularities of type $D_k^{\pm}$, $k = 4, 5$ of the distance squared functions and form closed regular curves at which the D_5 points are isolated. We observe that since the conformal maps preserve the spheres, they must preserve the contacts of the surfaces with the hyperspheres of $\mathbb{R}^4$. This implies that the semiumbilics curves, the strong principal curves and the ridges are conformally invariant. This fact allows to characterize the ridge points and the semiumbilic points of type D_5 of surfaces in $\mathbb{R}^4$, as the zeroes of conveniently defined conformally invariant forms respectively along the strong principal curves and the semiumbilic curves on such surfaces [91]. We must point out that the proof of the assertion that *any closed orientable surface generically immersed in* $\mathbb{R}^4$ *has at least two semiumbilic points of type* D_5, stablished in [91], has a mistake and therefore remains still as a conjecture.

It follows from the work of Montaldi [61] that there is a residual subset $\mathcal{E}$ of embeddings of $\mathbb{R}^m$ into $\mathbb{R}^n$ with the Whitney C^∞-topology such that for any f belonging to it the corresponding families $H(f)$ and $d(f)$ are generic families of functions on $\mathbb{R}^m$. For a detailed description of the term "generic family of functions" see [108]. This means, in particular, that these families are topologically stable, and for $n \leq 5$, smoothly stable too. The singularities of the different functions in such a generic family may have codimension at most $n - 1$ in the case of height functions and n in that of distance squared functions. These are well known for small enough values of n. For instance, for $n \leq 6$, they are all simple singularities and correspond to the extended list of catastrophe germs determined by V. I. Arnol'd [1]. A more complete classification, including all the possible singularities up to codimension 14 can be found in [2].

It is know that the inverse $\varphi : \mathbb{R}^n \to S^n$ of the stereographic projection determines a $\mathcal{K}$ equivalence between the family of the distance squared functions on an m-manifold M immersed in $\mathbb{R}^n$ and the family of height functions over the m-manifold $\varphi(M) \subset S^n \subset \mathbb{R}^{n+1}$ (see [85] or [99]). Therefore φ takes the singularities of a given type of distance squared functions on a k-codimension submanifold M of $\mathbb{R}^n$ into the singularities of the same type for height functions on the $(k+1)$ codimension submanifold $\varphi(M)$ of $\mathbb{R}^{n+1}$.

For a detailed study of the local geometry of surfaces in $\mathbb{R}^n$ for $n = 3, 4$ and 5 in terms of the singularities of these two families we refer to Chaps. 6, 7 and 8 in [46].

3 Maxwell Stratifications, Support Hyperplanes and Singularities of Gauss Maps

3.1 The Maxwell Stratification and the Convex Hull of a Hypersurface

A *stratification* of a subset S of a manifold M is a locally finite partition $\mathcal{S}$ of S into submanifolds of M, called *strata*. The pair $(S, \mathcal{S})$ is said to be a *stratified subset* of M. We observe that each stratum is locally closed, that is, given $x \in S \subseteq M$, there is a local neighbourhood U of x in M, such that $U \cap S$ is closed in U. This concept of stratification can be extended to a more general kind of spaces, such as the space of smooth maps $C^\infty(M, N)$ between two smooth manifolds M and N. E. Looijenga introduced in [50] a natural stratification for the space $C^\infty(M)$ of smooth functions on a manifold M (see [28]). This stratification takes into account all the singular points of each smooth function on M. We can construct an alternative stratification of the space $C^\infty(M)$ by just looking to the absolute minima of the function. This was first considered by R. Thom [104] who introduced the concept of *Maxwell subset* of $C^\infty(M)$. This is given by the complement of the set of functions that attain their absolute minimum at a unique non degenerate critical point. The complement of the Maxwell subset is open and dense in $C^\infty(M)$ and can be viewed as a union of open strata, which are determined by its connected path components. We say that two functions in the Maxwell subset belong to the same stratum provided their multijets at their absolute minima are equivalent in some sufficient multijet space. We can distinguish among the following types of strata in this stratification:

(1) *Morse strata*: Functions f, whose absolute minimum is attained at a unique point of type A_1.
(2) *Conflict strata*: Functions having exactly $k > 1$ absolute minima of h_v, each one of type A_1. They form codimension $k - 1$ strata in the Maxwell subset in $C^\infty(M)$.
(3) *Bifurcation strata*: Functions having a unique absolute minimum of type A_k, where $k > 1$ is an odd number. They give rise to a codimension $k - 1$ stratum of the Maxwell subset in $C^\infty(M)$.
(4) *Mixed strata*: Functions having $r > 1$ absolute minima of respective types $A_{k_1}, \dots, A_{k_r}$, where the k_i are odd numbers, at least one of them higher than 1. They form codimension $k_1 + \dots + k_r - 1$ strata in the Maxwell subset of $C^\infty(M)$.

The stratification of $C^\infty(M)$ defined by Looijenga is a refinement of this one, at which the singularity types of not only the absolute minima but also all the critical points of the functions are considered.

Given a stratification $\mathcal{S}$ of a subset S of a manifold M and a smooth map $f : N \to M$ transversal to $\mathcal{S}$ (i.e., f is transversal to all the strata of $\mathcal{S}$), we can consider the pull-back of $\mathcal{S}$ by f, that is the stratification $f^{-1}(\mathcal{S})$, whose strata are the subsets $f^{-1}(S_i)$, $\forall S_i \in \mathcal{S}$. In particular, given a family, $F : M \times C \to \mathbb{R}$, of functions on M with parameters in a manifold C, the pull back of the Maxwell stratification of $C^\infty(M)$ through the map $\Lambda : C \to C^\infty(M)$ that sends each parameter c to the function $F_c(x) = F(x, c)$ induces a stratification on C that we call *Maxwell stratification of C associated to F*. We can ensure, under appropriate (transversality) conditions on the family F, that the map Λ respects the incidence relations between the different strata in both stratifications. In such case, we say that F is a *generic family of functions*. We can apply this setting to the family of height functions on a submanifold M in $\mathbb{R}^n$,

$$\begin{array}{rcl} H : M \times S^{n-1} & \longrightarrow & \mathbb{R}^n \\ (x, v) & \longmapsto & h_v(x). \end{array}$$

in order to obtain some global geometrical properties.

Let us recall first some classical geometrical concepts and results.

The *convex hull* $\mathcal{H}(S)$ of a subset S in $\mathbb{R}^n$ is the intersection of all the convex subsets of $\mathbb{R}^n$ containing S, that is the minimal convex subset of $\mathbb{R}^n$ that contains S. In the case of a surface M in $\mathbb{R}^3$, it is classically known that $\mathcal{H}(M)$ is homeomorphic to a closed 3-disc and its boundary $H(M)$ is a C^1-surface of $\mathbb{R}^3$ C^1-diffeomorphic to the standard 2-sphere S^2. A closed surface M of $\mathbb{R}^3$ is said to be *convex* if and only M coincides with the boundary $H(M)$ of its convex hull $\mathcal{H}(M)$. It is also a well established geometrical property that a surface M is convex if and only if if its Gaussian curvature is non-negative. We can extend the criterium of convexity to curves in $\mathbb{R}^3$ as follows:

A closed curve γ in $\mathbb{R}^3$ is said to be *convex* if and only if the image of γ is contained in the boundary $H(\gamma)$ of its convex hull $\mathcal{H}(\gamma)$. We call $H(\gamma)$ the *convex envelope* of γ. In general, the points of a surface M that lie on $H(M)$ are called *external* or *exposed* points, the other being *internal* points.

It is not difficult to check that a point $x \in M$ is external if and only if x is an absolute minimum of some height function on M. In the case of a surface in $\mathbb{R}^3$, this height function is precisely defined, up to sign, by the unique normal direction to M at x.

Given a surface M in $\mathbb{R}^3$ whose associated height functions family is generic in the above sense, we can consider the Maxwell stratification on S^2 associated to this family. Then we have the following possibilities for the singularity type at the absolute minima of the different height functions on M:

(1) *Morse strata*: The absolute minimum of h_v is attained at a unique point of type A_1. The function h_v lies in the complement of the Maxwell subset in $C^\infty(M)$. We have that $v \in S^2$ is of type A_1. All the points of type A_1 form open regions in S^{n-1} whose complement is the Maxwell subset associated to H.
(2) *Conflict strata*: There are exactly $k = 2, 3$ absolute minima of h_v, each one of type A_1. Then h_v lies in a codimension $k - 1$ stratum of the Maxwell subset in $C^\infty(M)$ and so does v in S^2.
(3) *Bifurcation strata*: The absolute minimum of h_v is attained at a unique point of type A_3. The function h_v lies in a codimension 2 stratum of the Maxwell subset in $C^\infty(M)$ and so does v in S^2.

We can view the Maxwell stratification on S^2 associated to an embedding f of M into $\mathbb{R}^3$ as the pull-back of the Maxwell stratification of $C^\infty(M)$ through the map $\Lambda : S^2 \to C^\infty(M)$ given by $\Lambda(v) = h_v, \forall v \in S^2$, where $h_v(x) = \langle f(x), v\rangle$.

These arguments can be easily extended to hypersurfaces of $\mathbb{R}^n$, $n \geq 3$ and submanifolds of higher codimension (see [81, 83]). Let's denote

$$\mathcal{A}_{k_1,\dots,k_r} = \{v \in S^2 | \text{ the absolute min of } h_v \text{ has type } A_{k_1,\dots,k_r}\}.$$

Then for a generic immersion of a hypersurface M into $\mathbb{R}^n$, the subset $\mathcal{A}_{k_1,\dots,k_r}$ is a submanifold of codimension $r - 1 + \sum_{i=1}^r 2k_i$, made of a union of strata of the Maxwell stratification determined by the height functions family associated to this immersion. It is not difficult to see that the union of all the strata of codimension one or more coincides with the closure of the image of the Gauss map on the C^1-surface $H(M) \setminus M$. The behaviour of the Maxwell stratification in connection with the Gauss map and the convex hull structure of a hypersurface was studied with detail in [81, 83], where it was called *Core stratification*. In fact, it is shown that the strata of the Maxwell stratification of a generic hypersurface satisfies the following relation:

$$\chi(S^{n-1} - \mathcal{M}) + \sum_{j=0}^{n}(-1)^j(\chi(B_j) + \chi(M_j) + \chi(C_j)) = \begin{cases} 0 \text{ if } n \text{ even} \\ 2 \text{ if } n \text{ odd,} \end{cases}$$

where $\mathcal{M}$ is the union of strata of codimension at least one and B_j, M_j and C_j respectively represent the union of bifurcation, mixed and conflict strata of codimension j.

For a surface M generically embedded in $\mathbb{R}^3$, the Maxwell strata of the family of height functions on M determine a graph $\mathcal{M}$ on S^2 that we call the *Maxwell graph*. The extremal points of the Maxwell graph (bifurcation strata of type A_2) correspond to *external cusps* of the Gauss map (i.e., cusps of the Gauss map lying on the boundary of the convex hull of the surface). The other vertices of the graph (conflict strata of type $A_1 + A_1 + A_1$) are trivalent and correspond to the (isolated) tritangent support planes of the surface. On the other hand, the edges of the graph (conflict strata of type $A_1 + A_1$) are determined by the normal directions to the (1-parameter) family of support bitangent planes of the surface. By applying the equality above to this particular case, we obtain:

Corollary 3.1 ([83]) *Given a generic surface M in 3-space, the numbers C of external cusps of the Gauss map and T of tritangent support planes on M satisfy*

$$C - T = 4 - 2\chi(M \cap H(M)).$$

An interesting consequence is that *the existence of support tritangent planes implies the existence of (external) cusps of the Gauss map*. Moreover, if the Gauss map on a generic surface M has no external cusps, it follows that the internal part (= complement of the external part) of M must be non simply connected. Some other results in this direction for 3-manifolds immersed in $\mathbb{R}^4$ can be found in [81].

3.2 *Canal Surfaces of Curves in* $\mathbb{R}^3$

The canal hypersurface $C(\gamma)$ of a closed curve γ embedded in a $\mathbb{R}^3$ is given by

$$C(\gamma, \varepsilon) = \{\gamma(t) + \varepsilon\nu \,:\, t \in S^1,\ \nu \in (N_{\gamma(t)}M)_1\},$$

where $(N_{\gamma(t)}M)_1$ denotes the unit sphere in the normal subspace $N_{\gamma(t)}M$ of M at the point $\gamma(t)$. Clearly, for ε small enough, $C(\gamma, \varepsilon)$ is a torus embedded in $\mathbb{R}^3$. We can fix such a value of ε and just denote $C(\gamma)$. It is easy to see that we have the following relation between the singularities of the height functions on γ and on $C(\gamma)$ (respectively denoted h_v and $\bar{h}_v$):

(i) A point $x = \gamma(t)$ is a singular point of the height function h_v if and only if $(x, v) \in C(\gamma)$ is a singular point of $\bar{h}_v$. Furthermore, the $\mathcal{R}$-singularity type of h_v at x is the same as that of $\bar{h}_v$ at (x, v).
(ii) A point (x, v) is a degenerate singularity of the height function $\bar{h}_v$ on the surface $C(\gamma)$ if and only if (x, v) is a singular point of the Gauss map G on $C(\gamma)$.

It follows that the Maxwell graph of the (family of height functions on the) curve γ coincides with the Maxwell graph of its canal surface and a simple calculation shows the following:

(a) The cusps of the Gauss map on the canal surface correspond to the torsion zero points of the curve. More precisely, if $\gamma(t)$ is a torsion zero point of the curve γ, then the points $(\gamma(t), b(t))$, where $b(t)$ is the binormal vector at $\gamma(t)$, are cusps of the Gauss map of the canal surface of γ.
(b) Tritangent planes of $\gamma(t)$ correspond to pairs of tritangent planes on its canal surface. Therefore, the extremal vertices of the Maxwell graph defined by a curve in 3-space correspond to torsion zero points whose osculating plane is a support plane of the curve.
(c) The trivalent vertices correspond to support tritangent planes.
(d) The edges are given by the normal directions to the support bitangent planes family.

The local structure of the convex hull of a curve γ in $\mathbb{R}^3$ was described in [96]. The convex envelope of a closed curve substantially embedded in $\mathbb{R}^3$ is homeomorphic to a 2-sphere. For a conveniently defined (open and dense) subset of *generic curves* in $\mathbb{R}^3$, V. D. Sedykh proved that the open subset $H(\gamma) - \gamma$ is a C^1-surface, whereas the points in $\gamma \cap H(\gamma)$ are the C^0-singularities of $H(\gamma)$.

From a global viewpoint, the topological analysis of the Maxwell graphs in the case of generic curves leads to the following properties that were shown in [84]:

(i) *A generic closed curve is convex (i.e. lies on the boundary of its convex hull) if and only if its Maxwell graph has exactly two connected components. Otherwise, it has just one.*
(ii) *If a closed curve is convex, then each connected component of its Maxwell graph is contractible.*

All these considerations lead to the following:

Theorem 3.2 ([84]) *Given a closed curve γ generically immersed in $\mathbb{R}^3$, denote respectively by C, T and ρ the numbers of external torsion zero points, support tritangent planes and connected components of $\gamma - \gamma \cap H(\gamma)$. Then the following relation holds:*

$$C - T = 4 - 2\rho.$$

As an immediate consequence of this formula we get the following

4-**vertex theorem for closed curves in** 3-**space**: *A convex closed curve generically immersed in $\mathbb{R}^3$ has at least* 4 *torsion zero points.*

If we take into account that stereographic projection takes torsion zero points of curves in the 2-sphere to vertices of plane curves, we can view this result as a generalization of the well known 4-vertex theorem for closed curves in the plane. In fact, a similar argument, based in the fact that the vertices of a plane curve are the end points of its cut-locus, can be used in order to give a proof of the 4-vertex theorem for closed curves generically immersed in the plane (this proof is due to A. Weinstein, according to [104]). On the other hand, an extension to (non necessarily generic) convex closed curves in 3-space with no zero curvature points was obtained by V.D. Sedkh in [98]. A more general approach to the study of four vertex theorems (including vertices of curves in Minkowski planes) can be found [105, 107]. Some generalizations Theorem 3.2 to curves with possible isolated zero curvature points and/or singular points, respectively obtained in [12, 93] are the following:

(a) *Let $\alpha : S^1 \to \mathbb{R}^3$ be a C^3 simple closed convex curve, with possible isolated singular points and vanishing curvature at isolated points. Let S, K and V respectively denote the number of singular points, zero curvature points and vertices (zero torsion points) of α. Then the following relation holds,*

$$3S + 2K + V \geq 4.$$

(b) *Any regular simple closed C^4-curve $\alpha : S^1 \to \mathbb{R}^3$ with nowhere vanishing curvature, no bitangent lines, having finite total order contact with any plane and finitely many vanishing torsion points, satisfies*

$$C + 2\rho \geq 4 + P,$$

where C is the total number of points of the curve at which the osculating plane is of support, P is the total number of support planes whose contact points with the *curve are not all in the same line and ρ is the number of connected components of the set of points of the curve lying inside its convex hull.*

As a consequence of this we can assert:

Provided the torsion of such a curve never vanishes, the curve enters at least twice in the interior of its convex hull.

Given a closed spacial curve γ, we call the maximal arcs of γ lying on the boundary of its convex hull the *external segments* of γ. Then we say that a vertex of a closed spacial curve γ is *external* provided it entirely lies on the boundary of its convex hull. The following inequality was proven in [94]

Any C^3-embedded closed space curve γ with a finite number of vertices, having V external vertices, K zero-curvature points and d external segments satisfies the following relation

$$V + 2K + 2d \geq 4 + T.$$

Other relevant results of global type for space curves, which are obtained by similar methods, are the following:

1. *The number of tritangent planes of a curve in general position with no torsion zero points is even* [21].
2. *The number of bitangent osculating planes of a generic curve is even* [69, 97].
3. *The number of torsion zero points of generic closed curves without cross tangents and bitangent osculating planes is a multiple of* 4 [70].

Further results concerning the relation between the (indexed) number of torsion zero points and the number of tritangent planes were also obtained by Banchoff, Gaffney and McCrory [3] and Ozawa [74]. More recent results in this direction, due to M. Ghomi [26], provide inequalities involving the numbers of pairs of points with parallel tangent lines, inflections and vertices of closed space curves in 3-space. Also, an extension of the classical four vertex theorem to closed curves that bound simply connected compact surfaces of constant curvature can be found in [25].

4 The Geometry of Surfaces in $\mathbb{R}^n$ from a Viewpoint of Their Contacts with Models

Let M be a surface locally determined by a smooth immersion $f : \mathbb{R}^2 \to \mathbb{R}^n$, $n \geq 3$. The *curvature ellipse* of M at a point $x \in f(M)$ is constructed as follows [49]: Consider the unit circle in T_xM parametrized by the angle $\theta \in [0, 2\pi]$ and let γ_θ be the normal section of M in the direction θ and η_θ the curvature vector of γ_θ. The vector η_θ clearly lies in the normal space N_xM. As θ varies from 0 to 2π, the extremum of η_θ describes an ellipse in N_xM which is known as the *curvature ellipse* of M at x. A point $x \in M$ is said to be *hyperbolic*, *elliptic* or *parabolic* according to x lies outside, inside, or on the curvature ellipse of M at x. The curvature ellipse may degenerate to a segment a some points, called *semiumbilics*. An *inflection point* is a point x that is both, parabolic and semiumbilic, this means that the curvature ellipse degenerates to a segment at the point x, which lies on the affine span of this segment in $N_x M$. Inflection points may be *imaginary*, *real* or *flat* according to they lie off, inside or at one of the end points of the curvature segment. A direction $\theta \in T_x M$ is called *asymptotic* provided the vectors η_θ and $\frac{\partial \eta_\theta}{\partial \theta}$ are parallel. It is easy to see that there are exactly 2, 0 or 1 asymptotic directions respectively at hyperbolic, elliptic and parabolic points of M. In the case of inflection points, all the tangent directions are asymptotic. The vector joining the point x with the center of the curvature ellipse of M at x is the *mean curvature vector* of M at x. A point is said to be *minimal* if the mean curvature vector vanishes at it and a surface is called *minimal* if it is completely made of minimal points. It can be seen ([49], p. 28) that for most immersions the minimal points are isolated. Clearly, minimal surfaces are non generic in the sense that most small enough perturbations of them are non minimal. As an immediate consequence of the above definitions we can assert:

(1) Minimal surfaces in $\mathbb{R}^n$ do not have hyperbolic, nor parabolic points.
(2) Inflection points of minimal surfaces are all of real type.

An interesting question yet to be studied is the behavior of the set of inflection points at minimal surfaces (e.g., existence or local and global structure).

We shall see now how to interpret and analyze the behavior of the above geometrical concepts in terms of the singularities of height and distance squared functions on a surface M.

Consider the family of height functions on M in $\mathbb{R}^4$, where we suppose that M is (locally) given as the image of an embedding $f : \mathbb{R}^2 \to \mathbb{R}^n$, $n > 3$,

$$\begin{array}{rcl} H : M \times S^{n-1} & \longrightarrow & \mathbb{R}^n \\ (p, v) & \longmapsto & h_v(p) = \langle f(x), v \rangle . \end{array}$$

It can be seen [53] that

(1) A point is elliptic if and only if it is a nondegenerate singularity of any height function associated to a normal direction to M at x.
(2) A point is hyperbolic if and only if it is a degenerate critical point (generically of type A_3 along curves and A_4 at isolated points of M) of exactly two height functions on M. The corresponding normal vectors are said to be *binormals* of M at x.
(3) A point x is parabolic if and only if there is a unique height function having a degenerate critical point at x (generically of type A_2 and possibly of type A_3 or D_4 at isolated points).
(4) A parabolic point is an inflection point if and only if it is a singularity of type corank 2 (generically $D_4^{\pm}$) for the height function in the unique binormal direction of M at x. Flat inflections do not occur on surfaces generically immersed in $\mathbb{R}^4$.
(5) The set Δ of parabolic points of M is a (non necessarily connected) smooth curve with normal crossings at inflection points of real type, together with possible

isolated points corresponding to inflection points of imaginary type. If the surface is closed, then this curve is closed too.

(6) The curves of parabolic points separate M into hyperbolic and elliptic regions (respectively denoted M_h and M_e). The imaginary inflection points are isolated points lying in the interior of the hyperbolic region.
(7) The hyperbolic region of any closed surface in $\mathbb{R}^4$ is nonempty.

An interesting result, due to Dreibelbis [15] is the following: Consider the surface $B_M \subset S^3$, determined by all the binormal directions of a surface M immersed in $\mathbb{R}^4$, then provided the normal curvature of the M vanishes, the asymptotic curves on M lift to geodesics on B_M

We can also consider the family of distance squared functions on M,

$$\begin{array}{rcl} d : M \times S^{n-1} & \longrightarrow & \mathbb{R}^n \\ (x, a) & \longmapsto & ||f(x) - a||^2. \end{array}$$

Those points of $a \in \mathbb{R}^n$ for which the function d_a is non stable (i.e., non Morse) are known as the *focal centers* of M and form the *focal set* of M in R^n. They are the centers of hyperspheres of $\mathbb{R}^n$ with higher order of contact with M, i.e., the focal hyperspheres. The focal centers for which the function d_a has corank 2 are called *umbilical foci*. The generic contacts of surfaces with hyperspheres in $\mathbb{R}^4$ were first analyzed by Montaldi [59], who proved:

(i) The semiumbilic points of a surface M immersed in $\mathbb{R}^4$ are the singularities of corank 2 of the distance squared functions on M, i.e., the *umbilical foci*.
(ii) On a generically immersed surface, the semiumbilic points form immersed curves with no self intersections.

The semiumbilic points of surfaces immersed in higher codimension were studied in [63, 64], where it was shown that *the corank 2 singularities of distance squared functions on a surface immersed in $\mathbb{R}^n$, $n \geq 3$ coincide with the ν-umbilic points, for the different normal fields ν on the surface.*

The bitangencies between pairs of surfaces immersed in 4-space have been analyzed by Dreibelbis in [14], who proves that for a generic immersion $f : M \to \mathbb{R}^4$, the pair given by f and any small enough translation f_v of f along a generic direction $v \in S^3$, satisfies the following relations:

(i) $B + D = \nu(f)\nu(g)$,
(ii) $B + D + E_v + \frac{1}{2}P_v = \frac{1}{2}\nu(f)^2$,

where D and B are respectively the numbers of double points and bitangencies counted with sign; $\nu(f)$, $\nu(g)$ denote the normal Euler numbers of f and g, respectively; E_v is the number of elliptic points of f where v is a tangent vector and P_v is the number of parabolic points with a flecnodal normal perpendicular to v counted with sign. We can view the last result as a generalization of Fabricius-Bjerre's formula for a closed generic plane curve. Further extensions can be found in [16, 17].

In the case of a 3-manifold M immersed in $\mathbb{R}^n, n \geq 5$ the curvature ellipse, becomes a more sophisticated object, which is called *curvature Veronese*. This is obtained as a convenient projection (that depends on the coefficients of the second fundamental form) of the classical Veronese surface of order 2 in the normal space of M at each point. A complete description of its possible shapes for surfaces immersed into $\mathbb{R}^5$ and $\mathbb{R}^6$ is given in [6]. Some connections between the behaviour of the singularities of height functions and the shape of the curvature Veronese at each point can also be found in [6]. On the other hand, [18] provides a natural extension of the notions of asymptotic directions, parabolic curves and inflection points n-manifolds immersed in $\mathbb{R}^{2n}$, investigating some of their properties. The case of 3-manifolds immersed in $\mathbb{R}^6$ is analyzed with detail, describing the possible generic algebraic structures of the asymptotic vectors at parabolic and inflection points, as well as the generic topological structures of the parabolic surface.

4.1 Convexity and Existence of Inflection Points on Surfaces in $\mathbb{R}^4$

The asymptotic directions on the hyperbolic region of a surface immersed in $\mathbb{R}^4$ determine a pair of foliations, which are determined by convenient binary differential equations (see Sect. 7.3 in [46] for their definition and analysis)). The critical points of these foliations are the inflection points of the surface and their local behaviour at such points has been studied in [22] in the case of inflection points of imaginary type and in [9] at the inflection points of real type. We point out that the inflection points of imaginary type are the singular points (Darbouxian umbilics) of the fields of asymptotic directions and they have indexes $\frac{1}{2}$ (cases D_1 and D_2) or $-\frac{1}{2}$ (case D_3).

Since the inverse stereographic projection from $\mathbb{R}^3$ to S^3 maps curvature lines of surfaces in 3-space into asymptotic lines of their spherical images, considered as surfaces in 4-space, we can conclude easily that a closed surface immersed in $\mathbb{R}^4$ that lies a 3-sphere and is generic from the viewpoint of its contacts with hyperplanes of $\mathbb{R}^4$ has exactly two orthogonal asymptotic directions at each point, except at most at a finite number of inflection points of imaginary type. We shall see now that this property can be extended to a more general class of surfaces in $\mathbb{R}^4$. In fact, a surface M is said to be *locally convex* if it has a locally support hyperplane at each point. Notice that the orthogonal direction to a support hyperplane at some point $x \in M$ determines a height function that has either a (local) minimum or a (local) maximum at x. Then we have,

Proposition 4.1 ([53]) *A generic surface M in 4-space is locally convex if and only if M is made of hyperbolic and (isolated) inflection points.*

An immediate consequence is that *generic locally convex surfaces have two globally defined asymptotic directions fields whose critical points are isolated imaginary inflections.*

Remark 4.2 From a local viewpoint, we must point out the relation between the convexity and the existence of binormal/asymptotic directions at a given point. On the other hand, it was shown in [71] that a necessary and sufficient condition for the vanishing of the normal curvature at a point x of an m-submanifold of codimension 2 in Euclidean space is the existence of exactly m mutually orthogonal asymptotic directions at x.

Let M be a smooth compact surface (possibly with boundary) and suppose that V is a smooth line field on S with finitely many critical points, all contained in the interior of S, and finitely many (s) inner and (n) outer contact points with the boundary of S. Then the sum of the indices at the critical points of V is given by the following relation, known as the generalized Poincaré–Hopf formula [79],

$$Ind(V) = \chi(S) + \frac{s-n}{2}.$$

Observe that if $\partial S = \emptyset$ then $s = n = 0$, and this becomes the classical Poincaré–Hopf formula. Now, as a result of applying the Poincaré–Hopf formula to the asymptotic direction fields on generic locally convex surfaces, we obtain:

Theorem 4.3 ([22]) *A generic locally convex immersion of a closed surface M in $\mathbb{R}^4$ satisfies the following relation*

$$2|\chi(M)| \leq \sharp\{\textit{inflection points}\}.$$

An immediate consequence is the following:

Corollary 4.4 *Any generic locally convex immersion of a compact surface with nonvanishing Euler number in $\mathbb{R}^4$ has at least* 4 *inflection points.*

A particular case of convex surface in $\mathbb{R}^4$ is given by the image of a surface in 3-space trough the inverse of the stereographic projection ψ. The singular points (umbilics) of the principal curvature direction fields of a surface M in 3-space are taken by ψ to the singular points (inflections) of the asymptotic direction fields on $\psi(M)$ considered as a surface in 4-space. Consequently, the Corollary 4.4 generalizes the following result obtained by Feldman [20] for generic closed surfaces in 3-space:

Corollary 4.5 *Any* 2*-sphere generically immersed into $\mathbb{R}^3$ has at least* 4 *umbilic points.*

This result represents a generic version of the following more general statement,

Carathéodory conjecture: *Any closed convex and sufficiently smooth surface immersed in $\mathbb{R}^3$ has at least two umbilic points.*

A way to approach this question, which is based in the application of Poincaré–Hopf's formula and has been followed by several authors, consists in investigating the possible values for the indices of the umbilic points of surfaces immersed into $\mathbb{R}^3$.

It is known how to construct examples of local immersions of surfaces with umbilics of any index ≤ 1. Therefore, the following assertion can be considered as a local (stronger) form of the above conjecture.

Loewner conjecture: *The index of the principal directions field at an umbilic point of a surface immersed in* $\mathbb{R}^3$ *is at most* 1.

The proof of this last conjecture has a long history. According to D.J. Struik the first references go back to 1922, appearing in the works of Cohn-Vossen, Blaschke and Hamburger, who attributed the first conjecture to C. Carathéodory. The first attempt to prove it, which was due to Hamburger [41–43] considered the analytic case. Since then, several authors have searched a shorter and clearer proof [7, 47, 95, 106]. The proof, which is certainly very hard, seems to have some gaps in all these works. A good review on the state of the problem during the 20th century is provided in [40]. A more recent attempt that uses complex analysis techniques (analytic implicit functions, Weierstrass preparation theorem, Puiseux series, and circular root systems) is due to V.V. Ivanov [45]. Although there is so far some controversy on correctednes of the different proofs, the analytic version of the conjecture seems to be accepted as true by most authors. Some aspects of the smooth case were considered in [36, 101]. More recent attempts to prove the conjecture in the smooth case can be found in [31, 67] Also, an interesting related result reformulating the global conjecture for C^2-surfaces in terms of the existence of at least one umbilic point in the graphs of asymptotically constant functions on the plane has been obtained by M. Ghomi and R. Howard [27].

Now, in view of Corollary 4.4, it seems natural to propose the following:

Generalized Carathéodory conjecture: *Every closed convex immersion of the* 2*-sphere in* $\mathbb{R}^4$ *has at least two inflection points.*

Some facts giving support to this conjecture are the following:

(1) C. Gutiérrez and M. Ruas [37] proved that under very mild conditions, stated in terms of Newton polyhedra of the coordinate functions of the embedding, the index of an isolated inflection point of a locally strictly convex surface embedded in $\mathbb{R}^4$ is the same as the index of an umbilic point of a surface immersed in $\mathbb{R}^3$.

(2) J. Nuño Ballesteros [68] considered a particular case of locally convex surfaces immersed in $\mathbb{R}^4$ given by those that admit a local non-degenerate parallel normal field at each point. He proved that this condition is equivalent to asking that all the points are semiumbilic and that the Gaussian curvature of the surface does not vanish. In such case, it is possible to define a Gauss map $G_\nu : M \to S^3$, whose image defines an immersed submanifold M^ν in S^3. This submanifold has the following property: If M has a contact of a given type with a hyperplane, then M^ν has the same contact type with the translated hyperplane. But this means that M^ν has this same contact with the 2-sphere determined by the intersection of S^3 with the last hyperplane. By taking stereographic projection, this is transformed into the contact of a surface with some hypersphere in $\mathbb{R}^3$. In other words, this construction transforms diffeomorphically the asymptotic configuration on a surface M immersed in $\mathbb{R}^4$, into

the principal configuration of its image M^{ν} through the stereographic projection in $\mathbb{R}^3$. As consequence of this setting we can assert the following:

Loewner's and Carathéodory's conjectures on umbilic points of principal direction fields on surfaces in $\mathbb{R}^3$ hold if and only if they hold for inflection points of asymptotic configurations on surfaces with non vanishing Gaussian curvature in $\mathbb{R}^4$ totally made of semiumbilic points.

Here we observe that the asymptotic lines this surfaces coincide with their unique principal configuration. Moreover, in the particular case of surfaces contained in S^3, the inflection points become umbilics.

Remark 4.6 We point out that, as shown in [38], given any $n \in \mathbb{Z}$, there is an analytic immersion $f : \mathbb{R}^2 \to \mathbb{R}^4$ having a normal field ν and a ν-umbilic point x with index $\frac{n}{2}$. This implies that the Loewner conjecture does not hold for principal configurations associated to arbitrary normal fields on submanifolds in 4-space. So the above conjecture just concerns the asymptotic configurations, i.e., those associated to binormal fields on surfaces in 4-space.

We turn back now to the generic situation and describe how to extend the above results to non convex surfaces in $\mathbb{R}^4$. Observe that if we denote by $\mathcal{H}(M)$ the boundary of the convex hull of a surface M immersed in $\mathbb{R}^4$, we have that $M \cap \mathcal{H}(M) \neq \emptyset$. Then M is locally convex at any point $x \in M \cap \mathcal{H}(M)$ and hence x is a hyperbolic point of M, so we can assert:

The hyperbolic region of any closed surface M immersed in $\mathbb{R}^4$ is non empty.

Now, since the elliptic region of a non locally convex surface M is non empty, provided M has no real inflection points, the parabolic points form a smooth (non necessarily connected) closed curve separating the hyperbolic region M_h from the elliptic region M_e. The two asymptotic fields defined on the hyperbolic regions may become tangent to the parabolic curve at (generically) isolated points which coincide with those at which the curve of flat ridges meets the parabolic curve, so we call them *parabolic flat ridges*. On the other hand, the singularities of the asymptotic direction fields occur at isolated inflection points of imaginary type contained in the interior of the hyperbolic region. Therefore, as a consequence of the generalized Poincaré–Hopf formula [79], we obtain the following:

Theorem 4.7 ([22] *If M is a generically immersed closed surface (non necessarily orientable) in $\mathbb{R}^4$ with no inflection points of real type, then*

$$2|\chi(cl(M_h))| \leq \sharp\{inflectionpoints\} + \sharp\{parabolicflatridges\}.$$

Corollary 4.8 *If $\chi(cl(M_h)) \neq 0$, then M has either inflection points or parabolic flat ridges.*

We can also consider immersions in higher codimension. It can be shown that *a surface M generically immersed in $\mathbb{R}^n, n > 5$ does not have semiumbilic, nor inflexion, nor umbilic, nor minimal points* [30]. In such case, the curvature ellipse does not degenerate and defines a plane on the normal space at each point. Some specially relevant points are the *pseudo-umbilics*, i.e., the points at which the mean curvature vector $H(x)$, determined by the center of the curvature ellipse at x, is orthogonal to the plane determined by the curvature ellipse. It was also shown in [30] that such points coincide with the critical points of the relative mean curvature lines and are generically isolated. We consider now a natural generalization of the concept of *mean curvature lines* to the general case of surfaces immersed with codimension higher that two in Euclidean space: For a surface immersed into $\mathbb{R}^4$, the normal line in the direction $H(x)$ cuts the curvature ellipse in two opposite points (except at the special situations in which the ellipse degenerates into a radial segment, or if $H(x) = 0$). These two points determine a couple of orthogonal tangent directions known as the *mean curvature directions* at x. Such directions are characterized by the fact that the curvature vector of the normal section of the surface along them is parallel to the mean curvature vector $H(x)$. The generalization of this procedure to surfaces immersed in R^n with $n > 4$ embodies some problems due to the fact that the plane determined by the curvature ellipse does not pass through the origin of the normal space at a generic point x. This means that there are no tangent directions whose normal section's curvature vector is parallel to $H(x)$. To overcome this difficulty we observe that, from a qualitative viewpoint, all the principal configurations on the surface arise from normal vector fields parallel to the subspace determined by the curvature ellipse at every point (see [65]). In fact, any normal vector $v \in N_x M$ can be decomposed into a sum $v_1 + v_2$, where v_1 and v_2 are vectors respectively parallel and orthogonal to the plane determined by the curvature ellipse. Now, the shape operator associated to v_2 is a multiple of the identity and thus the eigenvectors of the shape operator associated to v coincide with those of the shape operator associated to v_1. This suggests the convenience of defining the *relative mean curvature directions* at a point x of a surface immersed in $\mathbb{R}^n$, $n \geq 4$ as those inducing normal sections whose curvature vector is parallel to the component of H lying on the ellipse plane at x. This procedure leads to two orthogonal foliations globally defined on the surface whose critical points are the semiumbilic and the pseudo-umbilic (with inflection points and minimal points considered as non-generic particular cases). The analysis of the behavior of these foliations together with Poincaré–Hopf's formula leads to the following,

Theorem 4.9 *(1) A generic immersion of a 2-sphere into $\mathbb{R}^5$ has either (isolated) semiumbilic points, or at least four points at which the mean curvature vector H is orthogonal to the normal subspace determined by the curvature ellipse.*

(2) A generic immersion of the 2-sphere into $\mathbb{R}^n$, $n \geq 5$, has at least four points at which the mean curvature vector H is orthogonal to the normal subspace determined by the curvature ellipse.

4.2 Existence of Semiumbilics, Semiumbilicity and Sphericity

Given a unit normal field ν on M, the ν-principal curvatures at a point $x \in M$ are the extremal values of the projection of the normal curvature vector $k(\theta)$ in the direction ν, for $\theta \in [0, 2\pi]$ and the principal directions are the corresponding angles $\theta_i, i = 1, 2$. So θ is a ν-principal direction if and only if

$$0 = \frac{d\langle \eta(\theta), \nu \rangle}{d\theta} = \langle \frac{d\eta(\theta)}{d\theta}, \nu \rangle.$$

These values of θ correspond to the points $\eta(\theta)$ at which ν is normal to the ellipse. In other words, the ν-principal curvatures at x are the maximum and minimum values of the height function

$$\begin{aligned} h_\nu : N_x M &\longrightarrow \mathbb{R} \\ w &\longmapsto \langle v, \nu \rangle. \end{aligned}$$

over the curvature ellipse in $N_x M$.

When x is a semiumbilic point, the extremal values of h_ν are reached at the end points of the curvature segment, for any normal direction ν. So the principal directions of any normal field ν on M coincide with the asymptotic directions at x and we thus have that *the asymptotic directions are orthogonal at the semiumbilic points of a surface immersed in* $\mathbb{R}^4$.

The semiumbilic points can also be characterized in terms of the behaviour of the normal fields on the surface. An umbilic point of a normal field ν on M is a point $x \in M$ at which the two ν-principal curvatures coincide. In such case, the shape operator $S_{\nu(x)}$ is a scalar multiple of the identity, the scalar being the (unique) principal curvature at x. We say that x is a ν-umbilic point. Suppose that x is a semiumbilic point of M, so the curvature ellipse is a segment and its span E_x is a 1-dimensional subspace of $N_x M$. Then it is not difficult to see that any normal field ν such that $\nu(x)$ lies in the orthogonal complement $E_x^\perp$ of E_x has x as an umbilic point, And conversely, if $\langle \eta(\theta), \nu \rangle$ is a constant for all θ at x, it follows that the curvature ellipse lies in $\nu^\perp$ and hence it is a segment, so x is a semiumbilic point and thus we have that

A point x of a surface M immersed in $\mathbb{R}^4$ is semiumbilic if and only if x is an ν-umbilic point of some normal field locally defined at x on M.

We quote from [49] some properties of the curvature ellipse of a surface immersed in 4-space that are be relevant in order to get some global results on the geometry of surfaces in 4-space:

The area of the curvature ellipse, which is a scalar invariant, at x coincides with $\frac{1}{2}|N|$, where N is the curvature of the normal bundle of M at x (see [49]) Then, provided the ellipse is not a circle, we can choose the tangent frame $\{e_1, e_2\}$ on the normal plane, so that the vector $\eta(\theta) - H$ coincides with the semimajor axis vector B for $\theta = 0$ and with the semiminor axis C for $\theta = \frac{\pi}{4}$. In the case of a circle we can work with any frame and we get the equality

$$\langle H, H\rangle - K = \langle B, B\rangle + \langle C, C\rangle,$$

where K is the Gaussian curvature and H is the mean curvature of M. Now, since the area of an ellipse with semiaxis B and C is given by $\pi|B||C|$, we get

$$|N| = 2|B||C|,$$

$$\langle H, H\rangle - K = \langle B, B\rangle + \langle C, C\rangle.$$

So the shape of the curvature ellipse is completely determined by the scalar invariants $|N|$ and $\langle H, H\rangle - K$. An immediate consequence is that *a point $x \in M$ is semiumbilic if and only if* $N(x) = 0$.

On the other hand, if θ_1 and θ_2 are the two asymptotic directions at a hyperbolic point x, we have the following formulae [109],

$$tan^2(\theta_1 - \theta_2) = \frac{\Delta}{N^2}, \tag{1}$$

$$tan^2\,\Omega = \frac{\Delta}{K^2}, \tag{2}$$

where Ω represents the angle at the origin subtended by the ellipse, that is, the angle determined by the normal vectors $\eta(\theta_1)$ and $\eta(\theta_2)$ in N_xM, which coincides in turn with the angle determined by the two binormals at x and Δ is a scalar invariant of the surface, defined in terms of the coefficients of the second fundamental form at each point, as follows:

$$\Delta = \frac{1}{4}\begin{vmatrix} a_1 & b_1 & c_1 & 0 \\ a_2 & b_2 & c_2 & 0 \\ 0 & a_1 & b_1 & c_1 \\ 0 & a_2 & b_2 & c_2 \end{vmatrix} \tag{3}$$
$$= \frac{1}{4}\left(4(a_1b_2 - a_2b_1)(b_1c_2 - b_2c_1) - (a_1c_2 - a_2c_1)^2\right).$$

An immediate consequence of the first formula is:

A point x is semiumbilic if and only if the two asymptotic directions are orthogonal at x.

We can summarize the above considerations on semiumbilic points in the following:

Theorem 4.10 *Given a surface M immersed in $\mathbb{R}^4$ and a point $x \in M$, the following assertions are equivalent:*

(1) x is semiumbilic.
(2) $N(x) = 0$.

(3) x *is* ν*-umbilic, for some normal field* ν *on* M.
(4) *There are two orthogonal asymptotic directions at* x.

It follows, in particular, that the critical points of the binormal fields are the inflection points of M and the umbilic points are critical points for all the principal configurations on M. Now, we can use the following:

Theorem 4.11 ([4]) *A surface embedded in* $\mathbb{R}^4$ *is orientable if and only if it admits some globally defined normal field.*

And as a consequence of the Poincaré–Hopf formula relating the Euler number and the index of a tangent direction field on a closed surface, we obtain the following result on the existence of semiumbilics:

Corollary 4.12 *Any embedding of an orientable closed surface with nonvanishing Euler number in* $\mathbb{R}^4$ *has semiumbilic points.*

Finally, since J. Little proved in [49] (page 291) that any embedding of the torus in $\mathbb{R}^4$ has semiumbilic points, we can state:

Corollary 4.13 *Any embedding of an orientable closed surface in* $\mathbb{R}^4$ *has semiumbilic points.*

Taking into account that the semiumbilicity condition is equivalent to the vanishing of the curvature of the normal bundle at the considered point, the above corollary amounts to say:

There are no closed orientable surfaces with never vanishing normal curvature immersed in $\mathbb{R}^4$.

A surface all whose points are semiumbilic is said to be a *totally semiumbilical* surface. A particular case of totally semiumbilic surfaces in $\mathbb{R}^4$ is made of those lying in a 3-sphere. It is thus quite natural to search for sufficient conditions on a totally semiumbilical surface to be hyperspherical.

Totally semiumbilic surface M have two binormal globally defined fields, b_1 and b_2, whose critical points are the inflection points (including the umbilic points as a particular case of inflection points) of M. The binormal fields are characterized by the fact that one of their two principal curvatures vanishes identically on M. We call the other principal curvatures the*binormal curvatures* on M. We denote by κ_1 and κ_2 the two binormal curvatures (respectively associated to b_1 and b_2) on a totally semiumbilical surface M. The following result provides necessary and sufficient conditions on a totally semiumbilical surface to lie on a 3-sphere.

Corollary 4.14 ([92]) *Suppose that* M *is a surface with isolated inflection points in* $\mathbb{R}^4$. *Then* M *is hyperspherical if and only if its asymptotic lines are globally defined and orthogonal and its binormal curvatures* $\{\kappa_i\}_{i=1,2}$ *satisfy the following relation*

$$\left(\frac{\kappa_1}{\kappa_2}+\frac{\kappa_2}{\kappa_1}+2cos\Omega\right)E = constant,$$

where Ω *is the angle between the two binormals at each point.*

Some of the above results can be generalized to submanifolds of codimension 2 in Euclidean space. In fact, the concept of asymptotic and binormal direction on submanifols of codimension 2 in $\mathbb{R}^n$ where introduced in [54] in a natural way in terms of singularities of height functions. It was there shown that:

(a) An $n-2$-*manifold immersed in* $\mathbb{R}^n$ *admits at most* $n-2$ *binormal directions at each point.*
(b) *A sufficient condition for the existence of at least one asymptotic direction at a point is the local convexity*. Here, we say that a submanifold is locally convex if it admits a locally support hyperplane at a point.

In [71] it was shown that *strict local convexity is a sufficient condition for the existence of the maximal number of binormal directions at a point*, where we say the M is strictly locally convex at a point x if the support hyperplane has non-degenerate contact with M (i.e., the orthogonal direction determines a non-degenerate elliptic function on M at x).

The concept of curvature locus at a point x of a submanifold M of dimension $n-2$ in $\mathbb{R}^n$ was considered in [71] as a natural generalization of the curvature ellipse of a surface in $\mathbb{R}^4$. This is a projection of the Veronese submanifold of dimension $n-3$ into the normal space of the submanifold at x and it has been shown that *it becomes a convex polygon at the points at which the curvature of the normal bundle (normal curvature) of* M *vanishes.* It is also proven that *submanifolds with everywhere vanishing normal curvature admit an orthogonal basis made of asymptotic directions at each point.*

For surfaces in $\mathbb{R}^4$, we have seen:

(1) M is totally semiumbilic if and only if M has everywhere vanishing normal curvature.
(2) M is strictly locally convex if and only if there exist 2 binormal fields over M.

On the other hand, in the case of higher dimensional submanifolds immersed with codimension 2 in $\mathbb{R}^n$, it can be seen [71]:

(1) If M is totally semiumbilic, then M has everywhere vanishing normal curvature. The last condition implies that there exist $n-2$ binormal fields over M.
(2) If M is totally semiumbilic, then M is strictly locally convex. The last condition implies that there exist $n-2$ binormal fields over M.

An example illustrating that the vanishing normal curvature does not imply strict local convexity is given by the 3-manifold embedded in $\mathbb{R}^5$ parameterized by $f: \mathbb{R}^3 \to \mathbb{R}^5$, where $f(u_{1,2}, u_3) = (u_1, u_2, u_3, u_1^2 - u_3^2, u_2^2 - u_3^2)$.

Clearly, for codimension 2 submanifold M in Euclidean space, hypersphericity implies semiumbilicity (and thus vanishing normal curvature), for the position vector provides an umbilic field globally defined on M. On the other hand, submanifolds of codimension 2 with vanishing normal curvature, or even semiumbilical $(n-2)$-

manifolds immersed in $\mathbb{R}^n$, do not need to lie in an $(n-1)$-sphere. Nevertheless, the following classical result due to Chen and Yano can help us to find a sufficient and necessary condition for hypersphericity in terms of the behaviour of the curvature ellipse.

Theorem 4.15 ([10]) *An $(n-2)$-submanifold lies in a hypersphere of $\mathbb{R}^n$ if and only if it is ν-umbilic for some parallel normal field ν.*

We observe now that the curvature locus is a segment at a point $x \in M$ if and only if x is ν-umbilic for some field ν, which is orthogonal to the direction defined by the segment in the normal plane of M at each point. Then, provided the umbilic points are isolated, we can apply this result to the open and dense submanifold determined by the complement of the umbilic points and extended it by continuity to the whole of M. So we can state,

Corollary 4.16 ([71]) *Let M be an $(n-2)$-manifold immersed in $\mathbb{R}^n$ with isolated umbilic points. Then M lies in a hypersphere if and only if the curvature locus at every point of M is a segment that defines a parallel field off the umbilic points of M.*

The concept of curvature locus can be naturally generalized to higher dimensional submanifolds. In the case of a submanifold immersed in codimension 2, it is a planar convex region (possibly degenerated into a segment or a point) some of whose properties have been analyzed in [71]. For 3-manifolds immersed in codimension 3 or more, the curvature locus is obtained as a linear projection of a Veronese submanifold of order 2 in the normal space of the manifold at each point and may have several shapes, whose singularities can be interpreted in terms of the behaviour of the principal directions of the normal fields on the submanifold [6]. Analogously to the case of surfaces immersed in $\mathbb{R}^n$, it is possible to connect several properties of the curvature locus with the behaviour of the height functions family on the considered submanifold. For instance, it can be seen that *a 3-manifold M immersed in $\mathbb{R}^{3+k}$ is strictly locally convex at a point x if and only if the origin of the normal plane is not contained in the convex hull of the curvature locus of M at x.* Some relations between the properties of the curvature locus, semiumbilicity, local convexity and existence of degenerate directions for height functions on submanifolds immersed with codimension higher than 2 have been described in [72].

4.3 Existence of 2-Regular Immersions

The concept of kth-regular immersion of a submanifold in Euclidean space was introduced in [19, 78] in terms of maps between osculating bundles. In the case of curves, the k-regularity condition means that the first k derivatives are linearly independent. Therefore, a 2-regular plane curve is a strictly convex (in the sense that it has never vanishing curvature) curve. The 2-regular space curves are the space curves with never vanishing curvature and the 3-regular space curves are the space curves with never vanishing curvature and torsion. An interesting question that arises in this context is the following:

Under what conditions can we ensure the existence of 2-regular immersions from a given closed submanifold into $\mathbb{R}^n$?

In the case of closed plane curves, ellipses and ovals provide a trivial positive answer for the existence 2-regular immersions. Yet in the case of space curves it can be shown by standard transversality techniques that the subspace of 2-regular closed simple curves is open and dense in $C^\infty(S^1, \mathbb{R}^3)$ (see for instance [84]). The existence of 3-regular embedded closed space curves was investigated by S.I.R. Costa [11], who determined necessary and sufficient conditions for a (p, q) toric curve to be 3-regular. On the other hand, as a consequence of Theorem 3.2, we can deduce that the convexity is an obstruction for the 3-regularity on embedded closed space curves.

For a surface M immersed in $\mathbb{R}^n$, $n \geq 4$ we have, in local coordinates, that a point $x \in M$ is *2-regular* if and only if the subset of vectors determined by the first and second derivatives of the immersion at x has maximal rank. E.A. Feldman [19] proved that *the set of 2-regular immersions of any closed surface M in $\mathbb{R}^n$ is open and dense for $n = 3$ and $n \geq 7$*. In the case of surfaces in $\mathbb{R}^4$, the 2-singular points coincide with the inflection points and it follows from Corollary 4.4 that the *local convexity is an obstruction for the 2-regularity of closed surfaces with nonvanishing Euler number.* An example of 2-regular convex surface with vanishing Euler characteristic is given by the Clifford torus obtained as the natural embedding of $S^1 \times S^1$ into in S^3.

The study of the existence of 2-regular immersions of surfaces in $\mathbb{R}^5$ appears to be more complicated and interesting. In this case, a 2-regular point x is characterized by the fact that the curvature ellipse is either a segment, or determines a plane passing through x. A classically known example of 2-regular immersion of S^2 into $\mathbb{R}^5$ is given by the *Veronese surface*. This surface is obtained as the restriction to S^2 of the Veronese map of order 2, $\xi : \mathbb{R}^3 \to \mathbb{R}^6$, which is given by

$$\xi(u, v, w) = \left(u^2, v^2, w^2, \sqrt{2}uv, \sqrt{2}uw, \sqrt{2}vw\right).$$

The subset $\xi(S^2)$ lies in a hyperplane, or more precisely, in a 4-sphere given by the intersection of a hyperplane with a hypersphere of $\mathbb{R}^6$. It is not difficult to check that $V|_{S^2}$ is a double covering of the Veronese surface. Examples of 2-regular immersions of closed orientable surfaces with non zero genus are not known so far.

We shall start by considering the particular case of a *surface M contained in S^4*. Given a surface immersed into $\mathbb{R}^5$, we say that a point $x \in M$ is of type M_k, $k = 1, 2, 3$, provided the second fundamental form has rank k at x. It is not difficult to see that the M_2 points of M coincide with the corank 2 singularities of height functions on M. Then, given a surface contained in $S^4 \subset \mathbb{R}^5$, since the stereographic projection $\phi : S^4 \to \mathbb{R}^4$ takes the corank 2 singularities of height functions on M to the corank 2 singularities of the distance squared functions on its image in $\mathbb{R}^4$, i.e. the semiumbilic points, of $\phi(M)$, as a consequence of Corollary 4.13 we can state the following.

Corollary 4.17 *There are no 2-regular embeddings of orientable closed surfaces in S^4.*

By a *4-dimensional ovaloid* we understand the image of a 4-sphere through an affine map. Taking into account that affine maps preserve contacts with hyperplanes and thus they preserve M_2 points, we obtain the following more general result.

Corollary 4.18 *There are no 2-regular embeddings of orientable closed surfaces into 4-dimensional ovaloids.*

We say that a surface M embedded in $\mathbb{R}^5$ is *convex* if it is contained in the boundary of its convex hull. It follows that any surface embedded in an ovaloid is convex in $\mathbb{R}^5$. Clearly, convex surfaces admit some (non necessarily unique) support hyperplane at every point. It seems now quite natural to propose the following,

Conjecture: *Orientable closed surfaces do not admit convex 2-regular embeddings in $\mathbb{R}^5$.*

A surface M immersed in $\mathbb{R}^5$ is said to be *strictly locally convex* at a point x provided it admits a locally support hyperplane with non-degenerate contact at x, or in other words, x is a non degenerate (Morse) local minimum of some height function on M. It is not difficult to see that a surface immersed in $\mathbb{R}^5$ is strictly locally convex at all the points at which the second fundamental form has maximum rank. The following more restrictive concept of local convexity for surfaces in $\mathbb{R}^5$ was first proposed in [64]: A surface M immersed in $\mathbb{R}^5$ is said to be *essentially convex* at x if there is some normal vector v lying in the normal plane determined by the curvature locus at x, such that the height h_v is a nondegenerate elliptic function. A surface that is essentially convex at each one of its points is said to be *essentially convex*. This definition of convexity is based in the following concept of *essential binormal and asymptotic directions* introduced in [64]: A binormal direction b at a non semiumbilic point x of a surface in $\mathbb{R}^5$ is said to be *essential* provided it lies in the subspace $E_x \subset N_xM$ spanned by the curvature ellipse at x. The tangent directions lying in the kernel of the Hessian of the height function h_b are said to be *essential asymptotic directions* at x. It can be shown that there are at most two essential binormals at those points $x \in M$ at which the second fundamental form has maximum rank 3. Such normal directions were said to be *essential* because all the possible principal configuration grids on the surface M are determined by them. In fact, any normal field ν on M can be decomposed at each point x as a sum of an umbilic component ν_1 (which is orthogonal to the linear span E_x of the curvature ellipse) plus an essential part ν_2 lying in E_x. It follows that the field ν_1 is umbilic and hence the ν_2-principal lines coincide with those of ν. The ν-principal curvatures are a linear combination of the umbilic curvature and the principal curvatures of the essential part at each point. So we may say that the essential normal fields, together with the umbilic field, which is generically unique, contain all the information relative to the principal configurations on the surface. Some interesting global consequences are the following,

(1) An essentially convex surface $M \subset \mathbb{R}^5$ without semiumbilic points admits two globally defined fields of essential binormal directions.

(2) Closed orientable essentially convex surfaces with non vanishing Euler number in $\mathbb{R}^5$ have semiumbilic points.

As a consequence, we can state the following,

Theorem 4.19 *An essentially convex immersion in $\mathbb{R}^5$ of a closed orientable surface with non vanishing Euler number cannot be 2-regular.*

Remark 4.20 Notice that (global) convexity does not imply essential convexity. In fact, any surface M immersed in S^4 is (globally) convex, but we have that M is essentially convex in $\mathbb{R}^5$ if and only if its image through the stereographic projection is (locally) convex in $\mathbb{R}^4$. So the inverse image of any non locally convex surface of $\mathbb{R}^4$ by stereographic projection provides an example of a convex surface in $\mathbb{R}^5$ that is non essentially convex.

Another obstruction to the 2-regularity of surfaces in $\mathbb{R}^5$ can be given in terms of the *umbilical curvature* of M. This is defined (up to sign) as the curvature associated to a unit vector which is orthogonal to the plane E_x determined by the curvature ellipse at each point $x \in M$. Clearly, the umbilic curvature of M does not vanish at the points at which the second fundamental form has maximal rank 3. Therefore, we get that the *2-regular surfaces have never vanishing umbilic curvature*. We observe that the embedding of the projective plane in $\mathbb{R}^5$, provided by the Veronese surface exhibited above as an example of 2-regular immersion of the 2-sphere, has constant non zero umbilic curvature. An interesting property of this immersion, that makes it very degenerate from a contact viewpoint, is the fact that all its points are flat ridges. Nevertheless, small enough perturbations of this immersion must provide generic examples of 2-regular (non necessarily injective) immersions of the 2-sphere in $\mathbb{R}^5$ that do not lie in a 4-sphere and do not have constant umbilic curvature.

References

1. Arnol'd, V.I., Guseĭn-Zade, S.M., Varchenko, A.N.: Singularities of Differentiable Maps. Vol. I. The Classification of Critical Points, Caustics and Wave Fronts. Monographs in Mathematics, vol. 82. Birkhuser (1985)
2. Arnold, V.I.: Singularity Theory. Selected Papers. LMS Lecture Notes Series, vol. 53. Cambridge University Press, Cambridge (1981)
3. Banchoff, T., Gaffney, T., McCrory, C.: Counting tritangent planes of space curves. Topology **24**, 15–23 (1985)
4. Banchoff, T., Farris, F.: Tangential and normal Euler numbers, complex points, and singularities of projections for oriented surfaces in four-space. Pac. J. Math. **161**, 1–24 (1993)
5. Banchoff, T., Gaffney, T., McCrory, C.: Cusps of Gauss Mappings. Research Notes in Mathematics, vol. 55. Pitman (Advanced Publishing Program), Boston (1982)
6. Binotto, R.R., Costa, S.I.R., Romero Fuster, M.C.: The curvature veronese of a 3-manifold immersed in Euclidean space. In: Real and Complex Singularities. Contemporary Maths. vol. 675, pp. 25–44. American Mathematical Society, Providence, RI (2016)
7. Bol, G.: Über Nabelpunkte auf einer Eifläche. Math. Z. **49**, 399–410 (1943/44)

8. Bruce, J.W., Giblin, P.J.: Curves and Singularities. Cambridge University Press, Cambridge (1992)
9. Bruce, J.W., Tari, F.: Families of surfaces in $\mathbb{R}^4$. Proc. Edinb. Math. Soc. (2) **45**(1), 181–203 (2002)
10. Chen, B.Y., Yano, K.: Umbilical submanifolds with respect to a nonparallel normal direction. J. Differ. Geom. **8**, 589–597 (1973)
11. Costa, S.I.R.: On closed twisted curves. Proc. Am. Math. Soc. **109**, 205–214 (1990)
12. Costa, S.I.R., Romero Fuster, M.C.: Nowhere vanishing torsion closed curves always hide twice. Geom. Dedicata **66**, 1–17 (1997)
13. Costa, S.I.R., Moraes, S.M., Romero-Fuster, M.C.: Geometric contacts of surfaces immersed in $\mathbb{R}^n$, $n \geq 5$. Differ. Geom. Appl. **27**, 442–454 (2009)
14. Dreibelbis, D.: Bitangencies on surfaces in four dimensions. Q. J. Math. **52**(2), 137–160 (2001)
15. Dreibelbis, D.: Singularities of the Gauss map and the binormal surface. Adv. Geom. **3**(4), 453–468 (2003)
16. Dreibelbis, D.: Invariance of the diagonal contribution in a bitangency formula. In: Real and Complex Singularities. Contemporary Mathematics, vol. 354, pp. 45–56. American Mathematical Society, Providence, RI (2004)
17. Dreibelbis, D.: Birth of bitangencies in a family of surfaces in R4. Differ. Geom. Appl. **24**(4), 321–331 (2006)
18. Dreibelbis, D.: Self-conjugate vectors of immersed 3-manifolds in $\mathbb{R}^6$. Topol. Appl. **159**, 450–456 (2012)
19. Feldman, E.A.: The geometry of immersions. I. Trans. Am. Math. Soc. **120**, 185–224 (1965)
20. Feldman, E.A.: On parabolic and umbilic points of immersed hypersurfaces. Trans. Am. Math. Soc. **127**, 1–28 (1967)
21. Freedman, M.H.: Planes triply tangent to curves with non-vanishing torsion. Topology **19**, 1–8 (1980)
22. Garcia, R., Mochida, D.K.H., Romero-Fuster, M.C., Ruas, M.A.S.: Inflection points and topology of surfaces in 4-space. Trans. Am. Math. Soc. **352**, 3029–3043 (2000)
23. Garcia, R., Sotomayor, J.: Lines of axial curvature on surfaces immersed in $\mathbb{R}^4$. Differ. Geom. Appl. **12**, 253–269 (2000)
24. Garcia, R., Mello, L.F., Sotomayor, J.: Principal mean curvature foliations on surfaces immersed in $\mathbb{R}^4$. EQUADIFF 2003 (2005), pp. 939–950. World Science Publisher, Hackensack, NJ
25. Ghomi, M.: Tangent lines, inflections and vertices of closed curves. Duke Math. J. **162**, 2691–2730 (2013)
26. Ghomi, M.: Vertices of closed curves in Riemannian surfaces. Comment. Math. Helv. **88**, 427–448 (2013)
27. Ghomi, M., Howard, R.: Normal curvatures of asymptotically constant graphs and Carathéodory's conjecture. Proc. Am. Math. Soc. **140**(12), 4323–4335 (2012)
28. Gibson, C.G., Wirthmuller, K., du Plessis, A.A., Looijenga, E.J.N.: Topological Stability of Smooth Mappings. Lecture Notes in Mathematics, vol. 552. Springer, Berlin (1976)
29. Golubitsky, M., Guillemin, V.: Stable Mappings and their Singularities. Springer GTM 14, Berlin (1973)
30. Gonçalves, R.A., Martínez Alfaro, J.A., Montesinos Amilibia, A., Romero-Fuster, M.C.: Relative mean curvature configurations for surfaces in $\mathbb{R}^n$, $n \geq 5$. Bull. Braz. Math. Soc., New Ser. **38**(2), 1–22 (2007)
31. Guilfoyle, B., Klingenberg, W.: Proof of the Carathéodory conjecture by mean curvature flow in the space of oriented affine lines. arXiv:0808.0851v1
32. Guíñez, V.: Positive quadratic differential forms and foliations with singularities on surfaces. Trans. Am. Math. Soc. **309**(2), 477–502 (1988)
33. Gutierrez, C., Guadalupe, I., Tribuzy, R., Guíñez.: Lines of curvature on surfaces immersed in $\mathbb{R}^4$. Bol. Soc. Brasil. Mat. (N.S.) **28**, 233–251 (1997)

34. Gutierrez, C., Guadalupe, I., Tribuzy, R., Guíñez, V.: A differential equation for lines of curvature on surfaces immersed in $\mathbb{R}^4$. Qual. Theory Dyn. Syst. **2**, 207–220 (2001)
35. Gutierrez, C., Guíñez, V.: Simple umbilic points on surfaces immersed in $\mathbb{R}^4$. Discret. Contin. Dyn. Syst. **9**, 877–900 (2003)
36. Gutiérrez, C., Mercuri, F., Sánchez-Bringas, F.: On a conjecture of Carathéodory: analyticity versus smoothness. Exp. Math. **5**(1), 33–37 (1996)
37. Gutiérrez, C., Ruas, M. A. S., Indices of Newton non-degenerate vector fields and a conjecture of Loewner for surfaces in $\mathbb{R}^4$. Real and complex singularities. Lecture Notes in Pure and Applied Mathematics, vol. 232, pp. 245–253. Dekker, New York (2003)
38. Gutierrez, C., Sánchez-Bringas, F.: On a Loewner umbilic-index conjecture for surfaces immersed in $\mathbb{R}^4$. J. Dyn. Control Syst. **4**, 127–136 (1998)
39. Gutiérrez, C., Sánchez-Bringas, F.: Planar vector field versions of Carathéodory's and Loewner's conjectures. In: Proceedings of the Symposium on Planar Vector Fields (Lleida, 1996), vol. 41, no. 1, pp. 169–179. Publ. Mat. (1997)
40. Gutierrez, C., Sotomayor, J.: Lines of curvature, umbilic points and Carathéodory conjecture. Resenhas **3**, 291–322 (1998)
41. Hamburger, H.L.: Beweis einer Caratheodoryschen Vermutung. Teil I. (German) Ann. Math. **41**(2), 63–86 (1940)
42. Hamburger, H.L.: Beweis einer Caratheodoryschen Vermutung. II. (German). Acta Math. **73**, 175–228 (1941)
43. Hamburger, H.L.: Beweis einer Caratheodoryschen Vermutung. III. (German). Acta Math. **73**, 229–332 (1941)
44. Haupt, O.: Vierscheitelsätze in der ebenen hyperbolischen Geometrie. Geometriae Dedicata **1**, 399–414 (1973)
45. Ivanov, V.V.: An analytic conjecture of Carathéodory. (Russian) Sibirsk. Mat. Zh. **43**(2), 314–405 (2002), ii; translation in Siberian Math. J. **43**(2), 251–322 (2002)
46. Izumiya, S., Romero Fuster, M.C., Ruas, M.A.S., Tari, F.: Differential Geometry from a Singularity Viewpoint. World Scientific, Singapore (2016)
47. Klotz, T.: On Bol's proof of Carathéodory's conjecture. Commun. Pure Appl. Math. **12**, 277–311 (1959)
48. Kommerell, K.: Riemannsche Flachen in ebenen Raum von vier Dimensionen. Math. Ann. **60**, 546–596 (1905)
49. Little, J.A.: On singularities of submanifolds of higher dimensional Euclidean spaces. Annali Mat. Pura Appl. **4A**(83), 261–336 (1969)
50. Looijenga, E.J.N.: Structural stability of smooth families of C^∞ functions, Thesis, University of Amsterdan (1974)
51. Mello, L.F.: Mean directionally curved lines on surfaces immersed in $\mathbb{R}^4$. Publ. Mat. **47**, 415–440 (2003)
52. Mochida, D.K.H.: Geometria generica de subvariedades em codimensao maior que 1. Ph.D thesis, University of São Paulo (1993)
53. Mochida, D.K.H., Romero Fuster, M.C., Ruas, M.A.S.: The geometry of surfaces in 4-space from a contact viewpoint. Geom. Dedicata. **54**, 323–332 (1995)
54. Mochida, D.K.H., Romero Fuster, M.C., Ruas, M.A.S.: Osculating hyperplanes and asymptotic directions of codimension two submanifolds of Euclidean spaces. Geom. Dedicata **77**, 305–315 (1999)
55. Mochida, D.K.H., Romero Fuster, M.C., Ruas, M.A.S.: Singularities and duality in the flat geometry of submanifolds of Euclidean spaces. Beiträge Algebra Geom. **42**, 137–148 (2001)
56. Mochida, D.K.H., Romero Fuster, M.C., Ruas, M.A.S.: Inflection points and nonsingular embeddings of surfaces in $\mathbb{R}^5$. Rocky Mountain J. Math. **33**, 995–1009 (2003)
57. Mond, D.M.Q.: On the classification of germs of maps from $\mathbb{R}^2$ to $\mathbb{R}^3$. Proc. Lond. Math. Soc. **50**, 333–369 (1985)
58. Mond, D.M.Q.: The Classification of Germs of Maps from Surfaces to 3-space, with Applications to the Differential Geometry of Immersions. Ph.D Thesis, Liverpool University (1982)

59. Montaldi, J.A.: Contact with applications to submanifolds. Ph.D Thesis, University of Liverpool (1983)
60. Montaldi, J.A.: On contact between submanifolds. Michigan Math. J. **33**, 195–199 (1986)
61. Montaldi, J.A.: On generic composites of maps. Bull. Lond. Math. Soc. **23**, 81–85 (1991)
62. Monera, M.G., Montesinos Amilibia, A., Sanabria Codesal, E.: The Taylor expansion of the exponential map and geometric applications. Rev. R. Acad. Cienc. Exactas F. Nat. Ser. A Math. RACSAM **108**(2), 881–906 (2014)
63. Moraes, S.M.: Elipses de curvatura no estudo de superficies imersas em $\mathbb{R}^n$, $n \geq 5$. Doctoral Thesis, UNICAMP (Brasil) (2002)
64. Moraes, S.M., Romero-Fuster, M.C.: Semiumbilic and 2-regular immersions of surfaces in Euclidean spaces. Rocky Mt. J. Math. **35**, 1327–1345 (2005)
65. Moraes, S.M., Romero-Fuster, M.C., Sanchez-Bringas, F.: Principal configurations and umbilicity of submanifolds in $\mathbb{R}^N$. Bull. Belg. Math. Soc. Simon Stevin **11**(2), 227–245 (2004)
66. Nabarro, A.C., Romero-Fuster, M.C.: 3-manifolds in Euclidean space from a contact viewpoint. Commun. Anal. Geom. **17**, 755–776 (2009)
67. Nikolaev, I.: Foliations on Surfaces, Ergebnisse der Mathematik und ihrer Grenzgebiete. 3. Folge A, Series of Modern Surveys in Mathematics. Springer, Berlin (2001)
68. Nuño Ballesteros, J.J.: Submanifolds with a non-degenerate parallel normal vector field in Euclidean spaces. In: Singularity Theory and its Applications. Advanced Studies in Pure Mathematics, vol. 43, pp. 311–332. Mathematical Society of Japan, Tokyo (2006)
69. Nuño Ballesteros, J.J., Romero Fuster, M.C.: A four vertex theorem for strictly convex space curves. J. Geom. **46**, 119–126 (1992)
70. Nuño Ballesteros, J.J., Romero Fuster, M.C.: Global bitangency properties of generic closed space curves. Math. Proc. Camb. Philos. Soc. **112**(3), 519–526 (1992)
71. Nuño Ballesteros, J.J., Romero-Fuster, M.C.: Contact properties of codimension 2 submanifolds with flat normal bundle. Rev. Mat. Iberoam. **26**, 799–824 (2010)
72. Nuño Ballesteros, J.J., Romero-Fuster, M.C., Sánchez Bringas, F.: Curvature locus and principal configurations of submanifolds in Euclidean space. To appear in Revista Iberoamericana de Matemáticas
73. Osserman, R.: The four-or-more vertex theorem. Am. Math. Mon. **92**(5), 332–337 (1985)
74. Ozawa, T.: The numbers of triple tangencies of smooth space curves. Topology **24**, 1–13 (1985)
75. Porteous, I.R.: The normal singularities of a submanifold. J. Differ. Geom. **5**, 543–564 (1971)
76. Porteous, I.R. (1983), The normal singularities of surfaces in $\mathbb{R}^3$. Singularities, Part 2 (Arcata, Calif., 1981). In: Proceedings of Symposia in Pure Mathematics, vol. 40, pp. 379–393. American Mathematical Society, Providence, RI (1983)
77. Porteous, I.R.: Ridges and umbilics of surfaces. The mathematics of surfaces, II (Cardiff, 1986). In: Institute of Mathematics and its Applications Conference Series, New Series, vol. 11, pp. 447–458. Oxford University Press, New York (1987)
78. Pohl, W.F.: Differential geometry of higher order. Topology **1**, 169–211 (1962)
79. Pugh, C.: A generalized Poincaré index formula. Topology **7**, 217–226 (1968)
80. Ramírez-Galarza, A.I., Sánchez-Bringas, F.: Lines of curvature near umbilical points on surfaces immersed in $\mathbb{R}^4$. Ann. Global Anal. Geom. **13**, 129–140 (1995)
81. Robertson, S.A., Romero Fuster, M.C.: The convex hull of a hypersurface. Proc. Lond. Math. Soc. **50**, 370–384 (1985)
82. Romero Fuster, M.C.: The convex hull of an immersion. Ph.D Thesis, University of Southampton (1981)
83. Romero Fuster, M.C.: Sphere stratifications and the Gauss map. Proc. Edinb. Math. Soc. **95A**, 115–136 (1983)
84. Romero Fuster, M.C.: Convexly generic curves in $\mathbb{R}^3$. Geom. Dedicata **28**, 7–29 (1988)
85. Romero Fuster, M.C.: Stereographic projections and geometric singularities. Workshop on real and complex singularities (São Carlos, 1996). Mat. Contemp. **12**, 167–182 (1997)

86. Romero Fuster, M.C.: Geometric contacts and 2-regularity of surfaces in Euclidean space. In: Singularity Theory, pp. 307–325. World Scientific Publishing, Hackensack, NJ (2007)
87. Romero Fuster, M.C. Ruas, M.A.S., Tari, F.: Asymptotic curves on surfaces in $\mathbb{R}^5$. Commun. Contemp. Math. **10**, 309–335 (2008)
88. Romero Fuster, M.C., Sanabria Codesal, E.: On the flat ridges of submanifolds of codimension 2 in $\mathbb{R}^n$. Proc. Edinb. Math. Soc. **132A**, 975–984 (2002)
89. Romero Fuster, M.C., Sanabria Codesal, E.: Lines of curvature, ridges and conformal invariants of hypersurfaces. Beiträge Algebra Geom. **45**, 615–635 (2004)
90. Romero Fuster, M.C., Sanabria Codesal, E.: Conformal invariants interpreted in de Sitter space. Mat. Contemp. **35**, 205–220 (2008)
91. Romero Fuster, M.C., Sanabria Codesal, E.: Conformal invariants and spherical contacts of surfaces in $\mathbb{R}^4$. Rev. Mat. Complut. **26**, 215–240 (2013)
92. Romero Fuster, M.C., Sánchez-Bringas F.: Umbilicity of surfaces with orthogonal asymptotic lines in $\mathbb{R}^4$. Differ. Geom. Appl. **16**, 213–224 (2002)
93. Romero Fuster, M.C., Sedykh, V.D.: On the number of singularities, zero curvature points and vertices of a simple convex space curve. J. Geom. **52**, 168–172 (1995)
94. Romero Fuster, M.C., Sedykh, V.D.: A lower estimate for the number of zero-torsion points of a space curve. Beitrage Algebra Geom. **38**, 183–192 (1997)
95. Scherbel, H.: A new proof of Hamburger's Index Theorem on umbilical points. Dissertation ETH No. 10281
96. Sedyh, V.D.: Structure of the convex hull of a space curve. (Russian). Trudy Sem. Petrovsk. **6**, 239–256 (1981)
97. Sedyh, V.D.: Double tangent planes to a space curve. (Russian). Sibirsk. Mat. Zh. **30**(1), 209–211 (1989); translation in Siberian Math. J. **30**(1), 161–162 (1989)
98. Sedykh, V.D.: The four-vertex theorem of a convex space curve. Funct. Anal. Appl. **26**, 28–32 (1992)
99. Sedykh, V.D.: A relationship between Lagrange and Legendre singularities in stereographic projection. (Russian). Mat. Sb. **185**(12), 123–130 (1994); translation in Russian Acad. Sci. Sb. Math. **83**(2), 533–540 (1995)
100. Sedykh, V.D.: On Euler characteristics of manifolds of singularities of wave fronts. (Russian). Funktsional. Anal. i Prilozhen. **46**, 92–96 (2012); translation in Funct. Anal. Appl. **46**, 77–80 (2012)
101. Smyth, B., Xavier, F.: A sharp geometry estimate for the index of an umbilic point on a smooth surface. Bull. Lond. Math. Soc. **24**, 176–180 (1992)
102. Sotomayor, J.: Historical Comments on Monge's Ellipsoid and the Configuration of Lines of Curvature on Surfaces Immersed in $\mathbb{R}^3$ (2004). arXiv:math/0411403v1
103. Sotomayor, J., Gutierrez, C.: Structurally stable configurations of lines of principal curvature. Bifurcation, ergodic theory and applications (Dijon, 1981). In: Astérisque, vol. 98–99, pp. 195–215. Soc. Math, France, Paris (1982)
104. Thom, R.: Sur le cut-locus d'une variété plongée. (French) Collection of articles dedicated to S.S. Chern and D.C. Spencer on their sixtieth birthdays. J. Differ. Geom. **6**, 577–586 (1972)
105. Thobergsson, G., Umehara, M.: A unified approach to the four vertex theorems. II. In: Differential and Symplectic Topology of Knots and Curves. American Mathematical Society Translations: Series 2, pp. 229–252. AMS, Providence, RI (1999)
106. Titus, C.J.: A proof of a conjecture of Carathéodory on umbilic points. Acta Math. **131**, 43–77 (1973)
107. Umehara, M.: A unified approach to the four vertex theorems I. Differential and symplectic topology of knots and curves. AMS Transl. **190**, 185–228 (1999)
108. Wegner, B.: A cyclographic approach to the vertices of plane curves. J. Geom. **50**, 186–201 (1944)
109. Wong, Y.C.: A new curvature theory for surfaces in a Euclidean 4-space. Comment. Math. Helv. **26**, 152–170 (1952)

Equisingularity

Lê Dũng Tráng

Abstract We give a quick survey of problems concerning Equisingularity.

Classifications 14B05 · 14B12 · 32S15

Introduction

A singularity is the germ of a complex analytic space (X, x). Equisingularity means the same singularity.

A naive view would be that two singularities are equisingular if they are analytically the same. It is known that two singularities (X, x) and (Y, y) are analytically the same if and only if the local rings $\mathcal{O}_{X,x}$ and $\mathcal{O}_{Y,y}$ are isomorphic (see e.g. [8]).

In the case of complex singularities of hypersurfaces, it seems that one may use analytic isomorphism to define equisingularity, since for the most "simple" singularities analytic isomorphisms and ambient homeomorphisms between the singularities are equivalent.

For instance two complex cusps of plane curves are equally locally homeomorphic in the local ambient space or analytically isomorphic. One expresses this property by saying that the moduli of a complex cusp singularity is reduced to a point. More generally the moduli of a space or a singularity is the parameter space of a deformation of complex analytic spaces or singularities having the same "topological features", but being analytically different.

However, one knows that the moduli of a singularity is in general not reduced to a point. In the case of plane curves the moduli of the germs of plane curves with one Puiseux pair (3, 7) (see e.g. [25] Sect. 1 p. 284) is not reduced to a point. There is the famous example of Riemann of a family of cubic curves having different analytic structures. The cones on these cubic curves define a family of complex singularities having different analytic structures.

L. D. Tráng (✉)
Université Aix-Marseille, Marseille, France
e-mail: ledt@ictp.it

R. N. Araújo dos Santos et al. (eds.), *Singularities and Foliations. Geometry, Topology and Applications*, Springer Proceedings in Mathematics & Statistics 222, https://doi.org/10.1007/978-3-319-73639-6_10

Therefore, the analytic isomorphism of germs does not seem to be the adequate answer for equisingularity, since the analytic type can change continuously on a complex analytic space.

Several mathematicians like H. Whitney, R. Thom, O. Zariski have tried to give a good definition of equisingularity.

H. Whitney (see [30]) introduced a partition of a general complex analytic space, called a Whitney stratification (see below). The article [30] is the fruit of several discussions that H. Whitney had with R. Thom. Later R. Thom and J. Mather (in [13, 22]) proved that locally along the strata of a Whitney stratification the analytic space is a topological product.

In a paper published in 1937, O. Zariski used an argument very similar to equisingularity to prove that the fundamental group of the complement of a complex projective hypersurface of dimension $n \geq 2$ is isomorphic to the fundamental group of the complement of the complex curve intersection of the hypersurface with a general complex plane in the general plane section (see "A theorem on the Poincaré group of an algebraic hypersurface", Ann. of Math. **38** (1937), 131–141).

Unfortunately, at that time O. Zariski did not have a clear definition of what should be a general plane section and a fortiori of what should be equisingular sections. O. Zariski used to call jokingly that paper his last italian paper.

The term of equisingularity appears in the papers of O. Zariski (see [32, 34]). The viewpoint changed somehow. One considers a partition of the analytic space X such that, for two points x_1, x_2 of a stratum of the partition, the germs (X, x_1) and (X, x_2) are equisingular. For Zariski, he considers algebraic varieties and he wants that the partition is defined by algebraic data. In the case of a complex hypersurface the big stratum is the stratum of non-singular points and the stratum of codimension one is the one such that transversal sections by a plane of dimension 2 give a germ of plane curve with the same Puiseux pairs, if the germ is a branch, or a germ of plane curve with a given topology in the case of several analytic branches.

He could characterize the codimension one stratum with a new concept called the saturation (see [36]).

In the period from 1965 to 1968, O. Zariski introduced the notion of saturation of a ring. Then, he published an algebraic understanding of what he called equisingularity in several papers [31–36]. Surprisingly these papers attract little attention of the community of algebraic geometers. One of the reasons of this attitude might be because the notion of equisingularity was not clearly defined but in the case of plane curves for which equisingular germs of plane curves are germs of plane curves with isomorphic saturation rings. Unfortunately this definition does not work in dimension ≥ 2.

When the hypersurface singularity is isolated, in [14] (1968) J. Milnor has introduced a multiplicity that we call the Milnor number of the isolated singularity which is a topological invariant of the embedded topology of the hypersurface (see e.g. [26] Proposition p. 261). However, two isolated hypersurface singularities having the same Milnor number may not be topologically equisingular: two plane curves with one Puiseux pair (p_1, q_1) and (p_2, q_2) such that $(p_1 - 1)(q_1 - 1) = (p_2 - 1)(q_2 - 1)$ have the same Milnor number:

$$\mu = (p_1 - 1)(q_1 - 1) = (p_2 - 1)(q_2 - 1)$$

but, if $p_1 \neq p_2$, are not topologically equisingular.

In 1968 in a seminar at IHES, H. Hironaka made a conjecture that in a family of plane curves with Milnor number constant, the local topology of the plane curve does not change. In 1970 I found a proof of this conjecture (see [24] published in 1971). In 1971 together with C.P. Ramanujam I extended this result to complex hypersurfaces of dimension ≥ 3 (see [27] published in 1976). The restriction on the dimension came from the use of the h-cobordism Theorem.

This topological result showed that equisingularity can be understood either topologically, or algebraically, as Zariski tried to do for plane curves. The different ways to define equisingularity should at least imply topological equisingularity. Furthermore one should be able to "stratify" an algebraic variety with equisingular germs along each strata. The case of plane curves which should correspond to strata of codimension one in a hypersurface would be the typical first example.

Finally, the concept of equisingularity, although vague, can be formulated in the following way:

Let X be a complex analytic space. There is an analytic partition $X = \coprod_{i \in I} X_i$, such that:

- The definition of the analytic partition should be given by algebraic conditions on the local ring $\mathcal{O}_{X,x}$;
- All the germs (X, x) with $x \in X_i$ should be equisingular, e.g. topologically equisingular in the case of hypersurfaces;
- Following Zariski (see [31]) the multiplicity of (X, x) should be constant along X_i.

In these notes, in a quick way we shall present most of the aspects of Equisingularity theory that is known nowadays, hoping that it will motivate younger mathematicians to make research in this direction.

1 Basic Notions

1.1 Germs

Let X be complex analytic spaces and let x be a point of X. One calls germ of X at the point x the pair (X, x). Let (X, x) and (Y, y) be complex analytic germs. The germ at x of morphism from (X, x) into (Y, y) is the equivalence class of complex analytic morphisms defined on a neighborhood of x in X into Y such that the image of x is y and two such morphisms coincide on a neighborhood of x in X.

Germs of complex analytic spaces with germs of complex analytic morphisms form a category that we shall call *German*. The objects of this category are germs of complex analytic spaces and the arrows from (X, x) into (Y, y) are the germs of morphisms of (X, x) into (Y, y).

Similarly complex analytic algebras isomorphic to the local ring $\mathcal{O}_{X,x}$ of germs of complex analytic functions of some complex analytic space X at x form a category *Algan* in which the objects are complex analytic algebras and the arrows are $\mathbb{C}$-homomorphisms of these algebras.

We have a natural functor:

$$\Phi : \text{German}^\circ \rightarrow \text{Algan}$$

where German° is the opposite category of German and such that $\Phi((X, x)) = \mathcal{O}_{X,x}$, where $\mathcal{O}_{X,x}$ is the local ring of germs at x of complex analytic functions on X.

It is known that (see [8] p. 13–02):

Theorem 1.1 *The functor* Φ : *German°→Algan is an equivalence of categories from the opposite of the category of complex analytic germs with the category of complex analytic local algebras.*

1.2 Analytic Equivalence

Of course one can classify singularities using analytic equivalence. Using Theorem 1.1 two singularities (X, x) and (Y, y) are analytically equivalent if the local analytic rings $\mathcal{O}_{X,x}$ and $\mathcal{O}_{Y,y}$ are isomorphic. However such a classification is too fine and if $x \in X$ the analytic structure of $\mathcal{O}_{X,x}$ can vary continuously.

For instance, the complex surface of $\mathbb{C}^3$ given by $X(X-Y)(X+Y)(X+TY) = 0$ has a continuous analytic structure at the points $(0, 0, t)$.

1.3 Topological Equivalence

In the case of hypersurfaces, one has an notion of topological equivalence.

The germs of hypersurfaces (X, x) and (Y, y) of $\mathbb{C}^n$ are topologically equivalent if there exists a germ of homeomorphisms φ of $(\mathbb{C}^n, x)$ into $(\mathbb{C}^n, y)$ such that the image of (X, x) is (Y, y). In what follows, we shall say that two topologically equivalent hypersurface singularities are topologically equisingular.

This notion of topological equivalence does not extend to codimension ≥ 2 analytic spaces. For instance, two analytically irreducible germs of curves of $\mathbb{C}^n$ are topologically equivalent if $n \geq 3$.

1.4 Plane Curves

The case of germs of complex plane curves is the test case where all the criteria for a good equivalence are working.

As Zariski did, we shall call analytic plane branch an analytically irreducible germ of reduced complex plane curve. Let us suppose that an analytic branch $(C, 0)$ is defined by the equation $f = 0$ where f is an irreducible germ of complex analytic function of $(\mathbb{C}^2, 0)$ at the origin 0. Let us suppose that the coordinates X, Y of $(\mathbb{C}^2, 0)$ satisfy the Weierstrass type condition:

$$f(0, Y) \neq 0.$$

One can define the Puiseux exponents relatively to the coordinates X, Y (see [25]). Those Puiseux exponents define the knot type of the intersection $\{f = 0\} \cap \mathbb{S}_\varepsilon(0)$ of C and a sufficiently small sphere $\mathbb{S}_\varepsilon(0)$ centered at 0 with radius ε (e.g. see [25] Sect. 1).

Puiseux Theorem shows that one can parametrize the branch $(C, 0)$, i.e. there exist a convergent series $\Phi(X^{1/n})$ in $X^{1/n}$ such that $f(X, \Phi(X^{1/n})) \equiv 0$ and n equals the valuation of $f(0, Y)$:

$$\phi\left(X^{\frac{1}{x}}\right) = \Sigma a_k X^{\frac{k}{n}}$$

Let us now define the Puiseux exponents relatively to the coordinates (X, Y).

If $n = 1$, the Puiseux expansion is a formal series with coefficients in $\mathbb{C}$. In this case, there are no Puiseux exponent.

If $n > 1$, the set $E_1 = \{k/n \notin \mathbb{N},\ a_k \neq 0\}$ is not empty, since n is the smallest integer ℓ, such that $\Phi(X^{1/n}) \in \mathbb{C}[[X^{1/\ell}]]$.

Define the first Puiseux exponent relatively to the coordinates (X, Y):

$$\frac{k_1}{n} = \inf\{k/n \notin \mathbb{N},\ a_k \neq 0\}.$$

Then, either (k_1, n) are relatively prime and there is only one Puiseux exponent, or

$$\frac{k_1}{n} = \frac{m_1}{n_1}$$

and $n_1 < n$. The set $E_1 = \{k/n \notin (1/n_1)\mathbb{N},\ a_k \neq 0,\ k > k_1\}$ is not empty, otherwise $\Phi(X^{1/n})$ belongs to $\mathbb{C}[[X^{1/n_1}]]$.

Define the second Puiseux exponent by:

$$\frac{k_2}{n} := \inf\{\frac{k}{n} \notin \frac{1}{n_1}\mathbb{N},\ a_k \neq 0,\ k > k_1\}.$$

There is a unique way to write:

$$\frac{k_2}{n} = \frac{m_2}{n_1 n_2}$$

in such a way that (m_2, n_2) are relatively prime.

Then, either $n_1 n_2 = n$ and there are only two Puiseux exponents, or $n_1 n_2 < n$ and the set:

$$E_2 = \{\frac{k}{n} \notin \frac{1}{n_1 n_2}\mathbb{N},\ a_k \neq 0,\ k > k_2\}$$

is not empty.

By induction, one defines $m_h/n_1 \dots n_h$, where (m_h, n_h) are relatively prime. Either, $n_1 \dots n_h = n$ and there are h Puiseux exponents, or $n_1 \dots n_h < n$ and the set:

$$E_h = \{\frac{k}{n} \notin \frac{1}{n_1 \dots n_h}\mathbb{N},\ a_k \neq 0,\ k > k_h\}$$

is not empty, in which case $\inf E_h = k_{h+1}/n = m_{h+1}/n_1 \dots n_{h+1}$, where (m_{h+1}, n_{h+1}) are relatively prime and unique.

The process has to end, since n has a finite number of divisors.

The pairs $(m_1, n_1), \dots, (m_h, n_h)$ are called the Puiseux pairs of $(C, 0)$ relatively to the coordinates (X, Y) and the exponents:

$$\frac{m_1}{n_1}, \dots, \frac{m_h}{n_1 \dots n_h}$$

are called the Puiseux exponents of $(C, 0)$ relatively to the coordinates (X, Y).

One can prove:

Theorem 1.2 *Two plane branches $(C_1, 0)$ and $(C_2, 0)$ are topologically equivalent if and only if, there are coordinates for which their Puiseux exponents are equal.*

In [36] O. Zariski introduced the notion of saturation $\tilde{\mathcal{O}}_{X,x}$ of the local ring $\mathcal{O}_{X,x}$ when (X, x) is a complex analytic plane branch (see below in Sect. 3.2). The saturation $\tilde{\mathcal{O}}_{X,x}$ is a local ring which contains the local ring $\mathcal{O}_{X,x}$ and is contained in the normalization $\bar{\mathcal{O}}_{X,x}$:

$$\mathcal{O}_{X,x} \subset \tilde{\mathcal{O}}_{X,x} \subset \bar{\mathcal{O}}_{X,x}.$$

It is known that the normalization $\bar{\mathcal{O}}_{X,x}$ is the ring of germs of meromorphic functions whose restriction to (X, x) are bounded (e.g. see [18] Chapter VI).

Similarly F. Pham and B. Teissier have proved that the saturation $\tilde{\mathcal{O}}_{X,x}$ is the ring of germs of meromorphic functions on (X, x) which are Lipschitz functions (see [19] or [7]).

1.5 Hypersurfaces

In the case of reduced hypersurfaces (X, x) we proved in [26] (Proposition of the Introduction) that the local monodromy of the local Milnor fibration of (X, x) (see [14] Sect. 4 for the definition and existence) is a topological invariant of (X, x). In particular the Milnor numbers of two topological equisingular hypersurfaces $(X_1, 0)$ and $(X_2, 0)$ are the same.

It is remarkable that, in a smooth family of complex hypersurfaces containing the origin 0 and having at 0 the same Milnor number, the hypersurfaces are topologically equisingular (see [27]). However the dimension n of the hypersurfaces is $\neq 2$ because the proof uses the h-cobordism Theorem.

Conjecture: It is natural to conjecture that this result holds also in dimension 2.

1.6 Whitney Stratifications

As we have mentioned in the introduction, in 1965 H. Whitney introduced the notion of Whitney condition (see [30]).

Let X be a reduced complex analytic space. Let Σ_X the subset of singular points of X. It is known that Σ_X is an complex analytic subspace of X. We have the partition of X:

$$X = (X - \Sigma_X) \coprod \Sigma_X.$$

Defining by induction $X_1 = \Sigma_X$ and, for $i \geq 1$, $X_{i+1} = \Sigma_{X_i}$, we have:

$$X = (X - \Sigma_X) \coprod (X_1 - X_2) \coprod (X_2 - X_3) \coprod \ldots$$

which has to be a finite partition since $X_i - X_{i+1}$ is a manifold and $\dim X_i > \dim X_{i+1}$ if $X_i \neq \emptyset$. It is called the partition by dimension of X.

The partition by the connected components of $X_i - X_{i+1}$ is called the *full partition* by dimension of X. If X is a complex analytic space, its full partition by dimension might not be finite but it is locally finite.

A complex analytic manifold Y contained in a complex analytic space X is *strict* if the closure $\bar{Y}$ of Y in X and the difference $\bar{Y} - Y$ are complex analytic subspace of X.

If Y is strict of dimension m, then Lemma 3.13 of [30] shows that the dimension of $\bar{Y}$ is m and the dimension of $\bar{Y} - Y$ is $< m$.

A *strict partition* of a complex analytic space X is a partition, which is locally finite, into strict manifolds. The elements of a strict partition are called the strata of the strict partition.

Lemma 18.2 of [30] states that the partition by dimension and the full partition by dimension of a complex analytic space X are strict partitions of X.

A strict partition $(X_i)_{i \in I}$ of a complex analytic space X satisfies *the frontier condition* if:

$$\forall i, j \in I, X_i \cap \bar{X}_j \neq \emptyset \Rightarrow X_i \subset \bar{X}_j \text{ and } \dim X_i < \dim X_j.$$

Definition 1.1 Let X be a complex analytic space. A stratification of X is a locally finite strict partition by connected strata which satisfies the frontier condition.

Now we can define the conditions of Whitney.

Definition 1.2 Let M and N be two complex analytic manifolds strict in the complex analytic space X. Assume that $N \subset \bar{M}$. Let $x \in N$. We may assume that locally at

x the are neighborhoods V of x in N and U of x in M such that $V \subset \bar{U} \subset \mathbb{C}^N$. One says that M satisfies the Whitney condition (a) at x along N if, for any sequence (x_n) of points of M converging x for which the sequence of tangent spaces $T_{x_n}(M)$ converge to T, we have $T_x(N) \subset T$.

Definition 1.3 Let M and N be two complex analytic manifolds strict in the complex analytic space X. Assume that $N \subset \bar{M}$. Let $x \in N$. We may assume that locally at x the are neighborhoods V of x in N and U of x in M such that $V \subset \bar{U} \subset \mathbb{C}^N$. One says that M satisfies the Whitney condition (b) at x along N if, for any sequence (x_n) of points of M and any sequence (y_n) of N converging x, for which the sequence of tangent spaces $T_{x_n}(M)$ converges to T and for which the sequence of lines $\overline{y_n x_n}$ converges to ℓ, we have $\ell \subset T$.

In [13] (Proposition 2.4) it is proven that If M satisfies the Whitney condition (b) at x along N, then it satisfies the condition of Whitney (a) at x along N.

We say that M satisfies Whitney condition (b) along N if it satisfies Whitney condition (b) at any point x of N along N.

Definition 1.4 A stratification $(S_i)_{i \in I}$ of the complex analytic space X is a Whitney stratification if, for any pair (S_i, S_j) of strata such that S_i is contained in the closure $\bar{S}_j$, the stratum S_j satisfies the condition of Whitney (b) at any point x of S_i along S_i.

In [30] (Theorem 19.2 p. 540, H. Whitney proved that any reduced complex analytic space has a Whitney stratification.

The remarkable result of Mather and Thom is that for any Whitney stratification of a complex analytic space X the topology of X along any strata is a local product. Namely let (S_i) be a Whitney stratification of X, for any point $x \in S_i$, there is a neighborhood U_x of x in X, such that U_x is homeomorphic to the product $(U_x \cap S_i) \times (N_x \cap U)$ where N_x is a slice of X transverse at x to S_i in a local smooth ambient space.

Since the strata of a Whitney stratification are pathwise connected, the topology of the germ of N_x at x does not depend on the point x in a stratum.

In fact, the notion of stratification as well as Whitney conditions can be extended to subanalytic spaces or even to definable spaces. We shall not consider this extension in these notes.

1.7 The Concept of Equisingularity

Equisingularity is up to now a rather vague concept. We shall try to fix some properties which should be satisfied by a proper definition of equisingularity.

As we said in the introduction, roughly speaking two germs of complex analytic spaces should be equisingular if their singularity are somehow the "same". We already mentioned that considering complex analytic equivalence is too strong, because in a family the analytic structure might change continuously.

We can give some basic features which should characterize Equisingularity:

1. It is an equivalence relation in the class of complex analytic germs;
2. Two equisingular hypersurfaces should be topologically equisingular;
3. If X is a complex analytic space, the disjoint subspaces:

$$\mathcal{S}_x = \{y \in X \mid (X, y) \text{ is equisingular to } (X, x)\}$$

   define a strict partition of X.
4. Two equisingular spaces (X, x) and (Y, y) should have the same multiplicity.
5. Equisingularity should be characterized algebraically.

In the paper of Zariski, "A theorem on the Poincaré group of an algebraic hypersurface", quoted in the introduction above, one of the main arguments of the proof is that two general hyperplane sections of a projective hypersurface and their embedding in their hypersurface are homeomorphic or equivalently the germs of their cones at the origin are topologically equisingular.

2 Whitney Equisingularity

A possible definition of Equisingularity is Whitney Equisingularity. Let X be a reduced complex analytic space. Let x and y be points of X.

Definition 2.1 The singularities (X, x) and (X, y) are Whitney equisingular if there is a Whitney stratification $(S_i)_{i\in I}$ of X such that x and y belong to the same stratum S_i.

We shall see that Whitney equisingularity satisfies the features mentioned above.

2.1 Topological Properties

Let X be a reduced complex analytic space. Let $\mathcal{S} = (S_i)_{i\in I}$ be a Whitney stratification of X. There is a local topological triviality of X along the strata of the stratification $\mathcal{S}$ in the following sense:

As we have said above, for any $x \in X$, let $S_{i(x)}$ be the stratum of the stratification $\mathcal{S}$ of X which contains x, then there exist an open neighborhood V of x in $S_{i(x)}$ and a *slice* $\mathcal{N}_x$, i.e. in a local embedding $(X, x) \subset (\mathbb{C}^N, x)$ the intersection of X with a linear subspace of $\mathbb{C}^N$ transverse to $S_{i(x)}$ at x in a neighborhood of x in X, such that a neighborhood of x in X is homeomorphic to the product $V \times (\mathcal{N}_x \cap V)$.

This result was announced by R. Thom in [22] and one can find a sketch of proof by J. Mather in [13].

As a consequence, using the tubular neighbourhoods of J. Mather (see p. 480 of [13]), we can prove that, for any point $x \in S_i$, the slices $\mathcal{N}_x$ are diffeomorphic.

In particular, if X is a hypersurface, if $x, y \in X$ are points of X, since they belong to the same Whitney stratum of some Whitney stratifications the germs (X, x) and

(X, y) are homeomorphic germs of hypersurface, so they are topologically equisingular.

2.2 Equimultiplicity

Let X be a reduced complex analytic space. Let $\mathcal{S} = (S_i)_{i \in I}$ be a Whitney stratification of X. In his paper [9] Corollary 6.2, H. Hironaka proves that for any points $x \in S_i$, the multiplicity of X is the same. Then, along its Whitney strata, a reduced analytic space is equimultiple.

2.3 Polar Varieties

Let X be an equidimensional reduced complex analytic space of dimension d and let x be a point of X. Consider the integers k, $2 \leq k \leq d+1$.

We may embed $(X, x) \subset (\mathbb{C}^N, x)$. In [28] (2.2.2) we show that the set of germs of projection:

$$p : (X, x) \to (\mathbb{C}^k, 0)$$

induced by surjective affine maps $(\mathbb{C}^N, x) \to (\mathbb{C}^k, 0)$ contains an Zariski dense subset Ω_k such that, for any $p \in \Omega_k$, the critical locus $C(p)$ of the restriction of p to the non-singular part $X \setminus \Sigma_X$ is a reduced complex analytic space and the multiplicity $m_k(X, x)$ of germ of the closure $\overline{C(p)}$ at the point x does not depend on $p \in \Omega_k$.

For $p \in \Omega_k$ the germ $(\overline{C(p)}, x)$ is called a polar variety $P^{k-1}(X, x)$ of (X, x) of dimension $k-1$. Beware that $P^{k-1}(X, x)$ can be empty in which case its multiplicity at x is 0.

Therefore, one can associate a d-uple $M(X, x) = (m_2(X, x), \ldots, m_{d+1}(X, x))$ to the germ (X, x). Notice that some $m_k(X, x)$ can be 0 and $m_{d+1}(X, x)$ is the multiplicity of X at x, because $P^d(X, x) = (X, x)$.

We have the following algebraic characterisation of Whitney stratification due to B. Teissier (see [21] Chapitre 5 Théorème 1.2) which gives somehow an algebraic characterisation of Whitney equisingularity:

Theorem 2.1 *Let X be a reduced equidimensional complex analytic space. Let $\mathcal{S} = (S_i)_{i \in I}$ be a stratification of X (see Definition 1.1). Suppose that, for any pair (S_i, S_j), such that $S_i \subset \overline{S}_j$, the* $\dim(S_j)$*-uple $M(\overline{S}_j, x)$ is constant for $x \in S_i$. Then, the stratification $\mathcal{S}$ is a Whitney stratification of X.*

2.4 Vanishing Euler Characteristics

Let X be a d-equidimensional reduced complex analytic space and x be a point of X. We may assume that $(X, x) \subset (\mathbb{C}^N, x)$. We have seen that, for any $p \in \Omega_k$, where Ω_k is a Zariski dense open subset of the space of projections of (X, x) onto $(\mathbb{C}^k, 0)$ induced by affine maps $(\mathbb{C}^N, x) \to (\mathbb{C}^k, 0)$, we can define Ω_k, such that the general local fiber $p^{-1}(u) \cap \mathbb{B}_\varepsilon$, where $\mathbb{B}_\varepsilon$ is the ball centered at x of radius ε in $\mathbb{C}^N$, $0 < \|u\| \ll \varepsilon$ and $u \in \mathbb{C}^k$ is a general point, of p at x has a homotopy type which does not depend on p (see 3.1.2 in [23]).

We call the general local fiber $p^{-1}(u) \cap \mathbb{B}_\varepsilon$ of p at x a *local vanishing fiber of* (X, x) *of dimension* $d - k$. When $k = d + 1$ the local vanishing finer is empty, so we put $\chi_{d+1}(X, x) = 0$ for the Euler characteristic of the empty fiber.

Definition 2.2 We call the Euler characteristic of the local vanishing fiber of $p : (X, x) \to (\mathbb{C}^k, 0)$ the vanishing Euler characteristic $\chi_k(X, x)$. The vanishing Euler characteristics of (X, x) is the dim X-uple:

$$\mathbb{K}(X, x) = (\chi_2(X, x), \dots, \chi_{\dim X+1}(X, x))$$

In [23] (Théorème (5.3.1)) we have the following characterization of Whitney stratification:

Theorem 2.2 *Let X be an equidimensional reduced complex analytic space and let $\mathcal{S} = (S_I)_{i \in I}$ be a stratification of X. Suppose that for any pair (S_i, S_j) of strata of $\mathcal{S}$, such that $S_i \subset \overline{S}_j$ we have that the vanishing Euler characteristics $\mathbb{K}(\overline{S}_j, x)$ is constant for $x \in S_i$, then the stratification $\mathcal{S}$ is a Whitney stratification.*

As it is noticed in [23] (5.3) this theorem can be understood as a converse of Thom-Mather first isotopy theorem.

In fact, Theorem 2.2 is a consequence of Teissier's Theorem 2.1 stated above by using Théorème 4.1.1 of [23].

2.5 Summary

All this results show that Whitney equisingularity satisfies the requirements of 1.7.

The Theorem 2.2 is given to show that a Whitney stratification can be characterized by topological data and leads naturally to the question:

Can a Whitney stratification on a real analytic space (or a subanalytic space) be characterized by a real version of Theorem 2.2?

3 Saturation

In this section we essentially follow O. Zariski in [36].

3.1 Definition

Let $\mathcal{O}$ be a ring with identity. Let K be its total ring of fractions and Let $L \subset K$ be s a subfield of K. We assume:

1. The ring has no divisor of zero $\neq 0$;
2. In view of the preceding hypothesis, the total ring of fractions K being noetherian, the ring K is the direct sum of finite number of fields:

$$K = K_1 \oplus \cdots \oplus K_r;$$

3. The field L contains the unit of K, or equivalently no element $\neq 0$ is a zero divisor of K;
4. Let ε_i be the unit of K_i. Then K_i is a finite separable extension of $L\varepsilon_i$;
5. If $R = \mathcal{O} \cap L$ then, the ring $\mathcal{O}$ is integral over R.

Let us fix an algebraic closure Ω of L. Consider L-homomorphisms of K into Ω. Let ψ such a homomorphism. Then, for some i, $\psi(\varepsilon_i) = 1$ and, for $j \neq i$, $\psi(\varepsilon_j) = 0$. Then, for $j \neq i$, $\psi(K_j) = 0$ while ψ induces an isomorphism of K_i onto its image and $\psi(\alpha\varepsilon_i) = \alpha$ for any $\alpha \in L$. According to the hypothesis 4 above the number of L-homomorphisms of K into Ω is finite.

For any given i,$1 \leq i \leq r$, the compositum K_i^* of the fields $\psi(K_i)$ as ψ varies, i.e. the smallest field of Ω which contains the $\psi(K_i)$'s, is a finite Galois extension of L. Similarly, the compositum K^* of the fields $\psi(K)$ is a finite Galois extension of L.

Following O. Zariski, we shall say for two elements ξ and η of K, ξ dominates η if for any pair of homomorphisms ψ_1 and ψ_2, either $\psi_1(\eta) \neq \psi_2(\eta)$ and the quotient:

$$\frac{\psi_1(\xi) - \psi_2(\xi)}{\psi_1(\eta) - \psi_2(\eta)}$$

is integral over R, while $\psi_1(\eta) = \psi_2(\eta)$ implies $\psi_1(\xi) = \psi_2(\xi)$.

Note that if. for some i, ψ_1 and ψ_1' are L-homomorphisms of K into Ω such that $\psi_1(\varepsilon_i) = \psi_1'(\varepsilon_i) = 1$, then there is a L-monomorphism ϕ_0 of $\psi_i(K)$ into Ω such that $\psi_1' = \phi_0\psi_1$. The monomorphism ϕ_0 can be extended to a L-automorphism of the compositum K^* of the fields $\psi(K)$. Thus, for any element η of K, the set of elements $\psi_1'(\eta) - \psi_2(\eta)$ is the set of ϕ-images of the elements $\psi_1(\eta) - \psi_2(\eta)$ (ψ_1' and ψ_1 being fixed as above).

It yields that, if one fixes for each $i = 1, \ldots, r$ (where r is the number of fields in the hypothesis 2 above) a L-homomorphism $\psi_1^{(i)}$ of K into Ω such that $\psi_1^{(i)}(\varepsilon_i) = 1$, then, in order to verify that ξ dominates η, it is sufficient to verify the conditions of domination only for the pairs (ψ_1, ψ_2) where ψ_1 ranges over the set $\{\psi_1^{(1)}, \ldots, \psi_1^{(r)}\}$ and ψ_2 is any L-homomorphism of K into Ω.

In particular if $\mathcal{O}$ is a domain of integrity K is a field and $r = 1$. So, we may assume that $K \subset \Omega$. Then, the compositum K^* in Ω is the smallest Galois extension of L containing K. One can take $\psi_1^{(1)}$ to be the injection map of K into Ω. Then, the definition of domination is the following:

The element ξ dominates η if, for any element σ of the Galois group of the compositum K^* over L, the following holds: if $\sigma.\eta \neq \eta$, then the quotient $(\sigma.\xi - \xi)/(\sigma.\eta - \eta)$ is integral over $R = \mathcal{O} \cap L$, while $\sigma.\eta - \eta = 0$ implies $\sigma.\xi - \xi = 0$.

Now, we can define:

Definition 3.1 Let $\overline{\mathcal{O}}$ be the integral closure of $\mathcal{O}$ in K. The ring $\mathcal{O}$ is said to be saturated with respect to the field L if it contains every element of $\overline{\mathcal{O}}$ which dominates an element of $\mathcal{O}$.

Since the integral closure $\overline{\mathcal{O}}$ is saturated with respect to the field L, the set of saturated rings with respect to L which contain $\mathcal{O}$ and are contained in $\overline{\mathcal{O}}$ is not empty.

The intersection of two rings saturated with respect to the field L which contain $\mathcal{O}$ and are contained in $\overline{\mathcal{O}}$ is also saturated with respect to L. It implies:

Proposition 3.1 *The set of saturated rings with respect to the filed L which contain $\mathcal{O}$ and are contained in $\overline{\mathcal{O}}$ has a smallest element for the order induced by inclusion.*

Definition 3.2 The smallest element of the set of saturated rings with respect to the field L which contain $\mathcal{O}$ and are contained in $\overline{\mathcal{O}}$ is called the saturation of $\mathcal{O}$ with respect to L and is denoted by $\tilde{\mathcal{O}}_L$.

3.2 Dimension 1

We shall be interested in complex analytic local rings, i.e. local rings isomorphic to quotients of a ring of convergent series by an ideal. It is known that a complex analytic local ring is noetherian (see e.g. [10]).

A complex analytic local ring $\mathcal{O}$ is isomorphic to the ring $\mathcal{O}_{X,x}$ of germs of complex analytic functions on complex analytic space X at a point x.

If $\mathcal{O}$ is reduced the normal closure of $\overline{\mathcal{O}}$ of $\mathcal{O}$ in its total ring of fractions is isomorphic to the germ of meromorphic functions which are bounded in a neighborhood of x in X.

In the case of a complex analytic local ring of dimension 1 a result of F. Pham and B. Teissier proves that the saturation $\tilde{\mathcal{O}}_{X,x}$ of $\mathcal{O}_{X,x}$ with respect to the quotient field of the ring of convergent series in a parameter u of $\mathcal{O}_{X,x}$ is the ring of meromorphic functions which are Lipschitz in a neighborhood of x in X (see [19]).

Then, they prove that two germs of plane branches are topologically equisingular if the saturations of their local rings with respect to the quotient field of the ring of convergent series in a general parameter are isomorphic.

In fact in [7] A. Fernandes proved a geometrical result:

Theorem 3.1 *Let $(X, 0)$ and $(X', 0)$ be germs of complex analytic curves in $\mathbb{C}^2$ with branches $X_i, i \in I$ and $X'_j, j \in J$:*

$$X = \cup_{i \in I} X_i \text{ and } X' = \cup_{j \in J} X'_j.$$

Then the following conditions are equivalent:

1. *There exists a germ of the subanalytic bi-Lipschitz map $F : (X, 0) \to (X', 0)$;*
2. *There exists a bijection $\sigma : I \to J$ such that $\beta(X_i) = \beta(X'_{\sigma(i)})$ for all $i \in I$, where $\beta(\Gamma)$ is the Puiseux exponents of the branch Γ at 0, and such that $(X_i, X_j)_0 = (X'_{\sigma(i)}, X'_{\sigma(j)})_0$, for all $i, j \in I$, where $(\bullet, \bullet)_0$ denotes the intersection multiplicity at the point 0;*
3. *$(X, 0)$ is topologically equivalent to $(X', 0)$;*
4. *There exist an integer d, a germ of the curve $(C, 0) \subset (\mathbb{C}^d, 0)$, and two linear projections $p, p' : \mathbb{C}^d \to \mathbb{C}^2$, both general for C at 0 and such that $p(C) = X$ and $p'(C) = X'$.*

In summary two germs of plane curves at 0 have isomorphic saturations with respect to the quotient field of the ring of series in a transversal parameter, i.e. a parameter whose valuation in the normalization of the local rings is equal to the multiplicity of the local rings, if and only if there is a bijection between the branches such that corresponding branches have the same topology and pairwise intersection numbers at 0 of branches and their corresponding branches are equal, i.e. if and only if the two germs of curves at 0 are topologically equisingular.

3.3 Zariski Equisingularity

In [35] O. Zariski introduced the notion of equisingularity in codimension one for an algebraic variety. It is easy to adapt his definition to define equisingularity in codimension one for a germ of complex analytic set.

Definition 3.3 Let (X, x) be a germ of a reduced equidimensional complex analytic space. Let Y be a codimension one complex analytic subspace of X which is smooth at x. We suppose that (X, x) is embedded in $\mathbb{C}^N$. We say that X is equisingular in codimension one if the intersections of X with smooth spaces S_v transverse to Y in $\mathbb{C}^N$ define germs of curves $(S_v \cap X, S_v \cap Y)$ which are equisingular for any point $S_v \cap Y$ is a neighborhood of x.

Here equisingularity is concerning curves and is taken in the sense of [34]. According to what is said above, equisingularity means that the saturations of the local rings $\mathcal{O}_{S_v \cap X, S_v \cap Y}$ with respect to the quotient field of the ring of convergent series of a general parameter of the local rings.

In the case the singular locus has not codimension one, e.g. in the case of a normal germ of complex analytic space, one cannot use equisaturation to define equisingularity since the saturation of a ring is contained in its normalization and, when the singular locus has codimension 2, the local ring might be normal.

This is why O. Zariski imagines to define equisingularity by induction (see [37] Definition 3 p. 589):

Let (X, x) be a germ of d-equidimensional reduced complex analytic space. Let (Y, x) the germ of a smooth subspace of (X, x) which is contained in the singular locus. Let $p : (X, x) \to (\mathbb{C}^d, 0)$ a general projection of (X, x) onto $(\mathbb{C}^d, 0)$. Then, p is finite and one can define the discriminant of p. Let $\Delta(p)$ be the discriminant of p. The reduced germ $(|\Delta(p)|, 0)$ contains $(p(Y), 0)$ which is smooth, since p is a general projection. Then, (X, x) is equisingular along (Y, x) at the point x if $(|\Delta(p)|, 0)$ is equisingular along $(p(Y), 0)$ at the point 0.

Then, Zariski equisingularity can be defined by induction on the dimension of the ambient space.

In the case of a hypersurface, if the germ $(Y, 0)$ has codimension one in (X, x), then $(p(Y), x) \subset (|\Delta(p)|, 0)$, and we know that (X, x) is topologically equisingular at x along (Y, x) if and only if it is equisaturated along Y at x, which means that the Milnor number of the plane curve, intersection of a plane transversal P to Y at $P \cap Y$, plus its multiplicty minus 1 is constant along Y in a neighborhood of x in Y which implies $(p(Y), x) = (|\Delta(p)|, 0)$.

In the case of a subspace (Y, x) of higher codimension little is known. Recently W. Neumann and A. Pichon have studied hypersurfaces of dimension 3 and have related Zariski equisingularity with Lipschitz equisingularity which we shall define in the following section.

4 Lipschitz Viewpoint

Although F. Pham and B. Teissier were the first to relate Lipschitz meromorphic function and Saturation of local rings (see [19]), T. Mostowski introduced Lipschitz equisingularity where instead of homeomorphisms, he considers Lipschitz homeomorphisms (see [15–17]).

In particular, T. Mostowski proved that any complex analytic space have a Lipschitz stratification (see [15]).

Little has been done about Lipschitz equisingularity. L. Birbrair and T. Mostowski have introduced the notion of normal embedding in [3]. For instance, suppose that the reduced complex analytic space $X \subset U \subset \mathbb{C}^N$, where U is an open set of $\mathbb{C}^N$, then X is endowed by two metrics: the outer and the inner metrics. The outer metric is the metric induced by the embedding $X \subset U$. The inner metric is the the metric defined by $d(x, y) = \inf l(\gamma)$ where γ is a piecewise C^1 continuous path and $l(\gamma)$ is the length of γ. It may happen that these two metrics are different. When, they are the same one says that the embedding $X \subset U$ is normal.

In fact, in [7] A. Fernandes proved that topological equisingularity and Lipschitz equisingularity are the same for germs of plane curves.

For the case of surfaces, there are several papers on the Lipschitz structure of a germ of surface which begins with [1] until [4].

In higher dimensions little is known about the Lipschitz structure.

A lot is to be done with Lipschitz viewpoint.

Let us cite the recent result of [2] where it is proved that a germ of d-equidimensional reduced complex analytic space (X, x) which is bi-Lipschitz homeomorphic by a subanalytic map to $(\mathbb{C}^d, 0)$ is non-singular. The results on the Lipschitz structure of germs of complex surfaces should encourage new results on germs of reduced complex analytic spaces of higher dimension.

5 Open Problems

In this section we shall list some open problems on equisingularity.

5.1 Zariski Multiplicity Conjecture

Among basic problems about equisingularity, there is a basic problem by O. Zariski (see [31] and [37] p. 483):

Conjecture 1 Let (X, x) and (X', x') be topologically equisingular hypersurfaces. Their multiplicity $e(X, x)$ and $e(X', x')$ are equal.

In fact, we can weaken this conjecture:

Conjecture 2 Let (X_t, x_t) be a complex analytic family of topologically equisingular hypersurfaces. The multiplicty $e(X_t, x_t)$ is constant.

Both of these conjectures are true for complex analytic plane curves.

In a natural way, one may ask the same conjecture in the case of Lipschitz singularities.

5.2 Do the Diverse Definitions of Equisingularity Satisfy the Conditions of 1.7?

Above in 1.7 we give some hints which should be satisfy by an notion of equisingularity on a given complex analytic space.

For instance:

Question 3 Does Lipschitz equisingularity have an algebraic definition?

We saw above that this conjecture has a positive answer for a hypersurface along a codimension one stratum of the singular locus where one has topological equisingularity.

Question 4 Does Lipschitz equisingularity imply equimultiplicity?

This is nearly proved by G. Comte in the case the constants of the bi-Lipschitz homeomorphism satisfy some inequality in relation with the multiplicities of the singularities (see [6]). His proof might imply Question 4 for an analytic family of reduced complex analytic spaces.

5.3 What Are the Relations Between the Diverse Equisingularites

It was asked by Zariski (see [37] p. 487):

"Does topological equisingularity implies differential equisingularity (i.e. Whitney equisingularity)?"

It was proved by J. Briançon and J.P. Speder that the answer is negative in [5].

However, one should investigate other relations between the diverse notions of equisingularity.

For instance, a result of R. Thom and J. Mather (see [22] and [13]) shows that Whitney equisingularity implies topological equisingularity on a reduced complex analytic space. By definition Lipschitz equisingularity implies topological equisingularity.

Recent results of W. Neumann and A. Pichon assert that Lipschitz equisingularity is equivalent to Zariski equisingularity in dimension ≤ 3. Of course, it remains to understand the general case.

5.4 Is There Any Other Type of Equisingularity?

Then, it remains to find if there are other types of equisingularity.

Since it could be required that equisingularity is defined by algebraic data, we should study algebraic invariants for some equisingularity. For instance, Teissier proved that Whitney equisingularity is defined by the constancy of the multiplicities of some Polar varieties (see [21]). One should investigate the meaning of the constancy of Lê numbers or Lê cycles introduced by D. Massey in [12] (see also [11]).

5.5 Real Case

As we have mentioned above, is there a result similar to Theorem 2.2 in the real case?

Is there a characterization of Whitney stratification in the real case?

References

1. Birbrair, L.: Local bi-Lipschitz classification of two-dimensional semialgebraic sets (Portuguese). Rev. Semin. Iberoam. Mat. Singul. Tordesillas **2**, 29–34 (1998)
2. Birbrair, L., Fernandes, A., Lê, D.T., Sampaio, J.E.: Lipschitz regular complex algebraic sets are smooth. Proc. Am. Math. Soc. **144**, 983–987 (2016)
3. Birbrair, L., Mostowski, T.: Normal embeddings of semialgebraic sets. Mich. Math. J. **47**, 125–132 (2000)
4. Birbrair, L., Neumann, W., Pichon, A.: The thick-thin decomposition and the bilipschitz classification of normal surface singularities. Acta Math. **212**, 199–256 (2014)
5. Briançon, J., Speder, J.P.: La trivialité topologique n'implique pas les conditions de Whitney. C. R. Acad. Sci. Paris Ser. A - B **280**(Aiii), A365–A367 (1975)
6. Comte, G.: Multiplicity of complex analytic sets and bi-Lipschitz maps. Real analytic and algebraic singularities (Nagoya/Sapporo/Hachioji, 1996). Pitman Research Notes in Mathematics Series, pp. 182–188. Longman, Harlow (1998)
7. Fernandes, A.: Topological equivalence of complex curves and bi-Lipschitz homeomorphisms. Mich. Math. J. **51**, 593–606 (2003)
8. Grothendieck, A.: Techniques de construction en Géométrie analytique VI: Etude locale des morphismes. Germes d'espaces analytiques, Platitude , Morphismes simples, Sém. Henri Cartan, 13 ème année, Exposé XIII, Secrétariat Math., 11 Rue Pierre Curie, 75005 (1962)
9. Hironaka, H.: Normal cones in analytic Whitney stratifications. Inst. Hautes Études Sci. Publ. Math. **36**, 127–138 (1969)
10. Houzel, C.:Géométrie analytique locale, I, in Séminaire H. Cartan 1960–1961, Familles d'espaces complexes, Exposé 18, Secrétariat Mathématique, IHP, 11 rue Pierre et Marie Curie, Paris (1962)
11. Massey, D.B.: Lê Cycles and Hypersurface Singularities. Lecture Notes in Mathematics, Xii+131 pp. Springer, Berlin (1995)
12. Massey, D.B.: The Lê varieties I. Invent. Math. **99**, 357–376 (1990); The Lê varieties II. Invent. Math. **104**, 113–148 (1991)
13. Mather, J.: Notes on topological stability. Bull. Am. Math. Soc. **49**, 475–506 (2012)
14. Milnor, J.: Singular points of complex hypersurfaces. Annals of Mathematics Studies **61** Princeton University Press, Princeton, New Jersey; University of Tokyo Press, Tokyo (1968) iii+122 pp
15. Mostowski, T.: Tangent cones and Lipschitz stratifications, Singularities (Warsaw, 1985), Banach Center Publ. vol. 20, pp. 303–322. PWN, Warsaw (1988)
16. Mostowski, T.: Lipschitz equisingularity. Diss. Math. (Rozprawy Mat.) **243**, 46 (1985)
17. Mostowski, T.: A criterion for Lipschitz equisingularity. Bull. Pol. Acad. Sci. Math. **37**(1–6), 109–116 (1989)
18. Narasimhan, R.: Introduction to the Theory of Analytic Spaces. SLN, vol. 25. Springer, Verlag (1966)
19. Pham F.: (work with B. Teissier), Fractions lipschitziennes et saturation de Zariski des algèbres analytiques complexes, Actes du Congrès International des Mathèmaticiens (Nice, 1970), Tome 2, pp. 649–654. Gauthier-Villars, Paris, (1971)
20. Teissier, B.: Cycles èvanescents, sections planes et conditions de Whitney, Singularitès á Cargèse (Rencontre Singularitès Géom. Anal., Inst. tudes Sci., Cargèse, 1972). Asterisque, Nos. 7 et 8, Société Mathématique de France, Paris, pp. 285–362. (1973)
21. Teissier, B.: Variétés polaires II. Multiplicités polaires, sections planes, et conditions de Whitney, Algebraic geometry (La Rábida, 1981), 314–491, Lecture Notes in Math. **961**, Springer, Berlin, 1982
22. Thom, R.: Ensembles et morphismes stratifiés. Bull. Am. Math. Soc. **75**, 240–284 (1969)
23. Lê, D.T., Teissier, B.: Cycles évanescents, sections planes et conditions de Whitney II, Singularities, Part 2 (Arcata, Calif., 1981). Proceedings of Symposia in Pure Mathematics, pp. 65–103. American Mathematical Society, Providence, RI (1983)

24. Lê, D.T.: Sur un critre d'équisingularité. C. R. Acad. Sci. Paris Ser. A - B **272**, A138–A140 (1971)
25. Lê, D.T.: Sur les nœuds algébriques. Compos. Math. **25**, 281–321 (1972)
26. Lê, D.T.: Calcul du nombre des cycles évanouissants d'une hypersurface complexe. Ann. Inst. Fourier **23**, 261–270 (1973)
27. Lê, D.T., Ramanujam, C.P.: The invariance of Milnor's number implies the invariance of the topological type. Am. J. Math. **98**, 67–78 (1976)
28. Lê, D.T., Teissier, B.: Variétés polaires locales et classes de Chern des variétés singulières. Ann. Math. **114**, 457–491 (1981)
29. Lê, D.T., Teissier, B.: Limites d'espaces tangents en géométrie analytique. Comment. Math. Helv. **63**, 540–578 (1988)
30. Whitney, H.: Tangents to an analytic variety. Ann. Math. **81**, 496–549 (1965)
31. Zariski, O.: Contributions to the problem of equisingularity, Questions on Algebraic Varieties (C.I.M.E., III Ciclo, Varenna, 1969), pp. 261–343. Edizioni Cremonese, Rome (1970)
32. Zariski, O.: Equisingular points on algebraic varieties (1965) in Seminari 1962/63 Anal. Alg. Geom. e Topol., Vol. 1, Ist. Naz. Alta Mat. pp. 164–177, Ediz. Cremonese, Rome
33. Zariski, O.: Equisingular points on algebraic varieties (1966). Some Recent Advances in the Basic Sciences In: Proceedings of the Annual Science Conference on Belfer Graduate School Science, Yeshiva University, New York, vol. 1, pp. 163–175. (1962–1964)
34. Zariski, O.: Studies in equisingularity. I. Equivalent singularities of plane algebroid curves. Am. J. Math. **87**, 507–536 (1965)
35. Zariski, O.: Studies in equisingularity. II. Equisingularity in codimension 1 (and characteristic zero). Am. J. Math. **87**, 972–1006 (1965)
36. Zariski, O.: Studies in equisingularity. III. Saturation of local rings and equisingularity. Am. J. Math. **90**, 961–1023 (1968)
37. Zariski, O.: Some open questions in the theory of singularities. Bull. Am. Math. Soc. **77**, 481–491 (1971)

On the Factorization of the Polar of a Plane Branch

A. Hefez, M. E. Hernandes and M. F. H. Iglesias

Abstract Irreducible complex plane curve germs with the same characteristic exponents form an equisingularity class. In this paper we determine the Zariski invariants that characterize the general polar of a general member of such an equisingularity class. More precisely, we will describe explicitly the characteristic exponents of the irreducible components of the polar and their mutual intersection multiplicities, allowing us in particular to describe completely the content of each of Merle's packages of the polar.

Keywords Polar curves · Polar decomposition · Equisingularity

1 Introduction

The polar of a complex plane curve is a classical object that was used in classical algebraic geometry for enumeration purposes, such as Plücker's formulas. The subject was resuscitated during the 70s in the work of B. Teissier [9] for the study of families of singular hypersurfaces. On the other hand, the description of the topological type or, equivalently, the equisingularity class of the general polar of a complex plane

---

A. Hefez and M. E. Hernandes were partially supported by the CNPq grants 307873/2016-1 and 303594/2014-4, respectively, while the third author was supported by a fellowship from CAPES/Fundação Araucária.

A. Hefez (✉)
Universidade Federal Fluminense, Av. Quintino Bocaiúva 187, Ap. 402,
São Francisco, Niterói, RJ 24.360-022, Brazil
e-mail: hefez@mat.uff.br

M. E. Hernandes
Universidade Estadual de Maringá, Av. Colombo 5790, Maringá, Paraná
87020-900, Brazil
e-mail: mehernandes@uem.br

M. F. H. Iglesias
Universidade Federal Fluminense, Rua Prof Marcos Waldemar de Freitas
S/N Bloco H 4o Andar Campus do Gragoatá, Niteroi, RJ 24.210-201, Brazil
e-mail: mhernandezi@pucp.pe

R. N. Araújo dos Santos et al. (eds.), *Singularities and Foliations. Geometry, Topology and Applications*, Springer Proceedings in Mathematics & Statistics 222, https://doi.org/10.1007/978-3-319-73639-6_11

branch in a given equisingularity class is an old and still open problem. F. Enriques and O. Chisini, in [4] described explicitly a cluster of infinitely near points obtained by means of the characteristic exponents of a Newton-Puiseux parametrization of the branch that should determine the equisingularity class of its general polar by requiring that it passes through it in a certain way. This is false as shown with a simple example in [8]. The equisingularity class of the general polar of the germ of a plane curve is actually an analytic invariant of the germ and not a topological invariant. In the papers [2, 3], E. Casas-Alvero showed that the answer by Enriques-Chisini was generically true, in the sense that it is true for a general branch in the given equisingularity class. The characterization of the equisingularity class of a curve given by a cluster, as realized by Enriques-Chisini and adopted by Casas-Alvero, was replaced later by the more explicit equivalent characterization given by Zariski in terms of the equisingularity characters: the number of branches of the curve, their characteristic exponents and their mutual intersection multiplicities. The aim of this paper is to show how to pass from the Enriques-Chisini-Casas description of the cluster of the general polar of the general member of an equisingularity class of plane branches, to its Zariski characters. This allows us to get by simple and elementary arithmetical operations all Zariski equisingularity characters of the general polar of a general member of an equisingularity class of branches in terms of the characteristic exponents of the branch.

It should be pointed out that there were two previous partial results in the direction of ours. On one hand, a rough decomposition of the polar into packages, not necessarily irreducible, as described in a theorem of Merle in [7], which we will make explicit later. This decomposition is, in some sense, the best result satisfied by all branches, not only by the general ones, in a given equisingularity class, but is not sufficient to describe the equisingularity class of the polar of the general branch. On the other hand, in the very particular case of branches with only one characteristic exponent, our result coincides with that of [2], the first attempt made by that author toward his solution of the general case of an arbitrary number of characteristic exponents reached by changing strategy with the use of the Enriques-Chisini approach. So, our result may be viewed as a generalization of [2].

Actually, in this paper, more precisely in Theorem 3.1, we describe all irreducible factors of the general polar of the general member of a given equisingularity class of plane branches, their characteristic exponents and their mutual intersection multiplicities, that is, Zariski characters of the equisingularity class of such polars. So, our result may also be viewed as a refinement of Merle's decomposition.

The content of the paper is as follows: in Sect. 2 we state some known results we will need, give a brief introduction to the theory of infinitely near points, recall the definition of a cluster, state Enriques' Theorem and describe the related diagrams. Section 3 contains our main result that consists of the description of the irreducible components of each of Merle's packages in a constructive and inductive way, the characteristic exponents of each component and their mutual intersection multiplicities. Our proof is by descent: we start determining the last Merle's package by using an appropriate infinitely near point. To obtain the next package, we consider a new cluster constructed by means of the original cluster, the package already obtained, the proximity relations and Noether's formula.

Finally, as an application, we use our analysis to classify all special equisingularity classes of irreducible plane germs such that their general members have general polar that admit only irreducible components with at most one less characteristic exponent than the branch, generalizing a result obtained in [6] in the case of curves with two characteristic exponents.

2 Classical Results

A germ of an analytic plane curve at the origin of $\mathbb{C}^2$ is a germ of set $C = C_f = \{(x, y) \in (\mathbb{C}^2, 0);\ f(x, y) = 0\}$, where $f \in \mathbb{C}\{x, y\}$ is a convergent complex power series in two variables at the origin. Two such germs will be considered analytically equivalent if there is a germ of analytic diffeomorphism φ of $(\mathbb{C}^2, 0)$, also called an analytic change of coordinates, such that $\varphi(C_f) = C_g$. When the above φ is just a homeomorphism, we say that C_f and C_g are topologically equivalent, or equisingular, writing, in this case, $C_f \equiv C_g$.

From now on, we will assume that f is an irreducible power series and call its associated curve $C = C_f$ a branch. After an analytic change of coordinates, if necessary, we may assume that $\mathrm{I}(f, x) = n < m = \mathrm{I}(f, y)$ and $n \nmid m$, where $\mathrm{I}(f, g)$ stands for the intersection multiplicity at the origin of the plane curve germs C_f and C_g. In this situation, n coincides with the multiplicity of C.

With such coordinates suitably chosen, it is well known that a branch C admits a Newton-Puiseux parametrization of the form (t^2, t^m), if $n = 2$, or $(t^n, \sum_{m \le i < c} a_i t^i)$, if $n > 2$, where c is some positive integer, called the conductor of C. Conversely, given such a parametrization, attached to it there is a well defined branch. It is also classically known that the topological, or equisingularity class of C is completely determined by n and the characteristic exponents $m_1, \ldots, m_r$, defined by

$$m_i = \min\{j;\ a_j \neq 0 \text{ and } e_{i-1} \nmid j\},$$

where $e_0 = n$ and, for $k > 0$, $e_k = \gcd(n, m_1, \ldots, m_k)$ and $e_r = 1$. The integer r is what is called the genus of C. This classical terminology, although confusing, has nothing to do with the genus of a curve in Riemann's sense. We also define the integers $d_0 = 1$ and $d_i = \frac{e_{i-1}}{e_i}$, for $i = 1, \ldots, r$.

When a germ of curve is not irreducible, but reduced, Zariski has shown that its equisingularity type is determined by the equisingularity type of its branches and by their mutual intersection multiplicities.

In what follows, we will consider the set $K(n, m_1, \ldots, m_r)$ that parametrizes all Newton-Puiseux finite expansions as above with multiplicity n and characteristic exponents $m_1, \ldots, m_r$.

Let f be a reduced power series. The germ of curve defined by $P_{(a:b)}(f) = af_x + bf_y = 0$ is the polar curve of f in the direction $(a : b) \in \mathbb{P}^1$. When $(a : b)$ is a general point of $\mathbb{P}^1$, we say that the associated polar $P_{(a:b)}(f) = 0$ is general and we denote it simply by $P(f)$.

In this paper we only consider the general polar of f and refer to it simply as the polar curve of f.

In general, the polar curve depends upon the equation f of C_f, however its topological type depends only upon the analytic type of C_f (see [1, Theorem 7.2.10]).

The next result due to M. Merle provides a rough decomposition of $P(f)$ in packages of curves, not necessarily irreducible, that gives partial information about the topology of $P(f)$.

Theorem 2.1 (Merle [7]) *Let C_f be a germ of an irreducible curve with multiplicity n and characteristic exponents $m_1, \ldots, m_r$. Then the general polar $P(f)$ has a decomposition of the form*

$$P(f) = \xi_1 \xi_2 \cdots \xi_r,$$

where each ξ_i, not necessarily. irreducible, satisfies the following conditions:

(i) *The multiplicity of ξ_i is given by $m(\xi_i) = d_0 d_1 d_2 \cdots d_{i-1}(d_i - 1)$;*
(ii) *Each irreducible factor $\xi^i_{j,k}$ of ξ_i satisfies*

$$\frac{\mathrm{I}(\xi^i_{j,k}, f)}{m(\xi^i_{j,k})} = \frac{1}{n} \sum_{w=1}^{i-1} (e_{w-1} - e_w) m_w + m_i.$$

When $i = 1$ we will denote $\xi^i_{j,k}$ by $\xi_{j,k}$.

Let us make some few remarks. Merle's Theorem does not describe completely the topology of $P(f)$, because it does not describe the branches inside each package ξ_i. Such branches depend upon the analytic type of f and not only upon its topological type. It also does not describe the intersection multiplicities among the branches of the polar. The terms in the second conclusion are the so-called polar quotients and the equality says that the branches $\xi^i_{j,k}$ have contact order with C_f equal to m_i, which implies that they have genus at least $i - 1$, but they may have greater genus.

On the other hand, Casas-Alvero in [3], determines the equisingularity class of $P(f)$, for an f corresponding to a general member of the set $K(n, m_1, \ldots, m_r)$ in terms of a certain weighted cluster obtained from the Enriques diagram attached to the resolution of C_f.

If $r = 1$, Casas-Alvero in [2] describes explicitly the factorization of $P(f)$ as follows:

Let n and m be two coprime natural numbers. Consider the euclidean GCD algorithm applied to the pair n, m:

$$\begin{aligned}
m &= h_0 n + n_1 \\
n &= h_1 n_1 + n_2 \\
n_1 &= h_2 n_2 + n_3 \\
&\vdots \\
n_{s-2} &= h_{s-1} n_{s-1} + 1 \\
n_{s-1} &= h_s 1.
\end{aligned}$$

We denote by $\frac{m}{n} = [h_0, \dots, h_s]$ the partial fraction decomposition of $\frac{m}{n}$, adjusted in such a way that s becomes even, say $s = 2t$ (for example, $[a_0, a_1] = [a_0, a_1 - 1, 1]$). Put $\frac{q_i}{p_i} = [h_0, \dots, h_i]$ in such a way that q_i and p_i are coprime. So, one has the following theorem:

Theorem 2.2 (Casas-Alvero [2]) *If f is a general member of $K(n, m)$ where $\gcd(n, m) = 1$, then $P(f)$ has branches $\xi_{i,j}$, $i = 1, \dots, t$, $j = 1, \dots, h_{2i}$, having multiplicity* $\mathrm{I}(\xi_{i,j}, X) = p_{2i-1}$ *and* $\mathrm{I}(\xi_{i,j}, Y) = q_{2i-1}$ *and such that*

$$\mathrm{I}(\xi_{i,j}, \xi_{i',j'}) = \min(p_{2i-1}q_{2i'-1}, p_{2i'-1}q_{2i-1}).$$

Remark 2.3 Notice that the branches of $P(f)$ for a general $f \in K(n, m)$ are all smooth if and only if $p_{2i-1} = 1$, for all i. But, since the p_i form an increasing sequence, this only may happen when $2t - 1 = 1$, that is, $t = 1$.

If $\frac{m}{n} = [h_0, h_1, h_2]$, then we have $m = h_0 n + n_1$; $n = h_1 n_1 + 1$; $n_1 = h_2 \cdot 1$. The condition that $\frac{q_1}{p_1} = [h_0, h_1]$ is an integer is equivalent to $h_1 = 1$ and $h_2 = n - 1$. Hence the fact that $P(f)$ has only smooth branches is equivalent to $m = (h_0 + 1)$ $n - 1$.

In the case where $\frac{m}{n} = [h_0, h_1 - 1, 1]$, so $\frac{q_1}{p_1} = [h_0, h_1 - 1]$. Now, the condition that $\frac{q_1}{p_1}$ is an integer is equivalent to $h_1 = 2$ and this in turn is equivalent to $n = 2$. Hence, the fact $P(f)$ has only smooth branches is equivalent to $m = h_0 \cdot 2 + 1 = (h_0 + 1)2 - 1$.

In conclusion, one has that $P(f)$, where f corresponds to a general member of $K(n, m)$, has only smooth branches, if and only if $m = \lambda n - 1$, where λ is some natural number greater than 1.

2.1 *The Infinitely Near Points*

Let $S_0 \subset \mathbb{C}^2$ be an open set containing the origin $0 = (0, 0)$. Let $\pi \colon S_1 \to S_0$ be the blow-up of S_0 centered at 0 and denote by $E_0 = \pi^{-1}(0)$ the exceptional divisor of π. We denote by $\mathcal{N}_0$ the set of infinitely near points to 0, which can be viewed as the disjoint union of 0 and all exceptional divisors obtained by successive blowing-ups above 0. The set of points on the exceptional divisor of the ith blow-up centered at a point $P \in S_{i-1}$ are called the first infinitesimal neighborhood of P and the ith infinitesimal neighborhood of 0. The set $\mathcal{N}_0$ is naturally endowed with an order relation defined by $P < Q$ if and only if $Q \in \mathcal{N}_P$.

Given $f \in \mathbb{C}\{x, y\}$ that defines a curve C and given P in the first infinitesimal neighborhood of 0, we denote by C_P the germ of curve at P defined via the strict transform $\widetilde{f_P}$ of f, which might be viewed as the germ at P of the closure of $\pi^{-1}(C \setminus \{0\})$. By induction we may obtain the strict transform of C at any point of $\mathcal{N}_0$.

The multiplicity of C_P at $P \in \mathcal{N}_0$ is $m_P(f) = m_P(\widetilde{f_P})$. We say that P lies on C, or belongs to it, if and only if $m_P(f) > 0$, and denote by $\mathcal{N}_0(f)$ the set of all such points. A point $P \in \mathcal{N}_0(f)$ is *simple* (resp. *multiple*) if and only if $m_P(f) = 1$ (resp.

$m_P(f) > 1$). Given two germs of curves C_f and C_g, their intersection multiplicity at 0 can be computed by means of Noether's formula as follows:

$$\mathrm{I}(f, g) = \sum_{P \in \mathcal{N}_0(f) \cap \mathcal{N}_0(g)} m_P(f) m_P(g). \tag{1}$$

Given $P, Q \in \mathcal{N}_0$ such that $P < Q$, we say that Q is *proximate* to P, written $Q \to P$. if and only if Q lies on the exceptional divisor E_P or in the strict transform of E_P. A point P is said to be *free* (resp. *satellite*) if it is proximate to exactly one point (resp. two points), and these are the only possibilities. Notice that $Q \to P$ implies $Q > P$, but not conversely.

An important formula due to Noether is the following:

$$m_P(f) = \sum_{Q \to P} m_Q(f).$$

A point $P \in \mathcal{N}_0(f)$ is *singular* if it is either multiple, or satellite, or precedes a satellite point on C_f, and it is *non-singular*, or *regular*, otherwise. Equivalently, P is non-singular if and only if it is free and there is no satellite point $Q > P$.

Let $C_f = \bigcup_{i=1}^{s} C_{f_i}$ be a reducible plane curve. The point $P_i \in \mathcal{N}_0(f)$ is the first regular point on C_{f_i}. We denote by

$$S(f) = \{Q \in \mathcal{N}_0(f);\ Q = P_i \text{ or } Q \text{ is singular}\}.$$

It is well known (see for instance [1, Theorem 3.8.6]) that two curves C_f and C_g are equisingular, if and only if there exists a bijection $\phi\colon S(f) \to S(g)$ such that both ϕ, ϕ^{-1} preserve the natural ordering and the proximity relations among their infinitely near points

Definition 2.4 A cluster $\mathcal{K}$ is a finite subset $K \subset \mathcal{N}_0$ such that if $P \in K$, then any other point $Q < P$ also belongs to K, together with a valuation $v_{\mathcal{K}}\colon K \longrightarrow \mathbb{Z}$. The set K is called the support of $\mathcal{K}$ and the number $v_{\mathcal{K}}(P)$ is the *virtual multiplicity* of P in $\mathcal{K}$.

We follow Casas-Alvero, representing a cluster by means of an Enriques diagram, which is a tree whose vertices are identified with the points in K (the root corresponds to the origin 0) and there is an edge between P and Q if and only if P lies on the first neighborhood of Q or vice-versa. Moreover, the edges are drawn according to the following rules:

(i) If Q is free and proximate to P, the edge joining P and Q is curved and if $P \neq 0$, it is tangent to the edge ending at P.
(ii) If P and Q (Q in the first neighborhood or P) have been represented, the other points proximate to P in successive neighborhoods of Q are represented on a straight half-line starting at Q and orthogonal to the edge ending at Q.

Definition 2.5 We will say that a curve C_f goes sharply through the cluster $\mathcal{K}$ if C_f goes through K with effective multiplicities equal to the virtual ones and has no singular points outside of K.

2.2 *Enriques' Theorem*

In what follows we will describe the *cluster of singularities* of a plane branch C_f, that is, the cluster $\mathcal{K}(f) = (S(f), v_{\mathcal{K}(f)})$, where $v_{\mathcal{K}(f)}(P) = m_P(f)$.

If C_f has multiplicity n and characteristic exponents $m_1, \ldots, m_r$, then C_f is analytically equivalent to a curve that admits a Newton-Puiseux parametrization of the form $x = t^n$, $y = \sum_{i \geq m_1} a_i t^i$ such that $a_{m_k} \neq 0$ for $k = 1, \ldots, r$ and $a_{m_1} = 1$.

Denoting $m_0 = 0$, $n_0^k = e_{k-1} = \gcd(n, m_1, \ldots, m_{k-1})$, $n_0^{k+1} = n_{s(k)}^k = e_k$, we consider the euclidean expansions

$$
\begin{aligned}
m_k - m_{k-1} &= h_0^k n_0^k + n_1^k \\
n_0^k &= h_1^k n_1^k + n_2^k \\
n_1^k &= h_2^k n_2^k + n_3^k \\
&\vdots \\
n_{s(k)-2}^k &= h_{s(k)-1}^k n_{s(k)-1}^k + n_{s(k)}^k \\
n_{s(k)-1}^k &= h_{s(k)}^k n_{s(k)}^k.
\end{aligned}
$$

When $k = 1$, we omit the index k in n_j^k, h_j^k and $s(k)$. With the previous notation we have following theorem (see [1, Theorem 5.5.1] or [4, IV.I]).

Theorem 2.6 (Enriques) *The cluster of C is composed by r blocks, which we describe below.*

The first block is composed as follows:

It starts with the point $P_{0,1} = O$, followed by points $P_{0,i} \in \mathcal{N}_0(f)$, $i = 2, \ldots, h_0$, each one in the first neighborhood of the preceeding one, all free with value n.

It continues with the point $P_{1,1}$, free in the fist neighborhood of P_{0,h_0}, followed by points $P_{1,i}$, $i = 2, \ldots, h_1$, not free and each in the first neighborhood of the preceeding one, with value n_1.

For $2 \leq j \leq s$, the point $P_{j,1}$ is proximate to $P_{j-2,h_{j-2}}$ and for $i = 2, \ldots, h_j$ we have $P_{j,i}$ proximate to $P_{j-1,h_{j-1}}$ in the first neighborhood of $P_{j,i-1}$ with value n_j.

For $1 < k \leq r$, we put $P^{k-1}_{s(k-1),h^{k-1}_{s(k-1)}} = P^k_{0,0}$. The points of the cluster in the kth block after $P^k_{0,0}$ are given by:

h^k_0 free points $P^k_{0,1}, \ldots, P^k_{0,h^k_0}$ with value n^k_0;

h^k_1 points $P^k_{1,1}, \ldots, P^k_{1,h^k_1}$ with value n^k_1 proximate to P^k_{0,h^k_0}.

For $2 \leq j \leq s(k)$, we have h^k_j points $P^k_{j,1}, \ldots, P^k_{j,h^k_j}$, where the first one is proximate to $P^k_{j-2,h^k_{j-2}}$ and for $i = 2, \ldots, h^k_j$, the point $P^k_{j,i}$ is proximate to $P^k_{j-1,h^k_{j-1}}$ and all of them have value n^k_j.

This yields the following Enriques diagrams:

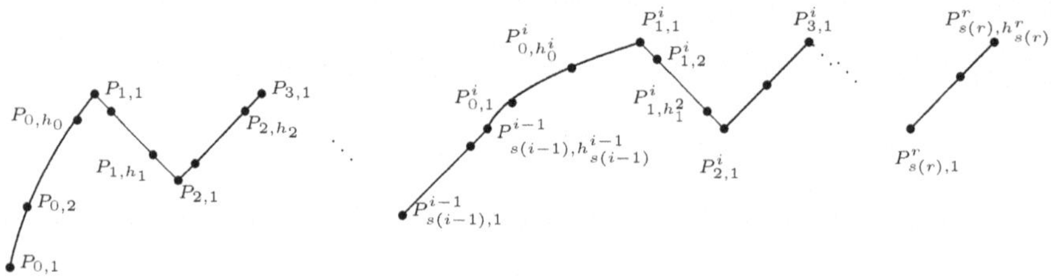

if $m_i - m_{i-1} > e_{i-1}$.

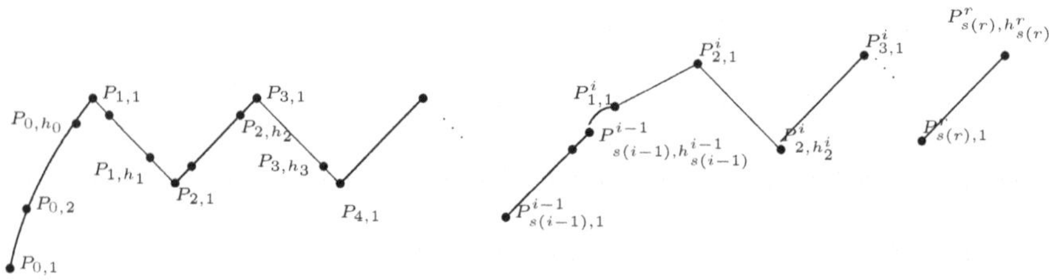

if $m_i - m_{i-1} < e_{i-1}$.

3 Description of the Packages in Merle's Theorem

By [3, Proposition 11.1], the cluster $\mathcal{K}^r$ of the polar of a branch corresponding to a general member of $K(n, m_1, \ldots, m_r)$ has the same support K^r as the cluster $\mathcal{K}(f)$ of the singularities of C_f, that is

$$K^r = \{P^k_{i,j}; 1 \le k \le r, 0 \le i \le s(k), 1 \le j \le h^i_{s(i)}\},$$

with valuation:

$$v_{K^r}(P^k_{i,j}) = \begin{cases} m_{P^k_{s(k),h^k_{s(k)}}}(f) - 1, & \text{if } (i,j) = (s(k), h^k_{s(k)}); \text{ otherwise,} \\ m_{P^k_{i,j}}(f) - 1, & \text{if } i \text{ is even,} \\ m_{P^k_{i,j}}(f), & \text{if } i \text{ is odd.} \end{cases} \tag{2}$$

To describe explicitly Merle's packages of such a polar, we firstly consider the cluster $\mathcal{K}'$ given as follows:

1. If $m_r - m_{r-1} > e_{r-1}$, then its support is $K' = \{P^r_{i,j}; 0 \le i \le s(r), 1 \le j \le h^r_{s(r)}\}$, with valuation:

$$v_{\mathcal{K}'}(P^r_{i,j}) = \begin{cases} 0, & \text{if } (i,j) = (s(r), h^r_{s(r)}); \text{ otherwise,} \\ n^r_i - 1, & \text{if } i \text{ is even,} \\ n^r_i, & \text{if } i \text{ is odd.} \end{cases}$$

Notice that $\mathcal{K}'$ represents the cluster of the polar of a general curve f_r in $K(e_{r-1}, m_r - m_{r-1})$ based at $P^r_{1,0}$.

2. If $m_r - m_{r-1} < e_{r-1}$, then its support is

$$K' = \{P^r_{i,j}; 1 \le i \le s(r), 1 \le j \le h^r_{s(r)}\} \cup \{P^{r-1}_{s(r-1),h^{r-1}_{s(r-1)}}\},$$

with same values as above on the first set and

$$v_{\mathcal{K}'}(P^{r-1}_{s(r-1),h^{r-1}_{s(r-1)}}) = e_{r-1} - 1.$$

Notice that this represents the cluster of the polar of a general curve f_r in $K(e_{r-1}, m_r - m_{r-1} + e_{r-1})$ based at $P^{r-1}_{s(r-1),h^{r-1}_{s(r-1)}}$.

Now, by Theorem 2.2, we have that:

$$P(f_r) = \prod_{i=1}^{[\frac{s(r)+1}{2}]} \prod_{j=1}^{h^r_{2i}} \gamma^r_{i,j}$$

where $\gamma^r_{i,j}$ is determined by $m_r - m_{r-1}$ and e_{r-1} according to the following cases:

1′. If $m_r - m_{r-1} > e_{r-1}$, writing $\frac{m_r - m_{r-1}}{e_{r-1}} = [h^r_0, \ldots, h^r_{s(r)}]$, one has that $\gamma^r_{i,j} \in K(p^r_{2i-1}, q^r_{2i-1})$, where $\frac{q^r_{2i-1}}{p^r_{2i-1}} = [h^r_0, \ldots, h^r_{2i-1}]$ and $\gcd(p^r_{2i-1}, q^r_{2i-1}) = 1$.

2′. If $m_r - m_{r-1} < e_{r-1}$, writing $\frac{m_r - m_{r-1} + e_{r-1}}{e_{r-1}} = [1, h^r_1, \ldots, h^r_{s(r)}]$, one has that $\gamma^r_{i,j} \in K(p^r_{2i-1}, p^r_{2i-1} + q^r_{2i-1})$, where $\frac{q^r_{2i-1}}{p^r_{2i-1}} = [0, h^r_1, \ldots, h^r_{2i-1}]$.

Now, by blowing down the branches $\gamma^r_{i,j}$ to the point $P_{0,1}$, with respect to the cluster of singularities of any element in $K(\widetilde{n}, \widetilde{m}_1, \ldots, \widetilde{m}_{r-1})$, where $\widetilde{n} = n/e_{r-1}$ and $\widetilde{m_i} = m_i/e_{r-1}$, $i = 1, \ldots, r-1$, we get branches $\xi^r_{i,j}$ that pass through the points $K^{r-1} \cup K^{'}_i$, where $K^{r-1} = \{P^k_{i,j};\ 1 \le k \le r-1, 0 \le i \le s(k), 1 \le j \le h^i_{s(i)}\}$, and $K^{'}_i = \{P^r_{0,1}, \ldots, P^r_{2i-1,h^r_{2i-1}}\}$, with multiplicities at the points of K^{r-1} given by

$$m_{P^k_{i,j}}(\xi^r_{i,j}) = \frac{m_{P^k_{i,j}}(f)}{e_{r-1}} p^r_{2i-1}, \quad k = 1, \ldots, r-1;$$

and the multiplicities at the points of $K^{'}$ given according to the following cases:
1″. For $m_r - m_{r-1} > e_{r-1}$ we have that the multiplicities of the $\xi^r_{i,j}$ at the points $P^r_{0,1}, \cdots, P^r_{2i-1,h^r_{2i-1}}$ are determined by $\frac{q^r_{2i-1}}{p^r_{2i-1}}$, then by definition of $\xi^r_{i,j}$ we have that the strict transform of the curve ξ_r in the point $P^r_{0,1}$ goes sharply through the cluster $\mathcal{K}^{'}$, since the strict transform of $\xi^r_{i,j}$ at the point $P^r_{0,1}$ coincides with $\gamma^r_{i,j}$.
2″. For $m_r - m_{r-1} < e_{r-1}$ we have that the multiplicities of the $\xi^r_{i,j}$ at the points $P^{r-1}_{s(r-1),h^{r-1}_{s(r-1)}}, P^r_{0,1}, \ldots, P^r_{2i-1,h^r_{2i-1}}$ are determined by $1 + \frac{q^r_{2i-1}}{p^r_{2i-1}}$ and, from the definition of $\xi^r_{i,j}$, the strict transform of curve ξ_r at the point $P^{r-1}_{s(r-1),h^{r-1}_{s(r-1)}}$ goes sharply through the cluster $\mathcal{K}^{'}$, for the same reason as above.

From the above analysis, one sees that

$$\xi_r = \prod_{i=1}^{[\frac{s(r)+1}{2}]} \prod_{j=1}^{h^r_{2i}} \xi^r_{i,j},$$

with

$$\xi^r_{i,j} \in K(p^r_{2i-1}\widetilde{n}, p^r_{2i-1}\widetilde{m}_1, \ldots, p^r_{2i-1}\widetilde{m}_{r-1}, p^r_{2i-1}\widetilde{m}_{r-1} + q^r_{2i-1}).$$

In order to describe the decomposition of the polar of f we consider the cluster $\overline{\mathcal{K}}$ whose support is the same as that of $\mathcal{K}(f)$ (or of $\mathcal{K}^r$), with valuation $v_{\overline{\mathcal{K}}}(P^k_{i,j}) = v_{\mathcal{K}^r}(P^k_{i,j}) - m_{P^k_{i,j}}(\xi_r)$.

In particular, we have that $v_{\overline{\mathcal{K}}}(P^r_{i,j}) = 0$ and if $m_r - m_{r-1} < e_{r-1}$, then

$$v_{\overline{\mathcal{K}}}(P^{r-1}_{r-1,s(r-1)}) = (e_{r-1} - 1) - (e_{r-1} - 1) = 0.$$

By a computation, using the proximity relations, one obtains

$$v_{\overline{\mathcal{K}}}(P^{r-1}_{i,j}) = \begin{cases} (\widetilde{n}^{r-1}_i e_{r-1} - 1) - (e_{r-1} - 1)\widetilde{n}^{r-1}_i = \widetilde{n}^{r-1}_i - 1, & \text{if } i \text{ is even,} \\ \widetilde{n}^{r-1}_i e_{r-1} - (e_{r-1} - 1)\widetilde{n}^{r-1}_i = \widetilde{n}^{r-1}_i, & \text{if } i \text{ is odd.} \end{cases}$$

where $\widetilde{n}^{r-1}_i = \frac{n^{r-1}_i}{e_{r-1}}$.

Using Noether's formulas and by a similar argument, it is possible to show that $v_{\overline{\mathcal{K}}}(P^k_{i,j}) = \widetilde{n}^k_i - 1$, if i is even, and $v_{\overline{\mathcal{K}}}(P^k_{i,j}) = \widetilde{n}^k_i$, if i is odd.

Finally, in any situation we have $v_{\overline{\mathcal{K}}}(P^k_{s(k),h^k_{s(k)}}) = \widetilde{n}^k_{s(k)} - 1$. In this way, the cluster $\overline{\mathcal{K}}$ represents the cluster of singularities of the polar curve of a generic branch g in $K(\widetilde{n}, \widetilde{m}_1, \cdots, \widetilde{m}_{r-1})$. Therefore,

$$P(f) = P(g)\xi_r.$$

Now, repeating the same procedure to $P(g)$, and so on, we obtain

$$P(f) = \xi_1 \cdots \xi_{r-1}\xi_r,$$

where $\xi_1 = P(f_1)$ and f_1 is a general member of $K(\frac{n}{e_1}, \frac{m_1}{e_1})$, which is explicitly described in Theorem 2.2. On the other hand,

$$\xi_{k+1} = \prod_{i=1}^{[\frac{s(k+1)+1}{2}]} \prod_{j=1}^{h_{2i}} \xi^{k+1}_{i,j}, \quad k = 1, \ldots, r-1, \tag{3}$$

where, if we write $\frac{m_{k+1}-m_k}{e_k} = [h^{k+1}_0, \ldots, h^{k+1}_{s(k+1)}]$ and define

$$\frac{q^{k+1}_{2i-1}}{p^{k+1}_{2i-1}} = [h^{k+1}_0, h^{k+1}_1, \ldots, h^{k+1}_{2i-1}], \text{ with } \gcd(p^{k+1}_{2i-1}, q^{k+1}_{2i-1}) = 1,$$

we have

$$\xi^{k+1}_{i,j} \in K\left(p^{k+1}_{2i-1}\frac{n}{e_k}, p^{k+1}_{2i-1}\frac{m_1}{e_k}, \ldots, p^{k+1}_{2i-1}\frac{m_k}{e_k}, p^{k+1}_{2i-1}\frac{m_k}{e_k} + q^{k+1}_{2i-1}\right).$$

Summarizing, we have proved part of the following result.

Theorem 3.1 *If f is a general branch in $K(n, m_1, \ldots, m_r)$, then the Merle decomposition of $P(f)$ is given by*

$$P(f) = \xi_1\xi_2 \cdots \xi_r,$$

where $\xi_1 = P(f_1)$ with f_1 a general member of $K(\frac{n}{e_1}, \frac{m_1}{e_1})$ and ξ_{k+1} is as in (3).

The intersection multiplicities of these branches are given by:

$\mathrm{I}(\xi^{k+1}_{i,j}, \xi^{k+1}_{u,v}) = p^{k+1}_{2i-1}p^{k+1}_{2u-1}\left(\frac{n}{e_k^2} + \sum_{w=1}^{k-1}\frac{e_w}{e_k^2}(m_{w+1} - m_w)\right) + q^{k+1}_{2u-1}p^{k+1}_{2i-1}$, *for* $i \leq u$.

$\mathrm{I}(\xi^{l+1}_{i,j}, \xi^{k+1}_{u,v}) = \frac{p^{l+1}_{2i-1}p^{k+1}_{2u-1}}{e_l e_k}\left(\sum_{w=1}^{l} m_w(e_{w-1} - e_w) + m_{l+1}e_l\right)$, *for* $k > l \geq 0$, *with the convention that* $\sum_{w=s}^{t} A_w = 0$, *if* $t < s$.

Proof It remains only to compute the intersection multiplicities.

By Noether's formula, we know that the intersection multiplicity of two branches is the sum of the products of the multiplicities in common points.

Case 1. The branches belong to the same package.

Suppose that $1 \le i \le u \le \left[\frac{s(k+1)+1}{2}\right]$, $1 \le j \le h_{2i}$ and $1 \le v \le h_{2u}$ and let

$$\xi_{i,j}^{k+1} \in K\left(p_{2i-1}^{k+1}\frac{n}{e_k}, p_{2i-1}^{k+1}\frac{m_1}{e_k}, \ldots, p_{2i-1}^{k+1}\frac{m_l}{e_k}, p_{2i-1}^{k+1}\frac{m_l}{e_k} + q_{2i-1}^{k+1}\right), \text{ and}$$

$$\xi_{u,v}^{k+1} \in K\left(p_{2u-1}^{k+1}\frac{n}{e_k}, p_{2u-1}^{k+1}\frac{m_1}{e_k}, \ldots, p_{2u-1}^{k+1}\frac{m_l}{e_k}, p_{2u-1}^{k+1}\frac{m_l}{e_k} + q_{2u-1}^{k+1}\right).$$

As $i \le u$, we have that the last common point of the two above branches is $P_{2i-1,h_{2i-1}^{k+1}}^{k+1}$. Using the clusters of both branches, we obtain that the sum of products of the multiplicities until the point $P_{s(k),h_{s(k)}^k}^k$ is

$$\left(\frac{e_1}{e_k^2}m_1 + \sum_{j=1}^{k-1}\frac{e_j}{e_k^2}(m_{j+1} - m_j)\right) p_{2i-1}^{k+1}p_{2u-1}^{k+1}.$$

On the other hand, since the branches at the point $P_{0,1}^{k+1}$ are the branches of the polar of a genus one curve, using Theorem 2.2, one gets

$$\mathrm{I}_{P_{0,1}^{k+1}}(\xi_{i,j}^{k+1}, \xi_{u,v}^{k+1}) = q_{2u-1}^{k+1}p_{2i-1}^{k+1}.$$

Summing up and using Noether's formula, one gets that

$$\mathrm{I}(\xi_{i,j}^{k+1}, \xi_{u,v}^{k+1}) = \left(\frac{e_j}{e_k^2} + \sum_{w=1}^{k-1}\frac{e_j}{e_k^2}(m_{w+1} - m_w)\right) p_{2i-1}^{k+1}p_{2u-1}^{k+1} + q_{2u-1}^{k+1}p_{2i-1}^{k+1}.$$

Case 2. The branches are in distinct packages.

Consider $\xi_{i,j}^{l+1}$ and $\xi_{u,v}^{k+1}$ where $0 \le l < k$, $1 \le i \le [\frac{s(l+1)+1}{2}]$, $1 \le u \le [\frac{s(k+1)+1}{2}]$, $1 \le j \le h_{2i}^{l+1}$ and $1 \le v \le h_{2u}^{k+1}$.

We have that the sum of products of the multiplicities until the point $P_{s(l),h_{s(l)}^l}^l$ is

$$\frac{p_{2i-1}^{l+1}p_{2u-1}^{k+1}}{e_le_k}\left(nm_1 + \sum_{w=1}^{l-1}e_w(m_{w+1} - m_w)\right),$$

while the sum of products of the multiplicities at the remaining points is

$$\frac{p_{2i-1}^{l+1}p_{2u-1}^{k+1}}{e_le_k}(m_{l+1} - m_l)e_l.$$

Therefore, if $e_0 = n$ and $m_0 = 0$, then

$$\begin{aligned} \mathrm{I}(\xi_{i,j}^{l+1}, \xi_{u,v}^{k+1}) &= \tfrac{p_{2i-1}^{l+1} p_{2u-1}^{k+1}}{e_l e_k} \left(n m_1 + \textstyle\sum_{w=1}^{l} e_w (m_{w+1} - m_w) \right) \\ &= \tfrac{p_{2i-1}^{l+1} p_{2u-1}^{k+1}}{e_l e_k} \left(\textstyle\sum_{w=1}^{l} m_w (e_{w-1} - e_w) + m_{l+1} e_l \right). \end{aligned}$$

By construction and by an analogous computation, we may show that

$$\frac{\mathrm{I}(\xi_{i,j}^{l+1}, f)}{m(\xi_{i,j}^{l+1})} = \frac{1}{n} \left(\sum_{w=1}^{l} m_w (e_{w-1} - e_w) + m_{l+1} e_l \right).$$

In this way, we see that ξ_u is precisely the uth package in Merle's Theorem. □

From the above theorem we get immediately the following result:

Corollary 3.2 *The number of branches of the* j*th package* ξ_j *in Merle's decomposition of the polar of a general member of* $K(n, m_1, \ldots, m_r)$ *is equal to*

$$\sum_{k=1}^{[\frac{s(j)+1}{2}]} h_{2k}^{j},$$

where the numbers that appear in the formula are obtained from the euclidean divisions described in (2.2).

Example 3.3 Let f be general member of $K(8, 12, 14, 15)$. The Euclidean divisions in this case are:

$m_1 = 12$ and $n = 8$	$m_2 - m_1 = 2$ and $e_1 = 4$	$m_3 - m_2 = 1$ and $e_2 = 2$
$12 = 1(8) + 4$	$2 = 0(4) + 2$	$1 = 0(2) + 1$
$8 = 2(4)$	$4 = 2(2)$	$2 = 2(1)$

In this way, we have

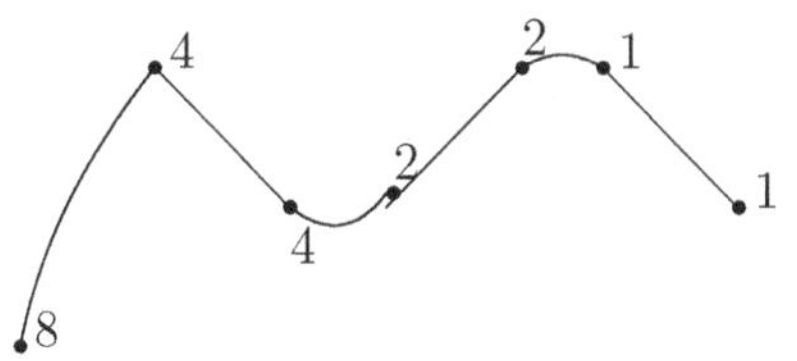

Enriques diagram of f

Since $h_0^3 = 0$, $h_1^3 = 1$ and $h_2^3 = 1$, according to Theorem 3.1, the third package ξ_3 of $P(f)$ has just one branch $\xi_{1,1}^3 \in K(4, 6, 7)$ whose Enriques diagram is

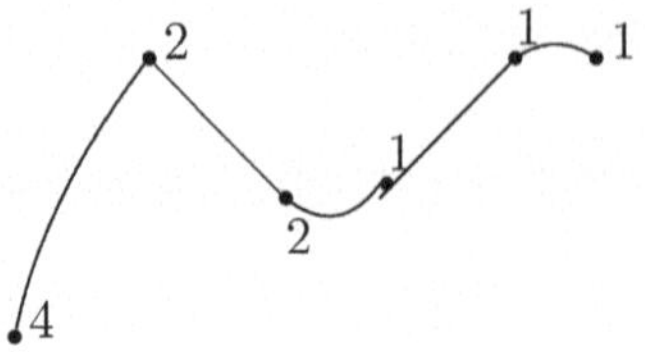

Now, since $h_0^2 = 0$, $h_1^2 = 1$ and $h_2^2 = 1$, the second package ξ_2 of $P(f)$ has just one branch $\xi_{1,1}^2 \in K(2,3)$, whose Enriques diagram is

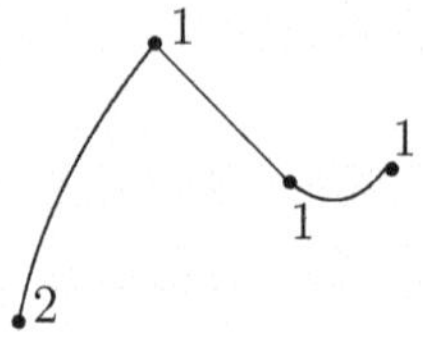

Finally, the first package is ξ_1, corresponding to the polar of a general member of $K(2,3)$, hence it has one smooth branch $\xi_{1,1}$, whose Enriques diagram is

It follows that Enriques diagram of $P(f) = \xi_1 \xi_2 \xi_3$ is

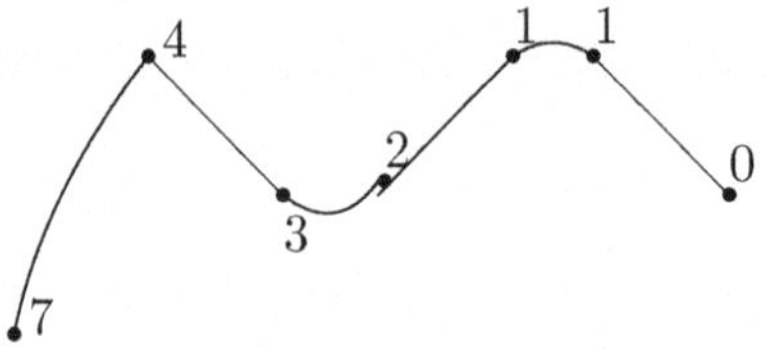

Enriques diagram of $P(f)$

For the intersection multiplicities of these branches, the theorem gives us

$$\mathrm{I}(\xi_{1,1}, \xi_{1,1}^2) = 3, \ \mathrm{I}(\xi_{1,1}, \xi_{1,1}^3) = 6, \ \mathrm{I}(\xi_{1,1}^2, \xi_{1,1}^3) = 13.$$

From Merle's Theorem it follows that each branch of the jth Merle's package ξ_j of the polar of an irreducible curve has genus at least $j - 1$. On the other hand, from the proof of Theorem 3.1 one may see that the genus of each component of ξ_j is less or equal than j, when the curve is general in its equisingularity class. This generality condition is a sufficient condition to guarantee the bound j from above for the genus of the components of ξ_j, as one may see in [5, Remark 2.1].

The problem we address now is to characterize the equisingularity classes $K(n, m_1, \ldots, m_r)$ for which the general member has its polar curve composed by branches with genus up to $r - 1$.

Corollary 3.4 *Let f be a power series corresponding to a general member of $K(n, m_1, \ldots, m_r)$. The polar of f has branches of genus at most $r - 1$, if and only if $m_r = m_{r-1} + \lambda e_{r-1} - 1$, for some integer $\lambda \geq 1$.*

Proof From Theorem 3.1, this happens if and only if the $\xi^r_{i,j}$ have genus $r - 1$. Since $\xi^r_{i,j} \in K(p^r_{2i-1}\frac{n}{e_{r-1}}, p^r_{2i-1}\frac{m_1}{e_{r-1}}, \ldots, p^r_{2i-1}\frac{m_{r-1}}{e_{r-1}}, p^r_{2i-1}\frac{m_{r-1}}{e_{r-1}} + q^r_{2i-1})$, this, in turn, happens if and only if $p^r_{2i-1} = 1$ for all $i = 1, \ldots, t(r)$, where $s(r) = 2t(r)$. Now, since the p^r_j form an increasing sequence, one must have $t(r) = 1$. We have two possibilities:

(1) $m_r - m_{r-1} = h^r_0 e_{r-1} + n^r_1$, $e_{r-1} = h^r_1 n^r_1 + 1$ and $n^r_1 = h^r_2 \cdot 1$. Now, since $\frac{q^r_1}{p^r_1} = [h^r_0, h^r_1]$ is an integer, we must have $h^r_1 = 1$. Therefore, the condition that $P(f)$ has branches of genus at most $r - 1$ is equivalent to

$$m_r - m_{r-1} = (h^r_0 + 1)e_{r-1} - 1.$$

(2) $m_r - m_{r-1} = h^r_0 e_{r-1} + 1$ and $e_{r-1} = (h^r_1 - 1) \cdot 1 + 1$. Since $\frac{q^r_1}{p^r_1} = [h^r_0, h^r_1 - 1]$ is an integer, then $h^r_1 = 2$. Which gives $e_{r-1} = 2$. Therefore, the condition that $P(f)$ has branches of genus at most $r - 1$ is equivalent to

$$m_r - m_{r-1} = (h^r_0 + 1)e_{r-1} - 1.$$

This concludes our proof. □

References

1. Casas-Alvero, E.: Singularities of Plane Curves. Cambridge University Press (2000)
2. Casas-Alvero, E.: On the singularities of polar curves. Manuscripta Math. **43**, 167–190 (1983)
3. Casas-Alvero, E.: Infinitely near imposed singularities and singularities of polar curves. Math. Ann. **287**, 429–454 (1990)
4. Enriques, F., Chisini, O.: Lezioni sulla teoria geometrica delle equazioni e delle funzioni algebriche. N. Zanichelli, Bologna (1915)
5. Hefez, A., Hernandes, M.E., Iglesias, M.F.H.: On polars of plane branches. In: Singularities in Geometry, Topology, Foliations and Dynamics. Trends in Mathematics, pp. 135–153. Birkhäuser (2017)
6. Hefez, A., Hernandes, M.E., Iglesias, M.F.H.: Plane branches with Newton nondegenerate polars. arxiv: 160.07522 [math.AG]
7. Merle, M.: Invariants polaires des courbes planes. Inventiones Mathematicae **41**, 103–111 (1977)

8. Pham, F.: Deformations equisingulières des idéaux jacobiens des courbes planes. In: Proceedings of Liverpool Symposium on Singularities II of Lecturer Notes in Mathematics, vol. 209, pp. Springer, New York (1971)
9. Teissier, B.: Variétés polaires I: Invariants polaires des singularités d'hypersurfaces. Inventiones Mathematicae **41**(3), 267–292 (1977)

Local Zeta Functions for Rational Functions and Newton Polyhedra

Miriam Bocardo–Gaspar and W. A. Zúñiga–Galindo

Abstract In this article, we introduce a notion of non-degeneracy, with respect to certain Newton polyhedra, for rational functions over non-Archimedean local fields of arbitrary characteristic. We study the local zeta functions attached to non-degenerate rational functions, we show the existence of meromorphic continuations for these zeta functions, as rational functions of q^{-s}, and give explicit formulas. In contrast with the classical local zeta functions, the meromorphic continuations of zeta functions for rational functions have poles with positive and negative real parts.

Keywords Igusa local zeta functions · Newton polyhedra · Non-degeneracy conditions

2000 Mathematics Subject Classification. Primary 14G10 · 11S40; Secondary 14M25

1 Introduction

The local zeta functions in the Archimedean setting, i.e. in $\mathbb{R}$ or $\mathbb{C}$, were introduced in the 50s by I. M. Gel'fand and G. E. Shilov [1]. An important motivation was that the meromorphic continuation for the local zeta functions implies the existence of fundamental solutions for differential operators with constant coefficients. The meromorphic continuation was established, independently, by M. Atiyah [2] and J. Bernstein [3]. On the other hand, by the middle of the 60s, A. Weil studied local

The second author was partially supported by Conacyt Grant No. 250845.

M. Bocardo–Gaspar · W. A. Zúñiga–Galindo (✉)
Centro de Investigacion y de Estudios Avanzados del I.P.N., Departamento de Matematicas, Av. Instituto Politecnico Nacional 2508 Col, C. P. 07360 San Pedro Zacatenco, Mexico D.F., Mexico
e-mail: wazuniga@math.cinvestav.edu.mx

M. Bocardo–Gaspar
e-mail: mbocardo@math.cinvestav.mx

R. N. Araújo dos Santos et al. (eds.), *Singularities and Foliations. Geometry, Topology and Applications*, Springer Proceedings in Mathematics & Statistics 222, https://doi.org/10.1007/978-3-319-73639-6_12

zeta functions, in the Archimedean and non-Archimedean settings, in connection with the Poisson–Siegel formula [4]. In the 70s, using Hironaka's resolution of singularities theorem, J.-I. Igusa developed a uniform theory for local zeta functions and oscillatory integrals attached to polynomials with coefficients in a field of characteristic zero [5, 6]. In the p-adic setting, local zeta functions are connected with the number of solutions of polynomial congruences mod p^m and with exponential sums mod p^m. In addition, there are many intriguing conjectures relating the poles of the local zeta functions with topology of complex singularities, see e.g. [6, 7]. More recently, J. Denef and F. Loeser introduced in [8] the motivic zeta functions which constitute a vast generalization of the p-adic local zeta functions.

In [9] W. Veys and W. A. Zúñiga–Galindo extended Igusa's theory to the case of rational functions, or, more generally, meromorphic functions f/g, with coefficients in a local field of characteristic zero. This generalization is far from being straightforward due to the fact that several new geometric phenomena appear. Also, the oscillatory integrals have two different asymptotic expansions: the 'usual' one when the norm of the parameter tends to infinity, and another one when the norm of the parameter tends to zero. The first asymptotic expansion is controlled by the poles (with negative real parts) of all the twisted local zeta functions associated to the meromorphic functions $f/g - c$, for certain special values c. The second expansion is controlled by the poles (with positive real parts) of all the twisted local zeta functions associated to f/g. There are several mathematical and physical motivations for studying these new local zeta functions. For instance, S. Gusein–Zade, I. Luengo and A. Melle–Hernández have studied the complex monodromy (and A'Campo zeta functions attached to it) of meromorphic functions, see e.g. [10–12]. This work drives naturally to ask about the existence of local zeta functions with poles related with the monodromies studied by the mentioned authors. From a physical perspective, the local zeta functions attached to meromorphic functions are very alike to parametric Feynman integrals and to p-adic string amplitudes, see e.g. [13–16]. For instance in [16, Sect. 3.15], M. Marcolli pointed out explicitly that the motivic Igusa zeta function constructed by J. Denef and F. Loeser may provide the right tool for a motivic formulation of the dimensionally regularized parametric Feynman integrals.

This article aims to study the local zeta functions attached to a rational function f/g with coefficients in a local field of arbitrary characteristic, when f/g is non-degenerate with respect to a certain Newton polyhedron. In [17] E. León–Cardenal and W. A. Zúñiga–Galindo studied similar matters. In this article, we present a more suitable and general notion of non-degeneracy which allows us to study the local zeta functions attached to much larger class of rational functions. Our article is organized as follows. In Sect. 2 we summarize some basic aspects about non-Archimedean local fields and compute some π-adic integrals that are needed in the article. In Sect. 3 we review some basic aspects about polyhedral subdivisions and Newton polyhedra, we also introduce a notion of non-degeneracy for polynomials mappings. It seems that our notion of non-degeneracy is a new one. In Sect. 4 we study the meromorphic continuation for multivariate local zeta functions attached to non-degenerate mappings. These local zeta functions were introduced by F. Loeser in [18]. We give a very general geometric description of the poles of the meromorphic continuation of these functions, see Theorem 1. Our results extend some of the well-known results due to Hoornaert and Denef [19], and Bories [20]. In Sect. 5 we study the local zeta functions attached to rational functions satisfying a suitable non-degeneracy condition. In Theorem 2, we give a geometric description of the poles of the meromorphic continuation of these functions. The real parts of the poles of the meromorphic continuation of these functions are positive and negative rational numbers. Finally, in Sect. 6, we describe the 'smallest positive and largest negative poles' appearing in the meromorphic continuation of these new local zeta functions, see Theorems 3 and 4.

2 Preliminaries

In this article we work with a non-Archimedean locally compact field K of arbitrary characteristic. We will say that a such field is a *non-Archimedean local field* of arbitrary characteristic. By a well-known classification theorem, a non-Archimedean local field is a finite extension of $\mathbb{Q}_p$, the field of p-adic numbers, or of *the field formal Laurent series* $\mathbb{F}_q((T))$ over a finite field $\mathbb{F}_q$. In the first case we say that K is *a p-adic field*. For further details the reader may consult [21, Chap. 1].

Let K be a non-Archimedean local field of arbitrary characteristic and let $\mathcal{O}_K$ be the ring of integers of K and let the residue field of K be $\mathbb{F}_q$, the finite field with $q = p^m$ elements, where p is a prime number. For $z \in K \setminus \{0\}$, let $ord(z) \in \mathbb{Z} \cup \{+\infty\}$ denote *the valuation* of z, let $|z|_K = q^{-ord(z)}$ denote the normalized *absolute value* (or *norm*), and let $ac(z) = z\pi^{-ord(z)}$ denote the *angular component*, where π is a fixed uniformizing parameter of K. We extend the norm $|\cdot|_K$ to K^n by taking $||(x_1, \ldots, x_n)||_K := \max\{|x_1|_K, \ldots, |x_n|_K\}$. Then $(K^n, ||\cdot||_K)$ is a complete metric space and the metric topology is equal to the product topology.

Along this paper, vectors will be written in boldface, so for instance we will write $\mathbf{b} := (b_1, \ldots, b_l)$ where l is a positive integer. For polynomials we will use $\boldsymbol{x} = (x_1, \ldots, x_n)$, thus $h(\boldsymbol{x}) = h(x_1, \ldots, x_n)$. For each n-tuple of natural numbers $\boldsymbol{k} = (k_1, \ldots, k_n) \in \mathbb{N}^n$, we will denote by $\sigma(\boldsymbol{k})$ the sum of all its components i.e. $\sigma(\boldsymbol{k}) = k_1 + k_2 + \ldots + k_n$. Furthermore, we will use the notation $|d\boldsymbol{x}|_K$ for the Haar measure on $(K^n, +)$ normalized so that the measure of $\mathcal{O}_K^n$ is equal to one. In dimension one, we will use the notation $|dx|_K$.

2.1 Multivariate Local Zeta Functions

We denote by $\mathcal{S}(K^n)$ the $\mathbb{C}$-vector space consisting of all $\mathbb{C}$-valued locally constant functions over K^n with compact support. An element of $\mathcal{S}(K^n)$ is called a *Bruhat-Schwartz function* or a *test function*. Along this article we work with a polynomial mapping $\boldsymbol{h} = (h_1, \ldots, h_r) : K^n \to K^r$ such that each $h_i(\boldsymbol{x})$ is a non-constant polynomial in $\mathcal{O}_K[x_1, \ldots, x_n] \backslash \pi\mathcal{O}_K[x_1, \ldots, x_n]$ and $r \leq n$. Let Φ a Bruhat–Schwartz function and let $\boldsymbol{s} = (s_1, \ldots, s_r) \in \mathbb{C}^r$. The local zeta function associated to Φ and $\boldsymbol{h}$ is defined as

$$Z_\Phi(\boldsymbol{s}, \boldsymbol{h}) = \int_{K^n \setminus D_K} \Phi(\boldsymbol{x}) \prod_{i=1}^{r} |h_i(\boldsymbol{x})|_K^{s_i} \, |d\boldsymbol{x}|_K$$

for $\mathrm{Re}(s_i) > 0$ for all i, where $D_K := \cup_{i\in\{1,\ldots,r\}} \{\boldsymbol{x} \in K^n; h_i(\boldsymbol{x}) = 0\}$. Notice that $Z_\Phi(\boldsymbol{s}, \boldsymbol{h})$ converges for $\mathrm{Re}(s_i) > 0$ for all $i = 1, \ldots, r$. If Φ is the characteristic function of $\mathcal{O}_K^n$ we use the notation $Z(\boldsymbol{s}, \boldsymbol{h})$ instead of $Z_\Phi(\boldsymbol{s}, \boldsymbol{h})$. In the case of polynomial mappings with coefficients in a local field of characteristic zero (not necessarily non-Archimedean and without the condition $r \le n$), the theory of local zeta functions of type $Z_\Phi(\boldsymbol{s}, \boldsymbol{h})$ was established by F. Loeser in [18].

Denote by $\overline{\boldsymbol{x}}$ the image of an element of $\mathcal{O}_K^n$ under the canonical homomorphism $\mathcal{O}_K^n \to \mathcal{O}_K^n/(\pi\mathcal{O}_K)^n \cong \mathbb{F}_q^n$, we call $\overline{\boldsymbol{x}}$ *the reduction modulo* π of $\boldsymbol{x}$. Given $h(\boldsymbol{x}) \in \mathcal{O}_K[x_1, \ldots, x_n]$, we denote by $\overline{h}(\boldsymbol{x})$ the polynomial obtained by reducing modulo π the coefficients of $h(\boldsymbol{x})$. Furthermore if $\boldsymbol{h} = (h_1, \ldots, h_r)$ is a polynomial mapping with $h_i \in \mathcal{O}_K[x_1, \ldots, x_n]$ for all i, then $\overline{\boldsymbol{h}} := (\overline{h}_1, \ldots, \overline{h}_r)$ denotes the polynomial mapping obtained by reducing modulo π all the components of $\boldsymbol{h}$.

2.2 Some π-Adic Integrals

Let $\boldsymbol{h} = (h_1, h_2, \ldots, h_r)$ be a polynomial mapping as above. For $\boldsymbol{a} \in (\mathcal{O}_K^\times)^n$, we set

$$J_{\boldsymbol{a}}(\boldsymbol{s}, \boldsymbol{h}) := \int_{\boldsymbol{a}+(\pi\mathcal{O}_K)^n \setminus D_K} \prod_{i=1}^{r} |h_i(\boldsymbol{x})|_K^{s_i} \, |d\boldsymbol{x}|_K, \tag{2.1}$$

$\boldsymbol{s} = (s_1, \ldots s_r) \in \mathbb{C}^r$ with $\mathrm{Re}(s_i) > 0$, $i = 1, \ldots, r$.

The Jacobian matrix of $\boldsymbol{h}$ at $\boldsymbol{a}$ is $Jac\,(\boldsymbol{h}, \boldsymbol{a}) = \left[\frac{\partial h_i}{\partial x_j}(\boldsymbol{a})\right]_{\substack{1\le i\le r\\ 1\le j\le n}}$ with $r \le n$. In a similar way we define the Jacobian matrix of $\overline{\boldsymbol{h}}$ at $\overline{\boldsymbol{a}}$. For every non-empty subset I of $\{1, \ldots, r\}$ we set $Jac\left(\overline{\boldsymbol{h}}_I, \overline{\boldsymbol{a}}\right) := \left[\frac{\partial \overline{h}_i}{\partial x_j}(\overline{\boldsymbol{a}})\right]_{\substack{i\in I\\ 1\le j\le n}}$.

Lemma 1 *Let I be the subset of $\{1, \ldots, r\}$ such that $\overline{h}_i(\overline{\boldsymbol{a}}) = 0 \Leftrightarrow i \in I$. Assume that $\boldsymbol{a} \notin D_K$ and that $Jac\left(\overline{\boldsymbol{h}}_I, \overline{\boldsymbol{a}}\right)$ has rank $m = Card(I)$ for $I \ne \varnothing$. Then $J_{\boldsymbol{a}}(\boldsymbol{s}, \boldsymbol{h})$ equals*

$$\begin{cases} q^{-n} & \text{if } I = \varnothing \\ q^{-n} \prod_{i\in I} \frac{(q-1)q^{-1-s_i}}{1-q^{-1-s_i}} & \text{if } I \ne \varnothing. \end{cases}$$

Proof By change of variables we get

$$J_a(s,h) = q^{-n} \int\limits_{\mathcal{O}_K^n \smallsetminus \cup_{i\in\{1,\ldots,r\}}\{x\in K^n;h_i(\pi x+a)=0\}} \prod_{i=1}^{r} |h_i(\pi x + a)|_K^{s_i} |dx|_K.$$

We first consider the case $I = \varnothing$. Then $h_i(a) \not\equiv 0 \bmod \pi$, thus $|h_i(\pi x + a)|_K = 1$, and $J_a(s,h) = q^{-n}$. In the case $I \neq \varnothing$, by reordering I (if necessary) we can suppose that $I = \{1,\ldots,m\}$ with $m \leq r$. Integral $J_a(s,h)$ is computed by changing variables as $y = \phi(x)$ with

$$y_i = \phi_i(x) := \begin{cases} \frac{h_i(a+\pi x)-h_i(a)}{\pi} & \text{if } i = 1,\ldots,m \\ x_i & \text{if } i \geq m+1. \end{cases}$$

By using that rank of $Jac(\overline{h}_I, \overline{a})$ is m we get that $det\left[\frac{\partial\phi_i}{\partial x_j}(\mathbf{0})\right]_{\substack{1\leq i\leq n\\1\leq j\leq n}} \not\equiv 0 \bmod \pi$,

which implies that $y = \phi(x)$ gives a measure-preserving map from $\mathcal{O}_K^n$ to itself (see e.g. [6, Lemma 7.4.3]), hence

$$J_a(s,h) = q^{-n} \prod_{i=1}^{m} \int\limits_{\mathcal{O}_K\setminus\{\pi y_i+h_i(a)=0\}} |\pi y_i + h_i(a)|_K^{s_i} |dy_i|_K =: q^{-n} \prod_{i=1}^{m} J'_a(y_i).$$

To prove the announced formula we compute integrals $J'_a(y_i)$. Now, since $h_i(a) \equiv 0 \bmod \pi$, by taking $z_i = \pi y_i + h_i(a)$ in $J'_a(y_i)$, we obtain

$$J'_a(y_i) = q^{-s_i} \int\limits_{\mathcal{O}_K\setminus\{0\}} |z_i|_K^{s_i} |dz_i|_K = \frac{(q-1)q^{-1-si}}{1-q^{-1-s_i}}.$$

Therefore

$$J_a(s,h) = \begin{cases} q^{-n} & I = \varnothing \\ q^{-n}\prod_{i\in I}\frac{(q-1)q^{-1-s_i}}{1-q^{-1-s_i}} & I \neq \varnothing. \end{cases} \tag{2.2}$$

□

Remark 1 If in integral (2.1), we replace $h_i(x)$ by $h_i(x) + \pi g_i(x)$, where each $g_i(x)$ is a polynomial with coefficients in $\mathcal{O}_K$, then the formulas given in Lemma 1 are valid.

For every subset $I \subseteq \{1,\ldots,r\}$ we set

$$\overline{V}_I := \left\{\overline{z} \in (\mathbb{F}_q^\times)^n;\ \overline{h}_i(\overline{z}) = 0 \Leftrightarrow i \in I\right\}. \tag{2.3}$$

To simplify the notation we will denote $\overline{V}_{\{1,\ldots,r\}}$ as $\overline{V}$.

Lemma 2 *Let $\boldsymbol{h} = (h_1, \ldots, h_r)$ with $r \leq n$, be as before. Assume that for all $I \neq \varnothing$ if $\overline{V}_I \neq \varnothing$, then*

$$rank_{\mathbb{F}_q}\left[\frac{\partial \overline{h}_i}{\partial x_j}(\overline{\boldsymbol{a}})\right]_{i\in I,\ j\in\{1,\ldots,n\}} = Card(I), \text{ for any } \overline{\boldsymbol{a}} \in \overline{V}_I.$$

Set

$$L(\boldsymbol{s}, \boldsymbol{h}) := \int_{(\mathcal{O}_K^{\times})^n \setminus D_K} \prod_{i=1}^{r} |h_i(\boldsymbol{x})|_K^{s_i} |d\boldsymbol{x}|_K, \ \boldsymbol{s} = (s_1, \ldots s_r) \in \mathbb{C}^r,$$

for Re $(s_i) > 0$ *for all i. Then, with the convention that $\prod_{i\in I} \frac{(q-1)q^{-1-s_i}}{1-q^{-1-s_i}} = 1$ when $I = \varnothing$, we have*

$$L(\boldsymbol{s}, \boldsymbol{h}) = q^{-n} \sum_{I\subseteq\{1,\ldots,r\}} Card(\overline{V}_I) \prod_{i\in I} \frac{(q-1)q^{-1-s_i}}{1-q^{-1-s_i}}.$$

Proof Note that $L(\boldsymbol{s}, \boldsymbol{h})$ can be expressed as a finite sum of integrals

$$J_{\boldsymbol{a}}(\boldsymbol{s}, \boldsymbol{h}) = \int_{\boldsymbol{a}+(\pi\mathcal{O}_K)^n \setminus D_K} \prod_{i=1}^{r} |h_i(\boldsymbol{x})|_K^{s_i} |d\boldsymbol{x}|_K,$$

where $\boldsymbol{a}$ runs through a fixed set of representatives $\mathcal{R}$ in $\left(\mathcal{O}_K^{\times}\right)^n$ of $(\mathbb{F}_q^{\times})^n$. Then $L(\boldsymbol{s}, \boldsymbol{h})$ is equals

$$\begin{aligned}
&\sum_{\overline{\boldsymbol{a}}\in\overline{V}_\varnothing} \int_{\boldsymbol{a}+(\pi\mathcal{O}_K)^n \setminus D_K} \prod_{i=1}^{r} |h_i(\boldsymbol{x})|_K^{s_i} |d\boldsymbol{x}|_K \\
&+ \sum_{\substack{I\subsetneq\{1,\ldots,r\}\\ I\neq\varnothing}} \sum_{\overline{\boldsymbol{a}}\in\overline{V}_I} \int_{\boldsymbol{a}+(\pi\mathcal{O}_K)^n \setminus D_K} \prod_{i=1}^{r} |h_i(\boldsymbol{x})|_K^{s_i} |d\boldsymbol{x}|_K \\
&+ \sum_{\overline{\boldsymbol{a}}\in\overline{V}} \int_{\boldsymbol{a}+(\pi\mathcal{O}_K)^n \setminus D_K} \prod_{i=1}^{r} |h_i(\boldsymbol{x})|_K^{s_i} |d\boldsymbol{x}|_K \\
&=: J(\boldsymbol{s}, \overline{V}_\varnothing) + \sum_{\substack{I\subsetneq\{1,\ldots,r\}\\ I\neq\varnothing}} J(\boldsymbol{s}, \overline{V}_I) + J(\boldsymbol{s}, \overline{V}),
\end{aligned}$$

with the convention that if $\overline{V}_I = \varnothing$, then $\sum_{\overline{\boldsymbol{a}}\in\overline{V}_I} \int_{\boldsymbol{a}+(\pi\mathcal{O}_K)^n \setminus D_K} \cdot = 0$. Notice that

$$J(\boldsymbol{s}, \overline{V}_\varnothing) = q^{-n} Card(\overline{V}_\varnothing). \tag{2.4}$$

Thus we may assume that $I \neq \varnothing$. In the calculation of $J(s, \overline{V}_I)$ we use the following result:

Claim.

$$\sum_{\overline{a} \in \overline{V}_I} \int_{a+(\pi\mathcal{O}_K)^n \setminus D_K} \prod_{i=1}^{r} |h_i(\boldsymbol{x})|_K^{s_i} |d\boldsymbol{x}|_K = \sum_{\substack{\overline{a} \in \overline{V}_I \\ a \notin D_K}} \int_{a+(\pi\mathcal{O}_K)^n \setminus D_K} \prod_{i=1}^{r} |h_i(\boldsymbol{x})|_K^{s_i} |d\boldsymbol{x}|_K.$$

The Claim follows from the following reasoning. The analytic mapping $h_1 \cdots h_r : \boldsymbol{a} + (\pi\mathcal{O}_K)^n \to K$ is not identically zero, otherwise by [6, Lemma 2.1.3], the polynomial $(h_1 \cdots h_r)(\boldsymbol{x})$ would be the constant polynomial zero contradicting the hypothesis that all the h_i's are non-constant polynomials. Hence there exists an element $\boldsymbol{b} \in \boldsymbol{a} + (\pi\mathcal{O}_K)^n$ such that $(h_1 \cdots h_r)(\boldsymbol{b}) \neq 0$. Finally, we use the fact that every point of a ball is its center, which implies that $\boldsymbol{a} + (\pi\mathcal{O}_K)^n = \boldsymbol{b} + (\pi\mathcal{O}_K)^n$.

By using Lemma 1,

$$J(s, \overline{V}_I) = q^{-n} Card(\overline{V}_I) \prod_{i \in I} \frac{(q-1)q^{-1-s_i}}{1 - q^{-1-s_i}}. \tag{2.5}$$

The formula for $J(s, \overline{V})$ is a special case of formula (2.5):

$$J(s, \overline{V}) = q^{-n} Card(\overline{V}) \prod_{i \in \{1, \ldots, r\}} \frac{(q-1)q^{-1-s_i}}{1 - q^{-1-s_i}}. \tag{2.6}$$

□

Remark 2 In integral $L(s, \boldsymbol{h})$ we can replace $\boldsymbol{h}$ by $\boldsymbol{h} + \pi g$, where g is a polynomial mapping over $\mathcal{O}_K$, and the formulas given in Lemma 2 remain valid.

3 Polyhedral Subdivisions of $\mathbb{R}_+^n$ and Non-degeneracy Conditions

In this section we review, without proofs, some well-known results about Newton polyhedra and non-degeneracy conditions that we will use along the article. Our presentation follows closely [22, 23].

3.1 Newton Polyhedra

We set $\mathbb{R}_+ := \{x \in \mathbb{R}; x \geqslant 0\}$. Let G be a non-empty subset of $\mathbb{N}^n$. The *Newton polyhedron* $\Gamma = \Gamma(G)$ associated to G is the convex hull in $\mathbb{R}_+^n$ of the set $\cup_{\boldsymbol{m} \in G} \left(\boldsymbol{m} + \mathbb{R}_+^n\right)$. For instance classically one associates a *Newton polyhedron*

$\Gamma(h)$ *(at the origin)* to a polynomial function $h(\boldsymbol{x}) = \sum_{\boldsymbol{m}} c_{\boldsymbol{m}} \boldsymbol{x}^{\boldsymbol{m}}$ ($\boldsymbol{x} = (x_1, \ldots, x_n)$, $h(\boldsymbol{0}) = \boldsymbol{0}$), where $G = \operatorname{supp}(h) := \{\boldsymbol{m} \in \mathbb{N}^n; c_{\boldsymbol{m}} \neq 0\}$. Further we will associate more generally a Newton polyhedron to a polynomial mapping.

We fix a Newton polyhedron Γ as above. We first collect some notions and results about Newton polyhedra that will be used in the next sections. Let $\langle \cdot, \cdot \rangle$ denote the usual inner product of $\mathbb{R}^n$, and identify the dual space of $\mathbb{R}^n$ with $\mathbb{R}^n$ itself by means of it.

Let H be the hyperplane $H = \{\boldsymbol{x} \in \mathbb{R}^n; \langle \boldsymbol{x}, \mathbf{b} \rangle = c\}$, H determines two closed half-spaces

$$H^+ = \left\{\boldsymbol{x} \in \mathbb{R}^n; \langle \boldsymbol{x}, \mathbf{b} \rangle \geq c\right\} \text{ and } H^- = \left\{\boldsymbol{x} \in \mathbb{R}^n; \langle \boldsymbol{x}, \mathbf{b} \rangle \leq c\right\}.$$

We say that H is a *supporting hyperplane* of $\Gamma(h)$ if $\Gamma(h) \cap H \neq \varnothing$ and $\Gamma(h)$ is contained in one of the two closed half-spaces determined by H. By a *proper face* τ of $\Gamma(h)$, we mean a non-empty convex set τ obtained by intersecting $\Gamma(h)$ with one of its supporting hyperplanes. By the *faces* of $\Gamma(h)$ we will mean the proper faces of $\Gamma(h)$ and the whole the polyhedron $\Gamma(h)$. By *dimension of a face* τ of $\Gamma(h)$ we mean the dimension of the affine hull of τ, and its *codimension* is $\operatorname{cod}(\tau) = n - \dim(\tau)$, where $\dim(\tau)$ denotes the dimension of τ. A face of codimension one is called a *facet*.

For $\boldsymbol{a} \in \mathbb{R}_+^n$, we define

$$d(\boldsymbol{a}, \Gamma) = \min_{x \in \Gamma} \langle \boldsymbol{a}, \boldsymbol{x} \rangle,$$

and *the first meet locus* $F(\boldsymbol{a}, \Gamma)$ of $\boldsymbol{a}$ as

$$F(\boldsymbol{a}, \Gamma) := \{\boldsymbol{x} \in \Gamma; \langle \boldsymbol{a}, \boldsymbol{x} \rangle = d(\boldsymbol{a}, \Gamma)\}.$$

The first meet locus is a face of Γ. Moreover, if $\boldsymbol{a} \neq \boldsymbol{0}$, $F(\boldsymbol{a}, \Gamma)$ is a proper face of Γ.

If $\Gamma = \Gamma(h)$, we define the *face function* $h_{\boldsymbol{a}}(\boldsymbol{x})$ *of* $h(\boldsymbol{x})$ *with respect to* $\boldsymbol{a}$ as

$$h_{\boldsymbol{a}}(\boldsymbol{x}) = h_{F(\boldsymbol{a},\Gamma)}(\boldsymbol{x}) = \sum_{\boldsymbol{m} \in F(\boldsymbol{a},\Gamma)} c_{\boldsymbol{m}} \boldsymbol{x}^{\boldsymbol{m}}.$$

In the case of functions having subindices, say $h_i(\boldsymbol{x})$, we will use the notation $h_{i,\boldsymbol{a}}(\boldsymbol{x})$ for the face function of $h_i(\boldsymbol{x})$ with respect to $\boldsymbol{a}$. Notice that $h_{\boldsymbol{0}}(\boldsymbol{x}) = h_{F(\boldsymbol{0},\Gamma)}(\boldsymbol{x}) = \sum_{\boldsymbol{m}} c_{\boldsymbol{m}} \boldsymbol{x}^{\boldsymbol{m}}$.

3.2 Polyhedral Subdivisions Subordinate to a Polyhedron

We define an equivalence relation in $\mathbb{R}_+^n$ by taking $\boldsymbol{a} \sim \boldsymbol{a}' \Leftrightarrow F(\boldsymbol{a}, \Gamma) = F(\boldsymbol{a}', \Gamma)$. The equivalence classes of $\sim$ are sets of the form

$$\Delta_\tau = \{\boldsymbol{a} \in \mathbb{R}_+^n;\ F(\boldsymbol{a}, \Gamma) = \tau\},$$

where τ is a face of Γ.

We recall that the *cone strictly spanned* by the vectors $\boldsymbol{a}_1, \dots, \boldsymbol{a}_l \in \mathbb{R}_+^n \setminus \{0\}$ is the set $\Delta = \{\lambda_1 \boldsymbol{a}_1 + \dots + \lambda_l \boldsymbol{a}_l;\ \lambda_i \in \mathbb{R}_+, \lambda_i > 0\}$. If $\boldsymbol{a}_1, \dots, \boldsymbol{a}_l$ are linearly independent over $\mathbb{R}$, Δ is called a *simplicial cone*. If $\boldsymbol{a}_1, \dots, \boldsymbol{a}_l \in \mathbb{Z}^n$, we say Δ is a *rational cone*. If $\{\boldsymbol{a}_1, \dots, \boldsymbol{a}_l\}$ is a subset of a basis of the $\mathbb{Z}$-module $\mathbb{Z}^n$, we call Δ a *simple cone*.

A precise description of the geometry of the equivalence classes modulo $\sim$ is as follows. Each *facet* γ of Γ has a unique vector $\boldsymbol{a}(\gamma) = (a_{\gamma,1}, \dots, a_{\gamma,n}) \in \mathbb{N}^n \setminus \{0\}$, whose nonzero coordinates are relatively prime, which is perpendicular to γ. We denote by $\mathfrak{D}(\Gamma)$ the set of such vectors. The equivalence classes are rational cones of the form

$$\Delta_\tau = \{\sum_{i=1}^{r} \lambda_i \boldsymbol{a}(\gamma_i);\ \lambda_i \in \mathbb{R}_+, \lambda_i > 0\},$$

where τ runs through the set of faces of Γ, and γ_i, $i = 1, \dots, r$ are the facets containing τ. We note that $\Delta_\tau = \{0\}$ if and only if $\tau = \Gamma$. The family $\{\Delta_\tau\}_\tau$, with τ running over the proper faces of Γ, is a partition of $\mathbb{R}_+^n \setminus \{0\}$; we call this partition a *polyhedral subdivision of* $\mathbb{R}_+^n$ *subordinate* to Γ. We call $\left\{\overline{\Delta_\tau}\right\}_\tau$, the family formed by the topological closures of the Δ_τ, a *fan subordinate* to Γ.

Each cone Δ_τ can be partitioned into a finite number of simplicial cones $\Delta_{\tau,i}$. In addition, the subdivision can be chosen such that each $\Delta_{\tau,i}$ is spanned by part of $\mathfrak{D}(\Gamma)$. Thus from the above considerations we have the following partition of $\mathbb{R}_+^n \setminus \{0\}$:

$$\mathbb{R}_+^n \setminus \{0\} = \bigcup_\tau \left(\bigcup_{i=1}^{l_\tau} \Delta_{\tau,i} \right),$$

where τ runs over the proper faces of Γ, and each $\Delta_{\tau,i}$ is a simplicial cone contained in Δ_τ. We will say that $\left\{\Delta_{\tau,i}\right\}$ is a *simplicial polyhedral subdivision of* $\mathbb{R}_+^n$ *subordinate* to Γ, and that $\left\{\overline{\Delta_{\tau,i}}\right\}$ is a *simplicial fan subordinate* to Γ.

By adding new rays, each simplicial cone can be partitioned further into a finite number of simple cones. In this way we obtain a *simple polyhedral subdivision* of $\mathbb{R}_+^n$ *subordinate* to Γ, and a *simple fan subordinate* to Γ (or a *complete regular fan*) (see e.g. [24]).

3.3 The Newton Polyhedron Associated to a Polynomial Mapping

Let $\boldsymbol{h} = (h_1, \dots, h_r)$, $\boldsymbol{h}(\boldsymbol{0}) = 0$, be a non-constant polynomial mapping. In this article we associate to $\boldsymbol{h}$ a Newton polyhedron $\Gamma(\boldsymbol{h}) := \Gamma\left(\prod_{i=1}^{r} h_i(\boldsymbol{x})\right)$. From a geometrical point of view, $\Gamma(\boldsymbol{h})$ is the Minkowski sum of the $\Gamma(h_i)$, for $i = 1, \cdots, r$, (see e.g. [22, 25]). By using the results previously presented, we can associate to $\Gamma(\boldsymbol{h})$ a simplicial polyhedral subdivision $\mathcal{F}(\boldsymbol{h})$ of $\mathbb{R}_+^n$ subordinate to $\Gamma(\boldsymbol{h})$.

Remark 3 A basic fact about the Minkowski sum operation is the additivity of the faces. From this fact follows:
(1) $F(\boldsymbol{a}, \Gamma(\boldsymbol{h})) = \sum_{j=1}^{r} F\left(\boldsymbol{a}, \Gamma\left(h_j\right)\right)$, for $\boldsymbol{a} \in \mathbb{R}_+^n$;
(2) $d(\boldsymbol{a}, \Gamma(\boldsymbol{h})) = \sum_{j=1}^{r} d\left(\boldsymbol{a}, \Gamma\left(h_j\right)\right)$, for $\boldsymbol{a} \in \mathbb{R}_+^n$;
(3) let τ be a proper face of $\Gamma(\boldsymbol{h})$, and let τ_j be proper face of $\Gamma\left(h_j\right)$, for $i = 1, \cdots, r$. If $\tau = \sum_{j=1}^{r} \tau_j$, then $\Delta_\tau \subseteq \overline{\Delta}_{\tau_j}$, for $i = 1, \cdots, r$.

Remark 4 Note that the equivalence relation,

$$\boldsymbol{a} \sim \boldsymbol{a}' \Leftrightarrow F(a, \Gamma(\boldsymbol{h})) = F(\boldsymbol{a}', \Gamma(\boldsymbol{h})),$$

used in the construction of a polyhedral subdivision of $\mathbb{R}_+^n$ subordinate to $\Gamma(\boldsymbol{h})$ can be equivalently defined in the following form:

$$\boldsymbol{a} \sim \boldsymbol{a}' \Leftrightarrow F(\boldsymbol{a}, \Gamma\left(h_j\right)) = F(\boldsymbol{a}', \Gamma\left(h_j\right)), \text{ for each } j = 1, \ldots, r.$$

This last definition is used in Oka's book [22].

3.4 Non-degeneracy Conditions

For $K = \mathbb{Q}_p$, Denef and Hoornaert in [19, Theorem 4.2] gave an explicit formula for $Z(\boldsymbol{s}, \boldsymbol{h})$, in the case $r = 1$ with $\boldsymbol{h}$ a non-degenerate polynomial with respect to its Newton polyhedron $\Gamma(\boldsymbol{h})$. This explicit formula can be generalized to the case $r \geq 1$ by using the condition of non-degeneracy for polynomial mappings introduced here.

Definition 1 Let $\boldsymbol{h} = (h_1, \ldots, h_r)$, $\boldsymbol{h}(\boldsymbol{0}) = 0$, be a polynomial mapping with $r \leq n$ as in Sect. 2.1 and let $\Gamma(\boldsymbol{h})$ be the Newton polyhedron of $\boldsymbol{h}$ at the origin. The mapping $\boldsymbol{h}$ is called *non-degenerate over* $\mathbb{F}_q$ *with respect to* $\Gamma(\boldsymbol{h})$, if for every vector $\boldsymbol{k} \in \mathbb{R}_+^n$ and for any non-empty subset $I \subseteq \{1, \ldots, r\}$, it verifies that

$$rank_{\mathbb{F}_q}\left[\frac{\partial \overline{h}_{i,\boldsymbol{k}}}{\partial x_j}(\overline{z})\right]_{i\in I,\ j\in\{1,\ldots,n\}} = Card(I) \tag{3.1}$$

for any

$$\overline{z} \in \left\{\overline{z} \in \left(\mathbb{F}_q^\times\right)^n ; \overline{h}_{i,\boldsymbol{k}}(\overline{z}) = 0 \Leftrightarrow i \in I\right\}. \tag{3.2}$$

We notice that above notion is different to the those introduced in [23, 26]. The notion introduced here is similar to the usual notion given by Khovansky, see

[22, 27]. For a discussion about the relation between Khovansky's non-degeneracy notion and other similar notions we refer the reader to [26].

Let Δ be a rational simplicial cone spanned by $\boldsymbol{w}_i$, $i = 1, \ldots, e_\Delta$. We define the *barycenter* of Δ as $b(\Delta) = \sum_{i=1}^{e_\Delta} \boldsymbol{w}_i$. Set $b(\{\boldsymbol{0}\}) := \boldsymbol{0}$.

Remark 5 (i)Let $\mathcal{F}(\boldsymbol{h})$ be a simplicial polyhedral subdivision of $\mathbb{R}_+^n$ subordinate to $\Gamma(\boldsymbol{h})$. Then, it is sufficient to verify the condition given in Definition 1 for $\boldsymbol{k} = b(\Delta)$ with $\Delta \in \mathcal{F}(\boldsymbol{h}) \cup \{\boldsymbol{0}\}$.

(ii) Notice that our notion of non-degeneracy agrees, in the case $K = \mathbb{Q}_p$, $r = 1$, with the corresponding notion in [19].

Example 1 Set $\boldsymbol{h} = (h_1, h_2)$ with $h_1(x, y) = x^2 - y$, $h_2(x, y) = x^2 y$ polynomials in $\mathcal{O}_K[x, y]$. Then a simplicial polyhedral subdivision subordinate to $\Gamma(\boldsymbol{h})$ is given by where $\mathbb{R}_{>0} := \mathbb{R}_+ \setminus \{\boldsymbol{0}\}$. Notice that for every $\boldsymbol{k} \in \mathbb{R}_+^n \setminus (\{\boldsymbol{0}\} \cup \Delta_3)$ and every

Cone	$h_{1,b(\Delta)}$	$h_{2,b(\Delta)}$
$\Delta_1 := (1,0)\mathbb{R}_{>0}$	y	x^2y
$\Delta_2 := (1,0)\mathbb{R}_{>0} + (1,2)\mathbb{R}_{>0}$	y	x^2y
$\Delta_3 := (1,2)\mathbb{R}_{>0}$	$x^2 - y$	x^2y
$\Delta_4 := (1,2)\mathbb{R}_{>0} + (0,1)\mathbb{R}_{>0}$	x^2	x^2y
$\Delta_5 := (0,1)\mathbb{R}_{>0}$	x^2	$x^2y,$

non-empty subset $I \subseteq \{1, 2\}$, the subset defined in (3.2) is empty, thus (3.1) is always satisfied. In the case $\boldsymbol{k} = \boldsymbol{0}$ and $\boldsymbol{k} \in \Delta_3$, $h_{1,\boldsymbol{k}} = x^2 - y$, $h_{2,\boldsymbol{k}} = x^2 y$, the conditions (3.2)–(3.1) are also verified. Hence $\boldsymbol{h}$ is non-degenerate over $\mathbb{F}_q$ with respect to $\Gamma(\boldsymbol{h})$.

Example 2 Let $\boldsymbol{h} = (h_1(\boldsymbol{x}), \ldots, h_r(\boldsymbol{x}))$ be a monomial mapping. In this case, $\Gamma(\boldsymbol{h}) = \boldsymbol{m}_0 + \mathbb{R}_+^n$ for some nonzero vector $\boldsymbol{m}_0$ in $\mathbb{N}^n$. Then for every vector $\boldsymbol{k} \in \mathbb{R}_+^n$ $h_{i,\boldsymbol{k}}(\boldsymbol{x}) = h_i(\boldsymbol{x})$ for $i = 1, \ldots, r$, and thus the subset in (3.2) is always empty, which implies that condition (3.1) is always satisfied. Therefore any monomial mapping (with $r \leq n$) is non-degenerate over $\mathbb{F}_q$ with respect to its Newton polyhedron.

Example 3 $f(\boldsymbol{x})$, $g(\boldsymbol{x}) \in \mathcal{O}_K[x_1, \ldots, x_n] \backslash \pi\mathcal{O}_K[x_1, \ldots, x_n]$ such that $g(\boldsymbol{x}) = \boldsymbol{x}^{\boldsymbol{m}_0}$, with $\boldsymbol{m}_0 \neq \boldsymbol{0}$, is a monomial. Suppose that f is non-degenerate with respect to $\Gamma(f)$ over $\mathbb{F}_q$. In this case, $\Gamma((f, g)) = \boldsymbol{m}_0 + \Gamma(f)$. Then the subset in (3.2) can take three different forms:

$$\text{(i) } \left\{\overline{z} \in \left(\mathbb{F}_q^\times\right)^n ; \overline{f}_{\boldsymbol{k}}(\overline{z}) = \overline{g}(\overline{z}) = 0\right\} = \varnothing, \text{ (ii) } \left\{\overline{z} \in \left(\mathbb{F}_q^\times\right)^n ; \overline{f}_{\boldsymbol{k}}(\overline{z}) = 0\right\},$$
$$\text{(iii) } \left\{\overline{z} \in \left(\mathbb{F}_q^\times\right)^n ; \overline{g}(\overline{z}) = 0, \overline{f}_{\boldsymbol{k}}(\overline{z}) \neq 0\right\} = \varnothing.$$

In the second case, conditions (3.2)–(3.1) are verified due to the hypothesis that f is non-degenerate with respect $\Gamma(f)$ over $\mathbb{F}_q$. Hence, (f, g) is a non-degenerate mapping over $\mathbb{F}_q$ with respect to $\Gamma((f, g))$ over $\mathbb{F}_q$.

4 Meromorphic Continuation of Multivariate Local Zeta Functions

Along this section, we work with a fix simplicial polyhedral subdivision $\mathcal{F}(\boldsymbol{h})$ subordinate to $\Gamma(\boldsymbol{h})$. Let $\Delta \in \mathcal{F}(\boldsymbol{h}) \cup \{\mathbf{0}\}$ and $I \subseteq \{1, \ldots, r\}$, we put

$$\overline{V}_{\Delta,I} := \left\{\overline{z} \in (\mathbb{F}_q^{\times})^n;\ \overline{h}_{i,b(\Delta)}(\overline{z}) = 0 \Leftrightarrow i \in I\right\}.$$

We use the convention $\overline{V}_{\Delta,\{1,\ldots,r\}} = \overline{V}_{\Delta}$. If $\Delta = \mathbf{0}$, then

$$\overline{V}_{\mathbf{0},I} = \left\{\overline{z} \in (\mathbb{F}_q^{\times})^n;\ \overline{h}_i(\overline{z}) = 0 \Leftrightarrow i \in I\right\} = \overline{V}_I,$$

where $\overline{V}_I$ is the set defined in (2.3). In particular, $\overline{V}_{\mathbf{0},\{1,\ldots,r\}} = \overline{V}$ and

$$\overline{V}_{\mathbf{0},\varnothing} = \left\{\overline{z} \in (\mathbb{F}_q^{\times})^n;\ \overline{h}_i(\overline{z}) \neq 0,\ i = 1, \ldots, r\right\} = \overline{V}_{\varnothing}.$$

If $\boldsymbol{h} = (h_1, \ldots, h_r)$ is non-degenerated polynomial mapping over $\mathbb{F}_q$ with respect to $\Gamma(\boldsymbol{h})$, then Lemma 2 is true for $\boldsymbol{h}_{b(\Delta)} = (h_{1,b(\Delta)}, \ldots, h_{r,b(\Delta)})$.

Theorem 1 *Assume that $\boldsymbol{h} = (h_1, \ldots, h_r)$ is non-degenerated polynomial mapping over $\mathbb{F}_q$ with respect to $\Gamma(\boldsymbol{h})$, with $r \leq n$ as before. Fix a simplicial polyhedral subdivision $\mathcal{F}(\boldsymbol{h})$ subordinate to $\Gamma(\boldsymbol{h})$. Then $Z(\mathbf{s}, \boldsymbol{h})$ has a meromorphic continuation to $\mathbb{C}^r$ as a rational function in the variables q^{-s_i}, $i = 1, \ldots, r$. In addition, the following explicit formula holds:*

$$Z(\boldsymbol{s}, \boldsymbol{h}) = L_{\{\mathbf{0}\}}(\boldsymbol{s}, \boldsymbol{h}) + \sum_{\Delta \in \mathcal{F}(\boldsymbol{h})} L_{\Delta}(\boldsymbol{s}, \boldsymbol{h}) S_{\Delta},$$

where

$$L_{\{\mathbf{0}\}} = q^{-n} \sum_{I \subseteq \{1,\ldots,r\}} Card(\overline{V}_I) \prod_{i \in I} \frac{(q-1)q^{-1-s_i}}{1 - q^{-1-s_i}},$$

$$L_{\Delta} = q^{-n} \sum_{I \subseteq \{1,\ldots,r\}} Card(\overline{V}_{\Delta,I}) \prod_{i \in I} \frac{(q-1)q^{-1-s_i}}{1 - q^{-1-s_i}},$$

with the convention that for $I = \varnothing$, $\prod_{i \in I} \frac{(q-1)q^{-1-s_i}}{1-q^{-1-s_i}} := 1$, and

$$S_{\Delta} = \sum_{\boldsymbol{k} \in \mathbb{N}^n \cap \Delta} q^{-\sigma(\boldsymbol{k}) - \sum_{i=1}^{r} d(\boldsymbol{k}, \Gamma(h_i)) s_i}.$$

Let Δ be the cone strictly positively generated by linearly independent vectors $\boldsymbol{w}_1, \ldots, \boldsymbol{w}_l \in \mathbb{N}^n \setminus \{\mathbf{0}\}$, then

$$S_\Delta = \frac{\sum_t q^{-\sigma(t)-\sum_{i=1}^r d(t,\Gamma(h_i))s_i}}{(1-q^{-\sigma(w_1)-\sum_{i=1}^r d(w_1,\Gamma(h_i))s_i})\cdots(1-q^{-\sigma(w_l)-\sum_{i=1}^r d(w_l,\Gamma(h_i))s_i})},$$

where t runs through the elements of the set

$$\mathbb{Z}^n \cap \left\{ \sum_{i=1}^{l} \lambda_i w_i;\, 0 < \lambda_i \le 1\, for\, i = 1, \ldots, l \right\}. \tag{4.1}$$

Proof By using the simplicial polyhedral subdivision $\mathcal{F}(h)$, we have

$$\mathbb{R}^n_+ = \{0\} \bigsqcup \bigsqcup_{\Delta \in \mathcal{F}(h)} \Delta.$$

We set for $k = (k_1, \ldots, k_n) \in \mathbb{N}^n$,

$$E_k := \left\{ (x_1, \ldots, x_n) \in \mathcal{O}_K^n;\, ord(x_i) = k_i, i = 1, \ldots, n \right\}.$$

Hence

$$Z(s,h) = \int_{(\mathcal{O}_K^\times)^n \setminus D_K} \prod_{i=1}^r |h_i(x)|_K^{s_i} |dx|_K + \sum_{\Delta \in \mathcal{F}(h)} \sum_{k \in \mathbb{N}^n \cap \Delta} \int_{E_k \setminus D_K} \prod_{i=1}^r |h_i(x)|_K^{s_i} |dx|_K.$$

For $\Delta \in \mathcal{F}(h)$, $k \in \mathbb{N}^n \cap \Delta$, and $x = (x_1, \ldots, x_n) \in E_k$, we put $x_j = \pi^{k_j} u_j$ with $u_j \in \mathcal{O}_K^\times$. Then

$$|dx|_K = q^{-\sigma(k)} |du|_K \text{ and } x^m = x_1^{m_1} \cdots x_n^{m_n} = \pi^{\langle k,m \rangle} u_1^{m_1} \cdots u_n^{m_n}.$$

Fix $i \in \{1, \ldots, r\}$ and k. For $m \in supp(h_i)$, the scalar product $\langle k, m \rangle$ attains its minimum $d(k, \Gamma(h_i))$ exactly when $m \in F(k, \Gamma(h_i))$, and thus $\langle k, m \rangle \ge d(k, \Gamma(h_i)) + 1$ for $m \in supp(h_i) \setminus supp(h_i) \cap F(k, \Gamma(h_i))$. This fact implies that

$$\begin{aligned} h_i(x) &= \pi^{d(k,\Gamma(h_i))}(h_{i,k}(u) + \pi \widetilde{h}_{i,k}(u)) \\ &= \pi^{d(k,\Gamma(h_i))}(h_{i,b(\Delta)}(u) + \pi \widetilde{h}_{i,k}(u)), \end{aligned}$$

where $\widetilde{h}_{i,k}(u)$ is a polynomial over $\mathcal{O}_K$ in the variables $u_1, \ldots, u_n$. Note that $h_{i,k}(u)$ does not depend on $k \in \Delta$, for this reason we take $h_{i,k}(u) = h_{i,b(\Delta)}(u)$. Therefore

$$Z(s,h) = L_{\{0\}}(s,h) + \sum_{\Delta \in \mathcal{F}(h)} L_\Delta(s,h) \sum_{k \in \mathbb{N}^n \cap \Delta} q^{-\sigma(k) - \sum_{i=1}^r d(k,\Gamma(h_i))s_i}$$

where

$$L_{\{0\}}(s,h) := \int_{(\mathcal{O}_K^\times)^n \setminus D_K} \prod_{i=1}^r |h_i(x)|_K^{s_i} |dx|_K,$$

$$L_{\Delta}(\boldsymbol{s},\boldsymbol{h}) := \int\limits_{(\mathcal{O}_K^{\times})^n\setminus D_{\Delta}} \prod_{i=1}^{r} |h_{i,b(\Delta)}(\boldsymbol{u}) + \pi\widetilde{h}_{i,\boldsymbol{k}}(\boldsymbol{u})|_K^{s_i} |d\boldsymbol{u}|_K$$

with $D_{\Delta} = \bigcup_{i=1}^{r} \left\{\boldsymbol{x} \in (\mathcal{O}_K^{\times})^n;\, h_{i,b(\Delta)}(\boldsymbol{u}) + \pi\widetilde{h}_{i,\boldsymbol{k}}(\boldsymbol{u}) = 0\right\}$. By using the non-degeneracy condition, integrals $L_{\{\boldsymbol{0}\}}(\boldsymbol{s},\boldsymbol{h})$, $L_{\Delta}(\boldsymbol{s},\boldsymbol{h})$ can be computed using Lemma 2 and Remarks 1 and 2.

Let Δ be the cone strictly positively generated by linearly independent vectors $\boldsymbol{w}_1,\ldots,\boldsymbol{w}_l \in \mathbb{N}^n\setminus\{\boldsymbol{0}\}$. If Δ is a simple cone then $\mathbb{N}^n \cap \Delta = (\mathbb{N}\setminus\{0\})\,\boldsymbol{w}_1 + \cdots + (\mathbb{N}\setminus\{0\})\,\boldsymbol{w}_l$. By using that the functions $d(\cdot,\Gamma(h_i))$ are linear over each cone Δ, and that

$$\sigma(\boldsymbol{w}_m) + \sum_{i=1}^{r} d(\boldsymbol{w}_m,\Gamma(h_i))\,\mathrm{Re}(s_i) > 0, m = 1,\ldots,l,$$

since $\mathrm{Re}(s_1),\ldots,\mathrm{Re}(s_r) > 0$, we obtain

$$S_{\Delta} = \sum_{\lambda_1,\ldots,\lambda_l\in\mathbb{N}\setminus\{0\}} q^{-\sigma(\lambda_1\boldsymbol{w}_1+\ldots+\lambda_l\boldsymbol{w}_l)-\sum_{i=1}^{r} d(\lambda_1\boldsymbol{w}_1+\ldots+\lambda_l\boldsymbol{w}_l,\Gamma(h_i))s_i}$$

$$= \sum_{\lambda_1=1}^{\infty}\left(q^{-\sigma(\boldsymbol{w}_1)-\sum_{i=1}^{r} d(\boldsymbol{w}_1,\Gamma(h_i))s_i}\right)^{\lambda_1}\cdots\sum_{\lambda_l=1}^{\infty}\left(q^{-\sigma(\boldsymbol{w}_l)-\sum_{i=1}^{r} d(\boldsymbol{w}_l,\Gamma(h_i))s_i}\right)^{\lambda_l}$$

$$S_{\Delta} = \frac{q^{-\sigma(\boldsymbol{w}_1)-\sum_{i=1}^{r} d(\boldsymbol{w}_1,\Gamma(h_i))s_i}}{1-q^{-\sigma(\boldsymbol{w}_1)-\sum_{i=1}^{r} d(\boldsymbol{w}_1,\Gamma(h_i))s_i}}\cdots\frac{q^{-\sigma(\boldsymbol{w}_l)-\sum_{i=1}^{r} d(\boldsymbol{w}_l,\Gamma(h_i))s_i}}{1-q^{-\sigma(\boldsymbol{w}_l)-\sum_{i=1}^{r} d(\boldsymbol{w}_l,\Gamma(h_i))s_i}}$$

$$= \frac{\sum_{\boldsymbol{t}} q^{-\sigma(\boldsymbol{t})-\sum_{i=1}^{r} d(\boldsymbol{t},\Gamma(h_i))s_i}}{(1-q^{-\sigma(\boldsymbol{w}_1)-\sum_{i=1}^{r} d(\boldsymbol{w}_1,\Gamma(h_i))s_i})\cdots(1-q^{-\sigma(\boldsymbol{w}_l)-\sum_{i=1}^{r} d(\boldsymbol{w}_l,\Gamma(h_i))s_i})},$$

where $\boldsymbol{t}$ runs through the elements of the set (4.1), which consists exactly of one element: $\boldsymbol{t} = \sum_{i=1}^{l} \boldsymbol{w}_i$. We now consider the case in which Δ is a simplicial cone. Note that $\mathbb{N}^n \cap \Delta$ is the disjoint union of the sets

$$\boldsymbol{t} + \mathbb{N}\boldsymbol{w}_1 + \cdots + \mathbb{N}\boldsymbol{w}_l,$$

where $\boldsymbol{t}$ runs through the elements of the set

$$\mathbb{Z}^n \cap \left\{\sum_{i=1}^{l} \lambda_i \boldsymbol{w}_i;\, 0 < \lambda_i \le 1 \text{ for } i = 1,\ldots,l\right\}.$$

Hence S_{Δ} equals

$$\sum_{\boldsymbol{t}} q^{-\sigma(\boldsymbol{t})-\sum_{i=1}^{r} d(\boldsymbol{t},\Gamma(h_i))s_i} \sum_{\lambda_1,\ldots,\lambda_l\in\mathbb{N}} q^{-\sigma(\sum_{j=1}^{l}\lambda_j\boldsymbol{w}_j)-\sum_{i=1}^{r} d(\lambda_1\boldsymbol{w}_1+\ldots+\lambda_l\boldsymbol{w}_l,\Gamma(h_i))s_i},$$

and since $\mathrm{Re}(s_1),\ldots,\mathrm{Re}(s_r) > 0$,

$$S_\Delta = \frac{\sum_t q^{-\sigma(t)-\sum_{i=1}^r d(t,\Gamma(h_i))s_i}}{(1-q^{-\sigma(w_1)-\sum_{i=1}^r d(w_1,\Gamma(h_i))s_i})\cdots(1-q^{-\sigma(w_l)-\sum_{i=1}^r d(w_l,\Gamma(h_i))s_i})}.$$

□

Remark 6 In the p-adic case, $K = \mathbb{Q}_p$, Theorem 1 is a generalization of Theorem 4.2 in [19] and Theorem 4.3 in [20].

5 Local Zeta Function for Rational Functions

From now on, we fix two non-constant polynomials $f(\boldsymbol{x}), g(\boldsymbol{x})$ in n variables, $n \geq 2$, with coefficients in $\mathcal{O}_K[x_1, \ldots, x_n] \backslash \pi\mathcal{O}_K[x_1, \ldots, x_n]$. From now on, we will assume that f and g are co-prime over $K[x_1, \ldots, x_n]$. Set $D_K := \{\boldsymbol{x} \in K^n; f(\boldsymbol{x}) = 0\} \cup \{\boldsymbol{x} \in K^n; g(\boldsymbol{x}) = 0\}$, and

$$\frac{f}{g} : K^n \setminus D_K \to K.$$

Furthermore, we define the *Newton polyhedron* $\Gamma\left(\frac{f}{g}\right)$ of $\frac{f}{g}$ to be $\Gamma(fg)$, and assume that the mapping $(f, g) : K^n \to K^2$ is non-degenerate over $\mathbb{F}_q$ with respect to $\Gamma\left(\frac{f}{g}\right)$ as before. In this case we will say that $\frac{f}{g}$ *is non-degenerate over* $\mathbb{F}_q$ *with respect to* $\Gamma\left(\frac{f}{g}\right)$. We fix a simplicial polyhedral subdivision $\mathcal{F}\left(\frac{f}{g}\right)$ of $\mathbb{R}_+^n$ subordinate to $\Gamma\left(\frac{f}{g}\right)$. For $\Delta \in \mathcal{F}\left(\frac{f}{g}\right) \cup \{\boldsymbol{0}\}$, we put

$$N_{\Delta,\{f\}} := Card\left\{\overline{\boldsymbol{a}} \in (\mathbb{F}_q^\times)^n; \overline{f}_{b(\Delta)}(\overline{\boldsymbol{a}}) = 0 \,\text{and}\, \overline{g}_{b(\Delta)}(\overline{\boldsymbol{a}}) \neq 0\right\},$$
$$N_{\Delta,\{g\}} := Card\left\{\overline{\boldsymbol{a}} \in (\mathbb{F}_q^\times)^n; \overline{f}_{b(\Delta)}(\overline{\boldsymbol{a}}) \neq 0 \,\text{and}\, \overline{g}_{b(\Delta)}(\overline{\boldsymbol{a}}) = 0\right\},$$
$$N_{\Delta,\{f,g\}} := Card\left\{\overline{\boldsymbol{a}} \in (\mathbb{F}_q^\times)^n; \overline{f}_{b(\Delta)}(\overline{\boldsymbol{a}}) = 0 \,\text{and}\, \overline{g}_{b(\Delta)}(\overline{\boldsymbol{a}}) = 0\right\},$$

with the convention that if $b(\Delta) = b(\boldsymbol{0}) = \boldsymbol{0}$, then $f_{b(\Delta)} = f$ and $g_{b(\Delta)} = g$. We also define $\mathfrak{D}\left(\frac{f}{g}\right) = \mathfrak{D}(f, g)$, which is the set of primitive vectors in $\mathbb{N}^n \setminus \{\boldsymbol{0}\}$ perpendicular to the facets of $\Gamma\left(\frac{f}{g}\right)$. We set

$$T_+ := \left\{\boldsymbol{w} \in \mathfrak{D}\left(\frac{f}{g}\right); d(\boldsymbol{w}, \Gamma(g)) - d(\boldsymbol{w}, \Gamma(f)) > 0\right\},$$
$$T_- := \left\{\boldsymbol{w} \in \mathfrak{D}\left(\frac{f}{g}\right); d(\boldsymbol{w}, \Gamma(g)) - d(\boldsymbol{w}, \Gamma(f)) < 0\right\},$$

$$\alpha := \alpha\left(\frac{f}{g}\right) = \begin{cases} \min_{\boldsymbol{w}\in T_+}\left\{\frac{\sigma(\boldsymbol{w})}{d(\boldsymbol{w},\Gamma(g))-d(\boldsymbol{w},\Gamma(f))}\right\} & \text{if } T_+ \neq \varnothing \\ +\infty & \text{if } T_+ = \varnothing, \end{cases}$$

$$\beta := \beta\left(\frac{f}{g}\right) = \begin{cases} \max_{\boldsymbol{w}\in T_-}\left\{\frac{\sigma(\boldsymbol{w})}{d(\boldsymbol{w},\Gamma(g))-d(\boldsymbol{w},\Gamma(f))}\right\} & \text{if } T_- \neq \varnothing \\ -\infty & \text{if } T_- = \varnothing, \end{cases}$$

and

$$\widetilde{\alpha} := \widetilde{\alpha}\left(\frac{f}{g}\right) = \min\{1, \alpha\},\ \widetilde{\beta} := \widetilde{\beta}\left(\frac{f}{g}\right) = \max\{-1, \beta\}.$$

Notice that $\alpha > 0$ and $\beta < 0$.

We define the local zeta function attached to $\frac{f}{g}$ as

$$Z\left(s, \frac{f}{g}\right) = Z(s, -s, f, g),\ s \in \mathbb{C},$$

where $Z(s_1, s_2, f, g)$ denotes the meromorphic continuation of the local zeta function attached to the polynomial mapping (f, g), see Theorem 1.

Theorem 2 *Assume that $\frac{f}{g}$ is non-degenerate over $\mathbb{F}_q$ with respect to $\Gamma\left(\frac{f}{g}\right)$, with $n \geq 2$ as before. We fix a simplicial polyhedral subdivision $\mathcal{F}\left(\frac{f}{g}\right)$ of $\mathbb{R}_+^n$ subordinate to $\Gamma\left(\frac{f}{g}\right)$. Then the following assertions hold:*

(i) $Z\left(s, \frac{f}{g}\right)$ has a meromorphic continuation to the whole complex plane as a rational function of q^{-s} and the following explicit formula holds:

$$Z\left(s, \frac{f}{g}\right) = \sum_{\Delta\in\mathcal{F}(\frac{f}{g})\cup\{\mathbf{0}\}} L_\Delta\left(s, \frac{f}{g}\right) S_\Delta(s),$$

where for $\Delta \in \mathcal{F}\left(\frac{f}{g}\right) \cup \{\mathbf{0}\}$,

$$L_\Delta(s, \frac{f}{g}) = q^{-n}\left[(q-1)^n - N_{\Delta,\{f\}}\frac{1-q^{-s}}{1-q^{-1-s}} - N_{\Delta,\{g\}}\frac{1-q^{s}}{1-q^{-1+s}}\right.$$
$$\left. -N_{\Delta,\{f,g\}}\frac{(1-q^{-s})(1-q^{s})}{q(1-q^{-1-s})(1-q^{-1+s})}\right]$$

and

$$S_\Delta(s) = \frac{\sum_t q^{-\sigma(t)-(d(t,\Gamma(f))-d(t,\Gamma(g)))s}}{\prod_{i=1}^{l}(1-q^{-\sigma(\boldsymbol{w}_i)-(d(\boldsymbol{w}_i,\Gamma(f))-d(\boldsymbol{w}_i,\Gamma(g)))s})},$$

for $\Delta \in \mathcal{F}\left(\frac{f}{g}\right)$ *a cone strictly positively generated by linearly independent vectors* $\boldsymbol{w}_1, \ldots, \boldsymbol{w}_l \in \mathfrak{D}\left(\frac{f}{g}\right)$, *and where* $\boldsymbol{t}$ *runs through the elements of the set*

$$\mathbb{Z}^n \cap \left\{ \sum_{i=1}^{l} \lambda_i \boldsymbol{w}_i ; 0 < \lambda_i \leq 1 \, for \, i = 1, \ldots, l \right\}.$$

By convention $S_{\mathbf{0}}(s) := 1$.
(ii) $Z\left(s, \frac{f}{g}\right)$ *is a holomorphic function on* $\widetilde{\beta} < \mathrm{Re}(s) < \widetilde{\alpha}$, *and on this band it verifies that*

$$Z\left(s, \frac{f}{g}\right) = \int\limits_{\mathcal{O}_K^n \setminus D_K} \left| \frac{f(x)}{g(x)} \right|^s |dx|; \tag{5.1}$$

(iii) the poles of the meromorphic continuation of $Z\left(s, \frac{f}{g}\right)$ *belong to the set*

$$\bigcup_{k \in \mathbb{Z}} \left\{ 1 + \frac{2\pi\sqrt{-1}k}{\ln q} \right\} \cup \bigcup_{k \in \mathbb{Z}} \left\{ -1 + \frac{2\pi\sqrt{-1}k}{\ln q} \right\} \cup$$
$$\bigcup_{\boldsymbol{w} \in \mathfrak{D}\left(\frac{f}{g}\right)} \bigcup_{k \in \mathbb{Z}} \left\{ \frac{\sigma(\boldsymbol{w})}{d(\boldsymbol{w}, \Gamma(g)) - d(\boldsymbol{w}, \Gamma(f))} + \frac{2\pi\sqrt{-1}k}{\{d(\boldsymbol{w}, \Gamma(g)) - d(\boldsymbol{w}, \Gamma(f))\} \ln q} \right\}.$$

Proof (i) The explicit formula for $Z(s, \frac{f}{g})$ follows from Theorem 1 as follows: we take $r = 2$, $s_1 = s$, $s_2 = -s$, $h_1 = f_{b(\Delta)}$ and $h_2 = g_{b(\Delta)}$ for $\Delta \in \mathcal{F}\left(\frac{f}{g}\right) \cup \{\mathbf{0}\}$, with the convention that if $b(\Delta) = b(\mathbf{0}) = \mathbf{0}$, then $h_1 = f$ and $h_2 = g$. Now

$$\overline{V}_\Delta = \left\{ \overline{z} \in \left(\mathbb{F}_q^\times\right)^n ; \overline{f}_{b(\Delta)}(\overline{z}) = \overline{g}_{b(\Delta)}(\overline{z}) = 0 \right\} \text{ for } \Delta \in \mathcal{F}\left(\frac{f}{g}\right) \cup \{\mathbf{0}\},$$

and thus $Card(\overline{V}_\Delta) = N_{\Delta,\{f,g\}}$. Now, with $I = \{1, 2\}$, by using (2.6), we have

$$J\left(s, -s, \overline{V}_\Delta\right) = \frac{q^{-n}\left(1 - q^{-1}\right)^2 N_{\Delta,\{f,g\}}}{\left(1 - q^{-1-s}\right)\left(1 - q^{-1+s}\right)}. \tag{5.2}$$

We now consider the case $I \neq \varnothing$, $I \subsetneq \{1, 2\}$, thus there are two cases: $I = \{1\}$ or $I = \{2\}$. Note that

$$\overline{V}_{\Delta,\{1\}} = \left\{ \overline{z} \in \left(\mathbb{F}_q^\times\right)^n ; \overline{f}_{b(\Delta)}(\overline{z}) = 0 \text{ and } \overline{g}_{b(\Delta)}(\overline{z}) \neq 0 \right\} \text{ for } \Delta \in \mathcal{F}\left(\frac{f}{g}\right) \cup \{\mathbf{0}\},$$

and that $Card\left(\overline{V}_{\Delta,\{1\}}\right) = N_{\Delta,\{f\}}$, with the convention that

$$\overline{V}_{\mathbf{0},\{1\}} = \left\{\overline{z} \in \left(\mathbb{F}_q^{\times}\right)^n ; \overline{f}\left(\overline{z}\right) = 0 \text{ and } \overline{g}\left(\overline{z}\right) \neq 0\right\}.$$

In this case, by using (2.5),

$$J\left(s, -s, \overline{V}_{\Delta,\{1\}}\right) = \frac{q^{-n-s}\left(1-q^{-1}\right) N_{\Delta,\{f\}}}{1-q^{-1-s}}. \tag{5.3}$$

Analogously,

$$J\left(s, -s, \overline{V}_{\Delta,\{2\}}\right) = \frac{q^{-n+s}\left(1-q^{-1}\right) N_{\Delta,\{g\}}}{1-q^{-1+s}}. \tag{5.4}$$

We now consider the case $I = \varnothing$, then

$$\overline{V}_{\Delta,\varnothing} = \left\{\overline{z} \in \left(\mathbb{F}_q^{\times}\right)^n ; \overline{f}_{b(\Delta)}\left(\overline{z}\right) \neq 0 \text{ and } \overline{g}_{b(\Delta)}\left(\overline{z}\right) \neq 0\right\} \text{ for } \Delta \in \mathcal{F}\left(\frac{f}{g}\right) \cup \{\mathbf{0}\},$$

with the convention that

$$\overline{V}_{\mathbf{0},\varnothing} = \left\{\overline{z} \in \left(\mathbb{F}_q^{\times}\right)^n ; \overline{f}\left(\overline{z}\right) \neq 0 \text{ and } \overline{g}\left(\overline{z}\right) \neq 0\right\}.$$

Notice that $Card(\overline{V}_{\Delta,\varnothing}) = (q-1)^n - N_{\Delta,\{f\}} - N_{\Delta,\{g\}} - N_{\Delta,\{f,g\}}$. Then, by using (2.4),

$$J\left(s, -s, \overline{V}_{\Delta,\varnothing}\right) = q^{-n} Card(\overline{V}_{\Delta,\varnothing}). \tag{5.5}$$

Then from Theorem 1 and (5.2)–(5.5), we get

$$L_{\Delta}(s, \frac{f}{g}) = \frac{q^{-n}\left(1-q^{-1}\right)^2 N_{\Delta,\{f,g\}}}{\left(1-q^{-1-s}\right)\left(1-q^{-1+s}\right)} + \frac{q^{-n-s}\left(1-q^{-1}\right) N_{\Delta,\{f\}}}{1-q^{-1-s}} +$$
$$\frac{q^{-n+s}\left(1-q^{-1}\right) N_{\Delta,\{g\}}}{1-q^{-1+s}} + q^{-n}\left\{(q-1)^n - N_{\Delta,\{f\}} - N_{\Delta,\{g\}} - N_{\Delta,\{f,g\}}\right\}.$$

The announced formula for $L_{\Delta}(s, \frac{f}{g})$ is obtained from the above formula after some simple algebraic manipulations.

(ii) Notice that for $\boldsymbol{w} \in \mathfrak{D}\left(\frac{f}{g}\right)$, $\frac{1}{1-q^{-\sigma(\boldsymbol{w})-(d(\boldsymbol{w},\Gamma(f))-d(\boldsymbol{w},\Gamma(g)))s}}$ is holomorphic on $\sigma(\boldsymbol{w}) + (d(\boldsymbol{w},\Gamma(f)) - d(\boldsymbol{w},\Gamma(g)))\,\mathrm{Re}(s) > 0$, and that $\frac{1}{1-q^{-1-s}}$ is holomorphic on $\mathrm{Re}(s) > -1$, and $\frac{1}{1-q^{-1+s}}$ is holomorphic on $\mathrm{Re}(s) < 1$, then, from the explicit formula for $Z(s, \frac{f}{g})$ given in (i) follows that it is holomorphic on the band $\widetilde{\beta} < \mathrm{Re}(s) < \widetilde{\alpha}$. Now, since $Z(s, \frac{f}{g}) = Z(s, -s, f, g)$, $Z(s, \frac{f}{g})$ is given by integral (5.1) because $Z(s_1, s_2, f, g)$ agrees with an integral on its natural domain.

(iii) It is a direct consequence of the explicit formula. $\square$

6 The Largest and Smallest Real Part of the Poles Of $Z(s, \frac{f}{g})$ (Different From −1 and 1, Respectively)

In this section we use all the notation introduced in Sect. 5. We work with a fix simplicial polyhedral subdivision $\mathcal{F}\left(\frac{f}{g}\right)$ of $\mathbb{R}_+^n$ subordinate to $\Gamma\left(\frac{f}{g}\right)$. We recall that in the case $T_- \neq \emptyset$,

$$\beta = \max_{w \in T_-} \left\{ \frac{\sigma(\boldsymbol{w})}{d(\boldsymbol{w}, \Gamma(g)) - d(\boldsymbol{w}, \Gamma(f))} \right\}$$

is the largest possible 'non-trivial' negative real part of the poles of $Z(s, \frac{f}{g})$. We set

$$\mathcal{P}(\beta) := \left\{ \boldsymbol{w} \in T_-;\ \frac{\sigma(\boldsymbol{w})}{d(\boldsymbol{w}, \Gamma(g)) - d(\boldsymbol{w}, \Gamma(f))} = \beta \right\},$$

and for $m \in \mathbb{N}$ with $1 \leq m \leq n$,

$$\mathcal{M}_m(\beta) := \left\{ \Delta \in \mathcal{F}\left(\frac{f}{g}\right);\ \Delta \text{ has exactly } m \text{ generators belonging to } \mathcal{P}(\beta) \right\},$$

and $\rho := \max\{m;\ \mathcal{M}_m(\beta) \neq \emptyset\}$.

Theorem 3 *Suppose that $\frac{f}{g}$ is non-degenerated over $\mathbb{F}_q$ with respect to $\Gamma(\frac{f}{g})$ and that $T_- \neq \emptyset$. If $\beta > -1$, then β is a pole of $Z(s, \frac{f}{g})$ of multiplicity ρ.*

Proof In order to prove that β is a pole of $Z(s, \frac{f}{g})$ of order ρ, it is sufficient to show that

$$\lim_{s \to \beta} (1 - q^{\beta - s})^{\rho} Z\left(s, \frac{f}{g}\right) > 0.$$

This assertion follows from the explicit formula for $Z(s, \frac{f}{g})$ given in Theorem 2, by the following claim:

Claim. $Res\,(\Delta, \beta) := \lim_{s \to \beta} (1 - q^{s-\beta})^{\rho} L_{\Delta}(s, \frac{f}{g}) S_{\Delta}(s) \geq 0$ for every cone $\Delta \in \mathcal{F}(\frac{f}{g})$. Furthermore, there exists a cone $\Delta_0 \in \mathcal{M}_{\rho}(\beta)$ such that $Res\,(\Delta_0, \beta) > 0$.

We show that for at least one cone Δ_0 in $\mathcal{M}_{\rho}(\beta)$, $Res\,(\Delta_0, \beta) > 0$, because for any cone $\Delta \notin \mathcal{M}_{\rho}(\beta)$, $Res\,(\Delta, \beta) = 0$. This last assertion can be verified by using the argument that we give for the cones in $\mathcal{M}_{\rho}(\beta)$. We first note that there exists at least one cone Δ_0 in $\mathcal{M}_{\rho}(\beta)$. Let $\boldsymbol{w}_1, \ldots, \boldsymbol{w}_{\rho}, \boldsymbol{w}_{\rho+1}, \ldots, \boldsymbol{w}_l$ its generators with $\boldsymbol{w}_i \in \mathcal{P}(\beta) \Leftrightarrow 1 \leq i \leq \rho$.

On the other hand,

$$\lim_{s \to \beta} L_{\Delta}\left(s, \frac{f}{g}\right) > 0 \tag{6.1}$$

for all cones $\Delta \in \mathcal{F}(\frac{f}{g}) \cup \{\boldsymbol{0}\}$. Inequality (6.1) follows from

$$L_\Delta\left(\beta, \frac{f}{g}\right) > q^{-n}((q-1)^n - N_{\Delta,\{f\}} - N_{\Delta,\{g\}} - N_{\Delta,\{f,g\}}) \geq 0$$

for all cones $\Delta \in \mathcal{F}(\frac{f}{g}) \cup \{\mathbf{0}\}$. We prove this last inequality in the case $N_{\Delta,\{f\}} > 0$, $N_{\Delta,\{g\}} > 0$, $N_{\Delta,\{f,g\}} > 0$ since the other cases are treated in similar form. In this case, the inequality follows from the formula for $L_\Delta(\beta, \frac{f}{g})$ given in Theorem 2 , by using that

$$N_{\Delta,\{f\}}\frac{1-q^{-\beta}}{1-q^{-1-\beta}} < N_{\Delta,\{f\}},\ N_{\Delta,\{g\}}\frac{1-q^{\beta}}{1-q^{-1+\beta}} < N_{\Delta,\{g\}},$$
$$N_{\Delta,\{f,g\}}\frac{(1-q^{-\beta})(1-q^{\beta})}{q(1-q^{-1-\beta})(1-q^{-1+\beta})} < N_{\Delta,\{f,g\}} \text{ when } \beta > -1.$$

We also notice that

$$\lim_{s\to\beta}\sum_{t} q^{-\sigma(\boldsymbol{t})-(d(\boldsymbol{t},\Gamma(f))-d(\boldsymbol{t},\Gamma(g)))s} > 0.$$

Hence in order to show that $Res\ (\Delta_0, \beta) > 0$, it is sufficient to show that

$$\lim_{s\to\beta}\frac{(1-q^{s-\beta})^\rho}{\prod_{i=1}^{l}(1-q^{-\sigma(\boldsymbol{w}_i)-(d(\boldsymbol{w}_i,\Gamma(f))-d(\boldsymbol{w}_i,\Gamma(g)))s})} > 0.$$

Now, notice that there are positive integer constants c_i such that

$$\prod_{i=1}^{\rho}(1-q^{-\sigma(\boldsymbol{w}_i)-(d(\boldsymbol{w}_i,\Gamma(f))-d(\boldsymbol{w}_i,\Gamma(g)))s}) = \prod_{i=1}^{\rho}(1-q^{(s-\beta)c_i})$$
$$= (1-q^{s-\beta})^\rho\prod_{i=1}^{\rho}\prod_{\varsigma^{c_i}=1,\varsigma\neq 1}\left(1-\varsigma q^{s-\beta}\right).$$

In addition, for $i = \rho+1,\ldots,l$,

$$1-q^{-\sigma(\boldsymbol{w}_i)-(d(\boldsymbol{w}_i,\Gamma(f))-d(\boldsymbol{w}_i,\Gamma(g)))\beta} > 0$$

because $-\sigma(\boldsymbol{w}_i) - (d(\boldsymbol{w}_i, \Gamma(f)) - d(\boldsymbol{w}_i, \Gamma(g)))\beta \leq 0$ for any $\boldsymbol{w}_i \in T_+ \cup T_-$ with $i = \rho+1,\ldots,l$. From these observations, we have

$$\lim_{s\to\beta}\frac{(1-q^{s-\beta})^\rho}{\prod_{i=1}^{l}(1-q^{-\sigma(\boldsymbol{w}_i)-(d(\boldsymbol{w}_i,\Gamma(f))-d(\boldsymbol{w}_i,\Gamma(g)))s})} =$$
$$\lim_{s\to\beta}\frac{(1-q^{s-\beta})^\rho}{(1-q^{s-\beta})^\rho\prod_{i=1}^{\rho}\prod_{\varsigma^{c_i}=1,\varsigma\neq 1}\left(1-\varsigma q^{s-\beta}\right)}\times$$

$$\lim_{s\to\beta}\frac{1}{\prod_{i=\rho+1}^{l}(1-q^{-\sigma(\boldsymbol{w}_i)-(d(\boldsymbol{w}_i,\Gamma(f))-d(\boldsymbol{w}_i,\Gamma(g)))s})}>0.$$

□

In the case $T_+ \neq \varnothing$,

$$\alpha=\min_{\boldsymbol{w}\in T_+}\left\{\frac{\sigma(\boldsymbol{w})}{d(\boldsymbol{w},\Gamma(g))-d(\boldsymbol{w},\Gamma(f))}\right\}.$$

is the smallest possible 'non-trivial' positive real part of the poles of $Z(s,\frac{f}{g})$. We set

$$\mathcal{P}(\alpha):=\left\{\boldsymbol{w}\in T_+;\ \frac{\sigma(\boldsymbol{w})}{d(\boldsymbol{w},\Gamma(g))-d(\boldsymbol{w},\Gamma(f))}=\alpha\right\},$$

and for $m\in\mathbb{N}$ with $1\le m\le n$,

$$\mathcal{M}_m(\alpha):=\left\{\Delta\in\mathcal{F}\left(\frac{f}{g}\right);\ \Delta \text{ has exactly } m \text{ generators belonging to } \mathcal{P}(\alpha)\right\},$$

and $\kappa:=\max\{m;\ \mathcal{M}_m(\alpha)\neq\varnothing\}$

The proof of the following result is similar to the proof of Theorem 3.

Theorem 4 *Suppose that $\frac{f}{g}$ is non-degenerated over $\mathbb{F}_q$ with respect to $\Gamma(\frac{f}{g})$ and that $T_+\neq\varnothing$. If $\alpha<1$, then α is a pole of $Z(s,\frac{f}{g})$ of multiplicity κ.*

Example 4 We compute the local zeta function for the rational function given in Example 1. With the notation of Theorem 2, one verifies that

Cone	L_Δ	S_Δ
$\{\boldsymbol{0}\}$	$q^{-2}((q-1)^2-(q-1)\frac{1-q^{-s}}{1-q^{-1-s}})$	1
Δ_1	$q^{-2}(q-1)^2$	$\frac{q^{-1+2s}}{1-q^{-1+2s}}$
Δ_2	$q^{-2}(q-1)^2$	$\frac{q^{-2+2s}+q^{-4+4s}}{(1-q^{-1+2s})(1-q^{-3+2s})}$
Δ_3	$q^{-2}((q-1)^2-(q-1)\frac{1-q^{-s}}{1-q^{-1-s}})$	$\frac{q^{-3+2s}}{1-q^{-3+2s}}$
Δ_4	$q^{-2}(q-1)^2$	$\frac{q^{-4+3s}}{(1-q^{-3+2s})(1-q^{-1+s})}$
Δ_5	$q^{-2}(q-1)^2$	$\frac{q^{-1+s}}{(1-q^{-1+s})}$.

Therefore

$$Z(s,\frac{f}{g})=\frac{\frac{(q-1)}{q^2}L(q^{-s})}{(1-q^{s-1})(1-q^{-1-s})(1-q^{2s-1})(1-q^{2s-3})},$$

where

$$L(q^{-s}) = q - q^{-1} - 2 - q^{2s-4} + q^{s-3} - q^{s-2} + q^{2s-2} + q^{3s-3} \\ + 2q^{2s-1} - q^{3s-2} - q^{3s-1} + q^{-s-1}.$$

Furthermore, $Z(s, \frac{f}{g})$ has poles with real parts belonging to $\{-1, 1/2, 1, 3/2\}$.

Acknowledgements The authors wish to thank the referee for his/her careful reading of the original manuscript.

References

1. Gel'fand, I.M., Shilov, G.E.: Generalized Functions, vol. 1. Academic Press, New York and London (1977)
2. Atiyah, M.F.: Resolution of singularities and division of distributions. Commun. Pure Appl. Math. **23**, 145–150 (1970)
3. Bernstein, I.N.: Modules over the ring of differential operators; the study of fundamental solutions of equations with constant coefficients. Funct. Anal. Appl. **5**(2), 1–16 (1972)
4. Weil, A.: Sur la formule de Siegel dans la théorie des groupes classiques. Acta Math. **113**, 1–87 (1965)
5. Igusa, J.-I.: Forms of higher degree. In: Tata Institute of Fundamental Research. Lectures on Mathematics and Physics, 59. Narosa Publishing House, New Delhi (1978)
6. Igusa, J.-I.: An Introduction to the Theory of Local Zeta Functions. AMS/IP Studies in Advanced Mathematics (2000)
7. Denef, J.: Report on Igusa's local zeta function. Séminarire Bourbaki 43 (1990–1991), exp 741; Astérique 201-202-203, 359–386 (1991)
8. Denef, J., Loeser, F.: Motivic Igusa zeta functions. J. Algebr. Geom. **7**, 505–537 (1998)
9. Veys, W., Zuñiga-Galindo, W.A.: Zeta functions and oscillatory integrals for meromorphic functions. Adv. Math. **311**, 295–337 (2017)
10. Gusein-Zade, S., Luengo, I., Melle Hernández, A.: Bifurcations and topology of meromorphic germs. New Developments in Singularity Theory (Cambridge, 2000). NATO Science Series II Mathematics, Physics and Chemistry, pp. 279–304. Kluwer Academic Publishers, Dordrecht (2001)
11. Gusein-Zade, S., Luengo I., Melle-Hernández, A.: On the topology of germs of meromorphic functions and its applications. (Russian) Algebra i Analiz **11**(5), 92–99 (1999); translation in St. Petersburg Math. J. **11**(5), 775–780(2000)
12. Gusein-Zade, S., Luengo, I., Melle-Hernández, A.: Zeta functions of germs of meromorphic functions, and the Newton diagram. (Russian) Funktsional. Anal. i Prilozhen. **32**(2), 26–35, 95 (1998); translation in Funct. Anal. Appl. **32**(2), 93–99 (1998)
13. Belkale, P., Brosnan, P.: Periods and Igusa local zeta functions. Int. Math. Res. Not. **49**, 2655–2670 (2003)
14. Bogner, C., Weinzierl, S.: Periods and Feynman integrals. J. Math. Phys. **50**(4), 042302, 16 (2009)
15. Brekke, L., Freund, P.G.O., Olson, M., Witten, E.: Non-archimedean string dynamics. Nucl. Phys. B **302**(3), 365–402 (1988)
16. Marcolli, M.: Feynman Motives. World Scientific Publishing Co. Pte. Ltd., Hackensack (2010)
17. León-Cardenal, E., Zuniga-Galindo, W.A.: Local zeta functions for non-degenerate Laurent polynomials over p-adic fields. J. Math. Sci. Univ. Tokyo **20**(4), 569–595 (2013)

18. Loeser, F.: Fonctions zêta locales d'Igusa à plusiers variables, intégration dans les fibres, et discriminants. Ann. Sc. Ec. Norm. Sup. **22**(3), 435–471 (1989)
19. Denef, J., Hoornaert, K.: Newton polyhedra and Igusa's local zeta function. J. Number Theory **89**, 31–64 (2001)
20. Bories, B.: Igusa's p-adic local zeta function associated to a polynomial mapping and a polynomial integration measure. Manuscripta Math. **138**(3), 395–417 (2012)
21. Weil, A.: Basic Number Theory. Springer, Berlin (1967)
22. Oka, M.: Actualités mathématiques. In: Non-degenerate Complete Intersection Singularity. Hermann, Paris (1997)
23. Zuñiga-Galindo, W.A.: Local zeta functions supported on analytic submanifolds and Newton polyhedra. Int. Math. Res. Not. IMRN. **2009**(15), 2855–2898 (2009)
24. Kempf, G., Knudsen, F., Mumford, D., Saint-Donat, B.: Toroidal Embeddings. Lectures Notes in Mathematics, vol. 339. Springer, Berlin (1973)
25. Sturmfels, B.: Gröbner Bases and Convex Polytopes. American Mathematical Society, Providence (1996)
26. Veys, W., Zúñiga-Galindo, W. A.: Zeta functions for analytic mappings, log-principalization of ideals, and Newton polyhedra. Trans. Am. Math. Soc. **360**(4), 2205–2227 (2008)
27. Khovanskii, A.G.: Newton polyhedra, and toroidal varieties. Funkcional. Anal. i Prilozhen. **11**(4), 56–64 (1977)

Symbolic Powers of Ideals

Hailong Dao, Alessandro De Stefani, Eloísa Grifo, Craig Huneke and Luis Núñez-Betancourt

Abstract We survey classical and recent results on symbolic powers of ideals. We focus on properties and problems of symbolic powers over regular rings, on the comparison of symbolic and regular powers, and on the combinatorics of the symbolic powers of monomial ideals. In addition, we present some new results on these aspects of the subject.

Keywords Symbolic powers · Differential operators · Uniform Symbolic Topologies · Packing problem

H. Dao—Partially supported by NSA Grant H98230-16-1-0012.
C. Huneke—Partially supported by the NSF Grant 1460638.
L. Núñez-Betancourt—Partially supported by the NSF Grant 1502282.

H. Dao
Department of Mathematics, University of Kansas, Lawrence, KS 66045-7523, USA
e-mail: hdao@math.ku.edu

A. De Stefani
Department of Mathematics, University of Nebraska, 203 Avery Hall, Lincoln, NE 68588, USA
e-mail: adestefani2@unl.edu

E. Grifo · C. Huneke
Department of Mathematics, University of Virginia, Charlottesville, VA 22904-4135, USA
e-mail: eloisa.grifo@virginia.edu

C. Huneke
e-mail: huneke@virginia.edu

L. Núñez-Betancourt (✉)
Centro de Investigación en Matemáticas, Guanajuato, Mexico
e-mail: luisnub@cimat.mx

R. N. Araújo dos Santos et al. (eds.), *Singularities and Foliations. Geometry, Topology and Applications*, Springer Proceedings in Mathematics & Statistics 222, https://doi.org/10.1007/978-3-319-73639-6_13

1 Introduction

In this survey, we discuss different algebraic, geometric, and combinatorial aspects of the symbolic powers. Specifically, we focus on the properties and problems of symbolic powers over regular rings, on the comparison of symbolic and regular powers, and on the combinatorics of the symbolic powers of monomial ideals.

Given an ideal I in a Noetherian domain R, its nth symbolic power is defined by

$$I^{(n)} = \bigcap_{\mathfrak{p} \in \mathrm{Ass}_R(R/I)} (I^n R_{\mathfrak{p}} \cap R).$$

For many purposes, one can focus on symbolic powers of prime ideals. If $\mathfrak{p}$ is a prime ideal, then $\mathfrak{p}^{(n)}$ is the $\mathfrak{p}$-primary component of $\mathfrak{p}^n$.

Symbolic powers do not, in general, coincide with the ordinary powers. From the definition it follows that $\mathfrak{p}^n \subseteq \mathfrak{p}^{(n)}$ for all n, but the converse may fail. The simplest such example can be constructed "generically" by letting $\mathfrak{p} = (x, y)$ in the hypersurface defined by $x^n - yz = 0$. It is easy to see that in this example y is in $\mathfrak{p}^{(n)}$ but is not even in $\mathfrak{p}^2$. It is a bit more subtle in a polynomial ring; however, for the prime ideal $\mathfrak{p} = (x^4 - yz, y^2 - xz, x^3y - z^2) \subseteq K[x, y, z]$, $\mathfrak{p}^{(2)} \neq \mathfrak{p}^2$.

The study and use of symbolic powers has a long history in commutative algebra. Krull's famous proof of his principal ideal theorem uses them in an essential way. Of course they first arose after primary decompositions were proved for Noetherian rings. Zariski used symbolic powers in his study of the analytic normality of algebraic varieties. Chevalley's famous lemma comparing topologies states that in a complete local domain the symbolic powers topology of any prime is finer than the $\mathfrak{m}$-adic topology. A crucial step in the vanishing theorem on local cohomology of Hartshorne and Lichtenbaum uses that for a prime $\mathfrak{p}$ defining a curve in a complete local domain, the powers of $\mathfrak{p}$ are cofinal with the symbolic powers of $\mathfrak{p}$. This important property of being cofinal was further developed by Schenzel in the 1970s, and is a critical point for much of this survey. Irena Swanson proved an important refinement which showed that when the symbolic power topology of a prime $\mathfrak{p}$ is cofinal with the usual powers, there is a *linear* relationship between the two: there exists a constant h such that for all n, $\mathfrak{p}^{(hn)} \subseteq \mathfrak{p}^n$.

In case the base ring is a polynomial ring over a field, one may interpret the nth symbolic power as the sheaf of all function germs over $X = \mathrm{Spec}(R)$ vanishing to order greater than or equal to n at $Z = \mathcal{V}(\mathfrak{p})$. Furthermore, if X is a smooth variety over a perfect field, then

$$\mathfrak{p}^{(n)} = \{f \in R \mid f \in \mathfrak{m}^n \text{ for every closed point } \mathfrak{m} \in Z\} \qquad (1.0.1)$$

by the Zariski–Nagata Theorem [69, 91]. The closed points in Eq. 1.0.1 can be taken only in the smooth locus of Z [22] (see Sect. 2.1 for details). In this survey, we present a characteristic-free proof of the Zariski–Nagata Theorem, which is based on the work of Zariski. Our approach uses the general definition of differential operators given by

Grothendieck [31]. As a consequence of this approach, we present a specific constant for a uniform version of the uniform Chevalley Theorem [52] for direct summands of polynomial rings (see Theorem 3.27 and Corollary 3.28). We point out that this result can be seen as a weaker version of the Zariski–Nagata Theorem which holds for complete local domains.

Equation 1.0.1 is used together with Euler's Formula to deduce $\mathfrak{p}^{(2)} \subseteq \mathfrak{m}\mathfrak{p}$ for a homogeneous prime ideal $\mathfrak{p}$ in a polynomial ring over a characteristic zero field, where $\mathfrak{m}$ is the maximal homogeneous ideal. This says that the minimal homogeneous generators of $\mathfrak{p}$ cannot have order at least 2 at all the closed points of the variety defined by $\mathfrak{p}$. Eisenbud and Mazur [23] conjectured that the same would hold for complete or affine local rings of equal characteristic zero. We devote Sect. 2.3 to this conjecture. As shown in [23], the fact that the symbolic square of a prime does not contain any minimal generators has close connections with the notion of evolutions. In Sect. 2.3, we define what an evolution is, and state what it means for it to be trivial. The existence of non-trivial evolutions for certain kinds of rings played a crucial role in Wiles's proof of Fermat's Last Theorem [89].

An important property of symbolic powers over regular rings containing fields, which was surprising at the time, is $\mathfrak{p}^{(dn)} \subseteq \mathfrak{p}^n$ for d-dimensional regular rings containing a field. This theorem was established by Ein, Lazarfeld, and Smith [20] in the geometric case in characteristic zero, Hochster and Huneke [41] in general for ring of equal characteristic, and Ma and Shwede [59] in the mixed characteristic case.

The fact that the constant d is uniform, so independent of $\mathfrak{p}$, is remarkable. This theorem motivated the following question:

Question 1.1 Let $(R, \mathfrak{m}, K)$ be a complete local domain. Does there exist a constant c, depending only on R, such that $\mathfrak{p}^{(cn)} \subseteq \mathfrak{p}^n$ for every prime ideal $\mathfrak{p}$?

A positive answer to this question would establish that $\mathfrak{p}$-adic and $\mathfrak{p}$-symbolic typologies are uniformly equivalent (see [48] for a survey on uniformity). In Sect. 3, we discuss the case where this question has a positive answer. In particular, we review the work of Huneke, Katz and Validashti on isolated singularities [51] and finite extensions [52]. We take advantage of this work to answer a question of Takagi about direct summands given by finite group actions (see Theorems 3.29 and Corollary 3.30).

In the final section of this survey we discuss the combinatorics that are encoded in the symbolic powers of monomial ideals. First, we discuss the work of Minh and Trung [66], and Varbaro [84] which characterizes the Cohen-Macaulayness property of the symbolic powers in terms of the corresponding simplicial complex (see Theorem 4.5).

We also discuss the equality $I^{(n)} = I^n$ for squarefree monomial ideals. This problem was related to a conjecture of Conforti and Cornuélos [13] on the max-flow and min-cut properties by Gitler, Valencia and Villarreal [28] and Gitler, Reyes, and Villarreal [27]. The Conforti–Cornuélos Conjecture is known in the context of symbolic powers as the Packing Problem (see Sect. 4.2), and it is a central problem in this theory. In Sect. 4.2, we give a relative version of the Packing Problem, and study

$I^{(n)} = I^n$ separately for each n. We discuss the theorem that establishes that if I_G is a monomial edge ideal associated to a graph G, then $I^{(n)} = I^n$ for every n if and only if G is bipartite.

We point out that the research and literature on this topic is vast, rich, and active. For this reason, we cannot cover all the aspects about symbolic powers in this survey. In particular, we do not cover the very large amount of material concerning symbolic powers and fat points. See [6] or [33] for material and references.

Although this article is largely expository, we present several results that are new, to the best of our knowledge. These include an example where radical ideals of extensions do not have uniform behavior (Example 3.24) and uniform bounds for direct summands of polynomial rings (Theorem 3.27, Corollary 3.28 and Theorem 3.29). Also in Theorem 4.18, we give an alternative proof for a characterization of k-packed edge ideals. In order to distinguish the new results from those which are already known, we make a citation explicitly after the numbering.

2 Symbolic Powers on Regular Rings

2.1 Zariski–Nagata Theorem

In this section, we discuss the Zariski-Nagata Theorem for regular rings [69, 91]. This fundamental theorem establishes that the nth symbolic power of an irreducible variety, or a prime ideal, consists precisely of the set of elements whose order is at least n in every closed point in the variety. Specifically,

$$\mathfrak{p}^{(n)} = \bigcap_{\substack{\mathfrak{m}\in\mathrm{Max}(R)\\ \mathfrak{p}\subseteq\mathfrak{m}}} \mathfrak{m}^n$$

whenever $\mathfrak{p}$ is a prime ideal in a polynomial ring over a field. The first step in proving this fact is showing that $\mathfrak{p}^{(n)} \subseteq \mathfrak{m}^n$ holds in a regular local ring, even in mixed characteristic. The previous fact is also known as the Zariski–Nagata Theorem. We present a proof of this fact based on rings of differential operators over polynomial rings over any ground field. We point out that this method also works for power series rings.

Definition 2.1 Let R be a finitely generated K-algebra. The K-linear differential operators of R of order n, $D^n_R \subseteq \mathrm{Hom}_K(R, R)$, are defined inductively as follows. The differential operators of order zero are $D^0_R = R \cong \mathrm{Hom}_R(R, R)$. We say that $\delta \in \mathrm{Hom}_K(R, R)$ is an operator of order less than or equal to n if $[\delta, r] = \delta r - r\delta$ is an operator of order less than or equal to $n - 1$ for all $r \in D^0_R$. The ring of K-linear differential operators is defined by $D_R = \bigcup_{n\in\mathbb{N}} D^n_R$. If R is clear from the context, we drop the subscript referring to the ring.

Definition 2.2 Let R be a finitely generated K-algebra. Let I be an ideal of R, and let n be a positive integer. We define the nth K-linear differential power of I by

$$I^{\langle n\rangle} = \{f \in R \mid \delta(f) \in I \text{ for all } \delta \in D_R^{n-1}\}.$$

Remark 2.3 Since $D_R^{n-1} \subseteq D_R^n$, it follows that $I^{\langle n+1\rangle} \subseteq I^{\langle n\rangle}$. If $I \subseteq J$, then $I^{\langle n\rangle} \subseteq J^{\langle n\rangle}$ for every $n \in \mathbb{N}$.

In order to prove the Zariski–Nagata Theorem, we need some properties of the differential powers.

Proposition 2.4 *Let R be a finitely generated K-algebra. Let I be an ideal of R, and n be a positive integer. Then, $I^{\langle n\rangle}$ is an ideal.*

Proof It is straightforward to verify that $f, g \in I^{\langle n\rangle}$ implies that $f + g \in I^{\langle n\rangle}$. It then suffices to show that $rf \in I^{\langle n\rangle}$ for $r \in R$ and $f \in I^{\langle n\rangle}$. Let $\delta \in D^{n-1}$. Then, $\delta(rf) = [\delta, r](f) + r\delta(f)$. Since $f \in I^{\langle n\rangle} \subseteq I^{\langle n-1\rangle}$, and $[\delta, r] \in D^{n-2}$, we have that $[\delta, r](f) \in I$. Since $f \in I^{\langle n\rangle}$, $\delta(f) \in I$. We conclude that $\delta(rf) \in I$. Hence, $rf \in I^{\langle n\rangle}$. □

Proposition 2.5 *Let R be a finitely generated K-algebra. Let I be an ideal of R, and n be a positive integer. Then, $I^n \subseteq I^{\langle n\rangle}$.*

Proof We proceed by induction on n.

$\underline{n = 1:}$ In this case, $I = I^{\langle n\rangle}$ because $D^0 = R$.

$\underline{n \Longrightarrow n + 1:}$ Let $f \in I$, $g \in I^n$ and $\delta \in D^n$. Then, $\delta(fg) = [\delta, f](g) + f\delta(g)$. Since $[\delta, r] \in D^{n-1}$ and $g \in I^n \subseteq I^{\langle n\rangle}$ by the induction hypothesis, we have that $[\delta, f](g) \in I$. Then, $\delta(fg) \in I$, and so, $fg \in I^{\langle n+1\rangle}$. Hence, $I^{n+1} \subseteq I^{\langle n+1\rangle}$. □

Proposition 2.6 *Let R be a finitely generated K-algebra. Let $\mathfrak{p}$ be a prime ideal of R, and n be a positive integer. Then, $\mathfrak{p}^{\langle n\rangle}$ is $\mathfrak{p}$-primary.*

Proof Once more, we use induction on n.

$\underline{n = 1:}$ In this case, $\mathfrak{p}^{\langle n\rangle} = \mathfrak{p}$ is a prime ideal.

$\underline{n \Longrightarrow n + 1:}$ Let $r \notin \mathfrak{p}$ and $f \in \mathfrak{p}$ such that $rf \in \mathfrak{p}^{\langle n+1\rangle}$. Let $\delta \in D^n$. Then, $\delta(rf) = [\delta, r](f) + r\delta(f) \in \mathfrak{p}$. Since $rf \in \mathfrak{p}^{\langle n+1\rangle} \subseteq \mathfrak{p}^{\langle n\rangle}$, we have that $f \in \mathfrak{p}^{\langle n\rangle}$ by the induction hypothesis. Then, $[\delta, r](f) \in \mathfrak{p}$, because $[\delta, r] \in D^{n-1}$. We conclude that $r\delta(f) = \delta(rf) - [\delta, r](f) \in \mathfrak{p}$. Then, $r\delta(f) \in \mathfrak{p}$, and so, $\delta(f) \in \mathfrak{p}$, because $\mathfrak{p}$ is a prime ideal and $r \notin \mathfrak{p}$. Hence, $f \in \mathfrak{p}^{\langle n+1\rangle}$. □

Remark 2.7 Let K be a field, R be either $K[x_1, \ldots, x_d]$ or $K[[x_1, \ldots, x_d]]$, and $\mathfrak{m} = (x_1, \ldots, x_d)$. In this case,

$$D_R^n = R\left\langle \frac{1}{\alpha_1!}\frac{\partial^{\alpha_1}}{\partial x_1^{\alpha_1}} \cdots \frac{1}{\alpha_d!}\frac{\partial^{\alpha_d}}{\partial x_d^{\alpha_d}} \,\middle|\, \alpha_1 + \cdots + \alpha_d \leqslant n \right\rangle$$

If $f \notin \mathfrak{m}^n$, then f has a monomial of the form $x_1^{\alpha_1} \cdots x_d^{\alpha_d}$, with nonzero coefficient $\lambda \in K$, which is minimal among all monomials appearing in f under the graded lexicographical order. Applying the differential operator $\frac{1}{\alpha_1!}\frac{\partial^{\alpha_1}}{\partial x_1^{\alpha_1}} \cdots \frac{1}{\alpha_d!}\frac{\partial^{\alpha_d}}{\partial x_d^{\alpha_d}}$ maps $\lambda x_1^{\alpha_1} \cdots x_d^{\alpha_d}$ to the nonzero element $\lambda \in K$, and any other monomial appearing in f either to a non constant monomial or to zero. Consequently, $f \notin \mathfrak{m}^{\langle n \rangle}$. Hence, $\mathfrak{m}^{\langle n \rangle} \subseteq \mathfrak{m}^n$. Since $\mathfrak{m}^n \subseteq \mathfrak{m}^{\langle n \rangle}$ by Proposition 2.5, we conclude that $\mathfrak{m}^{\langle n \rangle} = \mathfrak{m}^n$.

Exercise 2.8 Let K be a field of characteristic zero, R be either $K[x_1, \dots, x_d]$ or $K[[x_1, \dots, x_d]]$, and $\mathfrak{m} = (x_1, \dots, x_d)$. Then, $(\mathfrak{m}^t)^{\langle n \rangle} = \mathfrak{m}^{n+t-1}$.

Theorem 2.9 (Zariski–Nagata Theorem for polynomial and power series rings [91]) *Let K be a field, R be either $K[x_1, \dots, x_d]$ or $K[[x_1, \dots, x_d]]$, and $\mathfrak{m} = (x_1, \dots, x_d)$. Let $\mathfrak{p} \subseteq \mathfrak{m}$ be a prime ideal. Then, for any positive integer n, we have $\mathfrak{p}^{(n)} \subseteq \mathfrak{m}^n$. Furthermore, if* $\operatorname{char}(K) = 0$ *and $\mathfrak{p}$ is a prime ideal such that $\mathfrak{p} \subseteq \mathfrak{m}^t$, then $\mathfrak{p}^{(n)} \subseteq \mathfrak{m}^{n+t-1}$.*

Proof We have that $\mathfrak{m}^{\langle n \rangle} = \mathfrak{m}^n$ by Remark 2.7. Then, $\mathfrak{p}^{(n)} \subseteq \mathfrak{p}^{\langle n \rangle} \subseteq \mathfrak{m}^n$ by Proposition 2.6. The second claim follows from the fact that $\mathfrak{p}^{(n)} \subseteq \mathfrak{p}^{\langle n \rangle} \subseteq (\mathfrak{m}^t)^{\langle n \rangle} = \mathfrak{m}^{n+t-1}$ by Exercise 2.8. □

In order to show a more general version of Zariski–Nagata, we use the Hilbert–Samuel multiplicity:

Definition 2.10 Let $(R, \mathfrak{m}, K)$ be a d-dimensional local ring. The Hilbert–Samuel multiplicity of R is defined by

$$\mathrm{e}(R) = \lim_{n \to \infty} \frac{d!\,\lambda(R/\mathfrak{m}^n)}{n^d}.$$

The Hilbert–Samuel multiplicity is an important invariant which detects and measures singularities. For instance, under suitable hypotheses, $\mathrm{e}(R) = 1$ if and only if R is a regular ring. If R is a regular local ring and $f \in \mathfrak{m}$, then $\mathrm{e}(R/fR) = \operatorname{ord}(f)$, where $\operatorname{ord}(f) = \max\{t \in \mathbb{N} \mid f \in \mathfrak{m}^t\}$. Furthermore, under mild assumptions, we have that $\mathrm{e}(R_\mathfrak{p}) \leqslant \mathrm{e}(R)$ for every $\mathfrak{p} \in \operatorname{Spec}(R)$.

Theorem 2.11 (General version of Zariski–Nagata [69, 91]) *Let $(R, \mathfrak{m}, K)$ be a regular local ring, and $\mathfrak{p} \subseteq \mathfrak{m}$ be a prime ideal. Then, $\mathfrak{p}^{(n)} \subseteq \mathfrak{m}^n$.*

Proof For any element $f \in \mathfrak{m}$, we have that

$$\begin{aligned} \max\{t \in \mathbb{N} \mid f \in \mathfrak{p}^{(t)}\} &= \max\{t \in \mathbb{N} \mid f \in \mathfrak{p}^t R_\mathfrak{p}\} \\ &= \mathrm{e}((R/fR)_\mathfrak{p}) \leqslant \mathrm{e}(R/fR) = \max\{t \in \mathbb{N} \mid f \in \mathfrak{m}^t\}. \end{aligned}$$

As a consequence, we obtain that $f \in \mathfrak{p}^{(n)}$ implies that $f \in \mathfrak{m}^n$. □

We now present Eisenbud's and Hochster's [22] proof for Zariski's Main Lemma on Holomorphic Functions [91]. We point out that this result works in greater generality (cf. [22]).

Theorem 2.12 (Zariski [91], Eisenbud–Hochster [22]) *Let K be a field and R be a finitely generated regular K-algebra. Then,*

$$\mathfrak{p}^{(n)} = \bigcap_{\substack{\mathfrak{m}\in \operatorname{Max}(R)\\ \mathfrak{p}\subseteq\mathfrak{m}}} \mathfrak{m}^n$$

for every prime ideal $\mathfrak{p} \subseteq R$.

Proof The containment

$$\mathfrak{p}^{(n)} \subseteq \bigcap_{\substack{\mathfrak{m}\in \operatorname{Max}(R)\\ \mathfrak{p}\subseteq\mathfrak{m}}} \mathfrak{m}^n$$

follows from the Zariski–Nagata Theorem. In order to prove the other containment, it suffices to show that

$$\bigcap_{\substack{\mathfrak{m}\in \operatorname{Max}(R)\\ \mathfrak{p}\subseteq\mathfrak{m}}} \left(\mathfrak{m}^n \cdot R/\mathfrak{p}^{(n)}\right) = 0.$$

Let $M = R/\mathfrak{p}^{(n)}$ and $M_i = \mathfrak{p}^i \cdot R/\mathfrak{p}^{(n)}$. As an $R/\mathfrak{p}$-module, the locus where M_i/M_{i+1} is a free module is open. In addition, the regular locus of $R/\mathfrak{p}$ is open. Therefore, there exists $f \notin \mathfrak{p}$ such that $(M_i/M_{i+1})_f$ is a free $(R/\mathfrak{p})_f$-module and $(R/\mathfrak{p})_f$ is regular. We note that

$$\bigcap_{\substack{\mathfrak{m}\in \operatorname{Max}(R)\\ \mathfrak{p}\subseteq\mathfrak{m}}} \left(\mathfrak{m}^n \cdot R/\mathfrak{p}^{(n)}\right) \subseteq \bigcap_{\substack{\mathfrak{m}\in \operatorname{Max}(R)\\ \mathfrak{p}\subseteq\mathfrak{m}}} \left(\mathfrak{m}^n \cdot R/\mathfrak{p}^{(n)}\right)_f,$$

because $\mathfrak{m}^n \cdot R/\mathfrak{p}^{(n)} \subseteq (\mathfrak{m}^n \cdot R/\mathfrak{p}^{(n)})_f$, as $f \notin \mathfrak{p}$ and $\operatorname{Ass}_{R/\mathfrak{p}^{(n)}} = \{\mathfrak{p}\}$. Then we can replace R by R_f and assume that $R/\mathfrak{p}$ is regular and M_i/M_{i+1} is a free $R/\mathfrak{p}$-module.

We claim that

$$\mathfrak{m}^n M \cap M_i \subseteq \mathfrak{m} M_i \tag{2.1.1}$$

for every maximal ideal $\mathfrak{m}$ in R. If suffices to show this equality locally. If $\mathfrak{q} \neq \mathfrak{m}$, then $(\mathfrak{m}^n M \cap M_i)_\mathfrak{q} = (M_i)_\mathfrak{q} = (\mathfrak{m}M_i)_\mathfrak{q}$. Thus, it suffices to show our claim for $\mathfrak{q} = \mathfrak{m}$. Since $(R/\mathfrak{p})_\mathfrak{m}$ and $R_\mathfrak{m}$ are regular, there exists a regular sequence $x_1, \dots, x_d \in R_\mathfrak{m}$ such that $(x_1, \dots, x_c)R_\mathfrak{m} = \mathfrak{p}R_\mathfrak{m}$ and $(x_1, \dots, x_d)R_\mathfrak{m} = \mathfrak{m}R_m$. Furthermore, $\mathfrak{p}^{(n)}R_\mathfrak{m} = \mathfrak{p}^n R_\mathfrak{m}$. Then,

$$(\mathfrak{m}^n M \cap M_i)_\mathfrak{m} = \left(\frac{(x_1, \dots, x_d)^n}{(x_1, \dots, x_c)^n} R_\mathfrak{m}\right) \cap \left(\frac{(x_1, \dots, x_c)^i}{(x_1, \dots, x_c)^n} R_\mathfrak{m}\right) \subseteq \left(\frac{(x_1, \dots, x_d)(x_1, \dots, x_c)^i}{(x_1, \dots, x_c)^n} R_\mathfrak{m}\right),$$

which proves our claim.

We note that

$$\bigcap_{\substack{\mathfrak{m}\in \mathrm{Max}(R)\\ \mathfrak{p}\subseteq\mathfrak{m}}} (\mathfrak{m}(M_i/M_{i+1})) = 0, \tag{2.1.2}$$

because M_i/M_{i+1} is $R/\mathfrak{p}$-free and

$$\mathfrak{p} = \bigcap_{\substack{\mathfrak{m}\in \mathrm{Max}(R)\\ \mathfrak{p}\subseteq\mathfrak{m}}} \mathfrak{m},$$

as R is a Jacobson ring.

We consider an element

$$v \in \bigcap_{\substack{\mathfrak{m}\in \mathrm{Max}(R)\\ \mathfrak{p}\subseteq\mathfrak{m}}} \left(\mathfrak{m}^n \cdot R/\mathfrak{p}^{(n)}\right).$$

We want to show that $v = 0$. If $v \neq 0$, then we pick the largest integer i such that $v \in M_i$. From previous considerations, we deduce that

$$\begin{aligned} M_i \bigcap \left(\bigcap_{\substack{\mathfrak{m}\in \mathrm{Max}(R)\\ \mathfrak{p}\subseteq\mathfrak{m}}} \left(\mathfrak{m}^n \cdot R/\mathfrak{p}^{(n)}\right) \right) &= \bigcap_{\substack{\mathfrak{m}\in \mathrm{Max}(R)\\ \mathfrak{p}\subseteq\mathfrak{m}}} \left(M_i \cap (\mathfrak{m}^n \cdot R/\mathfrak{p}^{(n)})\right) \\ &\subseteq \bigcap_{\substack{\mathfrak{m}\in \mathrm{Max}(R)\\ \mathfrak{p}\subseteq\mathfrak{m}}} \mathfrak{m}M_i \text{ by Eq. 2.1.1;} \\ &\subseteq M_{i+1} \text{ by Eq. 2.1.2.} \end{aligned}$$

Then, $v \in M_{i+1}$, which contradicts our choice for i. Hence, $v = 0$. ☐

We now give an exercise that is helpful for the next theorem.

Exercise 2.13 Let $\{I_\alpha\}_{\alpha\in A}$ be an indexed family of ideals. Then,

$$\bigcap_{\alpha\in A} I_\alpha^{\langle n\rangle} = \left(\bigcap_{\alpha\in A} I_\alpha\right)^{\langle n\rangle}$$

for every positive integer n.

As a consequence of Theorem 2.12, we can show that differential powers and symbolic powers are the same for polynomial rings over any perfect field. This is usually presented only for fields of characteristic zero (see for instance [21, Theorem 3.14]).

Proposition 2.14 (Zariski–Nagata) *Let $R = K[x_1, \ldots, x_d]$ be a polynomial ring over K. If K is a perfect field and $\mathfrak{p} \subseteq R$ a prime ideal, then*

$$\mathfrak{p}^{(n)} = \mathfrak{p}^{\langle n \rangle}.$$

Proof Let $\mathfrak{m} \subseteq R$ be a maximal ideal. Using the fact that K is perfect and Remark 2.7, one can show that $\mathfrak{m}^{\langle n \rangle} = \mathfrak{m}^n$ for every positive integer n by going to $R \otimes_K \overline{K}$. Then,

$$\begin{aligned}
\mathfrak{p}^{(n)} &= \bigcap_{\substack{\mathfrak{m}\in \mathrm{Max}(R)\\ \mathfrak{p}\subseteq \mathfrak{m}}} \mathfrak{m}^n \text{ by Theorem 2.12;}\\
&= \bigcap_{\substack{\mathfrak{m}\in \mathrm{Max}(R)\\ \mathfrak{p}\subseteq \mathfrak{m}}} \mathfrak{m}^{\langle n \rangle} = \left(\bigcap_{\substack{\mathfrak{m}\in \mathrm{Max}(R)\\ \mathfrak{p}\subseteq \mathfrak{m}}} \mathfrak{m} \right)^{\langle n \rangle} \quad \text{by Exercise 2.13;}\\
&= \mathfrak{p}^{\langle n \rangle} \text{ because } R/\mathfrak{p} \text{ is a Jacobson ring.}
\end{aligned}$$

□

Exercise 2.15 Let $R = K[x_1, \ldots, x_d]$ be a polynomial ring over K, and K be a perfect field and $I \subseteq R$ be a radical ideal. Prove that $I^{(n)} = I^{\langle n \rangle}$. Show that this theorem does hold not if I is not radical (hint: find an example where $\mathrm{char}(K)$ is prime).

We can also characterize the symbolic powers in terms of join of ideals. This characterization is used to compute symbolic powers of prime ideals in polynomial rings. We start by recalling the definition of the join of two ideals.

Definition 2.16 Let $R = K[x_1, \ldots, x_d]$ be a polynomial ring over K. For ideals $I, J \subseteq R$, we consider the ideals $I[\underline{y}] \subseteq K[y_1, \ldots, y_d]$ and $J[\underline{z}] \subseteq K[z_1, \ldots, z_d]$, the ideals obtained from changing the variables in I and J. We define the join ideal of I and J by

$$I * J = (I[\underline{y}], J[\underline{z}], x_1 - y_1 - z_1, \ldots, x_d - y_d - z_d) \bigcap K[x_1, \ldots, x_d].$$

Suppose that $K = \mathbb{C}$. Let V and W denote the vanishing set of I and J in $\mathbb{C}^n$ respectively. Then, the vanishing set of $I * J$ is the Zariski closure of

$$\bigcup_{v\in V, w\in W} < v, w >,$$

where $< v, w >$ denotes the complex line that joins v and w.

Theorem 2.17 (Sullivant [77, Proposition 2.8]) *Let K be a perfect field. Let $R = K[x_1, \ldots, x_d]$ be a polynomial ring, and $\eta = (x_1, \ldots, x_n)$. Then,*

$$\mathfrak{p}^{(n)} = \mathfrak{p} * \eta^n$$

for every prime ideal $\mathfrak{p} \subseteq R$.

Proof Sketch: Let $K = \overline{K}$ denote the algebraic closure of K. We have that $\mathfrak{m} * \eta^n = \mathfrak{m}^n$ for every maximal ideal $\mathfrak{m} \subseteq R \otimes_K \overline{K}$. Since K is perfect, $\mathfrak{m} \otimes_K \overline{K}$ is the intersection of maximal ideals in $R \otimes_K \overline{K}$. Using this fact together with the faithful flatness of field extensions, we deduce that $\mathfrak{m} * \eta^n = \mathfrak{m}^n$ for every maximal ideal $\mathfrak{m} \subseteq R$. Then, we have

$$\mathfrak{p} * \eta^n = \left(\bigcap_{\substack{\mathfrak{m} \in \operatorname{Max}(R) \\ \mathfrak{p} \subseteq \mathfrak{m}}} \mathfrak{m} \right) * \eta^n = \bigcap_{\substack{\mathfrak{m} \in \operatorname{Max}(R) \\ \mathfrak{p} \subseteq \mathfrak{m}}} \mathfrak{m} * \eta^n = \bigcap_{\substack{\mathfrak{m} \in \operatorname{Max}(R) \\ \mathfrak{p} \subseteq \mathfrak{m}}} \mathfrak{m}^n = \mathfrak{p}^{(n)},$$

where the last equality follows from Theorem 2.12. □

Remark 2.18 The previous theorem gives an algorithm to compute the symbolic powers of radical ideals in a polynomial ring over a perfect field.

2.2 Uniform Bounds

The following striking result by Ein, Lazarsfeld and Smith shows that it is possible to find a uniform constant, c, that guarantees that $\mathfrak{p}^{(cn)} \subseteq \mathfrak{p}^n$ for smooth varieties over $\mathbb{C}$, as the following theorem makes explicit. We will revisit this theme of uniformity in Sect. 3.

Theorem 2.19 (Ein–Lazarsfeld–Smith [20]) *If* $\mathfrak{p}$ *is a prime ideal of codimension* h *in the coordinate ring of a smooth algebraic variety over* $\mathbb{C}$*, then* $\mathfrak{p}^{(hn)} \subseteq \mathfrak{p}^n$ *for all* $n \geqslant 1$.

Hochster and Huneke extended Ein–Lazarsfeld–Smithf Theorem to regular local rings containing a field, by reduction to characteristic $p > 0$ methods, and using tight closure arguments.

Theorem 2.20 (Hochster–Huneke [41]) *Let* $(R, \mathfrak{m})$ *be a regular local ring containing a field, let* $\mathfrak{p}$ *be a prime ideal, and let* h *be the height of* $\mathfrak{p}$*. Then* $\mathfrak{p}^{(hn)} \subseteq \mathfrak{p}^n$ *for all* $n \geqslant 1$.

Proof If $\mathfrak{p}$ is a maximal ideal, then symbolic powers coincide with regular powers, and the statement is clear. If $\dim(R) \leqslant 1$, then either $\mathfrak{p} = 0$ or $\mathfrak{p}$ is a maximal ideal, and the statement follows. Therefore, we assume that $\mathfrak{p}$ is neither maximal nor zero. We first assume that $\operatorname{char}(R) = p > 0$. Fix $n \geqslant 1$ and let $f \in \mathfrak{p}^{(hn)}$. For all $q = p^e$ write $q = an + r$ for some $a \in \mathbb{N}$ and $0 \leqslant r < n$. Then $f^a \in (\mathfrak{p}^{(hn)})^a \subseteq \mathfrak{p}^{(han)}$. Thus, $\mathfrak{p}^{hn} f^a \subseteq \mathfrak{p}^{hr} f^a \subseteq \mathfrak{p}^{(han+hr))} = \mathfrak{p}^{(hq)}$. We now want to show that $\mathfrak{p}^{(hq)} \subseteq \mathfrak{p}^{[q]}$ for all q.

Since $\mathrm{Ass}_R(R/\mathfrak{p}^{[q]}) = \{\mathfrak{p}\}$ by the flatness of Frobenius, we can check the containment just after localizing at $\mathfrak{p}$. Since $\mathfrak{p}R_\mathfrak{p}$ is the maximal ideal of the regular local ring $R_\mathfrak{p}$, it is generated by h elements. Furthermore, $\mathfrak{p}^{(hq)}R_\mathfrak{p} = \mathfrak{p}^{hq}R_\mathfrak{p}$. By the pigeonhole principle, $\mathfrak{p}^{hq}R_\mathfrak{p} \subseteq \mathfrak{p}^{[q]}R_\mathfrak{p}$, and this shows the containment in R as well. Taking nth powers yields $\mathfrak{p}^{n^2h}f^{an} \subseteq (\mathfrak{p}^n)^{[q]}$ for all q, and multiplying by f^r finally yields $\mathfrak{p}^{n^2h}f^q \subseteq (\mathfrak{p}^n)^{[q]}$. Choose any non-zero $c \in \mathfrak{p}^{n^2h}$, then $cf^q \in (\mathfrak{p}^n)^{[q]}$. Therefore

$$c \in \bigcap_q \left((\mathfrak{p}^n)^{[q]} : f^q\right) = \bigcap_q (\mathfrak{p}^n : f)^{[q]},$$

where the last equality follows from the flatness of Frobenius. Thus, either $f \in \mathfrak{p}^n$, and we are done, or $\mathfrak{p}^n : f \subseteq \mathfrak{m}$, so that $c \in \bigcap_q \mathfrak{m}^{[q]} \subseteq \bigcap_q \mathfrak{m}^q = (0)$, which is a contradiction. The result in characteristic zero follows by reduction to prime characteristic (see [39]). □

The previous two theorems can be restated for polynomial rings as follows. If $\mathfrak{p}$ is a prime ideal in $R = K[x_1, \ldots, x_d]$, then

$$\bigcap_{\substack{\mathfrak{m}\in\mathrm{Max}(R)\\ \mathfrak{p}\subseteq\mathfrak{m}}} \mathfrak{m}^{dn} \subseteq \left(\bigcap_{\substack{\mathfrak{m}\in\mathrm{Max}(R)\\ \mathfrak{p}\subseteq\mathfrak{m}}} \mathfrak{m}\right)^n.$$

This is a surprising fact, because the intersection is infinite.

The remarkable containment given by Theorems 2.20 and 2.19 might not, however, be the best possible. Given a prime ideal $\mathfrak{p}$ in a regular local ring and an integer a, one may ask what is the smallest $b \geqslant a$ such that $\mathfrak{p}^{(b)} \subseteq \mathfrak{p}^a$.

Question 2.21 (Huneke) Given a codimension 2 prime ideal $\mathfrak{p}$ in a regular local ring, does the containment

$$\mathfrak{p}^{(3)} \subseteq \mathfrak{p}^2$$

always hold?

Over the past decade, there has been a lot of work towards answering different versions of this question. If we consider the previous question for a radical ideal, I, the containment of the third symbolic power in the square has been shown to not hold in general [17], with an example later extended [34].[1] However, the containment does hold when I is a monomial ideal in a polynomial ring [35, 8.4.5], or an ideal defining a set of general points in $\mathbb{P}^2$ [33] and in $\mathbb{P}^3$ [16].

Harbourne has extended the question to higher powers [33, 35][2]:

---

[1] Akesseh [1] and Walker [87, 88] has also made recent progress regarding Question 2.21.

[2] The third and fourth authors [30] recently answered Question 2.21 affirmatively and proved Conjecture 2.22 for ideals defining F-pure rings.

Conjecture 2.22 (Harbourne) *Given a radical homogeneous ideal I in $k[\mathbb{P}^N]$, let h be the maximal height of an associated prime of I. Then for all $n \geqslant 1$,*

$$I^{(hn-h+1)} \subseteq I^n.$$

For a survey on the containment problem see [80].

Harbourne and Bocci introduced the resurgence of an ideal as an asymptotic measure of the best possible containment as the following definition makes explicit.

Definition 2.23 (*Harbourne–Bocci* [6, 7]) Let $I \subseteq K[x_1, \ldots, x_n]$ be an homogeneous ideal. The resurgence of I is defined by

$$\rho(I) = \sup\left\{\frac{n}{m} \mid I^{(n)} \not\subseteq I^m\right\}.$$

By Theorems 2.19 and 2.20, $\rho(I) \leqslant \dim(R)$. However, computing the resurgence of an ideal might be a very difficult task – instead, one may find bounds in terms of other invariants. One of these invariants is the Waldschmidt constant, which measures the asymptotic growth of the minimal degrees of the symbolic powers of the given ideal.

Definition 2.24 (*Waldschmidt* [85]) Let $I \subseteq K[x_1, \ldots, x_n]$ be an homogeneous ideal, and $\alpha(I) = \min\{t \mid I_t \neq 0\}$. The Waldschmidt constant of I is then defined to be

$$\widehat{\alpha}(I) = \lim_{m\to\infty} \frac{\alpha(I^{(n)})}{n}.$$

We point out that Waldschmidt showed that the previous limit exits. Bocci and Harbourne have showed that $\alpha(I)/\widehat{\alpha}(I) \leqslant \rho(I)$ [6, Theorem 1.2]. It is worth mentioning that the Zariski–Nagata Theorem (Theorem 2.11) guarantees that $1 \leqslant \widehat{\alpha}(I)$.

There are several cases where the Waldschmidt constant has been computed [4, 8, 19, 24, 32] or bounded [16, 18]. We point out that the function $\operatorname{reg}(R/I^{(n)})$ has also been studied [15, 36–38, 65].

2.3 *Eisenbud–Mazur Conjecture*

In this section we survey a famous conjecture of Eisenbud and Mazur, that can be stated in terms of containments involving symbolic powers. Given any ideal I inside a ring R, we always have an inclusion $I^{(2)} \subseteq I$. However, it is natural to ask whether something more precise can be said about the containment. Note that, if K is a field of characteristic zero, and $I \subseteq K[x_1, \ldots, x_n]$ is a homogeneous ideal, then for any homogeneous $f \in I^{(2)}$, say of degree D, we have

$$f = \frac{1}{D}\sum_{i=1}^{n} x_i \frac{\partial f}{\partial x_i} \in (x_1, \ldots, x_n)I,$$

since $\partial(f)/\partial x_i \in I$ for all i by Exercise 2.13 and Proposition 2.14. In other words, f is never a minimal generator of the ideal I, whenever $f \in I^{(2)}$. One can wonder whether this is true more generally.

Conjecture 2.25 (Eisenbud–Mazur [23]) *Let $(R, \mathfrak{m})$ be a localization of a polynomial ring $S = K[x_1, \ldots, x_n]$ over a field K of characteristic zero. If $I \subseteq R$ is a radical ideal, then $I^{(2)} \subseteq \mathfrak{m}I$.*

Conjecture 2.25 easily fails if the ambient ring is not regular. For example, if $R = \mathbb{C}[x, y, z]/(xy - z^2)$ and $I = (x, z)$, then $x \in I^{(2)} \smallsetminus \mathfrak{m}I$. The assumption on the characteristic is also needed. In fact, E. Kunz provided a counterexample to Conjecture 2.25 for any prime integer p:

Example 2.26 ([23]) Let p be a prime integer, and consider the polynomial $f = x_1^{p+1}x_2 - x_2^{p+1} - x_1x_3^p + x_4^p \in \mathbb{F}_p[x_1, x_2, x_3, x_4]$. Note that f is a quasi-homogeneous polynomial, and $f \notin (x_1, x_2, x_3, x_4)\sqrt{\mathrm{Jac}(f)}$. Let I be the kernel of the map

$$\begin{aligned} \mathbb{F}_p[x_1, x_2, x_3, x_4] &\longrightarrow \mathbb{F}_p[t] \\ x_1 &\longmapsto t^{p^2} \\ x_2 &\longmapsto t^{p(p+1)} \\ x_3 &\longmapsto t^{p^2+p+1} \\ x_4 &\longmapsto t^{(p+1)^2}. \end{aligned}$$

Then, $f \in I^{(2)} \smallsetminus (x_1, x_2, x_3, x_4)I$.

Conjecture 2.25 is open in most cases when the base field K has characteristic 0. The most recent results in this direction, to the best of our knowledge, are due to A. A. More, who proves Conjecture 2.25 for certain primes in power series rings [67].

Conjecture 2.25 is related to and motivated by the existence of non-trivial evolutions. In order to explain this connection, we first recall some basic facts about derivations and modules of Kähler differentials. For a more exhaustive and detailed treatment we refer the reader to [56].

Let K be a Noetherian ring, and let R be a K-algebra, essentially of finite type over K, and let M be a finitely generated R-module. A K-derivation $\partial : R \to M$ is a K-linear map that satisfies the Leibniz rule:

$$\partial(rs) = \partial(r)s + r\partial(s)$$

for all $r, s \in R$. The set of all derivations $\mathrm{Der}_K(R, M)$ is an R-module. When $M = R$, we denote $\mathrm{Der}_K(R) := \mathrm{Der}_K(R, R)$. As in the previous section, recall that D_R^n denotes the set of K-linear differential operators of R of order at most n.

Lemma 2.27 *Every element $\delta \in D^1_R$ can be written as the sum of a derivation $\partial \in \mathrm{Der}_K(R)$ and an operator $\mu_r \in D^0_R$, that is, multiplication by some element $r \in R$.*

Proof Note that multiplication by elements of R and derivations are differential operators of order zero and one, respectively. Now let $\delta \in D^1_R$. Let $\partial := \delta - \mu_{\delta(1)}$, where $\mu_{\delta(1)}(r) = \delta(1)r$ for all $r \in R$. Then ∂ is still a differential operator of order at most one, and it is clear that $\partial(\lambda) = 0$ for all $\lambda \in K$, by K-linearity of δ. Since ∂ has order one, for all $r, s \in R$ we have

$$\mu_t(s) = [\partial, r](s) = \partial(rs) - r\partial(s)$$

where μ_t is multiplication by some element $t \in R$. Applying this identity to $s = 1$, using that $\partial(1) = 0$, we obtain that $t = \partial(r)$. Therefore

$$\partial(rs) = \partial(r)s + r\partial(s)$$

for all $r, s \in R$, proving that $\partial \in \mathrm{Der}_K(S)$. □

Consider now the multiplication map $R \otimes_K R \to R$, and let $\mathcal{I}$ be its kernel. $\mathcal{I}$ is generated, both as a left and right R-module, by elements of the form $x \otimes 1 - 1 \otimes x$. In addition, one can show that $r(x \otimes 1 - 1 \otimes x) + \mathcal{I}^2 = (x \otimes 1 - 1 \otimes x)r + \mathcal{I}^2$, for all $r, x \in R$. We define

$$\Omega_{R/K} := \mathcal{I}/\mathcal{I}^2$$

which is an R-module (the actions on the left and on the right are the same, given the previous comment). The module of differentials comes equipped with a universal derivation $d_{R/K} : R \to \Omega_{R/K}$, which is the map that sends $r \in R$ to $r \otimes 1 - 1 \otimes r$. For any R-module M, we have an isomorphism

$$\mathrm{Der}_K(R, M) \cong \mathrm{Hom}_R(\Omega_{R/K}, M).$$

In fact, every derivation $\partial \in \mathrm{Der}_K(R, M)$ can be written as $\partial = \varphi \circ d_{R/K}$, for some R-linear homomorphism $\varphi \in \mathrm{Hom}_R(\Omega_{R/K}, M)$. Conversely, $\psi \circ d_{R/K} : R \to M$ is a K-derivation for any $\psi \in \mathrm{Hom}_R(\Omega_{R/K}, M)$. We now recall how to explicitly describe $\Omega_{R/K}$ when we have a presentation of R over K.

- If $R = K[x_1, \ldots, x_n]$ is a polynomial ring over K, then one can show that

$$\Omega_{R/K} \cong R dx_1 \oplus \cdots \oplus R dx_n$$

is a free R-module of rank n, with basis labeled by symbols dx_i. In this case the universal derivation $d_{R/K} : R \to \Omega_{R/K}$ turns out to be the standard differential. More explicitly, for $f \in K[x_1, \ldots, x_n]$, we have

$$d_{R/K}(f) = \sum_{i=1}^{n} \frac{\partial(f)}{\partial x_i} dx_i.$$

- If $R = S/I$, where $S = K[x_1, \dots, x_n]$ is a polynomial ring and $I \subseteq S$ is an ideal, then
$$\Omega_{R/K} \cong \frac{\Omega_{S/K}}{I\Omega_{S/K} + S \cdot d_{S/K}(I)},$$
where $d_{S/K}(I) = \{d_{S/K}(f) \mid f \in I\}$. The universal derivation is the map $d_{R/K} : R \to \Omega_{R/K}$ induced by $d_{S/K} : S \to \Omega_{S/K}$ on R.
- If $R = T_W$, where $T = K[x_1, \dots, x_n]/I$ and W is a multiplicatively closed set, we have
$$\Omega_{R/K} \cong \left(\Omega_{T/K}\right)_W$$
and $d_{R/K}$ obeys the classical quotient rule.

These rules are helpful to compute the module of differentials $\Omega_{R/K}$ when R is essentially of finite type over K. Given ring homomorphisms $K \to T \to R := T/I$, we obtain an exact sequence

$$I/I^2 \xrightarrow{\alpha} \Omega_{T/K} \otimes_T R \xrightarrow{\beta} \Omega_{R/K} \longrightarrow 0 \qquad (2.3.1)$$

where $\alpha(i + I^2) = d_{T/K}(i) + I\Omega_{T/K}$ and $\beta(d_{T/K}(t) \otimes r) = d_{R/K}(t) \cdot r$. Here we are using the same notation for elements in a module and classes in quotients of the same module.

We are now finally in a position to define evolutions.

Definition 2.28 Let R be a local K-algebra, where K is a field. An evolution of R is a surjective homomorphism $T \to R$ of K-algebras such that $R \otimes_T \Omega_{T/K} \to \Omega_{R/K}$ is an isomorphism. The evolution is said to be trivial if $T \to R$ is an isomorphism, and R is said to be evolutionary stable if every evolution of R is trivial.

Evolutions appear in the study of Hecke algebras and in the work of Wiles on Galois deformations, in relation with his proof of Fermat's Last Theorem. They have also been studied by Scheja–Storch [73] and Böger [9], under slightly different perspectives. Evolutions were formally introduced by Mazur [61] in relation with the work of Wiles on semistable curves [89]. This comes from a desire to compare some universal deformation ring with a particular quotient of it, which arises as a completion of a Hecke algebra. In many cases the induced quotient map is an evolution, and it was crucial to establish that it is trivial. When introducing evolutions, Mazur asked whether any ring arising in such a way is actually evolutionary stable. In this direction, further work of Wiles and Taylor-Wiles [81] showed that any such evolution is trivial. However, Mazur's more general question whether reduced algebras essentially of finite over fields of characteristic zero are evolutionary stable is still open. Some partial results have been established [23, 42–44, 49].

We start off by justifying the relation between evolutions and the Eisenbud–Mazur Conjecture 2.25. We closely follow the original arguments by Eisenbud and Mazur [23].

We first present a very useful criterion for existence of non-trivial evolutions, in terms of minimality, which is due to H. Lenstra.

Definition 2.29 Let T be a ring, and let $\phi : M \to N$ be an epimorphism of T-modules. We say that ϕ is minimal if there is no proper submodule $M' \subseteq M$ such that $\phi(M') = N$.

The following proposition actually works for local algebras essentially of finite type over any Noetherian ring. However, we just focus on algebras essentially of finite type over a field.

Proposition 2.30 (Lenstra [23, Proposition 1]) *Let R be a local K-algebra, essentially of finite type over K. Then R is evolutionary stable if and only if for some (equivalently all) presentations $R = S/I$, where S is a localization of a polynomial ring over K, the map*

$$\widetilde{\alpha} : I/I^2 \to \ker\left(R \otimes_S \Omega_{S/K} \to \Omega_{R/K}\right) \to 0$$

induced from the exact sequence (2.3.1) is minimal.

Proof We leave it to the reader to show that $\widetilde{\alpha}$ being minimal or not is independent of the chosen presentation. Let S/I be any presentation of R, with S a localization of a polynomial ring in finitely many variables over K. Let $J \subseteq I$ be an ideal, so that we have a surjection $A := S/J \to S/I = R \to 0$. A diagram chase on the sequences (2.3.1) induced by the surjections $S \to A \to R \to 0$ shows that $A \to R$ is an evolution if and only if the surjective map $\widetilde{\alpha} : I/I^2 \to \ker\left(\Omega_{S/K} \otimes_S R\right)$ carries $(J + I^2)/I^2$ onto the same image as I/I^2. In addition, by Nakayama's Lemma we have that $J = I$ if and only if $(J + I^2)/I^2 = I/I^2$. Therefore R has no non-trivial evolutions of the form S/J if and only if no proper submodule $(J + I^2)/I^2 \subsetneq I/I^2$ has the same image as I/I^2 via $\widetilde{\alpha}$, if and only if $\widetilde{\alpha}$ is minimal. □

Under certain assumptions, we can explicitly identify the kernel of the map α considered above.

Theorem 2.31 (Eisenbud–Mazur [23, Theorem 3]) *Let $(S, \mathfrak{m})$ be a localization of a polynomial ring in finitely many variables over K, and let I be an ideal of S. If $R := S/I$ is reduced and generically separable over K, then the kernel of $\alpha: I/I^2 \to R \otimes_K \Omega_{S/K}$ is $I^{(2)}/I^2$.*

As a consequence of Theorem 2.31, we can now fully justify the connection between Conjecture 2.25 and the existence of non-trivial evolutions.

Corollary 2.32 *Let $(S, \mathfrak{m})$ be a localization of a Noetherian polynomial ring over a field K, and let I be a radical ideal of S. If $R = S/I$ is generically separable over K, then R is evolutionary stable if and only if $I^{(2)} \subseteq \mathfrak{m}I$.*

Proof It is enough to observe that minimality of a surjective map $f: A \to B$ is equivalent to the fact that $\ker(f)$ does not contain any minimal generator of A. □

Over the complex numbers, Conjecture 2.25 can be equivalently restated even more explicitly.

Proposition 2.33 (Eisenbud–Mazur [23, Corollary 2]) *There exists a reduced local $\mathbb{C}$-algebra R of finite type whose localization at the origin is not evolutionary stable if and only if there exists a power series $f \in \mathbb{C}[[x_1, \dots, x_n]]$ without a constant term such that*

$$f \notin (x_1, \dots, x_n)\sqrt{(f, \partial_1(f), \dots, \partial_n(f))}.$$

We now present some results due to Hübl [42], that lead to new versions of the Eisenbud–Mazur conjecture. We closely follow his treatment of these topics.

Theorem 2.34 (Hübl [42, Theorem 1.1]) *Let K be a Noetherian ring, and let R be a local algebra essentially of finite type over K. The following conditions are equivalent:*

(1) R is evolutionary stable.
(2) Assume $(S, \mathfrak{m})$ is a local algebra, essentially of finite type and smooth over K, and $I \subseteq S$ is such $R = S/I$. If $f \in I$ and $\partial(f) \in I$ for all $\partial \in \mathrm{Der}_K(S)$, then $f \in \mathfrak{m}I$.

Proof Assume (1), and write $R = S/I$ for some radical ideal I. Let $f \in I$ be such that $\partial(f) \in I$ for all $\partial \in \mathrm{Der}_K(S)$. Assume, by way of contradiction, that $f \notin \mathfrak{m}I$. It follows that we can find $J \subseteq I$ such that $I = J + (f)$, so that $I/J \cong S/\mathfrak{m}$. Let $T := S/J$, and consider the surjection $T \to R$; we claim that this is an evolution, and this gives a contradiction. In fact, if we let $\mathcal{I} := I/J \subseteq T$, we have an exact sequence

$$\mathcal{I}/\mathcal{I}^2 \cong (f + J)/J \xrightarrow{\alpha} \Omega_{S/K}/\mathcal{I}\Omega_{S/K} \cong \Omega_{T/K} \otimes_T R \longrightarrow \Omega_{S/K} \longrightarrow 0.$$

If $\widetilde{f}$ is the image of f inside $\mathcal{I}/\mathcal{I}^2$, we have that $\alpha(\widetilde{f}) = df + \mathcal{I}\Omega_{S/K}$. Note that $\Omega_{S/K}$ is free over S, because S is essentially of finite type and smooth over K. Since $\partial(f) \in I$ for all $\partial \in \mathrm{Der}_K(S)$, and $\mathrm{Der}_K(S) \cong \mathrm{Hom}_S(\Omega_{S/K}, S)$, we have that $d(f) \in I\Omega_{S/K}$. We then have that $\alpha(\widetilde{f}) = 0$ in $\Omega_{S/K}/\mathcal{I}\Omega_{S/K}$, which shows that $T \to R$ is an evolution.

For the converse, assume (2), and consider any evolution $T = S/J \to S/I = R$ of R. By way of contradiction, assume that the evolution is non-trivial, so that $J \subsetneq I$. Without loss of generality, we can assume that $I = J + (f)$, for some $f \notin I$ such that $f\mathfrak{m} \subseteq J$. Clearly, $f \notin \mathfrak{m}I$, otherwise $I = J + (f) \subseteq J + \mathfrak{m}I \subseteq J$, contradicting the non-triviality of the evolution. We want to show that there exists $h \in I \smallsetminus \mathfrak{m}I$ such that $\partial(h) \in I$ for all $\partial \in \mathrm{Der}_K(S)$. To find such h, note that

$$\Omega_{R/K} \cong \frac{\Omega_{S/K}}{I\Omega_{S/K} + S \cdot d(I)} = \frac{\Omega_{S/K}}{J\Omega_{S/K} + f\Omega_{S/K} + S \cdot d(I)}.$$

On the other hand, since $T \to R$ is an evolution, we have

$$\Omega_{R/K} \cong \frac{\Omega_{T/K}}{f\Omega_{T/K}} \cong \frac{\Omega_{S/K}}{J\Omega_{S/K} + S \cdot d(J) + f\Omega_{S/K}}.$$

This shows that

$$d(f) \in d(I) \subseteq J\Omega_{S/K} + S \cdot d(J) + f\Omega_{S/K} = I\Omega_{S/K} + S \cdot d(J).$$

Let $g_1, \ldots, g_r \in J$ be such that $d(f) - \sum_i s_i d(g_i) = \eta \in I\Omega_{S/K}$, where $s_i \in S$. Set $h := f - \sum_i s_i g_i$, and note that

- $d(h) = d(f) - \sum_i s_i d(g_i) - \sum_i d(s_i)g_i = \eta - \sum_i d(s_i)g_i \in I\Omega_{S/K}$
- $I = J + (f) = J + (h)$; therefore, $h \in I$. In addition, $h \notin \mathfrak{m}I$; otherwise, $f \in \mathfrak{m}I$.

Since $d(h) \in I\Omega_{S/K}$, and $\operatorname{Der}_K(S) \cong \operatorname{Hom}_S(\Omega_{S/K}, S)$, we have that $\partial(h) \in I$ for all $\partial \in \operatorname{Der}_K(S)$. This concludes the proof. □

In light of Exercise 2.13 and Proposition 2.14, we see that, under the assumptions that $S = K[x_1, \ldots, x_n]$, $I \subseteq S$ is radical and K is perfect, we have that $I^{(2)} = I^{\langle 2 \rangle}$. Therefore Theorem 2.34 becomes just a restatement of Corollary 2.32.

Theorem 2.35 (Hübl [42, Theorem 1.2]) *Let K be a field of characteristic zero, and let S be a smooth algebra, essentially of finite type over K. For a radical ideal $I \subseteq S$ and $f \in I$, the following conditions are equivalent:*

(1) $\partial(f) \in I$ *for all* $\partial \in \operatorname{Der}_K(S)$*;*
(2) $f \in I^{(2)}$*;*
(3) $f^n \in I^{n+1}$ *for some* $n \in \mathbb{N}$.

Proof For simplicity, we prove only the case when K is algebraically closed, and $S = K[x_1, \ldots, x_n]$.

Assume (1). Let $\delta \in \mathcal{D}^1_K(S)$ be a differential operator of order at most one. By Lemma 2.27 we have that

$$\delta(f) = \partial(f) + \mu_{\delta(1)}(f) = \partial(f) + \delta(1) \cdot f \in I$$

for all $\delta \in \mathcal{D}^1_K(S)$. Therefore, we obtain that $f \in I^{\langle 2 \rangle}$. Given that K is a field of characteristic zero, and that S is a localization of a finite algebra over K, we obtain that $f \in I^{(2)}$ by Exercise 2.13 and Proposition 2.14, as desired.

Conversely, if $f \in I^{(2)} = I^{\langle 2 \rangle}$ we have that $\delta(f) \in I$ for all $\delta \in \mathcal{D}^2_K(S)$. Let $\partial \in \operatorname{Der}_K(S)$ be a derivation; then ∂ is, in particular, an element of $\mathcal{D}^2_K(S)$. In particular, $\partial(f) \in I$. Thus (1) and (2) are equivalent.

Now assume (1). Let Q be the field of fractions of S, and let $v_1, \ldots, v_t$ be the Rees valuations of I, with associated valuation rings $S \subseteq V_i \subseteq Q$. Note that V_i is essentially of finite type over S. Since $\partial(f) \in I$ for all $\partial \in \operatorname{Der}_K(S)$, and $\Omega_{S/K}$ is free over S by smoothness, we have that $d_{S/K}(f) \in I\Omega_{S/K}$. As the canonical map $\Omega_{S/K} \to \Omega_{V_i/K}$ is S-linear, we have that $d_{V_i/K}(f) \in I\Omega_{V_i/K}$. Similarly, we have that $d_{\widehat{V_i}/K}(f) \in I\widetilde{\Omega}_{\widehat{V_i}/K}$, after passing to completions (see [56] for a definition of the module of differentials in the complete case). Since $\widehat{V_i}$ is a DVR containing a field, there exists a parameter $t_i \in \widehat{V_i}$ and a field ℓ_i such that $\widehat{V_i} \cong \ell_i[[t_i]]$. Then

$$d_{\widehat{V_i}/K}(f) \in I\widetilde{\Omega}_{\widehat{V_i}/K} = I\widehat{V_i}dt_i,$$

and thus $\partial(f)/\partial(t_i) \in I \cdot \widehat{V_i}$. This means that

$$v_i(f) = v_i(\partial(f)/\partial(t_i)) + 1 \geqslant v_i(I) + 1, \tag{2.3.2}$$

for all Rees valuations v_i of I. Now write $I = (f, g_1, \ldots, g_t)$, for some elements $g_i \in I$, and let $\mathcal{R} = S[It]_{(ft)} = S\left[\frac{g_1}{f}, \ldots \frac{g_t}{f}\right]$ be the homogeneous localization of the Rees algebra at (ft). We claim that f is a unit in $\mathcal{R}$. If not, then $f \in \mathfrak{p}$ for some $\mathfrak{p}$ of height one. Let $\mathcal{S}$ be the integral closure of $\mathcal{R}$. Then $\mathcal{S}$ is finite over $\mathcal{R}$, since $\mathcal{R}$ is excellent. Then, there exists $Q \in \operatorname{Spec}(\mathcal{S})$ such that $Q \cap \mathcal{R} = \mathfrak{p}$. Then $\mathcal{S}_Q$ is a DVR, with valuation v. In particular, v is a Rees valuation of I, and $v(f) = v(I)$. This contradicts (2.3.2). Therefore $1/f \in \mathcal{R}$, which means that there exists a polynomial $F(T_1, \ldots, T_m)$, say of degree $n+1$, such that $1/f = F(g_1/f, \ldots, g_m/f)$. It follows that

$$f^n = f^{n+1} \cdot \frac{1}{f} = f^{n+1} \cdot F\left(\frac{g_1}{f}, \ldots, \frac{g_m}{f}\right) \in I^{n+1},$$

since the numerator of $F\left(\frac{g_1}{f}, \ldots, \frac{g_m}{f}\right)$ is a polynomial of degree $n+1$ in the g_i's. Finally, assume (3), so that $f + I^2$ is nilpotent in $G = \operatorname{gr}_I(S)$. This means that $f + I^2 \subseteq \mathcal{Q}$ for all $\mathcal{Q} \in \operatorname{Spec}(G)$. Let $\mathfrak{p}$ be a minimal prime over I, then $G_\mathfrak{p} = \operatorname{gr}_{\mathfrak{p}S_\mathfrak{p}}(S_\mathfrak{p})$, because $IS_\mathfrak{p} = \mathfrak{p}S_\mathfrak{p}$. Since $S_\mathfrak{p}$ is a regular local ring, with maximal ideal $\mathfrak{p}S_\mathfrak{p}$, we have that $G_\mathfrak{p}$ is a domain. As a consequence, $\ker(G \to G_\mathfrak{p})$ is a prime ideal of G, so that $f + I^2 \in \ker(G \to G_\mathfrak{p})$. Since this happens for all $\mathfrak{p} \in \min(I)$, we have that $f + I^2 \in \bigcap_{\mathfrak{p} \in \min(I)} \ker(I/I^2 \to (I/I^2)_\mathfrak{p}) = I^{(2)}/I^2$. □

Corollary 2.36 *Let K be a field of characteristic zero, and let S be a smooth algebra, essentially of finite type over K. Given a radical ideal $I \subseteq S$, there exists $N > 0$ such that $\left(I^{(2)}\right)^n \subseteq I^{n+1}$ for all $n \geqslant N$.*

Theorem 2.35 leads to the following new conjecture, which is equivalent to the Eisenbud–Mazur Conjecture 2.25 under the assumptions of Theorem 2.35.

Conjecture 2.37 (Hübl [42, Conjecture 1.3]) *Let $(R, \mathfrak{m})$ be a regular local ring, and let $I \subseteq R$ be a radical ideal. Let $f \in I$ be such that $f^n \in I^{n+1}$, for some $n \in \mathbb{N}$. Then $f \in \mathfrak{m}I$.*

While Conjecture 2.25 is known to be false for rings of positive characteristic, we are not aware of any counterexamples to Conjecture 2.37.

2.4 Intersection of Symbolic Powers and Serre's Intersection Multiplicity

In this section, we discuss an intriguing connection between symbolic powers and local intersection theory. Let $(R, \mathfrak{m})$ be a regular local ring and $\mathfrak{p}, \mathfrak{q} \in \operatorname{Spec} R$ such that $\sqrt{\mathfrak{p}+\mathfrak{q}} = \mathfrak{m}$. Most of the following materials follow from the work of Sather-Wagstaff [72]. Motivated by intersection multiplicities of two subvarieties inside an ambient affine or projective space, Serre proposed the following definition.

Definition 2.38 (*Serre* [76]) The intersection multiplicity of $R/\mathfrak{p}$ and $R/\mathfrak{q}$ is defined by

$$\chi^R(R/\mathfrak{p}, R/\mathfrak{q}) := \sum_{0}^{\dim R} \lambda(\operatorname{Tor}_i^R(R/\mathfrak{p}, R/\mathfrak{q})).$$

The definition makes sense since all the Tor modules have finite length. For unramified regular local rings (for instance, if R contains a field), Serre proved the following properties.

Theorem 2.39 (Serre [76]) *Let R be an unramified regular local ring and $\mathfrak{p}, \mathfrak{q} \in$ Spec R such that $\sqrt{\mathfrak{p}+\mathfrak{q}} = \mathfrak{m}$. Then we have:*

(1) (Dimension inequality) $\dim R/\mathfrak{p} + \dim R/\mathfrak{q} \leqslant \dim R$;
(2) (Non-negativity) $\chi^R(R/\mathfrak{p}, R/\mathfrak{q}) \geqslant 0$;
(3) (Vanishing) $\chi^R(R/\mathfrak{p}, R/\mathfrak{q}) = 0$ *if* $\dim R/\mathfrak{p} + \dim R/\mathfrak{q} < \dim R$;
(4) (Positivity) $\chi^R(R/\mathfrak{p}, R/\mathfrak{q}) > 0$ *if* $\dim R/\mathfrak{p} + \dim R/\mathfrak{q} = \dim R$.

These results and their potential extensions have been considered by many researchers over the last fifty years. For a comprehensive account we refer to P. Roberts' book on this topic [71].

The non-negativity property part of Serre's Theorem was extended to all regular local rings by Gabber [5]. In an attempt to extend Gabber's argument to prove the positivity property, which still remains a conjecture, Kurano and Roberts proved the following theorem.

Theorem 2.40 (Kurano–Roberts [57]) *Let $(R, \mathfrak{m})$ be ramified regular local ring and $\mathfrak{p}, \mathfrak{q} \in$ Spec R such that $\sqrt{\mathfrak{p}+\mathfrak{q}} = \mathfrak{m}$. Assume that the positivity property holds for R and* $\dim R/\mathfrak{p} + \dim R/\mathfrak{q} = \dim R$. *Then for each $n \geqslant 1$, $\mathfrak{p}^{(n)} \cap \mathfrak{q} \subseteq \mathfrak{m}^{n+1}$.*

Thus, the simple containment $\mathfrak{p}^{(n)} \cap \mathfrak{q} \subseteq \mathfrak{m}^{n+1}$ when $\dim R/\mathfrak{p} + \dim R/\mathfrak{q} = \dim R$ is a consequence of the positivity property. They conjectured that this holds for all regular local rings. Although this also stays open, a stronger statement was proved for regular local rings containing a field by Sather-Wagstaff.

Theorem 2.41 (Sather-Wagstaff [72, Theorem 1.6]) *Let $(R, \mathfrak{m})$ be a regular local ring containing a field. Let $\mathfrak{p}, \mathfrak{q} \in$ Spec R be such that $\sqrt{\mathfrak{p}+\mathfrak{q}} = \mathfrak{m}$ and* $\dim R/\mathfrak{p} + \dim R/\mathfrak{q} = \dim R$. *Then for $m, n > 0$, $\mathfrak{p}^{(m)} \cap \mathfrak{q}^{(n)} \subseteq \mathfrak{m}^{m+n}$.*

Note that without the condition $\dim R/\mathfrak{p} + \dim R/\mathfrak{q} = \dim R$, the conclusion fails easily. As an example, take $R = \mathbb{C}[[x, y, z]]$, $\mathfrak{p} = (x, y)$, $\mathfrak{q} = (y, z)$, then $y^n \in \mathfrak{p}^{(n)} \cap \mathfrak{q}^{(n)}$ for each n but $y^n \notin \mathfrak{m}^{2n}$. Also, taking $\mathfrak{p} = (x, y)$, $\mathfrak{q} = (z)$ we see that the exponent $m + n$ is sharp.

The proof of Theorem 2.41 relies on the following interesting result, which itself can be viewed as a generalization of Serre's dimension inequality above. For a local ring A, let $e(A)$ denote the Hilbert–Samuel multiplicity with respect to the maximal ideal.

Theorem 2.42 (Sather-Wagstaff [72, Theorem 1.7]) *Let $(A, \mathfrak{n})$ be a quasi-unmixed local ring containing a field and $P, Q \in \operatorname{Spec} A$ such that A/P, A/Q are analytically unramified. Suppose that $\sqrt{P+Q} = \mathfrak{n}$ and $\mathrm{e}(A) < \mathrm{e}(A_P) + \mathrm{e}(A_Q)$. Then $\dim A/P + \dim A/Q \leqslant \dim A$.*

Given the above Theorem, 2.41 follows readily. First we can pass to the completion of R using some minimal primes lying over $\mathfrak{p}, \mathfrak{q}$. Thus we may assume R is complete. Let $f \in \mathfrak{p}^{(n)} \cap \mathfrak{q}^{(n)}$. Suppose that $f \notin \mathfrak{m}^{m+n}$. Then set $A = R/fR$ and $P = \mathfrak{p}A$, $Q = \mathfrak{q}A$. It is clear that $\mathrm{e}(A_P) \geqslant m$, $\mathrm{e}(A_Q) \geqslant n$ and $\mathrm{e}(A) < m + n$. Thus by Theorem 2.42, we have

$$\dim R/\mathfrak{p} + \dim R/\mathfrak{q} = \dim A/P + \dim A/Q \leqslant \dim A < \dim R,$$

which gives a contradiction.

3 Uniform Symbolic Topologies Property

3.1 Background on Uniformity

In this subsection we present a few results on uniformity in commutative algebra. We refer to [48] for a survey on this topic.

In Sect. 2.2 we discussed an important uniformity result regarding symbolic powers of prime ideals in regular local rings (Theorems 2.19 and 2.20). The uniformity results we discuss in this section will be necessary to discuss results of the same flavor as Theorems 2.19 and 2.20 for more general classes of rings.

We start by recalling some assumptions under which the Uniform Briançon-Skoda Theorem and Uniform Artin–Rees Lemma hold.

Hypothesis 3.1 We consider a Noetherian reduced ring R satisfying one of the following conditions

(1) R is essentially of finite type over an excellent ring containing in a field;
(2) R is F-finite;
(3) R is essentially of finite type over $\mathbb{Z}$;

(4) R is an excellent Noetherian ring which is the homomorphic image of a regular ring R of finite Krull dimension such that for all P, R/P has a resolution of singularities obtained by blowing up an ideal.

The following results play an important role in the proof of many uniformity results about symbolic powers.

Theorem 3.2 (Uniform Briançon-Skoda Theorem, Huneke [46]) *Let R be a ring satisfying Hypothesis 3.1. Then, there exists $c = c(R)$ such that, for all ideals $I \subseteq R$ and all $n \geqslant 1$, $\overline{I^{n+c}} \subseteq I^n$.*

Theorem 3.3 (Uniform Artin–Rees Lemma, Huneke [46, 47]) *Let R be a ring satisfying Hypothesis 3.1, and let N and M be R-modules. Then, there exists c such that*

$$I^n M \cap N \subseteq I^{n-c} N$$

for every ideal $I \subseteq R$, and $n \geq c$.

The Uniform Artin–Rees Lemma is not effective, since we cannot explicitly compute the integer c in the previous theorem.

We now focus on a couple of results about associated primes, which are helpful while dealing with powers of ideals.

Exercise 3.4 (Matsumura [60, Proposition 9A]) Let $R \subseteq S$ be Noetherian rings, and M be an S-module. Then, $\mathrm{Ass}_R\, M = \{\mathfrak{q} \cap R \mid \mathfrak{q} \in \mathrm{Ass}_S\, M\}$. (Yassemi [90, Corollary 1.7])

Theorem 3.5 (Brodmann [10]) *Let R be a Noetherian ring, and I be an ideal. Then,*

$$A(I) := \bigcup_{n \in \mathbb{N}} \mathrm{Ass}_R\left(R/I^n\right)$$

is a finite set.

Proof Let $\mathcal{R} = \oplus_{n \in \mathbb{N}} I^n$ denote the Rees algebra associated to I. We know that $\mathcal{R}$ is a Noetherian algebra. Then, $\mathcal{R}/I\mathcal{R} = \oplus_{n \in \mathbb{N}} I^n/I^{n+1}$ is a finitely generated $\mathcal{R}$-module, and so, $\mathrm{Ass}_{\mathcal{R}}\left(\mathcal{R}/I\mathcal{R}\right)$ is finite. As a consequence, we have that $\mathrm{Ass}_R\left(\mathcal{R}/I\mathcal{R}\right) = \bigcup \mathrm{Ass}_R\left(I^n/I^{n+1}\right)$ is finite by Exercise 3.4. From the short exact sequences $0 \to I^n/I^{n+1} \to R/I^{n+1} \to R/I^n \to 0$, we obtain that

$$\mathrm{Ass}_R\left(R/I^{n+1}\right) \subseteq \mathrm{Ass}_R\left(R/I^n\right) \cup \mathrm{Ass}_R\left(I^n/I^{n+1}\right)$$

for every $n \in \mathbb{N}$. Then, $\bigcup_{n \in \mathbb{N}} \mathrm{Ass}_R(R/I^n) \subseteq \bigcup_{n \in \mathbb{N}} \mathrm{Ass}_R\left(I^n/I^{n+1}\right)$, and the claim follows. □

3.2 Linear Equivalence of Topologies

It is helpful to introduce slightly more general topologies. This generality is useful because under some standard operations such as completion, prime ideals do not necessarily stay prime. Completion cannot ultimately be avoided as Theorem 3.6 demonstrates.

Let A be a Noetherian ring and $I, J \subseteq A$ ideals. Write $I^n : \langle J \rangle := \bigcup_{m \geqslant 1}(I^n : J^m)$. This ideal is the saturation of I with respect to J. In terms of the primary decomposition of I, this saturation removes all components which contain J. Since the number of associated primes of powers of I are finite in number by Theorem 3.5, it follows that symbolic powers are always saturations with respect to a fixed suitable ideal.

Work byf Schenzel [74, 75], and Huckaba [45] investigates conditions under which the I-adic and $\{I^n : \langle J \rangle\}$ topologies are equivalent. Then, one requires for each $n \geqslant 1$ an integer $m \geqslant 1$ so that $I^m : \langle J \rangle \subseteq I^n$. In particular, a theorem of Schenzel [75, Theorem 3.2] says when certain ideal topologies are equivalent. In this section, we write $A(I)$ for the union over n of the associated primes of I^n, a finite set of prime ideals. We also write $\widehat{T}$ for the completion of the local ring T with respect to its maximal ideal. Here is Schenzel's Theorem:

Theorem 3.6 (Schenzel [75, Theorem 3.2]) *Let A be a Noetherian ring and $I, J \subseteq A$ two ideals. Then the following are equivalent.*

(1) The $\{I^n : \langle J \rangle\}$ topology is equivalent to the I-adic topology.
(2) $\dim(\widehat{R_\mathfrak{p}}/(I\widehat{R_\mathfrak{p}} + z)) > 0$, for all prime ideals $\mathfrak{p} \in A(I) \cap V(J)$, and prime ideals $z \in \operatorname{Ass}(\widehat{R_\mathfrak{p}})$.

We say that the topology determined by $\{I^n : \langle J \rangle\}$ is *linearly equivalent* to the topology determined by I^n if there is a constant h such that for all $n \geqslant 1$,

$$I^{hn} : \langle J \rangle \subseteq I^n.$$

This concept is a priori stronger than having equivalent topologies. However, Swanson [79] proved the following beautiful result relating the notions of equivalent and linearly equivalent topologies.

Theorem 3.7 (Swanson [79, Main Theorem 3.3.]) *Let A be a Noetherian ring and I, J ideals. Then the $\{I^n : \langle J \rangle\}$ and I-adic topologies are equivalent if and only if there exists $h \geqslant 1$ such that, for all $n \geqslant 1$, $I^{hn} : \langle J \rangle \subseteq I^n$.*

Two comments are in order regarding Theorem 3.7. The first is that h depends a priori on I. The second is that the theorem implies that if $\mathfrak{p} \subseteq A$ is a prime ideal, and the $\mathfrak{p}$-symbolic and $\mathfrak{p}$-adic topologies are equivalent, then there exists $h \geqslant 1$ so that $\mathfrak{p}^{(hn)} \subseteq \mathfrak{p}^n$, for all n. Swanson's Theorem sets the stage for uniform bounds for regular rings [20, 41] discussed in Sect. 2.2. In particular, if d is the dimension of the regular local ring R, then $\mathfrak{p}^{(dn)} \subseteq \mathfrak{p}^n$ for all $n \geqslant 1$ and all primes $\mathfrak{p}$. The point is that while h in Theorem 3.7 may depend on the ideal I, $h = d$ in Theorems 2.19 and 2.20 is independent of the ideal.

One could reasonably ask to do even better: could there be an integer t such that for every prime $\mathfrak{p}^{(n+t)} \subseteq \mathfrak{p}^n$? Even very simple examples show that this is hopeless. For example, in the hypersurface $x^2 - yz = 0$, the prime ideal $\mathfrak{p} = (x, y)$ has $2n^{\text{th}}$ symbolic power generated by y^n, which is not in $\mathfrak{p}^{2n-t}$ for any fixed constant t, as n gets large. Even the results of Theorems 2.19 and 2.20 cannot be improved in an asymptotic sense, as shown by Bocci and Harbourne [6]. All of these results lead to the following question:

Question 3.8 Let $(R, \mathfrak{m}, K)$ be a complete local domain. Does there exist a positive integer $b = b(R)$ such that $\mathfrak{p}^{(bn)} \subseteq \mathfrak{p}^n$, for all prime ideals $\mathfrak{p} \subseteq R$ and all $n \geqslant 1$?

We say that a ring satisfies the Uniform Symbolic Topologies Property, abbreviated as USTP, if the answer to the previous question is affirmative.

We now show that complete local domains have the property that the adic and symbolic topologies are linearly equivalent for all prime ideals. However, the constant c which gives the relation $\mathfrak{p}^{(cn)} \subseteq \mathfrak{p}^n$ depends a priori on the ideal $\mathfrak{p}$. We first make an observation needed for this result.

Remark 3.9 Let R be complete local domain. For ideals $I, J \subseteq R$, suppose there exist integers $d, t \geqslant 1$ such that $I^{dn} : \langle J \rangle \subseteq I^{n-t}$, for all $n \geq t$. An induction argument shows that for $c := d(t+1)$, $I^{cn} : \langle J \rangle \subseteq I^n$ for all $n \geqslant 1$.

We now show that Question 3.8 is well-posed.

Proposition 3.10 (Huneke–Katz–Validashti [51, Proposition 2.4]) *Let $(R, \mathfrak{m}, K)$ be a complete local domain and let $\mathfrak{p} \subseteq R$ be a prime ideal. Then the $\{\mathfrak{p}^{(n)}\}$ topology is linearly equivalent to the $\{\mathfrak{p}^n\}$ topology. In particular, there exists $c > 0$ (depending on $\mathfrak{p}$) such that $\mathfrak{p}^{(cn)} \subseteq \mathfrak{p}^n$, for all $n \geqslant 1$.*

Proof It suffices to prove the second statement. Let S denote the integral closure of R and set $I := \sqrt{\mathfrak{p}S}$. Since S is an excellent normal domain, it is locally analytically normal, so the completion of $S_{\mathfrak{q}}$ is a domain for all primes $\mathfrak{q}$. In particular, by Theorem 3.6, the $\{I^{(n)}\}$ topology is equivalent to the $\{I^n\}$ topology. Here we are writing $I^{(n)}$ for $I^n_U \cap S$, where $U := S \backslash \mathfrak{q}_1 \cup \cdots \cup \mathfrak{q}_r$, for $\mathfrak{q}_1, \ldots, \mathfrak{q}_r$ the primes in S lying over $\mathfrak{p}$, so that $I = \mathfrak{q}_1 \cap \cdots \cap \mathfrak{q}_r$. Thus, by 3.7, $\mathfrak{p}^{(kn)} \subseteq I^{(kn)} \subseteq I^n$ for some fixed k and all $n \geqslant 1$. On the other hand, there is an e such that $I^e \subseteq \mathfrak{p}S$, so that

$$\mathfrak{p}^{(ken)} \subseteq \mathfrak{p}^n S \cap R \subseteq \mathfrak{p}^{n-l},$$

for some l by Artin–Rees. By Remark 3.9, taking $c := ke(l+1)$ gives $\mathfrak{p}^{(cn)} \subseteq \mathfrak{p}^n$, for all n. □

It is worth noting that the previous result is motivated by the work of Swanson.

Theorem 3.11 (Swanson [78, Theorem 3.4]) *Let R be a Noetherian ring and $I \subseteq R$ be an ideal. Then, there exists a positive integer c and for each n an irredundant primary decomposition $\mathfrak{q}_{n,1} \cap \cdots \cap \mathfrak{q}_{n,s_n}$ of I^n so that for $\mathfrak{p}_j := \sqrt{(\mathfrak{q}_{n,j})}$, $\mathfrak{p}_j^{cn} \subseteq \mathfrak{q}_{n,j}$, for all $n \geqslant 1$ and $1 \leqslant j \leqslant s_n$.*

3.3 A Uniform Chevalley Theorem

In order to study the Uniform Symbolic Property, it is useful to have a general version of the Zariski–Nagata Theorem (Theorem 2.11). If $(R, \mathfrak{m}, K)$ is not regular, one cannot guarantee that $\mathfrak{p}^{(n)} \subseteq \mathfrak{m}^n$. Our goal is to show that for complete local domains there exists a constant h, not depending of $\mathfrak{p}$, such that $\mathfrak{p}^{(hn)} \subseteq \mathfrak{m}^n$, which can be seen as a uniform version of Chevalley's theorem.

Theorem 3.12 (Chevalley) *Let $(R, \mathfrak{m}, K)$ be a complete local ring, M be a finitely generated module, and let $\{M_i\}$ be a nonincreasing sequence of submodules. Under these conditions, $\bigcap\limits_{i\in\mathbb{N}} M_i = 0$ if and only if for every integer t there exists i such that $M_i \subseteq \mathfrak{m}^t M$.*

As a consequence, if $\{J_n\}_{n\geqslant 1}$ is a descending collection of ideals with $\bigcap_{n\geqslant 1} J_n = 0$, then the $\{J_n\}$ topology is finer than the $\mathfrak{m}$-adic topology. In other words, for all $n \geqslant 1$, there exists $t \geqslant 1$, such that $J_t \subseteq \mathfrak{m}^n$. Once again, we would like to understand when the symbolic topology is finer than the $\mathfrak{m}$-adic topology in not necessarily complete rings. To do so, it is convenient to introduce another type of saturation.

Notation 3.13 Let $S \subseteq R$ be a multiplicatively closed set and $L \subseteq R$ an ideal. We write $L : \langle S\rangle$ for $LR_S \cap R$.

Let I, J be ideals of R and take $s \in J$ with the following property: for all $\mathfrak{p} \in A(I)$, $s \in \mathfrak{p}$ if and only if $J \subseteq \mathfrak{p}$. Then $I^n : \langle S\rangle = I^n : \langle J\rangle$, for all $n \geqslant 1$. Consequently, any result about the $\{I^n : \langle S\rangle\}$ topology recovers the corresponding result about the $\{I^n : \langle J\rangle\}$ topology. Moreover, if we let S denote the complement of the union of the associated primes of I, then by definition, $I^{(n)} = I^n : \langle S\rangle$, for all n.

The following proposition is implicit in the work of McAdam [62] and Schenzel [75]. We present the proof given by Huneke, Katz and Validashti [51].

Proposition 3.14 (Huneke–Katz–Validashti [51, Proposition 2.2]) *Let $(R, \mathfrak{m}, K)$ be a local ring with completion $\widehat{R}$. Let $I \subseteq R$ be an ideal and $S \subseteq R$ a multiplicatively closed set. Write $\mathfrak{q}_1, \dots, \mathfrak{q}_s$ for the associated primes of $\widehat{R}$. Then the $\{I^n : \langle S\rangle\}$ topology is finer than the $\mathfrak{m}$-adic topology if and only if $(I\widehat{R} + \mathfrak{q}_i) \cap S = \emptyset$ for all $1 \leqslant i \leqslant s$.*

Proof Since $\widehat{R}$ is faithfully flat over R, the $\{I^n :_R \langle S\rangle\}$ topology is finer than the $\mathfrak{m}$-adic topology if and only if the $\{I^n\widehat{R} :_{\widehat{R}} \langle S\rangle\}$ topology is finer than the $\mathfrak{m}\widehat{R}$-adic topology. Thus, we may assume that R is a complete local ring with associated primes $\mathfrak{q}_1, \dots, \mathfrak{q}_s$. By Chevalley's Theorem, the $\{I^n : \langle S\rangle\}$ topology is *not* finer than the $\mathfrak{m}$-adic topology if and only if

$$\bigcap_{n\geqslant 1}(I^n : \langle S\rangle) \neq 0.$$

Suppose $0 \neq f \in \bigcap_{n \geqslant 1}(I^n : \langle S \rangle)$. Then for each $n \geqslant 1$, there exists $s \in S$ such that $s \in (I^n : f)$. By applying the Artin–Rees Lemma to $I^n \cap (f)$, we see that for n large, $s \in (0 : f) + I^{n-k}$, for some k. Taking $\mathfrak{q}_j$ so that $(0 : f) \subseteq \mathfrak{q}_j$, we have that $(I + \mathfrak{q}_j) \cap S \neq \emptyset$.

On the other hand, suppose that $(I + \mathfrak{q}_j) \cap S \neq \emptyset$, for some j. Then for all $n \geqslant 1$, $(I^n + \mathfrak{q}_j) \cap S \neq \emptyset$. Let $\mathfrak{q}_j = (0 : f)$. Then for each n, there exists $s \in S$ such that $s \in I^n + (0 : f)$, i.e., $sf \in I^n$. Thus, $0 \neq f \in \bigcap_{n \geqslant 1}(I^n : \langle S \rangle)$. □

Since R is analytically irreducible, $I = \mathfrak{p}$ is prime. By taking $S = R - \mathfrak{p}$ in the previous proposition, we obtain that the $\{\mathfrak{p}^{(n)}\}$ topology is finer that the $\mathfrak{m}$-adic topology.

We can now state and show the Uniform Chevalley Theorem.

Theorem 3.15 (Huneke–Katz–Validashti [51, Theorem 2.3]) *Let R be an analytically unramified local ring. Then there exists $h \geqslant 1$ with the following property: for all ideals $I \subseteq R$ and all multiplicatively closed sets $S \subseteq R$ such that the $\{I^n : \langle S \rangle\}$ topology is finer than the $\mathfrak{m}$-adic topology, $I^{hn} : \langle S \rangle \subseteq \mathfrak{m}^n$, for all $n \geqslant 1$.*

Proof Ideas: The statement reduces to the complete local domain case. In this case, we applied Rees' theory of Rees valuations and degree functions [70, Theorem 2.3] together with a theorem of Izumi [54] to obtain the result. □

As a corollary, we globalize the statement in Theorem 3.15.

Corollary 3.16 (Huneke–Katz–Validashti [51]) *Let R be a Noetherian ring and $J \subseteq R$ an ideal. Suppose that $R_{\mathfrak{p}}$ is analytically unramified for all $\mathfrak{p} \in A(J)$. Then there exists a positive integer h with the following property: for all ideals $I \subseteq R$ and multiplicatively closed sets S for which the $\{I^n : \langle S \rangle\}$ topology is finer than the J-adic topology, $I^{hn} : \langle S \rangle \subseteq J^n$, for all n.*

Proof The point of the proof is that we can combine Theorem 3.15 with Theorem 3.11. It follows from this that $\mathfrak{p}_j^{(cn)} \subseteq \mathfrak{q}_{n,j}$, for all n and all j.

On the other hand, if I, S are such that the $\{I^n : \langle S \rangle\}$ topology is finer than the J-adic topology, then the $\{I^n : \langle S \rangle\}$ topology is finer than the symbolic topology $\{\mathfrak{p}^{(n)}\}$ for any $\mathfrak{p} \in A(J)$. By our hypothesis on $A(J)$ and Theorem 3.15, there is a positive integer l such that $(I^{ln} : \langle S \rangle)_{\mathfrak{p}} \subseteq \mathfrak{p}_{\mathfrak{p}}^n$, for all n and all $\mathfrak{p} \in A(J)$. Therefore, $I^{ln} : \langle S \rangle \subseteq \mathfrak{p}^{(n)}$, for all $\mathfrak{p} \in A(J)$ and $n \geqslant 1$. Combining this with the previous paragraph and setting $h := cl$, it follows that $I^{hn} : \langle S \rangle \subseteq J^n$, for all I, S and $n \geqslant 1$. □

3.4 Isolated Singularities

The purpose of this subsection is to discuss a theorem of Huneke, Katz and Validashti [51] which proves that for a large class of isolated singularities, the symbolic topology defined by a prime ideal $\mathfrak{p}$ is uniformly linearly equivalent to the $\mathfrak{p}$-adic topology. We sketch their proof that for such isolated singularities R, there exists

$h \geqslant 1$, independent of $\mathfrak{p}$, such that for all primes $\mathfrak{p} \subseteq R$, $\mathfrak{p}^{(hn)} \subseteq \mathfrak{p}^n$, for all n. The following theorem is our focus:

Theorem 3.17 (Huneke–Katz–Validashti [51]) *Let R be an equicharacteristic local domain such that R is an isolated singularity. Assume that R is either essentially of finite type over a field of characteristic zero or R has positive characteristic and is F-finite. Then there exists $h \geqslant 1$ with the following property. For all ideals $I \subseteq R$ such that the symbolic topology of I is equivalent to the I-adic topology, $I^{(hn)} \subseteq I^n$, for all $n \geqslant 1$.*

There are three crucial ingredients in the proof of this theorem: the relation between the Jacobian ideal and symbolic powers established in [41], the uniform Artin–Rees Theorem (Theorem 3.3), and a uniform Chevalley Theorem (Theorem 3.15).

We now focus on a result needed to prove uniform bounds for isolated singularities. This result comes from the work of Hochster and Huneke [41], where the following property of the Jacobian ideal plays a crucial role in the main results concerning uniform linearity of symbolic powers over regular local rings:

Theorem 3.18 (Hochster–Huneke [41, Theorem 4.4]) *Let R be a an equidimensional local ring essentially of finite type over a field K of characteristic zero. Let J denote the square of the Jacobian ideal of R over K. Then there exists $k \geqslant 1$ such that*

$$J^n I^{(kn+ln)} \subseteq (I^{(l+1)})^n,$$

for all ideals I with positive grade and all $l, n \geqslant 1$.

Moreover, a parallel result is proved for F-finite local rings in characteristic p with isolated singularity [51]. In this case, J can be chosen to be a fixed $\mathfrak{m}$-primary ideal, though not necessarily the square of the Jacobian ideal.

The main point of the proof of the main result on isolated singularities is the following theorem.

Theorem 3.19 (Hunke–Katz–Validashti [51]) *Let R be a Noetherian ring in which the uniform Artin–Rees lemma holds and $J \subseteq R$ an ideal with positive grade. Assume that $R_\mathfrak{p}$ is analytically unramified for all $\mathfrak{p} \in A(J)$. Let $\mathcal{J}$ denote the collection of ideals $I \subseteq R$ for which the $\{I^{(n)}\}_{n\geqslant 1}$ topology is finer than the J-adic topology. Suppose there exists $k \geqslant 1$ with the following property: for all ideals $I \in \mathcal{J}$,*

$$J^n I^{(kn+ln)} \subseteq (I^{(l+1)})^n,$$

for all $l, n \geqslant 1$. Then there exists a positive integer h such that for all ideals $I \in \mathcal{J}$, $I^{(hn)} \subseteq I^n$, for all n.

Sketch of proof: By Corollary 3.16, we can choose $h_0 > k$ so that $I^{(h_0 n)} \subseteq J^n$, for all ideals $I \in \mathcal{J}$. Taking $l = 0$ in Theorem 3.18 gives $J^n I^{(kn)} \subseteq I^n$, for all n and all ideals $I \in \mathcal{J}$. Thus, $I^{(h_0 n)} I^{(kn)} \subseteq I^n$ for all n and all ideals $I \in \mathcal{J}$. On the other hand, if $n = 2$ and $l = h_0 r$ in Theorem 3.18, we get

$$J^2 I^{(2k+2h_0 r)} \subseteq (I^{(h_0 r+1)})^2 \subseteq I^{(h_0 r)} I^{(kr)} \subseteq I^r,$$

which holds for all r. Thus there exists a positive integer B such that $J^2 I^{(Br)} \subseteq I^r$ for all r. Choose a non-zerodivisor $c \in J^2$. Then $cI^{(Bn)} \subseteq I^n$, for all n. Thus, $I^{(Bn)} \subseteq (I^n : c)$ for all n and all I. By the uniform Artin–Rees Theorem (see Theorem 3.3), we find $q \geqslant 1$ with the property that

$$I^{(Bn+Bq)} \subseteq (I^{n+q} : c) \subseteq I^n,$$

for all n and all ideals $I \in \mathcal{J}$. Taking $h := B + Bq$ gives $I^{(hn)} \subseteq I^n$, for all $I \in \mathcal{J}$ and all $n \geqslant 1$, as required. □

We now can show the main result in this subsection.

Theorem 3.20 (Huneke–Katz–Validashti [51, Theorem 1.2]) *Let R be an equicharacteristic reduced local ring such that R is an isolated singularity. Assume either that R is equidimensional and essentially of finite type over a field of characteristic zero, or that R has positive characteristic and is F-finite. Then there exists $h \geqslant 1$ with the following property: for all ideals I with positive grade for which the I-symbolic and I-adic topologies are equivalent, $I^{(hn)} \subseteq I^n$, for all $n \geqslant 1$.*

Proof The ring R is excellent in both cases, and so R is analytically unramified (see [55] for the F-finite case). Let $d = \dim(R)$.

Suppose first that R is essentially of finite type over a field of characteristic zero. Let J denote the square of the Jacobian ideal. By Theorem 3.18, $J^n I^{(dn+ln)} \subseteq (I^{(l+1)})^n$ for all ideals I with positive grade and all $n \geqslant 1$ and all $l \geqslant 0$. Thus, since J is $\mathfrak{m}$-primary and R is analytically unramified, we may use Theorem 3.19 with $k = d$ to obtain the desired result.

The positive characteristic case follows similarly once an ideal J is constructed to play a similar role to the Jacobian ideal. □

As a consequence of the previous theorem, we obtain the Uniform Symbolic Topologies Property for isolated singularities.

Theorem 3.21 (Hunke–Katz–Validashti [51]) *Let R be an equicharacteristic local domain such that R is an isolated singularity. Assume that R is either essentially of finite type over a field of characteristic zero or R has positive characteristic, is F-finite and analytically irreducible. Then there exists $h \geqslant 1$ with the following property: for all prime ideals $\mathfrak{p} \neq \mathfrak{m}$, $\mathfrak{p}^{(hn)} \subseteq \mathfrak{p}^n$, for all n.*

3.5 Finite Extensions

We now present a uniform relation between the extension of an ideal and its radical in the case of a finite extension. This is a key ingredient in the proof that the USTP descends for finite extensions.

Lemma 3.22 (Hunke–Katz–Validashti [52]) *Let $R \subseteq S$ be a finite extension of domains, with R integrally closed. Let $e = [S : R]$, i.e. the degree of the quotient field of S over the quotient field of R. If $\mathfrak{q} \in \operatorname{Spec}(R)$, then $\left(\sqrt{\mathfrak{q}S}\right)^e \subseteq \overline{\mathfrak{q}S}$. Furthermore, if $e!$ is invertible, then $\left(\sqrt{\mathfrak{q}S}\right)^e \subseteq \mathfrak{q}S$.*

Proof Let $x \in \sqrt{\mathfrak{q}S}$. There is a polynomial $f(T) = T^e + r_1T^{e-1} + \cdots + r_e$ such that $f(x) = 0$, and with $r_i \in \mathfrak{q}$ for all i [3, Lemma 5.14 and Proposition 5.15]. As a consequence, $x^e \in \mathfrak{q}S$. If $e!$ is invertible, we have that $\left(\sqrt{\mathfrak{q}S}\right)^e = (x^e \mid x \in \sqrt{\mathfrak{q}S}) \subseteq \mathfrak{q}S$.

We now consider the case where $e!$ is not invertible. We have that elements of the form $x_1 \cdots x_e$, for $x_i \in \sqrt{\mathfrak{q}S}$, generate the ideal $\left(\sqrt{\mathfrak{q}S}\right)^e$. Then, $(x_1 \cdots x_e)^e = x_1^e \cdots x_e^e \in (\mathfrak{q}S)^e$. This implies that $x_1 \cdots x_e \in \overline{\mathfrak{q}S}$, and thus, $\left(\sqrt{\mathfrak{q}S}\right)^e \subseteq \overline{\mathfrak{q}S}$. □

Remark 3.23 If R satisfies Hypothesis 3.1 and $\mathfrak{q}^{(bn)} \subseteq \overline{\mathfrak{q}^n}$, then there exists t such that $\mathfrak{q}^{b(t+1)n} \subseteq \mathfrak{q}^n$ for all $n \geqslant 1$. This is a consequence of Theorem 3.2.

The following example shows that if the assumption that the extension is finite is dropped from Lemma 3.22, there does not necessarily exist a uniform c such that for all primes $\mathfrak{q}$ in R, $\left(\sqrt{\mathfrak{q}S}\right)^c \subseteq \mathfrak{q}S$. Computations using Macaulay2 [29] were crucial to find this example.

Example 3.24 Let $R = K[a,b,c,d]/(ad - bc)$, which includes in $S = K[x,y,u,v]$ via the map $h : R \longrightarrow S$ given by $h(a) = xy, h(b) = xu, h(c) = yv, h(d) = uv$.

For each integer A, let $\mathfrak{q}_A$ be the prime ideal in R given by the kernel of the map $f_A : R \longrightarrow k[t]$, where f_A is given by

$$f_A(a) = t^{4A},\ f_A(b) = t^{4A+1},\ f_A(c) = t^{8A+1},\ f_A(d) = t^{8A+2}.$$

Let $Q_A = h\left(q_A\right) S$, and $g_A = xu^{4A+1} - vy^{4A+1}$. Then $(g_A)^{4A} \in Q_A$, but $(g_A)^{4A-1} \notin Q_A$. As a consequence, $\left(\sqrt{Q_A}\right)^{4A-1} \not\subseteq Q_AS$.

To check this, fix A and write $\mathfrak{q} := \mathfrak{q}_A$, $Q := Q_A$ and $g := g_A$. We note that

(1) $\mathfrak{q} = \left(c - ab, d - b^2, b^{4A} - a^{4A+1}\right)$, so
(2) $Q = \left(y(v - x^2u), u(v - x^2u), x^{4A}\left(xy^{4A+1} - u^{4A}\right)\right)$.

The fact that $g^{4A} \in Q$, but $g^{4A-1} \notin Q$ follows because $\{y(v - x^2u), u(v - x^2u), x^{4A}\left(u^{4A} - y^{4A+1}\right)\}$ is a Gröbner basis for Q with respect to the lexicographical order induced by the following order on the variables: $v > u > x > y$. This can be checked applying the Buchberger's algorithm on the generating set

$$\{f_1 = yv - yux^2, f_2 = u(v - x^2u) = vu - u^2x^2, f_3 = u^{4A}x^{4A} - y^{4A+1}x^{4A+1}\}.$$

In this order, we now have that $\text{in}(Q) = \left(yv, uv, x^{4A}u^{4A}\right)$. Moreover,

$$g^n = \left(xu^{4A} - x^2y^{4A+1}\right)^n = \sum_{i=0}^{n} c_i x^i u^{4Ai} x^{2(n-i)} y^{(4A+1)(n-i)}, \text{ where } c_i \in \mathbb{Z},$$

so that g^n has leading term $x^n u^{4An}$. If $g^n \in Q$, then

$$\text{in}\left(g^n\right) = x^n u^{(4A+1)n} \in \left(yv, uv, x^{4A}u^{4A}\right).$$

This happens if and only if $n \geqslant 4A$, meaning that $g^n \notin Q$ for all $n < 4A$. On the other hand,

$$g^{4A} = \left(x\left(u^{4A} - xy^{4A+1}\right)\right)^{4A} = x^{4A}\left(u^{4A} - xy^{4A+1}\right)\left(u^{4A} - xy^{4A+1}\right)^{4A-1} \in Q.$$

We now present the main result in this subsection.

Theorem 3.25 (Hunke–Katz–Validashti [52, Corollary 3.4.]) *Let $R \subseteq S$ be a finite extension of domains, with R integrally closed, such that both rings satisfy Hypothesis 3.1. If S has USTP, then R has USTP.*

Proof We first show that there exists r such that for all $\mathfrak{p} \in \text{Spec}(S)$ the following is true: if $\mathfrak{p}^{(bn)} \subseteq \mathfrak{p}^n$ for all n and $\mathfrak{q} = \mathfrak{p} \cap R$, then $\mathfrak{q}^{(rbn)} \subseteq \mathfrak{q}^n$ for all n. Note that $\mathfrak{q}^{(n)} \subseteq \mathfrak{p}^{(n)}$. It suffices to show that there exists r, independent of $\mathfrak{p}$, such that $\mathfrak{p}^{rn} \cap R \subseteq \mathfrak{q}^n$. Indeed, this gives

$$\mathfrak{q}^{(rbn)} \subseteq \mathfrak{p}^{(rbn)} \cap R \subseteq \mathfrak{p}^{rn} \cap R \subseteq \mathfrak{q}^n.$$

By replacing S by $\overline{S}$, it is enough to show our claim for S integrally closed. It also suffices to show this separately for $R \subseteq T$ and $T \subseteq S$ separately, where T is the integral closure of R in some intermediate field E with $K \subseteq E \subseteq L$, where K is the fraction field of R and L is the fraction field of S.

We have two cases to consider:

(a) L is purely inseparable over K.
(b) L is separable over K.

Write $e = [S : R]$.

(a) L is purely inseparable over K.
For every element $u \in S$, $u^k \in R$ for some k, and thus $Q = \sqrt{\mathfrak{q}S}$.
If $u \in Q^{en} \cap R = (Q^e)^n \cap R$, then $u \in \overline{(\mathfrak{q}S)^n} \cap R \subseteq \overline{\mathfrak{q}^n}$. We have that $\overline{\mathfrak{q}^n} \subseteq \mathfrak{q}^{n-t}$ for t by Theorem 3.2. The claim now follows from Remark 3.23.
(b) By perhaps extending L, we can assume L is Galois over K. Write $\sqrt{\mathfrak{q}S} = \mathfrak{p}_1 \cap \cdots \cap \mathfrak{p}_k$, and notice that $k \leqslant e$, since the P_i are permuted by the Galois group.

Let $u \in Q^{e^2 n} \cap R$. Then

$$u^e \in \left(u^k\right) \subseteq \mathfrak{p}_1^{en} \dots \mathfrak{p}_k^{en} = (\mathfrak{p}_1 \dots \mathfrak{p}_k)^{en} \subseteq \left(\sqrt{\mathfrak{q}S}\right)^{en},$$

and by Lemma 3.22, $\left(\sqrt{\mathfrak{q}S}\right)^{en} \subseteq \left(\overline{\mathfrak{q}S}\right)^n \cap R$. Then,

$$u^e \in \left(\overline{\mathfrak{q}S}\right)^n \cap R \subseteq \overline{\mathfrak{q}^n} \subseteq \mathfrak{q}^{n-t}.$$

□

3.6 Direct Summands of Polynomial Rings

In this subsection we discuss uniform bounds for direct summands of polynomial rings. This includes affine toric normal rings [40], and rings on invariants. Rings corresponding to the cones of Grassmannian, Veronese and Segre varieties are also direct summands of a polynomial ring. If one assumes that the field has characteristic zero, then the ring associated to the $t \times t$ minors of an $n \times n$ generic matrix is also one of these rings. We continue with the strategy used to prove the Zariski–Nagata Theorem in Sect. 2.1. We start by recalling a property of direct summands that was used by Àlvarez-Montaner, the fourth, and fifth author in their study of D-modules.

Lemma 3.26 ([2]) *Let $R \subseteq S$ let be two finitely generated K-algebras. Let $\beta\colon S \to R$ be any R-linear morphism. Then, for every $\delta \in D^n_K(S)$, we have that $\tilde{\delta} := \beta \circ \delta_{|R} \in D^n_K(R)$.*

We now give a specific bound for the Chevalley Theorem for homogeneous ideals in a graded direct summand of a polynomial ring.

Theorem 3.27 *Let K be a field, $S = K[x_1, \dots, x_n]$, $\eta = (x_1, \dots, x_n)S$, $f_1, \dots, f_\ell \in S$ be homogeneous polynomials, $R = K[f_1, \dots, f_\ell]$, $\mathfrak{m} = (f_1, \dots, f_\ell)R$, and $B = \max\{\deg(f_1), \dots, \deg(f_\ell)\}$. Suppose that the inclusion, $R \subseteq S$, splits. Then,*

$$\mathfrak{q}^{(Bn)} \subseteq \mathfrak{m}^n$$

for every homogeneous prime ideal $\mathfrak{q} \subseteq R$.

Proof We first show that $\mathfrak{q}^{(Bn)} \subseteq \eta^{Bn} \cap R$. Since $\mathfrak{q}$ and $\eta^{Bn} \cap R$ are homogeneous ideals, it suffices to show that if a homogeneous element $f \notin \eta^{Bn} \cap R$, then $f \notin \mathfrak{q}^{(Bn)}$.

Since $\eta^{Bn} = \eta^{\langle Bn\rangle}$ and f is homogeneous, there exists an operator $\delta \in D^{n-1}_K(S)$ such that $\delta(f) = 1$. Then, $\beta \circ \delta(f) = 1$ by Lemma 3.26. Since $\beta \circ \delta_{|R} \in D^{n-1}_K(R)$, we have that $f \notin \mathfrak{q}^{\langle Bn\rangle}$. In particular, $f \notin \mathfrak{q}^{(Bn)}$, since $\mathfrak{q}^{(Bn)} \subseteq \mathfrak{q}^{\langle Bn\rangle}$.

We now show that $\eta^{Bn} \cap R \subseteq \mathfrak{m}^n$. Let $g \in \eta^{Bn} \cap R$. Then, g is a linear combination of products $f_1^{\alpha_1} \cdots f_\ell^{\alpha_\ell}$ such that

$$Bn \leqslant \deg(f) = \alpha_1 \deg(f_1) + \dots + \alpha_\ell \deg(f_\ell) \leqslant D(\alpha_1 + \dots + \alpha_\ell).$$

Then, $\alpha_1 + \cdots + \alpha_\ell \geqslant n$. Hence, $g \in \mathfrak{m}^n$. We conclude that

$$\mathfrak{q}^{(Bn)} \subseteq \eta^{Bn} \cap R \subseteq \mathfrak{m}^n.$$

□

As a corollary of the previous result, we find that 2 is a sufficient bound for determinantal rings.

Corollary 3.28 *Let X be a $n \times m$ generic matrix of variables, K a field, $R = K[X]/I_t(X)$, and $\mathfrak{m} = (x_{i,j})R$. If either $t = 2$ or $\operatorname{char}(K) = 0$, then*

$$\mathfrak{p}^{(2n)} \subseteq \mathfrak{m}^n$$

for every homogeneous prime ideal $\mathfrak{p} \subseteq \mathfrak{m}$.

Proof If either $t = 2$ or $\operatorname{char}(K) = 0$, then R is a direct summand of a polynomial ring, and it is generated by homogeneous polynomials of degree 2. The rest is a consequence of Theorem 3.27. □

We now focus on Question 3.8 for direct summands of polynomial rings. In case of direct summands whose extension is finite, we have specific values for the uniform bounds given in Theorem 3.25.

Theorem 3.29 *Let $S = K[x_1, \ldots, x_n]$ and $R \subseteq S$ a direct summand. Suppose that S is a finitely generated R-module. Let $\mathfrak{p} \subseteq R$ a prime ideal and $h = \operatorname{ht}(\mathfrak{p})$. If $k = [S : R]$, then*

$$\mathfrak{p}^{(khn)} \subseteq \mathfrak{p}^{n-d}$$

for every positive integer n. Furthermore, if $k!$ is invertible in R, then

$$\mathfrak{p}^{(khn)} \subseteq \mathfrak{p}^n.$$

Proof Let $\mathfrak{q}_1, \ldots, \mathfrak{q}_\ell$ be the minimal primes of $\mathfrak{p}S$. Since $R \subseteq S$ is an integral extension, and R is integrally closed, the going-up and going-down theorems apply. Then, $\mathfrak{q}_i \cap R = \mathfrak{p}$. Let $V = S \setminus (\mathfrak{q}_1 \cup \cdots \cup \mathfrak{q}_\ell)$. Then,

$$\begin{aligned} r \in R \setminus \mathfrak{p} &\Longrightarrow r \notin \mathfrak{q}_i \cap R \\ &\Longrightarrow r \notin \mathfrak{q}_i \text{ for all } i \\ &\Longrightarrow r \in S \setminus (\mathfrak{q}_1 \cup \cdots \cup \mathfrak{q}_\ell). \end{aligned}$$

Thus, $R \setminus \mathfrak{p} \subseteq V$.

We have that

$$\begin{aligned}
\mathfrak{p}^{(khn)} &= \mathfrak{p}^{knh} R_{\mathfrak{p}} \cap R \subseteq \mathfrak{p}^{khn} S_{\mathfrak{p}} \cap S; \\
&\subseteq \mathfrak{p}^{khn} V^{-1} S \cap S \text{ because } R \setminus \mathfrak{p} \subseteq V; \\
&\subseteq \sqrt{\mathfrak{p}S}^{khn} V^{-1} S \cap S = \sqrt{\mathfrak{p}S}^{(knh)}; \\
&\subseteq \sqrt{\mathfrak{p}S}^{kn} \text{ by Theorems 2.19 and 2.20}; \\
&= (\sqrt{\mathfrak{p}S}^{k})^n \subseteq \overline{\mathfrak{p}S}^{n} \text{ by applying Lemma 3.22}; \\
&\subseteq \mathfrak{p}^{n-d} S \text{ by Uniform Briançon-Skoda (Theorem 3.2)}.
\end{aligned}$$

Then, $\mathfrak{p}^{(khn)} \subseteq \mathfrak{p}^{n-d} S \cap R = \mathfrak{p}^{n-d}$.

If $k!$ is invertible, the claim follows the same lines as before, but we use the second part of Lemma 3.22. □

As a corollary of the previous result, we answer a question asked by Takagi to the fourth author.

Corollary 3.30 *Let $S = K[x_1, \dots, x_d]$, and G a finite group that acts on S. Let $R = S^G$ denote the ring of invariants. Let $\mathfrak{p} \subseteq R$ a prime ideal and $h = \operatorname{ht}(\mathfrak{p})$. If $k = |G|$ is invertible in K, then*

$$\mathfrak{p}^{(khn)} \subseteq \mathfrak{p}^{n-d}$$

for every positive integer n. Furthermore, if $k!$ is invertible in K, then

$$\mathfrak{p}^{(khn)} \subseteq \mathfrak{p}^{n}.$$

To the best of our knowledge, Question 3.8 is still open for direct summands R of polynomial rings such that the extension $R \to S$ is infinite. For recent progress on toric rings see the work of Walker [86].

4 Symbolic Powers of Monomial Ideals

4.1 Symbolic Powers, Monomial Ideals and Matroids

Let $S = K[x_1, \dots, x_n]$ denote the polynomial ring in n variables over a field K, and $\mathfrak{m} = (x_1, \dots, x_n)$. There is a bijection between the squarefree monomial ideals in S and simplicial complexes in n vertices, via the Stanley–Reisner correspondence. Some algebraic properties of such an ideal can be described via the combinatorial and topological properties of the corresponding simplicial complex, and vice-versa. Varbaro [84] and Minh and Trung [66] have independently shown that the property that all the symbolic powers of a Stanley–Reisner ideal are Cohen-Macaulay is equivalent to a combinatorial condition on the corresponding simplicial complex, namely that the simplicial complex is a matroid.

Definition 4.1 A *simplicial complex* on the set $[n] := \{1, \dots, n\}$ is a collection Δ of subsets of $[n]$, called *faces* of Δ, that satisfies the following property: given a face $\sigma \in \Delta$, if $\theta \subseteq \sigma$, then $\theta \in \Delta$. A *facet* is a face that is maximal under inclusion.

Given a simplicial complex on $[n]$, we can define a squarefree monomial ideal in S corresponding to Δ:

Definition 4.2 Given a simplicial complex Δ, the *Stanley–Reisner ideal* of Δ is the following squarefree monomial ideal:

$$I_\Delta = \left(x_{i_1} \cdots x_{i_s} : \{i_1, \dots, i_s\} \notin \Delta\right).$$

The quotient $K[\Delta] := S/I_\Delta$ is called the *Stanley–Reisner ring* of Δ.

On the other hand, given a squarefree monomial ideal, we can recover the simplicial complex associated to it, giving us a bijective correspondence:

Definition 4.3 Given a square free monomial ideal I in S, the *Stanley–Reisner complex* of I is given by

$$\Delta = \left\{\{i_1, \dots, i_s\} \subseteq [n] \mid x_{i_1} \dots x_{i_s} \notin I\right\}.$$

For a more details about Stanley–Reisner theory, we refer to [26, 63].

Definition 4.4 A simplicial complex Δ on $[n]$ is said to be a *matroid* if for all facets $F, G \in \Delta$ and all $i \in F$, there exists $j \in G$ such that $(F \setminus \{i\}) \cup \{j\} \in \Delta$ is still a facet.

This turns out to be precisely the combinatorial condition that corresponds to the following property of the symbolic powers of the Stanley–Reisner ideal:

Theorem 4.5 (Varbaro [84], Minh–Trung [66][3]) *Given a simplicial complex Δ on $[n]$, $S/I_\Delta^{(m)}$ is Cohen-Macaulay for all $m \geqslant 1$ if and only if Δ is a matroid.*

We point out that Terai and Trung [82] showed a more general result: If $S/I_\Delta^{(m)}$ is Cohen-Macaulay for some $m \geqslant 3$, then Δ is a matroid.

Example 4.6 The figure below represents the well-known Fano matroid, in 7 variables, where colinear points correspond to facets, considering the circle as a line. The Stanley–Reisner ideal of the Fano matroid in $\mathbb{F}_2[x_1, \dots, x_7]$ is given by

$$I = \begin{pmatrix} x_4x_2x_1, x_4x_3x_1, x_4x_3x_2, x_5x_2x_1, x_5x_3x_1, x_5x_3x_2, x_5x_4x_1, \\ x_5x_4x_2, x_6x_2x_1, x_6x_3x_1, x_6x_3x_2, x_6x_4x_1, x_6x_4x_3, x_6x_5x_2, \\ x_6x_5x_3, x_6x_5x_4, x_7x_2x_1, x_7x_3x_1, x_7x_3x_2, x_7x_4x_2, x_7x_4x_3, \\ x_7x_5x_1, x_7x_5x_3, x_7x_5x_4, x_7x_6x_1, x_7x_6x_2, x_7x_6x_4, x_7x_6x_5 \end{pmatrix}.$$

[3]See also [64].

By Theorem 4.5, we know that the symbolic powers of I are all Cohen-Macaulay. Using Macaulay2 [29] we can check that, for example, $I^{(2)} \neq I^2$ and $I^{(3)} \neq I^3$.

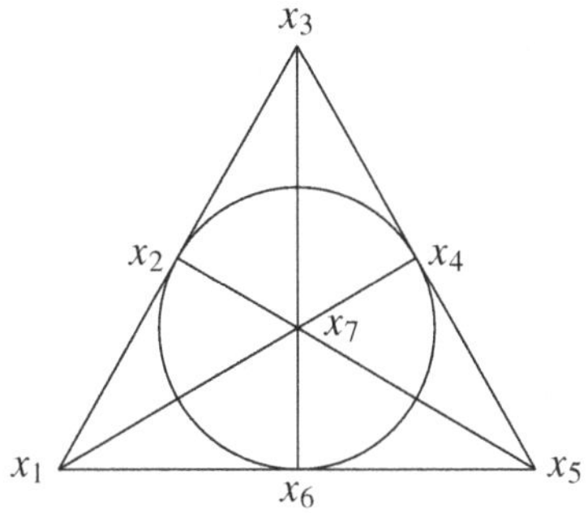

4.2 The Packing Problem

Although there are well-known instances in which the symbolic powers of an ideal are equal to its usual powers, for example complete intersections, there are essentially no theorems which give necessary and sufficient criteria for this equality to be true, except in a few cases. One of the most notable cases are the prime ideals defining curves which are licci [50]. In this latter case, being a complete intersection is both necessary and sufficient for the symbolic powers and regular powers to be the same. There is not even a good guess about what properties of an ideal are necessary and sufficient to guarantee the equality of the symbolic powers and usual powers. However, in the case of squarefree monomial ideals, there is a beautiful conjecture, first discovered by Conforti and Cornuéjols [13] in the context of max-flow min-cut properties, which was reworked by Gitler, Villarreal and others [27, 28] to place the conjecture within commutative algebra. We present this conjecture, and also introduce a new relative version of it. In addition, we give a proof of this relative version for graph ideals, and for the symbolic square. We first need to recall some definitions.

Definition 4.7 Let S be a polynomial ring over a field. A squarefree monomial ideal I of height c is *König* if there exists a regular sequence of monomials in I of length c. The ideal I is said to have the *packing property* if every ideal obtained from I by setting any number of variables equal to 0 or 1 is König.

The conjecture of Conforti and Cornuéjols can be restated in this language to say that the symbolic powers and usual powers of a squarefree monomial ideal coincide if and only if the ideal has the packing property. This conjecture has been the subject of much scrutiny [11, 12, 14, 25, 53, 68, 83]. If I is the edge ideal of a finite simple graph G, this conjecture is known [28]; in fact in this case $I^{(k)} = I^k$ for all k if and only if G is bipartite if and only if I has the packing property. In the graph case, the conjecture can also be reintrepreted in terms of well-known graph invariants. Namely, let G be a graph. Set $c(G)$ equal to the size of minimal vertex cover and $m(G)$ equal to the size of maximal set of disjoint edges. Then the height of I is $c(G)$, and the length of a maximal sequence in I of monomials is $m(G)$. Obviously, $c(G) \geqslant m(G)$. In this language, G has the König property if and only if $c(G) = m(G)$, while G is packed if and only if every minor of G has the König property.

One direction is not difficult. Assume that $I^{(n)} = I^n$ for all n. One can prove this property is preserved after setting variables equal to 0 or 1, so to prove that I is packed, one only needs to prove that I is König. Moreover, by setting variables which do not appear in the minimal generators of I to 0, one can further assume that every variable appears in some minimal prime of I. But if the ring is $K[x_1, \ldots, x_n]$, the monomial $m = x_1 \cdots x_n$ is in $\mathfrak{p}^c$ for every minimal prime $\mathfrak{p}$ of I, since I has height c. It follows that $m \in I^{(c)}$. By assumption, $m \in I^c$, and this implies that there are monomials $m_1, \ldots m_c \in I$ such that $m_1 \cdots m_c = m$. Since m is squarefree these monomials necessarily have disjoint support; then $m_1, \ldots, m_c$ form a regular sequence in I of maximal length c. The difficult direction is to prove that if I is packed then the symbolic and usual powers agree.

In this section we introduce a relative version of this conjecture. We make the following definition:

Definition 4.8 Let I be a squarefree monomial ideal. We say that I is k-König if there is a regular sequence of monomials in I of length at least $\min\{k, height(I)\}$. We say I is k-packed if every squarefree monomial ideal J obtained from I by setting variables equal to 0 or 1 is k-König.

With this language, a natural extension of the question of Conforti and Cornuéjols is:

Question 4.9 Is I k-packed if and only if $I^{(n)} = I^n$ for all $n \leqslant k$?

We prove that if I is the edge ideal of a graph, then Question 4.9 has a positive answer. To achieve this, we prove in our main theorem (see Theorem 4.13) that $I^{(k)} = I^k$ for $1 \leqslant k \leqslant n$ if and only if G contains no odd cycles of length at most $2n - 1$. In particular, if t is chosen to be the least integer such that $I^{(t)} \neq I^t$ (if such a t exists), then $2t - 1$ is the size of the smallest induced cycle of G. Another corollary of our main theorem is that if G is a finite graph with edge ideal I of height c, then $I^{(k)} = I^k$ for $1 \leqslant k \leqslant c$ implies that G is bipartite. We also observe that Question 4.9 has a positive answer if $k = 2$. We do not know whether the question has a positive answer if either I is generated by cubics, or for $k = 3$.

We begin our study of the graph ideal case by noting:

Proposition 4.10 *Let G be a finite simple graph.*

(1) Suppose G has an odd cycle $v_1, \ldots, v_{2n-1}$. Then $f = \prod v_i \in I_G^{(n)} \setminus I_G^n$.
(2) Assume that $c(G) > m(G) = n - 1$. Then G contains an odd cycle of length at most $2n - 1$.

Proof (1) We note that $f \notin I_G^n$ by degree reasons. But any minimal prime of I_G must contain n vertices of the cycle, so $f \in I_G^{(n)}$.

(2) Suppose G does not contain such cycle. Since any cycle of length at least $2n+1$ has a set of n disjoint edges, G must not contain any cycle of odd length. Thus G is bipartite, and König's theorem asserts that $c(G) = m(G)$, which gives a contradiction. □

Although our main concern in this section is with the edge ideals of graphs, our first reduction of the problem of the equality of powers and symbolic powers works equally well for any squarefree monomial ideal. We describe this reduction in the discussion and remark below.

Discussion 4.11 Let J be a squarefree monomial ideal, and fix a variable x. We let I_x denote $J : (x)$ and I be the ideal generated by the monomials in J not involving x. We have $J = I + xI_x$ and $I \subseteq I_x$. Suppose we know that $I_x^{(n)} = I_x^n$ and $I^{(n)} = I^n$ (for example, if $J = I_G$, this would be the case if we know by induction that the symbolic powers and usual powers of I_G are the same whenever we set variables equal to 0 or 1, provided we do so for at least one variable).

Clearly $J = I_x \cap (I, x)$. It follows that

$$\begin{aligned} J^{(n)} &= I_x^{(n)} \cap (I, x)^{(n)} \\ &= I_x^n \cap (I, x)^n \\ &= I^n + x(I^{n-1} \cap I_x^n) + x^2(I^{n-2} \cap I_x^n) + \cdots + x^n I_x^n. \end{aligned}$$

On the other hand, as $J = I + xI_x$:

$$J^n = I^n + xI^{n-1}I_x + x^2 I^{n-2} I_x^2 + \cdots + x^n I_x^n$$

As each term in this expression of J^n is inside the corresponding term of $J^{(n)}$, the equality $J^{(n)} = J^n$ is equivalent to

$$I^k \cap I_x^n = I^k I_x^{n-k} \quad \text{for} \quad 1 \leqslant k \leqslant n-1 \tag{4.2.1}$$

We can write $I_x = L + I$ where L is generated by monomials of $J : (x)$ which are not in I (in the case $J = I_G$, L is the ideal generated by the variables which correspond to neighbors of x).

The lefthand side of Eq. 4.2.1 can be written as:

$$\left(I^k \cap \sum_{0 \leqslant i \leqslant k-1} I^i L^{n-i} \right) + \sum_{j=k}^{n} I^j L^{n-j}.$$

We summarize this discussion in the following remark:

Remark 4.12 Let J be a squarefree monomial ideal, and let x be a variable. We let I_x denote $J : (x)$ and I be the ideal generated by monomials in J not involving x. Suppose we know that $I_x^{(n)} = I_x^n$ and $I^{(n)} = I^n$. Write $I_x = L + I$ where L is generated by monomials of $J : (x)$ which are not in I. Then $J^{(n)} = J^n$ if and only if for all k with $0 \leqslant i < k \leqslant n$

$$I^k \cap I^i L^{n-i} \subseteq \sum_{j=k}^{n} I^j L^{n-j}.$$

Note that the right hand side of this expression is precisely $I^k I_x^{n-k}$.

Before we begin the proof of main result in this subsection, we point out that this result also follows from the recent work of Lam and Trung [58, Corollary 4.5].

Theorem 4.13 *Let G be a finite simple graph, and let $I := I_G$, the edge ideal of G. Then $I^{(k)} = I^k$ for $1 \leqslant k \leqslant n$ if and only if G contains no odd cycles of length at most $2n - 1$. In particular, if t is chosen to be the least integer such that $I^{(t)} \neq I^t$ (if such a t exists), then $2t - 1$ is the size of the smallest induced cycle of G.*

Proof We first prove the last asserted statement, assuming the first. Let $2s - 1$ be the size of the smallest induced odd cycle. Hence, G has no odd cycles of length at most $2s - 3$, since any odd cycle contains an induced odd cycle of at most the same length. The first statement of the theorem then proves that $I^{(k)} = I^k$ for $1 \leqslant k \leqslant s - 1$. On the other hand, since G does contain an odd cycle of length $2s - 1$, the other direction of the first statement shows that $I^{(s)} \neq I^s$. Hence $t = s$, proving the second statement. We now prove the first statement.

Proposition 4.10 gives the "only if" direction of this theorem, so only the "if" direction needs to be proved. We shall prove this direction by induction on the number of vertices. It follows that we may assume that the symbolic and usual powers agree up to n for any ideal obtained from I by setting variables equal to 0 or 1, provided at least one variable is set equal to 0 or 1. By Remark 4.12, the proof is finished by proving the following lemma (we adopt the notation from the discussion and remark): □

Lemma 4.14 *Suppose that G has no odd cycles of length up to $2i + 3$. Then*

$$I^k \cap I^i L^{n-i} \subseteq \sum_{j=k}^{n} I^j L^{n-j} = I^n + I^{n-1} L + \cdots + I^k L^{n-k}$$

for any $i < k \leqslant n$.

Proof We use induction on n, then a backwards induction on k. For the second induction, note that if $k = n$, then the conclusion is satisfied trivially. By way of contradiction, consider any minimal degree monomial element $f \in I^k \cap I^i L^{n-i}$ such that $f \notin \sum_k^n I^j L^{n-j}$. We may write f as the product of k edges: $x_i y_i$ ($1 \leqslant i \leqslant a$,

$x_i \in L$), $u_i v_i$ ($1 \leqslant i \leqslant b$, $u_i, v_i \notin L$), and collect the rest of the variables dividing f into two sets, a set z_i ($1 \leqslant i \leqslant c$, $z_i \in L$), and possibly some extra variables, none of them in L. We denote this extra set of variables by F. Note that there might well be repetition among the variables. Also observe that no y_j can be in L, since this would give a 3-cycle in G. With this notation,

$$f = \prod_{1 \leqslant j \leqslant a} (x_j y_j) \cdot \prod_{1 \leqslant j \leqslant b} (u_j v_j) \cdot \prod_{1 \leqslant j \leqslant c} z_j \cdot \prod_{t \in F} t,$$

where $a + b \geqslant k$. We can assume that $a + b = k$, since if the sum is strictly greater, f would be in I^{k+1}, and we are done by the decreasing induction on k. We call this expression for f the *first expression* of f (it is not necessarily unique).

Since by assumption $f \in I^i L^{n-i}$, we may also write

$$f = \prod_{1 \leqslant j \leqslant i} m_j \cdot \prod_{1 \leqslant j \leqslant l} z'_j \cdot \prod_{w \in W} w,$$

where each m_j is a degree two monomial corresponding to an element in I, all $z'_j \in L$, and none of the extra variables w are in L. Finally, l is an integer such that $l \geqslant n - i$. We call this expression for f the *second expression* of f (again, it is not necessarily unique).

We make a general observation: since each m_j divides f, if for some j, $m_j = x_r y_r$ or $u_r v_r$ in the first representation of f, we can cancel m_j in the first representation so that $f/m_j \in I^{k-1} \cap I^{i-1} L^{n-i} = I^{n-1} + I^{n-2} L + \cdots + I^{k-1} L^{n-k}$ by the induction on n, and then $f \in \sum_k^n I^j L^{n-j}$, which gives a contradiction. Thus, without loss of generality we assume that m_j is not equal to any $x_r y_r$ or $u_r v_r$.

We claim that $c = 0$, and there are no extra variables F. If not, consider first the variable z_1. It must appear among the variables in the second expression for f as an element of $I^i L^{n-i}$. If z_1 is among the z'_j, we can cancel, and use the induction on n to reach a contradiction. If not, then z_1 divides one of the m_j, say $m_1 = z_1 s$. As s appears among the variables in the first expression, we simply combine it with z_1 and cancel m_1 from both sides. Notice that $f/m_1 \in I^{k-1}$, since the rearrangement affects at most one edge monomial in the first expression. The induction then gives a contradiction, as we observed above. Thus, no z_i can appear in the first expression for f. Now consider a variable $t \in F$. Using the same reasoning, t must appear in the second expression for f. It cannot be one of the z'_j, by assumption. If t appears in W, we can cancel it, and obtain that f/t is a smaller counterexample, which gives a contradiction. Thus t must divide some m_j. We can again recombine t in the first expression for f to replace possibly one edge monomial in I by m_j. After canceling m_j from each side and using induction, we reach a contradiction. We have reached the situation in which $\prod_{1 \leqslant j \leqslant i} m_j \cdot \prod_{1 \leqslant j \leqslant l} z'_j$ divides

$$f = \prod_{1 \leqslant j \leqslant a} (x_j y_j) \cdot \prod_{1 \leqslant j \leqslant b} (u_j v_j).$$

Since none of the u_j, v_j, y_j are in L, without loss of generality, $x_1 = z'_1, \ldots, x_l = z'_l$. After canceling these elements we obtain that $\prod_{1 \leqslant j \leqslant i} m_j$ divides

$$(y_1 y_2 \cdots y_l) \prod_{l+1 \leqslant j \leqslant a} (x_j y_j) \cdot \prod_{1 \leqslant j \leqslant b} (u_j v_j).$$

Note that $a - l + b = k - l \leqslant k - (n - i) = i + (k - n) < i$ (recall that we may assume $k < n$ now as the base case $k = n$ was handled at the beginning of the proof).

We next need to do the case $i = 0$ separately. In this case, our assumption is that G has no 3-cycles. In this case $l = n$, and since $x_1 = z'_1, \ldots, x_l = z'_l$, we must have n edge-monomials $x_1 y_1, \ldots, x_n y_n$ in the first expression for f. This forces $f \in I^n$ (after our reductions), which gives the necessary conclusion. Thus in the remainder of the proof, we may assume that $i \geqslant 1$. An important consequence of this reduction is that there cannot be an edge between any of the y_j, as this would give a 5-cycle.

We let D be the set of vertices $\{y_1, \ldots, y_l\}$. By Lemma 4.15 below applied with $e = a + b - l = k - l < i$ (see above) and $s_1 t_1, \ldots, s_e t_e$ being the edges $x_j y_j$ for $l + 1 \leqslant j \leqslant a$ together with all the $u_j v_j$, we can find an odd path $w_1, \ldots, w_l$ with $l \leqslant 2e + 1$ and $w_1, w_l \in F$ (so they are neighbors of some xs). Renumbering if necessary, we have an odd cycle $x, x_1, w_1, \ldots, w_l, x_2$ of length $l + 3$. Now there can be repetition, but any time we identify two non-adjacent vertices in that cycle, a new odd cycle appears. The proof is finished. □

We finish the proof of the main theorem by proving Lemma 4.15 below.

Lemma 4.15 *Let $e \geqslant 1$ be an integer. Suppose there are e edges $s_1 t_1, \cdots s_e t_e$ and a set of vertices D, together with an auxilliary set $\{m_1, \ldots m_{e+1}\}$ of edges (repetitions are allowed in all of these) such that $\prod_{1 \leqslant j \leqslant e+1} m_j$ divides $\prod_{1 \leqslant j \leqslant e} (s_j t_j) \cdot \prod_{w \in D} w$. Assume also that there are no edges between any of the elements in D. Then there is a path $w_1, \ldots, w_l$ with $l \leqslant 2e + 1$ even (so that the path length is odd) and $w_1, w_l \in D$.*

Proof We use induction on e. We claim that two of the edges m_is must contain exactly one from D. The reason is simply because there are no edges in D, so any m_i that contains some vertices in D contains exactly one, and there has to be at least two such edges, as the product of the m_is has degree $2e + 2$, while the degree of the product of $x_j y_j$ is $2e$.

We may write these edges as $m_1 = w_1 u$ and $m_2 = w_2 v$ where $w_1, w_2 \in D$. Consider u and v. If uv is an edge, then we have a length three path w_1, u, v, w_2 and we are done. Note that this settles the case $e = 1$, as in that situation uv has to be the edge $s_1 t_1$.

Hence without loss of generality we may assume that $e > 1$, $u = s_1$ and $v = s_2$. Now consider the set D' obtained from D by removing w_1, w_2 and including t_1, t_2, the $e - 2$ edges $s_3 t_3, \ldots, s_e t_e$, and the set of $e - 1$ auxilliary edges $m_3, \ldots, m_{e+1}$. We claim these sets satisfy the hypothesis of the lemma. We need to show that there are no edges between the elements of D'. If $t_1 t_2$ is an edge, then there is a path of length 5 from w_1 to w_2: $w_1, s_1, t_1, t_2, s_2, w_2$. If, on the other hand there is a $w_3 \in D$, apart from w_1, w_2 such that w_3t_1 is an edge, then we may construct a path of length three from w_1 to w_3, namely w_1, s_1, t_1, w_3, and again we are done. Hence there are no edges between the vertices in D'. By induction there is an odd length path of length at most $2(e-2)+1$ which begins and ends in vertices in D'. Now if either end of this path is t_1 or t_2, we may augment the path by two edges, e.g. by t_1s_1, s_1w_1, so that the beginning and end of the path are in D. This adds at most 4 edges to the path, so that the length, which is still odd, is at most $2(e-2)+1+4=2e+1$, as required. □

Proof This completes the proof of Theorem 4.13. □

Remark 4.16 We thank Susan Morey for pointing out that this result follows by putting together work in [11, 53, 68], although it is not explicitly stated. By combining the work in these papers one obtains the statement that a non-bipartite graph G whose smallest odd cycle has length $2t-1$ has the property that the associated primes of the powers of the edge ideal I of G is equal to the minimal primes of I for powers up to $t-1$, and the tth power of I has an embedded prime.

Corollary 4.17 *Let G be a finite graph and I be its edge ideal. Let $c = \text{height}(I)$. If $I^{(k)} = I^k$ for $1 \leqslant k \leqslant c$ then G is bipartite.*

Proof If G is not bipartite, there is a minimal odd cycle, say $x_1, x_2, \dots, x_{2i+1}$. The height of the edge ideal of this cycle is $i+1$ since the vertices are distinct by minimality. Hence $c \geqslant i+1$. By Theorem 4.13, G has no odd cycles of length at most $2c-1$. But $c \geqslant i+1$ implies that $2i+1 \leqslant 2(c-1)+1 = 2c-1$, a we reach a contradiction. □

We apply the work from above to give a positive answer to the relative version of the conjecture of Conforti and Cornuéjols which was given at the beginning of this subsection. Recall that Question 4.9 asks if I is k-packed if and only if $I^{(n)} = I^n$ for all $n \leqslant k$. In the case of quadrics, i.e., for edge ideals of simple graphs, we can give a positive answer:

Theorem 4.18 *Let I be the edge ideal of a graph, i.e., a squarefree ideal generated by quadrics. Then I is k-packed for some $k \geqslant 2$ if and only if $I^{(n)} = I^n$ for all $n \leqslant k$.*

Proof We first prove that if $I^{(n)} = I^n$ for all $n \leqslant k$ then I is k-packed. Since both of these properties are preserved by setting variables equal to 0 or 1, it suffices to prove that G is k-König. Suppose that the height of I is c, but that I does not contain a regular sequence of monomials of length at least the minimum of k and c. Let m denote the product of all the variables. Then $m \in I^{(c)}$, since at every minimal prime $\mathfrak{p}$ of I, $m \in \mathfrak{p}^c$. If $k \geqslant c$, then by assumption $m \in I^c$. But then m is a product of c elements of I which are necessarily a regular sequence as m is squarefree. This is a contradiction. If $c > k$, then $m \in I^{(c)} \subset I^{(k)} = I^k$, and again we reach a contradiction. Hence I is k-packed.

Conversely, assume that I is k-packed but that $I^{(n)} \neq I^n$ for some $n \leqslant k$. By decreasing k if necessary, we can assume that $n = k$. By Theorem 4.13, G must contain an odd cycle of length at most $2k+1$. Choose a minimal odd cycle in G, say $x_1, x_2, \ldots, x_{2j+1}$ of length $2j+1$, where $j \leqslant k$. We set all other variables equal to 0, and apply the assumption that the resulting ideal J is k-packed. Moreover, the height of J is at least the height of the edge ideal of this cycle, which has height $j+1$. Therefore there is a regular sequence of length at least j in J. This is clearly impossible since there are only $2j+1$ variables, and such a regular sequence would require at least $2j$ variables. □

There are a few easy cases one can also do, which we record as a remark;

Remark 4.19 We observe the following: if I is k-packed, and $J^{(n)} = J^n$ for all $n \leqslant k$ and for all ideals J which are obtained from I by setting at least one variable equal to 0 or 1, and I has height k, then $I^{(n)} = I^n$ for all $n \leqslant k$. To see this suppose not and let f be a minimal monomial which is in some $I^{(n)} \setminus I^n$ for some $n \leqslant k$. We may assume that f contains every variable, since if not, we can set a variable equal to 0 without changing the fact that f is in $I^{(n)} \setminus I^n$, contradicting our assumption. We may also assume that f is squarefree. This is because the product of all the variables is clearly in $\mathfrak{p}^k$ for all primes $\mathfrak{p}$ containing I, since k is the height of I. It follows that f must be the product of all the variables. But since I is k-packed there is a regular sequence of monomials in I of length k, and the product of them must divide f since they are disjoint. Thus $f \in I^k$, contradiction.

A consequence of the above Remark is that Question 4.9 has a positive answer for $k = 2$.

Corollary 4.20 *I is 2-packed if and only if $I^{(2)} = I^2$.*

Proof The case in which I has height one is trivial. If I has height two, the previous remark applies. □

Acknowledgements We thank Jeff Mermin for many helpful conversations concerning the packing problem, and in particular for discussions leading to Remark 4.19 and Corollary 4.20. We thank Jonathan Montaño, Andrew Conner, Jack Jeffries, and Robert Walker for helpful comments. Part of this work was done when the second and fifth authors were at the University of Virginia. They wish to thank this institution for its hospitality. Finally, the fifth author thanks the organizing committee for the 'Brazil-Mexico 2nd meeting on Singularities' in Salvador, Bahia, Brazil, where this project was initiated.

References

1. Akesseh, S.: Ideal containments under flat extensions. J. Algebra. **492**, 44–51 (2017)
2. Àlvarez Montaner, J., Huneke, C., Núñez-Betancourt, L.: D-modules, Bernstein–Sato polynomials, and F-invariants of direct summands. Adv. Math. **321**, 298–325 (2017). arXiv:1611.04412
3. Atiyah, M.F., Macdonald, I.G.: Introduction to Commutative Algebra. Addison-Wesley Publishing Co., Reading (1969)

4. Baczyńska, M., Dumnicki, M., Habura, A., Malara, G., Pokora, P., Szemberg, T., Szpond, J., Tutaj-Gasińska, H.: Points fattening on $\mathbb{P}^1 \times \mathbb{P}^1$ and symbolic powers of bi-homogeneous ideals. J. Pure Appl. Algebra **218**(8), 1555–1562 (2014)
5. Berthelot, P.: Altérations de variétés algébriques (d'après A. J. de Jong). Astérisque **241**, Exp. No. 815, 5, 273–311 (1997); Séminaire Bourbaki, Vol. 1995/96
6. Bocci, C., Harbourne, B.: Comparing powers and symbolic powers of ideals. J. Algebr. Geom. **19**(3), 399–417 (2010)
7. Bocci, C., Harbourne, B.: The resurgence of ideals of points and the containment problem. Proc. Am. Math. Soc. **138**(4), 175–1190 (2010)
8. Bocci, C., Cooper, S., Guardo, E., Harbourne, B., Janssen, M., Nagel, U., Seceleanu, A., Van Tuyl, A., Thanh, V.: The Waldschmidt constant for squarefree monomial ideals. J. Algebr. Combin. **44**(4), 875–904 (2016)
9. Böger, E.: Differentielle und ganz-algebraische Abhängigkeit bei Idealen analytischer Algebren. Math. Z. **121**, 188–189 (1971)
10. Brodmann, M.: Asymptotic stability of $\mathrm{Ass}(M/I^nM)$. Proc. Am. Math. Soc. **74**(1), 16–18 (1979)
11. Chen, J., Morey, S., Sung, A.: The stable set of associated primes of the ideal of a graph. Rocky Mt. J. Math. **32**(1), 71–89 (2002)
12. Cornuéjols, G.: Combinatorial Optimization: Packing and Covering. CBMS-NSF Regional Conference Series in Applied Mathematics, vol. 74. Society for Industrial and Applied Mathematics (SIAM), Philadelphia (2001)
13. Cornuéjols, G., Conforti, M.: A decomposition theorem for balanced matrices. Integer Program. Comb. Optim. **74**, 147–169 (1990)
14. Cornuéjols, G., Guenin, B., Margot, F.: The packing property. Math. Program. **89**(1, Ser. A), 113–126 (2000)
15. Dale Cutkosky, S.: Irrational asymptotic behaviour of Castelnuovo–Mumford regularity. J. Reine Angew. Math. **522**, 93–103 (2000)
16. Dumnicki, M.: Containments of symbolic powers of ideals of generic points in $\mathbb{P}^3$. Proc. Am. Math. Soc. **143**(2), 513–530 (2015)
17. Dumnicki, M., Szemberg, T., Tutaj-Gasińska, H.: Counterexamples to the $I^{(3)} \subseteq I^2$ containment. J. Algebra **393**, 24–29 (2013)
18. Dumnicki, M., Harbourne, B., Szemberg, T., Tutaj-Gasińska, H.: Linear subspaces, symbolic powers and Nagata type conjectures. Adv. Math. **252**, 471–491 (2014)
19. Dumnicki, M., Harbourne, B., Nagel, U., Seceleanu, A., Szemberg, T., Tutaj-Gasińska, H.: Resurgences for ideals of special point configurations in $\mathbf{P}^N$ coming from hyperplane arrangements. J. Algebra **443**, 383–394 (2015)
20. Ein, L., Lazarsfeld, R., Smith, K.E.: Uniform bounds and symbolic powers on smooth varieties. Invent. Math. **144**(2), 241–252 (2001)
21. Eisenbud, D.: Commutative Algebra: With a View Toward Algebraic Geometry. Graduate Texts in Mathematics, vol. 150. Springer, New York (1995)
22. Eisenbud, D., Hochster, M.: A Nullstellensatz with nilpotents and Zariski's main lemma on holomorphic functions. J. Algebra **58**(1), 157–161 (1979)
23. Eisenbud, D., Mazur, B.: Evolutions, symbolic squares, and fitting ideals. J. Reine Angew. Math. **488**, 189–201 (1997)
24. Fatabbi, G., Harbourne, B., Lorenzini, A.: Inductively computable unions of fat linear subspaces. J. Pure Appl. Algebra **219**(12), 5413–5425 (2015)
25. Francisco, C.A., Hà, H.T., Van Tuyl, A.: Colorings of hypergraphs, perfect graphs, and associated primes of powers of monomial ideals. J. Algebra **331**, 224–242 (2011)
26. Francisco, C.A., Mermin, J., Schweig, J.: A survey of Stanley–Reisner theory. Connections Between Algebra, Combinatorics, and Geometry, pp. 209–234. Springer, Berlin (2014)
27. Gitler, I., Reyes, E., Villarreal, R.H.: Blowup algebras of ideals of vertex covers of bipartite graphs. Algebraic Structures and their Representations. Contemporary Mathematics, vol. 376, pp. 273–279. American Mathematical Society, Providence (2005)

28. Gitler, I., Valencia, C., Villarreal, R.H.: A note on the Rees algebra of a bipartite graph. J. Pure Appl. Algebra **201**(1–3), 17–24 (2005)
29. Grayson, D.R., Stillman, M.E.: Macaulay2, a software system for research in algebraic geometry
30. Grifo, E., Huneke, C.: Symbolic powers of ideals defining F-pure and strongly F-regular rings. Int. Mat. Res. Not. (2017). http://doi.org/10.1093/imrn/rnx213
31. Grothendieck, A.: Éléments de géométrie algébrique. IV. Étude locale des schémas et des morphismes de schémas IV. Inst. Hautes Études Sci. Publ. Math. **32**, 361 (1967)
32. Guardo, E., Harbourne, B., Van Tuyl, A.: Symbolic powers versus regular powers of ideals of general points in $\mathbb{P}^1 \times \mathbb{P}^1$. Canad. J. Math. **65**(4), 823–842 (2013)
33. Harbourne, B., Huneke, C.: Are symbolic powers highly evolved? J. Ramanujan Math. Soc. **28A**, 247–266 (2013)
34. Harbourne, B., Seceleanu, A.: Containment counterexamples for ideals of various configurations of points in $\mathbf{P}^N$. J. Pure Appl. Algebra **219**(4), 1062–1072 (2015)
35. Harbourne, B., Kapustka, M., Knutsen, A., Syzdek, W., Bauer, T., Di Rocco, S., Szemberg, T.: A primer on Seshadri constants. Contemp. Math. **496**, 39–70 (2009)
36. Herzog, J., Tuân Hoa, L., Trung, N.V.: Asymptotic linear bounds for the Castelnuovo–Mumford regularity. Trans. Am. Math. Soc. **354**(5), 1793–1809 (2002)
37. Hoa, L.T., Trung, T.N.: Partial Castelnuovo–Mumford regularities of sums and intersections of powers of monomial ideals. Math. Proc. Camb. Philos. Soc. **149**(2), 229–246 (2010)
38. Hoa, L.T., Trung, T.N.: Castelnuovo–Mumford regularity of symbolic powers of two-dimensional squarefree monomial ideals. J. Commut. Algebra **8**(1), 77–88 (2016)
39. Hochster, M., Huneke, C.: Tight closure in equal characteristic zero (1999)
40. Hochster, M.: Rings of invariants of tori, Cohen–Macaulay rings generated by monomials, and polytopes. Ann. Math. **2**(96), 318–337 (1972)
41. Hochster, M., Huneke, C.: Comparison of symbolic and ordinary powers of ideals. Invent. Math. **147**(2), 349–369 (2002)
42. Hübl, R.: Evolutions and valuations associated to an ideal. J. Reine Angew. Math. **517**, 81–101 (1999)
43. Hübl, R.: Powers of elements and monomial ideals. Commun. Algebra **33**(10), 3771–3781 (2005)
44. Hübl, R., Huneke, C.: Fiber cones and the integral closure of ideals. Collect. Math. **52**(1), 85–100 (2001)
45. Huckaba, S.: Symbolic powers of prime ideals with applications to hypersurface rings. Nagoya Math. J. **113**, 161–172 (1989)
46. Huneke, C.: Uniform bounds in Noetherian rings. Invent. Math. **107**(1), 203–223 (1992)
47. Huneke, C.: Desingularizations and the uniform Artin–Rees theorem. J. Lond. Math. Soc. (2) **62**(3), 740–756 (2000)
48. Huneke, C., Raicu, C.: Introduction to uniformity in commutative algebra. Commutative Algebra and Noncommutative Algebraic Geometry. Vol. I. Mathematical Sciences Research Institute Publications, vol. 67, pp. 163–190. Cambridge University Press, New York (2015)
49. Huneke, C., Ribbe, J.: Symbolic squares in regular local rings. Math. Z. **229**(1), 31–44 (1998)
50. Huneke, C., Ulrich, B.: Powers of licci ideals. Commutative Algebra (Berkeley, CA, 1987). Mathematical Sciences Research Institute Publications, vol. 15, pp. 339–346. Springer, New York (1989)
51. Huneke, C., Katz, D., Validashti, J.: Uniform equivalente of symbolic and adic topologies. Ill. J. Math. **53**(1), 325–338 (2009)
52. Huneke, C., Katz, D., Validashti, J.: Uniform symbolic topologies and finite extensions. J. Pure Appl. Algebra **219**(3), 543–550 (2015)
53. Huy Tài Hà and Susan Morey: Embedded associated primes of powers of squarefree monomial ideals. J. Pure Appl. Algebra **214**(4), 301–308 (2010)
54. Izumi, S.: A measure of integrity for local analytic algebras. Publ. Res. Inst. Math. Sci. **21**(4), 719–735 (1985)
55. Kunz, E.: On Noetherian rings of characteristic p. Am. J. Math. **98**(4), 999–1013 (1976)

56. Kunz, E.: Kähler Differentials. Advanced Lectures in Mathematics. Friedr. Vieweg & Sohn, Braunschweig (1986)
57. Kurano, K., Roberts, P.C.: The positivity of intersection multiplicities and symbolic powers of prime ideals. Compos. Math. **122**(2), 165–182 (2000)
58. Lam, H.M., Trung, N.V.: Associated primes of powers of edge ideals and ear decompositions of graphs (2015). arXiv:1506.01483
59. Ma, L., Schwede, K.: Perfectoid multiplier/test ideals in regular rings and bounds on symboic powers (2017). arxiv:1705.02300
60. Matsumura, H. Commutative algebra. 2nd edn. Mathematics Lecture note series, vol. 56, pp. Xv+313. Benjamin/Cummings Publishing Co., Inc., Reading, Mass., (1980). ISBN: 0-8053-7026-9
61. Mazur, B.: Deformations of Galois representations and Hecke algebras. Harvard Course Notes (1994)
62. McAdam, S.: Quintasymptotic primes and four results of Schenzel. J. Pure Appl. Algebra **47**(3), 283–298 (1987)
63. Miller, E., Sturmfels, B.: Combinatorial Commutative Algebra. Graduate Texts in Mathematics, vol. 227. Springer, New York (2005)
64. Minh, N.C., Trung, N.V.: Corrigendum to Cohen-Macaulayness of monomial ideals and symbolic powers of Stanley–Reisner ideals [Adv. Math. **226**(2), 1285–1306 (2011)] [mr2737785]. Adv. Math. **228**(5), 2982–2983 (2011)
65. Minh, N.C., Trung, N.V.: Regularity of symbolic powers and Arboricity of matroids (2017). arXiv:1702.04491
66. Minh, N.C., Trung, N.V.: Cohen-Macaulayness of monomial ideals and symbolic powers of Stanley–Reisner ideals. Adv. Math. **226**(2), 1285–1306 (2011)
67. More, A.A.: A note on the Eisenbud–Mazur conjecture. J. Pure Appl. Algebra **219**(7), 2740–2755 (2015)
68. Morey, S., Villarreal, R.H.: Edge ideals: algebraic and combinatorial properties. Progress in Commutative Algebra 1, pp. 85–126. de Gruyter, Berlin (2012)
69. Nagata, M.: Local Rings. Interscience (1962)
70. Rees, D.: Degree functions in local rings. Proc. Camb. Philos. Soc. **57**, 1–7 (1961)
71. Roberts, P.C.: Multiplicities and Chern classes in local algebra. Cambridge Tracts in Mathematics, vol. 133. Cambridge University Press, Cambridge (1998)
72. Sather-Wagstaff, S.: Intersections of symbolic powers of prime ideals. J. Lond. Math. Soc. (2) **65**(3), 560–574 (2002)
73. Scheja, G., Storch, U.: Über differentielle Abhängigkeit bei Idealen analytischer Algebren. Math. Z. **114**, 101–112 (1970)
74. Schenzel, P.: Symbolic powers of prime ideals and their topology. Proc. Am. Math. Soc. **93**(1), 15–20 (1985)
75. Schenzel, P.: Finiteness of relative Rees rings and asymptotic prime divisors. Math. Nachr. **129**, 123–148 (1986)
76. Serre, J.-P.: Local Algebra. Springer Monographs in Mathematics. Springer, Berlin (2000). (Translated from the French by CheeWhye Chin and revised by the author)
77. Sullivant, S.: Combinatorial symbolic powers. J. Algebra **319**(1), 115–142 (2008)
78. Swanson, I.: Powers of ideals. Primary decompositions, Artin–Rees lemma and regularity. Math. Ann. **307**(2), 299–313 (1997)
79. Swanson, I.: Linear equivalence of topologies. Math. Zeitschrift **234**, 755–775 (2000)
80. Szemberg, T., Szpond, J.: On the containment problem (2016). arXiv:1601.01308
81. Taylor, R., Wiles, A.: Ring-theoretic properties of certain Hecke algebras. Ann. Math. (2) **141**(3), 553–572 (1995)
82. Terai, N., Trung, N.V.: Cohen-Macaulayness of large powers of Stanley–Reisner ideals. Adv. Math. **229**(2), 711–730 (2012)
83. Trung, N.V., Tuan, T.M.: Equality of ordinary and symbolic powers of Stanley–Reisner ideals. J. Algebra **328**, 77–93 (2011)
84. Varbaro, M.: Symbolic powers and matroids. Proc. Am. Math. Soc. **139**, 2357–2366 (2011)

85. Waldschmidt, M.: Propriétés arithmétiques de fonctions de plusieurs variables. II. Séminaire Pierre Lelong (Analyse) année 1975/76. Lecture Notes in Mathematics, vol. 578, pp. 108–135. Springer, Berlin (1977)
86. Walker, R.M.: Rational singularities and uniform symbolic topologies. Ill. J. Math. **60**(2), 541–550 (2016)
87. Walker, R.M.: Uniform Harbourne–Huneke bounds via flat extensions (2016). arXiv:1608.02320
88. Walker, R.M.: Uniform symbolic topologies via multinomial expansions (2017). arXiv:1703.04530
89. Wiles, A.: Modular elliptic curves and Fermat's last theorem. Ann. Math. (2) **141**(3), 443–551 (1995)
90. Yassemi, S.: Weakly associated primes under change of rings. Commun. Algebra **26**(6), 2007–2018 (1998)
91. Zariski, O.: A fundamental lemma from the theory of holomorphic functions on an algebraic variety. Ann. Mat. Pura Appl. **4**(29), 187–198 (1949)

Some Open Problems in Complex Singularities

José Seade

Abstract We discuss some open problems and questions related with five different topics in complex singularities. These are: (i) Topological and holomorphic ranks of an isolated singularity germ and the Zariski-Lipman conjecture; (ii) Graph manifolds and links of surface singularities. (iii) Milnor's fibration for complex singularities and the topology of analytic foliations near an isolated singularity. (iv) Rochlin's signature theorem and Gorenstein surface singularities. (v) The index of a vector field on a singular variety. These are all topics on which I have been interested for a long time.

Keywords Zariski-Lipman conjecture · Graph manifolds · Gorenstein singularities · Milnor fibrations · Foliations · Rochlin signature theorem · Indices of vector fields

In this article I discuss some open problems and questions related with complex singularities. Some of these have arisen from work either by myself or with co-authors, and others are somehow *folklore* problems on which I have been interested for years. The purpose of this note is inspiring others to think on these problems, or possibly on others that may spring along the way.

I have grouped this discussion in five topics:

(i) Topological and holomorphic ranks of an isolated singularity germ. The Zariski-Lipman conjecture.

(ii) Graph manifolds and links of surface singularities.

(iii) Milnor's fibration for complex singularities and the topology of analytic foliations near an isolated singularity.

(iv) Rochlin's signature theorem and Gorenstein surface singularities.

(v) The index of a vector field on a singular variety.

Thanks: Partially supported by CONACYT, Mexico, and PAPIIT-UNAM project IN 101114

J. Seade (✉)
Instituto de Matemáticas, Universidad Nacional Autónoma de México, Mexico City, Mexico
e-mail: jseade@matcuer.unam.mx

R. N. Araújo dos Santos et al. (eds.), *Singularities and Foliations. Geometry, Topology and Applications*, Springer Proceedings in Mathematics & Statistics 222, https://doi.org/10.1007/978-3-319-73639-6_14

Hence this paper has five sections, one for each of these topics. Each section is self-contained and provides an introduction to the subject, together with some bibliography and references. In each of these topics I have stated explicitly some open questions, but I know these are subjects on which there is a lot more to be said, and I am sure that anyone that spends some time thinking on the matter, will come out with her/his own questions, and hopefully also with some good ideas and interesting results.

In Sect. 1 we consider the problem of finding the maximal number of either continuous or holomorphic vector fields on an isolated singularity germ, or on a smoothing of it, which are linearly independent at all points. This naturally springs from the -still open- celebrated Zariski-Lipman conjecture, as well as from an old conjecture stated by A. Durfee [9, 1.6], that I proved long ago, in [36, 37]. This states that the tangent bundle of every smoothing of a normal Gorenstein surface singularity is topologically trivial.

In Sect. 2 we look at the 3-manifolds which are links of isolated complex surface singularities, and discuss properties of the dual graphs of their resolutions. The deep (*folklore*) question here is studying which conditions are imposed on a plumbing graph by the fact that this corresponds to the resolution of a hypersurface singularity, or to an ICIS germ, or to a Gorenstein germ.

Section 3 springs from the study of the topology of linear actions of $\mathbb{C}^n$ in $\mathbb{C}^m$, $m \gg n$, when we look at them "a la Milnor", namely by considering a small sphere centered at a singular point, then look at the intersection of the sphere with the leaves of the corresponding foliation, and see what happens as we make the sphere smaller (see [6, 24]). This all is much indebted to a remarkable work by René Thom [43], full of ideas about using Morse Theory to studying foliated manifolds. Some of those ideas were adapted in [24] to study the topology of holomorphic foliations near an isolated singularity, and this brings us to an elementary open question, that we discuss below, regarding the topology of a holomorphic map-germ $\mathbb{C}^n \to \mathbb{C}$.

Section 4 springs from an old paper by H. Esnault, E. Viehweg and myself [10]. We re-visit a classical theorem in 4-dimensional manifolds theory, namely Rochlin's signature theorem. In the case of compact complex surfaces this theorem is equivalent to stating that the parity of the Todd genus is measured by an invariant in $\mathbb{Z}_2$ which equals the mod (2) index of the Dirac operator associated to a smoothing of the canonical divisor. We discuss an integral lifting of this congruence to an equality over the integers. We briefly discuss too how this is related with Laufer's formula for the Milnor number and the geometric genus of surface singularities.

Section 5 concerns the GSV-index of vector fields on hypersurface germs, a concept that extends the notion of the local Poincaré-Hopf index. The question we rise here is about describing what is, or should be, the correct notion of a generic holomorphic vector field on an isolated hypersurface germ.

1 Topological and Holomorphic Ranks of an Isolated Singularity Germ. The Zariski-Lipman Conjecture

Let M be a complex manifold of dimension n and $X_1, \ldots, X_k$ vector fields on M. We say that they are linearly independent (briefly L.I.) if for all $x \in M$ the vectors $X_1(x), \ldots, X_k(x)$ are linearly independent over $\mathbb{C}$. The topological rank of M, denoted by $\mathrm{Rank}_{\mathrm{top}}(M)$, is the largest number of continuous L.I. vector fields on M. The holomorphic rank $\mathrm{Rank}_{\mathrm{hol}}(M)$ is defined similarly. So, for instance, $\mathrm{Rank}_{\mathrm{hol}}(M) = 1$ if there exists a holomorphic vector field on M with no singularities, but there are not two such vector fields which are L. I. everywhere.

Now consider a normal complex isolated singularity germ $(V, \underline{0})$ of dimension n in some complex space $\mathbb{C}^N$. Set $V^* = V \setminus \{\underline{0}\}$. Define the topological (or holomorphic) rank of V to be the maximal number of continuous (or holomorphic) vector fields on V which are linearly independent at every point in V^*, for some representative of the germ.

We emphazise that when we speak of the topological and holomorphic ranks, we are speaking of linearly independent sections over $\mathbb{C}$. For instance, given an arbitrary isolated normal singularity germ $(V, \underline{0})$ of complex dimension two, the complex manifold V^* is always an orientable manifold of real dimension 4, diffeomorphic to the product $L_V \times \mathbb{R}$, where L_V is the link. Since every oriented 3-manifold has trivial tangent bundle, it follows that as a real vector bundle TV^* is always trivial, isomorphic to $V^* \times \mathbb{R}^4$. However, if the singularity is not numerically Gorenstein (see next section), TV^* is not topologically trivial as a complex bundle. For instance, the singularities one gets by considering a holomorphic line bundle with Chern class < -2 over the Riemann sphere, by blowing down the zero section to a point, are never numerically Gorenstein.

1.1 The Hypersurface and ICIS Cases

Assume V is defined by a holomorphic map-germ $(\mathbb{C}^{n+1}, \underline{0}) \xrightarrow{f} (\mathbb{C}, 0)$ with a unique critical point at $\underline{0}$, with $V = f^{-1}(0)$ and $V^* = V \setminus \{\underline{0}\}$. In this case it is an exercise to show that the complex bundle TV^* is topologically trivial, so $\mathrm{Rank}_{\mathrm{top}}(V) = n$. This follows because the normal bundle is trivialized by the gradient vector field $\overline{\nabla} f$, and this implies that TV^* is trivial by standard obstruction theory arguments (basically because $H^{2n}(V^*, \mathbb{Z}) = 0$). Similarly, and by the same reason, if $F = f^{-1}(t) \cap \mathbb{B}_\varepsilon$ is the local Milnor fibre, one has $\mathrm{Rank}_{\mathrm{top}}(F) = n$.

The same statements regarding the topological rank hold for ICIS germs.

Now consider the holomorphic rank. One has that the space of germs of holomorphic vector fields in $\mathbb{C}^{n+1}$ with an isolated singularity at $\underline{0}$ and tangent to V is infinite dimensional. Surprisingly one has:

Theorem 1.1 (A. Lins Neto [25]) *Let V be a germ at* $\underline{0} \in \mathbb{C}^{n+1}$ *of a hypersurface with an isolated singularity at* $\underline{0}$*. Then* $1 \leq \mathrm{Rank_{hol}}(V) \leq 2$*. Moreover, if V is quasi-homogeneous or if* $n \leq 3$*, then* $\mathrm{Rank_{hol}}(V) = 1$.

That the holomorphic rank is always ≥ 1 actually holds in a more general setting (see [4] or Theorem 5.1 below). The new part here is that this is at most 2.

Notice that this theorem gives a proof of the Zariski-Lipman conjecture for hypersurface and quasi-homogeneous singularities. This conjecture, which springs from [26], is one of the main open problems in singularity theory. It asserts that a complex variety with locally free tangent sheaf is necessarily non-singular. This was proved by G. Scheja and U. Storch in [35] for hypersurfaces. In [17] it was proved for quasi-homogeneous singularities. This conjecture has also been settled in various other cases, see for instance [3, 17, 19].

Question 1.2 What about the holomorphic rank for the Milnor fibre F?

It follows easily from [12] (see also [5, Chapter III]) that $\mathrm{Rank_{hol}}(F) \geq 1$. As far as I know, nothing else is known about this question.

As noticed in [25], it is natural to ask:

Question 1.3 How does Theorem (1.1) generalize to ICIS germs?

1.2 On Gorenstein Singularities

A normal singularity germ $(V, \underline{0})$ is Gorenstein if its local ring is Cohen-Macauley and its dualizing sheaf is free at $\underline{0}$.

When V has complex dimension 2, this all amounts to saying that the canonical bundle $\mathcal{K}_{V^*}$ of $V^* = V \setminus \{\underline{0}\}$ is holomorphically trivial, *i.e.*, there exists a nowhere vanishing holomorphic 2-form on V^*. The singularity is numerically Gorenstein if the complex bundle $\mathcal{K}_{V^*}$ is topological trivial.

An n-dimensional isolated singularity germ $(V, \underline{0})$ is smoothable if there exists a complex analytic germ (Z, p) of dimension $n+1$ and a flat morphism $\mathcal{F} : (Z, p) \to (\mathbb{C}, 0)$ such that $\mathcal{F}^{-1}(0)$ is $(V, \underline{0})$ and $\mathcal{F}^{-1}(t)$ is non-singular for all $t \neq 0$ sufficiently close to 0. In this case, given a small ball $\mathbb{B}_\varepsilon$ in the ambient space whose boundary $\mathbb{S}_\varepsilon$ is a Milnor sphere for both Z and V, one has Milnor type fibration, with fibers $F_t := \mathcal{F}^{-1}(t) \cap \mathbb{B}_\varepsilon$ with $\varepsilon \gg |t| > 0$. The manifold F_t is called a smoothing of $(V, \underline{0})$.

The following theorem was conjectured by A. Durfee in [9] and proved in [36, 37].

Theorem 1.4 *Let* $(V, \underline{0})$ *be a smoothable normal Gorenstein surface singularity and let F be a smoothing. Then the complex bundle TF is topologically trivial and therefore* $\mathrm{Rank_{top}}(F) = 2$.

This rises several questions. Some of them are:

Question 1.5 If $(V, \underline{0})$ is a smoothable normal numerically Gorenstein surface singularity and F is a smoothing, is the complex bundle *TF* topologically trivial?

Question 1.6 What about the topological and holomorphic ranks for smoothings of normal isolated Gorenstein singularities in higher dimensions?

2 Graph Manifolds and Links of Surface Singularities

Let $(V, \underline{0})$ be a normal surface singularity germ, and consider a good resolution $\pi : \widetilde{V} \to V$. That is,

(i) $\widetilde{V}$ is a non-singular surface and π is a proper analytic map.

(ii) $E := \pi^{-1}(\underline{0})$ is a divisor and $\pi : \widetilde{V} \setminus E \longrightarrow V \setminus \{\underline{0}\}$ is a biholomorphism.

(iii) The irreducible components $E_1, \ldots, E_r$ of E are compact Riemann surfaces that meet transversally and every three of them have empty intersection.

The complex manifold $\widetilde{V}$ has E as a deformation retract.

For a good resolution $\widetilde{V}$ we associate a dual graph $\Gamma = \Gamma(\widetilde{V})$ in the classical way (see for instance [39, Chapter IV]: We assign a vertex to each irreducible componente E_i; two vertices v_i, v_j are joined by as many edges as points in which they meet. And each vertex is decorated with two integers (w_i, g_i). The first is the self-intersection number E_i^2, which equals the Chern class of the normal bundle of E_i, and g_i is the genus.

It is well-known that the resolution can be constructed by plumbing according to the graph: a construction introduced by Milnor in the 1950s for producing exotic spheres. It is an exercise to see that the topology of the resolution is fully described by the graph (Fig. 1).

It follows that the link $M := V \cap \mathbb{S}_\varepsilon$ is by definition a graph manifold, since it can be regarded as the boundary of the resolution, which can be constructed by plumbing.

One side of the following important theorem is due to Mumford [30], and also Du Val. The other side follows easily from the work of Grauert [15]:

Theorem 2.1 *The link of every normal surface singularity is a graph manifold with negative definite intersection matrix. And conversely, every graph 3-manifold with negative definite intersection matrix is orientation preserving homeomorphic to the link of a normal surface singularity.*

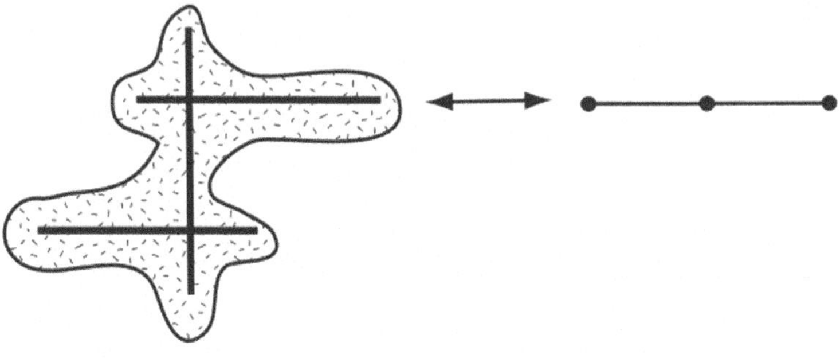

Fig. 1 Dual graph of a resolution

There are several natural questions, all of which have been studied by various authors. We now say a few words about some of these.

As said before, a normal surface singularity germ $(V, \underline{0})$ is Gorenstein if away from $\underline{0}$ its canonical bundle $\mathcal{K} := \wedge^2 T^*(V \setminus \underline{0})$ is holomorphically trivial. The singularity germ is numerically Gorenstein (briefly n-Gorenstein) if the complex bundle $\mathcal{K}$ is topologically trivial.

One has natural inclusions:

$$\{\text{hypersurface germs}\} \subset \{\text{ICIS}\} \subset \{\text{Gorenstein germs}\} \subset \{\text{n-Gorenstein germs}\,.\}$$

That is, every hypersuface singularity is a complete intersection (an ICIS for short), every ICIS is Gorenstein and these are obviously n-Gorenstein.

Being Gorenstein is analytic, while being n-Gorenstein is a topological condition, fully determined by the graph. To explain this we recall that a good resolution has a canonical class K; by definition K is the unique homology class in $H_2(\widetilde{V}; \mathbb{Q})$ such that

$$\mathcal{A} \cdot K = (2g_1 - 2, \cdots, 2g_r - 2) \in \mathbb{Q}^r\,,$$

where $\mathcal{A}$ is the isomorphism in $H_2(\widetilde{V}; \mathbb{Q})$ determined by the intersection matrix $A := ((E_i \cdot E_j))$. That is,

$$K = \mathcal{A}^{-1}(2g_1 - 2, \cdots, 2g_r - 2) \in \mathbb{Q}^r\,.$$

Then n-Gorenstein is equivalent to asking K to be integral (see [9, 20]). This latter condition depends only on the matrix A and the genera g_i attached to the vertices. Hence we can speak of n-Gorenstein graphs.

Yet, a remarkable theorem from [33] says that every n-Gorenstein graph is the dual graph of a resolution of some Gorenstein singularity. So the question is:

Question 2.2 Which weighted graphs correspond to n-Gorenstein singularities?

As far as I know, a full characterization of such graphs is not known. Yet, below are a couple of results in that direction.

Theorem 2.3 (Larrión, Seade [20]) *Given a finite graph Γ with weights w_i, there are infinitely many choices of genera g_i for the vertices, such that the corresponding singularity is n-Gorenstein.*

Theorem 2.4 (Popescu-Pampu, Seade [32]) *Given a finite graph Γ with genera g_i, if Γ is not a loop, then there are finitely many weights that make it n-Gorenstein.*

If the graph Γ is a loop with genera for the vertices, the equivalent statement is more complicated, but it can be decided under which conditions there are finitely many weights that make it n-Gorenstein (see [32]).

Example 2.5 Assume Γ has only one vertex with weight $w < 0$ and genus $g \geq 0$; so E consists of a single irreducible component. Then:

$$K = \left(\frac{2g-2}{w} - 1\right) \cdot E\,.$$

Hence the singularity is numerically Gorenstein if and only if $(2g - 2)$ is a multiple of the weight w. Thus:

(a) If we fix the weight $w < 0$, for all genera $g \geq 0$ of the form $g = tw + 1$ with $t \in \mathbb{Z}$, the graph is n-Gorenstein.

(b) On the other hand, if we fix genus $g \neq 1$, then only finitely many possibilities of weights w for which graph is n-Gorenstein.

For instance, if $g = 0$ only $w = -1, -2$ make it n-Gorenstein. But if $g = 1$ then $K = -E$ independently of w, so these are all n-Gorenstein.

The following theorem partially summarizes the previous discussion.

Theorem 2.6 *Let Γ be a finite graph with no loops and n vertices. Then for almost all sets of negative weights $w = (w_1, \ldots, w_n)$ for the vertices, the corresponding intersection matrix A is negative definite, and for each such vector of weights, there are infinitely many vectors of genera $g = (g_1, \ldots, g_n)$, $g_i \geq 0$, such that the corresponding plumbing graph (Γ, w, g) is the dual graph of a resolution of some Gorenstein singularity.*

Here is an elementary question taken from [32]:

Question 2.7 Given an arbitrary finite graph Γ with a genus $g_i \geq 0$ attached to each vertex, is the set of weights that make the graph minimal and n-Gorenstein always non-empty?

Minimal here means that there is no vertex with genus 0 and weight -1.

To conclude this section, here is a rather subtle question:

Question 2.8 Which weighted graphs appear as graphs of a hypersurface singularity in $\mathbb{C}^3$?

This is very interesting open problem.

Similarly, one may ask:

Question 2.9 What conditions on the graph are imposed by the fact that there is an ICIS with such a graph?

This is (also an open problem) studied in [7], where the authors classify the weighted graphs such that all singularities that have this as dual graph of some resolution are ICIS.

3 Milnor's Fibration for Complex Singularities and the Topology of Holomorphic Foliations

It is well known that if $f : (\mathbb{C}^{n+1}, \underline{0}) \to (\mathbb{C}, 0)$ is a holomorphic map-germ with an isolated singularity at the origin, then the non-critical levels $V_t := f^{-1}(t)$, $t \neq 0$,

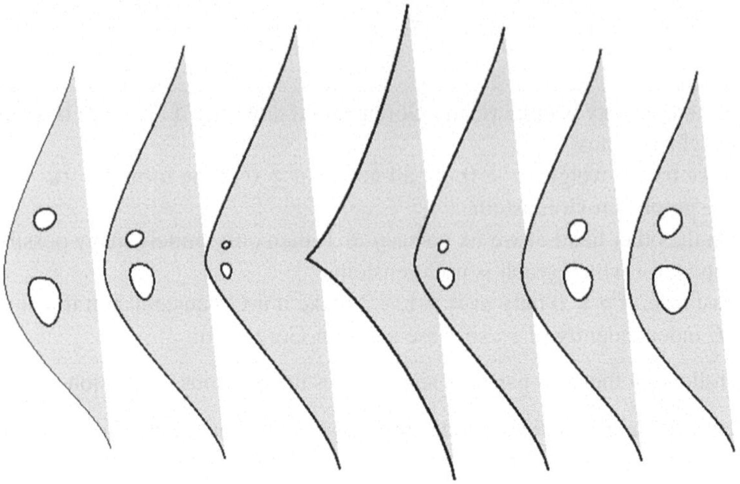

Fig. 2 The Milnor fibration

determine a locally trivial fibration in a tubular neighborhood of the special fibre $V := f^{-1}(0)$. This is the celebrated Milnor fibration (Fig. 2).

More precisely, we know that there exists $\varepsilon > 0$ small enough so that for every $\varepsilon' > 0$ which is $\leq \varepsilon$ we have that the sphere $\mathbb{S}_{\varepsilon'}$ of radius ε' and center $\underline{0}$ meets V transversally. Any such a sphere is called a Milnor sphere for the germ and the intersection $L_V := V \cap \mathbb{S}_\varepsilon$ is the link (see [29]). Now fix such an ε. Then there exists $\delta > 0$ sufficiently small such that every non-critical level V_t with $|t| \leq \delta$ meets the sphere $\mathbb{S}_\varepsilon$ transversally. [In the case we envisage here, where f has an isolated critical point, this follows from the implicit function theorem. The statement still is true for general f, but the proof is much harder and follows from a deep theorem of Hironaka, proving that every $\mathbb{C}$-valued holomorphic map-germ on a complex analytic space, has the Thom a_f-property.]

By definition, a Milnor tube for f is a set of the form

$$N(\varepsilon, \delta) := f^{-1}(\mathbb{D}_\delta \setminus \{0\}) \cap \mathbb{B}_\varepsilon\,,$$

for ε and δ as above, where $\mathbb{D}_\delta$ is the disc in $\mathbb{C}$ of all points with $|t| \leq \delta$. Then Milnor's fibration theorem is essentially equivalent to saying that:

$$f : N(\varepsilon, \delta) \longrightarrow \mathbb{D}_\delta \setminus \{0\}\,,$$

is a fibre bundle.

Notice that on $f^{-1}(\mathbb{D}_\delta) \cap \mathbb{S}_\varepsilon$ we have a codimension 2, real analytic foliation; indeed a locally trivial fibration with fibre the link L_V (this follows easily either from Ehresmann's fibration lemma or from the Thom-Mather isotopy theorem).

If we fix the sphere $\mathbb{S}_\varepsilon$ but allow continuously larger values of $|t|$, a time comes when the non-critical level $f^{-1}(t)$ has points of non-transversality with the sphere $\mathbb{S}_\varepsilon$, points where the tangent space to $f^{-1}(t)$ actually is contained in the space tangent to $\mathbb{S}_\varepsilon$. We thus get, by taking all fibers $f^{-1}(t)$ and their intersections with the sphere $\mathbb{S}_\varepsilon$, a codimension 2 real analytic foliation $\mathcal{F}_\varepsilon$ on the sphere, with singular set Σ_ε being the points where the fibers of f are tangent to the sphere, and the leaves close to L_V being all ambient isotopic to the link.

Question 3.1 Does the topology of this foliation depend on the choice of ε for sufficiently small spheres. Or more precisely, can we choose an appropriate metric in $\mathbb{C}^{n+1}$ such that for all sufficiently small spheres (for the metric), the corresponding foliations are topologically equivalent? (*i.e.*, there exists a homeomorphism taking one sphere into the other, that preserves the singular sets and carries leaves into leaves).

I believe that the answer is positive, though I have not succeeded in proving it. This is a particular case of a problem studied in [24] in larger generality, with a positive answer in some particular cases that do not occur in the setting describe above. Let us say a few words about the general setting.

In 1964 René Thom wrote a beautiful article [43] explaining how the classical ideas of Morse theory can be adapted to studying the topology of non-singular foliations on smooth manifolds. In [24] the authors adapted some of those ideas to studying the topology of holomorphic foliations near an isolated singular point. The main ideas are as follows.

Consider a holomorphic foliation $\mathcal{F}$ of dimension $d \geq 1$ defined on an open neighbourhood $\mathcal{U}$ of the origin $\underline{0} \in \mathbb{C}^n$, with a unique singular point at $\underline{0}$. To study its topology, we consider a function g on $\mathcal{U}$ with a Morse critical point at $\underline{0}$ of index 0, so that its non-critical levels are diffeomorphic to spheres; call these simply "spheres". Look at the restriction of g to the leaves of $\mathcal{F}$. The critical points of $g|_{\mathcal{L}}$ are the points of contact of $\mathcal{F}$ with the fibres of g, where the leaves of $\mathcal{F}$ are tangent to the fibres of g. The set of all such points is the *polar variety* of $\mathcal{F}$ relative to g, in analogy with this classical notion in algebraic geometry, where the role of g is usually played by linear forms.

Inspired by the way how J. Milnor studied the topology of complex singularities in [29], the authors study the topology of $\mathcal{F}$ near $\underline{0}$ by looking at the intersection of the leaves of $\mathcal{F}$ with a sufficiently small sphere $\mathbb{S}_\varepsilon$, and the way how these intersections change as we make the radius of the sphere tend to 0. In fact this point of view for studying holomorphic foliations already appears in [1, 6, 14].

Of course the contacts of the leaves of $\mathcal{F}$ with the fibres of g can be degenerate. If the contacts are all non-degenerate, then we say that $\mathcal{F}$ *carries a Morse structure compatible with* g, or simply that M carries the Morse structure of g, for short.

Theorem 1 in [24] is the analogous in our setting of a classical result for polar varieties (see for instance [22]) saying that the restriction of g to the leaves of $\mathcal{F}$ is a Morse function on each leaf if and only if away from $\underline{0}$, the polar variety $M^* = M \setminus \{\underline{0}\}$ is a smooth, reduced, submanifold of $\mathbb{C}^n$ of codimension $2d$ and M^* is everywhere transversal to $\mathcal{F}$.

The topology of holomorphic foliations near a singular point can be rather complicated. In order to have a certain understanding of the behaviour of $\mathcal{F}$ near $\underline{0}$ we ask for the condition that the polar set $M = M(\mathcal{F}, g)$ be real analytic at $\underline{0}$. This happens in many interesting families, including all foliations of dimension or codimension one. In this case we say that the foliation is *contact-analytic* with respect to the Morse function g. The Theorem 2 in [24] says that if $\mathcal{F}$ is contact-analytic and carries the Morse structure of g, then the corresponding polar variety has finitely many irreducible components, say $M_1, \ldots, M_r$, all of them pairwise disjoint away from $\underline{0}$; each of these is smooth away from $\underline{0}$, of real codimension $2d$, transversal to the foliation and consists of points where the contacts have all the same Morse index.

Using this we can give the following topological picture of $\mathcal{F}$ near $\underline{0}$: we equip $\mathcal{F}$ with *the gradient flow on the leaves*, *i.e.*, the flow $\mathcal{G}_{\mathcal{F}}$ of the vector field on $\mathcal{U}$ obtained by projecting at each point the gradient of g to the tangent space of $\mathcal{F}$. Let $\mathbb{B}_\varepsilon$ be a sufficiently small g-ball centred at $\underline{0}$. This ball splits in two disjoint $\mathcal{G}_{\mathcal{F}}$-invariant sets: the saturated $\widehat{M}$ of $M \cap \mathbb{B}_\varepsilon$ by $\mathcal{F}$ and its complement $K := \mathbb{B}_\varepsilon \setminus \widehat{M}$. On K the topology is somehow simple and the dynamics can be rich, while on $\widehat{M}$ the topology is rich and the dynamics is often simpler.

Each leaf $\mathcal{L}$ in K is homeomorphic to a product $(\mathcal{L} \cap \mathbb{S}_\varepsilon) \times \mathbb{R}$, immersed in $\mathbb{C}^n$ so that it is transversal to each g-sphere around $\underline{0}$ and the α-limit of each orbit in K of the gradient flow $\mathcal{G}_{\mathcal{F}}$ is the origin $\underline{0}$. Thus, for each positive number $\varepsilon' < \varepsilon$, the leaves of $\mathcal{F}$ in K meet the g-sphere $\mathbb{S}_{\varepsilon'}$ transversally and define a real analytic foliation on it, which can have very rich dynamics. For instance [6], if $\mathcal{F}$ is defined by a linear vector field in the Poincaré domain with generic eigenvalues, then the foliation on $\mathbb{S}_{\varepsilon'}$ actually is defined by a flow which is Morse-Smale.

The picture on $\widehat{M}$ is rather different: the *α-limit* of $\mathcal{G}_{\mathcal{F}}$ of each leaf in $\widehat{M}$ is the set of points where the corresponding leaf meets the polar variety M. The intersection of the leaves of $\mathcal{F}$ with M is always transverse, and the topology of every leaf $\mathcal{L} \subset \widehat{M}$ is determined by its intersection with the boundary sphere, $\mathcal{L} \cap \mathbb{S}_\varepsilon$, and the points where $\mathcal{L}$ meets M: each such intersection point comes with a Morse index, that tells us what type of handle we must attach to the leaf when passing through that point. In particular, if $\mathcal{L}$ is compact, then its Euler-Poincaré characteristic $\chi(\mathcal{L})$ equals the number of intersection points in $\mathcal{L} \cap M$, counted with sign. The sign is negative when the corresponding Morse index is odd, and positive otherwise.

For instance, for generic linear actions of $\mathbb{C}^m$ in $\mathbb{C}^n$ in the Siegel domain (see [6, 24, 28]), $m << n$, each Siegel leaf $\mathcal{L}$ is a copy of $\mathbb{C}^m$ embedded in $\mathbb{C}^n$ with a unique point of minimal distance to $\underline{0}$. The gradient flow of the quadratic form $(z_1, \cdots, z_n) \mapsto |z_1|^2 + \cdots + |z_n|^2$ describes each Siegel leaf as a cylinder over its intersection with a sphere, $\mathcal{L} \cap \mathbb{S}^{2n-1} \cong \mathbb{S}^{2m-1}$, to which we attach a handle of Morse index 0, *i.e.*, a $2m$-disc.

We remark that the topology of the polar varieties that arise in this way can be rather interesting. In fact these are all manifolds with a canonical complex structure determined by a foliated atlas for $\mathcal{F}$, and with a rich geometry. For instance, when $\mathcal{F}$ is the foliation of a linear flow in the Siegel domain with generic eigenvalues and g is the usual metric, the manifold $M_\varepsilon = M \cap \mathbb{S}_\varepsilon$ is the space of Siegel leaves of the flow and it has very interesting topology [6, 27]. This type of manifolds has been studied by several authors giving rise to the theory of *LVM manifolds*, see for instance [39, Chapter VI] or [18].

Remark 3.2 (Polar weighted and mixed singularities) As noted in [39, VI.3] the above discussion can be extended to germs of holomorphic vector fields in general. The study of the corresponding polar varieties gives rise to interesting real analytic singularities. Let us say a few words about this (see [39, Chapter VII]). Consider for instance the non-linear vector fields in $\mathbb{C}^n$ of the form $F(z) = (\lambda_1 z_{\sigma(1)}^{a_1}, \cdots, \lambda_n z_{\sigma(n)}^{a_n})$, where the λ_i are non-zero complex numbers, $a_i \geq 2$, and σ is a permutation of the set $\{1, \cdots, n\}$. One has a real analytic map, $\psi_F : \mathbb{C}^n \to \mathbb{C}$, defined by $\psi_F(z) = \langle F(z), z\rangle := \sum_{i=1}^n \lambda_i z_{\sigma(i)}^{a_i} \cdot \overline{z}_i$. The zero-set of ψ_F is the variety of contacts of the solutions of F with the spheres around the origin. It is proved in [34, 39] that all these singularities admit a "good action"of the group $\mathbb{S}^1 \times \mathbb{C}^*$, and that having this action implies that these singularities admit a local Milnor fibration. These were called in [39] *Twisted Pham-Brieskorn singularities* because when the permutation σ is the identity, these singularities are equivalent to the classical Pham–Brieskorn singularities. From these considerations spring the concepts of *polar actions* and *polar weighted singularities* introduced by Cisneros-Molina in [8], as well as the concept of *mixed singularities* developed by Oka in various articles, see for instance [31].

As an example, consider the vector field in $\mathbb{C}^2$ defined by

$$F(z_1, z_2) = (q z_2^{q-1}, p z_1^{p-1}),$$

for some integers $p, q > 2$. Its integral lines are the level curves of the map $(z_1, z_2) \overset{f}{\mapsto} z_1^p - z_2^q$. It is easy to see that in these examples the square of the distance function to $\underline{0}$ is a Morse function restricted to each leaf. There is one separatrix, $V := \{z_1^p = z_2^q\}$, which is transversal to all the spheres around 0, and all other fibres meet the z_1-axis at p points and they have q points on the z_2-axis. A straightforward computation shows that these are all points with Morse index 0. There is a third component of the polar variety M, which is a real analytic surface with an isolated singularity at 0:

$$M^- := \{(z_1, z_2) \in \mathbb{C}^2 \,|\, q z_2^{q-1} \overline{z}_1 = -p z_1^{p-1} \overline{z}_2 \,;\, z_1, z_2 \neq 0\}.$$

Each leaf has pq saddles in this surface. As a matter of fact, one can easily check that M^- is ambient homeomorphic to the separatrix V. That is, there is a homeomorphism of $\mathbb{C}^2$ carrying M^- into V.

Question 3.3 Is this a coincidence or a special case of a general theorem?

I do not know the answer.

Coming back to the general setting, notice that if $\mathbb{S}_\varepsilon$ is a small g-sphere in $\mathcal{U}$ around $\underline{0}$, then the intersection of the leaves of $\mathcal{F}$ with $\mathbb{S}_\varepsilon$ defines a real analytic foliation $\mathcal{F}_\varepsilon$ on the sphere, which is singular at $M_\varepsilon := \mathbb{S}_\varepsilon \cap M$, where M is the polar variety. Away from M_ε the leaves of $\mathcal{F}_\varepsilon$ have real dimension $2d - 1$. We may then address the problem of studying how the topology of the foliation $\mathcal{F}_\varepsilon$ varies as $\varepsilon \to 0$. For instance if $M = \{\underline{0}\}$, as it happens for linear vector fields in the Poincaré domain, then the topology of $\mathcal{F}_\varepsilon$ is independent of ε. This happens also for generic flows in the Siegel domain, and more generally for generic linear $\mathbb{C}^m$-actions on $\mathbb{C}^n$ in the Siegel domain. In [14] it was proved that if $\mathcal{F}$ has dimension 1 and all the points in M have Morse index 0, *i.e.*, they correspond to local minimal points in their leaves, then the topology of $\mathcal{F}_\varepsilon$ is independent of ε. In [24, Theorem 3] it is proved that the same statement is true for foliations of arbitrary dimension.

Alas the previous results do not ever apply to the setting on which we started this section, where the foliation is given by the fibers of a holomorphic map-germ $f : (\mathbb{C}^{n+1}, \underline{0}) \to (\mathbb{C}, 0)$ with an isolated singularity at the origin, because the fibers always have points of contact with the spheres which are either degenerate or local saddles.

We may close this section with another question which I am yet unable to answer:

Question 3.4 Given a map-germ $f : (\mathbb{C}^{n+1}, \underline{0}) \to (\mathbb{C}, 0)$ as above, is it always possible to find a Hermitian metric in a neighborhood of $\underline{0}$ such that the contacts of the fibers with the spheres are all non-degenerate?

4 Rochlin's Signature Theorem and Gorenstein Surface Singularities

Recall that if X is a closed oriented 4-manifold, the cup product determines a non-degenerate bilinear form:

$$H^2(X; \mathbb{R}) \cup H^2(X; \mathbb{R}) \longrightarrow H^4(X; \mathbb{R}) \cong \mathbb{R}\,.$$

By definition the signature of X, denoted $\sigma(X) \in \mathbb{Z}$, is the signature of this quadratic form, *i.e.*, the number of positive eigenvalues minus the number of negative ones.

For the sequel we need to speak of *Spin* and $Spin^c$ structures.

The spin group $Spin(n)$ is the non-trivial double cover of the special orthogonal group $SO(n) = SO(n, R)$, such that there exists a short exact sequence of Lie groups

$$1 \to \mathbb{Z}_2 \to Spin(n) \to SO(n) \to 1\,.$$

For $n > 2$, $Spin(n)$ is simply connected and so coincides with the universal cover of $SO(n)$. Hence $Spin(3)$ is $SU(2) \cong \mathbb{S}^3 \cong Sp(1)$ while $Spin(4)$ is $SU(2) \times SU(2)$.

The 4-manifold X admits a *Spin* structure if the classifying map of its tangent bundle, $X \to B(SO(4))$, has a lifting to $B(Spin(4))$. It is an exercise to see that such a lifting exists if and only if the 2nd Stiefel-Whitney class $\omega_2(X)$ vanishes, and if this happens, then a spin structure on X means the homotopy of one such lifting. The spin structures on X are classified by $H^1(X; \mathbb{Z}_2)$.

If X is a complex surface, then its anti-canonical class $-K_X$ is dual to Chern class $c_1(M)$. Since for complex bundles the reduction modulo 2 of the 1st Chern class is the 2nd Stiefel-Whitney class, it follows that a complex manifold is spin if and only if its canonical class is even.

There is also a "complex spin group" $Spin^c(n)$. This is defined by the exact sequence:

$$1 \to Z_2 \to Spin^c(n) \to SO(n) \times U(1) \to 1\,,$$

which has important applications to low dimensional manifolds, particularly for the Seiberg-Witten invariants.

As before, an oriented manifold X admits a $Spin^c$ structure if the classifying map of its tangent bundle, $X \to B(SO(4))$, has a lifting to $B(Spin^c(4))$. Such a lifting exists if and only if there exists an integral homology class in $H_2(X;\mathbb{Z})$ whose reduction modulo 2 is $\omega_2(X)$, the 2nd Stiefel-Whitney class. The homotopy classes of such liftings are the different $Spin^c$ structures on X. These are classified by the homology classes in $H_2(X;\mathbb{Z})$ whose reduction modulo 2 is dual to $\omega_2(X)$.

It follows that every $Spin$ manifold is canonically $Spin^c$, and so is every complex manifold with a canonical $Spin^c$ structure determined by its canonical bundle.

Definition 4.1 (*Rochlin, 1970s*) Let W be an oriented 2-submanifold of a closed oriented manifold X^4. Then W is a characteristic submanifold if $[W] \in H_2(X;\mathbb{Z})$ reduced modulo 2 is the dual of $\omega_2(X)$.

If X is $Spin$ then the empty manifold $\emptyset$ is characteristic. Notice that if X is a complex surface, then K_M, being the divisor of a holomorphic bundle, can always be smoothed C^∞ by considering an approximation defined by a differentiable section of the canonical bundle, which is transversal to the zero-section. Then the smoothing is a characteristic submanifold.

The classical Rochlin's signature theorem states that the signature of a closed oriented 4-manifold X is always divisible by 16. That theorem was later improved by Kervaire and Milnor, proving that if there is a 2-sphere S in X which is characteristic, then one has:

$$\sigma(X) - S^2 \equiv 0 \mod (16)\,,$$

where S^2 is the self-intersection number, which essentially equals the Euler class of its normal bundle. This theorem was later strengthen by Rohlin himself in the following way:

Theorem 4.2 *Let W be a characteristic submanifold of X, then*

$$\sigma(M) - W^2 \equiv 8\mathrm{Arf}\, W \mod (16)$$

where Arf $W \in \{0,1\}$ *is an Arf invariant associated to a quadratic form on* $H_1(W;\mathbb{Z}_2)$.

Recall that the non-degenerate quadratic forms on finite dimensional vector spaces over $\mathbb{Z}_2$ are classified by their Arf invariant and the dimension of the vector space.

The Arf invariant of such a form is defined to be 0 if and only if the form carries more elements to 0 than to 1.

There is an elegant self-contained proof of Theorem 4.2 in Freedman and Kirby's paper [11]. On the other hand, the Arf invariant in Theorem 4.2 can be interpreted in the following way. One knows (see for instance in [23]) that if W is characteristic in X, then W is naturally equipped with a spin structure. A $Spin$-manifold has an associated Spin bundle and a corresponding Dirac Operator. It was noticed in [2] that in the case of Riemann surfaces equipped with a $Spin$ structure, one has a mod (2) index of the Dirac operator, with several interesting properties. In particular it is a $Spin$-cobordism invariant. That is, two Riemann surfaces equipped with $Spin$ structures represent the same element in the $Spin$ cobordism group Ω_2^{Spin} if and only if the mod two index of the corresponding Dirac operators coincides. Furthermore, by [23] this invariant coincides with the Arf invariant in Theorem 4.2. Hence:

$$\text{Arf } W = 0 \quad \Leftrightarrow W \quad \text{is a spin boundary}$$

In the case when the manifold X in Theorem 4.2 is a complex surface, Rochlin's signature theorem has a nice re-interpretation. To explain this, recall first that Hirzebruch signature theorem implies:

$$\sigma(M) = \frac{1}{3}\, p_1(M)[M]\,,$$

where p_1 is the Pontryagin class. For compact (almost) complex surfaces one has

$$p_1 = c_1^2 - 2c_2 \quad , \quad c_1(M) = -K_M \quad \text{and} \quad c_2(M)[M] = \chi(M).$$

Recalling that the 2nd Todd polynomial is $\frac{1}{12}(c_1^2 + c_2)$. We may re-state Rochlin's theorem for (almost) complex manifolds as:

$$\sigma(M) - K_M^2 = -8Td(M)[M]\,.$$

Thus, if $\widetilde{K}$ is a C^∞ smoothing of the canonical divisor K, then Rohlin's theorem can be re-stated as:

$$Td(M)[M] \equiv \text{Arf } \widetilde{K} \quad (24)\,.$$

Furthermore, by Hirzebruch-Riemann-Roch's theorem, the Todd genus equals the analytic Euler characteristic:

$$Td(M)[M] = \chi(M, \mathcal{O}_M)\,.$$

So we get the following reformulation of Rohlin's theorem:

Theorem 4.3 *For a compact complex surface M, the parity of its analytic Euler characteristic is that of the Arf invariant* Arf $\widetilde{K}_M$.

We want a similar expression in algebraic geometry, not with a topological smoothing of K_M but with the actual canonical divisor.

Everything I say from now on is contained in [10].

Definition 4.4 A characteristic divisor W of M is an effective divisor of a bundle $\mathcal{L}$ of the form $\mathcal{L} = \mathcal{K}_M \otimes \mathcal{D}^{-2}$.

Notice that such $W = K_M - 2D$ represents a homology class whose reduction modulo 2 coincides with that of K

Definition 4.5 Let W be a characteristic divisor of M. Define its mod (2)-index by:

$$\mathfrak{h}(W) = \dim\, H^0(W, \mathcal{D}|_W) \mod 2\,.$$

If W is non-singular, this is the mod (2) index of the Dirac operator considered in [2, 23]. Hence in that case this index coincides with the Arf invariant in Rochlin's theorem.

Notice that for the anti-canonical class $-K = -K_M$ one has $\mathcal{D} = \mathcal{K}_M$:

$$\mathfrak{h}(-K) = \dim\, H^0(-K, \mathcal{K}_M|_K) \mod 2\,.$$

We have:

Theorem 4.6 (Esnault-Seade-Viehweg) *The parity of the analytic Euler characteristic coincides with the mod (2) index* $\mathfrak{h}(-K)$*:*

$$\mathfrak{h}(-K) = \chi(M, \mathcal{O}_M) \mod (2)\ .$$

More generally, let $W = K_M - 2D$ *be a characteristic divisor, where* D *is a divisor of some holomorphic bundle* $\mathcal{D}$*. Then:*

$$\dim\, H^0(W, \mathcal{D}|_W) \equiv \chi(M, \mathcal{D}) \mod 2\,,$$

where $\chi(M, \mathcal{D}) = \sum_{i=0}^{2}(-1)^i h^i(M, \mathcal{D})$ *is the analytic Euler characteristic of* M *with coefficients in* $\mathcal{D}$*.*

It is natural to ask:

Question 4.7 Is this congruence modulo (2) the reduction of some equality over the integers?

This would provide an integral lifting of Rochlin's theorem in the case of complex manifolds. An answer is provided by the next theorem from [40], but it is not yet satisfactory:

Theorem 4.8 *Let* M *and* $W = K_M - 2D$ *be as above. If* $W \neq 0$*, then:*

$$\chi(M, \mathcal{D}) = h^0(W; \mathcal{D}|_W) - R\,,$$

with R an even integer associated to the divisor W:

$$R = h^1(M; \mathcal{D}) - 2h^2(M; \mathcal{D}) + \dim \operatorname{Ker}(\hat{\beta})\,,$$

where $\hat{\beta}$ is a skew symmetric bilinear form on $H^1(M; D)$.

The problem now is understanding R, perhaps relating it with more recent invariants of low dimensional manifolds. Of course this can be regarded from the viewpoint of the Atiyah-Singer index theorem.

This all springs from [10] where the authors use it to study the case of Gorenstein singularities, defined above in Sect. 2. We refer to [40] for details on what follows.

Consider again a hypersurface germ $(V, \underline{0})$ defined by a map-germ $f : (\mathbb{C}^3, \underline{0}) \to (\mathbb{C}, 0)$ with an isolated critical point at $\underline{0}$.

Laufer in [21] proved an intriguing formula that expresses the Milnor number $\mu(V)$ in terms of invariants associated to a resolution of $(V, \underline{0})$:

Theorem 4.9 *Assume the germ $(V, \underline{0})$ is a hypersurface germ. Then*

$$\mu(V) + 1 = \chi(\widetilde{V}) + K^2 + 12\rho_g(V)\,,$$

where $\chi(\widetilde{V})$ is the Euler characteristic of a resolution, K^2 is the self-intersection number of the canonical class (defined in Sect. 2), and $\rho_g(V)$ is the geometric genus.

In fact the same statement, with essentially the same proof, holds for all ICIS germs.

It is natural to ask how this formula extends to more general singularities. Observe that the left hand side in Laufer's formula has no *a priori* meaning when the singularity is not an ICIS. Yet, the right hand side is an integer defined always for all normal numerically Gorenstein surface singularities, and it is an invariant of $(V, \underline{0})$, independent of the choice of resolution.

Definition 4.10 Let $(V, \underline{0})$ be a numerically Gorenstein normal surface singularity germ. We call:

$$\operatorname{La}(V, \underline{0}) := \chi(\widetilde{V}) + K^2 + 12\rho_g(V)$$

the Laufer invariant of $(V, \underline{0})$.

In [16] Greuel and Steenbrink proved that if $(V, \underline{0})$ is a normal Gorenstein surface singularity which is further smoothable, then the first Betti number of every smoothing vanishes, while the second Betti number $b_2(F)$ is independent of the choice of smoothing F. Hence one has in this setting a well-defined notion of the Milnor number $\mu_{GS}(V)$. Laufer's formula was generalized to this setting by Steenbrink [42]:

Theorem 4.11 *Let $(V, \underline{0})$ be a smoothable Gorenstein normal surface singularity germ. Then*

$$\mu_{GS}(V) + 1 = \chi(\widetilde{V}) + K^2 + 12\rho_g(V)\,.$$

So a natural question is:

Question 4.12 How does this formula generalize to numerically Gorenstein singularities?

The difference between Gorenstein and numerically Gorenstein singularities is that in the first case the canonical bundle is holomorphically trivial, while in the second case this bundle is only topologically trivial. As noted in [36] Gorenstein implies that the topological rank (as defined above) of every smoothing F is 2. This implies that the first Chern class of the tangent bundle TF vanishes. What happens when the singularity is only numerically Gorenstein?

A way how one may try to generalize Theorem 4.11, or to have new understandings about it, is to define an invariant associated to the link L_V of the singularity, which can be computed in terms of compact non-singular manifolds with some appropriate structure and boundary the link L_V. This is done in [36] to some extent, by using a cobordism invariant associated to Gorenstein singularities, that can be computed in terms of either a resolution of the singularity or a smoothing of it. Alas this only yields an invariant in the cyclic group of order 24, and one gets:

$$\mu_{GS}(V) + 1 \equiv \chi(\widetilde{V}) + K^2 + 12Arf(\widetilde{K}) \mod (24),$$

where the latter term is the Arf invariant appearing in Laufer's theorem, which is defined only in $\mathbb{Z}_2$. It is proved in [10] that in this setting the invariant $Arf(\widetilde{K})$ actually lifts to an invariant in $\mathbb{Z}$ and coincides with the geometric genus. So it is natural to ask whether this point of view can yield to an alternative proof of Theorem 4.11 that may throw some new light on this topic. This is explored in [41].

5 Indices of Vector Fields on Singular Varieties

Given a germ of a (say continuous) vector field v on a smooth manifold M, we know that its most basic invariant at an isolated singularity p of M is its local Poincaré-Hopf index $\mathrm{Ind}_{\mathrm{PH}}(v; p)$. This has several remarkable properties. Some of these are:

1. It is stable under perturbations. More precisely, if we make a small perturbation of v near p, the singularity of v at p may split into several singularities, each having its own local index $\mathrm{Ind}_{\mathrm{PH}}(v; p)$, the sum of all of these indices equals that of v at p. Moreover: take an arbitrarily small closed disc $\mathbb{B}_\varepsilon$ centered at p, and let ζ be another vector field on M defined in a neighborhood of p, which has isolated singularities in $\mathbb{B}_\varepsilon$ and coincides with v on the boundary sphere $\partial\mathbb{B}_\varepsilon$. Then $\mathrm{Ind}_{\mathrm{PH}}(v; p)$ equals the sum of the indices $\mathrm{Ind}_{\mathrm{PH}}(\zeta; p_i)$ at the singularities of ζ in $\mathbb{B}_\varepsilon$.
2. If M is compact, say connected and orientable, and v is defined and non-singular on its boundary, then v can be extended to the interior of M with finitely many singularities, and the total index $\mathrm{Ind}(v; M)$ of v in M is independent of the choice of

the extension. In particular, if v is everywhere transversal to the boundary, pointing outwards, then $\mathrm{Ind}(v; M)$ equals the Euler characteristic of M independently of the choice of the extension. This is Poincaré-Hopf's theorem for manifolds with boundary. Of course this applies also in the special case when the boundary is empty.

3. If M is a compact complex manifold, then $\mathrm{Ind}(v; M)$ is the Poincaré dual of the top Chern class of M.
4. If M is a complex n-manifold, and v is holomorphic, then choosing a coordinate chart for M so that $v = (v_1, ..., v_n)$ where the v_i are the components of v, we may express $\mathrm{Ind}_{\mathrm{PH}}(v; (M, p))$ as the intersection number
$$\mathrm{Ind}_{\mathrm{PH}}(v; p) = \dim_{\mathbb{C}} \frac{\mathcal{O}_{M,p}}{(v_1, ..., v_n)},$$
where $(v_1, ..., v_n)$ denotes here the ideal generated by these functions. In particular $\mathrm{Ind}_{\mathrm{PH}}(v; p) \geq 0$ and this is 0 if and only if v actually is non-singular at p.

What about all of this when we consider vector fields on singular varieties? The point is that there is not a unique way of extending the notion of the local Poincaré-Hopf index $\mathrm{Ind}_{\mathrm{PH}}(v; p)$ to the case of singular varieties. There are rather, several possible such extensions, each having its own properties and characteristics. The extension we want somehow depends on the kind of property of the local index that we want to preserve. This is also much related with asking who plays the role of the tangent bundle at the singular points of the ambient space. We refer to [5] for a thorough account of the subject.

Here we focus on an specific question regarding the GSV-index, originally defined in [13, 38], which is an extension of the local Poincaré-Hopf index to the case of vector fields on hypersurface singularities. This arises when we search for an index that preserves the stability property (5.1) above with respect to perturbations of both, the vector field and the defining function. Let us make this precise.

Consider a holomorphic map-germ $(\mathbb{C}^{n+1}, \underline{0}) \xrightarrow{f} (\mathbb{C}, 0)$ with an isolated critical point at $\underline{0}$, and the Milnor fibration, where the fibers are given by the non-critical levels $V_t = f^{-1}(t)$. We thus have a family of complex manifolds V_t that degenerate to the special fiber $V = f^{-1}(0)$.

Now consider a continuous vector field v on V, singular only at $\underline{0}$. This means a continuous section of the bundle $T\mathbb{C}^{n+1}|_V$ which is tangent to V at each point $x \neq \underline{0}$. As noticed in Sect. 3, on $f^{-1}(\mathbb{D}_\delta) \cap \mathbb{S}_\varepsilon$ we have a codimension 2 locally trivial fibration with fibre the link L_V, where $\mathbb{D}_\delta$ is a small enough disc in $\mathbb{C}$ centered at 0. We may thus move, by an isotopy, a neighborhood of the link in $V \cap \mathbb{S}_\varepsilon$ to the boundary ∂F of a Milnor fibre $F := f^{-1}(t) \cap \mathbb{S}_\varepsilon$. Hence we can think of v as being a vector field on a neighborhood of ∂F in F. By the Property (2) of the index stated above, we can extend it to a vector field on F with finitely many singularities $p_1, ..., p_r$. Then define the GSV index of v on V at $\underline{0}$ to be:

Fig. 3 The GSV index

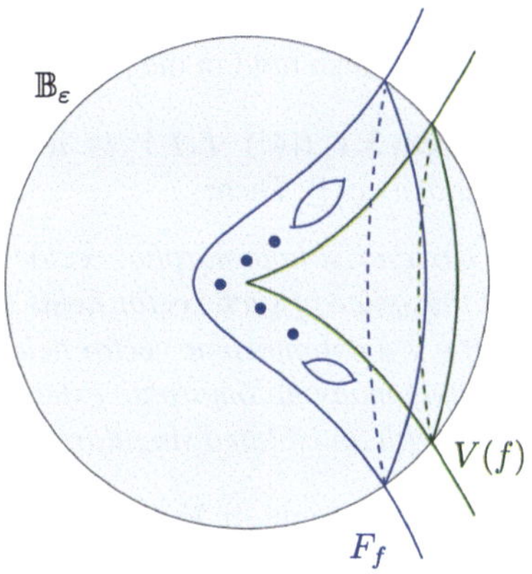

$$\mathrm{Ind}_{\mathrm{GSV}}(v; (V, \underline{0})) = \sum_{i=1}^{r} \mathrm{Ind}_{\mathrm{PH}}(v; (F, p_i)) \,.$$

Alternatively, just as we think of V as being a "limit" of the non-singular levels V_t, we may think that we have a continuous vector field v_t on each V_t with finite singularities, and depending continuously on the parameter t. As t tends to 0 the manifolds V_t degenerate to V and the vector fields v_t degenerate to v. Then the GSV-index of v is the sum of the Poincaré-Hopf indices of each v_t, before its singularities merge into $\underline{0}$ (Fig. 3).

We notice that for each $x \in V^* = V \setminus \{\underline{0}\}$, the tangent space $T_x V^*$ consists of all vectors in $T_x \mathbb{C}^{n+1}$ which are mapped to 0 by the derivative of f:

$$T_x V^* \;=\; \{\zeta \in T_x \mathbb{C}^{n+1} \,\big|\, df_x(\zeta) = 0\} \,.$$

For example, if f is the polynomial map in $\mathbb{C}^2$ defined by $f(z_1, z_2) = z_1^2 + z_2^3$, then the line tangent to $V = f^{-1}(0)$ at a point (z_1, z_2), other than the origin, is spanned by the vector $\widetilde{\zeta}(z_1, z_2) = (-3z_2^2, 2z_1)$. To see this notice one has

$$df_z = 2\, z_1 dz_1 + 3\, z_2^3 dz_2 \,.$$

Hence:

$$df_z(\widetilde{\zeta}) = df_z(-3z_2^2, 2z_1) \;=\; 0 \,.$$

Now, a vector field v on V can be thought of as being a continuous map $(V, \underline{0}) \stackrel{v}{\rightarrow} (\mathbb{C}^{n+1}, \underline{0})$ which is non-zero on V^* and whose image is contained in the linear space tangent to V at each given point. Since V is a closed subset of $\mathbb{B}_\varepsilon$, this map extends to a neighborhood of V in $\mathbb{C}^{n+1}$. Geometrically this means that the vector field v on V can always be extended to the ambient space, or equivalently that v can always be considered as the restriction to V of a vector field in the ambient space.

However the extension of v to V is by no means unique. Furthermore, all these statements also hold in the holomorphic category:

Theorem 5.1 ([4]) *Let V be a complex analytic variety in $\mathbb{C}^m$ with an isolated singularity at $\underline{0}$. Then:*

(1) *There exist holomorphic vector fields on V with an isolated singularity at $\underline{0}$. In fact the space of such vector fields is infinite-dimensional.*
(2) *If v is a holomorphic vector field on V with an isolated singularity, then there are infinitely many holomorphic extensions of v to a neighborhood of $\underline{0}$ in the ambient space with an isolated singularity.*

As an example, if V is defined in $\mathbb{C}^2$ by a map $f : (\mathbb{C}^2, \underline{0}) \to (\mathbb{C}, 0)$, then the Hamiltonian vector field $\widetilde{\zeta}(z_1, z_2) = (-\frac{\partial f}{\partial z_2}, \frac{\partial f}{\partial z_1})$ is tangent to V and it is zero only at the origin. Notice that this vector field is actually tangent to all the fibers $f^{-1}(t)$. Let ζ be the restriction of $\widetilde{\zeta}$ to V. Notice that ζ can be extended to $\mathbb{C}^2$ in many other ways; for example, if g is a holomorphic function on $\mathbb{C}^2$ that vanishes exactly on V and represents a non-zero element in the local ring $\mathcal{O}_{(\mathbb{C}^2,\underline{0})}$, then

$$\xi = \left(g - \frac{\partial f}{\partial z_2}, g + \frac{\partial f}{\partial z_1}\right),$$

coincides with ζ on V and is no longer tangent to the fibers of f; choosing g appropriately we can also assure that ξ has an isolated singularity at 0. Hence the GSV-index of this vector field is 0.

In $\mathbb{C}^3$ one has the following example from [12]. Let $f : (\mathbb{C}^3, \underline{0}) \to (\mathbb{C}, 0)$ have an isolated critical point at $\underline{0}$, set $V = f^{-1}(0)$ and choose the coordinates (z_1, z_2, z_3) so that V meets only at $\underline{0}$ the analytic set where the partial derivatives of f with respect to z_2 and z_3 vanish, *i.e.*,

$$V \cap \left\{\frac{\partial f}{\partial z_2} = \frac{\partial f}{\partial z_3} = 0\right\} = \{\underline{0}\}.$$

Define a holomorphic vector field in $\mathbb{C}^3$ by

$$\widetilde{\zeta} = \left(f, \frac{\partial f}{\partial z_3}, -\frac{\partial f}{\partial z_2}\right),$$

Notice $\widetilde{\zeta}$ has an isolated singularity at 0 and

$$df(\widetilde{\zeta}) = f\frac{\partial f}{\partial z_1},$$

hence $df(\widetilde{\zeta})$ vanishes at the points where f vanishes, so $\widetilde{\zeta}$ is tangent to V. If we set $\zeta = \widetilde{\zeta}|_V$, then we have a holomorphic vector field on V with an isolated singularity at the origin, and an extension $\widetilde{\zeta}$ of it to $\mathbb{C}^3$ which also has an isolated singularity.

Notice however that, unlike the previous example, $\widetilde{\zeta}$ is no longer tangent to the fibers of f. Yet, we may forget we are given $\widetilde{\zeta}$ and just consider the vector field ζ on V. Since f vanishes exactly on V, ζ takes the form $\zeta = (0, \frac{\partial f}{\partial z_3}, -\frac{\partial f}{\partial z_2})$ and we can extend it to a holomorphic vector field $\widetilde{\xi}$ on $\mathbb{C}^3$ defined by:

$$\widetilde{\xi} = \left(0, \frac{\partial f}{\partial z_3}, -\frac{\partial f}{\partial z_2}\right).$$

This is tangent to all the non-singular hypersurfaces $f^{-1}(t)$, $t \neq 0$. The singular set of $\widetilde{\xi}$ is the complete intersection curve defined by the ideal $(\frac{\partial f}{\partial z_2}, \frac{\partial f}{\partial z_3})$, which meets each non-singular fiber $f^{-1}(t)$ at finitely many points, whose total sum (counting multiplicities) is constant. This constant is an index that depends only on ζ and the way V is embedded in $\mathbb{C}^3$. This is the GSV-index.

By Property (4) above, the index of a holomorphic vector field in $\mathbb{C}^n$ at an isolated singularity is positive. Moreover, we know that every holomorphic vector field can be approximated, in a neighborhood of a singular point, by vector fields having only Morse singularities. In the holomorphic case, Morse singularities have all index 1. Thus we have:

In the space of germs of holomorphic vector fields in $\mathbb{C}^n$ with an isolated singularity at $\underline{0}$, there is a dense open subset of vector fields with a Morse singularity, and these have Poincaré-Hopf index 1.

What about holomorphic vector fields on singular varieties?

Example 5.2 Let f be a homogeneous polynomial $f(z_1, ..., z_n) = z_1^k + \cdots + z_n^k$, $n, k \geq 2$. It is an exercise to show that the radial vector field $v_{rad}(z_1, ..., z_n) = \sum_{i=1}^{n} \frac{\partial}{\partial z_i}$ is tangent to $V = f^{-1}(0)$ at each point in V. It is clear that when we move v_{rad} to the boundary of a Milnor fibre by an isotopy, it becomes transversal to the boundary, pointing outwards everywhere. Hence, by Property (2) of the index, its total Poincaré-Hopf index on F is equal to the Euler characteristic $\chi(F)$. And by [29] we have:

$$\chi(F) = 1 + (-1)^{n-1}\mu = 1 + (-1)^{n-1}(k-1)^n,$$

where μ is the Milnor number. Hence if n is even and $k > 2$, then

$$\mathrm{Ind}_{\mathrm{GSV}}(v_{rad}; (V, \underline{0})) < 0,$$

and v_{rad} obviously is holomorphic.

By the way, we leave it as an exercise to see what happens when $k = 2$.

So in the singular case, the GSV-index can be negative. Yet the following theorem says that it cannot be arbitrarily negative: it is bounded by below. This theorem follows immediately from [4, Theorem 2.2].

Theorem 5.3 *Let $(V, \underline{0})$ be a hypersurface singularity germ. Then there is a (possibly negative) largest $K > -\infty$, such that every germ of a holomorphic vector field v on V satisfies:*

$$Ind_{\mathrm{GSV}}(v_{rad}; (V, \underline{0})) \geq K .$$

Furthermore, the holomorphic vector fields on V with the smallest index form a dense open subset in the space of germs of vector fields on $(V, \underline{0})$.

The integer K obviously is an analytic invariant of V. This rises some natural questions:

Question 5.4 What is a "generic" vector field on $(V, \underline{0})$? And what is K?

I believe that $K = 1 + (-1)^n \mu$ where μ is the Milnor number and n is the dimension of V.

References

1. Arnold, V.I.: Remarks on singularities of finite codimension in complex dynamical systems. Funct. Anal. Appl. **3**, 1–5 (1969)
2. Atiyah, M.F.: Riemann Surfaces and Spin Structures. Ann. Scient. Ec. Norm. Sup. **4**, 47–62 (1971)
3. Biswas, I., Gurjar, R.V., Kolte, S.U.: On the Zariski-Lipman conjecture for normal algebraic surfaces. J. Lond. Math. Soc., II. Ser. **90**(1), 270–286 (2014)
4. Bonatti, Ch., Gómez-Mont, X.: The index of a holomorphic vector field on a singular variety. Astérisque **222**, 9–35 (1994)
5. Brasselet, J.P., Seade, J., Suwa, T.: Vector Fields on Singular Varieties. Lecture notes in mathematics # 1987. Springer, Berlin (2009)
6. Camacho, C., Kuiper, N., Palis, J.: The topology of holomorphic flows with singularity. Inst. Hautes Et. Sci. Publ. Math. **48**, 5–38 (1978)
7. Chung, F., Xu, Y.-J., Yau, S.T.: Classification of weighted dual graphs with only complete intersection singularities structures. Trans. AMS **261**, 3535–3596 (2009)
8. Cisneros-Molina, J.L.: Join theorem for polar weighted homogeneous singularities. In Singularities II. Geometric and topological aspects. Proceedings Internatinal Conference School and workshop on the geometry and topology of singularities. A. M. S. Contemporary Mathematics **475**, 43–59 (2008)
9. Durfee, A.H.: The signature of smoothings of complex surface singularities. Math. Ann. **232**, 85–98 (1978)
10. Esnault, H., Seade, J., Viehweg, E.: Characteristic divisors on complex manifolds. J. Reine Angew. Math. **424**, 17–30 (1992)
11. Freedman, M., Kirby, R.: A geometric proof of Rochlins theorem. In: Algebraic and Geometric Topology. Proceedings of Symposia Pure Mathematics, vol. XXXII, pp. 85–97. A.M.S (1978)
12. Gómez-Mont, X.: An algebraic formula for the index of a vector field on a hypersurface with an isolated singularity. J. Algebraic Geom. **7**, 731–752 (1998)
13. Gómez-Mont, X., Seade, J., Verjovsky, A.: The index of a holomorphic flow with an isolated singularity. Math. Ann. **291**, 737–751 (1991)
14. Gómez-Mont, X., Seade, J., Verjovsky, A.: Topology of a holomorphic vector field around an isolated singularity. Funct. Anal. Appl. **27**, 97–103 (1993)

15. Grauert, H.: Über Modifikationen und exzeptionnelle analytische Mengen. Math. Ann. **146**, 331–368 (1962)
16. Greuel, G.-M., Steenbrink, J.: On the Topology of Smoothable Singularities. Proceedings of sympsia pure mathematics Part 1, pp. 535–545. A.M.S. (1983)
17. Hochster, M.: The Zariski-Lipman conjecture in the graded case. J. Algebra **47**(2), 411–424 (1977)
18. Ishida, H.: Torus invariant transverse Kähler foliations. Trans. A. M. S. **369**, 5137–5155 (2017)
19. Källström, R.: The Zariski-Lipman conjecture for complete intersections. J. Algebra **337**, 169–180 (2011)
20. Larrión, F., Seade, J.: Complex surface singularities from the combinatorial point of view. Topol. Appl. **66**, 251–265 (1995)
21. Laufer, H.B.: On μ for surface singularities. In: Several Complex Variables Proceedings of SymposIa Pure Mathematics Part 1, vol. XXX, pp. 45–49. AMS (1977)
22. Lê, D.T.: Computation of the milnor number of an isolated singularity of a complete intersection. Funct. Anal. Appl. **8**, 127–131 (1974)
23. Libgober, A.: Theta characteristics on singular curves, spin structures and Rokhlin theorem. Ann. Sci. Éc. Norm. Supér. **4(21)**(4), 623–635 (1988)
24. Limón, B., Seade, J.: Morse theory and the topology of holomorphic foliations near an isolated singulariry. J. Topol. **4**(3), 667–686 (2011)
25. Lins Neto, A.: Holomorphic Rank of Hypersurfaces with an Isolated Singularity. Bol. Soc. Bras. Mat. **29**(1), 145–161 (1998)
26. Lipman, J.: Free derivation modules on algebraic varieties. Am. J. Math. **87**, 874–898 (1965)
27. López de Medrano, S.: Singularities of real homogeneous quadratic mappings Revista de la Real Academia de Ciencias Exactas, Fisicas y Naturales. Serie A. Matemáticas, pp. 1–18 (2012)
28. Meersseman, L.: A new geometric construction of compact complex manifolds in any dimension. Math. Ann. **317**, 79–115 (2000)
29. Milnor, J.: Singular Points of Complex Hypersurfaces. Annals of mathematics studies. Princeton University Press, New York (1968)
30. Mumford, D.: The topology of normal singularities of an algebraic surface and a criterion for simplicity. Publ. Math. IHES **9**, 229–246 (1961)
31. Oka, M.: Non-degenerate mixed functions. Kodai Math. J. **33**, 1–62 (2010)
32. Popescu-Pampu, P., Seade, J.: A finitness theorem for dual graphs of surface singularities. Int. J. Maths. **20**(8), 1057–1068 (2009)
33. Popescu-Pampu, P.: Numerically Gorenstein surface singularities are homeomorphic to Gorenstein ones. Duke Math. J. **159**(3), 539–559 (2011)
34. Ruas, M.A.S., Seade, J., Verjovsky, A.: On real singularities with a Milnor fibration. In: Libgober, A., Tibăr, M. (eds.) Trends in Singularities, pp. 191–213 (2002)
35. Scheja, G., Storch, U.: Differentielle Eigenschaften der Lokalisierungen analytischer Algebren. Math. Ann. **197**, 137–170 (1972)
36. Seade, J.: A cobordism invariant for surface singularities. Proceedings Symposia Pure Mathematics part 2, vol. 40, pp. 479–484 (1983)
37. Seade, J.: Vector fields on smoothings of complex singularities. In Ramírez de Arellano, E., Sundararaman, D. (eds.) Topics in Several Complex Variables, Research Notes in Mathematics, vol. 112, pp. 152–157. Pitman Advanced Publishing Program (1985)
38. Seade, J.: The index of a vector field on a complex surface with singularities. In: Verjovsky, A. (ed.) The Lefschetz Centennial Conference Contemporary Mathematics Part III, American Mathematical Society, vol. 58, pp. 225–232 (1987)
39. Seade, J.: On the Topology of Isolated Singularities in Analytic Spaces. Progress in mathematics. Birkhauser, Basel (2006)
40. Seade, J.: Remarks on Laufer's formula for the Milnor number, Rochlin's signature theorem and the analytic Euler characteristic of compact complex manifolds. Meth. Appl. Anal. **24**, 105–123. Special issue in honor of Henry Laufer's 70th Birthday (2017)
41. Seade, J.: A note on 3-manifolds and complex surface singularities. Preprint 2018, to be published

42. Steenbrink, J.: Mixed hodge structures associated with isolated singularities. Proceedings of Symposia Pure Mathematics, In: Singularities Part 2 (Arcata, Calif., 1981), vol. 40, pp. 513–536 (1983)
43. Thom, R.: Généralization de la théorie de Morse et variétés feuilletées. Ann. Inst. Fourier (Grenoble) **14**, 173–190 (1964)

Part III
Selected Papers in Foliations and Singularity Theory

A Comprehensive Approach to the Moduli Space of Quasi-homogeneous Singularities

Leonardo M. Câmara and Bruno Scárdua

Abstract We study the relationship between singular holomorphic foliations at $(\mathbb{C}^2, 0)$ and their separatrices. Under mild conditions we describe a complete set of analytic invariants characterizing foliations with quasi-homogeneous separatrices. Further, we give the full moduli space of quasi-homogeneous plane curves. This paper has an expository character in order to make it accessible also to non-specialists.

Keywords Singular holomorphic foliations · Holonomy groups · Quasi-homogeneous curves

2010 Mathematics Subject Classification 37F75 · 32S65 · 32S45

1 Introduction

In this paper we deal with the classification of germs of curves and germs of holomorphic foliations at $(\mathbb{C}^2, 0)$ (cf. Theorems A and B). The problem of the classification of germs of analytic plane curves has been addressed by several authors since the XVII th century with different methods (see for instance [2, 3, 21, 34]). In the first part of the present work, we study the problem of the analytic classification of germs of singular curves with many branches from an algebro-geometric viewpoint. We establish pre-normal forms for quasi-homogeneous polynomials, then we use the standard resolution of these singularities in order to stratify them and thus identify

This work was partially supported by CAPES-PROCAD grant no. 0007056.

L. M. Câmara (✉)
Departamento de Matemática, Universidade Federal do Espírito Santo,
Av. Fernando Ferrari 514, Vitória-ES 29075-910, Brazil
e-mail: leonardo.camara@ufes.br

B. Scárdua
Intituto de Matemática, Universidade Federal do Rio de Janeiro, Caixa Postal 68530,
Rio de Janeiro-RJ 21945-970, Brazil
e-mail: scardua@im.ufrj.br

R. N. Araújo dos Santos et al. (eds.), *Singularities and Foliations. Geometry, Topology and Applications*, Springer Proceedings in Mathematics & Statistics 222, https://doi.org/10.1007/978-3-319-73639-6_15

the moduli space of each stratum. As a consequence, our method provides an effective way to identify if two quasi-homogeneous curves are equivalent. Notice that the analytic type of a quasi-homogeneous curve is one of the invariants determining the analytic type of a foliation having such a curve as separatrix set (cf. Theorem B). Therefore, the present classification completes the classification of such germs of complex analytic foliations.

On the other hand, the problem of local classification of differential equations of the form $Adx + Bdy = 0$ in two variables has been studied by various mathematicians — since the end of the nineteenth century — as C. A. Briot, J. C. Bouquet, H. Dulac, H. Poincaré, I. Bendixson, G. D. Birkhoff, C. L. Siegel, A. D. Brjuno et al. In the middle 1970s R. Thom restored the interest in this question with a series of talks at IHES. In fact, he conjectured that a germ of foliation $\mathcal{F}$ at $(\mathbb{C}^2, 0)$ with a finite number of separatrices, i.e. a finite number of analytic invariant curves through the origin, has its analytic type characterized by its holonomy with respect to the separatrix set (cf. [13], pp. 162, 163). In [26–28] it is proved that the conjecture has an affirmative answer if the linear part of the vector field defining the foliation is non-nilpotent. In [29] it is proved that the conjecture is not true in general with the introduction of an analytic invariant called vanishing holonomy. Further, in [5] it is proved that any germ of singular holomorphic foliation at $(\mathbb{C}^2, 0)$ has a nonempty separatrix set, which is denoted by $\text{Sep}(\mathcal{F})$. Since this time, the problem of finding a complete set of analytic invariants determining the analytic type of a germ of foliation at $(\mathbb{C}^2, 0)$ having a finite number of separatrices is known as Thom's problem (cf. [16], pp. 60, 98). In [13] the results of [29] are generalized, classifying a Zariski open subset of the nilpotent singularities in terms of the vanishing holonomy (now called projective holonomy). Other contributions have been given by many authors such as [4, 16, 32], etc.

In [25] the problem of moduli space is studied from the deformation viewpoint. There it is proved that the moduli space of local unfoldings of quasi-homogeneous foliations is determined by the conjugacy class of the projective holonomy and the analytic type of its separatrix set for a generic class of foliations called quasi-hyperbolic (cf. [25], Definition 1.1, p. 255; Theorem B, p. 256; and Definition 6.8, p. 273). Namely, a germ of foliation $\mathcal{F}$ is called *quasi-hyperbolic generic* provided that the following conditions are satisfied: (i) its resolution $\widetilde{\mathcal{F}}$ has at least one non-solvable projective holonomy; (ii) $\widetilde{\mathcal{F}}$ has no saddle-nodes and the ratio between the eigenvalues of each of its singular points is not a positive real number. Using this result, in [17] it is proved that any two quasi-hyperbolic generic quasi-homogeneous foliations can be linked by such kind of unfoldings, classifying the quasi-hyperbolic generic quasi-homogeneous foliations. Unfortunately, the author of [17] forgot to mention the above hypotheses in his work.

From a quite different viewpoint, we show in the second part of our work an analogous result with less restrictive hypotheses on the foliation $\mathcal{F}$ (cf. Theorem B), using a geometric and much simpler proof. In fact, this geometrical approach leads also to the classification of curves.

We would like to remark that one of the main sources of inspiration for this work was the relationship between singular holonomies (cf. e.g. [7–10]) and the analytic type of a foliation near their Hopf components (see definition below). Furthermore, our approach can be used to understand the moduli space of more general germs of singular foliations, for instance, in the presence of saddle nodes.

The plan of the article is as follows. First we determine normal forms for quasi-homogeneous algebraic curves obtaining some geometric properties for the resolution of the separatrix set. With this geometric features at hand, we determine the moduli space in terms of the moduli space of punctured Riemann spheres. In the sequel, we study the semilocal invariants of resolved foliation determining the analytic type of each Hopf component of the foliation. Then we introduce natural cocycles that measure the obstruction for two analytically componentwise equivalent foliations to be really analytically equivalent. Finally we use the geometric description of the separatrix set in order to trivialize these cocycles and construct an explicit conjugation between two foliations with the same quasi-homogeneous curve and analytically conjugate projective holonomies.

We would like to thank Maria Aparecida Ruas for calling our attention to the fact that the analytic classification of germs of quasi-homogeneous curves was given by Kang in [22]. His classification is obtained in terms of equivalence classes between the coefficients of the homogeneous polynomials. The strategy used here allows to study the moduli space of bi-Lipschitz and analytically equivalent function-germs. This is the subject of a forthcoming paper of the first author in collaboration with M. A. Ruas.

Finally, we would like to thank the referee for the careful reading of the manuscript and for many helpful comments, corrections, and suggestions.

Part 1. Classification of Curves

2 Preliminaries

In this section we introduce the main notions for the first part of the paper. Let C be a singular curve and $\pi : (\mathcal{M}, D) \longrightarrow (\mathbb{C}^2, 0)$ its standard resolution, i.e. the minimal resolution of C whose *strict transform* $\widetilde{C} := \pi^{-1}(C)\backslash D$ is transversal to the exceptional divisor $D = \pi^{-1}(0)$. A germ of holomorphic function $f \in \mathbb{C}\{x, y\}$ is said to be *quasi-homogeneous* if there is a local system of coordinates in which f can be represented by a quasi-homogeneous polynomial, i.e. $f(x, y) = \sum_{ai+bj=d} a_{ij}x^i y^j$, where $a, b, d \in \mathbb{N}$. Let M be a manifold and $M_\Delta(n) := \{(x_1, \cdots, x_n) \in M^n : x_i \neq x_j \text{ for all } i \neq j\}$. Let S_n denote the group of permutations of n elements and consider its action in $M_\Delta(n)$ given by $(\sigma, \lambda) \mapsto \sigma \cdot \lambda = (\lambda_{\sigma(1)}, \cdots, \lambda_{\sigma(n)})$. The quotient space induced by this action is denoted by $\mathrm{Symm}(M_\Delta(n))$. Now suppose a Lie group G acts in M and let G act in $M_\Delta(n)$ in the natural way $(g, \lambda) = (g \cdot \lambda_1, \cdots, g \cdot \lambda_n)$ for every $\lambda \in M_\Delta(n)$. Then the actions of G and S_n in $M_\Delta(n)$ commute. Thus one obtains a natural action of G in $\mathrm{Symm}(M_\Delta(n))$. Given $\lambda \in M_\Delta(n)$, denote its equivalence class in $\mathrm{Symm}(M_\Delta(n))/G$ by $[\lambda]$.

Let C be a quasi-homogeneous curve determined by $f = 0$, where f is a reduced polynomial. Then Lemma 3.3 says that f can be (uniquely) written in the form

$$f(x, y) = x^m y^k \prod_{j=1}^{n} (y^p - \lambda_j x^q)$$

where $m, k \in \mathbb{Z}_2$, $p, q \in \mathbb{Z}_+$, $p \leq q$, $\gcd(p, q) = 1$, and $\lambda_j \in \mathbb{C}^*$ are pairwise distinct. In particular C has $n + m + k$ distinct branches. Since the exceptional divisor of the standard resolution and the number of irreducible components are analytic invariants of a germ of curve, then Lemmas 3.4 and 3.5 ensure that the triple (p, q, n) is an analytic invariant of the curve. Thus we have to consider the following three distinct cases:

(i) $f(x, y) = x^m \prod_{j=1}^{n} (y - \lambda_j x)$, where $m \in \mathbb{Z}_2$, and $\lambda_j \in \mathbb{C}$.

(ii) $f(x, y) = x^m \prod_{j=1}^{n} (y - \lambda_j x^q)$, where $m \in \mathbb{Z}_2$, $q \in \mathbb{Z}_+$, $q \geq 2$ and $\lambda_j \in \mathbb{C}$.

(iii) $f(x, y) = x^m y^k \prod_{j=1}^{n} (y^p - \lambda_j x^q)$, where $m, k \in \mathbb{Z}_2$, $p, q \in \mathbb{Z}_+$, $2 \leq p < q$, $\gcd(p, q) = 1$, and $\lambda_j \in \mathbb{C}^*$.

A quasi-homogeneous curve is said to be of *type* $(1, 1, n)$, $(1, q, n)$, and (p, q, n) respectively in cases (i), (ii), and (iii).

Theorem A *The analytic moduli space of germs of quasi-homogeneous curves of type* (p, q, n) *are given respectively by*

(i) $\frac{Symm(\mathbb{P}^1_\Delta(n))}{PSL(2,\mathbb{C})}$ *if* $(p, q) = (1, 1)$*;*
(ii) $\mathbb{Z}_2 \times \frac{Symm(\mathbb{C}_\Delta(n))}{Aff(\mathbb{C})}$ *if* $p = 1$ *and* $q > 1$*;*
(iii) $\mathbb{Z}_2 \times \mathbb{Z}_2 \times \frac{Symm(\mathbb{C}^*_\Delta(n))}{GL(1,\mathbb{C})}$ *if* $1 < p < q$.

3 Quasi-homogeneous Polynomials

In this section we describe the main algebro-geometric features of quasi-homogeneous polynomials.

3.1 Normal Forms

A quasi-homogeneous polynomial $f \in \mathbb{C}[x, y]$ is called *commode* if its Newton polygon intersects both coordinate axes. Further, notice that a polynomial in two variables $P \in \mathbb{C}[x, y]$ may be considered as a polynomial in the variable y with coefficients in $\mathbb{C}[x]$, i.e. $P \in (\mathbb{C}[x])[y]$. Let $\text{ord}_y P$ be the order of P as a polynomial in $(\mathbb{C}[x])[y]$. Similarly let $\text{ord}_x P$ be the order of P as an element of $(\mathbb{C}[y])[x]$. Therefore, a quasi-homogeneous polynomial $P \in \mathbb{C}[x, y]$ is commode if and only if $\text{ord}_x P = \text{ord}_y P = 0$. Next, we recall the general behavior of a quasi-homogeneous polynomial.

Lemma 3.1 *Let $P \in \mathbb{C}[x, y]$ be a quasi-homogeneous polynomial, then it has a unique decomposition in the form*

$$P(x, y) = x^m y^n P_0(x, y)$$

where $m, n \in \mathbb{N}$, $\lambda \in \mathbb{C}$, and P_0 is a commode quasi-homogeneous polynomial.

Proof Let $m := \text{ord}_x P$ and $n := \text{ord}_y P$. Clearly, both x^m and y^n divide P. Hence P can be written in the form $P(x, y) = \sum_{ai+bj=d} a_{ij} x^i y^j$, where $i \geq m$ and $j \geq n$. Thus $P(x, y) = x^m y^n P_0(x, y)$, where $P_0(x, y) = \sum_{ai'+bj'=d'} a_{i'+m, j'+n} x^{i'} y^{j'}$ and $d' := d - am - bn$. Since $m = \text{ord}_x P$ and $n = \text{ord}_y P$, then $\text{ord}_x P_0 = 0 = \text{ord}_y P_0$. The result then follows directly from the above remark. □

Definition 3.1 A commode polynomial $P \in \mathbb{C}[x, y]$ is called monic in y if it is a monic polynomial in $(\mathbb{C}[x])[y]$.

Lemma 3.2 *Let $P \in \mathbb{C}[x, y]$ be a commode quasi-homogeneous polynomial, which is monic in y. Then P can be written uniquely as*

$$P(x, y) = \prod_{\ell=1}^{k} (y^p - \lambda_\ell x^q),$$

where $\gcd(p, q) = 1$ and $\lambda_\ell \in \mathbb{C}^*$.

Proof First remark that any quasi-homogeneous polynomial can be written in the form $P(x, y) = \sum_{pi+qj=m} a_{ij} x^i y^j$, where $p, q, m \in \mathbb{N}$ and $\gcd(p, q) = 1$. Since P is commode, there are $i_0, j_0 \in \mathbb{N}$ such that $qj_0 = m$ and $pi_0 = m$; in particular $k := m/pq \in \mathbb{N}$. Therefore $pi + qj = pqk$. Since $\gcd(p, q) = 1$, then q divides i and p divides j. If we let $i = qi'$ and $j = pj'$, then $pqi' + qpj' = pqk$. Thus P can be written in the form $P(x, y) = \sum_{i+j=k} a_{qi,pj} x^{qi} y^{pj}$. Let $y = tx^{\frac{q}{p}}$, then the above equation assumes the form $P(x, tx^{q/p}) = x^{qk} \sum_{i+j=k} a_{qi,pj} t^{pj}$. Now let $\{\lambda_j\}_{j=1}^k$ be the roots of the polynomial $g(z) = \sum_{i+j=k} a_{qi,pj} z^j$, then

$$\begin{aligned} P(x, y) &= x^{qk} \prod_{\ell=1}^{k} (t^p - \lambda_l) = x^{qk} \prod_{\ell=1}^{k} (\frac{y^p}{x^q} - \lambda_l) \\ &= \prod_{\ell=1}^{k} (y^p - \lambda_l x^q). \end{aligned}$$

□

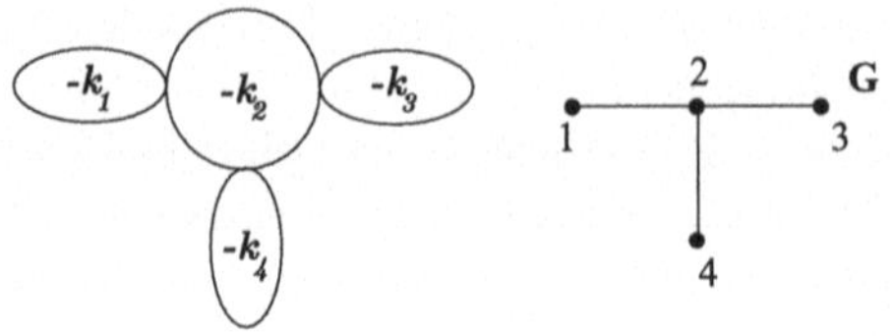

Fig. 1 The dual graph

Lemma 3.3 *Let $P \in \mathbb{C}[x, y]$ be a quasi-homogeneous polynomial. Then P can be written, uniquely, in the form*

$$P(x, y) = \mu x^m y^n \prod_{\ell=1}^{k} (y^p - \lambda_\ell x^q)$$

where $m, n, p, q \in \mathbb{N}$, $\mu, \lambda_\ell \in \mathbb{C}^*$, and $\gcd(p, q) = 1$.

Proof In view of Lemmas 3.1 and 3.2, it is enough to remark that any commode quasi-homogeneous polynomial $P \in \mathbb{C}[x, y]$ can be written uniquely as $P = \mu P_0$, where P_0 is monic in y. □

3.2 Resolution

We recall the geometry of the exceptional divisor of the minimal resolution of a germ of quasi-homogeneous curve.

A *tree of projective lines* is an embedding of a connected and simply connected chain of projective lines intersecting transversely in a complex surface (two dimensional complex analytic manifold) with two projective lines in each intersection. In fact, it consists of a pasting of Hopf bundles whose zero sections are the projective lines themselves. A *tree of points* is any tree of projective lines in which a finite number of points is discriminated. The above nomenclature has a natural motivation. In fact, as well know, we can assign to each projective line a point together with its respective self-intersection number and to each intersection an edge in order to form the *weighted dual graph*. Two trees of projective lines are called *isomorphic* if their weighted dual graph are isomorphic (as graphs). It is well known that any germ of analytic curve C at $(\mathbb{C}^2, 0)$ has a standard resolution, which we denote by $\widetilde{C}$. If the exceptional divisor of $\widetilde{C}$ has just one projective line intersecting three or more branches of $\widetilde{C}$, then it is called the *principal projective line* of $\widetilde{C}$ and denoted by $D_{\mathrm{pr}(\widetilde{C})}$. A tree of projective lines is called a *linear chain* if each of its projective lines intersects at most other two projective lines of the tree (Fig. 1). A projective line of a linear chain is called an *end* if it intersects only one of the projective lines of the chain.

Lemma 3.4 *Let C be a commode quasi-homogeneous curve. Then its standard resolution tree is a linear chain and its standard resolution $\widetilde{C}$ intersects only one projective line of D, i.e. C has one of the following diagrams of resolution:*

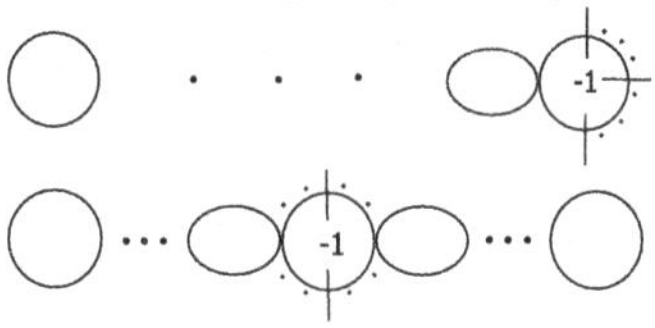

Proof From Lemma 3.2, there is a local system of coordinates (x, y) such that $C = f^{-1}(0)$, where $f(x, y) = \prod_{l=1}^{k}(y^p - \lambda_j x^q)$ with $p < q$ and $\gcd(p, q) = 1$. The result then follows immediately from the standard resolution of each irreducible curve $y^p - \lambda_l x^q = 0$. □

Let $\#\mathrm{irred}(\widetilde{C})$ denote the number of irreducible components of $\widetilde{C}$.

Lemma 3.5 *Let C be a non-commode quasi-homogeneous curve. Then its minimal resolution tree is a linear chain having a principal projective line such that $\#irred(\widetilde{C}) - 2 \leq \#(\widetilde{C} \cap D_{pr(\widetilde{C})}) \leq \#irred(\widetilde{C})$. Furthermore $\widetilde{C} \cap D_j = \emptyset$ whenever D_j is neither the principal projective line nor an end; i.e. C has one of the following diagrams of resolution:*

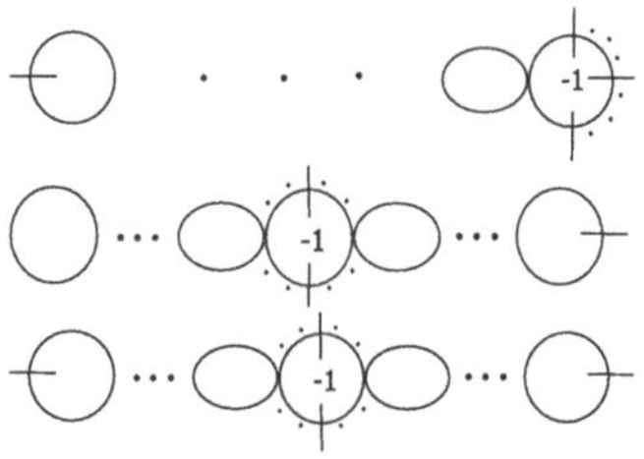

Proof From Lemma 3.3, there is a local system of coordinates (x, y) such that $C = f^{-1}(0)$, where $f(x, y) = \mu x^m y^n \prod_{l=1}^{k}(y^p - \lambda_j x^q)$, $p < q$, and $\gcd(p, q) = 1$. Since $\mu x^m y^n$ is resolved after one blowup, then $f(x, y)$ is resolved together with the fibration $\frac{y^p}{x^q} \equiv \text{const}$, as before. Then the result follows from Lemma 3.4. □

4 Quasi-homogeneous Curves

In this section we consider types (i), (ii), and (iii) separately and prove Theorem A in a series of lemmas.

4.1 Curves of Type $(1, 1, n)$

In this case the curve is given as the zero set of a polynomial of the form $f(x, y) = x^m \prod_{j=1}^{n}(y - \lambda_j x)$, where $m \in \mathbb{Z}_2$, and $\lambda_j \in \mathbb{C}$; in particular it is resolved after one blowup. Thus, given $\lambda = (\lambda_1, \cdots, \lambda_n) \in \mathbb{P}^1_\Delta(n)$ we define $f_\lambda(x, y) = x \prod_{j \neq i}(y - \lambda_j x)$ if $\lambda_i = \infty$ or $f_\lambda(x, y) = \prod_{j=1}^{n}(y - \lambda_j x)$ if $\lambda_j \neq \infty$ for all $j = 1, \ldots, k$. We denote the curve $f_\lambda = 0$ by C_λ. Recall that the natural action of $\mathrm{PSL}(2, \mathbb{C})$ in $\mathbb{P}^1$ as the group of homographies induces a natural action of $\mathrm{PSL}(2, \mathbb{C})$ in $\mathrm{Symm}(\mathbb{P}^1_\Delta(n))$. Moreover, recall that the equivalence class of $\lambda \in \mathbb{P}^1_\Delta(n)$ in $\mathrm{Symm}(\mathbb{P}^1_\Delta(n))/\mathrm{PSL}(2, \mathbb{C})$ is denoted by $[\lambda]$.

Lemma 4.1 *Two homogeneous curves C_λ and C_μ are analytically equivalent if and only if $[\lambda] = [\mu] \in Symm(\mathbb{P}^1_\Delta(n))/PSL(2, \mathbb{C})$.*

Proof Suppose C_λ and C_μ are analytically equivalent and let $\Phi \in \mathrm{Diff}(\mathbb{C}^2, 0)$ taking C_λ into C_μ. Let $\widetilde{\Phi}$ be the blowup of Φ, then it takes the strict transform of C_λ into the strict transform of C_μ. Blowing up f_λ and f_μ we obtain at once that the first tangent cones of C_λ and C_μ are respectively given by $\{\lambda_1, \cdots, \lambda_n\}$ and $\{\mu_1, \cdots, \mu_n\}$. Therefore, there is $\sigma \in S_n$ such that the Möbius transformation $\varphi = \widetilde{\Phi}|_{\mathbb{P}^1}$ satisfies $\mu_{\sigma(j)} = \varphi(\lambda_j)$ for all $j = 1, \ldots, n$. In other words, $[\lambda] = [\mu]$. Conversely, suppose $[\lambda] = [\mu]$. Reordering the indexes of $\{\mu_1, \cdots, \mu_n\}$ we may suppose, without loss of generality, that there is a Möbius transformation $\varphi(z) = \frac{az+b}{cz+d}$, with $ad - bc = 1$, such that $\mu_j = \varphi(\lambda_j)$ for all $j = 1, \ldots, n$. Now consider the linear transformation $T(x, y) = (dx + cy, bx + ay)$ with inverse $T^{-1}(x, y) = (ax - cy, -bx + dy)$. Then a straightforward calculation shows that $f_\lambda = \alpha \cdot T^* f_\mu$, where $\alpha \in \mathbb{C}^*$. Thus C_λ is analytically equivalent to C_μ, as desired. □

Remark 4.1 Recall that for any three distinct points $\{\lambda_1, \lambda_2, \lambda_3\} \subset \mathbb{P}^1$ there is a Möbius transformation φ such that $\varphi(0) = \lambda_1$, $\varphi(1) = \lambda_2$ and $\varphi(\infty) = \lambda_3$.

As a straightforward consequence of Lemma 4.1 and Remark 4.1 one has:

Corollary 4.2 *Let $\lambda, \mu \in \mathbb{P}^1_\Delta(n)$ with $n \leq 3$. Then C_λ and C_μ are analytically equivalent.*

4.2 Curves of Type $(1, q, n)$, $q \geq 2$

In this case, the curve is given as the zero set of a polynomial of the form $f_{m,\lambda}(x, y) = x^m \prod_{j=1}^{n}(y - \lambda_j x^q)$, where $m \in \mathbb{Z}_2, q \in \mathbb{Z}_+, q \geq 2$, and $\lambda_j \in \mathbb{C}$. Given $m \in \mathbb{Z}_2$ and $\lambda = (\lambda_1, \cdots, \lambda_n) \in \mathbb{C}_\Delta(n)$, we denote a curve of type $(1, q, n)$ by $C_{m,\lambda}$ if it is given as the zero set of $f_{m,\lambda}$. Recall that the group of affine transformations of $\mathbb{C}$, denoted by $\mathrm{Aff}(\mathbb{C})$, acts in a natural way in $\mathrm{Symm}(\mathbb{C}_\Delta(n))$. As before, denote by $[\lambda]$ the equivalence class of $\lambda \in \mathbb{C}_\Delta(n)$ in $\mathrm{Symm}(\mathbb{C}_\Delta(n))/\mathrm{Aff}(\mathbb{C})$.

Lemma 4.3 *Two homogeneous curves $C_{m,\lambda}$ and $C_{m,\mu}$ are analytically equivalent if and only if $[\lambda] = [\mu] \in Symm(\mathbb{C}_\Delta(n))/Aff(\mathbb{C})$.*

Proof Suppose $\Phi \in \text{Diff}(\mathbb{C}^2, 0)$ is an equivalence between $C_{m,\lambda}$ and $C_{m,\mu}$. From the proof of Lemma 3.4, both curves are resolved after q blowups. Furthermore, after $q-1$ blowups Φ will be lifted to a local conjugacy $\Phi^{(q-1)}$ between the germs of curves given in local coordinates (x, y) respectively by $p_\lambda(x, y) = x\prod_{j=1}^{n}(y - \lambda_j x)$ and $p_\mu(x, y) = x\prod_{j=1}^{n}(y - \mu_j x)$, where $(x = 0)$ is the local equation of the exceptional divisor $D^{(q-1)}$. Let π denote an additional blowup given in local coordinates by $\pi(t, x) = (x, tx)$ and $\pi(u, y) = (u, uy)$, and $\Phi^{(q)}$ be the map obtained by the lifting of $\Phi^{(q-1)}$ by π. Moreover, let $\varphi = \Phi^{(q)}|_{D_q}$, where $D_q = \pi^{-1}(0)$. Since $\Phi^{(q)}$ preserves the irreducible components of $\pi^*(D^{(q-1)})$, then $\varphi(t) = \Phi^{(q)}(t, 0)$ is a homography fixing ∞ and conjugating the first tangent cones of $p_\lambda = 0$ and $p_\mu = 0$ respectively. Thus $[\lambda] = [\mu] \in \text{Symm}(\mathbb{C}_\Delta(n))/\text{Aff}(\mathbb{C})$. Conversely, reordering the indexes of μ, if necessary, suppose there is $\varphi(z) = az + b \in \text{Aff}(\mathbb{C})$ such that $\mu_j = \varphi(\lambda_j)$ for all $j = 1, \ldots, n$, and let $T(x, y) = (x, ay + bx^q)$. Then a straightforward calculation shows that $f_{m,\lambda} = \alpha \cdot T^* f_{m,\mu}$, where $\alpha \in \mathbb{C}^*$. Thus $C_{m,\lambda}$ and $C_{m,\mu}$ are analytically equivalent, as desired. □

As a straightforward consequence of Lemma 4.3 and Remark 4.1 one has:

Corollary 4.4 *Let $\lambda, \mu \in \mathbb{C}_\Delta(n)$ with $n \leq 2$. Then $C_{m,\lambda}$ and $C_{m,\mu}$ are analytically equivalent.*

4.3 Curves of Type (p, q, n), $2 \leq P < q$

In this case, the curve is given as the zero set of a polynomial of the form $f_{m,k,\lambda}(x, y) = x^m y^k \prod_{j=1}^{n}(y^p - \lambda_j x^q)$, where $m, k = 0, 1$, $p, q \in \mathbb{Z}_+$, $2 \leq p < q$, and $\lambda_j \in \mathbb{C}^*$. Given $\lambda = (\lambda_1, \cdots, \lambda_n) \in \mathbb{C}^*_\Delta(n)$ we denote a curve of type (p, q, n) by $C_{m,k,\lambda}$ if it is given as the zero set of $f_{m,k,\lambda}(x, y)$. Recall that the group of linear transformations of $\mathbb{C}$, denoted by $\text{GL}(1, \mathbb{C})$, acts in a natural way in $\text{Symm}(\mathbb{C}^*_\Delta(n))$. Recall that the equivalence class of $\lambda \in \mathbb{C}^*_\Delta(n)$ in $\text{Symm}(\mathbb{C}^*_\Delta(n))/\text{GL}(1, \mathbb{C})$ is denoted by $[\lambda]$.

Lemma 4.5 *Two homogeneous curves $C_{m,k,\lambda}$ and $C_{m,k,\mu}$ are analytically equivalent if and only if $[\lambda] = [\mu] \in Symm(\mathbb{C}^*_\Delta(n))/GL(1, \mathbb{C})$.*

Proof First recall from the proof of Lemma 3.4 that $C_{m,k,\lambda}$ is resolved after N blowups, where N depends on the Euclid's division algorithm between q and p. Further, at the (N-1)th step we have to blowup a singularity given in local coordinates (x, y) as the zero set of the polynomial $g_\lambda(x, y) = xy\prod_{j=1}^{n}(y - \lambda_j x)$. Therefore, if $\Phi \in \text{Diff}(\mathbb{C}^2, 0)$ is an equivalence between $C_{m,k,\lambda}$ and $C_{m,k,\mu}$ and $\Phi^{(N-1)}$ is its lifting to the $(N-1)^{th}$ step of the resolution, then it conjugates the germs of curves given in local coordinates (x, y) respectively by $p_\lambda(x, y) = xy\prod_{j=1}^{n}(y - \lambda_j x)$ and $p_\mu(x, y) = xy\prod_{j=1}^{n}(y - \mu_j x)$, where $(x = 0)$ and $(y = 0)$ are local equations for the exceptional divisor $D^{(N-1)}$. Let π denote the final blowup of the resolution given in local coordinates by $\pi(t, x) = (x, tx)$ and $\pi(u, y) = (u, uy)$, and $\Phi^{(N)}$ be the map obtained by the lifting of $\Phi^{(N-1)}$ by π. Further let $\varphi = \Phi^{(N)}\big|_{D_N}$, where $D_N = \pi^{-1}(0)$. Since $\Phi^{(N)}$ preserves the irreducible components of $\pi^*(D^{(q-1)})$, then $\varphi(t) = \Phi^{(q)}(t, 0)$ is a homography fixing 0 and ∞, and conjugating the first tangent cones of $p_\lambda = 0$ and $p_\mu = 0$ respectively. Thus $[\lambda] = [\mu] \in \mathrm{Symm}(\mathbb{C}^*_\Delta(n))/\mathrm{GL}(1, \mathbb{C})$. Conversely, reordering the indexes of μ, if necessary, suppose there is $\varphi(z) = az \in \mathrm{GL}(1, \mathbb{C})$ such that $\mu_j = \varphi(\lambda_j)$ for all $j = 1, \ldots, n$, and let $T(x, y) = (x, \sqrt[p]{a}y)$. Then a straightforward calculation shows that $f_{m,\lambda} = \alpha \cdot T^* f_{m,\mu}$, where $\alpha \in \mathbb{C}^*$. Thus $C_{m,\lambda}$ and $C_{m,\mu}$ are analytically equivalent, as desired. □

As a straightforward consequence of Lemma 4.5 and Remark 4.1 one has:

Corollary 4.6 *Let $\lambda, \mu \in \mathbb{C}^*_\Delta(1)$, then $C_{m,k,\lambda}$ and $C_{m,k,\mu}$ are analytically equivalent.*

5 Resolution and Factorization

In this section we study the relationship between the resolution tree and the factorization of a quasi-homogeneous polynomial. We use the resolution in order to study the equivalence between two quasi-homogeneous polynomials.

First recall that a quasi-homogeneous polynomial splits uniquely in the form $P = x^m y^n P_0$, where P_0 is a commode quasi-homogeneous polynomial. In particular P and P_0 share the same resolution process.

Corollary 5.1 *Let $P \in \mathbb{C}[x, y]$ be a commode quasi-homogeneous polynomial with weights (p, q), where $\gcd(p, q) = 1$. Let $q_j = s_j p_j + r_j$, $j = 1, \ldots, \ell$, be given by Euclid's algorithm of (p, q), where $q_1 := q$, $p_1 := p$, $q_{j+1} := p_j$, and $p_{j+1} := r_j$ for all $j = 1, \ldots, \ell - 1$. Then the exceptional divisor of its minimal resolution is given by a linear chain of projective lines, namely $D = \cup_{j=1}^{\nu} D_j$, whose self-intersection numbers are given as follows:*

1. *If $\ell = 2\alpha - 1$, then*

$$D_j \cdot D_j = \begin{cases} -(s_{2k} + 2) & \text{if } j = s_1 + \cdots + s_{2k-1},\ k = 1, \ldots, \alpha - 1; \\ -1 & \text{if } j = s_1 + \cdots + s_{2\alpha-1}; \\ -(s_{2k+1} + 2) & \text{if } j = \ell - (s_2 + \cdots + s_{2k-2}) + 1,\ k = 1, \ldots, \alpha - 1; \\ -(s_{2\alpha-1} + 1) & \text{if } j = \ell - (s_1 + \cdots + s_{2\alpha-2}) + 1; \\ -2 & \text{otherwise.} \end{cases}$$

2. *If $\ell = 2\alpha$, then*

$$D_j \cdot D_j = \begin{cases} -(s_{2k} + 2) & \text{if } j = s_1 + \cdots + s_{2k-1},\ k = 1, \ldots, \alpha - 1; \\ -(s_{2\alpha} + 1) & \text{if } j = s_1 + \cdots + s_{2\alpha-1}; \\ -(s_{2k+1} + 2) & \text{if } j = \ell - (s_2 + \cdots + s_{2k-2}) + 1,\ k = 1, \ldots, \alpha - 1; \\ -1 & \text{if } j = \ell - (s_1 + \cdots + s_{2\alpha-2}) + 1; \\ -2 & \text{otherwise.} \end{cases}$$

Finally, if C is given by $f = 0$, where $f(x, y) = x^m y^n \prod_{j=1}^{k}(y^p - \lambda_j x^q)$, then a representative of $[\lambda]$ is determined by the intersection of the strict transform of C with the exceptional divisor D.

Proof The proof shall be performed by induction on ℓ, the length of the Euclidean algorithm. In order to better understand the arguments, the reader have to keep in mind the proof of Lemma 3.4. From Lemma 3.2, we may suppose, without loss of generality, that P can be written in the form $P(x, y) = \prod_{j=1}^{k}(y^p - \lambda_j x^q)$. First remark that if $\ell = 1$ then $p = 1$. Thus we prove the statement for $\ell = 1$ by induction on q. For $q = 1$ the result is easily verified after one blowup. Now suppose the result is true for all $q \leq q_0 - 1$. Then after one blowup $\pi(t, x) = (x, tx)$, $\pi(u, y) = (uy, y)$, P is transformed into $\pi^* P(t, x) = x \prod_{j=1}^{k}(t - \lambda_j x^{q-1})$. Thus the result follows for $\ell = 1$ by induction on q. Suppose the result is true for all polynomials whose pair of weights has Euclid's algorithm length less than ℓ, and let (p, q) with length ℓ. Since $p_j = s_j q_j + r_j$, $j = 1, \dots, \ell$, is the Euclid's algorithm of (p, q), then $p_j = s_j q_j + r_j$, $j = 2, \dots, \ell$, is the Euclid's algorithm of (p_2, q_2). In particular the Euclid's algorithm of (p_2, q_2) has length $\ell - 1$. Reasoning in a similar way as in the case $\ell = 1$, we have after s_1 blowups a linear chain of projective lines $\cup_{j=1}^{s_1} D_j^{(1)}$ such that $D_j^{(1)} \cdot D_j^{(1)} = -2$ for all $j = 1, \dots, s_1 - 1$ and $D_{s_1}^{(1)} \cdot D_{s_1}^{(1)} = -1$. Besides, the strict transform of $P = 0$ is given by the zero set of the polynomial $\widetilde{P}(t, x) = \prod_{j=1}^{k}(t^{p_1} - \lambda_j x^{r_1}) = \lambda_1 \cdots \lambda_k \prod_{j=1}^{k}(x^{p_2} - \lambda_j t^{q_2})$, where the local equation for $D_{s_1}^{(1)}$ is $(x = 0)$. The first statement thus follows from the induction hypothesis. The last statement comes immediately from the above reasoning. For the above induction arguments ensure that the strict transform of P assume the form $\widetilde{P} = 0$, with $\widetilde{P}(x, y) = \prod_{j=1}^{k}(y - \lambda_j x)$, just before the last blowup. □

The above Corollary gives an easy way to compute the relatively prime weights of a quasi-homogeneous polynomials from the dual weighted tree of its minimal resolution. Also it shows that the minimal resolution can be used both to split a quasi-homogeneous polynomial into irreducible components and also to determine its analytic type.

Part 2. Classification of Foliations

6 Preliminaries

In this section we introduce the main notions for the second part of the paper. A germ of singular foliation $(\mathcal{F} : \omega = 0)$ at $(\mathbb{C}^2, 0)$ of codimension 1 is, roughly speaking, the set of integral curves of a given germ of 1-form $\omega \in \Omega^1(\mathbb{C}^2, 0)$, which may be assumed to have just an isolated singularity at the origin. Let $\mathrm{Diff}(\mathbb{C}^k, 0)$ be the group of germs of analytic diffeomorphisms of $(\mathbb{C}^k, 0)$ fixing the origin. Two germs of foliations $(\mathcal{F}_j : \omega_j = 0)$ at $(\mathbb{C}^2, 0)$, $j = 1, 2$, are analytically equivalent if there is $\Phi \in \mathrm{Diff}(\mathbb{C}^2, 0)$ sending leaves of $\mathcal{F}_1$ into leaves of $\mathcal{F}_2$. One says that

$h_1, h_2 \in \mathrm{Diff}(\mathbb{C}, 0)$ are analytically conjugate if there is $\phi \in \mathrm{Diff}(\mathbb{C}, 0)$ such that $Ad_\phi(h_1) := \phi \circ h_1 \circ \phi^{-1} = h_2$. We denote the *Hopf bundle of order* k (see Definition 7.1) by $p_{(k)} : \mathcal{H}(-k) \to D$, where $D \simeq \mathbb{CP}(1)$, or just by its total space $\mathbb{H}(-k)$.

The study of germs of singular holomorphic foliation at $(\mathbb{C}^2, 0)$ is given by the resolution process (cf. [31]). Essentially, giving a germ of singular foliation $\mathcal{F}$ at $(\mathbb{C}^2, 0)$, it says that after the composition of a finite number of blowing-ups one obtains a map $\pi : (\widetilde{X}, D) \longrightarrow (\mathbb{C}^2, 0)$ such that the singularities appearing along the exceptional divisor $D = \pi^{-1}(0)$ of the strict transform $\widetilde{\mathcal{F}}$ are isolated and assume, in local coordinates (x, y), one of the following forms:

(1). $\widetilde{\omega}(x, y) = \lambda x dy + \mu y dx + \cdots$ with $\lambda \cdot \mu \neq 0$ and $\lambda/\mu \notin \mathbb{Q}_-$;
(2). $\widetilde{\omega}(x, y) = \lambda y dx + \cdots, \lambda \neq 0$;

where the dots mean higher order terms. By definition $\widetilde{\mathcal{F}}$ is the unique extension of $\pi^*(\mathcal{F})$ whose singular set has codimension greater or equal to 2 (cf. [6]). Notice that these are minimal models in the sense that they are stable under new blowing-ups. These singularities are called *reduced* or *simple* singularities. A reduced singularity is called *non-degenerate* if it's of type (1), otherwise it's called a *saddle node* (cf. [6]). Furthermore, if $D = \cup D_j$ is the decomposition of the exceptional divisor into irreducible components, where D_j has self-intersection number equal to $-k_j$, $j = 1, \ldots, n$, then recall from the theory of algebraic curves that a suitable neighborhood of D in $\widetilde{X}$ results from pasting together suitable neighborhoods of the zero sections of $\mathbb{H}(-k_j)$. For each Hopf bundle $p_j : \mathcal{H}_j \to D_j$ of a given resolution, we denote by $\widetilde{\mathcal{F}}_j$ the germ of foliation in $(\mathcal{H}_j, D_j)$ induced by the restriction of $\widetilde{\mathcal{F}}$ and call it the jth *Hopf component* of the resolution. The singular points of the exceptional divisor, namely $c_{ij} := D_i \cap D_j$, are called *corners* and the singularities on such points are called *corner singularities* (or just corners) and denoted by $\widetilde{\mathcal{F}}_{ij}$. The "strict transform" of $\mathrm{Sep}(\mathcal{F})$ at $D_j \subset \mathcal{H}_j$, i.e. the set of local separatrices of $\widetilde{\mathcal{F}}_j$, namely $\mathrm{Sep}(\widetilde{\mathcal{F}}_j) = \overline{(\pi^*\mathrm{Sep}(\mathcal{F}))\,|_{\mathcal{H}_j} \backslash D_j}$, is called the jth *Hopf component* of $\pi^*(\mathrm{Sep}(\mathcal{F}))$. Two foliations having analytically equivalent Hopf components are called *analytically componentwise equivalent*.

Let $p : \mathcal{H} \to D$ be a Hopf bundle and $\mathcal{F}$ a germ of foliation defined in $(\mathcal{H}, D)$. Then $\mathcal{F}$ is called *non-dicritical* if D is an invariant set of $\mathcal{F}$, and *dicritical* otherwise. In the former case the holonomy of $\mathcal{F}$ with respect to D evaluated at a transversal section Σ is called *projective holonomy* of $\mathcal{F}$ and denoted by $\mathrm{Hol}_\Sigma(\mathcal{F}, D)$. One says that $\mathcal{F}$ is *resolved* if it has just *reduced* singularities (cf. [6]). Let $\widetilde{\mathcal{F}}^1$ and $\widetilde{\mathcal{F}}^2$ be two germs of non-dicritical singular foliations about $D \subset \mathcal{H}$ without saddle-nodes and having the same singular set, say $\{t_j\}_{j=1}^n$. Let $t_0 \in D$ be a regular point of $\widetilde{\mathcal{F}}^1$ and denote by h^i_γ the holonomy of a path $\gamma \in \pi_1(D \backslash \{t_j\}_{j=1}^n, t_0)$ with respect to D evaluated at a transversal section $\Sigma_0 := p^{-1}(t_0)$. Then one says that the projective holonomies of these foliations are *analytically conjugate* if there is $\phi \in \mathrm{Diff}(\mathbb{C}, 0)$ such that $Ad_\phi(h^1_\gamma) = h^2_\gamma$ for every $\gamma \in \pi_1(D \backslash \{t_j\}_{j=1}^n, t_0)$.

A *generalized curve foliation* is a germ of singular foliation at $(\mathbb{C}^2, 0)$ that has no saddle-nodes along its minimal resolution (cf. [6]). Naturally, a non dicritical germ of generalized curve foliation has a finite number of irreducible separatrices.

A germ of holomorphic function $f \in \mathbb{C}\{x, y\}$ is said to be *quasi-homogeneous* if there is a local system of coordinates in which f can be represented by a quasi-homogeneous polynomial, i.e. $f(x, y) = \sum_{ai+bj=d} a_{ij} x^i y^j$, where $a, b, d \in \mathbb{N}$. The separatrix set of a germ of foliation $\mathcal{F}$ at $(\mathbb{C}^2, 0)$ is said to be quasi-homogeneous if $\mathrm{Sep}(\mathcal{F}) = f^{-1}(0)$, where f is a quasi-homogeneous function. The set of generalized curve foliations at $(\mathbb{C}^2, 0)$ with quasi-homogeneous separatrix set is denoted by $\mathcal{QHS}$; in particular, if $\mathrm{Sep}(\mathcal{F})$ is commode, then $\mathcal{F}$ is called a *commode* $\mathcal{QHS}$ foliation.

Recall, from [31], that any germ of holomorphic foliation $\mathcal{F}$ at $(\mathbb{C}^2, 0)$ has a minimal resolution. We denote it by $\widetilde{\mathcal{F}}$ and its ambient surface by $M_{\widetilde{\mathcal{F}}}$. If the exceptional divisor of $\widetilde{\mathcal{F}}$ has just one projective line containing three or more singular points of $\widetilde{\mathcal{F}}$, then it is called the *principal projective line* of $\widetilde{\mathcal{F}}$ and denoted by $D_{\mathrm{pr}(\widetilde{\mathcal{F}})}$ (see definition in Sect. 9.1.1). If $\widetilde{\mathcal{F}}$ has a principal projective line, then the projective holonomy of its principal projective line is called the *projective holonomy of the foliation* $\mathcal{F}$. Later on, we will see that any $\mathcal{QHS}$ foliation has a principal projective line. Then one says that $\mathcal{F} \in \mathcal{QHS}$ is *generic* if the corner singularities of $\widetilde{\mathcal{F}}$ in $D_{\mathrm{pr}(\widetilde{\mathcal{F}})}$ are in the Poincaré domain or in the Siegel domain with the quotient of eingenvalues satisfying the Yoccoz-Bryuno condition (cf. [1]).

Now in are in a position to state the main result of this work.

Theorem B *Let $\mathcal{F}$ and $\mathcal{F}'$ be two $\mathcal{QHS}$ germs of foliations with the same separatrix set. Suppose that $\mathcal{F}$ and $\mathcal{F}'$ are both commode or generic. Then $\mathcal{F}$ and $\mathcal{F}'$ are analytically equivalent if and only if their projective holonomies are analytically conjugate.*

7 Hopf Bundles and Projective Holonomy

In this section we consider germs of reduced foliations with only non degenerate singularities leaving invariant the zero section of a Hopf bundle. Under some natural geometric conditions, we describe the invariants that determine their analytic type.

First recall the definition of a Hopf bundle.

Definition 7.1 Let $k \in \mathbb{Z}_+$ and consider two copies of $\mathbb{C}^2$ with coordinates given respectively by (t, x) and (u, y). Then the line bundle over $\mathbb{CP}(1)$ given by the transition maps

$$\begin{cases} y = t^k x \\ u = 1/t \end{cases}$$

for all $t \neq 0$ is called the Hopf bundle of order k and denoted by $p_{(k)} : \mathbb{H}(-k) \to \mathbb{CP}(1)$ or just by its total space $\mathbb{H}(-k)$.

Clearly, two analytically equivalent singularities have isomorphic weighted dual trees of singular points along their minimal resolution. Thus, if we consider analytically equivalent Hopf components, it is clear that isomorphic points have the same linear part and that their local holonomy generators are conjugated by a global map. To clarify the ideas, we need the following

Definition 7.2 Let M be a complex surface and $S \subset M$ a smooth curve invariant by the germ of holomorphic foliation $\mathcal{F}$ in (M, S) such that $\mathrm{Sing}(\mathcal{F}) \subset S$ has only non-degenerate reduced singularities. Then we say that a germ of holomorphic map $f : (M, S) \to S$ is a fibration transversal to $\mathcal{F}$ if it satisfies:

(1) f is a retraction, i.e. f is a submersion and $f|_S = \mathrm{id}|_S$;
(2) the fiber $f^{-1}(t_j)$ is a separatrix of F for each $t_j \in \mathrm{Sing}(\mathcal{F})$;
(3) $f^{-1}(t)$ is transversal to F for every (regular) point $t \in S \backslash \mathrm{Sing}(\mathcal{F})$.

Let $\mathcal{F}$ be a germ of reduced holomorphic foliation having only non degenerate singularities defined in a neighborhood of the zero section of the Hopf bundle $p : \mathcal{H} \to D$, $f : (\mathcal{H}, D) \to D$ be a fibration transversal to $\mathcal{F}$, and $t_0 \in D\backslash \mathrm{Sing}(\mathcal{F})$ be a regular point of $\mathcal{F}$. Hence the path lifting construction ensures that the projective holonomy $\mathrm{Hol}_{f^{-1}(t)}(\mathcal{F}, D)$ is completely determined by $\mathrm{Hol}_{f^{-1}(t_o)}(\mathcal{F}, D)$ for any $t, t_0 \in D\backslash \mathrm{Sing}(\mathcal{F})$. Such a holonomy is called *projective holonomy* of $\mathcal{F}$ with respect to f. If there is no doubt about the fibration, we only talk about the projective holonomy of the foliation and denote it by $\mathrm{Hol}(\mathcal{F}, D)$.

Definition 7.3 Let $\mathcal{F}$ and $\mathcal{F}_o$ be germs of reduced foliations with only non degenerate singularities leaving invariant the zero section of the Hopf bundle $p : \mathcal{H} \to D$ with the same singular set S. Then we set

$$\mathrm{Diff}_{\mathcal{F},\mathcal{F}_o}(\mathcal{H}, D) := \{\Phi \in \mathrm{Diff}(\mathcal{H}, D) : \Phi_*(\mathcal{F}) = \mathcal{F}_o \text{ and } \Phi|_S = \mathrm{id}\}$$

and call

$$\mathrm{Aut}(\mathcal{F}_o) := \{\Phi \in \mathrm{Diff}_{\mathcal{F}_o,\mathcal{F}_o}(\mathcal{H}, D) : \Phi|_S = \mathrm{id}\}$$

the group of automorphisms of $\mathcal{F}_o$. Furthermore, if $f : (\mathcal{H}, D) \to D$ is a fibration transversal to $\mathcal{F}_o$, then the set of elements of $\mathrm{Aut}(\mathcal{F}_o)$ preserving f is denoted by $\mathrm{Aut}(\mathcal{F}_o, f)$.

Proposition 7.1 *Let $\mathcal{F}^i$, $i = 1, 2$, be two germs of reduced foliations with only non degenerate singularities leaving invariant the zero section of the Hopf bundle $p : \mathcal{H} \to D$. Suppose that $Sep(\mathcal{F}^1) = Sep(\mathcal{F}^2)$ and that there is a fibration $f_i : (\mathcal{H}, D) \longrightarrow D$ transversal to $\mathcal{F}^i$. Then $\mathcal{F}^1$ and $\mathcal{F}^2$ are analytically equivalent if and only if their projective holonomies are analytically conjugate.*

Proof As already remarked, the necessary part is straightforward. Let us treat the sufficient part. Since the separatrices of $\mathcal{F}^1$ and $\mathcal{F}^2$ coincide, then their singular sets also coincide. Let $\mathrm{Sing}(\mathcal{F}^i) = \{t_j\}_{j=1}^n$ and $t_0 \in D$ be a regular point. Suppose there is $\phi \in \mathrm{Diff}(\mathbb{C}, 0)$ such that $\phi \circ (h_j^1) \circ \phi^{-1} = h_j^2$ for all $j = 1, \ldots, n$. Then define the map $\Phi : \mathcal{F}\backslash \bigcup_{j=1}^n f_1^{-1}(t_j) \longrightarrow \mathcal{F}'\backslash \bigcup_{j=1}^n f_2^{-1}(t_j)$ by

$$\Phi(t, x) := \Phi_t(x) := h_t^2 \circ \phi \circ (h_t^1)^{-1}(x),$$

where $x \in f_1^{-1}(t)$ and $h_t^i : f_i^{-1}(t_0) \longrightarrow f_i^{-1}(t)$ are the holonomy maps obtained by path lifting a curve connecting t_0 to t along the leaves of $\mathcal{F}^i$. Note that this map does not depend on the chosen base curves, since ϕ conjugates the elements of the respective projective holonomies of $\mathcal{F}^1$ and $\mathcal{F}^2$. Since Φ is holomorphic in each variable separately, then (complex) ODE theory and Hartogs' theorem ensure that Φ is holomorphic. Finally, since $\mathcal{F}^1$ has just reduced singularities, then [26, 28] ensure that the union of the saturated of $\Sigma_0 := f_1^{-1}(t_0)$ along the leaves of $\mathcal{F}^1$ and the local separatrices $\mathrm{Sep}(\mathcal{F}^1) = \bigcup_{j=1}^n f_1^{-1}(t_j)$ gives rise to a neighborhood of D. Thus we can use Riemann's extension theorem in order to extend Φ to $\mathrm{Sep}(\mathcal{F}^1)$ in a neighborhood of D. □

8 Analytic Invariants

In this section we consider germs of foliations at $(\mathbb{C}^2, 0)$ and use the weighted dual trees of their minimal resolutions, the first jet of each singularity of these resolutions, and the projective holonomies of their Hopf components in order to determine analytic componentwise equivalence. Next, we identify some analytical cocycles that appear as obstructions to extend the analytically componentwise isomorphisms introduced in Proposition 7.1. Finally, we relate these obstructions with the analytic classification of the foliations.

8.1 Componentwise Equivalence and Realization

In this section we find conditions to determine whether two $\mathcal{QHS}$ foliations with the same quasi-homogeneous separatrix set are componentwise equivalent. In order to do that, we split $\mathcal{QHS}$ into subclasses with increasing degree of information about the analytic type of foliations therein.

Let $\mathcal{QHS}_f$ denote the set of $\mathcal{QHS}$ foliations with the same separatrix set $f = 0$. This means that the separatrix of any $\mathcal{F} \in \mathcal{QHS}_f$ is given by the same curve $\mathrm{Sep}(\mathcal{F}) = (f = 0)$.

Remark 8.1 The resolution of a non dicritical generalized curve foliation coincides with the resolution of its separatrix [6]. In particular, any foliations in $\mathcal{QHS}_f$ has its weighted dual graph automatically determined by the separatrix $f = 0$.

Now notice that to each branch of $f = 0$ there corresponds one singular point of $\widetilde{\mathcal{F}}$. Since the first jet of this singularity is an analytic invariant of the foliation, up to multiplication by a nonvanishing complex number, then we define a subclass of $\mathcal{QHS}_f$ by attaching to each singular point of $\mathcal{F} \in \mathcal{QHS}_f$ this new analytic invariant as follows: Let $\mathcal{F}, \mathcal{F}' \in \mathcal{QHS}_f$ and $\widetilde{\mathcal{F}}, \widetilde{\mathcal{F}}'$ be respectively their minimal resolutions. Let $\mathrm{Sing}(\widetilde{\mathcal{F}}) = \bigcup_{i=1}^k \mathrm{Sing}(\widetilde{\mathcal{F}}_i)$, where $\mathrm{Sing}(\widetilde{\mathcal{F}}_i) = \{p_{i,j_i} : j_i = 1, \ldots, n_i\}$, k is the number of Hopf components of $\widetilde{\mathcal{F}}$, and $n_i := \#\text{Sing}(\widetilde{\mathcal{F}}_i)$. Let $\omega_{i,j_i} = 0$ and $\omega'_{i,j_i} = 0$ determine the germs of $\widetilde{\mathcal{F}}$ and $\widetilde{\mathcal{F}}'$ at p_{i,j_i}. Then one says that $\widetilde{\mathcal{F}}'$ is *analytically componentwise equivalent* to $\widetilde{\mathcal{F}}$ *up to first order* if $J^1(\omega_{i,j_i}) = J^1(\omega'_{i,j_i})$ (i.e. if they have the same linear part) for all $i = 1, \dots, k$ and $j_i = 1, \dots, n_i$. The set of $\mathcal{QHS}_f$ foliations analytically componentwise equivalent up to first order to $(\mathcal{F} : \omega = 0)$ is denoted by $\mathcal{QHS}^{c,1}_{\omega,f}$.

Finally, denote the set of $\mathcal{QHS}$ (respect. $\mathcal{QHS}_f$) foliations analytically componentwise equivalent to $(\mathcal{F} : \omega = 0)$ by $\mathcal{QHS}^c_\omega$ (respect. $\mathcal{QHS}^c_{\omega,f}$).

We determine now the moduli space $\mathcal{QHS}^{c,1}_\omega / \mathcal{QHS}^c_\omega$. The following result is a straightforward consequence of Proposition 7.1.

Proposition 8.1 *Let $\mathcal{F}$ and $\mathcal{F}'$ belonging to the same equivalence class in $\mathcal{QHS}^{c,1}_\omega$. Then they belong to the same equivalence class in $\mathcal{QHS}^c_\omega$ if and only if their projective holonomies are analytically conjugate.*

Given two germs of foliations in $\mathcal{QHS}^c_\omega$, we want to verify under what conditions they are in fact globally holomorphically conjugate. For this sake, we need the following realization data.

Definition 8.1 A complex surface is called resolution-like if it is obtained by a holomorphic pasting of Hopf bundles with negative Chern classes, in such a way that the union of their zero sections become a tree of projective lines isomorphic to the exceptional divisor of a composition of a finite numbers of blowups applied to $(\mathbb{C}^2, 0)$.

Clearly, this definition is given in such a way that every resolution surface of some singularity is automatically resolution-like. In fact, any resolution-like surface is biholomorphic to the resolution surface of some singularity as shown in

Proposition 8.2 ([12]) *Let M be a resolution-like surface with tree of projective lines D. Then (M, D) can be realized as a neighborhood of the exceptional divisor of a composition of a finite number of blowups applied to* $(\mathbb{C}^2, 0)$.

In order to prove this proposition, we need the following results about complex line bundles.

Theorem 8.3 (Grauert [18]) *Let S be a complex surface and $C \subset S$ be a rational curve with negative self-intersection number. Then there are neighborhoods U and V of C, respectively in S and $N(C; S)$ (the normal bundle of C in S), and a biholomorphism $\Psi : U \to V$ sending C in the zero section of $N(C; S)$.*

Theorem 8.4 (Grothendieck [19]) *Two complex line bundles over the Riemann sphere have the same Chern class if and only if they are biholomorphic.*

Proof of Proposition 8.2 The proof is performed by induction on the number of projective lines in the chain. If the chain is composed by just one projective line, the result follows immediately from the theorems of Grauert and Grothendieck. Suppose the result is true for all chains composed by $n \geq 1$ projective lines and let

D have $n+1$ projective lines. From the hypothesis, D has a projective line with self intersection -1, namely C^1. Now consider a projective line C^2 intersecting C^1 with self-intersection number $-k$, $k \geq 2$. Applying Grauert's and Grothendieck's theorems, we obtain that a neighborhood of each curve is biholomorphic to a neighborhood of the zero section of the Hopf bundle with Chern classes given by their self-intersection numbers. Thus we can blow down a neighborhood of the curve C^1 obtaining yet an analytic surface defined in a neighborhood of a Riemann sphere, say $\pi(C^2)$ — where π stands for the blow down. Since $\pi(C^2)$ is smooth, it is well known that its self-intersection number is $-k-1$ (cf. e.g. [23, 24, 30]). The result now follows from the induction hypothesis. □

Remark 8.2 Although the resolutions of two foliations in $\mathcal{QHS}_\omega^c$ are not necessarily defined in the same ambient surface, they all can be modeled by $(\mathcal{F} : \omega = 0)$ in the sense that they are analytically componentwise equivalent to $\widetilde{\mathcal{F}}$. Anyway, the ambient surfaces of their resolutions will be automatically equivalent whenever they have equivalent cocycles (definition found below).

8.2 Analytic Cocycles

We construct some cocycles associated with analytically componentwise equivalent foliations. In some sense, these cocycles measure how far two analytically componentwise equivalent foliations are from being analytically equivalent.

Let $\mathcal{F}^o \in \mathcal{QHS}$, $\widetilde{\mathcal{F}}^o$ its minimal resolution, and $M^o = M_{\widetilde{\mathcal{F}}^o}$ the ambient surface, where $\widetilde{\mathcal{F}}^o$ is defined. Let Pseudo(M^o) denote the pseudogroup of transformations of M^o and Aut$(\widetilde{\mathcal{F}}^o)$ denote its subset given by those $\phi \in$ Pseudo(M^o) satisfying the following properties:

(a) $\phi : U \longrightarrow \phi(U)$ preserves the Hopf components of the exceptional divisor, i.e. $\phi(U \cap D_j) = \phi(U) \cap D_j$;
(b) ϕ fixes the singularities of $\widetilde{\mathcal{F}}^o$, i.e. $\phi|_{\mathrm{Sing}(\widetilde{\mathcal{F}}^o)} = \mathrm{id}|_{\mathrm{Sing}(\widetilde{\mathcal{F}}^o)}$;
(c) ϕ preserves the leaves of $\widetilde{\mathcal{F}}_j^o$, i.e. $\phi^*(\widetilde{\mathcal{F}}_j^o|_{\phi(U)}) = \widetilde{\mathcal{F}}_j^o|_U$.

At this point some comments about the above definition are worthwhile. First, notice that all conditions can be verified explicitly. The first two are quite obvious and the third can be achieved with the aid of the path lifting procedure. In fact, choose a section Σ transversal to D_j and pick an element $\psi : \phi(\Sigma) \longrightarrow \Sigma$ of the classical holonomy pseudogroup of $\widetilde{\mathcal{F}}_j^o|_U$ with respect to D_j. Since the holonomy characterizes $\widetilde{\mathcal{F}}_j^o|_U$ (cf. Proposition 7.1, [26, 28]), it is enough to verify that $\psi \circ \phi \in$ Diff(Σ) commutes with the generators of $\mathrm{Hol}_\Sigma(\widetilde{\mathcal{F}}_j^o|_U, D_j)$. Moreover, note that we decided to deal with just local and semilocal leaves (i.e. those determined by the holonomies of $\widetilde{\mathcal{F}}_j^o|_U$) avoiding, for the time being, questions related to Dulac maps (cf. [10, 11]) that are very difficult to handle concretely in the global sense. This task will be performed by the pasting cocycles we define next.

Definition 8.2 Let $(\mathcal{F} : \omega = 0)$ be a germ of foliation at $(\mathbb{C}^2, 0)$. Then the set

$$\mathrm{Aut}(\mathcal{F}) = \{\phi \in \mathrm{Diff}(\mathbb{C}^2, 0) : \phi^*\omega \wedge \omega = 0\}$$

is called group of automorphisms of $\mathcal{F}$. Besides, if $f : (M, S) \to S$ is a fibration transversal to $\mathcal{F}$, then $\mathrm{Aut}(\mathcal{F}, f)$ denote the subgroup determined by elements of $\mathrm{Aut}(\mathcal{F})$ preserving f.

Let $(\mathcal{F} : \omega = 0)$ be a non dicritical generalized curve foliation and pick $\mathcal{F}^o$ analytically componentwise equivalent to $\mathcal{F}$. Consider a resolution $\widetilde{\mathcal{F}}^o$ of $\mathcal{F}^o$ such that $\mathrm{Sep}(\widetilde{\mathcal{F}}^o_j)$ consists of fibers of a fibration $f_j : (\mathcal{H}_j, D_j) \longrightarrow D_j$ transversal to $\widetilde{\mathcal{F}}^o_j$ (the existence of such a resolution, called rectifier, is proved in [12], Proposition 1, p. 4). Then $\mathcal{F}^o$ is called a *fixed model* for $\mathcal{F}$ and a map $\Phi_j \in \mathrm{Diff}(\widetilde{\mathcal{F}}_j, \widetilde{\mathcal{F}}^o_j)$ is called a *projective chart* for $\widetilde{\mathcal{F}}$ with respect to $\widetilde{\mathcal{F}}^o$. From what we have done before, it is straightforward that:

Lemma 8.5 *For each $\widetilde{\mathcal{F}}_j = \widetilde{\mathcal{F}}|_{(\mathcal{H}_j, D_j)}$ and each fixed model component $\widetilde{\mathcal{F}}^o_j$, there exists only one projective chart up to left composition with an element of $\mathrm{Aut}(\widetilde{\mathcal{F}}^o_j)$.*

Let $D = \cup D_j$ be the exceptional divisor of $\widetilde{\mathcal{F}}^o$. One says that $\mathcal{U} := \cup U_j$ is a *good covering* for D if each U_j is a simply-connected neighborhood of $D_j \subset \mathcal{H}_j$ and each intersection $U_i \cap U_j$ is simply-connected. For each good covering $\mathcal{U}$ and each foliation $\mathcal{F}$ one can associate a cocycle $\Phi(\mathcal{F}) := (\Phi_{i,j})$ given by $\Phi_{i,j} := \Phi_i \circ \Phi_j^{-1}$, where each Φ_i is a projective chart for $\widetilde{\mathcal{F}}$ with respect to $\widetilde{\mathcal{F}}^o$. Note that $(\Phi_{i,j})$ does not depend neither on the fixed models nor on the chosen (good) covering up to analytical componentwise equivalence.

Proposition 8.6 *Two analytically componentwise equivalent non dicritical generalized curve foliations $\mathcal{F}$ and $\mathcal{G}$ are analytically equivalent if and only if $\Phi(\mathcal{F}) = \Phi(\mathcal{G})$.*

Proof Let $\Phi(\mathcal{F}) = (\Phi_1 \circ \Phi_2^{-1}, \cdots, \Phi_{k-1} \circ \Phi_k^{-1})$ and $\Phi(\mathcal{G}) = (\Psi_1 \circ \Psi_2^{-1}, \cdots, \Psi_{k-1} \circ \Psi_k^{-1})$. First, let us verify the necessary part. Suppose H is a global conjugation between $\mathcal{F}$ and $\mathcal{G}$, i.e. $H^*(\mathcal{G}) = \mathcal{F}$. From Lemma 8.5, there is $\Xi_j \in \mathrm{Aut}(\widetilde{\mathcal{F}}^o_j)$ such that $\Psi_j = \Xi_j \circ \Phi_j \circ H$. Therefore

$$\begin{aligned}\Psi_{j-1} \circ \Psi_j^{-1} &= \Xi_{j-1} \circ \Phi_{j-1} \circ H \circ H^{-1} \circ \Phi_j^{-1} \circ \Xi_j^{-1} \\ &= \Xi_{j-1} \circ \Phi_{j-1} \circ \Phi_j^{-1} \circ \Xi_j^{-1}.\end{aligned}$$

Now let us verify the sufficient part. Notice that $\mathcal{F}$ and $\mathcal{G}$ have the same fixed model. Hence, if $(\Phi_1 \circ \Phi_2^{-1}, \cdots, \Phi_{k-1} \circ \Phi_k^{-1}) = \Phi(\mathcal{F}) = \Phi(\mathcal{G}) = (\Psi_1 \circ \Psi_2^{-1}, \cdots, \Psi_{k-1} \circ \Psi_k^{-1})$, there is a collection $(\Xi_j) \subset \mathrm{Aut}(\widetilde{\mathcal{F}}^o_j)$ such that $\Psi_{j-1} \circ \Psi_j^{-1} = \Xi_{j-1} \circ \Phi_{j-1} \circ \Phi_j^{-1} \circ \Xi_j^{-1}$. Therefore $\left(\Xi_{j-1} \circ \Phi_{j-1}\right)^{-1} \circ \Psi_{j-1} = \left(\Xi_j \circ \Phi_j\right)^{-1} \circ \Psi_j$. Thus we can define a global conjugation between them just by letting $H := \left(\Xi_j \circ \Phi_j\right)^{-1} \circ \Psi_j$ for all $j = 1, \ldots, k$. □

Remark 8.3 It is not difficult to verify that $\mathrm{Aut}(\widetilde{\mathcal{F}}^o)$ is itself a pseudogroup of transformations of M^o. Therefore the sheaf of germs of elements of $\mathrm{Aut}(\widetilde{\mathcal{F}}^o)$, generated by inductive limit, is a sheaf of groupoids over the exceptional divisor D^o of $\widetilde{\mathcal{F}}^o$ (cf. [20]). We denote this sheaf by $\mathfrak{Aut}_{\widetilde{\mathcal{F}}^o}$. Consider the first cohomology set $H^1(\mathcal{U}, \mathfrak{Aut}_{\widetilde{\mathcal{F}}^o})$, and let $H^1(D, \mathfrak{Aut}_{\widetilde{\mathcal{F}}^o})$ be the inductive limit of $H^1(\mathcal{U}, \mathrm{Aut}_{\widetilde{\mathcal{F}}^o})$ for all good coverings of D. Then Proposition 8.2 ensures that the map

$$\begin{array}{ccc} \mathcal{QHS}^c_\omega & \xrightarrow{\Phi} & Z^1(D, \mathrm{Aut}_{\widetilde{\mathcal{F}}^o}) \\ \mathcal{F} & \mapsto & (\Phi_{i,j}) := \Phi_i \circ \Phi_j^{-1} \end{array}$$

is well defined and onto $H^1(D, \mathrm{Aut}_{\widetilde{\mathcal{F}}^o})$. Since $\Phi(\mathcal{F})$ does not depend on the fixed models up to componentwise equivalence class, it determines a characteristic class for non dicritical generalized curve foliations appearing as obstruction for the global pasting of analytically componentwise isomorphisms. For the reader not acquainted with groupoids and the cohomology of their sheaves, we refer to [14, 15, 20].

9 Trivializing Cocycles

In this section we use the algebraic and geometric features of the separatrix set in order to construct an auxiliary fibration that helps us trivialize the cocycles. For this sake, we have first to introduce the concept of leaf preserving automorphism. Besides, we use the geometry of the divisors of both the foliation and the fibration in order to provide a method for trivializing $\Phi(\mathcal{F})$.

9.1 Quasi-homogeneous Polynomials and Companion Fibrations

In order to prove Theorem B, we need to perform an accurate geometric analysis of the interplay between the foliation $\mathcal{F}$ and its companion fibration $\mathcal{G}$.

9.1.1 Multivalued First Integrals and the Branches of $\mathcal{F}$

Let $\mathcal{F} \in \mathcal{QHS}^c_{\omega,f}$, where

$$f(x, y) = \mu y^m x^n \prod_{j=1}^{d} (y^p - \lambda_j x^q), \tag{9.1}$$

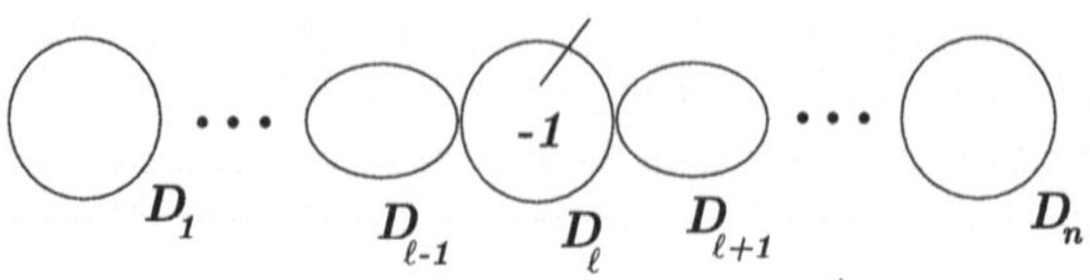

Fig. 2 The principal projective line for $p \neq 1$

$1 \leq p < q, m, n \in \mathbb{N}^*$, $\gcd(p,q) = 1$, and $\lambda_j, \mu \in \mathbb{C}^*$. Then we order the first projective line to arise in the course of the resolution process with 1, the next one intersecting it with 2, and so on (see Lemma 3.4 and Fig. 2), until we reach the last projective line in the minimal resolution. The principal projective line is denoted by $D_{\mathrm{pr}(\widetilde{\mathcal{F}})}$, where $\mathrm{pr}(\widetilde{\mathcal{F}})$ denotes the index of the principal projective line in the above definition. For the sake of simplicity, we set $\ell := \mathrm{pr}(\widetilde{\mathcal{F}})$ and call the subset of $\widetilde{\mathcal{F}}$ given by $\mathcal{B}_+\mathcal{F} := \cup_{j>\ell}\widetilde{\mathcal{F}}_j$ (respect. $\mathcal{B}_-\mathcal{F} := \cup_{j<\ell}\widetilde{\mathcal{F}}_j$) the *positive* (respect. *negative) branch* of $\widetilde{\mathcal{F}}$.

Recall from the geometry of the exceptional divisor of $\widetilde{\mathcal{F}}$ that each $\widetilde{\mathcal{F}}_j$ has only two singularities given in affine charts (t_j, x_j) by a 1-form $\widetilde{\omega}_j(t_j, x_j) = \nu_j x_j dt_j + \mu_j t_j dx_j + \cdots$, where $\nu_j, \mu_j \in \mathbb{C}^*$, for all $j \neq \mathrm{pr}(\widetilde{\mathcal{F}})$.

Remark 9.1 In case $\mathcal{F}$ is commode each end of its resolution $\widetilde{\mathcal{F}}$ has only one singular point, thus the respective holonomy is trivial. As a consequence, each Hopf component $\widetilde{\mathcal{F}}_j$ with $j \neq \mathrm{pr}(\widetilde{\mathcal{F}})$ admits a holomorphic first integral. Therefore, in this case the corner singularities of the principal divisor are automatically linearizable.

Now we are in a position to state the following geometric characterization of the branches of $\widetilde{\mathcal{F}}$ for a commode or generic foliation $\mathcal{F} \in \mathcal{QHS}^c_{\omega,f}$.

Lemma 9.1 *$\widetilde{\mathcal{F}}_j$ is linearizable for each $j \neq pr(\widetilde{\mathcal{F}})$. In particular, it has a multivalued first integral. More precisely, there is $\Phi_j \in Diff_{\widetilde{\mathcal{F}}_j, \widetilde{\mathcal{F}}_j^{lin}}(\mathcal{H}_j, D_j)$, where $(\widetilde{\mathcal{F}}_j^{lin} : d\widetilde{f}_j^{lin} = 0)$ is given by the (global) multivalued first integral*

$$\begin{cases} \widetilde{f}_j^{lin}(t_j, x_j) = t_j^{\nu_j} x_j^{\mu_j}, \\ \widetilde{f}_j^{lin}(u_j, y_j) = u_j^{k_j\mu_j - \nu_j} y_j^{\mu_j}, \end{cases}$$

where $\nu_j, \mu_j \in \mathbb{C}^*$ are non-resonant and $-k_j$ is the first Chern class of $\mathbb{H}_j$ for all $j \neq \ell$.

Proof Since $\mathcal{F}$ is commode or generic, then the corner singularities of $\widetilde{\mathcal{F}}_{\mathrm{pr}(\widetilde{\mathcal{F}})}$ are linearizable (cf. [33]). But Lemmas 3.4 and 3.5 ensure that $\widetilde{\mathcal{F}}_j$ has at most two singularities for all $j \neq \mathrm{pr}(\widetilde{\mathcal{F}})$, thus both singularities share the same holonomy with respect to D_j. Recall from [26] that a reduced and non-degenerate (i.e. a non saddle-node) singularity is linearizable if and only if its holonomy is linearizable. Thus Proposition 7.1 ensures that $\widetilde{\mathcal{F}}_j$ is linearizable whenever one of its singularities is linearizable. □

If (x, y) is a system of coordinates about the origin in which $\mathrm{Sep}(\mathcal{F})$ assumes the form (9.1), then it induces canonical affine coordinates for $M := \cup_{j=1}^{n}\mathbb{H}_j(-k_j)$, denoted by

$$\mathcal{A} := \{(t_j, x_j), (u_j, y_j) : u_j = 1/t_j, y_j = t_j^{k_j}x_j, y_j = t_{j+1}, u_j = x_{j+1}\}. \quad (9.2)$$

Now we prove that $\mathcal{B}_+\mathcal{F}$ (respect. $\mathcal{B}_-\mathcal{F}$) has a multivalued first integral and describe its feature in this system of coordinates. Let $\mathbb{D}_r$ denote the disk centered at the origin with radius r.

Lemma 9.2 *$\mathcal{B}_+\mathcal{F}$ (respect. $\mathcal{B}_-\mathcal{F}$) has a multivalued first integral denoted by $\widetilde{f}_+$ (respect. $\widetilde{f}_-$). More precisely, $\widetilde{f}_+$ (respect. $\widetilde{f}_-$) is given in the system of coordinates $\mathcal{A}$ by $\widetilde{f}_+ = \widetilde{f}_j$, where*

$$\begin{cases} \widetilde{f}_j(t_j, x_j) = t_j^{\nu_j}x_j^{\mu_j}U_j(t_j, x_j), \\ \widetilde{f}_j(u_j, y_j) = u_j^{k_j\mu_j-\nu_j}y_j^{\mu_j}V_j(u_j, y_j), \end{cases}$$

with $U_j, V_j \in \mathcal{O}^*(\mathbb{D}_{1+\epsilon} \times \mathbb{D}_\epsilon)$ for some $\epsilon > 0$ and all $j = 1, \ldots, \ell - 1$ (respect. $j = \ell + 1, \ldots, n - 1$).

Proof We prove the statement for the positive branch case, the other one being completely analogous. Pick $\Phi_{\ell+1} \in \mathrm{Diff}_{\widetilde{\mathcal{F}}_{\ell+1}, \widetilde{\mathcal{F}}^{lin}_{\ell+1}}(\mathcal{H}_{\ell+1}, D_{\ell+1})$ and let $\widetilde{f}_{\ell+1} := \Phi^*_{\ell+1}\widetilde{f}^{lin}_{\ell+1}$. Let p be a regular point of $D_{\ell+2}$ near the corner $c_{\ell+1,\ell+2} := D_{\ell+1} \cap D_{\ell+2}$ and Σ_p be the fiber of $\mathbb{H}_{\ell+2}$ over p. Recall that $\Phi_{\ell+1}$ induces a bijective map between the spaces of leaves of $\widetilde{\mathcal{F}}_{\ell+1}$ and $\widetilde{\mathcal{F}}^{lin}_{\ell+1}$ which can be realized as $\phi_{\ell+2} \in \mathrm{Diff}(\Sigma_p, p)$. In particular, $\phi_{\ell+2}$ takes $\mathrm{Hol}_{\Sigma_p}(\widetilde{\mathcal{F}}_{\ell+2}, D_{\ell+2})$ in $\mathrm{Hol}_{\Sigma_p}(\widetilde{\mathcal{F}}^{lin}_{\ell+2}, D_{\ell+2})$. Since $\widetilde{\mathcal{F}}_{\ell+2}$ has just two singularities, then Proposition 7.1 ensures that one can extend $\phi_{\ell+2}$ to $\Phi_{\ell+2} \in \mathrm{Diff}_{\widetilde{\mathcal{F}}_{\ell+2}, \widetilde{\mathcal{F}}^{lin}_{\ell+2}}(\mathcal{H}_{\ell+2}, D_{\ell+2})$ by classical path lifting arguments along the fibers of $\mathbb{H}_{\ell+2}$ (just use the same arguments in the proof of Lemma 9.1). Now recall that $\widetilde{\mathcal{F}}_{\ell+1,\ell+2}$ is the germ of the foliation at the corner $c_{\ell+1,\ell+2}$ induced by $\widetilde{\mathcal{F}}$. Since $\Phi_{\ell+1}$ and $\Phi_{\ell+2}$ induce the same bijective map between the spaces of leaves of $\widetilde{\mathcal{F}}_{\ell+1,\ell+2}$ and $\widetilde{\mathcal{F}}^{lin}_{\ell+1,\ell+2}$, then $\Phi_{\ell+2} \circ \Phi^{-1}_{\ell+1}$ fixes the leaves of $\widetilde{\mathcal{F}}^{lin}_{\ell+1,\ell+2}$. Therefore, if we set $\widetilde{f}_{\ell+2} := \Phi^*_{\ell+2}\widetilde{f}^{lin}_{\ell+2}$, then $\widetilde{f}_{\ell+2} = \widetilde{f}_{\ell+1}$ about $c_{\ell+1,\ell+2}$. Proceeding by induction on $j > \ell$ we obtain a multivalued first integral for $\mathcal{B}_+\widetilde{\mathcal{F}}$. Finally, let us verify that $\widetilde{f}_+$ has the desired form. Since $\widetilde{f}^{lin}_j(t_j, x_j) = t_j^{\nu_j}x_j^{\mu_j}$ and Φ_j is of the form $\Phi_j(t_j, x_j) = (t_j, \alpha_j x_j + x_j a_j(t_j, x_j))$, with $\alpha_j \in \mathbb{C}^*$ and $a_j \in \mathfrak{m}_2$ (where $\mathfrak{m}_2$ denotes the maximal ideal of $\mathcal{O}_2$), then a straightforward calculation shows that $\widetilde{f}_j(t_j, x_j) = t_j^{\nu_j}x_j^{\mu_j}U_j(t_j, x_j)$, where $U_j(t_j, x_j) = [\alpha_j + a_j(t_j, x_j)]^{\mu_j} \in \mathcal{O}^*(\mathbb{D}_{1+\epsilon} \times \mathbb{D}_\epsilon)$ for some $\epsilon > 0$. Similarly $\widetilde{f}^{lin}_j(u_j, y_j) = u_j^{k_j\mu_j-\nu_j}y_j^{\mu_j}$ and $\Phi_j(u_j, y_j) = (u_j, \beta_j y_j + y_j b_j(u_j, y_j))$, with $\beta_j \in \mathbb{C}^*$ and $b_j \in \mathfrak{m}_2$. Thus $\widetilde{f}_j(u_j, y_j) = u_j^{k_j\mu_j-\nu_j}y_j^{\mu_j}V_j(u_j, y_j)$, where $V_j(u_j, y_j) = [\beta_j + b_j(u_j, y_j)]^{\mu_j} \in \mathcal{O}^*(\mathbb{D}_{1+\epsilon} \times \mathbb{D}_\epsilon)$ for some $\epsilon > 0$. □

9.1.2 Holomorphic First Integrals and the Geometry of Sing($\mathcal{G}$)

The arguments used in the proof of Lemma 3.4 ensure that $\mathcal{F}$ is resolved together with any "generic" fiber of the *companion fibration* $\frac{y^p}{x^q} \equiv$ const, i.e. $(\mathcal{G} : \eta = 0)$ given by $\eta(x, y) = pxdy - qydx$. In other words, $\mathcal{F}$ and $\mathcal{G}$ are resolved by the same sequence of blowups. In particular, the minimal resolution of $\mathcal{G}$ has the same tree of projective lines of the minimal resolution of any element of $\mathcal{QHS}^{c,1}_{\omega,f}$ and contains its separatrices as fibers. Furthermore, for each $j \neq \text{pr}(\widetilde{\mathcal{F}})$ the foliation $\widetilde{\mathcal{G}}_j$ has a (global) holomorphic first integral of the form

$$\begin{cases} \widetilde{\eta}(t_j, x_j) = d(t_j^{r_j} x_j^{s_j}), \\ \widetilde{\eta}(u_j, y_j) = d(u_j^{k_j s_j - r_j} y_j^{s_j}), \end{cases}$$

where $r_j, s_j \in \mathbb{N}$ are relatively prime. Since $\widetilde{\mathcal{G}}_{\text{pr}(\widetilde{\mathcal{F}})}$ is a radial fibration, then $\widetilde{\mathcal{G}}_{\text{pr}(\widetilde{\mathcal{F}})-1}$ has just one singularity (cf. Fig. 3). More precisely, the corners of the principal divisor are regular points of $\widetilde{\mathcal{G}}$.

9.1.3 Comparing the Indexes of $\mathcal{F}$ and $\mathcal{G}$

First, recall the smooth version of the celebrated Camacho–Sad's index theorem. Let S be a complex surface, $C \subset S$ a smooth analytic curve, and $\mathcal{F}$ a germ of singular foliation defined in a neighborhood of C with just isolated singularities. For each singular point p of $\mathcal{F}$ in S, the Camacho–Sad's index is defined as follows: choose local coordinates for S around p such that C is given by $(y = 0)$. Let $\mathcal{F}$ be given by $\omega = 0$, where $\omega(x, y) = a(x, y)dx + b(x, y)dy$. Then $CS_p(\mathcal{F}, S) = \text{Res}_{x=0} \frac{\partial}{\partial y}(\frac{a}{b}(x, y)|_{y=0})dx$. In particular, if $\omega(x, y) = \mu y(1 + \cdots)dx + \nu x(1 + \cdots)dy$, where $\mu, \nu \neq 0$, then $CS_0(\mathcal{F}, S) = \frac{\mu}{\nu}$. A straightforward calculation shows that this index does not depend on the coordinates.

Theorem 9.3 (Camacho–Sad [5]) *Let S be a complex surface, $C \subset M$ a smooth analytic curve, and $\mathcal{F}$ a germ of singular foliation defined in a neighborhood of S with just isolated singularities. Then*

$$\sum_{p \in Sing(\mathcal{F})} CS_p(\mathcal{F}, S) = C \cdot C$$

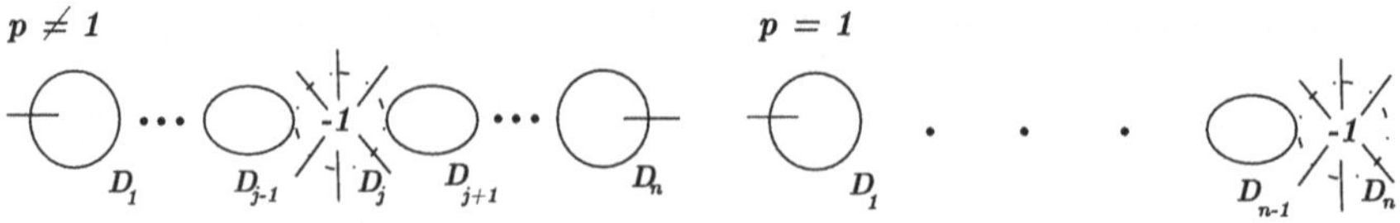

Fig. 3 The resolution tree of $\mathcal{G} : (\frac{x^p}{y^q} = const.)$

where $C \cdot C$ is the self-intersection number of C in S.

As mentioned above, the resolution of the companion fibration $\mathcal{G}$ is regular at the corners of the principal divisor $D_{\mathrm{pr}(\widetilde{\mathcal{F}})}$. Therefore, comparing the Camacho–Sad's indices of $\widetilde{\mathcal{F}}_j$ and $\widetilde{\mathcal{G}}_j$ (going from $\mathrm{pr}(\widetilde{\mathcal{F}})-1$ to 1 and from $\mathrm{pr}(\widetilde{\mathcal{F}})+1$ to n), we conclude form the Camacho–Sad's index theorem that

$$\nu_j s_j - \mu_j r_j \neq 0 \text{ for all } j \neq \mathrm{pr}(\widetilde{\mathcal{F}}). \tag{9.3}$$

Remark 9.2 If $\mathrm{Sep}(\mathcal{F})$ is commode, then $\mathcal{F}$ is automatically generic. In fact, Lemma 3.4 ensures that any Hopf components of $\widetilde{\mathcal{F}}$ about an end of D has just one singularity. Therefore, with arguments similar to that used for $\mathcal{G}$, one can verify that each Hopf component $\widetilde{\mathcal{F}}_j$ has linear and periodic holonomy for all $j \neq \mathrm{pr}(\widetilde{\mathcal{F}})$. Thus it is linearizable and has a holomorphic first integral (cf. [26]).

9.2 *Cocycles Fixing the Leaves of $\mathcal{F}$ and $\mathcal{G}$*

In this section we show how to trivialize $\Phi(\mathcal{F})$ and prove Theorem B.

9.2.1 Fixing Leaves Locally

We first introduce some notation in order to clarify the ideas. Let $\mathcal{F}$ be a germ of reduced singular foliation at $(\mathbb{C}^2, 0)$. Since it is characterized by its (local) holonomy group (cf. [26, 28]), then given a smooth fibration f it is classical to identify the space of leaves of $\mathcal{F}$ with the the quotient of $(\mathbb{C}^2, 0)$ by the action of the unique fibre preserving suspension of this holonomy in $\mathrm{Aut}(\mathcal{F}, f)$. Therefore, we say that $\phi \in \mathrm{Aut}(\mathcal{F})$ fixes the leaves of $\mathcal{F}$ if its action in the space of leaves of $\mathcal{F}$ is trivial. We denote the set of such automorphisms by $\mathrm{Fix}(\mathcal{F})$. As before, this condition can be verified explicitly by path lifting arguments. In particular, if U is an open neighborhood of some point in the exceptional divisor of $\mathcal{B}_+\mathcal{F}$ (respect. $\mathcal{B}_-\mathcal{F}$) and $\phi \in \mathrm{Diff}(U)$, then we say that ϕ fixes the leaves of $\mathcal{B}_+\mathcal{F}$ (respect. $\mathcal{B}_-\mathcal{F}$), denoting it just by $\phi \in \mathrm{Fix}(\mathcal{B}_+\mathcal{F})$ (respect. $\mathcal{B}_-\mathcal{F}$), if ϕ preserves the level sets of the first integrals introduced in Lemma 9.2.

Let $\mathcal{QHS}_\omega$ denote the set of $\mathcal{QHS}$ foliations that are analytically equivalent to $(\mathcal{F}_\omega : \omega = 0)$, and $f = 0$ be the separatrix set of $\mathcal{F}_\omega$. From the discussion in Sect. 8.2, in order to determine the moduli space $\mathcal{QHS}^c_{\omega,f}/\mathcal{QHS}_\omega$, we have to pick a fixed model $\mathcal{F}^o \in \mathcal{QHS}^c_{\omega,f}$ and a collection of projective charts (Φ_j) for any $\mathcal{F} \in \mathcal{QHS}^c_{\omega,f}$ (with respect to $\mathcal{F}^o$) preserving $f = 0$. In order to simplify the expression of (Φ_j), it is natural to ask it to preserve not just $f = 0$ but the whole companion fibration $\mathcal{G}$. On the other hand, it is not difficult to see that the geometry of the exceptional divisor of $\mathcal{F}$ allows us to simplify inductively the transversal structure of $\Phi(\mathcal{F})$ in such a way that each $\Phi_{i,j}$ fixes (locally) the leaves of $\mathcal{F}$. This, of course, will also simplify the expression of $\Phi(\mathcal{F})$. In the next sections we shall see that it is possible to do both at the same time, simplifying a lot the expression of $\Phi(\mathcal{F})$.

9.2.2 Projective Charts and First Integrals Adapted to a Fixed Model

In each componentwise equivalence class pick a model $(\mathcal{F}^o : \omega^o = 0)$ and fix first integrals f_+^o and f_-^o for $\mathcal{B}_+\mathcal{F}^o$ and $\mathcal{B}_-\mathcal{F}^o$ as in Lemma 9.2.

Now, for any $\mathcal{F} \in \mathcal{QHS}^c_{\omega^o,f}$, we shall construct first integrals for $\mathcal{B}_+\mathcal{F}$ and $\mathcal{B}_-\mathcal{F}$ and a collection of projective charts taking the level sets of the first integral of $\mathcal{B}_+\mathcal{F}$ (respect. $\mathcal{B}_-\mathcal{F}$) in the level set of f_+^o (respect. f_-^o). In order to be more precise, let us first introduce some notions: one says that a collection of projective charts (Φ_j) for $\mathcal{F} \in \mathcal{QHS}^c_{\omega^o,f}$ with respect to $\mathcal{F}^o$ and first integrals f_+ for $\mathcal{B}_+\mathcal{F}$ and f_- for $\mathcal{B}_-\mathcal{F}$ are *adapted* to $(\mathcal{F}^o, f_+^o, f_-^o)$ if each Φ_i takes $(f_- = c)$ in $(f_-^o = c)$ and Φ_j takes $(f_+ = c)$ in $(f_+^o = c)$ for all $i = 1, \dots, \ell$, $j = \ell, \dots, n$, and all $c \in \mathbb{C}$ sufficiently close to zero.

Lemma 9.4 *For each $\mathcal{F} \in \mathcal{QHS}^c_{\omega^o,f}$ there is a collection of projective charts (Φ_j) for $\mathcal{F}$ with respect to $\mathcal{F}^o$ and first integrals f_+ for $\mathcal{B}_+\mathcal{F}$ and f_- for $\mathcal{B}_-\mathcal{F}$ adapted to $(\mathcal{F}^o, f_+^o, f_-^o)$.*

Proof We prove the statement for the positive branch case, the other one being completely analogous. Pick $\Phi_\ell \in \mathrm{Diff}_{\tilde{\mathcal{F}}_\ell, \tilde{\mathcal{F}}^o_\ell}(\mathcal{H}_\ell, D_\ell)$ and let $\tilde{f}_{\ell,\ell+1} := \Phi_\ell^* \tilde{f}^o_{\ell,\ell+1}$, where $\tilde{f}^o_{\ell,\ell+1}$ is the germ of $\tilde{f}^o_{\ell+1}$ at the corner $c_{\ell,\ell+1} := D_\ell \cap D_{\ell+1}$. Let p be a regular point of $D_{\ell+1}$ near the corner $c_{\ell,\ell+1}$ and Σ_p be the fiber of $\mathbb{H}_{\ell+1}$ over p. Recall that Φ_ℓ induces a bijective map between the spaces of leaves of $\tilde{\mathcal{F}}_{\ell,\ell+1}$ and $\tilde{\mathcal{F}}^o_{\ell,\ell+1}$ which can be realized as $\phi_{\ell+1} \in \mathrm{Diff}(\Sigma_p, p)$. In particular, $\phi_{\ell+1}$ takes $\mathrm{Hol}_{\Sigma_p}(\tilde{\mathcal{F}}_{\ell,\ell+1}, D_{\ell+1})$ onto $\mathrm{Hol}_{\Sigma_p}(\tilde{\mathcal{F}}^o_{\ell,\ell+1}, D_{\ell+1})$. Since $\tilde{\mathcal{F}}_{\ell+1}$ has just two singularities, then the spaces of leaves of $\tilde{\mathcal{F}}_{\ell,\ell+1}$ and $\tilde{\mathcal{F}}_{\ell+1}$ coincide as the spaces of leaves of $\tilde{\mathcal{F}}^o_{\ell,\ell+1}$ and $\tilde{\mathcal{F}}^o_{\ell+1}$. Therefore Proposition 7.1 ensures that one can extend $\phi_{\ell+1}$ to $\Phi_{\ell+1} \in \mathrm{Diff}_{\tilde{\mathcal{F}}_{\ell+1}, \tilde{\mathcal{F}}^o_{\ell+1}}(\mathcal{H}_{\ell+1}, D_{\ell+1})$ along the fibers of $\mathbb{H}_{\ell+1}$ by classical path lifting arguments (just use the same arguments in the proof of Lemma 9.1). Since Φ_ℓ and $\Phi_{\ell+1}$ induce the same bijective map between the spaces of leaves of $\tilde{\mathcal{F}}_{\ell,\ell+1}$ and $\tilde{\mathcal{F}}^o_{\ell,\ell+1}$, then $\Phi_\ell \circ \Phi_{\ell+1}^{-1}$ fixes the leaves of $\tilde{\mathcal{F}}^o_{\ell,\ell+1}$. Therefore, if we let $\tilde{f}_{\ell+1} := \Phi^*_{\ell+1} \tilde{f}^o_{\ell+1}$, then $\tilde{f}_{\ell+1} = \tilde{f}_{\ell,\ell+1}$ about $c_{\ell,\ell+1}$. Proceeding by induction on $j > \ell + 1$ we obtain a multivalued first integral for $\mathcal{B}_+\tilde{\mathcal{F}}$ and the collection of projective charts with the desired properties. □

In order to give a better understanding of the proof of the next lemma, let us make a brief digression about the simultaneous linearization of two transversal non-singular foliations. As it is well known, two germs of non-singular holomorphic foliations $\mathcal{F}$ and $\mathcal{G}$ can be simultaneously linearized. In fact the problem can be easily reduced to the following: given the germs of holomorphic functions $f(x, y) = yU(x, y)$, $f^o(x, y) = y$ and $g(x, y) = x$ about the origin, where $U \in \mathcal{O}_2^*$, find out $\Phi \in \mathrm{Diff}(\mathbb{C}^2, 0)$ such that $\Phi^* f^o = f$ and $\Phi^* g = g$. If we let

$\Phi(x, y) = (a(x, y), b(x, y))$, then the problem reduces to the following system of equations

$$\begin{cases} b(x, y) = yU(x, y); \\ a(x, y) = x. \end{cases}$$

whose solution is evident. The core of the proof of the following result is analogous (cf. (9.4)).

Lemma 9.5 *For each $\mathcal{F} \in \mathcal{QHS}^c_{\omega^o, f}$ and each $j = 1, \ldots, \ell - 1$ (respect. $j = \ell + 1, \ldots, n$) there is $\Psi_j \in Fix(\widetilde{\mathcal{G}}_j)$ such that $f_-^{lin} = \Psi_{j*} f_-$ (respect. $f_+^{lin} = \Psi_{j*} f_+$).*

Proof We prove the result for the positive branch, the negative one being completely analogous. In view of the second part of Lemma 9.2, one just has to find a solution $\Phi_j := \Psi_j^{-1} := (a_j(t_j, x_j), b_j(t_j, x_j))$ to the system of equations

$$\begin{cases} \Phi_j^* \widetilde{f}^{lin}(t_j, x_j) = \widetilde{f}^o(t_j, x_j) \\ \Phi_j^* \widetilde{g}(t_j, x_j) = \widetilde{g}(t_j, x_j) \end{cases} \Leftrightarrow \begin{cases} a_j(t_j, x_j)^{\nu_j} b_j(t_j, x_j)_j^{\mu_j} = t_j^{\nu_j} x_j^{\mu_j} U(t_j, x_j) \\ a_j(t_j, x_j)^{r_j} b_j(t_j, x_j)^{s_j} = t_j^{r_j} x_j^{s_j} \end{cases} \tag{9.4}$$

But this can be given in the affine charts (t_j, x_j) by

$$\begin{cases} a_j(t_j, x_j) = t_j [U(t_j, x_j)]^{\frac{s_j}{\nu_j s_j - \mu_j r_j}}, \\ b_j(t_j, x_j) = x_j [U(t_j, x_j)]^{\frac{r_j}{\mu_j r_j - \nu_j s_j}}, \end{cases}$$

which is well defined by (9.3). A straightforward calculation shows that the expression of Φ_j in the affine chart (u_j, y_j) is given by

$$\Phi_j(u_j, y_j) = (u_j [V(u_j, y_j)]^{\frac{s_j}{\mu_j r_j - \nu_j s_j}}, y_j [V(u_j, y_j)]^{\frac{r_j - k_j s_j}{\mu_j r_j - \nu_j s_j}})$$

where $V(u_j, y_j) := U(1/u_j, u_j^{k_j} y_j) \in \mathcal{O}^*(\mathbb{D}_{1+\epsilon}, \mathbb{D}_\epsilon)$. □

Remark 9.3 As a straightforward consequence of the above lemma, there is a system of coordinates $\widetilde{\mathcal{A}}_j := \{(\widetilde{t}_j, \widetilde{x}_j), (\widetilde{u}_j, \widetilde{y}_j) \in \mathbb{C}^2 : \widetilde{u}_j = 1/\widetilde{t}_j, \widetilde{y}_j = \widetilde{t}_j^{k_j} \widetilde{x}_j\}$ for $\mathbb{H}_j(-k_j)$ such that the first integrals of $\widetilde{\mathcal{F}}_j$ and $\widetilde{\mathcal{G}}_j$ are given respectively by $\widetilde{t}_j^{\nu_j} x_j^{\mu_j}, \widetilde{u}_j^{k_j \mu_j - \nu_j} \widetilde{y}_1^{\mu_j}$ and $\widetilde{t}_j^{r_j} \widetilde{x}_j^{s_j}, \widetilde{u}_j^{k_j s_j - r_j} \widetilde{y}_j^{s_j}$ for all $j \neq \ell$.

Now we enrich a little bit the structure preserved by the cocyles.

Lemma 9.6 *Let $\mathcal{F} \in \mathcal{QHS}^c_{\omega^o, f}$, then there is a collection of projective charts (Φ_j) with respect to $\mathcal{F}^o$ such that $\Phi_j \in Fix(\widetilde{\mathcal{G}}_j)$ for all $j = 1, \ldots, n$, $\Phi_j \circ \Phi_{j+1}^{-1} \in Fix(\mathcal{B}_+ \mathcal{F}^o)$ for all $j = \ell, \ldots, n - 1$ and $\Phi_j \circ \Phi_{j+1}^{-1} \in Fix(\mathcal{B}_- \mathcal{F}^o)$ for all $j = 1, \ldots, \ell - 1$.*

Proof From Lemma 9.4 one knows that there is a collection of projective charts (Υ_j) for $\mathcal{F}$ with respect to $\mathcal{F}^o$ and first integrals f_+ for $\mathcal{B}_+ \mathcal{F}$ and f_- for $\mathcal{B}_- \mathcal{F}$ adapted to $(\mathcal{F}^o, f_+^o, f_-^o)$, where $\Upsilon_\ell \in \mathrm{Fix}(\mathcal{G}_\ell)$; thus we let $\Phi_\ell := \Upsilon_\ell$. Now we construct Φ_j for $j \neq \ell$. From Lemma 9.5 there are $\Psi_j, \Xi_j \in \mathrm{Diff}_{\tilde{\mathcal{F}}_j, \tilde{\mathcal{F}}_j^{lin}}(\mathcal{H}_j, D_j)$ such that $\Xi_{j*}(f_+) = f_+^{lin}$ (respect. $\Xi_{j*}(f_-) = f_-^{lin}$), $\Psi_{j*}(f_+^o) = f_+^{lin}$ (respect. $\Psi_{j*}(f_-^o) = f_-^{lin}$) and $\Xi_j, \Psi_j \in \mathrm{Fix}(\mathcal{G}_j)$. Then define $\Phi_j := \Psi_j^{-1} \circ \Xi_j$ in order to obtain the following commutative diagram

$$\begin{array}{ccc} & \tilde{\mathcal{F}}_j & \\ {}^{\Phi_j}\swarrow & \circlearrowleft & \searrow^{\Xi_j} \\ \tilde{\mathcal{F}}_j^o & \xrightarrow{\Psi_j} & \tilde{\mathcal{F}}_j^{lin} \end{array} \tag{9.5}$$

□

9.2.3 Trivializing Cocycles

Here we follow the program outlined in Sect. 9.2.1 in order to trivialize the cocycles associated with a given fixed model. Recall that f_+^o (respect. f_-^o) is the multivalued first integral for $\mathcal{B}_+\mathcal{F}^o$ (respect. $\mathcal{B}_-\mathcal{F}^o$).

Lemma 9.7 *Let* $\Psi_{j,j+1} \in Fix(\tilde{\mathcal{F}}_{j,j+1}^{lin}) \cap Fix(\tilde{\mathcal{G}}_{j,j+1})$ *for* $j = 1, \ldots, n-1$. *Then* $\Psi_{j,j+1}$ *has a unique extension to* $\Psi_{j+1} \in Fix(\tilde{\mathcal{F}}_{j+1}^{lin}) \cap Fix(\tilde{\mathcal{G}}_{j+1})$ *for all* $j \geq \ell$. *Analogously,* $\Psi_{j,j+1}$ *has a unique extension to* $\Psi_j \in Fix(\tilde{\mathcal{F}}_j^{lin}) \cap Fix(\tilde{\mathcal{G}}_j)$ *for all* $j < \ell$.

Proof We prove the first part of the Lemma, the second one being completely analogous. We adopt the coordinate system $\mathcal{A}$ introduced in (9.2). Notice that the corner $c_{j,j+1} = D_j \cap D_{j+1}$ is represented by the origin in the affine chart (t_{j+1}, x_{j+1}) for $\mathcal{H}_{j+1}$, thus $\Phi_{j,j+1}(t_{j+1}, x_{j+1}) = (a_{j+1}(t_{j+1}, x_{j+1}), b_{j+1}(t_{j+1}, x_{j+1}))$, where a_{j+1}, $b_{j+1} \in \mathcal{O}(\mathbb{D}_{\epsilon_1} \times \mathbb{D}_{\epsilon_2})$. Since $\Phi_{j,j+1} \in \mathrm{Fix}(\tilde{\mathcal{F}}_{j,j+1}^{lin}) \cap \mathrm{Fix}(\tilde{\mathcal{G}}_{j,j+1})$, then (denoting $i := j+1$ for simplicity) a_i and b_i satisfy the following system of equations

$$\begin{cases} a_i(t_i, x_i)^{\nu_i} b_i(t_i, x_i)^{\mu_i} = t_i^{\nu_i} x_i^{\mu_i} \\ a_i(t_i, x_i)^{r_i} b_i(t_i, x_i)^{s_i} = t_i^{r_i} x_i^{s_i} \end{cases}$$

whose solutions are of the form $a_i(t_i, x_i) = \alpha t_i$ and $b_i(t_i, x_i) = \beta x_i$, where $\alpha, \frac{1}{\beta}$ are $(\nu_i s_i - \mu_i r_i)$-roots of unity. The uniqueness is straightforward since both $\Phi_{j,j+1}$ and its extension Φ_{j+1} are holomorphic. □

Now we are in a position to show that the cocycles generated by generic elements of $\mathcal{QHS}^c_{\omega^o, f}$ are in fact trivial.

Lemma 9.8 *Let* $\Phi_{j,j+1} \in Fix(\tilde{\mathcal{F}}_{j,j+1}^o) \cap Fix(\tilde{\mathcal{G}}_{j,j+1})$ *for* $j = 1, \ldots, n-1$. *Then* $\Phi_{j,j+1}$ *has a unique extension to* $\Phi_{j+1} \in Fix(\tilde{\mathcal{F}}_{j+1}^o) \cap Fix(\tilde{\mathcal{G}}_{j+1})$ *for all* $j \geq \ell$. *Analogously,* $\Phi_{j,j+1}$ *has a unique extension to* $\Phi_j \in Fix(\tilde{\mathcal{F}}_j^o) \cap Fix(\tilde{\mathcal{G}}_j)$ *for all* $j < \ell$.

Proof We prove the first part of the Lemma, since the second one is completely analogous. Let $(\Psi_j) \in \text{Fix}(\widetilde{\mathcal{G}}_j)$, $j = 1, \ldots, n$, be the collection of maps introduced in Lemma 9.5 and $\overline{\Phi}_{j,j+1} := \Psi_{j+1} \circ \Phi_{j,j+1} \circ (\Psi_{j+1})^{-1}$. Since $\Psi_{j*} f_+^o = f_+^{lin}$, then $\overline{\Phi}_{j,j+1} \in \text{Fix}(\mathcal{F}_{j,j+1}^{lin}) \cap \text{Fix}(\widetilde{\mathcal{G}}_{j,j+1})$ for all $j = \ell, \ldots, n-1$ (cf. (9.5)). Hence Lemma 9.7 assures that $\overline{\Phi}_{j,j+1}$ can be extended to $\overline{\Phi}_{j+1} \in \text{Fix}(\widetilde{\mathcal{F}}_{j+1}^{lin}) \cap \text{Fix}(\widetilde{\mathcal{G}}_{j+1})$ for all $j = \ell, \ldots, n-1$. Therefore, $\Phi_{j+1} := (\Psi_{j+1})^{-1} \circ \overline{\Phi}_{j+1} \circ \Psi_{j+1} \in \text{Fix}(\widetilde{\mathcal{F}}_{j+1}^o) \cap \text{Fix}(\widetilde{\mathcal{G}}_{j+1})$ extends $\Phi_{j,j+1}$. A similar reasoning works for all $j < \ell$. □

9.2.4 Extending Semi-local Conjugations

Here we use all the machinery developed above in order to prove Theorem B. In fact, we show that the vanishing of the cocycles in the positive (respect. negative) branch means that we can extend to the positive (respect. negative) branch any conjugation from $\widetilde{\mathcal{F}}_\ell$ to $\widetilde{\mathcal{F}}_\ell^o$.

Proof of Theorem B Let $\mathcal{F}^o \in \mathcal{QHS}_{\omega^o,f}^c$, where $(\mathcal{F}^o : \omega^o = 0)$ is a fixed model. Let (Φ_j) be a collection of projective charts given by Lemma 9.6 and $\Phi_{i,j} := \Phi_i \circ \Phi_j^{-1}$. Then Lemma 9.8 ensures that there is $\Xi_{\ell+1} \in \text{Fix}(\widetilde{\mathcal{F}}_{\ell+1}^o) \cap \text{Fix}(\widetilde{\mathcal{G}}_{\ell+1})$ such that $\Xi_{\ell+1} = \Phi_{\ell,\ell+1}$. Let $(\Phi_j^{(1)})$ be given by $\Phi_j^{(1)} := \Phi_j$ for all $j \neq \ell+1$ and $\Phi_{\ell+1}^{(1)} := \Xi_{\ell+1} \circ \Phi_{\ell+1}$. Then $(\Phi_j^{(1)})$ is a collection of projective charts such that $\Phi_{j,j+1}^{(1)} \in \text{Fix}(\widetilde{\mathcal{F}}_{j,j+1}^o) \cap \text{Fix}(\widetilde{\mathcal{G}}_{j,j+1})$ and $\Phi_{\ell,\ell+1}^{(1)} = \text{id}$. Repeating inductively the same arguments for $j > \ell+1$ we obtain a collection of projective charts $(\Phi_j^{(n-\ell)})$ such that $\Phi_{j,j+1}^{(n-\ell)} \in \text{Fix}(\widetilde{\mathcal{F}}_{j,j+1}^o) \cap \text{Fix}(\widetilde{\mathcal{G}}_{j,j+1})$ for all $j = 1, \ldots, n-1$ and $\Phi_{j,j+1}^{(n-\ell)} = \text{id}$ for all $j \geq \ell$. An analogous reasoning works for all $j < \ell$, generating a collection of projective charts $(\Phi_j^{(n-1)})$ such that $\Phi_{j,j+1}^{(n-1)} = \text{id}$ for all $j = 1, \ldots, n-1$. In particular, this family paste together in order to define a map $\Phi \in \text{Diff}(M, D)$ such that $\Phi_* \widetilde{\mathcal{F}} = \widetilde{\mathcal{F}}^o$, as desired. □

References

1. Arnold, V.I.: Chapitres supplémentaires de la théorie des équations différentielles ordinaires, edns., pp. 324. Mir, Moscou (1980). MR 0626685 (83a:34003)
2. Berzolari, L.: Allgemeine Theorie der höheren ebenen algebraischen Kurven, Enzyklopädie der math. Wissenschaften, Bd. III 2, 1, 313–455 (1906)
3. Brieskorn, E.: Beispiele zur Differentietopologie von Singularitäten. Invent. Math. **2**, 1–14 (1966). MR 0206972 (34 #6788)
4. Berthier, M., Meziani, R., Sad, P.: On the classification of nilpotent singularities. Bull. Sci. Math. **123**(5)351–370 (1999). MR 1703202 (2000j:32052)
5. Camacho, C., Sad, P.: Invariant varieties through singularities of holomorphic vector fields. Ann. Math. (2) **115**(3), 579–595 (1982). MR 0657239 (83m:58062)
6. Camacho, C., Lins-Neto, A., Sad, P.: Topological invariants and equidesingularization for holomorphic vector fields. J. Differ. Geom. **20**(1), 143–174 (1984). MR 0772129 (86d:58080)

7. Camacho, C., Scárdua, B.A.: The transcendence of the solutions of a complex differential equation. In: Mozo, J. (ed.) Ecuationes Diferenciales y Singularidades, Publicaciones de la Universidad de Valladolid (1997)
8. Camacho, C., Scárdua, B.A.: Beyond Liouvillian transcendence. Math. Res. Lett. **6**(1), 31–41 (1999). MR 1682717 (2000b:32057)
9. Camacho, C., Scárdua, B.A.: Complex foliations with algebraic limit sets, Géométrie Complexe et Systémes Dynamiques (Orsay, 1995), Astérisque (2000), no. 261, xi, 57–88.MR 1755437 (2001d:32045)
10. Camacho, C., Scárdua, B.A.: Holomorphic foliations with Liouvillian first integrals, Ergodic Theory Dynam. Systems **21**(3), 717–756 (2001). MR 1836428 (2002k:37080)
11. Camacho, C., Scárdua, B.A.: Erratum to: "Holomorphic foliations with Liouvillian first integrals" [Ergodic Theory Dynam. Systems **21**(3), 717–756 (2001); MR 1836428 (2002k:37080)]. Ergodic Theory Dynam. Systems **23**(3) 985–987 (2003). MR 1992674 (2004g:37062)
12. Câmara, L.M.: Non-linear analytic differential equations and their invariants, IMPA preprint Série C 5/2001. http://preprint.impa.br/visualizar?id=5636
13. Cerveau, D., Moussu, R.: Groups d'automorphismes de $(\mathbb{C}, 0)$ et équations différentielles $y\,dy + \cdots = 0$. Bull. Soc. Math. France **116**(4), 459–488 (1988). MR 1005391 (90m:58192)
14. Ehresmann, C.: Catégories différentiable et géométrie différentielle, Université de Montréal, Montréal (1961). MR 0149401 (26 #6890)
15. Ehresmann, C.: Catégories et structures. Dunod, Paris (1965). MR 0213410 (35 #4274)
16. Elizarov, P.M., Il'yashenko, Yu.S., Shcherbakov, A.A., Voronin, S.M.: Finitely generated groups of germs of one-dimensional conformal mappings, and invariants for complex singular points of analytic foliations of the complex plane. Nonlinear Stokes phenomena, pp. 57–105, Advances in Soviet Mathematics, 14, American Mathematical Society, Providence, RI (1993). MR 1206042 (94e:32055)
17. Genzmer, Y.: Analytical and formal classifications of quasi-homogeneous foliations in $(\mathbb{C}^2, 0)$. J. Differ. Equ. **245**(6), 1656–1680 (2008). MR 2436456 (2009j:37075)
18. Grauert, H.: Über Modifikationen und exzeptionelle analytische Mengen. Math. Ann. **146**, 331–368 (1962). MR 0137127 (25 #583)
19. Grothendieck, A.: Sur la classification des fibrés holomorphe sur la sphere de Riemann. Am. J. Math. **79**, 121–138 (1957). MR 0087176 (19,315b)
20. Haefliger, A.: Structures feuilletées et cohomologie à valeur dans un faisceux de groupoïdes. Comment. Math. Helv. **32** (1958), 248–329. MR 0100269 (20 #6702)
21. Hefez, A., Hernandes, M.E.: The analytic classification of plane branches. Bull. Lond. Math. Soc. **43**(2), 289–298 (2011). MR 2781209 (2012c:14065)
22. Kang, C.: Analytic types of plane curve singularities defined by weighted homogeneous polynomials. Trans. Am. Math. Soc. **352**(9), 3995–4006 (2000). MR 1661266 (2000m:32035)
23. Laufer, H.B.: Normal two dimensional singularities. In: Annals of Mathematics Studies, vol. 71. Princeton University Press, Princeton, N.J.; University of Tokyo Press, Tokyo (1971). MR 0320365 (47 #8904)
24. Lins-Neto, A.: Construction of singular holomorphic vector fields and foliations in dimension two. J. Diff. Geom. **26**(1), 3–31 (1987). MR 0892029 (88f:32047)
25. Mattei, J.-F.: Quasihomogénéité et equireducibilité de feuilletages holomorphes em dimenson deux, Géométrie complexe et systèmes dynamiques (Orsay, 1995), Astérisque No. 261 (2000), xix, 253–276. MR 1755444 (2001g:37069)
26. Mattei, J.-F., Moussu, R.: Holonomie et intégrales premières. Ann. Scient. Ec. Norm. Sup., 4e série. **13**, 469–523 (1980). MR 0608290 (83b:58005)
27. Martinet, J., Ramis, J.-P.: Problèmes de modules pour des équations différentielles non-linéaires du premier ordre. Inst. Hautes Études Sci. Publ. Math. **55**, 63–164 (1982). MR 0672182 (84k:34011)
28. Martinet, J., Ramis, J.-P.: Classification analytique des équations différentielles non-linéaires résonnantes du premier ordre. Ann. Sci. École Norm. Sup. **16**(4), 571–621 (1983, 1984). MR 0740592 (86k:34034)

29. Moussu, R.: Holonomie évanescente des équations différentielles dégénérées transverses. In: Singularities and Dynamical Systems (Iráklion, 1983), North-Holland Mathematics Studies, vol. 103, pp. 161–173. North-Holland, Amsterdam (1985). MR 0806187 (87d:58008)
30. Poincaré, H.: Note sur les propriétés des fonctions définies par des équations différentielles. J. Ec. Pol. **45**^{e} cahier, 13–26 (1878)
31. Seindenberg, A.: Reduction of the singularities of the differentiable equation $Ady = Bdx$. Am. J. Math **90**, 248–269 (1968). MR 0220710 (36 #3762)
32. Strozyna, E.: Orbital formal normal forms for general Bogdanov-Takens singularity. J. Differ. Equ. **193**(1) 239–259 (2003). MR 1994066 (2004e:37074)
33. Yoccoz, J.-C.: Linearisation des germes de diffeomorphismes holomorphes de $(\mathbf{C}, 0)$. C. R. Acad. Sci. Paris Sér. I Math. **306**(1), 55–58 (1988). MR 0929279 (89i:58123)
34. Zariski, O.: On the topology of algebroid singularities. Am. J. Math. **54**(3), 453–465 (1932). MR 1507926

On the Roots of an Extended Lens Equation and an Application

Mutsuo Oka

Abstract We consider zero points of a generalized Lens equation $L(z, \bar{z}) = \bar{z}^m - p(z)/q(z)$ and also harmonically splitting Lens type equation $L^{hs}(z, \bar{z}) = r(\bar{z}) - p(z)/q(z)$ with deg $q(z) = n$, deg $p(z) \leq n$ whose numerator is a mixed polynomials, say $f(z, \bar{z})$, of degree $(n + m; n, m)$. To such a polynomial, we associate a strongly mixed weighted homogeneous polynomial $F(\mathbf{z}, \bar{\mathbf{z}})$ of two variables and we show the topology of Milnor fibration of F is described by the number of roots of $f(z, \bar{z}) = 0$.

Keywords Lens equation · Mixed curves · Link components

2000 Mathematics Subject Classification 14P05 · 14N99

1 Introduction

Consider a mixed polynomial of one variable $f(z, \bar{z}) = \sum_{\nu,\mu} a_{\nu,\mu} z^\nu \bar{z}^\mu$. We denote the set of roots of f by $V(f)$. Assume that $z = \alpha$ is an isolated zero of $f = 0$. Put $f(z, \bar{z}) = g(x, y) + ih(x, y)$ with $z = x + iy$. A root α is called *simple* if the Jacobian $J(g, h)$ is not vanishing at $z = \alpha$. We call α an orientation preserving or positive root (respectively orientation reversing, or negative), if the Jacobian $J(g, h)$ is positive (resp. negative) at $z = \alpha$. There are two basic questions.

(1) Determine the number of roots with sign.
(2) Determine the number of roots without sign.

M. Oka (✉)
Department of Mathematics, Tokyo University of Science, 1-3 Kagurazaka, Shinjuku-ku, Tokyo 162-8601, Japan
e-mail: oka@rs.kagu.tus.ac.jp

R. N. Araújo dos Santos et al. (eds.), *Singularities and Foliations. Geometry, Topology and Applications*, Springer Proceedings in Mathematics & Statistics 222, https://doi.org/10.1007/978-3-319-73639-6_16

1.1 Number of Roots with Sign

Let C be a mixed projective curve of polar degree d defined by a strongly mixed homogeneous polynomial $F(\mathbf{z}, \bar{\mathbf{z}})$, $\mathbf{z} = (z_1, z_2, z_3)$ of radial degree $d_r = d + 2s$ and let $L = \{z_3 = 0\}$ be a line in $\mathbb{P}^2$. Recall that $F(\mathbf{z}, \bar{\mathbf{z}})$ is strongly mixed weighted homogeneous if it is polar and radial weighted homogeneous with respect to the same weight vector [9]. We assume that L intersects C transversely.

Proposition 1 (Theorem 4.1, [9]) *With the hypothesis above, the fundamental class $[C]$ is mapped to $d[\mathbb{P}^1]$ and thus the intersection number $[C] \cdot [L]$ is given by d. This is also given by the number of the roots of $F(z_1, z_2, 0) = 0$ in $\mathbb{P}^1$ counted with sign.*

We assume that the point at infinity $z_2 = 0$ is not in the intersection $C \cap L$ and use the affine coordinate $z = z_1/z_2$. Then $C \cap L$ is described by the roots of the mixed polynomial $f(z) := F(z, 1, 0)$ which is written as

$$f(z) = z^{d+s}\bar{z}^s + (\text{lower terms}) = 0$$

with respect to the mixed degree. The second term is a linear combination of monomials $z^a\bar{z}^b$ with $a + b < d + 2s$, $a \le d + s$, $b \le s$.

Generic mixed polynomials do not come from mixed projective curves through a holomorphic line section as above. The following is useful to compute the number of zeros with sign of such polynomials. Let $f(z, \bar{z})$ be a given mixed polynomial of one variable, we consider the filtration by the degree:

$$f(z, \bar{z}) = f_d(z, \bar{z}) + f_{d-1}(z, \bar{z}) + \cdots + f_0(z, \bar{z}).$$

Here $f_\ell(z, \bar{z}) := \sum_{\nu+\mu=\ell} c_{\nu,\mu} z^\nu \bar{z}^\mu$. Note that we have a unique factorization of f_d as follows.

$$\begin{aligned} f_d(z, \bar{z}) &= c z^p \bar{z}^q \textstyle\prod_{j=1}^s (z + \gamma_j \bar{z})^{\nu_j}, \\ & p + q + \textstyle\sum_{j=1}^s \nu_j = d,\ c \in \mathbb{C}^*. \end{aligned}$$

where $\gamma_1, \ldots, \gamma_s$ are mutually distinct non-zero complex numbers. We say that $f(z, \bar{z})$ is *admissible at infinity* if $|\gamma_j| \ne 1$ for $j = 1, \ldots, s$. For non-zero complex number ξ, we put

$$\varepsilon(\xi) = \begin{cases} 1 & |\xi| < 1 \\ -1 & |\xi| > 1 \end{cases}$$

and we consider the following integer:

$$\beta(f) := p - q + \sum_{j=1}^{s} \varepsilon(\gamma_j)\nu_j,$$

The following equality holds.

Theorem 2 ([10]) *Assume that $f(z, \bar{z})$ is an admissible mixed polynomial at infinity. Then the total number of roots with sign is equal to $\beta(f)$.*

Remark 3 Here if α is a non-simple root, we count the number with multiplicity. The multiplicity is defined by the local rotation number at α of the normalized Gauss mapping $S_\varepsilon(\alpha) \to S^1, z \mapsto f(z)/|f(z)|$.

1.2 Number of Roots Ignoring the Sign

In this paper, we are interested in the total number of $V(f)$ which we denote by $\rho(f)$, the cardinality of $\sharp V(f)$ for particular classes of mixed polynomials ignoring the sign. The notion of the multiplicity is not well defined for a root without sign. Thus we assume that roots are all simple. The problem is that $\rho(f)$ is not described by the highest degree part f_d, which was the case for the number of roots with sign $\beta(f)$. We will give an example of mixed polynomial below $\rho(f) = n^2$. Another example is known by Wilmshurst [13].

Example 4 Let us consider the Chebycheff polynomial $T_n(x)$. It has two critical values 1 and -1 and the roots of $T_n(x) = 0$ are in the interval $(-1, 1)$. Consider a polynomial

$$F(x, y) = (y - T_n(x) + i(x - aT_n(by)), \quad a, b \gg 1.$$

By the assumption $a, b \gg 1$, $F = 0$ has n^2 roots in $(-1, 1) \times (-1, 1)$. Consider F as a mixed polynomial by substituting x, y by $x = (z + \bar{z})/2$, $y = -i(z - \bar{z})/2$. This example gives an extreme case for which the possible complex roots (by Bezout theorem) of $\Re F = \Im F = 0$ are all real roots (Fig. 1).

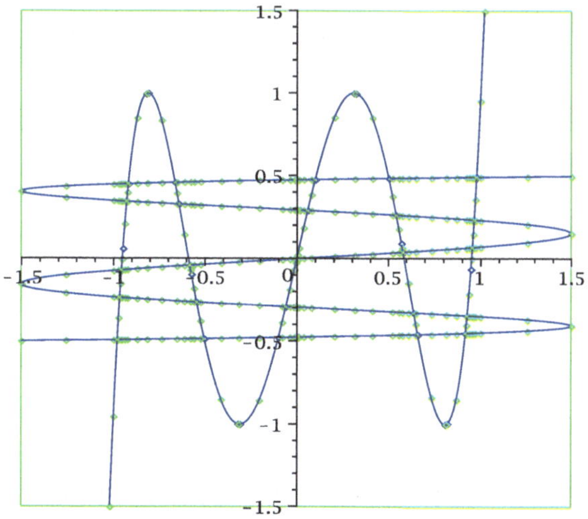

Fig. 1 Roots of $F(x, y) = 0, n = 5, a = 3/2, b = 2$

The above example shows implicitly that the behavior of the number of roots without sign behaves very violently if we do not assume any assumption on f.

Consider a mixed polynomial of one variable $f(z,\bar{z}) = \sum_{\nu,\mu} a_{\nu,\mu} z^{\nu}\bar{z}^{\mu}$. Put

$$\begin{aligned}
\deg_z f &:= \max\{\nu \mid a_{\nu,\mu} \neq 0\} \\
\deg_{\bar{z}} f &:= \max\{\mu \mid a_{\nu,\mu} \neq 0\} \\
\deg f &:= \max\{\mu + \nu \mid a_{\nu,\mu} \neq 0\}
\end{aligned}$$

We call $\deg_z f$, $\deg_{\bar{z}} f$, $\deg f$ *the holomorphic degree*, *the anti-holomorphic degree* and *the mixed degree* of f respectively. We consider the following subclasses of mixed polynomials:

$$\begin{aligned}
L(n+m;n,m) &:= \{\bar{z}^m q(z) - p(z) \mid \deg_z q(z) = n,\ \deg_z p(z) \le n\}, \\
L^{hs}(n+m;n,m) &:= \{r(\bar{z})q(z) - p(z) \mid \deg_{\bar{z}} r(\bar{z}) = m, \\
&\qquad\qquad \deg_z q(z) = n,\ \deg_z p(z) \le n\}, \\
M(n+m;n,m) &:= \{f(z,\bar{z}) \mid \deg f = n+m,\ \deg_z f = n,\ \deg_{\bar{z}} f = m\}.
\end{aligned}$$

where $p(z), q(z) \in \mathbb{C}[z]$, $r(\bar{z}) \in \mathbb{C}[\bar{z}]$. We have canonical inclusions:

$$L(n+m;n,m) \subset L^{hs}(n+m;n,m) \subset M(n+m;n,m).$$

The class $L(n+m;n,m)$, $L^{hs}(n+m;n,m)$ come from harmonic functions

$$\bar{z}^m - \frac{p(z)}{q(z)},\quad r(\bar{z}) - \frac{p(z)}{q(z)}$$

as their numerators. Especially $L(n+1;n,1)$ corresponds to the lens equation. We call $\bar{z}^m - \frac{p(z)}{q(z)} = 0$ *a generalized lens equation* and $r(\bar{z}) - \frac{p(z)}{q(z)} = 0$ *a harmonically splitting lens type equation* respectively. The corresponding numerators are called a generalized lens polynomial and a harmonically splitting lens type polynomial respectively. The polynomials which attracted us in this paper are these classes. We thank to A. Galligo for sending us their paper where we learned this problem [3].

1.3 Lens Equation

The following equation is known as the lens equation.

$$L(z,\bar{z}) = \bar{z} - \sum_{i=1}^{n} \frac{\sigma_i}{z - \alpha_i} = 0, \quad \sigma_i, \alpha_i \in \mathbb{C}^*. \tag{1}$$

We identify the left side rational function with the mixed polynomial given by its numerator

$$L(z, \bar{z}) \prod_{i=1}^{n} (z - \alpha_i) \in M(n+1; n, 1).$$

throughout this paper. The real and imaginary part of this polynomial are polynomials of x, y of degree $n + 1$. Unlike the previous example, $\rho(f)$ is much more smaller than $(n + 1)^2$. This type of equation is studied by astrophysicists. For more explanation from astrophysical viewpoint, see for example Petters–Werner [11]. The lens equation can be written as

$$L(z, \bar{z}) := \bar{z} - \varphi(z), \quad \varphi(z) = \frac{p(z)}{q(z)} \neq 0, \qquad (2)$$
$$\deg\, p(z) \leq n,\ \deg\, q(z) = n.$$

A slightly simpler equation is

$$L'(z, \bar{z}) := \bar{z} - p(z), \quad \deg_z\, p = n. \qquad (3)$$

Both equations are studied using complex dynamics. Consider the function $r : \mathbb{P}^1 \to \mathbb{P}^1$ defined $r(z) = \overline{\varphi(\overline{\varphi(z)})}$. It is easy to see that r is a rational mapping of degree n^2. Observe that if z is a root of $L(z) = 0$, then z is a fixed point of $r(z)$, that is $z = r(z)$. It is known that

Proposition 5 *The number of zeros $\rho(L')$ of L', is bounded by $3n - 2$ Khavinson–Świątęk [5] and the number of zeros $\rho(L)$ of L is bounded by $5n - 5$ by Khavinson–Neumann [4].*

Bleher–Homma–Ji–Roeder have determined the exact range of $\rho(L)$:

Theorem 6 (Theorem 1.2, [2]) *Suppose that the lens equation has only simple solutions. Then the set of possible numbers of solutions is equal to*

$$\{n - 1 + 2k \mid 0 \leq k \leq 2n - 2\} = \{n - 1, n + 1, \ldots, 5n - 7, 5n - 5\}.$$

The estimation in Proposition 5 are optimal. Rhie gave an explicit example of f which satisfies $\rho(f) = 5n - 5$ (See Rhie [12], Bleher–Homma–Ji–Roeder [2], and also Theorem 21 below). Thus the inequality $\rho(f) \leq 5(n - 1)$ is optimal. The minimum of ρ is $n - 1$ and it can be obtained for example by $\bar{z}z^n - 1$.

In the proof of Proposition 5, the following principle in complex dynamics plays a key role.

Lemma 7 *Let r be an rational function on $\mathbb{P}^1$. If z_0 is an attracting or rationally neutral fixed point, then z_0 attracts some critical point of r.*

Elkadi and Galligo studied this problem from computational point of view to construct such a mixed polynomial explicitly and proposed the similar problem for generalized lens polynomials $L(n+m; n, m)$ [3].

2 Relation of Strongly Polar Weighted Homogeneous Polynomials and Number of Zeros Without Sign

Consider a strongly mixed weighted homogeneous polynomial $F(\mathbf{z}, \bar{\mathbf{z}})$ of two variables $\mathbf{z} = (z_1, z_2)$ with polar weight $P = {}^t(p, q)$, $\gcd(p, q) = 1$ and let d_p, d_r be the polar and radial degrees respectively. See [7, 9] for definition. This is equivalent to the equality (5) holds. Let

$$\mathbb{C}^* \times \mathbb{C}^2 \to \mathbb{C}^2, \quad (\rho, (z_1, z_2)) \mapsto \rho \underset{P}{\circ} (z_1, z_2) := (z_1\rho^p, z_2\rho^q) \tag{4}$$

be the associated $\mathbb{C}^*$-action. Recall that F satisfies the Euler equality:

$$F(r\exp(\theta i) \underset{P}{\circ} (\mathbf{z}, \bar{\mathbf{z}})) = r^{d_r}\exp(d_p\theta i)F(\mathbf{z}, \bar{\mathbf{z}}). \tag{5}$$

A strongly mixed homogeneous polynomial is the case where the weight is the canonical weight $\mathbf{1} := {}^t(1, 1)$. Consider the global Milnor fibration $F : \mathbb{C}^2 \setminus F^{-1}(\mathbf{0}) \to \mathbb{C}^*$ and let $M = \{\mathbf{z} \in \mathbb{C}^2 | F(\mathbf{z}, \bar{\mathbf{z}}) = 1\}$ be the Milnor fiber.

We assume further that F is convenient. Namely $F|_{z_1=0}$, $F|_{z_2=0}$ are not identically zero. By the convenience assumption and the strong mixed weighted homogenuity, we can find some integers n, r such that

$$d_p = npq, \quad d_r = (n+2r)pq$$

and we can write $F(z_1, \bar{z}_1, z_2, \bar{z}_2)$ as a linear combination of monomials $z_1^{\nu_1} z_2^{\nu_2} \bar{z}_1^{\mu_1} \bar{z}_2^{\mu_2}$ where the summation satisfies the equality

$$(\nu_1 + \mu_1)p + (\nu_2 + \mu_2)q = d_r \quad (\nu_1 - \mu_1)p + (\nu_2 - \mu_2)q = d_p.$$

In particular, we see that the coefficients of $z_1^{(n+r)q}\bar{z}_1^{rq}$ and $z_2^{(n+r)p}\bar{z}_2^{rp}$ are non-zero and any other monomials satisfies

$$\nu_1, \mu_1 \equiv 0 \mod q, \quad \nu_2, \mu_2 \equiv 0 \mod p.$$

The monodromy mapping $h : M \to M$ is defined by

$$h : M \to M, \quad \mathbf{z} \mapsto \exp(2\pi i/npq) \underset{P}{\circ} \mathbf{z} = (\exp(2\pi i/nq)z_1, \exp(2\pi i/np)z_2).$$

Thus there exists a strongly mixed homogeneous polynomial $G(\mathbf{w}, \bar{\mathbf{w}})$, $\mathbf{w} = (w_1, w_2)$ of polar degree n and radial degree $(n + 2r)$ such that

$$F(\mathbf{z}, \bar{\mathbf{z}}) = G(z_1^q, \bar{z}_1^q, z_2^p, \bar{z}_2^p).$$

The curve $F = 0$ is invariant under the $\mathbb{C}^*$-action given by (4). Let $\mathbb{P}^1(P)$ be the weighted projective line which is the quotient space of $\mathbb{C}^2 \setminus \{\mathbf{0}\}$ by the action (4). It has two singular points $A = [0, 1]$ and $B = [0, 1]$ (if $p, q \geq 2$) and the complement $U := \mathbb{P}^1(P) \setminus \{A, B\}$ is isomorphic to $\mathbb{C}^*$ with coordinate $z := z_1^q / z_2^p$. Note that z is well defined on $z_2 \neq 0$. The zero locus $V(F)$ of F in $\mathbb{P}^1(P)$ does not contain A, B and it is defined on U by the mixed polynomial $f(z, \bar{z}) = 0$ where f is defined by the equality:

$$\begin{aligned} f(z, \bar{z}) &:= F(\mathbf{z}, \bar{\mathbf{z}}) / (z_2^{(n+r)p} \bar{z}_2^{rp}) \\ &= c\, z^{n+r} \bar{z}^r + \sum_{i,j} a_{i,j} z^i \bar{z}^j \end{aligned}$$

where the summation is taken for $i \leq n + r$, $j \leq r$ and $i + j < n + 2r$ and $c \neq 0$ is the coefficient of $z_1^{(n+r)q} \bar{z}_1^{rq}$ in F. Note also that $g(z) = f(z)$ where

$$g(w) := G(w_1, \bar{w}_1, w_2, \bar{w}_2) / (w_2^{n+r} \bar{w}_2^r), \quad w = w_1 / w_2.$$

Thus in these affine coordinates z, w, we have

$$\begin{aligned} & z_1^{qn_1} \bar{z}_1^{qn_2} z_2^{pm_1} \bar{z}_2^{pm_2} / (z_2^{(n+r)p} \bar{z}_2^{rp}) = z^{n_1} \bar{z}^{n_2}, \\ & w_1^{n_1} \bar{w}_1^{n_2} w_2^{m_1} \bar{w}_2^{m_2} / (w_2^{n+r} \bar{w}_2^r) = w^{n_1} \bar{w}^{n_2} \end{aligned}$$

This implies that $f(z) = g(z)$, the number of points of $V(f(z))$ and $V(g(w))$ are equal in their respective projective spaces and

$$f = g \in M(n + 2r; n + r, r).$$

The associated $\mathbb{C}^*$-action to $G(\mathbf{w}, \bar{\mathbf{w}})$ is the canonical linear action and we simply denote it as $\rho \circ \mathbf{w}$ instead of $\rho \underset{1}{\circ} \mathbf{w}$. Let $M(G)$ be the Milnor fiber of G and let $\mathbb{P}^1$ be the usual projective line. The monodromy mapping $h_G : M(G) \to M(G)$ of G is given by $h_G(\mathbf{w}) = \exp(2\pi i / n) \circ \mathbf{w}$. Then we have a canonical diagram

$$\begin{array}{ccc} \mathbb{C}^2 & \xrightarrow{\varphi_{q,p}} & \mathbb{C}^2 \\ \uparrow & & \uparrow \\ M & \xrightarrow{\varphi_{q,p}} & M(G) \\ \downarrow \pi & & \downarrow \pi' \\ \mathbb{P}^1(P) \setminus V(f) & \xrightarrow{\bar{\varphi}_{q,p}} & \mathbb{P}^1 \setminus V(g) \end{array}$$

π is a $\mathbb{Z}/d_p\mathbb{Z}$-cyclic covering branched over $\{A, B\}$,($d_p = npq$) while π' is a $\mathbb{Z}/n\mathbb{Z}$-cyclic covering without any branch locus. $\varphi_{q,p}$ is defined $\varphi_{q,p}(z_1, z_2) = (z_1^q, z_2^p)$ which satisfies $\varphi_{q,p}(\rho \underset{P}{\circ} \mathbf{z}) = \rho^{pq} \circ (z_1^q, z_2^p)$, $\rho \in S^1$ and thus $\varphi_{q,p} \circ h = h_G \circ \varphi_{q,p}$ as we have

$$\begin{aligned}\varphi_{q,p}(h(\mathbf{z})) &= \varphi((\exp(2\pi i/nq)z_1, \exp(2\pi i/np)z_2) \\ &= (\exp(2\pi i/n)z_1^q, \exp(2\pi i/n)z_2^p) = h_G(\varphi_{q,p}(\mathbf{z})).\end{aligned}$$

The mapping $\bar{\varphi}_{q,p}$ is canonically induced by $\varphi_{q,p}$ and we observe that $\bar{\varphi}_{q,p}$ gives a bijection of

$$\bar{\varphi}_{q,p} : \mathbb{P}^1(P) \setminus \{A, B\} \to \mathbb{P}^1 \setminus \{\bar{A}, \bar{B}\}$$

and it induces an bijection between $V(f)$ and $V(g)$. Here $\bar{A} = [1 : 0]$ and $\bar{B} = [0 : 1]$. Recall that by [1, 8], we have

Proposition 8 *(1)* $\chi(M(G)) = n(2 - \rho(g))$.
(2) $\chi(M) = -npq\rho(f) + n(p + q)$.
(3) The links $K_F := F^{-1}(0) \cap S^3$ *and* $K_G := G^{-1}(0) \cap S^3$ *have the same number of components and it is given by* $\rho(f)$.

Proof The assertion follows from a simple calculation of Euler characteristics. (1) is an immediate result that $M(G) \xrightarrow{\pi'} \mathbb{P}^1 \setminus V(G)$ is an n-fold cyclic covering. (2) follows from the following.

$$\pi : M \cap \mathbb{C}^{*2} \to \mathbb{P}^1(P) \setminus (\{A, B\} \cup V(F))$$

is an npq-cyclic covering while $M \cap \{z_1 = 0\}$ and $M \cap \{z_2 = 0\}$ are np and nq points respectively. Thus

$$\begin{aligned}\chi(M) &= \chi(M \cap \mathbb{C}^{*2}) + \chi(M \cap \{z_1 = 0\}) + \chi(M \cap \{z_2 = 0\}) \\ &= npq(-\rho(f)) + np + nq\end{aligned}$$

The link components of K_F and K_G are S^1 invariant and the assertion (3) follows from this observation. □

The correspondence $F(\mathbf{z}, \bar{\mathbf{z}}) \mapsto f(z)$ is reversible. Namely we have

Proposition 9 *For a given* $f(z, \bar{z}) \in M(n + m; n, m)$ *and any weight vector* $P = {}^t(p, q)$, *we can define a strongly mixed weighted homogeneous polynomial of two variables* $\mathbf{z} = (z_1, z_2)$ *with weight* P *by*

$$F(\mathbf{z}, \bar{\mathbf{z}}) := f(z_1^q/z_2^p, \bar{z}_1^q/\bar{z}_2^p)z_2^{pn}\bar{z}_2^{pm}.$$

The polar degree and the radial degree of F *are* $(n - m)pq$ *and* $(n + m)pq$ *respectively. The coefficient of* $z_1^{nq}\bar{z}_1^{qm}$ *in* F *is the same as that of* $z^n\bar{z}^m$.

If f has non-zero constant term, F is convenient polynomial. The correspondence

$$F(\mathbf{z},\bar{\mathbf{z}}) \mapsto f(z,\bar{z}), \quad f(z,\bar{z}) \mapsto F(\mathbf{z},\bar{\mathbf{z}})$$

are inverse of the other.

Proof In fact, the monomial $z^i\bar{z}^j$, $i+j \le n+m$, $i \le n$, $j \le m$ changes into $z_1^{qi}\bar{z}_1^{jq}z_2^{p(n-i)}\bar{z}_2^{p(m-j)}$. In particular,

$$z^n\bar{z}^m \mapsto z_1^{qn}\bar{z}_1^{qm}, \quad 1 \mapsto z_2^{pn}\bar{z}_2^{pm}.$$

□

It is well-known that the Milnor fibration of a weighted homogeneous polynomial $h(\mathbf{z}) \in \mathbb{C}[z_1,\dots,z_n]$ with an isolated singularity at the origin is described by the weight and the degree by Orlik–Milnor [6]. This assertion is not true for a mixed weighted homogeneous polynomials.

Let

$$\tilde{M}(n+m;n,m;P),\ \tilde{L}^{hs}(n+m;n,m;P),\ \tilde{L}(n+m;n,m;P)$$

be the space of strongly mixed weighted homogeneous convenient polynomials of two variables with weight $P=(p,q)$, $\gcd(p,q)=1$ and with isolated singularity at the origin which corresponds to $M(n+m;n,m)$, $L^{hs}(n+m;n,m)$, $L(n+m;n,m)$ respectively through Propositions 8 and 9. For $P=(1,1)$, we simply write as

$$\tilde{M}(n+m;n,m),\ \tilde{L}^{hs}(n+m;n,m),\ \tilde{L}(n+m;n,m)$$

Proposition 10 *The moduli spaces $\tilde{M}(n+m;n,m;P)$, $\tilde{L}^{hs}(n+m;n,m;P)$, $\tilde{L}(n+m;n,m;P)$ are isomorphic to the moduli spaces $M(n+m;n,m)$, $L^{hs}(n+m;n,m)$, $L(n+m;n,m)$ respectively.*

As the above moduli spaces do not depend on the weight P (up to isomorphism), we only consider hereafter strongly mixed homogeneous polynomials. Assume that two polynomials F_1, F_2 are in a same connected component. Then their Milnor fibrations, are equivalent. Thus

Corollary 11 *Assume that $F_1, F_2 \in \tilde{M}(n+m;n,m)$ have different number of link components $\rho(f_1)$, $\rho(f_2)$. Then they belongs to different connected components of $\tilde{M}(n+m;n,m)$. In particular, the number of the connected components of $\tilde{M}(n+m;n,m)$ is not smaller than the number of $\{\rho(f) \mid f \in M(n+m;n,m)\}$.*

Remark 12 For a fixed number ρ of link components, we do not know if the subspace of the moduli space with link number ρ is connected or not.

Example 13 Consider a strongly mixed homogeneous polynomial F of polar degree 1 and radial degree 3. Namely $f \in M(3; 2, 1)$. Its possible link components are 1, 3, 5. 1 and 3 are given in Example 59, [8]. An example of 5 components are given by Bleher–Homma–Ji–Roeder [2]. For example, we can take

$$f(z) = \bar{z}(z^2 - 1/2) - z + 1/30$$
$$F(\mathbf{z}, \bar{\mathbf{z}}) = \bar{z}_1(z_1^2 - z_2^2/2) - (z_1 z_2 - z_2^2/30)\bar{z}_2$$

3 Extended Lens Equation

3.1 Extended Lens Equation

One of the main purposes of this paper is to study the number of zeros of the following extended lens equation for a given $m \geq 1$ and its perturbation.

$$L(z, \bar{z}) = \bar{z}^m - \frac{p(z)}{q(z)}, \quad \deg\, q = n,\ \deg\, p \leq n.$$

The corresponding mixed polynomial is in $L(n + m; n, m) \subset M(n + m; n, m)$. We will construct a mixed polynomial for which the example of Rhie is extended. However a simple generalization of Proposition 5 seems not possible. The reason is the following. Consider the function

$$\varphi := \sqrt[m]{\overline{\frac{p(z)}{q(z)}}}$$

and the composition $\psi := \varphi \circ \varphi$. ψ is a locally holomorphic function but the point is that φ and ψ are multi-valued functions, not single valued if $m \geq 2$. Thus we do not know any meaningful upper bound of $\rho(L)$.

3.2 A Symmetric Case

Here is one special case where we can say more. Suppose that m divide n and put $n_0 = n/m$. Assume that $p(z)/q(z)$ is m-symmetric, in the sense that there exists polynomials $p_0(z), q_0(z)$ so that $p(z) = p_0(z^m)$ and $q(z) = q_0(z^m)$. We assume that $p_0(0) \neq 0$. In this case, we can consider the lens equation

$$L_0(z, \bar{z}) := \bar{z} - \varphi_0(z), \quad \varphi_0(z) = \frac{p_0(z)}{q_0(z)}, \tag{6}$$

$$L(z, \bar{z}) := \bar{z}^m - \varphi(z), \quad \varphi(z) = \frac{p(z)}{q(z)}. \tag{7}$$

As $L(z, \bar{z}) = L_0(z^m, \bar{z}^m)$, there is $m : 1$ correspondence between the non-zero roots of L and L_0. Thus by Proposition 5, we have

$$\rho(L) = m\rho(L_0) \leq m(5n_0 - 5) = 5n - 5m.$$

Corollary 14 *Suppose that* $n = 2m$ *and let* $f(z, \bar{z}) = \bar{z} - \frac{z-1/30}{z^2-1/2}$ *as in Example 13. Put* $f_{2m}(z) = f(z^m, \bar{z}^m)$. *Then* $\rho(f_{2m}) = 5m$ *and the corresponding strongly mixed homogeneous polynomial* F_{2m} *is contained in* $\tilde{L}(3m; 2m, m)$.

3.3 Generalization of the Rhie's Example

So we will try to generalize the example of Rhie for the case $m \geq 2$ without assuming $n \equiv 0 \mod m$. First we consider the following extended Lens equation:

$$\ell_{n,m}(z, \bar{z}) = \bar{z}^m - \frac{z^{n-m}}{z^n - a^n} = 0, \quad n > m > 0, \ a \in \mathbb{R}_+ \tag{8}$$

Hereafter by abuse of notation, we also denote the corresponding mixed polynomial (i.e., the numerator) by the same $\ell_{n,m}(z, \bar{z})$. For the study of $V(\ell_{n,m}) \setminus \{0\}$, we may consider equivalently the following:

$$|z|^{2m} - \frac{z^n}{z^n - a^n} = 0. \tag{9}$$

This can be rewritten as

$$z^n(|z|^{2m} - 1) = |z|^{2m} a^n.$$

Thus we have

Proposition 15 *Take a non-zero root* z *of* $\ell_{n,m} = 0$. *If* $a > 0$ *and* $z \neq 0$, *then* $|z| \neq 1$ *and* z^n *is a real number. Thus* z^{2n} *is a positive real number.*

Let us consider the half lines

$$\mathbb{R}_+(\theta) := \{re^{i\theta} \mid r \geq 0\}$$

and lines L_θ which are the union of two half lines:

$$L_\theta = \mathbb{R}_+(\theta) \cup \mathbb{R}_+(\theta + \pi).$$

Put

$$L(n) := \bigcup_{j=0}^{n-1} L_{2\pi j/n}, \quad L(n)' := \bigcup_{j=0}^{n-1} L_{(2j+1)\pi/n}$$

$$\mathcal{L}(2n) := \bigcup_{j=0}^{2n-1} \mathbb{R}_+(j\pi/n) = \{z \in \mathbb{C} \,|\, z^{2n} \geq 0\}.$$

Observation 16 *(1) If n is odd, $L_{2\pi j/n} = L_{(2j+n)\pi/n}$ and thus $L(n) = L(n)'$ and they consists of n lines and $L(n) = \mathcal{L}(2n)$.*
(2) If n is even, $L(n) \cap L(n)' = \{0\}$, $\mathcal{L}(2n) = L(n) \cup L'(n)$ and lines of $L(n)$ and $L(n)'$ are doubled. That is, each half line $\mathbb{R}_+(2\pi j/n)$ and $\mathbb{R}_+(\pi(2j+1)/n)$ appear twice in $L(n)$ and in $L(n)'$ respectively.

We identify $\mathbb{Z}/n\mathbb{Z}$ with complex numbers which are nth root of unity and we consider the canonical action of $\mathbb{Z}/n\mathbb{Z} \subset \mathbb{C}^*$ on $\mathbb{C}$ by multiplication. Thus it is easy to observe that

Lemma 17 *$V(\ell_{n,m})$ is a subset of $\mathcal{L}(2n)$ and $V(\ell_{n,m}) \cap L(n)$ and $V(\ell_{n,m}) \cap L(n)'$ are stable by the action of $\mathbb{Z}/n\mathbb{Z}$.*

For non-zero real number solutions of (8) are given by the roots of the following equation:

$$\ell_{n,m}(z, \bar{z}) = z^{2m} - \frac{z^n}{z^n - a^n} = 0, \quad z \in \mathbb{R}^*. \tag{10}$$

Equivalently

$$\begin{cases} z^n - a^n - z^{n-2m} = 0, & n > 2m \\ z^{2m-n}(z^n - a^n) - 1 = 0, & n \leq 2m. \end{cases}$$

Note that for n odd, $V(\ell_{n,m}) \subset L(n)$ and the generator $e^{2\pi i/n}$ of $\mathbb{Z}/n\mathbb{Z}$ acts cyclicly as

$$L_{2\pi j/n} \cap V(\ell_{n,m}) \mapsto L_{2\pi(j+1)/n} \cap V(\ell_{n,m})$$
$$L_{\pi(2j+1)/n} \cap V(\ell_{n,m}) \mapsto L_{\pi(2j+3)/n} \cap V(\ell_{n,m})$$

For n even, $V(\ell_{n,m}) \subset L(n) \cup L(n)'$. To consider the roots on $L(n)'$, we put $z = \exp(\pi(2j+1)i/n) \circ u$ with $u \in \mathbb{R}^*$. Then by (9) u satisfies

$$u^{2m} - \frac{-u^n}{-u^n - a^n} = 0, \text{ if } u \in \mathbb{R}. \tag{11}$$

This is equivalent to

$$\begin{cases} u^n + a^n - u^{n-2m} = 0, & n > 2m \\ u^{2m-n}(u^n + a^n) - 1 = 0, & n \leq 2m. \end{cases} \tag{12}$$

3.4 Preliminary Result Before a Bifurcation

The first preliminary result is the following (Lemmas 18, 19).

Lemma 18 *If $n > 2m$, for a sufficiently small $a > 0$, $\rho^*(\ell_{n,m}) = 3n$. Here $\rho^*(f)$ is the number of roots in $\mathbb{C}^*$.*

Proof The proof is parallel to that of Rhie ([2, 12]. We know that roots are on $L(n)$ or $L(n)'$ by Proposition 15. Consider non-zero real roots of $\ell_{n,m}(z, \bar{z}) = 0$. It satisfies the equality:

$$z^m - \frac{z^{n-m}}{z^n - a^n} = 0 \iff z^n - z^{n-2m} - a^n = 0, \quad z \in \mathbb{R} \setminus \{0\}. \tag{13}$$

(1) Assume that n is odd. Then the function $w = z^n - z^{n-2m}$ has three real points on the real axis, $(-1, 0)$, $(0, 0)$, $(1, 0)$ and the graph looks like Fig. 2. As we see in the Figure, they have one relative maximum $\alpha > 0$ and one relative minimum $-\alpha$. Thus the horizontal line $w = a^n$ intersects with this graph at three points if $a^n < \alpha$.

Thus (13) has three real roots for a sufficiently small a. Now we consider the action of $\mathbb{Z}/n\mathbb{Z}$ on $V(\ell_{n,m})$, we have $3n$ solutions on $V(\ell_{n,m}) \cap L(n)$.

(2) Assume that n is even. In this case, we have to notice that the action of $\mathbb{Z}/n\mathbb{Z}$ on $V(f) \cap L(n)$ is $2:1$ off the origin.

In this case, the graph of $y = t^n - t^{n-2m}$ looks like Fig. 3. Thus for a sufficiently small $a > 0$, $t^n - t^{n-2m} - a^n = 0$ has two real roots. Thus by the above remark, it gives $2n/2 = n$ roots on $V(\ell_{n,m}) \cap L(n)$. Now we consider the roots on the line $L_{(2j+1)\pi/n}$, $j = 0, \ldots, n-1$. Putting $z = u\zeta$, $\zeta^n = -1$ with u being real, from (12), we get the equality:

$$-u^n + u^{n-2m} - a^n = 0. \tag{14}$$

The graph of $y = -t^n + t^{n-2m}$ is the mirror image of Fig. 3 with respect to t-axis. Thus $-t^n + t^{n-2m} - a^n = 0$ has 4 real roots. Counting all the roots on the lines in $L(n)'$, it gives 4n/2 = 2n roots. Thus altogether, we get $3n$ roots. □

Now we consider the case $2m \geq n$.

Lemma 19 *If $2m \geq n > m$, $\rho^*(\ell_{n,m}) = 2n$ for a sufficiently small $a > 0$.*

Proof (1a) Assume that $2m > n$ and n is odd. The equation of the real solutions of (8) reduces to

$$z^{2m-n}(z^n - a^n) = 1. \tag{15}$$

It is easy to see that there are two real solutions (one positive and one negative). See Fig. 4. Considering other solutions of the argument $2\pi j/n$, $j = 0, \ldots, n-1$, we get $2n$ solutions.

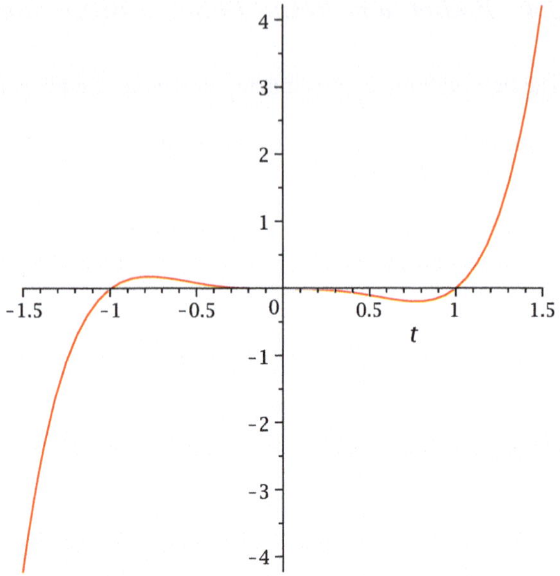

Fig. 2 Graph of $y = t^n - t^{n-2m}$, $n = 5, m = 1$

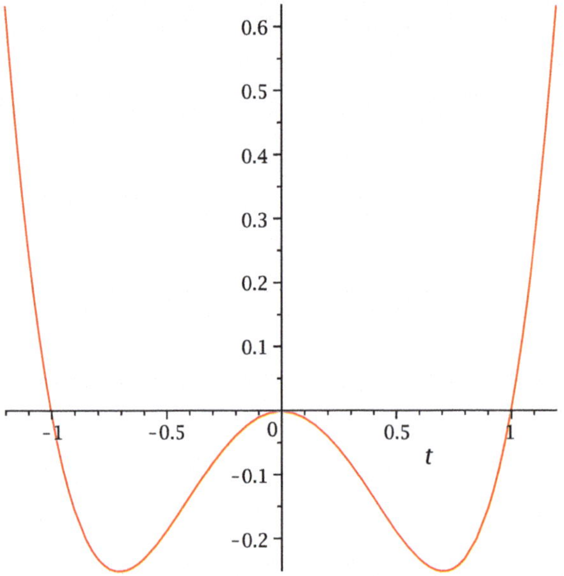

Fig. 3 Graph of $y = t^n - t^{n-2m}, n = 4, m = 1$

(1b) Assume that $2m > n$ and n is even. The equation for the real solutions is

$$z^{2m-n}(z^n - a^n) = 1$$

Fig. 4 Graph of $y = z^{2m-n}(z^n - a^n)$, $n = 5, m = 4, a = 1$

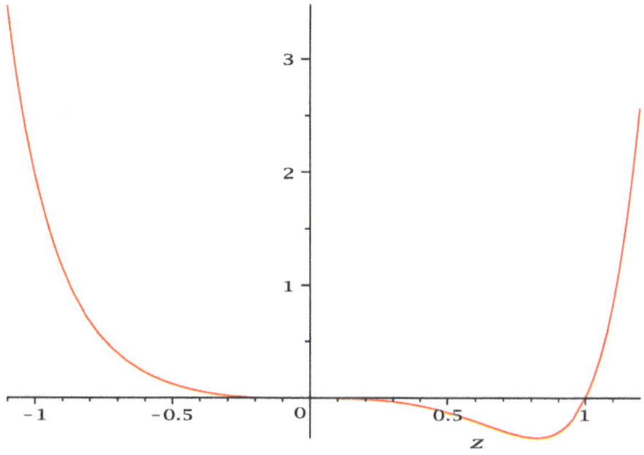

Fig. 5 Graph of $y = z^{2m-n}(z^n - a^n)$, $n = 6, m = 4, a = 1$

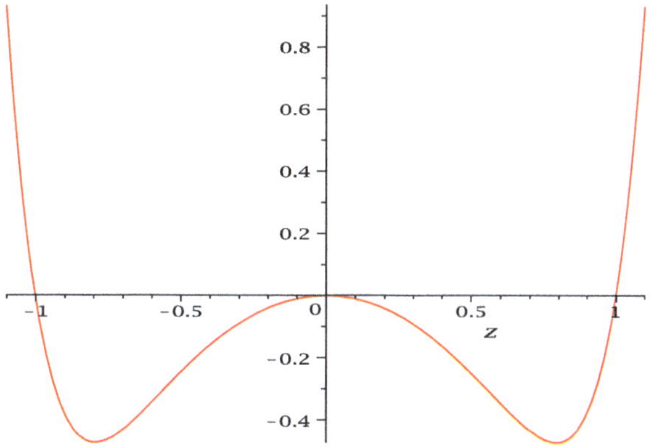

and it has two real solutions. Thus on the lines $L_{2j\pi/n}$, $2n/2 = n$ solutions. See Fig. 5. On the real lines $L_{(2j+1)\pi/n}$, the equation reduces to

$$u^{2m-n}(u^n + a^n) = 1.$$

Thus it has $2n/2 = n$ solutions on these lines and altogether, we gave $2n$ solutions.
(2) Assume that $n = 2m$. Then (15) reduces to

$$z^{2m} - a^{2m} = 1.$$

This has two real roots on L_0 and thus we get $2n/2 = n$ roots on the lines $L(n)$. On the lines $\arg z = (2j + 1)\pi/n$, putting $z = u\exp(\pi/n)$, the equation is given by $u^n + a^n = 1$. This has two roots provided $a < 1$ and thus n roots on $L(n)'$. Thus altogether, we get $2n$ roots. □

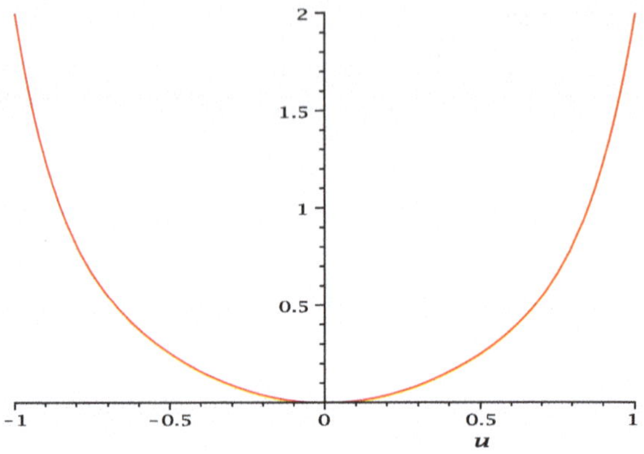

Fig. 6 Graph of $y = z^{2m-n}(z^n + a^n)$, $n = 6, m = 4, a = 1$

4 Bifurcation of the Root and the Main Result

We considered the extended lens equation for a fixed $a > 0$ as in Lemma 18. Note that $z = 0$ is a root with multiplicity. We want to change these roots into $2n$ regular roots using a small bifurcation.

$$\ell^{\varepsilon}_{n,m} := \bar{z}^m - \frac{z^{n-m}}{z^n - a^n} - \frac{\varepsilon}{z^m}, \quad \varepsilon > 0. \tag{16}$$

Note that the mixed polynomial, given by the numerator of $\ell^{\varepsilon}_{n,m}$ (by abuse of the notation, we denote this numerator also by the same notation) satisfies (Fig. 6)

$$\ell^{\varepsilon}_{n,m} \in L(n_1 + m; n_1, m) \subset M(n_1 + m; n_1, m)$$
$$\text{where} \qquad n_1 := n + m.$$

First we observe (16) implies

$$z^n(|z|^{2m} - 1 - \varepsilon) = -\varepsilon a^n + |z|^{2m} a^n$$

which implies that z^{2n} is a positive real number as the situation before the bifurcation. We observe that

Proposition 20 *$V(\ell^{\varepsilon}_{n,m})$ is also a subset of $\mathcal{L}(2n)$ and it is $\mathbb{Z}/n\mathbb{Z}$-invariant.*

4.1 *The Case m Is Not So Big*

Assume that $n > 2m$ or $n_1 > 3m$. The following is our main result which generalize the result of Rhie for the case $m = 1$.

Theorem 21 *(1) Assume that $n > 2m$ i.e., $n_1 > 3m$. For a sufficiently small positive ε, $\rho(\ell^{\varepsilon}_{n,m}) = 5(n_1 - m)$.*
(2) For the case $n = 2m$, let f_{2m} be as in Corollary 14. Then $f_{2m} \in L(3m; 2m, m)$ and $\rho(f_{2m}) = 5m$.

Remark 22 For the modified polynomial $\ell^{\varepsilon}_{2m,m}$, we have $\rho(\ell^{\varepsilon}_{2m,m}) = 3m$.

Proof We prove the assertion for the case $n > 2m$, the assertion for $n = 2m$ is in Corollary 14. First observe that $3n$ roots of $\ell_{n,m}$ are all simple. Put them $\xi_1, \dots, \xi_{3n}$. Take a small radius r so that the disks $D_r(\xi_j)$, $j = 1, \dots, 3n$ of radius r centered at ξ_j are disjoint each other and they do not contain 0 and the Jacobian of $\Re\ell_{n,m}$, $\Im\,\ell_{n,m}$ has rank two everywhere on $D_r(\xi_j)$. Then for any sufficiently small $\varepsilon > 0$, there exists a single simple root in $D_r(\xi_j)$ for $\ell^{\varepsilon}_{n,m} = 0$.

First consider the case n being odd. The real root of $\ell^{\varepsilon}_{n,m} = 0$ satisfies the equation

$$\bar{z}^m = \frac{z^{n-m}}{z^n - a^n} + \frac{\varepsilon}{z^m} \quad \text{or} \tag{17}$$

$$f_\varepsilon := |z|^{2m}(z^n - a^n) - z^n - \varepsilon(z^n - a^n) = 0. \tag{18}$$

We consider the possible roots which bifurcate from $z = 0$. The second equation (18) is written as

$$z^n|z|^{2m} - z^n(1+\varepsilon) - a^n(|z|^{2m} - \varepsilon) = 0 \tag{19}$$

$(z^{2m} - \varepsilon) = 0$ has two real roots $\alpha_{0+} > 0 > \alpha_{0-}$. Take a sufficiently small $s > 0$ and consider the disk $B_{0\pm}$ centered at $\alpha_{0\pm}$ of radius $s\varepsilon^{1/2m}$ so that they do not contain zero.

Note that $|z^{2m} - \alpha_{0\pm}| \ge s^{2m}\varepsilon$ on $\partial B_{0\pm}$. As the other terms of f_ε (that is, $z^n|z|^{2m} - z^n(1+\varepsilon)$) are of order greater than or equal to $\varepsilon^{n/2m} \ll \varepsilon$. Thus taking ε small enough, we may assume that $f_\varepsilon = 0$ has a simple root inside the disk B_{0+} and B_{0-}.

Here is another slightly better argument. We consider the scale change $z = w\varepsilon^{1/2m}$ and put

$$\begin{aligned}\tilde{f}_\varepsilon(w) :=& \frac{1}{\varepsilon} f(w\varepsilon^{1/2m})\\ =& -a^n(|w|^{2m} - 1) + \varepsilon^{n/2m} w^n|w|^{2m} - \varepsilon^{(n-2m)/2m} w^n(1+\varepsilon).\end{aligned}$$

In this coordinate, $3n$ roots ξ_j are far from the origin and we see clearly there are two roots near $w = \pm 1$ as long as ε is sufficiently small.

We consider now roots on $L(n)$. By the $\mathbb{Z}/n\mathbb{Z}$-invariance, we have also two roots on each $L_{2\pi j/n}$ and thus we get $2n$ simple roots which are bifurcating from $z = 0$. Thus altogether, we get $5n = 5(n_1 - m)$ roots.

We consider now the case n being even. Then every root of (19) on $\arg z = 2j\pi/n$ are counted twice. Thus we have n roots on these real line. In this case, there are also roots on the real lines $\arg z = (2j+1)\pi/n$. In fact, put $z = u \exp \pi i/n$ in (19).

Then the equation in u takes the form:

$$-u^n|u|^{2m} + u^n(1+\varepsilon) - a^n(|u|^{2m} - \varepsilon) = 0 \tag{20}$$

This has two real roots. Thus we found another $2n/2 = n$ roots. Therefore there are $2n$ simple roots which bifurcate from $z = 0$. Thus we have $5n$ roots for $\ell^{\varepsilon}_{n,m}$ in any case. □

4.2 Rhie's Equations

Applying Theorem 21, we get lens equation with maximal number of zeros $5(n-1)$ in the form:

$$\bar{z} = f_n(z), \quad f_n(z) = \frac{z^{n-2}}{z^{n-1} - a^{n-1}} + \frac{\varepsilon}{z}, \; 0 < \varepsilon \ll a \ll 1$$

for $n \geq 4$. For example, for $n = 4$, we can take for example

$$f_4(z) = \frac{z^2}{z^3 - 1/5} + \frac{1/800}{z}.$$

For $n = 2, 3$, the previous construction does not work and we need a special care. In fact, we can take f_n for $n = 2, 3$ as follows (Compare with [2]):

$$f_2(z) = \frac{z - 1/30}{z^2 - 1/2}, \quad f_3(z) = \frac{z^2 - 1/1000}{z^3 - 1/8}.$$

In Fig. 7, the red curve is $\Im(numerator(\bar{z} - f_3(z)) = 0$ and the green curve is the zero set of $\Re(numerator(\bar{z} - f_3(z)) = 0$. The 10 intersections of green and red curves are zeros of $\bar{z} - f_3(z) = 0$. Graph is lifted -1 vertically.

4.3 The Case m Is Big

Assume that $2m \geq n > m$. In this case, we have the following result.

Theorem 23 *Assume that $2m \geq n > m$. Then for a sufficiently small $\varepsilon > 0$, $\rho(\ell^{\varepsilon}_{n,m}) \geq 3n = 3(n_1 - m)$.*

Proof We have shown in Lemma 19 that $\ell_{n,m}$ has $2n$ simple roots. Thus we need to show under the bifurcation equation $\ell^{\varepsilon}_{n,m}$, we get n further roots.

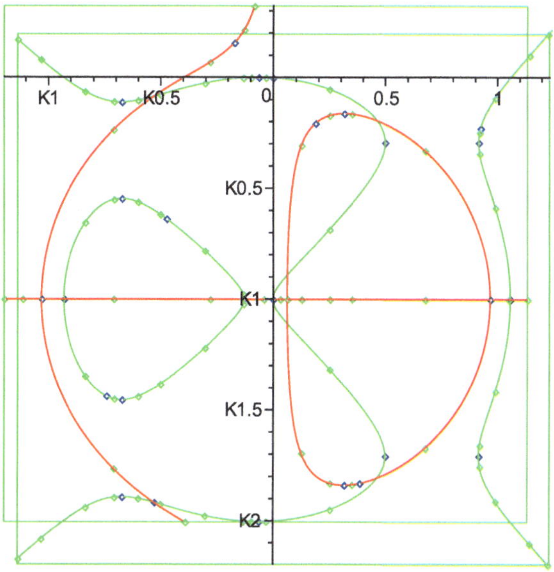

Fig. 7 Graph of f_3

$$\bar{z}^m - \frac{z^{n-m}}{z^n - a^n} - \frac{\varepsilon}{z^m} = 0$$

is equivalent to

$$|z|^{2m}(z^n - a^n) - z^n - \varepsilon(z^n - a^n) = 0 \text{ or} \tag{21}$$

$$|z|^{2m}(z^n - a^n) - z^n(1 + \varepsilon) + a^n \varepsilon = 0 \tag{22}$$

(I-1) We first consider the case $2m > n$ and n is odd. $-z^n(1 + \varepsilon) + a^n\varepsilon = 0$ has one positive root $z = \beta = a\sqrt[n]{\varepsilon/(1 + \varepsilon)}$. By a similar argument as in the previous section, (22) has a simple root near β. Thus by the $\mathbb{Z}/n\mathbb{Z}$-action stability, we have n simple bifurcating roots and altogether, we have $2n + n = 3n$ simple roots.
(I-2) Assume that $2m > n$ and n is even. $-z^n(1 + \varepsilon) + a^n\varepsilon = 0$ has one positive and one negative roots. Then by the stability (22) has $2n/2 = n$ simple roots. To see the roots on $L(n)'$, put $z = u\exp(i\pi/n)$. Then (22) is reduced to

$$|u|^{2m}(-u^n - a^n) + u^n(1 + \varepsilon) + a^n\varepsilon = 0$$

We see this has no real root. Thus altogether, we have $3(n_1 - m)$ roots.
(II) Assume that $n = 2m$. Then (22) can be written as

$$|z|^{2m} z^{2m} - z^{2m}(a^{2m} + 1 + \varepsilon) + a^{2m}\varepsilon = 0 \tag{23}$$

if z real. This has two real roots and adding all roots in the lines of $L(n)$, we get $2n/2 = n$ roots.

Consider other roots on $L(n)'$. Then we can write $z = u\zeta$ with $\zeta^n = -1$ and $u \in \mathbb{R}$ and

$$-|u|^{2m}u^{2m} + u^{2m}(1 - a^{2m} + \varepsilon) + a^{2m}\varepsilon = 0.$$

This has no zeros near the origin, as we have assumed $0 < a < 1$ to have 2n zeros in $\ell_{n,m} = 0$. Thus the above bifurcation equation has no real root. Thus altogether we conclude $\rho(\ell^{\varepsilon}_{n,m}) = 3n = 3(n_1 - m)$. □

4.4 Application

4.4.1 $L^{hs}(n+m; n, m)$

The space of harmonically splitting Lens type polynomials apparently can take bigger number of zeros than generalized lens polynomials. To show this, we start from arbitrary lens equation

$$\ell_n(z) := \bar{z} - \frac{p(z)}{q(z)}, \quad \deg_z q = n,\ \deg_z p \le n,$$

Put $k = \rho(\ell_n)$. We assume that 0 is not a root of ℓ_n for simplicity and $q(z)$ has coefficient 1 for z^n. We consider its small perturbation in $L^{hs}(n+m; n, m)$:

$$\phi_t(z) := -t\bar{z}^m + \ell_n(z) = -t\bar{z}^m + \bar{z} - \frac{p(z)}{q(z)}, \quad 1 \gg t > 0.$$

We assert

Theorem 24 *For sufficiently small $t > 0$, $\rho(\phi_t) = k + m - 1$.*

Proof As before, we identify ϕ_t, ℓ_n with their numerators. For sufficiently small t and for each zero root α of ℓ_n, there exists a zero α' of ϕ_t in a neighborhood of α which has the same orientation as α. For $t \ne 0$, we know that $\beta(\phi_t) = n - m$ and $\beta(\ell_n) = n - 1$. Here $\beta(f)$ is the number of zeros of f with sign. See Theorem 2. By the assumption, ℓ_n has k zeros, say $\alpha_1, \ldots, \alpha_k$ and $\beta(\ell_n) = n - 1$. First we choose a positive number R so that $1/R < |\alpha_j| < R$ for $j = 1, \ldots, 5n - 5$. Thus it is clear that ϕ_t has k zeros near each $\alpha_j(\varepsilon)$ with the same sign as α_j in the original equation $\ell_n = 0$. On the other hand, $\beta(\phi_t) = n - m,\ t \ne 0$. Thus ϕ_ε has at least $m - 1$ new negative zeros.

We assert that ϕ_t obtains exactly $m - 1$ new negative zeros near infinity. To see this near infinity, we change the coordinate $u = 1/z$ and consider the numerator: $((-t/\bar{u}^m - 1/\bar{u})q(1/u) - p(1/u))\,\bar{u}^m u^n$. This takes the form

$$\Phi_t = (-t + \bar{u}^{m-1})\tilde{q}(u) - \bar{u}^m \tilde{p}(u)$$

where $\tilde{q}$, $\tilde{p}$ are polynomials defined as $\tilde{q}(u) = u^n q(1/u)$, $\tilde{p}(u) = u^n p(1/u)$. By assumption we can write

$$\tilde{q}(u) = 1 + \sum_{i=1}^{n} b_i u^i$$

$$\tilde{p}(u) = \sum_{i=0}^{n} c_i u^i.$$

We will prove that for a sufficiently small $t > 0$, there exist exactly $m-1$ zeros $u(t)$ which converges to 0 as $t \to 0$. The zero set

$$\{(u, t) \in \mathbb{C} \times \mathbb{R} \mid \Phi_t(u) = 0\}$$

in $\mathbb{C} \times \mathbb{R}$ is a real algebraic set. Thus we need only check the components which intersect with $t = 0$. We use the Curve selection lemma. Suppose that

$$\Phi_{t(s)}(u(s)) \equiv 0, \ \ t(s) = s^a, \tag{24}$$

$$u(s) = \sum_{j=p}^{\infty} d_j s^j, \ \ d_p \neq 0. \tag{25}$$

Note that the possible lowest order of $(-t(s) + \bar{u}(s)^{m-1})\tilde{q}(u(s))$ is $\min(a, p(m-1))$, while the lowest order of the second term $\bar{u}(s)^m \tilde{p}(u(s))$ is pm. Thus (24) says

$$a = p(m-1), \quad -1 + \overline{d_p}^{\,m-1} = 0.$$

Thus we can write

$$d_p = \exp(2\pi j i/(m-1)), \quad \exists j, \ 0 \le j \le m-2. \tag{26}$$

We assert that

Assertion 25 *For a fixed j, there exist a unique $u(s)$ which satisfies (24) and (26).*

We prove the coefficients d_j of $u(s)$ are uniquely determined by induction. Put

$$(-t(s) + \bar{u}^{m-1}(s)\tilde{q}(u(s)) = \sum_{\nu=p(m-1)}^{\infty} \gamma_\nu s^\nu$$

$$\bar{u}(s)^m \tilde{p}(u(s)) = \sum_{\nu=pm}^{\infty} \delta_\nu s^\nu.$$

We have shown $\gamma_{p(m-1)} = 0$ as $d_p = \exp(2\pi ji/(m-1))$. Suppose that $d_j,\ p \le j \le \mu - 1$ are uniquely determined. We consider the coefficient of $s^{p(m-2)+\mu}$ in (24). We need to have

$$\gamma_{p(m-2)+\mu} = \delta_{p(m-2)+\mu}.$$

Observe that

$$\gamma_{p(m-2)+\mu} = (m-1)\overline{d}_p^{m-1}\overline{d}_\mu + r'$$

where r' is a polynomial of coefficients $\{\overline{d}_j,\ j \le \mu - 1\} \cup \{\underline{b}_j,\ j = 1, \dots, n\}$. On the other hand, $\delta_{p(m-2)+\mu}$ is a polynomial of coefficients $\{\overline{d}_j,\ j \le \mu - 1\} \cup \{c_j,\ j = 0, \dots, n\}$. Thus d_μ is uniquely determined by the equality $\gamma_{p(m-2)+\mu} = \delta_{p(m-2)+\mu}$. □

As $\Phi_t \in L^{hs}(n+m; n, m)$, combining with Theorem 6, we obtain the following.

Corollary 26 *The set of the number of zeros $\rho(f)$ of harmonically splitting lens type polynomials $f \in L^{hs}(n+m; n, m)$ includes $\{n+m-2, n+m, \cdots, 5n+m-6\}$.*

4.4.2 The Moduli Space $M(n+m; n, m)$

Now we consider the bigger class of polynomials $M(n+m; n, m) \supset L^{hs}(n+m; n, m)$. As $\beta(F) = n - m$ for $F \in M(n+m; n, m)$, the lowest possible number of zeros of a polynomial in $M(n+m; n, m)$ is $n - m$. In fact we assert

Corollary 27 *The set $\{\rho(f) \mid f \in M(n+m; n, m)\}$ includes $\{n-m, n-m+2, \dots, n+m-2, \dots, 5n+m-6\}$.*

Proof By Corollary 26, it is enough to show that any of $\{n-m, n-m+2, \dots, n+m-4\}$ can be ρ of some $f \in M(n+m; n, m)$. Let $j = n - m + 2a,\ 0 \le a \le m - 2$. Consider the polynomial

$$f_a(z) = (z^{n-a}\bar{z}^{m-a} - 1)(z^a - 2)(\bar{z}^a - 3)$$

Then we see that $\rho(f_a) = n - m + 2a$ and $f_a \in M(n+m; n, m)$. □

Example 28 Consider $M(5; 3, 2)$. The possible ρ are $\{1, 3, \dots, 11\}$. For $\rho = 1, 3, 5$, we can take for example mixed polynomials associated with thefollowing polynomials

$$f(z) = z^3\bar{z}^2 - 1,\ (z^2\bar{z} - 1)(z-2)(\bar{z}-3),\ (z-1)(z^2-2)(\bar{z}^2-3).$$

The higher values $\{7, 9, 11\}$ are given by

$$\varepsilon\bar{z}^2 + \bar{z} - \frac{p(z)}{q(z)},\ \deg\ p(z) \le 3,\ \deg\ q(z) = 3,\ \varepsilon \ll 1$$

where $\bar{z} - \frac{p(z)}{q(z)} = 0$ is a lens type equation with $\rho = 6, 8, 10$.

4.5 Further Remark

4.5.1 $L(n+2m; n+m, m)$ with $2m < n$

We observe that in Theorem 21, $\rho(\ell^{\varepsilon}_{n,m}) = 5(n_1 - m)$ with $n_1 = n + m$ which is exactly the optimal upper bound for $m = 1$. Thus we may expect that the number $5(n_1 - m)$ might be optimal upper bound for the polynomials in $L(n_1 + m; n_1, m)$. However in the proof for $m = 1$, a result about an attracting or rationally neutral fixed points in complex dynamics played an important role and the the argument there does not apply directly in our case.

4.5.2 $L(2m + m; 2m, m)$

Our polynomial $\ell^{\varepsilon}_{2m,m}$ is not good enough. We have seen in Corollary 14 that the mixed polynomial f_{2m} has $5m$ zeros, while our polynomial $\ell^{\varepsilon}_{m,m}$ has only $3m$ zeros.

Problem 29 • *Determine the upper bound of* ρ *for* $L(n+m; n, m)$ *for* $n > 3m$.

- *Determine the possible number of* ρ *for* $L(n+m; n, m)$. *Is it* $\{n - m + 2k \mid 0 \le k \le 2n - 2m\}$?
- *Determine the upper bound of* ρ *for* $L^{hs}(n+m; n, m)$ *or* $M(n+m; n, m)$.
- *Are the subspaces of the moduli* $L(n+m; n, m)$, $L^{hs}(n+m; n, m)$, $M(n+m; n, m)$ *with a fixed* ρ *connected? If not, give an example.*

References

1. Blanloeil, V., Oka, M.: Topology of strongly polar weighted homogeneous links. SUT J. Math. **51**(1), 119–128 (2015)
2. Bleher, P., Homma, Y., Ji, L., Roeder, R.: Counting zeros of harmonic rational functions and its application to gravitational lensing. Int. Math. Res. Not. IMRN **8**, 2245–2264
3. Elkadi, M., Galligo, A.: Exploring univariate mixed polynomials of bidegree. In: Proceeding of SNC'2014, pp. 50–58 (2014)
4. Khavinson, D., Neumann, G.: On the number of zeros of certain rational harmonic functions. Proc. Am. Math. Soc. **134**(6), 666–675 (2008)
5. Khavinson, D., Świątek, G.: On the number of zeros of certain harmonic polynomials. Proc. Am. Math. Soc. **131**(2), 409–414 (2003)
6. Milnor, J., Orlik, P.: Isolated singularities defined by weighted homogeneous polynomials. Topology **9**, 385–393 (1970)
7. Oka, M.: Topology of polar weighted homogeneous hypersurfaces. Kodai Math. J. **31**(2), 163–182 (2008)
8. Oka, M.: Non-degenerate mixed functions. Kodai Math. J. **33**(1), 1–62 (2010)
9. Oka, M.: On mixed projective curves. Singularities in Geometry and Topology. IRMA Lectures in Mathematics and Theoretical Physics, vol. 20, pp. 133–147. European Mathematical Society, Zürich (2012)
10. Oka, M.: Intersection theory on mixed curves. Kodai Math. J. **35**(2), 248–267 (2012)
11. Petters, A.O., Werner, M.C.: Mathematics of gravitational lensing: multiple imaging and magnification. Gen. Rel. Gravit. **42**(9), 2011–2046 (2010)
12. Rhie, S.H.: n-point gravitational lenses with $5(n-1)$ images (2003). arXiv:astro-ph/0305166
13. Wilmshurst, A.S.: The valence of harmonic polynomials. Proc. Am. Math. Soc. **126**(7), 2077–2081 (1998)

A Lefschetz Coincidence Theorem for Singular Varieties

J.-P. Brasselet, A. K. M. Libardi, T. F. M. Monis, E. C. Rizziolli and M. J. Saia

Abstract This article provides a survey concerning Lefschetz fixed point formula and Lefschetz coincidence formula in the smooth and singular cases, moreover we show a Lefschetz type formula for the Coincidence number of two maps. As a consequence we obtain a relation with correspondences, and we provide some examples.

Keywords Coincidence · Lefschetz fixed point theorem · Intersection homology · Singular varieties

---

The first named author thanks CNPq, Bolsa Pesquisador Visitante Especial Process: 400580/2012-8 and FAPESP Process: 2015/06697-9. The second named author thanks FAPESP Process: 2012/24454-8. The fourth and fifth named authors thank FAPESP Process: 2014/00304-2 and CNPq Processes: 482183/2013-6 and 300733/2009-7.

J.-P. Brasselet
I2M Aix-Marseille University, Marseille, France
e-mail: jean-aul.brasselet@univ-amu.fr

J.-P. Brasselet
IBILCE-UNESP, São José do Rio Preto, S.P., Brasil

A. K. M. Libardi · T. F. M. Monis · E. C. Rizziolli
IGCE-UNESP, Rio Claro, S.P., Brasil
e-mail: alicekml@rc.unesp.br

T. F. M. Monis
e-mail: tfmonis@rc.unesp.br

E. C. Rizziolli
e-mail: eliris@rc.unesp.br

M. J. Saia (✉)
ICMC-USP, São Carlos, S.P., Brasil
e-mail: mjsaia@icmc.usp.br

R. N. Araújo dos Santos et al. (eds.), *Singularities and Foliations. Geometry, Topology and Applications*, Springer Proceedings in Mathematics & Statistics 222, https://doi.org/10.1007/978-3-319-73639-6_17

1 Introduction

The present paper is the result of discussions between the authors, during the Brasil-Mexico 2nd Meeting on Singularities and the 3rd Singularity Theory Meeting of the Northeast region conferences yield in Salvador de Bahia in July 2015. The aim of discussions was to understand the Goresky–MacPherson paper on Lefschetz fixed point formula in the case of singular spaces, to write explicitly the case of coincidences and to compute some examples. This article provides a survey concerning Lefschetz fixed point formula and Lefschetz coincidence formula in the smooth and singular cases, moreover we show in the Theorem 3.23 a Lefschetz type formula for the *Coincidence number* of maps (f, g) determined by the intersection of the canonical homology classes of the graphs $[G(f)]$ and $[G(g)]$. As a consequence we obtain a relation with correspondences, as well as providing some examples.

2 The Smooth Case

2.1 The Lefschetz Number - Smooth Case

The story starts with the notion of index of a vector field at a singular point, in the Poincaré viewpoint: we will denote by M a compact, oriented, n-dimensional smooth manifold. Let us consider a vector field tangent to M, with finitely many isolated singularities. In such a singular point a, the vector field v vanishes and the index $I(v, a)$ is well defined. The Poincaré–Hopf Theorem, first proved by Poincaré for surfaces and by Hopf in general dimensions, says that the Euler–Poincaré characteristic of M is equal to

$$\chi(M) = \sum_{a_i \in \mathrm{Sing}(v)} I(v, a_i).$$

Later on, Brouwer considers maps $f : M \to M$ with isolated fixed points a_j, i.e. points such that $f(a_j) = a_j$. In such a point a_j, let us consider a neighborhood $U \cong \mathbb{D}^n$ of a_j in M, such that $f(U) \subset U$. One defines the degree $I(f, a_j)$ of f at a_j as the degree of the induced map in homology:

$$f_* : H_{n-1}(\mathbb{D}^n \setminus \{a_j\}; \mathbb{Z}) \cong \mathbb{Z} \longrightarrow H_{n-1}(\mathbb{D}^n \setminus \{a_j\}; \mathbb{Z}) \cong \mathbb{Z}.$$

The first definition of the Lefschetz index is:

$$L(f) = \sum_{a_j} I(f, a_j) \tag{2.1}$$

where a_j describes the fixed points of f.

Let us consider now, for any $0 \le k \le n$, the induced map in homology

$$f_k : H_k(M;\mathbb{Q}) \to H_k(M;\mathbb{Q}).$$

Lefschetz gave a second definition of the Lefschetz number (known as Lefschetz fixed point formula [9]):

$$L(f) = \sum_{k=0}^{n} (-1)^k \mathrm{Trace}(f_k). \tag{2.2}$$

Let us denote by $G(f) \subset M \times M$ the graph of f. In general, $G(f)$ is not transverse to the diagonal Δ_M. However, one can find a map $f' : M \to M$ homotopic to f and the graph of which $G(f')$ is transverse to Δ_M. The (oriented) cycles $G(f')$ and Δ_M are transverse and complementary dimensional in $M \times M$. Moreover, there are finitely many intersection points $b_j \in G(f') \cap \Delta_M$. In such a point, the intersection number $I(G(f'), \Delta_M; b_j)$ is well defined. One has the third definition of the Lefschetz number:

$$L(f) = \sum_{b_j} I(G(f'), \Delta_M; b_j). \tag{2.3}$$

The number does not depend on the map f' homotopic with f and such that $G(f')$ is transverse to Δ_M.

To provide the fourth definition of the Lefschetz number, let us consider the following diagram with rational coefficients:

$$\begin{array}{ccccc}
H^n(M \times M, M \times M \setminus \Delta_M) & \xrightarrow{\ j^*\ } & H^n(M \times M) & \xrightarrow{(f,\mathrm{id}_M)^*} & H^n(M) \\
{\scriptstyle A_{M\times M}}\downarrow\cong & & {\scriptstyle P_{M\times M}}\downarrow\cong & & {\scriptstyle P_M}\downarrow\cong \\
H_n(\Delta_M) & \xrightarrow{\ i_*\ } & H_n(M \times M) & \overset{\tilde{f}}{\dashrightarrow} & H_0(M)
\end{array}$$

where j^* is the map defined in the long exact sequence of a pair in cohomology, i is the inclusion $i : \Delta_M \hookrightarrow M \times M$, $P_{M\times M}$ and P_M are Poincaré isomorphisms, $\tilde{f}$ is defined by commutativity of the right square and $A_{M\times M}$ is the Alexander isomorphism that makes commutative the following diagram (see [4]):

$$\begin{array}{ccc}
H^n(M \times M, M \times M \setminus \Delta_M) & \xrightarrow[\cong]{\ Thom\ } & H^0(\Delta_M) \\
{\scriptstyle A_{M\times M}}\downarrow\cong & \swarrow{\scriptstyle P_{\Delta_M}}\ \cong & \\
H_n(\Delta_M) & &
\end{array}$$

and so that the left square is commutative.

By Alexander isomorphism, the Thom class $U_{\Delta_M} \in H^n(M \times M, M \times M \setminus \Delta_M)$ corresponds to the fundamental class $[\Delta_M] \in H_n(\Delta_M)$. One has:

$$j^*(U_{\Delta_M}) = P^{-1}_{M\times M}(i_*[\Delta_M]).$$

Then the Lefschetz number is defined by:

$$L(f) = P_M \circ (f, \mathrm{id}_M)^* \circ j^*(U_{\Delta_M}). \tag{2.4}$$

Theorem 2.5 *The four definitions of Lefschetz fixed point number coincide.*

Main properties of the Lefschetz number are the following: If $L(f) \neq 0$, then f admits fixed points. If $f = \mathrm{id}_M$ then $L(f) = \chi(M)$. If f and g are two homotopic maps from M to M, then $L(f) = L(g)$.

2.2 *The Coincidence Number - Smooth Case*

Let us consider two maps $f : M \to N$ and $g : M \to N$ where M and N are compact oriented smooth manifolds without boundaries.

Definition 2.6 The coincidence set $C(f, g)$ is defined as

$$C(f, g) = \{x \in M \mid f(x) = g(x)\}.$$

Let us suppose that $\dim M = \dim N = n$ and consider the graphs $G(f)$ and $G(g)$ both n-dimensional (oriented) cycles in $M \times N$. As in (2.3), one can assume that, up to homotopy, the graphs $G(f)$ and $G(g)$ are transverse. Then the intersection $G(f) \cap G(g)$ is a finite number of points c_k at which the intersection number $I(G(f), G(g), c_k)$ is well defined.

The first definition of the Lefschetz coincidence index is given by

$$L(f, g) = \sum_{c_k \in G(f)\cap G(g)} I(G(f), G(g), c_k).$$

The second definition generalises (2.2). More precisely, for any $0 \leq k \leq n$, let us consider the commutative diagram:

$$\begin{array}{ccc} H_k(M;\mathbb{Q}) & \xrightarrow{f_k} & H_k(N;\mathbb{Q}) \\ P_M \uparrow \cong & & P_N \uparrow \cong \\ H^{n-k}(M;\mathbb{Q}) & \xleftarrow{g^{n-k}} & H^{n-k}(N;\mathbb{Q}) \end{array}$$

One defines the maps $\Theta_k : H_k(M; \mathbb{Q}) \to H_k(M; \mathbb{Q})$ by

$$\Theta_k = P_M \circ g^{n-k} \circ P_N^{-1} \circ f_k$$

and one defines

$$L(f, g) = \sum_{k=0}^{n} (-1)^k \mathrm{Trace}(\Theta_k).$$

One can also consider the maps

$$\Theta^{n-k} = P_N^{-1} \circ f_k \circ P_M \circ g^{n-k} : H^{n-k}(N; \mathbb{Q}) \to H^{n-k}(N; \mathbb{Q}).$$

One has $\mathrm{Trace}(\Theta_k) = \mathrm{Trace}(\Theta^{n-k})$ and then:

$$L(f, g) = \sum_{k=0}^{n} (-1)^k \mathrm{Trace}(\Theta^{n-k}).$$

If one defines

$$\widehat{L}(f, g) = \sum_{k=0}^{n} (-1)^k \mathrm{Trace}(\Theta^k)$$

then $\widehat{L}(f, g) = (-1)^n L(f, g)$.

The third definition of the Lefschetz coincidence number is a generalisation of (2.4). Let us consider the commutative diagram, with rational coefficients:

$$\begin{array}{ccccc}
H^n(N \times N, N \times N \setminus \Delta_N) & \xrightarrow{j^*} & H^n(N \times N) & \xrightarrow{(f,g)^*} & H^n(M) \\
{\scriptstyle A_{N\times N}} \downarrow \cong & & {\scriptstyle P_{N\times N}} \downarrow \cong & & {\scriptstyle P_M} \downarrow \cong \\
H_n(\Delta_N) & \xrightarrow{i_*} & H_n(N \times N) & & H_0(M).
\end{array}$$

One defines the Lefschetz coincidence number of the two maps f and g from M to N as:

$$L(f, g) = P_M \circ (f, g)^* \circ j^*(U_{\Delta_N}) = P_M \circ (f, g)^* \circ P_{N\times N}^{-1}([\Delta_N]). \qquad (2.7)$$

Theorem 2.8 *The three definitions of the Lefschetz coincidence number coincide.*

3 The Singular Case

Goresky and MacPherson proved in [7] the Lefschetz fixed point theorem in the context of "placid" self maps of singular spaces, by using intersection homology. Let us remind the main definitions.

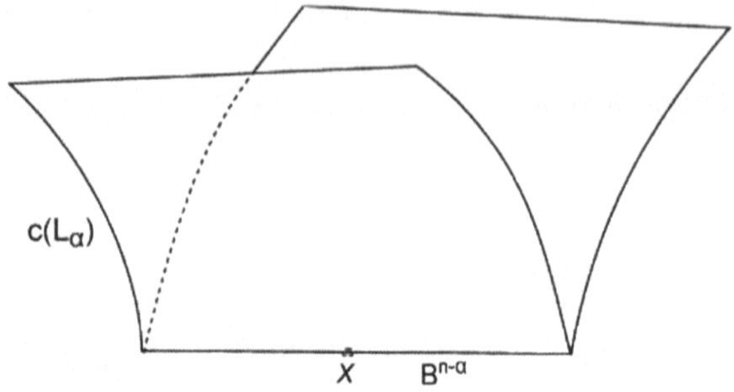

Fig. 1 The neighborhood U_x of a point x is $\mathbb{B}^{n-\alpha} \times \mathrm{c}(L_\alpha) \cong c(L_x)$, where L_α consists of two points and L_x is the boundary of U_x

3.1 Intersection Homology

Reference for the entire section are the Goresky–MacPherson original papers [5, 7] that we use (and abuse).

The singular varieties we consider are pseudomanifolds:

Definition 3.1 ([5]) An n-dimensional pseudomanifold X (without boundary) is a purely n-dimensional piecewise linear (P.L. for short) polyhedron which admits a triangulation such that each $(n-1)$ simplex is a face of exactly two n-simplices.

A pseudomanifold admits a piecewise linear stratification [2, I.1.4], which is a filtration by closed subspaces $\emptyset \subset X_0 \subset X_1 \subset \ldots \subset X_{n-2} \subset X_n = X$, with the singular part $\Sigma(X)$ of X being (included in) the element X_{n-2} of the filtration and such that for each point $x \in X_{n-\alpha} - X_{n-\alpha-1}$ there is a neighborhood U_x and a P.L. stratum preserving homeomorphism between U_x and $\mathbb{B}^{n-\alpha} \times \mathrm{c}(L_\alpha)$, where $\mathbb{B}^{n-\alpha}$ is an open ball of dimension $n-\alpha$ and $\mathrm{c}(L_\alpha)$ denotes the (open) cone on the link L_α of the stratum $X_{n-\alpha} - X_{n-\alpha-1}$. The link L_α is itself a stratified pseudomanifold of dimension $(\alpha - 1)$ (see [2, I.1.1]). If $X_{n-\alpha} - X_{n-\alpha-1}$ is non empty, it is a (non necessarily connected) manifold of dimension $n-\alpha$, and is called the $(n-\alpha)$-dimensional stratum of the stratification.

The neighborhood U_x of the point x is homeomorphic to a cone $c(L_x)$ over the link of the point x. Notice that the link of the point x is different from the link L_α of the stratum containing x. One has (see Fig. 1):

$$L_x = \left(\mathbb{B}^{n-\alpha} \times L_\alpha \bigcup \partial\mathbb{B}^{n-\alpha} \times \mathrm{c}(L_\alpha)\right).$$

The notion of perversity is fundamental for the definition of intersection homology and cohomology. A perversity $\overline{p}$ is a multi-index sequence of integers $(p(2), p(3), \ldots)$ such that $p(2) = 0$ and $p(\alpha) \leq p(\alpha+1) \leq p(\alpha)+1$, for $\alpha \geq 2$. Any perversity $\overline{p}$ lies between the zero perversity $\overline{0} = (0, 0, 0, \ldots)$ and the total perversity $\overline{t} = (0, 1, 2, 3, \ldots)$. In particular, one has the lower middle perversity, denoted $\overline{m}$ and the upper middle perversity, denoted $\overline{n}$, such that

$$m(\alpha) = \left[\frac{\alpha-2}{2}\right] \quad \text{and} \quad n(\alpha) = \left[\frac{\alpha-1}{2}\right], \qquad \text{for } \alpha \geq 2.$$

Let X be an n-dimensional pseudomanifold and $\overline{p}$ a perversity. The intersection homology groups (with compact support), denoted $IH_i^{\overline{p}}(X)$, are the homology groups of the chain complex

$$IC_i^{\overline{p}}(X) = \left\{ \xi \in C_i(X) \mid \begin{array}{ll} \dim(|\xi| \cap X_{n-\alpha}) \leq & i - \alpha + p(\alpha) \text{ and} \\ \dim(|\partial\xi| \cap X_{n-\alpha}) \leq i - 1 - \alpha + p(\alpha), \ \forall \alpha \geq 2. \end{array} \right\}, \tag{3.2}$$

where $C_i(X)$ denotes the group of compact i-dimensional P.L. chains ξ of X and $|\xi|$ denotes the support of ξ.

Intersection homology groups are the good ones to generalize main duality theorems for singular varieties, in particular Poincaré duality (see [5]). They appear to be the good ones also to define Lefschetz fixed points indices and numbers.

3.2 *The Lefschetz Number - Singular Case*

Goresky and MacPherson showed in [7] that one can prove a Lefschetz formula in intersection homology, for placid self maps of singular spaces.

A subanalytic map $f : X \to Y$ between two subanalytic pseudo-manifolds is called placid if there exists a subanalytic stratification of Y such that for each stratum S in Y we have

$$\operatorname{codim}_X f^{-1}(S) \geq \operatorname{codim}_Y(S).$$

Goresky and MacPherson proved that:

Proposition 3.3 ([7, Proposition 4.1]) *Let us assume that $f : X \to Y$ is a placid map. Then pushforward of chains and pullback of generic chains induce homomorphisms on intersection homology,*

$$f_i : IH_i^{\bar{m}}(X) \to IH_i^{\bar{m}}(Y) \qquad f^i : IH_i^{\bar{m}}(Y) \to IH_{i+dim(X)-dim(Y)}^{\bar{m}}(X), \tag{3.4}$$

where $\bar{m}$ is the (lower) middle perversity.

Definition 3.5 ([7, Sect. 4 Definition]) The intersection homology Lefschetz number of a placid self-map $f : X \to X$ is given by the formula:

$$IL(f) = \sum_{i=0}^{\dim X} (-1)^i Trace(f_i : IH_i^{\bar{m}}(X;\mathbb{Z}) \to IH_i^{\bar{m}}(X;\mathbb{Z})). \tag{3.6}$$

Goresky and MacPherson provide the intersection version of the Lefschetz fixed point Theorem in the context of Witt spaces:

Definition 3.7 ([6, 11]) A stratified pseudomanifold X is a $\mathbb{Q}$-Witt space if, for each stratum of odd codimension $\alpha = 2k + 1$, then $IH_k^{\bar{m}}(L_\alpha; \mathbb{Q}) = 0$, where L_α is the link of the stratum.

If X is a $\mathbb{Q}$-Witt space, then intersection homology of the two middle perversities coincide:

$$IH_*^{\bar{m}}(X; \mathbb{Q}) \cong IH_*^{\bar{n}}(X; \mathbb{Q}).$$

Proposition 3.8 ([7], Proposition 4.2) *If $f : X \to Y$ is a placid map between two compact oriented $\mathbb{Q}$-Witt spaces, with $n = \dim X$, then the graph of f determines a canonical homology class $[G(f)] \in IH_n^{\bar{m}}(X \times Y; \mathbb{Q})$.*

For a placid self map of a Witt space, both the graph of f and the diagonal carry fundamental classes in the intersection homology of $X \times X$, more precisely one has:

Theorem 3.9 ([7], Theorem I) *Let $f : X \to X$ be a placid self map of an n-dimensional Witt space. Let $[G(f)]$ and $[\Delta]$ be the homology classes of the graph of f and of the diagonal in $IH_n^{\bar{m}}(X \times X; \mathbb{Q})$. Then the Lefschetz number $IL(f)$ is given by*

$$IL(f) = [G(f)] \bullet [\Delta]$$

where $\bullet$ denotes the intersection product of cycles in intersection homology.

3.3 Local Lefschetz Numbers

In [7, Sects. 7–12], Goresky and MacPherson developed two notions of local Lefschetz numbers at isolated fixed points of a placid map $f : X \to X$. The first one is the "local contribution at x of the Lefschetz number", denoted by $\mu_\Delta([G_L(f)])$ in [7, Sects. 8 and 9], the second one is the "local trace of f at x" [7, Sect. 10].

On the one hand, the intersection homology Lefschetz number $IL(f)$ is the sum of the local Lefschetz numbers at fixed points. On the other hand, in the case of so-called "contracting" isolated fixed points, the two notions coincide ([7, Theorem III]).

In concrete examples, it appears that the second notion is easier to compute than the first one. That is what we will do in examples. Another way to define local Lefschetz numbers is developped in [1] using Čech-de Rham theory. The coincidence of this later notion with Goresky and MacPherson ones is shown in [3].

In order to illustrate the notion of local contribution, we will recall the two definitions by Goresky and MacPherson.

3.3.1 Local Contribution of the Lefschetz Number

In the same way that each point of a n-pseudomanifold admits a neighborhood U_x homeomorphic to $c(L_x)$ where L_x is the link of the point x, the main tool for the definition of local contribution of the Lefschetz number is the notion of the link $\mathcal{L}$ of a point (x, x) in $X \times X$. The corresponding conical neighborhood of (x, x) in $X \times X$ will be cone$(\mathcal{L})$.

In a more general way, one defines the link of a point (x_0, y_0) in a product $X \times Y$ of pseudomanifolds. Let us denote by L_1 the link of x_0 in X and by $h_1 : c(L_1) \to U_1$ a stratum preserving homeomorphism between the cone on L_1 and a neighborhood U_1 of x_0 such that x_0 is image of the vertex of the cone and $h_1|_{L_1}$ is identity. One defines a "radial distance" function $|\ \ | : U_1 \to [0, 1]$ by $|h_1(l, t)| = t$ for $l \in L_1$ and $t \in [0, 1]$.

One makes a similar choice of homeomorphism $h_2 : c(L_2) \to U_2$ between the cone on L_2 and a neighborhood U_2 of y_0.

Let us denote by $\mathcal{L} = L_1 * L_2$ the join of the links L_1 and L_2, that is the set of triples (a, s, b) where $a \in L_1, s \in [0, 1], b \in L_2$ and where we identify $(a, 0, b) \sim (a', 0, b)$ or $(a, 1, b) \sim (a, 1, b')$.

The previous construction provides an embedding $H : \text{cone}(\mathcal{L}) \to X \times Y$ which is a homeomorphism onto the conical neighborhood

$$V = \{(x, y) \in U_1 \times U_2 |\ |x| + |y| \leq 1\}$$

of (x_0, y_0) in $X \times Y$, by writing $H((l_1, s, l_2), t) = (h_1(l_1, t(1 - s)), h_2(l_2, st))$.

The join $\mathcal{L} = L_1 * L_2$ is the boundary of the conical neighborhood V of (x, y) in $X \times Y$:

$$\mathcal{L} = \partial V = \{(x, y) \in X \times Y |\ |x| + |y| = 1\},$$

so that $\mathcal{L}$ is the link of the point $\{(x, y)\}$ in $X \times Y$.

Let $f : X \to X$ be a placid map with isolated fixed points. For each fixed point x choose a sufficiently small neighborhood U_x which contains no other fixed points than x and consider the link $\mathcal{L}$ of the point (x, x) in $X \times X$ by the previous construction. The graph $G(f)$ of f and the diagonal Δ intersect the link $\mathcal{L}$ in disjoint $(n - 1)$-cycles $G_L(f) = G(f) \cap \mathcal{L}$ and $\Delta_L = \Delta \cap \mathcal{L}$. Notice that the full construction may be performed with respect of orientations.

Remark 3.10 In the following construction, given by by Goresky and Mac-Pherson, the diagonal Δ appears to have a prefered role, however the proof in [7] shows that the roles of Δ and $G(f)$ could be interchanged. That is N could be a small neighborhood of $G(f)$, or on other words Δ could be written as $G_L(Id_X)$ where Id_X is the identity map $Id : X \to X$.

This fact is important because it shows that the following construction holds for coincidence, replacing the pair $(\Delta, G(f)) = (G(Id_X), G(f))$ by the pair $(G(f), G(g))$.

Let N be a small regular neighborhood in $\mathcal{L}$ of Δ_L. By Alexander duality, one defines a nondegenerate linking pairing

$$\mu : IH_{n-1}^{\bar{m}}(N) \otimes IH_{n-1}^{\bar{m}}(\mathcal{L} - N) \to \mathbb{Q}$$

such that $\mu(a \times b) = \partial_*^{-1}(a) \bullet b$ where ∂_* is the connecting homomorphism in the long exact sequence of a pair (see [7, Sect. 8])

$$0 = IH_n^{\bar{m}}(\mathcal{L}) \longrightarrow IH_n^{\bar{m}}(\mathcal{L}, N) \xrightarrow{\partial_*} IH_{n-1}^{\bar{m}}(N) \longrightarrow IH_{n-1}^{\bar{m}}(\mathcal{L}) = 0.$$

On the one hand, again according to [7], Δ_L determines an unique class $[\Delta_L] \in IH_{n-1}^{\bar{m}}(N)$ and, if X is normal that group is one-dimensional. On the other hand, by the choice of the conical neighborhood U_x of x in X, one has the corresponding conical neighborhood $c(\mathcal{L})$ of (x, x) in $X \times X$. The neighborhood U_x being sufficiently small and containing no other fixed point of f, the graph $G(f)$ of f is transverse to $\mathcal{L}$. The intersection $G_L(f) = G(f) \cap \mathcal{L}$, oriented with the product orientation, determines a homology class [7, Sect. 9]

$$[G_L(f)] \in IH_{n-1}^{\bar{m}}(\mathcal{L} - N).$$

One can define:

Definition 3.11 ([7, Sect. 8]) The local contribution of the Lefschetz number at x is the linking number

$$\mu([\Delta_L] \otimes [G_L(f)]) \in \mathbb{Q}.$$

Theorem 3.12 ([7, Theorem II]) *The intersection Lefschetz number $IL(f) = [G(f)] \bullet [\Delta]$ is the sum of the local contributions taken over all the fixed points.*

3.3.2 Local Trace

We repeat here mainly Sect. 10 in [7].

Definition 3.13 ([8, 2.3]) An n-pseudomanifold X with boundary ∂X is an n-dimensional compact P-L space such that $X - \partial X$ is a pseudomanifold and ∂X is a compact $(n-1)$-dimensional P.L. subspace of X which has a collared neighborhood W in X, i.e. there is a P.L. homeomorphism $\varphi : W \cong \partial W \times [0, 1)$ such that the restriction $\varphi|_{\partial X}$ is the identity map.

Let us recall the Poincaré duality "with boundary", particular case of the relative Poincaré duality theorem:

Theorem 3.14 *Let X be a compact n-dimensional pseudomanifold with boundary ∂X, the intersection pairing*

$$IH_i^{\bar{p}}(X - \partial X) \otimes IH_{n-i}^{\bar{q}}(X, \partial X) \to \mathbb{Q}$$

where $\bar{q}$ is the complementary perversity of $\bar{p}$, is nondegenerate.

Let X be a compact n-dimensional pseudomanifold and let U be a conical neighborhood of x with boundary ∂U. Let L_x denote the link of the point x, then one has

$$IH_i(U) = \begin{cases} 0 & \text{for } i \geq \left[\frac{n+1}{2}\right] \\ IH_i(L_x) & \text{for } i \leq \left[\frac{n-1}{2}\right] \end{cases} \qquad IH_{n-i}(U, \partial U) = \begin{cases} 0 & \text{for } i \geq \left[\frac{n+1}{2}\right] \\ IH_{n-i-1}(L_x) & \text{for } i \leq \left[\frac{n-1}{2}\right] \end{cases}$$

Let $f : X \to X$ be a placid map with isolated fixed points. Let U_1 and U_2 be conical neighborhood of x with boundaries ∂U_1 and ∂U_2 such that $U_1 \subset f^{-1}(U_2 \setminus \partial U_2)$. If ξ is a compactly supported cycle in $IC_i(U_1)$ then $f_*(\xi)$ is a compactly supported cycle in $IC_i(U_2)$. One determines a local homomorphism

$$(f_*^x)_i : IH_i^{\bar{m}}(U_1) \to IH_i^{\bar{m}}(U_2). \tag{3.15}$$

The adjoint to f_* is a homomorphism

$$(f_x^*)_{n-i} : IH_{n-i}(U_2, \partial U_2) \to IH_{n-i}(U_1, \partial U_1) \tag{3.16}$$

which may be interpreted geometrically as assigning to almost every relative cycle $\xi \in IC_{n-i}(U_2, \partial U_2)$ the (appropriately oriented) relative cycle $f^{-1}(\xi) \cap U_1 \in IC_{n-i}(U_1, \partial U_1)$.

Definition 3.17 ([7, Sect. 10]) The local trace of f at x is the sum

$$\mathrm{Tr}_x(f) = \sum_{i=0}^{n} (-1)^i \,\mathrm{Trace}(f_*^x)_i = (-1)^n \sum_{i=0}^{n} (-1)^i \,\mathrm{Trace}(f_x^*)_i.$$

3.3.3 Contracting Fixed Points [7], Sect. 12

Definition 3.18 A fixed point $x \in X$ of a placid map $f : X \to X$ is contracting if there exists a (canonical) distinguished neighborhood U of x which contains no other fixed points and such that $\overline{U} \subset f^{-1}(\text{interior } (U))$.

Theorem 3.19 ([7, Theorem III]) *Suppose $x \in X$ is an isolated contracting fixed point of f. Then the local contribution at x to the Lefschetz number of f and the local trace of f at x coincide.*

Corollary 3.20 *In the previous situation, the intersection Lefschetz number $IL(f)$ is the sum of the local traces at fixed points.*

This result leads us naturally to the question of coincidence, already pointed out in [7] through the study of correspondences, although not explicitly. Our main goal is to explicit the formula of the Lefschetz coincidence number for placid maps $f, g : X \to Y$ between oriented compact $\mathbb{Q}$-Witt spaces of dimension n and to prove the Lefschetz coincidence theorem in this setting.

3.4 *The Coincidence Number - Singular Case*

Definition 3.21 Given $f, g : X \to Y$ placid maps between n-dimensional oriented compact $\mathbb{Q}$-Witt spaces, the Lefschetz coincidence number is defined by

$$IL(f, g) = \sum_i (-1)^i \mathrm{Trace}(g^i f_i), \tag{3.22}$$

where $f_i : IH_i^{\bar{m}}(X) \to IH_i^{\bar{m}}(Y)$ and $g^i : IH_i^{\bar{m}}(Y) \to IH_i^{\bar{m}}(X)$ are defined for the lower middle perversity $\bar{m}$ (Proposition 3.3).

Theorem 3.23 (Main Theorem). *The Lefschetz coincidence number of (f, g) is determined by the intersection of the canonical homology classes of the graphs, $[G(f)]$ and $[G(g)]$:*

$$IL(f, g) = (-1)^n [G(f)] \bullet [G(g)].$$

Corollary 3.24 *If $IL(f, g) \neq 0$ then there is $x \in X$ such that $f(x) = g(x)$.*

As a particular case of the Main Theorem, we recover the Lefschetz fixed point theorem of Goresky–MacPherson, [7, Theorem I]. In fact, considering the identity map $Id_X : X \to X$ and a placid self map $f : X \to X$, the coincidence number $IL(f, Id_X)$ is the Lefschetz number of f, denoted $IL(f)$ as defined by Goresky–MacPherson. Thus,

Corollary 3.25 *If $IL(f) \neq 0$ then there is $x \in X$ such that $f(x) = x$.*

4 Proof of the Main Theorem

From now on, X and Y denote n-dimensional oriented compact $\mathbb{Q}$-Witt spaces, $f : X \to Y$ a placid map and we denote by $IH_*^{\bar{m}}(X; \mathbb{Q})$ the intersection homology of X for middle perversity $\bar{m}$, with coefficients in $\mathbb{Q}$.

Let $u_1, \ldots, u_\alpha$ be a basis of $IH_*^{\bar{m}}(X; \mathbb{Q})$ with dual basis $u_1^*, \ldots, u_\alpha^*$. We also consider $v_1, \ldots, v_\beta$ be a basis of $IH_*^{\bar{m}}(Y; \mathbb{Q})$ with dual basis $v_1^*, \ldots, v_\beta^*$. Recall that these dual bases are obtained by Poincaré duality satisfying $u_i^* \otimes u_i = 1$ and $v_j^* \otimes v_j = 1$ for all $i = 1, \ldots \alpha$ and $j = 1, \ldots, \beta$.

Proposition 4.1 ([7], Proposition 6.2). *The homology class of the graph, $[G(f)]$, is given by*

$$[G(f)] = \sum_i \sum_j (-1)^{|u_i|(n-|u_i|)} F_{ij} u_i^* \otimes v_j,$$

where

$$f_*(u_i) = \sum_j F_{ij} v_j, \ i = 1, \ldots, \alpha.$$

Lemma 4.2 *If* $g_*(u_i^*) = \sum_j G_{ij} v_j^*$ *then* $g^*(v_j) = \sum_i G_{ij} u_i$.

Proof Let us write $g^*(v_j) = \sum_i G'_{ji} u_i$. Let u_i and v_j be such that $|u_i| = |v_j|$. From the formula 3 [7], one has

$$g_*(g^*(v_j) \otimes u_i^*) = v_j \otimes g_*(u_i^*)$$

and then it follows that

$$(-1)^{|u_i|(n-|u_i|)} G'_{ji} = (-1)^{|v_j|(n-|v_j|)} G_{ij}.$$

Since $|u_i| = |v_j|$ we conclude that $G'_{ji} = G_{ij}$. □

Proof of Theorem 3.23. Let us write

$$f_*(u_i) = \sum_j F_{ij} v_j \ \text{ and } \ g_*(u_i^*) = \sum_j G_{ij} v_j^*.$$

Following ([7], Proposition 6.2), the fundamental classes of the graph of f and g can be written, respectively, as

$$[G(f)] = \sum_i \sum_j (-1)^{|u_i|(n-|u_i|)} F_{ij} u_i^* \otimes v_j \ \text{ and } \ [G(g)] = \sum_k \sum_l G_{kl} u_k \otimes v_l^*.$$

Therefore,

$$\begin{aligned}
[G(f)] \bullet [G(g)] &= \sum_i \sum_j (-1)^{|u_i|(n-|u_i|)} (-1)^{(n-|v_j|)(n-|u_i|)} F_{ij} G_{ij} ((u_i^* \otimes u_i) \bullet (v_j \otimes v_j^*)) \\
&= \sum_i \sum_j (-1)^{n-|u_i|} F_{ij} G_{ij} \\
&= (-1)^n \sum_i \sum_j (-1)^{|u_i|} F_{ij} G_{ij} \\
&\overset{\star}{=} (-1)^n \sum_i (-1)^i \operatorname{Trace}(g^* f_*) \\
&= (-1)^n IL(f, g),
\end{aligned}$$

where the equality $\star$ follows from the fact that the matrix of g^* with respect to the basis $\{v_1, \ldots, v_\beta\}$ and $\{u_1, \ldots, u_\alpha\}$ is the matrix of g_* with respect to the basis $\{u_1^*, \ldots, u_\alpha^*\}$ and $\{v_1^*, \ldots, v_\beta^*\}$, see Lemma 4.2. Furthermore,

$$g^* f_*(u_i) = g^*(\sum_j F_{ij} v_j) = \sum_j F_{ij} g^*(v_j) = \sum_i \sum_j F_{ij} G_{ij} u_i \qquad \square$$

5 Relation with Correspondences

In [7], a placid correspondence C between n-dimensional Witt spaces X and Y is defined as being an n-dimensional compact oriented pseudomanifold $C \subset X \times Y$ such that the projections $\pi_X : C \to X$ and $\pi_Y : C \to Y$ are placid maps. A correspondence C determines a canonical homology class $[C] \in IH_n^{\overline{m}}(X \times Y; \mathbb{Q})$.

Theorem 5.1 ([7, Theorem I']) *The Lefschetz number $IL(C_1, C_2)$ of two correspondences is equal to the intersection product $[C_1] \bullet [C_2]$ of the intersection homology classes represented by C_1 and C_2.*

Given a pair of placid maps $f, g : X \to Y$ let us consider the respective graphs $C_1 := G(f)$ and $C_2 := G(g)$ and the projections maps

$$\pi_X^1 : G(f) \to X, \quad \pi_Y^1 : G(f) \to Y,$$

$$\pi_X^2 : G(g) \to X, \quad \pi_Y^2 : G(g) \to Y.$$

Corollary 5.2 ([7, Sects. 14 and 16]) *If $f, g : X \to Y$ are placid maps then $IL(f, g) = IL(G(f), G(g))$.*

Proof Observe that the projection maps induce homomorphisms in intersection homology and intersection cohomology groups:

$$\left(\pi_X^1\right)^* : IH_i^{\overline{m}}(X; \mathbb{Q}) \to IH_i^{\overline{m}}(G(f); \mathbb{Q}) \text{ and } \left(\pi_X^2\right)_* : IH_i^{\overline{m}}(G(g); \mathbb{Q}) \to IH_i^{\overline{m}}(X; \mathbb{Q}).$$

$$\left(\pi_Y^1\right)_* : IH_i^{\overline{m}}(G(f); \mathbb{Q}) \to IH_i^{\overline{m}}(Y; \mathbb{Q}) \text{ and } \left(\pi_Y^2\right)^* : IH_i^{\overline{m}}(Y; \mathbb{Q}) \to IH_i^{\overline{m}}(G(g); \mathbb{Q}).$$

Let us consider α a basis of $IH_i^{\overline{m}}(X; \mathbb{Q})$ and α_1 the corresponding basis of $IH_i^{\overline{m}}(G(f); \mathbb{Q})$ and recalling that $\pi_X^1 : G(f) \to X$ is a homeomorphism one has the identity matrix as the matrix of $(\pi_X^1)^*$ in the basis α and α_1. The same holds for $(\pi_X^2)_*$. Also we consider

$$X \xrightarrow{\phi_1} G(f) \xrightarrow{\pi_Y^1} Y$$

and

$$X \xrightarrow{\phi_2} G(g) \xrightarrow{\pi_Y^2} Y,$$

where $\phi_1(x) = (x, f(x))$ and $\phi_2(x) = (x, g(x))$. We point out that the induced maps by these compositions are the same as the ones induced by f and g.

Since f and g are placid maps, $G(f)$ and $G(g)$ are placid correspondences. Therefore by [7, Theorem I'] one has $IL(G(f), G(g)) = [G(f)] \bullet [G(g)]$.

By the following composition:

$$IH_i^{\overline{m}}(X;\mathbb{Q}) \xrightarrow{(\pi_X^1)^*} IH_i^{\overline{m}}(G(f);\mathbb{Q}) \xrightarrow{(\pi_Y^1)_*} IH_i^{\overline{m}}(Y;\mathbb{Q}) \xrightarrow{(\pi_Y^2)^*} IH_i^{\overline{m}}(G(g);\mathbb{Q}) \xrightarrow{(\pi_X^2)_*} IH_i^{\overline{m}}(X;\mathbb{Q})$$

we conclude that the alternating sum of traces of the $\left(\pi_X^2\right)_* \left(\pi_Y^2\right)^* \left(\pi_Y^1\right)_* \left(\pi_X^1\right)^*$ is equal (up to sign) to the alternating sum of traces of $g^* f_*$ [7, S16, formulas 7 and 9]:

$$IL(G(f), G(g)) = \sum_i (-1)^i \mathrm{Trace}(\left(\pi_X^2\right)_* \left(\pi_Y^2\right)^* \left(\pi_Y^1\right)_* \left(\pi_X^1\right)^*) =$$

$$\sum_i (-1)^i \mathrm{Trace}(g^* f_*) = IL(f, g).$$

□

5.1 Local Contribution

Goresky and MacPherson show that the Lefschetz number $IL(C_1, C_2)$ of two correspondences (Theorem 5.1) is sum of local contributions:

Theorem 5.3 ([7, Theorem II']) *Suppose the correspondences C_1 and C_2 intersect in finitely point s $(x_1, y_1), \ldots (x_k, y_k)$. Let $\mathcal{L}_i$ the full link (in $X \times Y$) of (x_i, y_i) and let $C_j^i = C_j \cap \mathcal{L}_i$ be the intersection of C_i with this link. Then, by Alexander duality, there is a well defined linking number $\mu(C_1^i, C_2^i)$ which is the local contribution of (x_i, y_i) (see [7, Sect. 8], here Definition 3.11) and one has:*

$$IL(C_1, C_2) = \sum_{i=1}^{k} \mu(C_1^i, C_2^i).$$

6 Local Coincidence Numbers

6.1 *Local Contribution*

Let $f, g : X \to Y$ be placid maps between n-dimensional oriented compact $\mathbb{Q}$-Witt spaces, with a finite number of coincidence points $\{x_i\} \in C(f, g)$. Let $\mathcal{L}_i$ be the link (in $X \times Y$) of $(x_i, f(x_i)) = (x_i, g(x_i))$. Let us denote $C^i_f = G(f) \cap \mathcal{L}_i$ and $C^i_g = G(g) \cap \mathcal{L}_i$.

As a corollary of Theorem 5.3 (see also Remark 3.10), one has:

Corollary 6.1 *Let $f, g : X \to Y$ be placid maps between n-dimensional oriented compact $\mathbb{Q}$-Witt spaces, the Lefschetz coincidence number is equal to:*

$$IL(f, g) = \sum_{x_i \in C(f,g)} \mu(C^i_f, C^i_g).$$

6.2 *Local Trace*

Let $f, g : X \to Y$ be a pair of placid maps between n-dimensional oriented compact $\mathbb{Q}$-Witt spaces, with isolated coincidence points. Let x be such a coincidence point. Let U_1 and U_2 be conical neighborhoods of x and $f(x) = g(x)$ respectively with boundaries ∂U_1 and ∂U_2 such that $U_1 \subset f^{-1}(U_2 \setminus \partial U_2) \cap g^{-1}(U_2 \setminus \partial U_2)$. If ξ is a compactly supported cycle in $IC_i(U_1)$ then $f_*(\xi)$ is a compactly supported cycle in $IC_i(U_2)$. In the same way than (3.15), (3.16), one determines local homomorphisms

$$(f^x_*)_i : IH^{\bar{m}}_i(U_1) \to IH^{\bar{m}}_i(U_2).$$

and

$$(g^*_x)_{n-i} : IH_{n-i}(U_2, \partial U_2) \to IH_{n-i}(U_1, \partial U_1).$$

One has a commutative diagram (middle perversity and rational coefficients):

$$\begin{array}{ccc} IH_i(U_1) & \xrightarrow{(f^x_*)_i} & IH_i(U_2) \\ P_1 \uparrow \cong & & P_2 \uparrow \cong \\ IH_{n-i}(U_1, \partial U_1) & \xleftarrow{(g^*_x)_{n-i}} & IH_{n-i}(U_2, \partial U_2) \end{array}$$

where vertical arrows are Poincaré–Lefschetz isomorphisms and where all elements are zero if $i > \left[\frac{n-1}{2}\right]$. It induces a diagram:

$$\begin{array}{ccc} IH_i^{\bar{m}}(L_1) & \longrightarrow & IH_i^{\bar{m}}(L_2) \\ \uparrow \cong & & \uparrow \cong \\ IH_{n-i-1}(L_1) & \longleftarrow & IH_{n-i-1}(L_2) \end{array}$$

where L_1 and L_2 are the links of x and $f(x) = g(x)$ in X and Y respectively.

Definition 6.2 ([7, Sect. 10]) The local trace of (f, g) at the isolated coincidence point x is the sum

$$\mathrm{Tr}_x(f, g) = \sum_{i=0}^{n}(-1)^i \ \mathrm{Trace}(P_1 \circ (g_x^*)_{n-i} \circ P_2^{-1} \circ (f_*^x)_i).$$

One obtains, in the case of coincidence, a similar result than the one for the Lefschetz intersection number. The proof is a copy of the proof provided in [7], Sect. 12.

Theorem 6.3 *The Lefschetz coincidence number $IL(f, g)$ is the sum of the local traces at coincidence points.*

$$IL(f, g) = \sum_{x \in C(f,g)} \mathrm{Tr}_x(f, g). \tag{6.4}$$

7 Examples

7.1 The Pinched Torus

Let us consider the "pinched torus", which has one singular point N (see Fig. 2). That is a pseudomanifold naturally stratified in: $X \setminus \{N\}, \{N\}$. There is only one possible perversity, the 0-perversity. An i-dimensional chain ξ containing the singular point N is allowable if and only if (see Eq. 3.2):

$$0 = \dim(|\xi| \cap \{N\} \leq i - \alpha + p(\alpha) = i - 2 + 0.$$

That is $i \geq 2$. The computation of intersection homology groups is then the following:
• $IH_0^{\bar{0}}(X) = \mathbb{Z}_x$ In fact all points in $X \setminus \{N\}$ are homologous in intersection homology.
• $IH_1^{\bar{0}}(X) = 0$. The possible cycles candidates as generators of $IH_1^{\bar{0}}(X)$ are the cycle G containing the points A and B and a "great" cycle passing through N. The first one is an allowable boundary (the boundary of a cone with vertex N and basis G), i.e. half of X, then gives 0 in intersection homology. The second one is not allowable, because 1-dimensional and containing N.

Fig. 2 The "pinched torus"

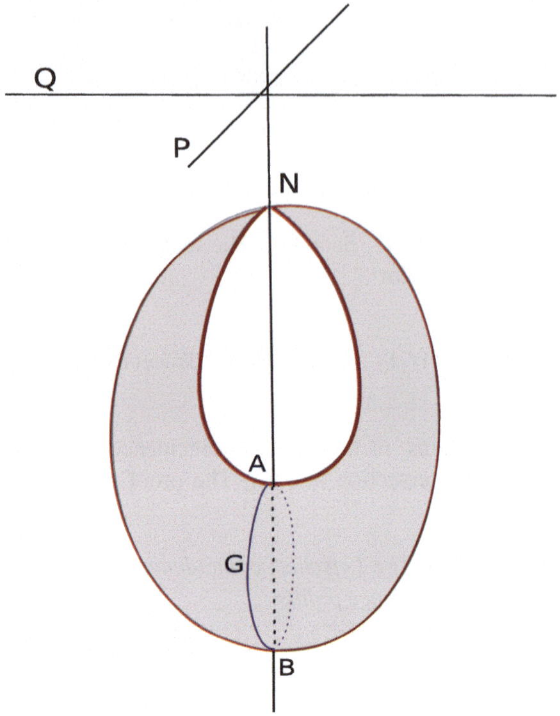

- $IH_2^{\overline{0}}(X) = \mathbb{Z}_{[X]}$ The generator is the fundamental class of X, denoted by $[X]$.

Let us consider now the following applications $X \to X$ (see Fig. 2 for notations):
r_E the vertical rotation of angle π around the line E,
s_P the symmetry relatively to the plane P,
s_Q the symmetry relatively to the plane Q.

The matrices M_0 and M_2 of the maps in intersection homology, induced by these maps and relatively to the basis $\{x\}$ of $IH_0^{\overline{0}}(X)$ and $[X]$ of $IH_2^{\overline{0}}(X)$ are the following:

$$r_E \qquad M_0 = \begin{pmatrix}1\end{pmatrix} \quad M_2 = \begin{pmatrix}1\end{pmatrix}$$

$$s_P \qquad M_0 = \begin{pmatrix}1\end{pmatrix} \quad M_2 = \begin{pmatrix}-1\end{pmatrix}$$

$$s_Q \qquad M_0 = \begin{pmatrix}1\end{pmatrix} \quad M_2 = \begin{pmatrix}-1\end{pmatrix}$$

The intersection Lefschetz numbers of these maps can be computed either globally, or locally.

In order to compute the local Lefschetz numbers, one can use one of the interpretations given by Goresky and MacPherson: either the local linking numbers denoted by $\mu_\Delta([G_L(f)])$ (see [7, Sects. 8 and 9]) or the local trace (see [7, §10]). Both are

equal at an isolated contracting fixed point ([7, Theorem III]) that is the case of all isolated fixed points considered.
The rotation r_E - Global calculus.

The fixed points of the rotation r_E are the pinched point N and the points A and B (see Fig. 2). Using the formula (3.6), one obtains the Lefschetz number $IL(r_E) = 2$.
The rotation r_E - Local calculus.

Let us compute the local trace at each fixed point: One consider firstly the fixed point N. According to Goresky–MacPherson, one has to consider a conical neighborhood U_N of N, that is union of two open cones with vertex N. Intersection homology with compact supports of U_N is equal to:

$$IH_0^{\bar{0}}(U_N) = \mathbb{Z}_{x_1} \oplus \mathbb{Z}_{x_2}$$

where x_1 and x_2 are points in each of the two cones. One remarks that these two points are not homologous in intersection homology.

$$IH_1^{\bar{0}}(U_N) = IH_2^{\bar{0}}(X) = 0$$

One remarks also that there is no fundamental class for the open neighborhood U_N in intersection homology with compact support (there is one in homology with closed support).

The matrix M_0 of the map r_E in intersection homology, relatively to the basis $\{x_1\}, \{x_2\}$ of $IH_0^{\bar{0}}(U_N)$ is the following:

$$r_E \quad M_0 = \begin{pmatrix} 0 & 1 \\ 1 & 0 \end{pmatrix}$$

The matrices M_1 and M_2 are obviously zero. One concludes that the local trace of r_E at N is zero, that is the local index of r_E at N.

Let us consider now the point A (the calculus for B will be the same). At the point A, which is a regular point of X, the local Lefschetz index can be computed by classical ways. One can also consider a neighborhood U_A of A, and local intersection homology (in that case equal to ordinary homology). One has:

$$IH_0^{\bar{0}}(U_A) = \mathbb{Z}_x$$

where x is any point in U_A, and

$$IH_1^{\bar{0}}(U_A) = IH_2^{\bar{0}}(X) = 0,$$

for the same reasons than above.

The matrix M_0 of the map r_E in intersection homology, relatively to the basis $\{x\}$ of $IH_0^{\bar{0}}(U_A)$ is the following:

$$r_E \quad M_0 = (1)$$

The matrices M_1 and M_2 are obviously zero. One concludes that the local trace of r_E at A is 1, that is the local index of r_E at A.

We have the same calculus at B and one obtains that $IL(r_E) = 2$, either globally or as the sum of local contributions.

The symmetry relatively to the plane P.

The fixed points of the symmetry relatively to the plane P are the pinched point N and the circle G (see Fig. 2). Using the formula (3.6), one obtains the intersection Lefschetz number $IL(s_P) = 0$.

In fact, the local Lefschetz number of the symmetry s_P at N is 0, with the same calculus than above. It is possible to deform the symmetry in a homotopic map without other fixed point, that is another way to get the result.

The symmetry relatively to the plane Q.

The fixed points of the symmetry relatively to the plane Q are the great circles passing through N and A and passing through N and B. (see Fig. 2). Using the formula (3.6), one obtains the intersection Lefschetz number $IL(s_Q) = 0$.

7.2 The "Bipinched Torus"

Let us consider the "bipinched torus", that is the suspension of two circles (cf. [10, Sect. 1.2, Example 1]) which has two singular points (see Fig. 3). The bipinched torus is not a pseudomanifold, in the sense that the regular part is not connected. However, one can compute some intersection Lefschetz fixed points numbers and intersection Lefschetz coincidence numbers, deciding orientations of the smooth components.

The bipinched torus has two singular points, the "north pole" N and the "south pole" S. The stratification is given by $X \setminus \{N, S\}, \{N, S\},$. Let us denote by X_1 and X_2 the two connected components of $X \setminus \{N, S\}$.

There is only one possible pervesity, the 0-perversity. The computation of intersection homology groups is the following:

• $IH_0^{\bar{0}}(X) = \mathbb{Z}_{x_1} \oplus \mathbb{Z}_{x_2}$,
where each point x_i belongs to one of the two connected components of $X \setminus \{N, S\}$. The two points x_1 and x_2 are not homologous. In fact, if γ is an arc joining the two points, γ must pass through one of the poles N or S. This arc is not an allowable chain because the following inequality is not verified:

$$\dim(|\gamma| \cap \{N\}) \leq 1 - 2 + 0.$$

• $IH_1^{\bar{0}}(X) = 0$

The three candidate cycles to be the generators of $IH_1^{\bar{0}}(X)$ are: the two circles C_1, C_2 and a cycle Γ going through the points A, N, D, S (see Fig. 3). The two first are allowable cycles and allowable boundaries: They are boundaries of the allowable cones $c(C_i)$ with vertex N and base C_1 or C_2:

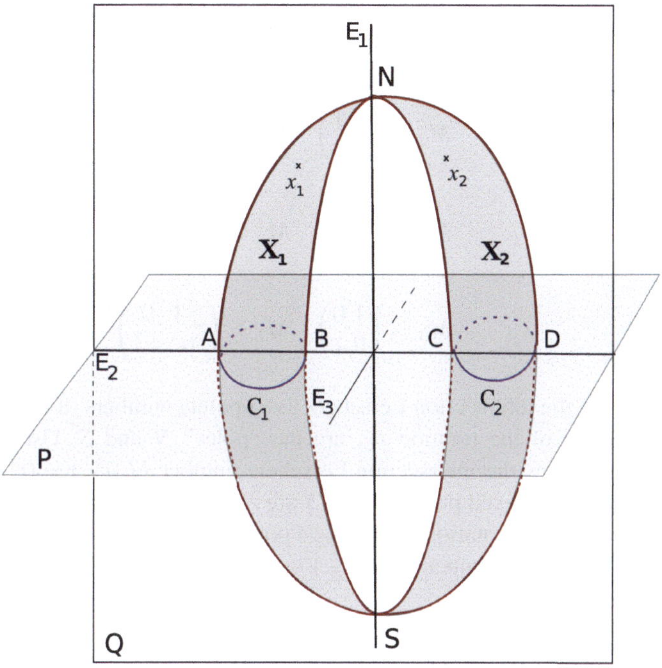

Fig. 3 The "bipinched torus"

$$\dim(|c(C_i)| \cap \{N\}) \leq 2 - 2 + 0.$$

The last cycle Γ is not allowable for the same reason than the chain γ.

• $IH_2^{\overline{0}}(X) = \mathbb{Z}_{[X_1]} \oplus \mathbb{Z}_{[X_2]}$,

where $[X_1]$ and $[X_2]$ are the fundamental classes of X_1 and X_2 for the chosen orientations.

Let us consider now the following maps $X \to X$:

r_{E_1} the vertical rotation of angle π around the line E_1,
r_{E_2} the horizontal rotation of angle π around the line E_2,
r_{E_3} the horizontal rotation of angle π around the line E_3,
s_P the symmetry relatively to the plane P containing E_2 and E_3,
s_Q the symmetry relatively to the plane Q containing E_1 and E_2.

The matrices M_0 and M_2 of the maps in intersection homology, induced by these maps and relatively to the basis $\{x_1, x_2\}$ of $IH_0^{\overline{0}}(X)$ and $[X_1], [X_2]$ of $IH_2^{\overline{0}}(X)$ are the following:

$$r_{E_1} \quad M_0 = \begin{pmatrix} 0 & 1 \\ 1 & 0 \end{pmatrix} \quad M_2 = \begin{pmatrix} 0 & 1 \\ 1 & 0 \end{pmatrix}$$

$$r_{E_2} \quad M_0 = \begin{pmatrix} 1 & 0 \\ 0 & 1 \end{pmatrix} \quad M_2 = \begin{pmatrix} 1 & 0 \\ 0 & 1 \end{pmatrix}$$

$$r_{E_3} \quad M_0 = \begin{pmatrix} 0 & 1 \\ 1 & 0 \end{pmatrix} \quad M_2 = \begin{pmatrix} 0 & 1 \\ 1 & 0 \end{pmatrix}$$

$$s_P \quad M_0 = \begin{pmatrix} 1 & 0 \\ 0 & 1 \end{pmatrix} \quad M_2 = \begin{pmatrix} -1 & 0 \\ 0 & -1 \end{pmatrix}$$

$$s_Q \quad M_0 = \begin{pmatrix} 1 & 0 \\ 0 & 1 \end{pmatrix} \quad M_2 = \begin{pmatrix} -1 & 0 \\ 0 & -1 \end{pmatrix}$$

Let us compute the intersection Lefschetz fixed points numbers of these maps.

The fixed points of the rotation r_{E_1} are the "poles" N and S. Using the formula (3.6), one obtains the intersection Lefschetz number $IL(r_{E_1}) = 0$. The local Lefschetz indices at the fixed points N and S are zero (see previous example).

The fixed points of the rotation r_{E_2} are the 4 points: A, B, C and D. Formula (3.6) shows that the Lefschetz number is $IL(r_{E_2}) = +4$. The Lefschetz index in each of the points A, B, C and D is $+1$.

The rotation r_{E_3} has no fixed point. One has $IL(r_{E_3}) = 0$.

The fixed points of the symmetry relatively to the plane P are the two circles C_1 and C_2. Formula (3.6) shows that $IL(s_P) = 0$. The graph of the symmetry s_P is not transverse to the diagonal Δ_X in $X \times X$. One may consider a map $\varphi : X \to X$ homotopic to s_P and whose graph is transverse to Δ_X. Then intersection points are 4 points (corresponding to intersection of a deformation of each of the two C_i with C_i). The sum of intersection indices is then 0.

The fixed points of the symmetry relatively to the plane Q are points of the intersection $X \cap Q$. Formula (3.6) shows that $IL(s_Q) = 0$.

Let us now compute some Lefschetz coincidence numbers for these maps, using Definition 3.21:

$IL(r_{E_1}, r_{E_2}) = 0$. There is no coincidence point.

$IL(r_{E_2}, s_P) = 0$. There are six coincidence points: N, S, A, B, C, D. Coincidence numbers are two by two opposite signs. That is confirmed by application of formula (6.4).

$IL(r_{E_1}, r_{E_3}) = 0$. The coincidence points are the two circles C_1 and C_2. Formula (3.22). provides the result. One computes the coincidence number in the same way than the Lefschetz fixed points number of s_P.

$IL(s_P, s_Q) = +4$. There are four coincidence points: A, B, C, D. The coincidence number in each of them is equal to $+1$. That is confirmed by application of formula (6.4).

References

1. Bisi, C., Bracci, F., Izawa, T., Suwa, T.: Localized intersection of currents and the Lefschetz coincidence point theorem. arXiv:1406.0607v1
2. Borel, A., et al.: Intersection Cohomology. Progress in Mathematics, vol. 50. Swiss seminars, Boston (1984). ISBN 0-8176-3274-3
3. Brasselet, J.-P., Suwa, T.: Quelle coïncidence ! Preprint 2016
4. Brasselet, J.-P.: Définition combinatoire des homomorphismes de Poincaré, Alexander et Thom pour une pseudo-variété, Astérisque, pp. 82–83 (1981)
5. Goresky, M., MacPherson, R.: Intersection homology theory. Topology **19**(2), 135–162 (1980)
6. Goresky, M., MacPherson, R.: Intersection homology II. Invent. Math. **72**(1), 77–129 (1983)
7. Goresky, M., MacPherson, R.: Letschetz fixed point theorem for intersection homology. Comment. Math. Helvetici **60**, 366–391 (1985)
8. Goresky, M., Pardon, W.: Wu numbers of singular spaces. Topology **28**, 325–367 (1989)
9. Lefschetz, S.: Continuous transformations of manifolds Proc. Acad. Sci. USA **9**, 90–93 (1923)
10. MacPherson, R.: Intersection homology and perverse sheaves Report, December 15 1990, Colloquium Lecture notes distributed by AMS (1991)
11. Siegel, P.: Witt spaces: a geometric cycle theory for KO-homology at odd primes. Am. J. Math. **105**(5), 1067–1105 (1983)

Preservation of Immersed or Injective Properties by Composing Generic Generalized Distance-Squared Mappings

Shunsuke Ichiki and Takashi Nishimura

Abstract Any generalized distance-squared mapping of equidimensional case has singularities, and their singularity types are wrapped into mystery in higher dimensional cases. Any generalized distance-squared mapping of equidimensional case is not injective. Nevertheless, in this paper, it is shown that the immersed property or the injective property of a mapping is preserved by composing a generic generalized distance-squared mapping of equidimensional case.

Keywords Generalized distance-squared mapping · Immersion · Injective Embedding · Transverse

2010 Mathematics Subject Classification 57R35 · 57R40 · 57R42

1 Introduction

Throughout this paper, i, j, ℓ, m and n stand for positive integers. In this paper, unless otherwise stated, all manifolds and mappings belong to class C^∞ and all manifolds are without boundary. Let $p_i = (p_{i1}, p_{i2}, \ldots, p_{im})$ $(1 \leq i \leq \ell)$ (resp., $A = (a_{ij})_{1\leq i\leq \ell, 1\leq j\leq m}$) be a point of $\mathbb{R}^m$ (resp., an $\ell \times m$ matrix with non-zero entries). Set $p = (p_1, p_2, \ldots, p_\ell) \in (\mathbb{R}^m)^\ell$. Let $G_{(p,A)} : \mathbb{R}^m \to \mathbb{R}^\ell$ be the mapping defined by

The first author is Research Fellow DC1 of Japan Society for the Promotion of Science.

S. Ichiki
Graduate School of Environment and Information Sciences, Yokohama National University, Yokohama 240-8501, Japan
e-mail: ichiki-shunsuke-jb@ynu.jp

T. Nishimura (✉)
Research Group of Mathematical Sciences, Research Institute of Environment and Information Sciences, Yokohama National University, Yokohama 240-8501, Japan
e-mail: nishimura-takashi-yx@ynu.jp

R. N. Araújo dos Santos et al. (eds.), *Singularities and Foliations. Geometry, Topology and Applications*, Springer Proceedings in Mathematics & Statistics 222, https://doi.org/10.1007/978-3-319-73639-6_18

$$G_{(p,A)}(x) = \left(\sum_{j=1}^{m} a_{1j}(x_j - p_{1j})^2, \sum_{j=1}^{m} a_{2j}(x_j - p_{2j})^2, \ldots, \sum_{j=1}^{m} a_{\ell j}(x_j - p_{\ell j})^2 \right),$$

where $x = (x_1, x_2, \ldots, x_m) \in \mathbb{R}^m$. The mapping $G_{(p,A)}$ is called a *generalized distance-squared mapping*, and the ℓ-tuple of points $p = (p_1, \ldots, p_\ell) \in (\mathbb{R}^m)^\ell$ is called the *central point* of the generalized distance-squared mapping $G_{(p,A)}$. A *distance-squared mapping D_p* (resp., *Lorentzian distance-squared mapping L_p*) is the mapping $G_{(p,A)}$ satisfying that each entry of A is 1 (resp., $a_{i1} = -1$ and $a_{ij} = 1$ $(j \neq 1)$).

In [4] (resp., [5]), a classification result on distance-squared mappings D_p (resp., Lorentzian distance-squared mappings L_p) is given.

In [7], a classification result on generalized distance-squared mappings of the plane into the plane is given. If the rank of A is two, a generalized distance-squared mapping having a generic central point is a mapping of which any singular point is a fold point except one cusp point. The singular set is a rectangular hyperbola. If the rank of A is one, a generalized distance-squared mapping having a generic central point is $\mathcal{A}$-equivalent to the normal form of fold singularity $(x_1, x_2) \mapsto (x_1, x_2^2)$.

In [6], a classification result on generalized distance-squared mappings of $\mathbb{R}^{m+1}$ into $\mathbb{R}^{2m+1}$ is given. If the rank of A is $m+1$, a generalized distance-squared mapping having a generic central point is $\mathcal{A}$-equivalent to the normal form of Whitney umbrella $(x_1, \ldots, x_{m+1}) \mapsto (x_1^2, x_1x_2, \ldots, x_1x_{m+1}, x_2, \ldots, x_{m+1})$. If the rank of A is less than $m+1$, a generalized distance-squared mapping having a generic central point is $\mathcal{A}$-equivalent to the inclusion $(x_1, \ldots, x_{m+1}) \mapsto (x_1, \ldots, x_{m+1}, 0, \ldots, 0)$.

As described above, in [6, 7], the properties of generalized distance-squared mappings having a generic central point are investigated. On the other hand, in this paper, the property of the compositions of a given immersion (resp., a given injection) and generalized distance-squared mappings having a generic central point is investigated (see Theorem 1 (resp., Theorem 2)).

We have the following original motivation. Height functions and distance-squared functions have been investigated in detail so far, and they are a useful tool in the applications of singularity theory to differential geometry (for instance, see [2]). The mapping in which each component is a height function is nothing but a projection. In [8], compositions of generic projections and embeddings are investigated.

On the other hand, the mapping in which each component is a distance-squared function is a distance-squared mapping. And, the notion of generalized distance-squared mapping is an extension of the distance-squared mappings. Therefore, it is natural to investigate generalized distance-squared mappings as well as projections.

Any generalized distance-squared mapping of equidimensional case $G_{(p,A)} : \mathbb{R}^m \to \mathbb{R}^m$ has singularities (see Lemma 5.1 in Appendix). Nevertheless, in Theorem 1, it is shown that the immersed property of a mapping is preserved by composing a generic generalized distance-squared mapping of equidimensional case.

Theorem 1 *Let N be an n-dimensional manifold, and let $f : N \to \mathbb{R}^m$ be an immersion $(m \geq 2n)$. Then, there exists a subset Σ of $(\mathbb{R}^m)^m$ with Lebesgue measure* *zero such that for any* $p \in (\mathbb{R}^m)^m - \Sigma$*, the composition* $G_{(p,A)} \circ f : N \to \mathbb{R}^m$ *is an immersion.*

Any generalized distance-squared mapping of equidimensional case $G_{(p,A)} : \mathbb{R}^m \to \mathbb{R}^m$ is not injective (see Lemma 5.2 in Appendix). Nevertheless, in Theorem 2, it is shown that the injective property of a mapping is preserved by composing a generic generalized distance-squared mapping of equidimensional case.

Theorem 2 *Let* N *be an* n*-dimensional manifold, and let* $f : N \to \mathbb{R}^m$ *be injective* ($m \geq 2n + 1$)*. Then, there exists a subset* Σ *of* $(\mathbb{R}^m)^m$ *with Lebesgue measure zero such that for any* $p \in (\mathbb{R}^m)^m - \Sigma$*, the composition* $G_{(p,A)} \circ f : N \to \mathbb{R}^m$ *is injective.*

By combining Theorems 1 and 2, we have the following proposition.

Proposition 1 *Let* N *be an* n*-dimensional manifold, and let* $f : N \to \mathbb{R}^m$ *be an injective immersion* ($m \geq 2n + 1$)*. Then, there exists a subset* Σ *of* $(\mathbb{R}^m)^m$ *with Lebesgue measure zero such that for any* $p \in (\mathbb{R}^m)^m - \Sigma$*, the composition* $G_{(p,A)} \circ f : N \to \mathbb{R}^m$ *is an injective immersion.*

1.1 Remark

Suppose that the mapping $G_{(p,A)} \circ f : N \to \mathbb{R}^m$ is proper in Proposition 1. Then, the injective immersion of $G_{(p,A)} \circ f$ implies the embedding of it (see [3], p.11). Hence, we have the following as a corollary of Proposition 1.

Corollary 1 *Let* N *be an* n*-dimensional compact manifold, and let* $f : N \to \mathbb{R}^m$ *be an embedding* ($m \geq 2n + 1$)*. Then, there exists a subset* Σ *of* $(\mathbb{R}^m)^m$ *with Lebesgue measure zero such that for any* $p \in (\mathbb{R}^m)^m - \Sigma$*, the composition* $G_{(p,A)} \circ f : N \to \mathbb{R}^m$ *is an embedding.*

In Sect. 2, it is reviewed some of standard definitions, and an important lemma for the proofs of Theorems 1 and 2 is given. Section 3 (resp., Sect. 4) devotes the proof of Theorem 1 (resp., Theorem 2). Finally, in Sect. 5.1 (resp., Sect. 5.2), for the readers' convenience, it is given the proof that any generalized distance-squared mapping of equidimensional case has singularities (resp., the proof that any generalized distance-squared mapping of equidimensional case is not injective).

2 Preliminaries

Let $J^r(n, p)$ be the set of all r-jets of map-germs $(\mathbb{R}^n, 0) \to (\mathbb{R}^p, 0)$. Let N and P be manifolds and let $J^r(N, P)$ be the space of r-jets of mappings of N into P. For a given mapping $g : N \to P$, the mapping $j^r g : N \to J^r(N, P)$ is defined by $q \mapsto j^r g(q)$

(for details on the space $J^r(N, P)$ or the mapping $j^r g : N \to J^r(N, P)$, see for example [3]).

Next, we recall the definition of transversality.

Definition 1 Let W be a submanifold of P. For a given mapping $g : N \to P$, we say that $g : N \to P$ is *transverse* to W if for any $q \in N$, $g(q) \notin W$ or in the case of $g(q) \in W$, the following holds:

$$dg_q(T_q N) + T_{g(q)} W = T_{g(q)} P.$$

For the proofs of Theorems 1 and 2, the following lemma is important.

Lemma 2.1 *([1, 8]) Let N, P, Z be manifolds, and let W be a submanifold of P. Let $\Gamma : N \times Z \to P$ be a mapping. If Γ is transverse to W, then there exists a subset Σ of Z with Lebesgue measure zero such that for any $p \in Z - \Sigma$, $\Gamma_p : N \to P$ is transverse to W, where $\Gamma_p(q) = \Gamma(q, p)$.*

3 Proof of Theorem 1

Let $\{(U_\lambda, \varphi_\lambda)\}_{\lambda \in \Lambda}$ be a coordinate neighborhood system of N. Let $\pi : J^1(N, \mathbb{R}^m) \to N \times \mathbb{R}^m$ be the natural projection defined by $\pi(j^1 g(q)) = (q, g(q))$. Let $\Phi_\lambda : \pi^{-1}(U_\lambda \times \mathbb{R}^m) \to \varphi_\lambda(U_\lambda) \times \mathbb{R}^m \times J^1(n, m)$ be the homeomorphism defined by

$$\Phi_\lambda \left(j^1 g(q)\right) = \left(\varphi_\lambda(q), g(q), j^1(\psi_\lambda \circ g \circ \varphi_\lambda^{-1} \circ \widetilde{\varphi}_\lambda)(0)\right),$$

where $\widetilde{\varphi}_\lambda : \mathbb{R}^n \to \mathbb{R}^n$ (resp., $\psi_\lambda : \mathbb{R}^m \to \mathbb{R}^m$) is the translation defined by $\widetilde{\varphi}_\lambda(0) = \varphi_\lambda(q)$ (resp., $\psi_\lambda(g(q)) = 0$). Then, $\{(\pi^{-1}(U_\lambda \times \mathbb{R}^m), \Phi_\lambda)\}_{\lambda \in \Lambda}$ is a coordinate neighborhood system of $J^1(N, \mathbb{R}^m)$. For any k ($k = 1, \ldots, n$), set

$$\Sigma^k = \left\{j^1 g(0) \in J^1(n, m) \mid \dim \mathrm{Ker} J g(0) = k\right\}.$$

For any k ($k = 1, \ldots, n$), set

$$\Sigma^k(N, \mathbb{R}^m) = \bigcup_{\lambda \in \Lambda} \Phi_\lambda^{-1} \left(\varphi_\lambda(U_\lambda) \times \mathbb{R}^m \times \Sigma^k\right).$$

Then, the set $\Sigma^k(N, \mathbb{R}^m)$ is a subfiber-bundle of $J^1(N, \mathbb{R}^m)$ such that

$$\begin{aligned} \mathrm{codim}\ \Sigma^k(N, \mathbb{R}^m) &= \dim J^1(N, \mathbb{R}^m) - \dim \Sigma^k(N, \mathbb{R}^m) \\ &= k(m - n + k). \end{aligned}$$

(for details on $\Sigma^k(N, \mathbb{R}^m)$, see for example [3], pp.60–61).

Now, let $\Gamma : N \times (\mathbb{R}^m)^m \to J^1(N, \mathbb{R}^m)$ be the mapping defined by

$$\Gamma(q, p) = j^1(G_{(p,A)} \circ f)(q).$$

We will show first that the mapping Γ is transverse to the submanifold $\Sigma^k(N, \mathbb{R}^m)$ for any k ($k = 1, \ldots, n$). It is sufficient to show that if $\Gamma(\widetilde{q}, \widetilde{p}) \in \Sigma^k(N, \mathbb{R}^m)$, then the following $(*)$ holds.

$$d\Gamma_{(\widetilde{q},\widetilde{p})}(T_{(\widetilde{q},\widetilde{p})}(N \times (\mathbb{R}^m)^m)) + T_{\Gamma(\widetilde{q},\widetilde{p})}\Sigma^k(N, \mathbb{R}^m) = T_{\Gamma(\widetilde{q},\widetilde{p})}J^1(N, \mathbb{R}^m). \qquad (*)$$

There exists a coordinate neighborhood $\left(U_{\widetilde{\lambda}} \times (\mathbb{R}^m)^m, \varphi_{\widetilde{\lambda}} \times id\right)$ containing the point $(\widetilde{q}, \widetilde{p})$ of $N \times (\mathbb{R}^m)^m$, where id is the identity mapping of $(\mathbb{R}^m)^m$ into $(\mathbb{R}^m)^m$, and the mapping $\varphi_{\widetilde{\lambda}} \times id : U_{\widetilde{\lambda}} \times (\mathbb{R}^m)^m \to \mathbb{R}^n \times (\mathbb{R}^m)^m$ is defined by $\left(\varphi_{\widetilde{\lambda}} \times id\right)(q, p) = \left(\varphi_{\widetilde{\lambda}}(q), id(p)\right)$. There exists a coordinate neighborhood $\left(\pi^{-1}(U_{\widetilde{\lambda}} \times \mathbb{R}^m), \Phi_{\widetilde{\lambda}}\right)$ containing the point $\Gamma(\widetilde{q}, \widetilde{p})$ of $J^1(N, \mathbb{R}^m)$. Let $t = (t_1, \ldots, t_n) \in \mathbb{R}^n$ be a local coordinate containing $\varphi_{\widetilde{\lambda}}(\widetilde{q})$. Then, the mapping Γ is locally given by the following:

$$\begin{aligned}
&(\Phi_{\widetilde{\lambda}} \circ \Gamma \circ (\varphi_{\widetilde{\lambda}} \times id)^{-1})(t, p)\\
&= (\Phi_{\widetilde{\lambda}} \circ \Gamma \circ (\varphi_{\widetilde{\lambda}}^{-1} \times id^{-1}))(t, p)\\
&= (\Phi_{\widetilde{\lambda}} \circ \Gamma)(\varphi_{\widetilde{\lambda}}^{-1}(t), p)\\
&= \Phi_{\widetilde{\lambda}}(\Gamma(\varphi_{\widetilde{\lambda}}^{-1}(t), p))\\
&= \Phi_{\widetilde{\lambda}}(j^1(G_{(p,A)} \circ f)(\varphi_{\widetilde{\lambda}}^{-1}(t)))\\
&= (\Phi_{\widetilde{\lambda}} \circ j^1(G_{(p,A)} \circ f) \circ \varphi_{\widetilde{\lambda}}^{-1})(t)\\
&= \Bigg(t, (G_{(p,A)} \circ f \circ \varphi_{\widetilde{\lambda}}^{-1})(t),\\
&\quad \frac{\partial(G_1 \circ f \circ \varphi_{\widetilde{\lambda}}^{-1})}{\partial t_1}(t), \ldots, \frac{\partial(G_1 \circ f \circ \varphi_{\widetilde{\lambda}}^{-1})}{\partial t_n}(t),\\
&\qquad \cdots\cdots\cdots,\\
&\quad \frac{\partial(G_m \circ f \circ \varphi_{\widetilde{\lambda}}^{-1})}{\partial t_1}(t), \ldots, \frac{\partial(G_m \circ f \circ \varphi_{\widetilde{\lambda}}^{-1})}{\partial t_n}(t)\Bigg)\\
&= \Bigg(t, (G_{(p,A)} \circ f \circ \varphi_{\widetilde{\lambda}}^{-1})(t),\\
&\quad 2\sum_{j=1}^{m} a_{1j}(\widetilde{f}_j(t) - p_{1j})\frac{\partial \widetilde{f}_j}{\partial t_1}(t), \ldots, 2\sum_{j=1}^{m} a_{1j}(\widetilde{f}_j(t) - p_{1j})\frac{\partial \widetilde{f}_j}{\partial t_n}(t),\\
&\qquad \cdots\cdots\cdots,\\
&\quad 2\sum_{j=1}^{m} a_{mj}(\widetilde{f}_j(t) - p_{mj})\frac{\partial \widetilde{f}_j}{\partial t_1}(t), \ldots, 2\sum_{j=1}^{m} a_{mj}(\widetilde{f}_j(t) - p_{mj})\frac{\partial \widetilde{f}_j}{\partial t_n}(t)\Bigg),
\end{aligned}$$

where $p = (p_{11}, \ldots, p_{1m}, \ldots, p_{m1}, \ldots, p_{mm})$, $f = (f_1, \ldots, f_m)$, $G_{(p,A)} = (G_1, \ldots, G_m)$, and $\widetilde{f_j} = f_j \circ \varphi_{\widetilde{\lambda}}^{-1}$ $(1 \le j \le m)$. The Jacobian matrix of the mapping Γ at $(\widetilde{q}, \widetilde{p})$ is the following:

$$J\Gamma_{(\widetilde{q},\widetilde{p})} = \left(\begin{array}{c|cccc} E_n & 0 & \cdots\cdots & & 0 \\ \hline & * & \cdots\cdots & & * \\ & B_1 & & 0 & \\ ** & & B_2 & & \\ & & 0 & \ddots & \\ & & & & B_m \end{array}\right)_{(\varphi_{\widetilde{\lambda}}(\widetilde{q}),\widetilde{p})},$$

where E_n is the $n \times n$ unit matrix and B_i $(1 \le i \le m)$ is the following $n \times m$ matrix.

$$B_i = \begin{pmatrix} -2a_{i1}\frac{\partial \widetilde{f_1}}{\partial t_1}(t) & \cdots & -2a_{im}\frac{\partial \widetilde{f_m}}{\partial t_1}(t) \\ \vdots & \ddots & \vdots \\ -2a_{i1}\frac{\partial \widetilde{f_1}}{\partial t_n}(t) & \cdots & -2a_{im}\frac{\partial \widetilde{f_m}}{\partial t_n}(t) \end{pmatrix}_{t=\varphi_{\widetilde{\lambda}}(\widetilde{q})}.$$

In the matrix $J\Gamma_{(\widetilde{q},\widetilde{p})}$, remark that the matrix $**$ is an $(m + nm) \times n$ matrix and each of the matrices $*$ is an $m \times m$ matrix.

Since $\Sigma^k(N, \mathbb{R}^m)$ is a subfiber-bundle of $J^1(N, \mathbb{R}^m)$ with fiber Σ^k, in order to show $(*)$, it is clearly seen that the rank of the following matrix C is $n + m + nm$.

$$C = \left(\begin{array}{c|cccc} E_{n+m} & * & \cdots\cdots & & * \\ \hline & B_1 & & 0 & \\ 0 & & B_2 & & \\ & & 0 & \ddots & \\ & & & & B_m \end{array}\right)_{(\varphi_{\widetilde{\lambda}}(\widetilde{q}),\widetilde{p})},$$

where E_{n+m} is the $(n + m) \times (n + m)$ unit matrix. Notice that for any i $(1 \le i \le m^2)$, the $(n + m + i)$th column vector of C is the $(n + i)$th column vector of $J\Gamma_{(\widetilde{q},\widetilde{p})}$. Let $Jf_{\widetilde{q}}$ be the Jacobian matrix of the mapping f at $\widetilde{q}$. Since $a_{ij} \neq 0$ for any i, j $(1 \le i, j \le m)$, there exists an $m \times m$ regular matrix R_i such that $B_i R_i = {}^t(Jf_{\widetilde{q}})$ for any i $(1 \le i \le m)$, where ${}^tX$ means the transposed matrix of X. Hence, there exists an $(n + m + m^2) \times (n + m + m^2)$ regular matrix $\widetilde{R}$ such that

$$C\widetilde{R} = \left(\begin{array}{c|cccc} E_{n+m} & * & \cdots & \cdots & * \\ \hline & {}^t(Jf_{\widetilde{q}}) & & 0 & \\ 0 & & {}^t(Jf_{\widetilde{q}}) & & \\ & & 0 & \ddots & \\ & & & & {}^t(Jf_{\widetilde{q}}) \end{array}\right)_{(\varphi_{\widetilde{\lambda}}(\widetilde{q}),\widetilde{p})}.$$

Since the mapping f is an immersion ($n \leq m$), we have that the rank of the matrix $C\widetilde{R}$ is $n+m+nm$. Therefore, the rank of the matrix C must be $n+m+nm$. Hence, we have $(*)$. Thus, the mapping Γ is transverse to the submanifold $\Sigma^k(N, \mathbb{R}^m)$.

By Lemma 2.1, for any k ($k = 1, \ldots, n$), there exists a subset $\widetilde{\Sigma}^k$ of $(\mathbb{R}^m)^m$ with Lebesgue measure zero such that for any $p \in (\mathbb{R}^m)^m - \widetilde{\Sigma}^k$, the mapping $\Gamma_p : N \to J^1(N, \mathbb{R}^m)$ is transverse to the submanifold $\Sigma^k(N, \mathbb{R}^m)$. Set $\Sigma = \bigcup_{k=1}^n \widetilde{\Sigma}^k$. Notice that Σ is a subset of $(\mathbb{R}^m)^m$ with Lebesgue measure zero. Then, for any $p \in (\mathbb{R}^m)^m - \Sigma$, the mapping $\Gamma_p : N \to J^1(N, \mathbb{R}^m)$ is transverse to the submanifold $\Sigma^k(N, \mathbb{R}^m)$ for any k ($k = 1, \ldots, n$).

In order to show that for any $p \in (\mathbb{R}^m)^m - \Sigma$, the mapping $G_{(p,A)} \circ f : N \to \mathbb{R}^m$ is an immersion, it is sufficient to show that for any $p \in (\mathbb{R}^m)^m - \Sigma$, it follows that $\Gamma_p(N) \bigcap \bigcup_{k=1}^n \Sigma^k(N, \mathbb{R}^m) = \emptyset$.

Suppose that there exists an element $p_0 \in (\mathbb{R}^m)^m - \Sigma$ such that there exists an element $q_0 \in N$ such that $\Gamma_{p_0}(q_0) \in \bigcup_{k=1}^n \Sigma^k(N, \mathbb{R}^m)$. Then, there exists a natural number k' ($1 \leq k' \leq n$) such that $\Gamma_{p_0}(q_0) \in \Sigma^{k'}(N, \mathbb{R}^m)$. Since Γ_{p_0} is transverse to $\Sigma^{k'}(N, \mathbb{R}^m)$, we have the following:

$$d(\Gamma_{p_0})_{q_0}(T_{q_0}N) + T_{\Gamma_{p_0}(q_0)}\Sigma^{k'}(N, \mathbb{R}^m) = T_{\Gamma_{p_0}(q_0)}J^1(N, \mathbb{R}^m).$$

Hence, we have

$$\begin{aligned} \dim d(\Gamma_{p_0})_{q_0}(T_{q_0}N) &\geq \dim T_{\Gamma_{p_0}(q_0)}J^1(N, \mathbb{R}^m) - \dim T_{\Gamma_{p_0}(q_0)}\Sigma^{k'}(N, \mathbb{R}^m) \\ &= \operatorname{codim} T_{\Gamma_{p_0}(q_0)}\Sigma^{k'}(N, \mathbb{R}^m). \end{aligned}$$

Thus, we have $n \geq k'(m - n + k')$. This contradicts the assumptions $m \geq 2n$ and $k' \geq 1$. □

4 Proof of Theorem 2

Let Δ be the subset of $\mathbb{R}^{2m}$ defined by $\Delta = \{(y, y) \mid y \in \mathbb{R}^m\}$. It is clearly seen that Δ is a submanifold of $\mathbb{R}^{2m}$ such that

$$\operatorname{codim} \Delta = \dim \mathbb{R}^{2m} - \dim \Delta = m.$$

Set $N^{(2)} = \{(q, q') \in N^2 \mid q \neq q'\}$. Notice that $N^{(2)}$ is an open submanifold of N^2.

Now, let $\Gamma : N^{(2)} \times (\mathbb{R}^m)^m \to \mathbb{R}^{2m}$ be the mapping defined by

$$\Gamma(q, q', p) = \left((G_{(p,A)} \circ f)(q), (G_{(p,A)} \circ f)(q')\right).$$

We will show first that the mapping Γ is transverse to the submanifold Δ. It is sufficient to show that if $\Gamma(\widetilde{q}, \widetilde{q}', \widetilde{p}) \in \Delta$, then the following $(**)$ holds.

$$d\Gamma_{(\widetilde{q},\widetilde{q}',\widetilde{p})}(T_{(\widetilde{q},\widetilde{q}',\widetilde{p})}(N^{(2)} \times (\mathbb{R}^m)^m)) + T_{\Gamma(\widetilde{q},\widetilde{q}',\widetilde{p})}\Delta = T_{\Gamma(\widetilde{q},\widetilde{q}',\widetilde{p})}\mathbb{R}^{2m}. \qquad (**)$$

Let $\{(U_\lambda, \varphi_\lambda)\}_{\lambda\in\Lambda}$ be a coordinate neighborhood system of N. There exists a coordinate neighborhood $\left(U_{\widetilde{\lambda}} \times U_{\widetilde{\lambda}'} \times (\mathbb{R}^m)^m, \varphi_{\widetilde{\lambda}} \times \varphi_{\widetilde{\lambda}'} \times id\right)$ containing the point $(\widetilde{q}, \widetilde{q}', \widetilde{p})$ of $N^{(2)} \times (\mathbb{R}^m)^m$, where id is the identity mapping of $(\mathbb{R}^m)^m$ into $(\mathbb{R}^m)^m$, and the mapping $\varphi_{\widetilde{\lambda}} \times \varphi_{\widetilde{\lambda}'} \times id : U_{\widetilde{\lambda}} \times U_{\widetilde{\lambda}'} \times (\mathbb{R}^m)^m \to \mathbb{R}^n \times \mathbb{R}^n \times (\mathbb{R}^m)^m$ is defined by $\left(\varphi_{\widetilde{\lambda}} \times \varphi_{\widetilde{\lambda}'} \times id\right)(q, q', p) = \left(\varphi_{\widetilde{\lambda}}(q), \varphi_{\widetilde{\lambda}'}(q'), id(p)\right)$. Let $t = (t_1, \ldots, t_n)$ be a local coordinate containing $\varphi_{\widetilde{\lambda}}(\widetilde{q})$, and let $t' = (t_1' \ldots, t_n')$ be a local coordinate containing $\varphi_{\widetilde{\lambda}'}(\widetilde{q}')$. Then, the mapping Γ is locally given by the following:

$$\begin{aligned}
&\Gamma \circ \left(\varphi_{\widetilde{\lambda}} \times \varphi_{\widetilde{\lambda}'} \times id\right)^{-1}(t, t', p) \\
&= \Gamma \circ \left(\varphi_{\widetilde{\lambda}}^{-1} \times \varphi_{\widetilde{\lambda}'}^{-1} \times id^{-1}\right)(t, t', p) \\
&= \Gamma\left(\varphi_{\widetilde{\lambda}}^{-1}(t), \varphi_{\widetilde{\lambda}'}^{-1}(t'), p\right) \\
&= \left((G_{(p,A)} \circ f \circ \varphi_{\widetilde{\lambda}}^{-1})(t), (G_{(p,A)} \circ f \circ \varphi_{\widetilde{\lambda}'}^{-1})(t')\right) \\
&= \left(\sum_{j=1}^m a_{1j}(\widetilde{f}_j(t) - p_{1j})^2, \ldots, \sum_{j=1}^m a_{mj}(\widetilde{f}_j(t) - p_{mj})^2,\right. \\
&\qquad \left.\sum_{j=1}^m a_{1j}(\widetilde{f}_j'(t') - p_{1j})^2, \ldots, \sum_{j=1}^m a_{mj}(\widetilde{f}_j'(t') - p_{mj})^2\right),
\end{aligned}$$

where $p = (p_{11}, \ldots, p_{1m}, \ldots, p_{m1}, \ldots, p_{mm})$, $f = (f_1, \ldots, f_m)$, $\widetilde{f}_j = f_j \circ \varphi_{\widetilde{\lambda}}^{-1}$, and $\widetilde{f}_j' = f_j \circ \varphi_{\widetilde{\lambda}'}^{-1}$ $(1 \leq j \leq m)$. The Jacobian matrix of the mapping Γ at $(\widetilde{q}, \widetilde{q}', \widetilde{p})$ is the following:

$$J\Gamma_{(\widetilde{q},\widetilde{q}',\widetilde{p})} = \left(\begin{array}{c|cccc}
 & \mathbf{b}_1 & & & 0 \\
 & & \mathbf{b}_2 & & \\
 & & & \ddots & \\
 & 0 & & & \mathbf{b}_m \\
* & \mathbf{b}'_1 & & & 0 \\
 & & \mathbf{b}'_2 & & \\
 & & & \ddots & \\
 & 0 & & & \mathbf{b}'_m
\end{array}\right)_{(\varphi_{\widetilde{\lambda}}(\widetilde{q}),\varphi_{\widetilde{\lambda}'}(\widetilde{q}'),\widetilde{p})},$$

where

$$\begin{aligned}
\mathbf{b}_i &= -2\left(a_{i1}(\widetilde{f}_1(t) - p_{i1}), \ldots, a_{im}(\widetilde{f}_m(t) - p_{im})\right), \\
\mathbf{b}'_i &= -2\left(a_{i1}(\widetilde{f}_1'(t') - p_{i1}), \ldots, a_{im}(\widetilde{f}_m'(t') - p_{im})\right).
\end{aligned}$$

By seeing the construction of $T_{\Gamma(\widetilde{q},\widetilde{q}',\widetilde{p})}\Delta$, in order to show (∗∗), it is sufficient to show that the rank of the following matrix D is $2m$.

$$D = \left(\begin{array}{c|cccc} & \mathbf{b}_1 & & & 0 \\ E_m & & \mathbf{b}_2 & & \\ & & & \ddots & \\ & 0 & & & \mathbf{b}_m \\ \hline & \mathbf{b}'_1 & & & 0 \\ E_m & & \mathbf{b}'_2 & & \\ & & & \ddots & \\ & 0 & & & \mathbf{b}'_m \end{array}\right)_{(\varphi_{\widetilde{X}}(\widetilde{q}),\varphi_{\widetilde{X}'}(\widetilde{q}'),\widetilde{p})},$$

where E_m is the $m \times m$ unit matrix. Notice that for any i $(1 \le i \le m^2)$, the $(m+i)$th column vector of D is the $(2n+i)$th column vector of $J\Gamma_{(\widetilde{q},\widetilde{q}',\widetilde{p})}$.

By $a_{ij} \neq 0$, there exist an $(m+m^2) \times (m+m^2)$ regular matrix Q_1 such that the following holds:

$$DQ_1 = \left(\begin{array}{c|ccccc} & \widetilde{f}_1(t) - p_{11} \cdots \widetilde{f}_m(t) - p_{1m} & & & & \\ E_m & 0 & & \ddots & & 0 \\ & & & & & \widetilde{f}_1(t) - p_{m1} \cdots \widetilde{f}_m(t) - p_{mm} \\ \hline & \widetilde{f}'_1(t') - p_{11} \cdots \widetilde{f}'_m(t') - p_{1m} & & & & \\ E_m & 0 & & \ddots & & 0 \\ & & & & & \widetilde{f}'_1(t') - p_{m1} \cdots \widetilde{f}'_m(t') - p_{mm} \end{array}\right)_{(t,t',p)},$$

where $(t, t', p) = (\varphi_{\widetilde{X}}(\widetilde{q}), \varphi_{\widetilde{X}'}(\widetilde{q}'), \widetilde{p})$. It is clearly seen that there exist a $2m \times 2m$ regular matrix Q_2 and an $(m+m^2) \times (m+m^2)$ regular matrix Q_3 such that the following holds:

$$Q_2DQ_1Q_3 = \left(\begin{array}{c|ccccc} E_m & & & 0 & & \\ \hline & \widetilde{f}'_1(t') - \widetilde{f}_1(t) \cdots \widetilde{f}'_m(t') - \widetilde{f}_m(t) & & & & \\ 0 & 0 & & \ddots & & 0 \\ & & & & & \widetilde{f}'_1(t') - \widetilde{f}_1(t) \cdots \widetilde{f}'_m(t') - \widetilde{f}_m(t) \end{array}\right)_{(t,t',p)},$$

where $(t, t', p) = (\varphi_{\widetilde{X}}(\widetilde{q}), \varphi_{\widetilde{X}'}(\widetilde{q}'), \widetilde{p})$. Since f is injective, there exists a natural number j $(1 \le j \le m)$ such that $\widetilde{f}'_j(t') - \widetilde{f}_j(t) \neq 0$. Hence, we have that the rank of $Q_2DQ_1Q_3$ is $2m$. Therefore, the rank of the matrix D must be $2m$. Hence, we have (∗∗). Thus, the mapping Γ is transverse to the submanifold Δ.

By Lemma 2.1, there exists a subset Σ of $(\mathbb{R}^m)^m$ with Lebesgue measure zero such that for any $p \in (\mathbb{R}^m)^m - \Sigma$, the mapping $\Gamma_p : N^{(2)} \to \mathbb{R}^{2m}$ is transverse to the submanifold Δ.

In order to prove that for any $p \in (\mathbb{R}^m)^m - \Sigma$, the mapping $G_{(p,A)} \circ f$ is injective, it is sufficient to show that for any $p \in (\mathbb{R}^m)^m - \Sigma$, it follows that $\Gamma_p(N^{(2)}) \cap \Delta = \emptyset$. Suppose that there exists an element $p_0 \in (\mathbb{R}^m)^m - \Sigma$ such that there exists an element $(q_0, q_0') \in N^{(2)}$ such that $\Gamma_{p_0}(q_0, q_0') \in \Delta$. Since Γ_{p_0} is transverse to Δ, we have the following:

$$d(\Gamma_{p_0})_{(q_0,q_0')}(T_{(q_0,q_0')}N^{(2)}) + T_{\Gamma_{p_0}(q_0,q_0')}\Delta = T_{\Gamma_{p_0}(q_0,q_0')}\mathbb{R}^{2m}.$$

Hence, we have

$$\begin{aligned}\dim d(\Gamma_{p_0})_{(q_0,q_0')}(T_{(q_0,q_0')}N^{(2)}) &\geq \dim T_{\Gamma_{p_0}(q_0,q_0')}\mathbb{R}^{2m} - \dim T_{\Gamma_{p_0}(q_0,q_0')}\Delta \\ &= \operatorname{codim} T_{\Gamma_{p_0}(q_0,q_0')}\Delta.\end{aligned}$$

Thus, we have $2n \geq m$. This contradicts the assumption $m \geq 2n + 1$. □

5 Appendix

The proofs of the following Lemmas 5.1 and 5.2 are given in Sects. 5.1 and 5.2, respectively.

Lemma 5.1 *Any generalized distance-squared mapping of equidimensional case* $G_{(p,A)} : \mathbb{R}^m \to \mathbb{R}^m$ *has singularities.*

Lemma 5.2 *Any generalized distance-squared mapping of equidimensional case* $G_{(p,A)} : \mathbb{R}^m \to \mathbb{R}^m$ *is not injective.*

5.1 Proof of Lemma 5.1

Let $J(G_{(p,A)})_x$ be the Jacobian matrix of the mapping $G_{(p,A)}$ at x.

$$J(G_{(p,A)})_x = 2\begin{pmatrix} a_{11}(x_1 - p_{11}) & \cdots & a_{1m}(x_m - p_{1m}) \\ \vdots & \ddots & \vdots \\ a_{m1}(x_1 - p_{m1}) & \cdots & a_{mm}(x_m - p_{mm}) \end{pmatrix}_x.$$

If $x = p_i$ $(1 \leq i \leq m)$, then we have that $\operatorname{rank} J(G_{(p,A)})_{p_i} \leq m - 1$. □

5.2 *Proof of Lemma 5.2*

Set $G_{(p,A)} = (G_1, \dots, G_m)$. It is clear that $G_{(p,A)}^{-1}(\{0\} \times \mathbb{R}^{m-1}) = G_1^{-1}(0)$. Since G_1 has the form $G_1(x) = \sum_{j=1}^{m} a_{1j}(x_j - p_{1j})^2$ $(a_{1j} \neq 0)$, it is easy to see that $G_1^{-1}(0) = \{p_1\}$ or $G_1^{-1}(0) - \{p_1\}$ is homotopy equivalent to $S^k \times S^{m-2-k}$ where k is an integer such that $0 \leq k \leq m-2$. Hence, it follows that the set-germ $(\{0\} \times \mathbb{R}^{m-1}, G_{(p,A)}(p_1))$ is not homeomorphic to the set-germ $(G_1^{-1}(0), p_1)$.

On the other hand, suppose that $G_{(p,A)}$ is injective. Then, by the invariance of domain theorem ([9]), $G_{(p,A)}^{-1} : G_{(p,A)}(\mathbb{R}^m) \to \mathbb{R}^m$ must be a homeomorphism. It follows that the set-germ $(\{0\} \times \mathbb{R}^{m-1}, G_{(p,A)}(p_1))$ is homeomorphic to the set-germ $(G_1^{-1}(0), p_1)$, which is a contradiction. Therefore, $G_{(p,A)}$ is not injective. □

Acknowledgements The authors are most grateful to the anonymous reviewer for his/her careful reading of the first manuscript of this paper and invaluable suggestions. The first author is supported by JSPS KAKENHI Grant Number 16J06911.

References

1. Abraham, R.: Transversality in manifolds of mappings. Bull. Am. Math. Soc. **69**, 470–474 (1963)
2. Bruce, J.W., Giblin, P.J.: Curves and Singularities, 2nd edn. Cambridge University Press, Cambridge (1992)
3. Golubitsky, M., Guillemin, V.: Stable Mappings and Their Singularities. Graduate texts in mathematics, vol. 14. Springer, New York (1973)
4. Ichiki, S., Nishimura, T.: Distance-squared mappings. Topol. Appl. **160**, 1005–1016 (2013)
5. Ichiki, S., Nishimura, T.: Recognizable classification of Lorentzian distance-squared mappings. J. Geom. Phys. **81**, 62–71 (2014)
6. Ichiki, S., Nishimura, T.: Generalized distance-squared mappings of $\mathbb{R}^{n+1}$ into $\mathbb{R}^{2n+1}$ Contemporary Mathematics. Am. Math. Soc. Providence RI, **675**, 121–132 (2016)
7. Ichiki, S., Nishimura, T., Sinha, R.O., Ruas, M.A.S.: Generalized distance-squared mappings of the plane into the plane. Adv. Geom. **16**, 189–198 (2016)
8. Mather, J.N.: Generic projections. Ann. Math. **98**(2), 226–245 (1973)
9. Spanier, E.H.: Algebraic Topology. McGraw-Hill Book Company, New York (1966)

Arc Criterion of Normal Embedding

Lev Birbrair and Rodrigo Mendes

Abstract We present a criterion of local normal embedding of a semialgebraic (or definable in a polynomially bounded o-minimal structure) germ contained in $\mathbb{R}^n$ in terms of orders of contact of arcs. Namely, we prove that a semialgebraic germ is normally embedded if and only if for any pair of arcs, coming to this point the inner order of contact is equal to the outer order of contact.

Keywords Normal embedding · Singularities

1991 Mathematics Subject Classification 14B05 · 32S50

1 Introduction

A connected subset of Euclidean space is called normally embedded if the two natural metrics, outer and the inner metric, are bi-Lipschitz equivalent and the bi-Lipschitz homeomorphism is given by the identity map. The notion of Normal Embedding (or in other words Lipschitz Normal Embedding) became rather popular in recent development of Singularity Theory. Is used in Metric Homology Theory of Birbrair and Brasselet [1], in Vanishing Homology of Valette [11], in Lipschitz Regularity

L. Birbrair—partially supported by CAPES-COFECUB and by CNPq-Brazil, grants no. 302655/2014-0.
R. Mendes—partially supported by Capes.

L. Birbrair
Departamento of Matemática, Universidade Federal do Ceará,
Campus do Pici Bloco 914, Fortaleza, CE 60455-760, Brazil
e-mail: lev.birbrair@gmail.com

R. Mendes (✉)
Instituto de Ciências Exatas e da Natureza, Universidade de Integração Internacional
da Lusofonia Afro-Brasileira (unilab), Campus dos Palmares, Acarape, CE
62785-000, Brazil
e-mail: rodrigomendes@unilab.edu.br

R. N. Araújo dos Santos et al. (eds.), *Singularities and Foliations. Geometry, Topology and Applications*, Springer Proceedings in Mathematics & Statistics 222, https://doi.org/10.1007/978-3-319-73639-6_19

theorem [2]. Several authors are investigating some special algebraic and semialgebraic sets in the spirit of their normal embedding. See, for example, the recent works [3, 7, 9, 10]. In this note we present an arc criterion of Normal Embedding that, we hope, it can be useful in these studies. The criterion is based on the arc selection lemma, an extremely important tool of Real Algebraic Geometry.

2 Normally Embedded Sets

Let $X \subset \mathbb{R}^n$ be a connected semialgebraic set. We define an inner metric on X as follows: Let $x, y \in X$. The inner distance $d_X(x, y)$ is defined as the infimum of lengths of rectifiable arcs $\gamma : [0, 1] \to X$ such that $\gamma(0) = x$ and $\gamma(1) = y$. Notice that for connected semialgebraic sets the inner metric is well-defined.

Definition 2.1 A semialgebraic subset $X \subset \mathbb{R}^n$ is called *normally embedded* if there exists $\lambda > 0$ such that

$$d_X(x_1, x_2) \leq \lambda \|x_1 - x_2\|,$$

for all $x_1, x_2 \in X$.

X is called normally embedded at x_0 if for sufficiently small $\epsilon > 0$, $X \cap B_{x_0,\epsilon}$ is normally embedded. We may also say that the germ of X at x_0 is normally embedded.

Considering real or complex cusps $x^2 = y^3$, one can see that the inner metric is not bi-Lipschitz equivalent to the Euclidean metric. On the other hand, the smooth compact semialgebraic sets are normally embedded.

By the results of Kurdyka and Orro [8] (see also Birbrair and Mostowski [4]), there exists a semialgebraic metric

$$d_P : X \times X \to \mathbb{R},$$

such that (X, d_X) and (X, d_P) are bi-Lipschitz equivalent and the identity map is bi-Lipschitz for d_X and d_P.

An *arc* γ with initial point at x_0 is a continuous semialgebraic map $\gamma : [0, \epsilon) \to \mathbb{R}^n$ such that $\gamma(0) = x_0$. When it does not lead to a confusion, we use the same notation for an arc and its image in $\mathbb{R}^n$.

For a semialgebraic function of one variable $f(t)$, where $f(0) = 0$, we have $f(t) = a_1 t^\alpha + o(t^\alpha)$, for some $a_1 \in \mathbb{R}$ and $\alpha \in \mathbb{Q}_+$. The number α is called the order of f at 0. We use the notation $ord_t f$.

We can define the *outer order of tangency* in the following way:

$$tord(\gamma_1, \gamma_2) = ord_t \|\gamma_1(t) - \gamma_2(t)\|,$$

where the arcs γ_1 and γ_2 are parametrized by the outer distance to the singular point, i.e., $\|\gamma_i(t) - x_0\| = t$, $i = 1, 2$. Given two arcs γ_1, γ_2 contained in X, we may define the *inner order of tangency* by

$$tord_{inn}(\gamma_1, \gamma_2) = ord_t(d_P(\gamma_1(t), \gamma_2(t))),$$

where the arcs γ_1 and γ_2 are again parametrized by the outer distance to the singular point.

Theorem 2.2 (Criterion of Normal embedding) *Let $X \subset \mathbb{R}^n$ be a closed semialgebraic germ at x_0. Then the following assertions are equivalent:*

- *The germ of X at x_0 is normally embedded;*
- *There exists a constant $k > 0$ such that for any pair of arcs $\gamma_1, \gamma_2 \subset X$ and parametrized by the distance at x_0, ($\gamma_i(0) = x_0$) we have*

$$d_X(\gamma_1(t), \gamma_2(t)) \le k\|\gamma_1(t) - \gamma_2(t))\|;$$

- *For any pair of arcs γ_1, γ_2 parametrized by the distance to x_0 one has:*

$$tord(\gamma_1, \gamma_2) = tord_{inn}(\gamma_1, \gamma_2).$$

Remark 2.3 The theorem is formulated in the semialgebraic category, but the result is true for polynomially bounded o-minimal structures. Actually, all the ingredients of the proof work in that case.

Proof If X is normally embedded at x_0, the inequality above follows from the definition. Assume now that X is not normally embedded at x_0. Consider a map $\psi : X \times X \to \mathbb{R}^2$, defined as follows: $\psi(x_1, x_2) = (\|x_1 - x_2\|, d_P(x_1, x_2))$. This map is semialgebraic. Since the distance functions $\|x_1 - x_2\|$ and $d_P(x_1, x_2)$ are continuous, then the image $\psi(X \times X)$ is closed and locally connected at $\psi(x_0, x_0) = 0 \in \mathbb{R}^2$, $x_0 \in X$. Moreover, this set is semialgebraic, according to Tarski-Seidenberg theorem. Since X is not normally embedded at x_0, the set $\psi(X \times X)$ must be locally bounded near $0 \in \mathbb{R}^2$ by an arc $\beta \subset \psi(X \times X)$ such that β is tangent to the y-axis. Taking a arc $(\tilde{\gamma}_1, \tilde{\gamma}_2)$ belonging to the inverse image of β, we obtain that

$$ord_t d_P(\tilde{\gamma}_1(t), \tilde{\gamma}_2(t)) < ord_t(\|\tilde{\gamma}_1(t) - \tilde{\gamma}_2(t)\|). \tag{1}$$

We may suppose that the arc $\tilde{\gamma}_1$ is parametrized by the distance to the singular point x_0. But, we cannot suppose that the other arc is also parametrized the same way. That is why we need the order of comparison lemma.

Remark 2.4 (Non-archimedean property) (see for example [6]). Let γ_1, γ_2 and γ_3 be three different semialgebraic arcs of X, $\gamma_i(0) = x_0$ $(i = 1, 2, 3)$. Let α_{12}, α_{23} and α_{13} be outer orders of tangency between the pairs (γ_1, γ_2), (γ_2, γ_3) and (γ_1, γ_3). Suppose that $\alpha_{12} \le \alpha_{23} \le \alpha_{13}$. Then $\alpha_{12} = \alpha_{23}$.

Proof $\|\gamma_1(t) - \gamma_2(t)\| \le \|\gamma_1(t) - \gamma_3(t)\| + \|\gamma_3(t) - \gamma_2(t)\| \Rightarrow \alpha_{12} \ge \alpha_{23}$. □

Observe that the function $d_P(\gamma_1(t), \gamma_2)$ given by

$$d_P(\gamma_1(t), \gamma_2) = inf\{d_P(\gamma_1(t), y); y \in \gamma_2\} \tag{2}$$

is a semialgebraic function and $d_P(\gamma_1(0), \gamma_2) = 0$. Then, $ord_t d_P(\gamma_1(t), \gamma_2)$ is well defined.

Lemma 2.5 (Inner order comparison lemma) $tord_{inn}(\gamma_1, \gamma_2) = ord_t(d_P(\gamma_1(t), \gamma_2))$.

Proof Consider a pancake decomposition $\{X_i\}$ of X, where the metric d_P corresponds to this decomposition (see [4]). By definition of metric d_P, we choose semialgebraic arcs $\tilde{\beta}_1, \ldots \tilde{\beta}_N$, $\tilde{\beta}_i(0) = x_0$, $i = 1, \ldots, N$, such that all the pairs $(\gamma_1, \tilde{\beta}_1), (\tilde{\beta}_1, \tilde{\beta}_2), \ldots, (\tilde{\beta}_N, \gamma_2)$ belong to the same "pancake", i.e., a normally embedded subset of X (see [4]). For $t \in [0, \delta)$, δ sufficient small, we have

$$d_P(\gamma_1(t), \gamma_2) = \|\gamma_1(t) - \tilde{\beta}_1(t)\| + \|\tilde{\beta}_1(t) - \tilde{\beta}_2(t)\| + \ldots + \|\tilde{\beta}_N(t) - \tilde{\gamma}_2(t)\|.$$

Notice that arcs $\tilde{\beta}_i$ are not necessarily parametrized by the distance to x_0. Moreover, we have

$$\|\gamma_1(t) - \tilde{\beta}_S(t)\| \geq d_{outer}(\gamma_1(t), \tilde{\beta}_S), \forall S, \tag{3}$$

where $d_{outer}(\gamma_1(t), \tilde{\beta}_S)$ is defined as in (2), considering the euclidean distance. Otherwise,

$$d_P(\gamma_1(t), \gamma_2(t)) \leq \|\gamma_1(t) - \beta_1(t)\| + \|\beta_1(t) - \beta_2(t)\| + \ldots + \|\beta_N(t) - \gamma_2(t)\|,$$

where now $\beta_i(t)$ and $\gamma_2(t)$ is a parametrization of $\tilde{\beta}_i$ and γ_2 by distance to the origin. By the non-archimedean property, we have

$$\|\gamma_1(t) - \beta_1(t)\| + \|\beta_1(t) - \beta_2(t)\| + \ldots + \|\beta_N(t) - \gamma_2(t)\| \simeq \|\beta_S(t) - \beta_{S-1}(t)\|,$$

for some $S \in \{1, \ldots, N+1\}$, where $\beta_0(t)$ is $\gamma_1(t)$ and $\beta_{N+1}(t)$ is $\gamma_2(t)$ and

$$\|\beta_S(t) - \beta_{S-1}(t)\| \simeq \|\gamma_1(t) - \beta_S(t)\|,$$

where $f(t) \simeq g(t)$ means that the functions have the same order. Now, the outer order comparison lemma of [6] says:

$$\|\gamma_1(t) - \beta_S(t)\| \simeq d_{outer}(\gamma_1(t), \tilde{\beta}_S).$$

So, there exists constant $C_2 > 0$ such that

$$d_P(\gamma_1(t), \gamma_2(t)) \leq C_2 d_{outer}(\gamma_1(t), \tilde{\beta}_S). \tag{4}$$

Hence, by (3) and (4) the lemma is proved. □

End of the proof of Theorem 2.2:

Since d_X is bi-Lipschitz equivalent to the d_P, using the previous lemma and the inequality (1), we obtain that

$$\lim_{t\to 0^+} \frac{d_X(\gamma_1(t), \gamma_2(t))}{\|\gamma_1(t) - \gamma_2(t)\|} = +\infty,$$

or, in other words, $tord(\gamma_1, \gamma_2) > tord_{inn}(\gamma_1, \gamma_2)$, where $\gamma_1,\ \gamma_2$ can be considered parametrized by the distance to the point x_0. □

Acknowledgements We would like to thank Alexandre Fernandes, Edson Sampaio, Anne Pichon and Walter Neumann for useful discussions. We would like also to thank the anonymous referee for his patience and extremely useful comments and corrections.

References

1. Birbrair, L., Brasselet, J.-P.: Metric homology. Commun. Pure Appl. Math. **LIII**, 1434–1447 (2000)
2. Birbrair, L., Fernandes, A., Le, D.T., Sampaio, J.E.: Lipschitz regular complex algebraic sets are smooth. Proc. Amer. Math. Soc. **144**(3), 983–987 (2016)
3. Birbrair, L., Fernandes, A., Neumann, W.D.: On normal embedding of complex algebraic surfaces. Real and complex singularities. Lond. Math. Soc. Lecture Note Ser. (380), 17–22. Cambridge University Press, Cambridge (2010)
4. Birbrair, L., Mostowski, T.: Normal embeddings of semialgebraic sets. Michigan Math. J. **47**, 125–132 (2000)
5. Bröcker, L.: Families of semialgebraic sets and limits. In: Real Algebraic Geometry, Lect. Notes. Math. (1524) pp. 145–162. Springer, Berlin (1992)
6. Birbrair, L., Fernandes, A.: Metric theory of semialgebraic curves. Rev. Mat. Complut. **13**(2), 369–382 (2000)
7. Katz, K., Katz, M., Kerner, D., Liokumovich, Y., Solomon, J.: Determinantal variety and bilipschitz equivalence. J. Topol. Anal. **10**, 27 (2018)
8. Kurdyka, K., Orro, P.: Distance géodésique sur un sous-analytique. Revista Mat. Univ. Complutense de Madrid **10**, 173–182 (1997)
9. Neumann, W.D., Pedersen, H.M., Pichon, A.: Minimal surface singularities are Lipschitz normally embedded (2015). arXiv:1503.03301
10. Pedersen, H.M.: Maria Aparecida Soares Ruas. Lipschitz Normal Embeddings and Determinantal Singularities (2016). arXiv:1607.07746
11. Valette, G.: Vanishing homology. Selecta Math. (N.S.) **16**(2), 267–296 (2010)

The manufacturer's authorised representative in the EU is Springer Nature Customer Service Centre GmbH, Europaplatz 3, 69115 Heidelberg, Germany. If you have any concerns regarding our products, please contact ProductSafety@springernature.com

Printed and bound by CPI Group (UK) Ltd, Croydon, CR0 4YY
01/10/2026
02226615-0001